Signals and Systems

PRENTICE-HALL SIGNAL PROCESSING SERIES

Alan V. Oppenheim, Editor

Prentice-Hall, Inc., Englewood Cliffs, New Jersey 07632

ALAN V. OPPENHEIM
ALAN S. WILLSKY

with
IAN T. YOUNG

SIGNALS and SYSTEMS

Library of Congress Cataloging in Publication Data

OPPENHEIM, ALAN V.   (date)
  Signals and systems.

    (Prentice-Hall signal processing series)
    Includes index.
    1. System analysis. 2. Signal theory
(Telecommunication) I. Willsky, Alan S.
II. Young, Ian T. III. Title. IV. Series.
QA402.063   003   81–22652
ISBN 0–13–809731–3   AACR2

Editorial production/supervision
   by Gretchen K. Chenenko
Chapter opening design
   by Dawn Stanley
Cover executed by Judy Matz
Manufacturing buyers: Joyce Levatino
   and Anthony Caruso

Printed in the United States of America

10 9 8 7 6 5 4 3

ISBN 0-13-809731-3

Prentice-Hall International, Inc., *London*
Prentice-Hall of Australia Pty. Limited, *Sydney*
Editora Prentice-Hall do Brasil, Ltda., *Rio de Janeiro*
Prentice-Hall Canada Inc., *Toronto*
Prentice-Hall of India Private Limited, *New Delhi*
Prentice-Hall of Japan, Inc., *Tokyo*
Prentice-Hall of Southeast Asia Pte. Ltd., *Singapore*
Whitehall Books Limited, *Wellington, New Zealand*

To our families

Contents

3

Linear Time-Invariant Systems   69

4

Fourier Analysis for Continuous-Time Signals and Systems   *161*

5
Fourier Analysis for Discrete-Time Signals and Systems   291

6
Filtering   *397*

Contents

7

Modulation   *447*

8

Sampling   *513*

9

The Laplace Transform   *573*

10

The z-Transform   629

11

Linear Feedback Systems   685

Contents

Appendix   Partial Fraction Expansion   *767*

Bibliography   *777*

Index   *783*

Preface

This book is designed as a text for an undergraduate course in signals and systems. While such courses are frequently found in electrical engineering curricula, the concepts and techniques that form the core of the subject are of fundamental importance in all engineering disciplines. In fact the scope of potential and actual applications of the methods of signal and system analysis continues to expand as engineers are confronted with new challenges involving the synthesis or analysis of complex processes. For these reasons we feel that a course in signals and systems not only is an essential element in an engineering program but also can be one of the most rewarding, exciting, and useful courses that engineering students take during their undergraduate education.

Our treatment of the subject of signals and systems is based on lecture notes that were developed in teaching a first course on this topic in the Department of Electrical Engineering and Computer Science at M.I.T. Our overall approach to the topic has been guided by the fact that with the recent and anticipated developments in technologies for signal and system design and implementation, the importance of having equal familiarity with techniques suitable for analyzing and synthesizing both continuous-time and discrete-time systems has increased dramatically. To achieve this goal we have chosen to develop in parallel the methods of analysis for continuous-time and discrete-time signals and systems. This approach also offers a distinct and extremely important pedagogical advantage. Specifically, we are able to draw on the similarities between continuous- and discrete-time methods in order to share insights and intuition developed in each domain. Similarly, we can exploit the differences between them to sharpen an understanding of the distinct properties of each.

In organizing the material, we have also considered it essential to introduce the student to some of the important uses of the basic methods that are developed in the book. Not only does this provide the student with an appreciation for the range of applications of the techniques being learned and for directions of further study, but it also helps to deepen understanding of the subject. To achieve this goal we have included introductory treatments on the subjects of filtering, modulation, sampling, discrete-time processing of continuous-time signals, and feedback. In addition, we have included a bibliography at the end of the book in order to assist the student who is interested in pursuing additional and more advanced studies of the methods and applications of signal and system analysis.

The book's organization also reflects our conviction that full mastery of a subject of this nature cannot be accomplished without a significant amount of practice in using and applying the basic tools that are developed. Consequently, we have included a collection of more than 350 end-of-chapter homework problems of several types. Many, of course, provide drill on the basic methods developed in the chapter. There are also numerous problems that require the student to apply these methods to problems of practical importance. Others require the student to delve into extensions of the concepts developed in the text. This variety and quantity will hopefully provide instructors with considerable flexibility in putting together homework sets that are tailored to the specific needs of their students. Solutions to the problems are available to instructors through the publisher. In addition, a self-study course consisting of a set of video-tape lectures and a study guide will be available to accompany this text.

Students using this book are assumed to have a basic background in calculus as well as some experience in manipulating complex numbers and some exposure to differential equations. With this background, the book is self-contained. In particular, no prior experience with system analysis, convolution, Fourier analysis, or Laplace and z-transforms is assumed. Prior to learning the subject of signals and systems most students will have had a course such as basic circuit theory for electrical engineers or fundamentals of dynamics for mechanical engineers. Such subjects touch on some of the basic ideas that are developed more fully in this text. This background can clearly be of great value to students in providing additional perspective as they proceed through the book.

A brief introductory chapter provides motivation and perspective for the subject of signals and systems in general and our treatment of it in particular. We begin Chapter 2 by introducing some of the elementary ideas related to the mathematical representation of signals and systems. In particular we discuss transformations (such as time shifts and scaling) of the independent variable of a signal. We also introduce some of the most important and basic continuous-time and discrete-time signals, namely real and complex exponentials and the continuous-time and discrete-time unit step and unit impulse. Chapter 2 also introduces block diagram representations of interconnections of systems and discusses several basic system properties ranging from causality to linearity and time-invariance. In Chapter 3 we build on these last two properties, together with the sifting property of unit impulses to develop the convolution sum representation for discrete-time linear, time-invariant (LTI) systems and the convolution integral representation for continuous-time LTI systems. In this treatment we use the intuition gained from our development of the discrete-time case as an aid in deriving and understanding its continuous-time counterpart. We then turn to a discussion of systems characterized by linear constant-coefficient differential and difference equations. In this introductory discussion we review the basic ideas involved in solving linear differential equations (to which most students will have had some previous exposure), and we also provide a discussion of analogous methods for linear difference equations. However, the primary focus of our development in Chapter 3 is not on methods of solution, since more convenient approaches are developed later using transform methods. Instead, in this first look, our intent is to provide the student with some appreciation for these extremely important classes of systems, which will be encountered often in subsequent chapters. Included in this discussion is the introduction of block diagram representations of LTI systems described by difference equations and differential equations using adders, coefficient multipliers, and delay elements (discrete-time) or integrators (continuous-time). In later chapters we return to this theme in developing cascade and parallel structures with the aid of transform methods. The inclusion of these representations provides the student not only with a way in which to visualize these systems but also with a concrete example of the implications (in terms of suggesting alternative and distinctly different structures for implementation) of some of the mathematical properties of LTI systems. Finally, Chapter 3 concludes with a brief discussion of singularity functions—steps, impulses, doublets, and so forth—in the context of their role in the description and analysis of continuous-time LTI systems. In particular, we stress the interpretation of these signals in terms of how they are defined under convolution—for example, in terms of the responses of LTI systems to these idealized signals.

Chapter 4 contains a thorough and self-contained development of Fourier analysis for continuous-time signals and systems, while Chapter 5 deals in a parallel fashion with the discrete-time case. We have included some historical information about the development of Fourier analysis at the beginning of Chapters 4 and 5, and at several points in their development to provide the student with a feel for the range of disciplines in which these tools have been used and to provide perspective on some of the mathematics of Fourier analysis. We begin the technical discussions in both chapters by emphasizing and illustrating the two fundamental reasons for the important role Fourier analysis plays in the study of signals and systems: (1) extremely broad classes of signals can be represented as weighted sums or integrals of complex exponentials; and (2) the response of an LTI system to a complex exponential input is simply the same exponential multiplied by a complex number characteristic of the system. Following this, in each chapter we first develop the Fourier series representation of periodic signals and then derive the Fourier transform representation of aperiodic signals as the limit of the Fourier series for a signal whose period becomes arbitrarily large. This perspective emphasizes the close relationship between Fourier series and transforms, which we develop further in subsequent sections. In both chapters we have included a discussion of the many important properties of Fourier transforms and series, with special emphasis placed on the convolution and modulation properties. These two specific properties, of course, form the basis for filtering, modulation, and sampling, topics that are developed in detail in later chapters. The last two sections in Chapters 4 and 5 deal with the use of transform methods to analyze LTI systems characterized by differential and difference equations. To supplement these discussions (and later treatments of Laplace and z-transforms) we have included an Appendix at the end of the book that contains a description of the method of partial fraction expansion. We use this method in several examples in Chapters 4 and 5 to illustrate how the response of LTI systems described by differential and difference equations can be calculated with relative ease. We also introduce the cascade and parallel-form realizations of such systems and use this as a natural lead-in to an examination of the basic building blocks for these systems—namely, first- and second-order systems.

Our treatment of Fourier analysis in these two chapters is characteristic of the nature of the parallel treatment we have developed. Specifically, in our discussion in Chapter 5, we are able to build on much of the insight developed in Chapter 4 for the continuous-time case, and toward the end of Chapter 5, we emphasize the complete duality in continuous-time and discrete-time Fourier representations. In addition, we bring the special nature of each domain into sharper focus by contrasting the differences between continuous- and discrete-time Fourier analysis.

Chapters 6, 7, and 8 deal with the topics of filtering, modulation, and sampling, respectively. The treatments of these subjects are intended not only to introduce the student to some of the important uses of the techniques of Fourier analysis but also to help reinforce the understanding of and intuition about frequency domain methods. In Chapter 6 we present an introduction to filtering in both continuous-time and discrete-time. Included in this chapter are a discussion of ideal frequency-selective filters, examples of filters described by differential and difference equations, and an introduction, through examples such as an automobile suspension system and the class of Butterworth filters, to a number of the qualitative and quantitative issues and tradeoffs that arise in filter design. Numerous other aspects of filtering are explored in the problems at the end of the chapter.

Our treatment of modulation in Chapter 7 includes an in-depth discussion of continuous-time sinusoidal amplitude modulation (AM), which begins with the most straightforward application of the modulation property to describe the effect of modulation in the frequency domain and to suggest how the original modulating signal can be recovered. Following this, we develop a number of additional issues and applications based on the modulation property such as: synchronous and asynchronous demodulation, implementation of frequency-selective filters with variable center frequencies, frequency-division multiplexing, and single-sideband modulation. Many other examples and applications are described in the problems. Three additional topics are covered in Chapter 7. The first of these is pulse-amplitude modulation and time-division multiplexing, which forms a natural bridge to the topic of sampling in Chapter 8. The second topic, discrete-time amplitude modulation, is readily developed based on our previous treatment of the continuous-time case. A variety of other discrete-time applications of modulation are developed in the problems. The third and final topic, frequency modulation (FM), provides the reader with a look at a non-linear modulation problem. Although the analysis of FM systems is not as straightforward as for the AM case, our introductory treatment indicates how frequency domain methods can be used to gain a significant amount of insight into the characteristics of FM signals and systems.

Our treatment of sampling in Chapter 8 is concerned primarily with the sampling theorem and its implications. However, to place this subject in perspective we begin by discussing the general concepts of representing a continuous-time signal in terms of its samples and the reconstruction of signals using interpolation. After having used frequency domain methods to derive the sampling theorem, we use both the frequency and time domains to provide intuition concerning the phenomenon of aliasing resulting from undersampling. One of the very important uses of sampling is in the discrete-time processing of continuous-time signals, a topic that we explore at some length in this chapter. We conclude our discussion of continuous-time sampling with the dual problem of sampling in the frequency domain. Following this, we turn to the sampling of discrete-time signals. The basic result underlying discrete-time sampling is developed in a manner that exactly parallels that used in continuous time, and the application of this result to problems of decimation, interpolation, and transmodulation are described. Again a variety of other applications, in both continuous- and discrete-time, are addressed in the problems.

Chapters 9 and 10 treat the Laplace and z-transforms, respectively. For the most part, we focus on the bilateral versions of these transforms, although we briefly discuss unilateral transforms and their use in solving differential and difference equations with nonzero initial conditions. Both chapters include discussions on: the close relationship between these transforms and Fourier transforms; the class of rational transforms and the notion of poles and zeroes; the region of convergence of a Laplace or z-transform and its relationship to properties of the signal with which it is associated; inverse transforms using partial fraction expansion; the geometric evaluation of system functions and frequency responses from pole-zero plots; and basic transform properties. In addition, in each chapter we examine the properties and uses of system functions for LTI systems. Included in these discussions are the determination of system functions for systems characterized by differential and difference equations, and the use of system function algebra for interconnections of LTI systems. Finally, Chapter 10 uses the techniques of Laplace and z-transforms to discuss transformations for mapping continuous-time systems with rational system functions into discrete-time systems with rational system functions. Three important examples of such transformations are described and their utility and properties are investigated.

The tools of Laplace and z-transforms form the basis for our examination of linear feedback systems in Chapter 11. We begin in this chapter by describing a number of the important uses and properties of feedback systems, including stabilizing unstable systems, designing tracking systems, and reducing system sensitivity. In subsequent sections we use the tools that we have developed in previous chapters to examine three topics that are of importance for both continuous-time and discrete-time feedback systems. These are root locus analysis, Nyquist plots and the Nyquist criterion, and log magnitude/phase plots and the concepts of phase and gain margins for stable feedback systems.

The subject of signals and systems is an extraordinarily rich one, and a variety of approaches can be taken in designing an introductory course. We have written this book in order to provide instructors with a great deal of flexibility in structuring their presentations of the subject. To obtain this flexibility and to maximize the usefulness of this book for instructors, we have chosen to present thorough, in-depth treatments of a cohesive set of topics that forms the core of most introductory courses on signals and systems. In achieving this depth we have of necessity omitted the introductions to topics such as descriptions of random signals and state space models that are sometimes included in first courses on signals and systems. Traditionally, at many schools, including M.I.T., such topics are not included in introductory courses but rather are developed in far more depth in courses explicitly devoted to their investigation. For example, thorough treatments of state space methods are usually carried out in the more general context of multi-input/multi-output and time-varying systems, and this generality is often best treated after a firm foundation is developed in the topics in this book. However, whereas we have not included an introduction to state space in the book, instructors of introductory courses can easily incorporate it into the treatments of differential and difference equations in Chapters 2-5.

A typical one-semester course at the sophomore-junior level using this book would cover Chapters 2, 3, 4, and 5 in reasonable depth (although various topics in each chapter can be omitted at the discretion of the instructor) with selected topics chosen from the remaining chapters. For example, one possibility is to present several of the basic topics in Chapters 6, 7, and 8 together with a treatment of Laplace and z-transforms and perhaps a brief introduction to the use of system function concepts to analyze feedback systems. A variety of alternate formats are possible, including one that incorporates an introduction to state space or one in which more focus is placed on continuous-time systems (by deemphasizing Chapters 5 and 10 and the discrete-time topics in Chapters 6, 7, 8, and 11). We have also found it useful to introduce some of the applications described in Chapters 6, 7, and 8 during our development of the basic material on Fourier analysis. This can be of great value in helping to build the student's intuition and appreciation for the subject at an earlier stage of the course.

In addition to these course formats this book can be used as the basic text for a thorough, two-semester sequence on linear systems. Alternatively, the portions of the book not used in a first course on signals and systems, together with other sources can form the basis for a senior elective course. For example, much of the material in this book forms a direct bridge to the subject of digital signal processing as treated in the book by Oppenheim and Schafer.† Consequently, a senior course can be constructed that uses the advanced material on discrete-time systems as a lead-in to a course on digital signal processing. In addition to or in place of such a focus is one that leads into state space methods for describing and analyzing linear systems.

As we developed the material that comprises this book, we have been fortunate to have received assistance, suggestions, and support from numerous colleagues, students, and friends. The ideas and perspectives that form the heart of this book were formulated and developed over a period of ten years while teaching our M.I.T. course on signals and systems, and the many colleagues and students who taught the course with us had a significant influence on the evolution of the course notes on which this book is based. We also wish to thank Jon Delatizky and Thomas Slezak for their help in generating many of the figure sketches, Hamid Nawab and Naveed Malik for preparing the problem solutions that accompany the text, and Carey Bunks and David Rossi for helping us to assemble the bibliography included at the end of the book. In addition the assistance of the many students who devoted a significant number of

---

†A. V. Oppenheim and R. W. Schafer, *Digital Signal Processing* (Englewood Cliffs, N.J. Prentice-Hall, Inc., 1975).

hours to the reading and checking of the galley and page proofs is gratefully acknowledged.

We wish to thank M.I.T. for providing support and an invigorating environment in which we could develop our ideas. In addition, some of the original course notes and subsequent drafts of parts of this book were written by A.V.O. while holding a chair provided to M.I.T. by Cecil H. Green; by A.S.W. first at Imperial College of Science and Technology under a Senior Visiting Fellowship from the United Kingdom's Science Research Council and subsequently at Le Laboratoire des Signaux et Systèmes, Gif-sur-Yvette, France, and L'Université de Paris-Sud; and by I.T.Y. at the Technical University Delft, The Netherlands under fellowships from the Cornelius Geldermanfonds and the Nederlandse organisatie voor zuiver-wetenschappelijk onderzoek (Z.W.O.). We would like to express our thanks to Ms. Monica Edelman Dove, Ms. Fifa Monserrate, Ms. Nina Lyall, Ms. Margaret Flaherty, Ms. Susanna Natti, and Ms. Helene George for typing various drafts of the book and to Mr. Arthur Giordani for drafting numerous versions of the figures for our course notes and the book. The encouragement, patience, technical support, and enthusiasm provided by Prentice-Hall, and in particular by Hank Kennedy and Bernard Goodwin, have been important in bringing this project to fruition.

The concepts of signals and systems arise in an extremely wide variety of fields, and the ideas and techniques associated with these concepts play an important role in such diverse areas of science and technology as communications, aeronautics and astronautics, circuit design, acoustics, seismology, biomedical engineering, energy generation and distribution systems, chemical process control, and speech processing. Although the physical nature of the signals and systems that arise in these various disciplines may be drastically different, they all have two very basic features in common. The signals are functions of one or more independent variables and typically contain information about the behavior or nature of some phenomenon, whereas the systems respond to particular signals by producing other signals. Voltages and currents as a function of time in an electrical circuit are examples of signals, and a circuit is itself an example of a system, which in this case responds to applied voltages and currents. As another example, when an automobile driver depresses the accelerator pedal, the automobile responds by increasing the speed of the vehicle. In this case, the system is the automobile, the pressure on the accelerator pedal the input to the system, and the automobile speed the response. A computer program for the automated diagnosis of electrocardiograms can be viewed as a system which has as its input a digitized electrocardiogram and which produces estimates of parameters such as heart rate as outputs. A camera is a system that receives light from different sources and reflected from objects and produces a photograph.

In the many contexts in which signals and systems arise, there are a variety of problems and questions that we may consider. In some cases, we are presented with a specific system and are interested in characterizing it in detail to understand how it will respond to various inputs. One example is the long and extensive history of

Introduction

1

research directed at gaining an understanding of the human auditory system. Another example is the development of an understanding and a characterization of the economic system in a particular geographical area in order to be better able to predict what its response will be to potential or unanticipated inputs, such as crop failures, new oil discoveries, and so on.

In other contexts of signal and system analysis, rather than analyzing existing systems, our interest may be focused on the problem of designing systems to process signals in particular ways. Economic forecasting represents one very common example of such a situation. We may, for example, have the history of an economic time series, such as a set of stock market averages, and it would be clearly advantageous to be able to predict the future behavior based on the past history of the signal. Many systems, typically in the form of computer programs, have been developed and refined to carry out detailed analysis of stock market averages and to carry out other kinds of economic forecasting. Although most such signals are not totally predictable, it is an interesting and important fact that from the past history of many of these signals, their future behavior is somewhat predictable; in other words, they can at least be approximately extrapolated.

A second very common set of applications is in the restoration of signals that have been degraded in some way. One situation in which this often arises is in speech communication when a significant amount of background noise is present. For example, when a pilot is communicating with an air traffic control tower, the communication can be degraded by the high level of background noise in the cockpit. In this and many similar cases, it is possible to design systems that will retain the desired signal, in this case the pilot's voice, and reject (at least approximately) the unwanted signal, i.e. the noise. Another example in which it has been useful to design a system for restoration of a degraded signal is in restoring old recordings. In acoustic recording a system is used to produce a pattern of grooves on a record from an input signal that is the recording artist's voice. In the early days of acoustic recording a mechanical recording horn was typically used and the resulting system introduced considerable distortion in the result. Given a set of old recordings, it is of interest to restore these to a quality that might be consistent with modern recording techniques. With the appropriate design of a signal processing system, it is possible to significantly enhance old recordings.

A third application in which it is of interest to design a system to process signals in a certain way is the general area of image restoration and image enhancement. In receiving images from deep space probes, the image is typically a degraded version of the scene being photographed because of limitations on the imaging equipment, possible atmospheric effects, and perhaps errors in signal transmission in returning the images to earth. Consequently, images returned from space are routinely processed by a system to compensate for some of these degradations. In addition, such images are usually processed to enhance certain features, such as lines (corresponding, for example, to river beds or faults) or regional boundaries in which there are sharp contrasts in color or darkness. The development of systems to perform this processing then becomes an issue of system design.

Another very important class of applications in which the concepts and techniques of signal and system analysis arise are those in which we wish to modify the characteristics of a given system, perhaps through the choice of specific input signals or by combining the system with other systems. Illustrative of this kind of application is the control of chemical plants, a general area typically referred to as process control. In this class of applications, sensors might typically measure physical signals, such as temperature, humidity, chemical ratios, and so on, and on the basis of these measurement signals, a regulating system would generate control signals to regulate the ongoing chemical process. A second example is related to the fact that some very high performance aircraft represent inherently unstable physical systems, in other words, their aerodynamic characteristics are such that in the absence of carefully designed control signals, they would be unflyable. In both this case and in the previous example of process control, an important concept, referred to as feedback, plays a major role, and this concept is one of the important topics treated in this text.

The examples described above are only a few of an extraordinarily wide variety of applications for the concepts of signals and systems. The importance of these concepts stems not only from the diversity of phenomena and processes in which they arise, but also from the collection of ideas, analytical techniques, and methodologies that have been and are being developed and used to solve problems involving signals and systems. The history of this development extends back over many centuries, and although most of this work was motivated by specific problems, many of these ideas have proven to be of central importance to problems in a far larger variety of applications than those for which they were originally intended. For example, the tools of Fourier analysis, which form the basis for the frequency-domain analysis of signals and systems, and which we will develop in some detail in this book, can be traced from problems of astronomy studied by the ancient Babylonians to the development of mathematical physics in the eighteenth and nineteenth centuries. More recently, these concepts and techniques have been applied to problems ranging from the design of AM and FM transmitters and receivers to the computer-aided restoration of images. From work on problems such as these has emerged a framework and some extremely powerful mathematical tools for the representation, analysis, and synthesis of signals and systems.

In some of the examples that we have mentioned, the signals vary continuously in time, whereas in others, their evolution is described only at discrete points in time. For example, in the restoration of old recordings we are concerned with audio signals that vary continuously. On the other hand, the daily closing stock market average is by its very nature a signal that evolves at discrete points in time (i.e., at the close of each day). Rather than a curve as a function of a continuous variable, then, the closing stock average is a sequence of numbers associated with the discrete time instants at which it is specified. This distinction in the basic description of the evolution of signals and of the systems that respond to or process these signals leads naturally to two parallel frameworks for signal and system analysis, one for phenomena and processes that are described in *continuous time* and one for those that are described in *discrete time*. The concepts and techniques associated both with continuous-time signals and systems and with discrete-time signals and systems have a rich history and are conceptually closely related. Historically, however, because their applications have in the past been sufficiently different, they have for the most part been studied and developed somewhat separately. Continuous-time signals and systems have very strong roots in problems associated with physics and, in the more recent past, with electrical circuits and communications. The techniques of discrete-time signals and systems have strong roots in numerical analysis, statistics, and time-series analysis associated with such applications as the analysis of economic and demographic data. Over the past several decades the disciplines of continuous-time and discrete-time signals and systems have become increasingly entwined and the applications have become highly interrelated. A strong motivation for this interrelationship has been the dramatic advances in technology for the implementation of systems and for the generation of signals. Specifically, the incredibly rapid development of high-speed digital computers, integrated circuits, and sophisticated high-density device fabrication techniques has made it increasingly advantageous to consider processing continuous-time signals by representing them by equally spaced time samples (i.e., by converting them to discrete-time signals). As we develop in detail in Chapter 8, it is a remarkable fact that under relatively mild restrictions a continuous-time signal can be represented totally by such a set of samples.

Because of the growing interrelationship between continuous-time signals and systems and discrete-time signals and systems and because of the close relationship among the concepts and techniques associated with each, we have chosen in this text to develop the concepts of continuous-time and discrete-time signals and systems in parallel. Because many of the concepts are similar (but not identical), by treating them in parallel, insight and intuition can be shared and both the similarities and differences become better focused. Furthermore, as will be evident as we proceed through the material, there are some concepts that are inherently easier to understand in one framework than the other and, once understood, the insight is easily transferable.

As we have so far described them, the notions of signals and systems are extremely general concepts. At this level of generality, however, only the most sweeping statements can be made about the nature of signals and systems, and their properties can be discussed only in the most elementary terms. On the other hand, an important and fundamental notion in dealing with signals and systems is that by carefully choosing subclasses of each with particular properties that can then be exploited, we can analyze and characterize these signals and systems in great depth. The principal focus in this book is a particular class of systems which we will refer to as linear time-invariant systems. The properties of linearity and time invariance that define this class lead to a remarkable set of concepts and techniques which are not only of major practical importance, but also intellectually satisfying.

As we have indicated in this introduction, signal and system analysis has a long history out of which have emerged some basic techniques and fundamental principles which have extremely broad areas of application. Also, as exemplified by the continuing development of integrated-circuit technology and its applications, signal and system analysis is constantly evolving and developing in response to new problems, techniques, and opportunities. We fully expect this development to accelerate in pace as improved technology makes possible the implementation of increasingly complex systems and signal processing techniques. In the future we will see the tools and concepts of signal and system analysis applied to an expanding scope of applications. In some of these areas the techniques of signal and system analysis are proving to have direct and immediate application, whereas in other fields that extend far beyond those that are classically considered to be within the domain of science and engineering, it is the set of ideas embodied in these techniques more than the specific techniques themselves that are proving to be of value in approaching and analyzing complex problems. For these reasons, we feel that the topic of signal and system analysis represents a body of knowledge that is of essential concern to the scientist and engineer. We have chosen the set of topics presented in this book, the organization of the presentation, and the problems in each chapter in a way that we feel will most help the reader to obtain a solid foundation in the fundamentals of signal and system analysis; to gain an understanding of some of the very important and basic applications of these fundamentals to problems in filtering, modulation, sampling, and feedback system analysis; and to develop some perspective into an extremely powerful approach to formulating and solving problems as well as some appreciation of the wide variety of actual and potential applications of this approach.

2.0 INTRODUCTION

To develop the techniques of signal and system analysis it is necessary to establish an analytical framework that captures the intuitive notions of signals and systems given in Chapter 1. In this chapter we begin our development of this framework by introducing the mathematical description and representation of signals and systems and by using these representations to develop some of the basic concepts behind signal and system analysis. In this way we will gain some insight and intuition into the properties of signals and systems and our representations of them.

2.1 SIGNALS

As we discussed in Chapter 1, signals may describe an extremely wide variety of physical phenomena. Although signals can be represented in many ways, in all cases the information in a signal is contained in a pattern of variations of some form. For example, the human vocal mechanism produces speech by creating fluctuations in acoustic pressure. Figure 2.1 is an illustration of a recording of such a speech signal, obtained by using a microphone to sense variations in acoustic pressure which are then converted into an electrical signal. As can be seen in the figure, different sounds correspond to different patterns in the variations of acoustic pressure, and the human vocal system produces intelligible speech by generating particular sequences of these patterns. As another example, consider the monochromatic picture, shown in Figure 2.2. In this case it is the pattern of variations in brightness that is important.

Signals and Systems

2

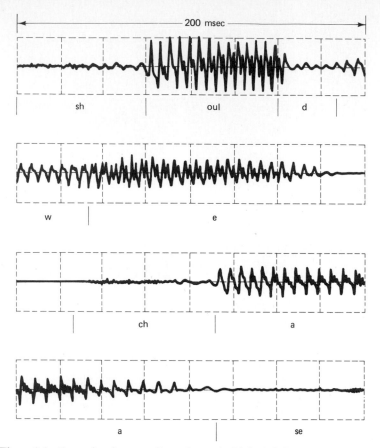

Figure 2.1 Example of a recording of speech. [Adapted from *Applications of Digital Signal Processing*, A. V. Oppenheim, ed. (Englewood Cliffs, N.J.: Prentice-Hall, Inc., 1978), p. 121.] The signal represents acoustic pressure variations as a function of time for the spoken words "should we chase." The top line of the figure corresponds to the word "should," the second line to the word "we," and the last two to the word "chase" (we have indicated the approximate beginnings and endings of each successive sound in each word).

Signals are represented mathematically as functions of one or more independent variables. For example, a speech signal would be represented mathematically by acoustic pressure as a function of time, and a picture is represented as a brightness function of two spatial variables. In this book we focus attention on signals involving a single independent variable. For convenience we will generally refer to the independent variable as time, although it may not in fact represent time in specific applications. For example, signals representing variations with depth of physical quantities such as density, porosity, and electrical resistivity are used in geophysics to study the structure of the earth. Also, knowledge of the variations of air pressure, temperature, and wind speed with altitude are extremely important in meteorological investigations. Figure 2.3 depicts a typical example of annual average vertical wind profile as a function of height. The measured variations of wind speed with height are used in examining

Figure 2.2 A monochromatic picture.

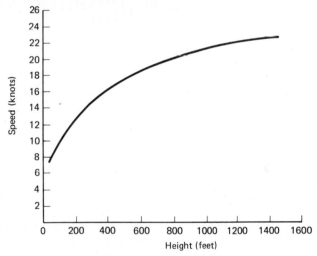

Figure 2.3 Typical annual average vertical wind profile. (Adapted from Crawford and Hudson, National Severe Storms Laboratory Report, ESSA ERLTM-NSSL 48, August 1970.)

weather patterns as well as wind conditions that may affect an aircraft during final approach and landing.

In Chapter 1 we indicated that there are two basic types of signals, continuous-time signals and discrete-time signals. In the case of continuous-time signals the independent variable is continuous, and thus these signals are defined for a continuum of values of the independent variable. On the other hand, discrete-time signals are only defined at discrete times, and consequently for these signals the independent variable takes on only a discrete set of values. A speech signal as a function of time and atmospheric pressure as a function of altitude are examples of continuous-time signals. The weekly Dow Jones stock market index is an example of a discrete-time signal and is illustrated in Figure 2.4. Other examples of discrete-time signals can be found in demographic studies of population in which various attributes, such as average income, crime rate, or pounds of fish caught, are tabulated versus such discrete variables as years of schooling, total population, or type of fishing vessel, respectively. In Figure 2.5 we have illustrated another discrete-time signal, which in this case is an example of the type of species-abundance relation used in ecological studies. Here the independent variable is the number of individuals corresponding to any particular species, and the dependent variable is the number of species in the ecological community under investigation that have a particular number of individuals.

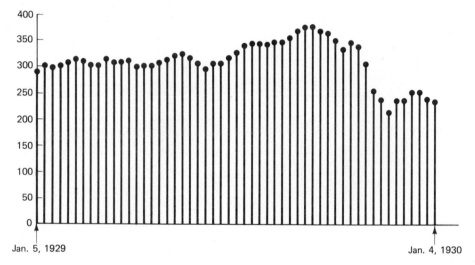

Figure 2.4 An example of a discrete-time signal: the weekly Dow-Jones stock market index from January 5, 1929 to January 4, 1930.

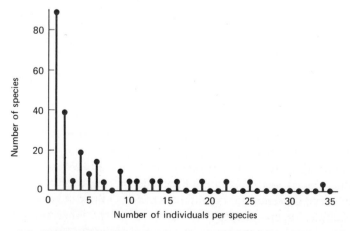

Figure 2.5 Signal representing the species-abundance relation of an ecological community. [Adapted from E. C. Pielou, *An Introduction to Mathematical Ecology* (New York: Wiley, 1969).]

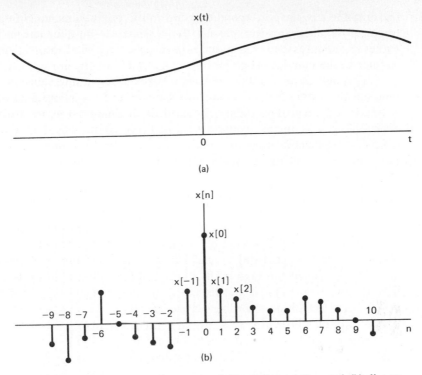

Figure 2.6 Graphical representations of (a) continuous-time and (b) discrete-time signals.

The nature of the signal shown in Figure 2.5 is quite typical in that there are several abundant species and many rare ones with only a few representatives.

To distinguish between continuous-time and discrete-time signals we will use the symbol t to denote the continuous-time variable and n for the discrete-time variable. In addition, for continuous-time signals we will enclose the independent variable in parentheses (·), whereas for discrete-time signals we will use brackets [·] to enclose the independent variable. We will also have frequent occasions when it will be useful to represent signals graphically. Illustrations of a continuous-time signal $x(t)$ and of a discrete-time signal $x[n]$ are shown in Figure 2.6. It is important to note that the discrete-time signal $x[n]$ is defined *only* for integer values of the independent variable. Our choice of graphical representation for $x[n]$ emphasizes this fact, and for further emphasis we will on occasion refer to $x[n]$ as a discrete-time *sequence*.

A discrete-time signal $x[n]$ may represent a phenomenon for which the independent variable is inherently discrete. Signals such as species-abundance relations or demographic data such as those mentioned previously are examples of this. On the other hand, a discrete-time signal $x[n]$ may represent successive *samples* of an underlying phenomenon for which the independent variable is continuous. For example, the processing of speech on a digital computer requires the use of a discrete-time sequence representing the values of the continuous-time speech signal at discrete points in time. Also, pictures in newspapers, or in this book for that matter, actually consist of a very fine grid of points, and each of these points represents a sample of the brightness of the corresponding point in the original image. No matter what the origin of the data, however, the signal $x[n]$ is defined only for integer values of n. It makes no more sense to refer to the $3\frac{1}{2}$th sample of a digital speech signal than it does to refer to the number of species having $4\frac{1}{3}$ representatives.

Throughout most of this book we will treat discrete-time signals and continuous-time signals separately but in parallel so that we can draw on insights developed in one setting to aid our understanding of the other. In Chapter 8 we return to the question of sampling, and in that context we will bring continuous-time and discrete-time concepts together in order to examine the relationship between a continuous-time signal and a discrete-time signal obtained from it by sampling.

2.2 TRANSFORMATIONS OF THE INDEPENDENT VARIABLE

In many situations it is important to consider signals related by a modification of the independent variable. For example, as illustrated in Figure 2.7, the signal $x[-n]$ is obtained from the signal $x[n]$ by a reflection about $n = 0$ (i.e. by reversing the signal). Similarly, as depicted in Figure 2.8, $x(-t)$ is obtained from the signal $x(t)$ by a reflection about $t = 0$. Thus, if $x(t)$ represents an audio signal on a tape recorder, then $x(-t)$ is the same tape recording played backward. As a second example, in Figure 2.9 we have illustrated three signals, $x(t)$, $x(2t)$, and $x(t/2)$, that are related by linear scale changes in the independent variable. If we again think of the example of $x(t)$ as a tape recording, then $x(2t)$ is that recording played at twice the speed, and $x(t/2)$ is the recording played at half-speed.

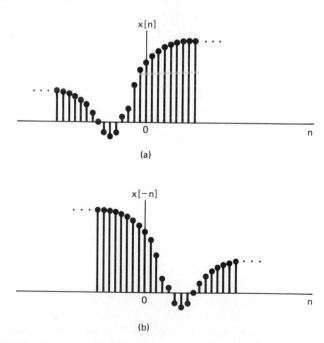

(a)

(b)

Figure 2.7 (a) A discrete-time signal $x[n]$; (b) its reflection, $x[-n]$, about $n = 0$.

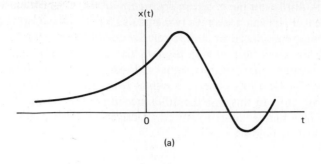

(a)

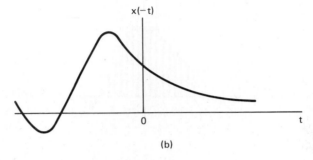

(b)

Figure 2.8 (a) A continuous-time signal $x(t)$; (b) its reflection, $x(-t)$, about $t = 0$.

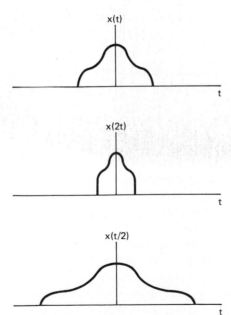

Figure 2.9 Continuous-time signals related by time scaling.

A third example of a transformation of the independent variable is illustrated in Figure 2.10, in which we have two signals $x[n]$ and $x[n - n_0]$ that are identical in shape but that are displaced or shifted relative to each other. Similarly, $x(t - t_0)$ represents a time-shifted version of $x(t)$. Signals that are related in this fashion arise in applications such as sonar, seismic signal processing, and radar, in which several receivers at different locations observe a signal being transmitted through a medium (water, rock, air, etc.). In this case the difference in propagation time from the point of origin of the transmitted signal to any two receivers results in a time shift between the signals measured by the two receivers.

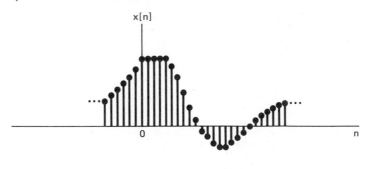

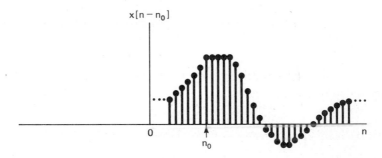

Figure 2.10 Discrete-time signals related by a time shift.

In addition to their use in representing physical phenomena such as the time shift in a sonar signal and the reversal of an audio tape, transformations of the independent variable are extremely useful in examining some of the important properties that signals may possess. In the remainder of this section we discuss these properties, and later in this chapter and in Chapter 3 we use transformations of the independent variable as we analyze the properties of systems.

A signal $x(t)$ or $x[n]$ is referred to as an *even* signal if it is identical with its reflection about the origin, that is, in continuous time if

$$x(-t) = x(t) \tag{2.1a}$$

or in discrete time if

$$x[-n] = x[n] \tag{2.1b}$$

A signal is referred to as *odd* if

$$x(-t) = -x(t) \tag{2.2a}$$

$$x[-n] = -x[n] \tag{2.2b}$$

Note that an odd signal must necessarily be 0 at $t = 0$ or $n = 0$. Examples of even and odd continuous-time signals are shown in Figure 2.11.

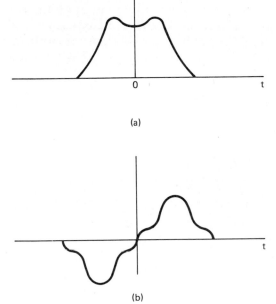

(a)

(b)

Figure 2.11 (a) An even continuous-time signal; (b) an odd continuous-time signal.

An important fact is that any signal can be broken into a sum of two signals, one of which is even and one of which is odd. To see this, consider the signal

$$\mathcal{E}v\{x(t)\} = \tfrac{1}{2}[x(t) + x(-t)] \tag{2.3}$$

which is referred to as the *even part* of $x(t)$. Similarly, the *odd part* of $x(t)$ is given by

$$\mathcal{O}d\{x(t)\} = \tfrac{1}{2}[x(t) - x(-t)] \tag{2.4}$$

It is a simple exercise to check that the even part is in fact even, that the odd part is odd, and that $x(t)$ is the sum of the two. Exactly analogous definitions hold in the discrete-time case, and an example of the even–odd decomposition of a discrete-time signal is given in Figure 2.12.

Throughout our discussion of signals and systems we will have occasion to refer to *periodic* signals, both in continuous time and in discrete time. A periodic continuous-time signal $x(t)$ has the property that there is a positive value of T for which

$$x(t) = x(t + T) \qquad \text{for all } t \tag{2.5}$$

In this case we say that $x(t)$ is *periodic with period* T. An example of such a signal is given in Figure 2.13. From the figure or from eq. (2.5) we can readily deduce that if $x(t)$ is periodic with period T, then $x(t) = x(t + mT)$ for all t and for any integer m. Thus, $x(t)$ is also periodic with period $2T, 3T, 4T, \ldots$. The *fundamental period* T_0 of $x(t)$ is the smallest positive value of T for which eq. (2.5) holds. Note that this definition of the fundamental period works except if $x(t)$ is a constant. In this case the fundamental period is undefined since $x(t)$ is periodic for *any* choice of T (so

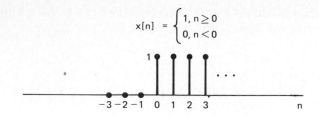

$$x[n] = \begin{cases} 1, n \geq 0 \\ 0, n < 0 \end{cases}$$

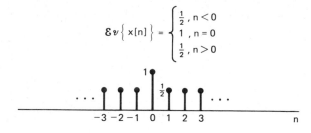

$$\mathcal{E}v\{x[n]\} = \begin{cases} \frac{1}{2}, n < 0 \\ 1, n = 0 \\ \frac{1}{2}, n > 0 \end{cases}$$

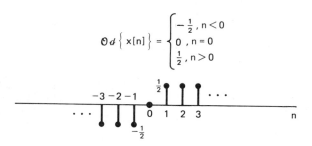

$$\mathcal{O}d\{x[n]\} = \begin{cases} -\frac{1}{2}, n < 0 \\ 0, n = 0 \\ \frac{1}{2}, n > 0 \end{cases}$$

Figure 2.12 The even–odd decomposition of a discrete-time signal.

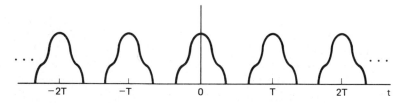

Figure 2.13 Continuous-time periodic signal.

there is no smallest positive value). Finally, a signal $x(t)$ that is not periodic will be referred to as an *aperiodic* signal.

Periodic signals are defined analogously in discrete time. Specifically, a discrete-time signal $x[n]$ is periodic with period N, where N is a positive integer, if

$$x[n] = x[n + N] \qquad \text{for all } n \qquad (2.6)$$

If eq. (2.6) holds, then $x[n]$ is also periodic with period $2N, 3N, \ldots$, and the *fundamental period* N_0 is the smallest positive value of N for which eq. (2.6) holds.

2.3 BASIC CONTINUOUS-TIME SIGNALS

In this section we introduce several particularly important continuous-time signals. Not only do these signals occur frequently in nature, but they also serve as basic building blocks from which we can construct many other signals. In this and subsequent chapters we will find that constructing signals in this way will allow us to examine and understand more deeply the properties of both signals and systems.

2.3.1 Continuous-Time Complex Exponential and Sinusoidal Signals

The continuous-time *complex exponential signal* is of the form

$$x(t) = Ce^{at} \tag{2.7}$$

where C and a are, in general, complex numbers. Depending upon the values of these parameters, the complex exponential can take on several different characteristics. As illustrated in Figure 2.14, if C and a are real [in which case $x(t)$ is called a *real exponential*], there are basically two types of behavior. If a is positive, then as t increases $x(t)$ is a growing exponential, a form that is used in describing a wide variety of phe-

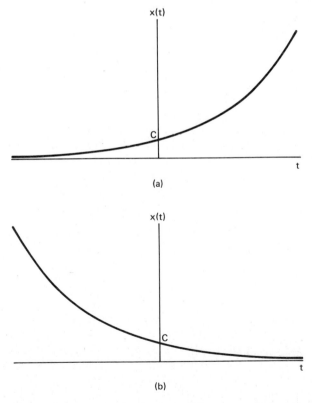

Figure 2.14 Continuous-time real exponential $x(t) = Ce^{at}$: (a) $a > 0$; (b) $a < 0$.

nomena, including chain reactions in atomic explosions or complex chemical reactions and the uninhibited growth of populations such as in bacterial cultures. If a is negative, then $x(t)$ is a decaying exponential. Such signals also find wide use in describing radioactive decay, the responses of RC circuits and damped mechanical systems, and many other physical processes. Finally, we note that for $a = 0$, $x(t)$ is constant.

A second important class of complex exponentials is obtained by constraining a to be purely imaginary. Specifically, consider

$$x(t) = e^{j\omega_0 t} \tag{2.8}$$

An important property of this signal is that it is periodic. To verify this, we recall from eq. (2.5) that $x(t)$ will be periodic with period T if

$$e^{j\omega_0 t} = e^{j\omega_0(t+T)} \tag{2.9}$$

or, since

$$e^{j\omega_0(t+T)} = e^{j\omega_0 t} e^{j\omega_0 T}$$

we must have that

$$e^{j\omega_0 T} = 1 \tag{2.10}$$

If $\omega_0 = 0$, then $x(t) = 1$, which is periodic for any value of T. If $\omega_0 \neq 0$, then the fundamental period T_0 of $x(t)$, that is, the smallest positive value of T for which eq. (2.10) holds, is given by

$$T_0 = \frac{2\pi}{|\omega_0|} \tag{2.11}$$

Thus, the signals $e^{j\omega_0 t}$ and $e^{-j\omega_0 t}$ both have the same fundamental period.

A signal closely related to the periodic complex exponential is the *sinusoidal signal*

$$x(t) = A \cos(\omega_0 t + \phi) \tag{2.12}$$

as shown in Figure 2.15. With the units of t as seconds, the units of ϕ and ω_0 are radians and radians per second, respectively. It is also common to write $\omega_0 = 2\pi f_0$, where f_0 has the units of cycles per second or Hertz (Hz). The sinusoidal signal is also periodic with fundamental period T_0 given by eq. (2.11). Sinusoidal and periodic complex

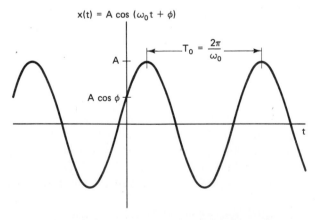

Figure 2.15 Continuous-time sinusoidal signal.

exponential signals are also used to describe the characteristics of many physical processes. The response of an *LC* circuit is sinusoidal, as is the simple harmonic motion of a mechanical system consisting of a mass connected by a spring to a stationary support. The acoustic pressure variations corresponding to a single musical note are also sinusoidal.

By using Euler's relation,† the complex exponential in eq. (2.8) can be written in terms of sinusoidal signals with the same fundamental period:

$$e^{j\omega_0 t} = \cos \omega_0 t + j \sin \omega_0 t \tag{2.13}$$

Similarly, the sinusoidal signal of eq. (2.12) can be written in terms of periodic complex exponentials, again with the same fundamental period:

$$A \cos (\omega_0 t + \phi) = \frac{A}{2} e^{j\phi} e^{j\omega_0 t} + \frac{A}{2} e^{-j\phi} e^{-j\omega_0 t} \tag{2.14}$$

Note that the two exponentials in eq. (2.14) have complex amplitudes. Alternatively, we can express a sinusoid in terms of the complex exponential signal as

$$A \cos (\omega_0 t + \phi) = A \, \mathcal{R}e\{e^{j(\omega_0 t + \phi)}\} \tag{2.15}$$

where if c is a complex number, $\mathcal{R}e\{c\}$ denotes its real part. We will also use the notation $\mathcal{I}m\{c\}$ for the imaginary part of c.

From eq. (2.11) we see that the fundamental period T_0 of a continuous-time sinusoidal signal or a periodic complex exponential is inversely proportional to $|\omega_0|$, which we will refer to as the *fundamental frequency*. From Figure 2.16 we see graphically what this means. If we decrease the magnitude of ω_0, we slow down the rate of oscillation and therefore increase the period. Exactly the opposite effects occur if we increase the magnitude of ω_0. Consider now the case $\omega_0 = 0$. In this case, as we mentioned earlier, $x(t)$ is constant and therefore is periodic with period T for any positive value of T. Thus, the fundamental period of a constant signal is undefined. On the other hand, there is no ambiguity in defining the fundamental frequency of a constant signal to be zero. That is, a constant signal has a zero rate of oscillation.

Periodic complex exponentials will play a central role in a substantial part of our treatment of signals and systems. On several occasions we will find it useful to consider the notion of *harmonically related* complex exponentials, that is, sets of periodic exponentials with fundamental frequencies that are all multiples of a single positive frequency ω_0:

$$\phi_k(t) = e^{jk\omega_0 t}, \qquad k = 0, \pm 1, \pm 2, \ldots \tag{2.16}$$

For $k = 0$, $\phi_k(t)$ is a constant, while for any other value of k, $\phi_k(t)$ is periodic with fundamental period $2\pi/(|k|\omega_0)$ or fundamental frequency $|k|\omega_0$. Since a signal that is periodic with period T is also periodic with period mT for any positive integer m, we see that all of the $\phi_k(t)$ have a common period of $2\pi/\omega_0$. Our use of the term "harmonic" is consistent with its use in music, where it refers to tones resulting from variations in acoustic pressure at frequencies which are harmonically related.

†Euler's relation and other basic ideas related to the manipulation of complex numbers and exponentials are reviewed in the first few problems at the end of the chapter.

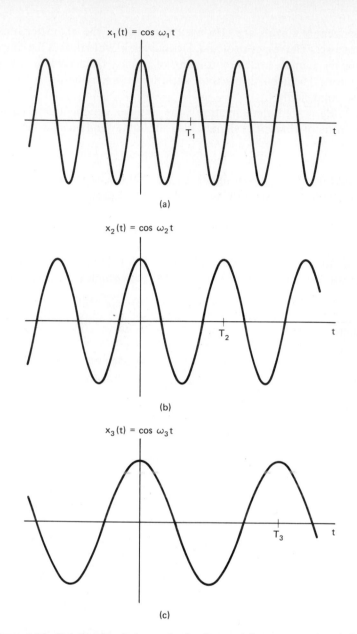

Figure 2.16 Relationship between the fundamental frequency and period for continuous-time sinusoidal signals; here $\omega_1 > \omega_2 > \omega_3$, which implies that $T_1 < T_2 < T_3$.

The most general case of a complex exponential can be expressed and interpreted in terms of the two cases we have examined so far: the real exponential and the periodic complex exponential. Specifically, consider a complex exponential Ce^{at}, where C is expressed in polar form and a in rectangular form. That is,

$$C = |C|e^{j\theta}$$

and
$$a = r + j\omega_0$$

Then
$$Ce^{at} = |C|e^{j\theta}e^{(r+j\omega_0)t} = |C|e^{rt}e^{j(\omega_0 t+\theta)} \qquad (2.17a)$$

Using Euler's relation we can expand this further as
$$Ce^{at} = |C|e^{rt}\cos(\omega_0 t + \theta) + j|C|e^{rt}\sin(\omega_0 t + \theta)$$
$$= |C|e^{rt}\cos(\omega_0 t + \theta) + j|C|e^{rt}\cos\left(\omega_0 t + \theta - \frac{\pi}{2}\right) \qquad (2.17b)$$

Thus, for $r = 0$ the real and imaginary parts of a complex exponential are sinusoidal. For $r > 0$ they correspond to sinusoidal signals multiplied by a growing exponential, and for $r < 0$ they correspond to sinusoidal signals multiplied by a decaying exponential. These two cases are shown in Figure 2.17. The dashed lines in Figure 2.17 correspond to the functions $\pm|C|e^{rt}$. From eq. (2.17a) we see that $|C|e^{rt}$ is the magnitude of the complex exponential. Thus, the dashed curves act as an *envelope* for the oscillatory curve in Figure 2.17 in that the peaks of the oscillations just reach these curves, and in this way the envelope provides us with a convenient way in which to visualize the general trend in the amplitude of the oscillations. Sinusoidal signals multiplied by decaying exponentials are commonly referred to as *damped sinusoids*. Examples of such signals arise in the response of *RLC* circuits and in mechanical systems containing both damping and restoring forces, such as automotive suspension systems.

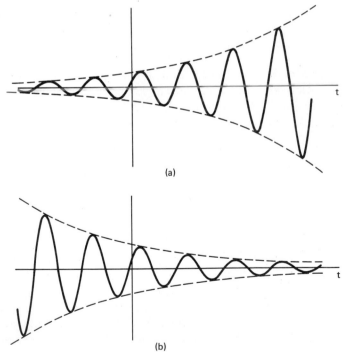

Figure 2.17 (a) Growing sinusoidal signal $x(t) = Ce^{rt}\cos(\omega_0 t + \theta)$, $r > 0$; (b) decaying sinusoid $x(t) = Ce^{rt}\cos(\omega_0 t + \theta)$, $r < 0$.

2.3.2 The Continuous-Time Unit Step
and Unit Impulse Functions

Another basic continuous-time signal is the *unit step function*

$$u(t) = \begin{cases} 0, & t < 0 \\ 1, & t > 0 \end{cases}$$

(2.18)

which is shown in Figure 2.18. Note that it is discontinuous at $t = 0$. As with the

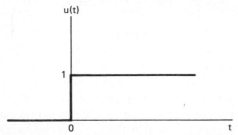

Figure 2.18 Continuous-time unit step function.

complex exponential, the unit step function will be very important in our examination of the properties of systems. Another signal that we will find to be quite useful is the continuous-time *unit impulse function* $\delta(t)$, which is related to the unit step by the equation

$$u(t) = \int_{-\infty}^{t} \delta(\tau) \, d\tau$$

(2.19)

That is, $u(t)$ is the *running integral* of the unit impulse function. This suggests that

$$\delta(t) = \frac{du(t)}{dt}$$

(2.20)

There is obviously some formal difficulty with this as a definition of the unit impulse function since $u(t)$ is discontinuous at $t = 0$ and consequently is formally not differentiable. We can, however, interpret eq. (2.20) by considering $u(t)$ as the limit of a continuous function. Thus, let us define $u_\Delta(t)$ as indicated in Figure 2.19, so that $u(t)$ equals the limit of $u_\Delta(t)$ as $\Delta \longrightarrow 0$, and let us define $\delta_\Delta(t)$ as

$$\delta_\Delta(t) = \frac{du_\Delta(t)}{dt}$$

(2.21)

as shown in Figure 2.20.

We observe that $\delta_\Delta(t)$ has unity area for any value of Δ and is zero outside the interval $0 \leq t \leq \Delta$. As $\Delta \longrightarrow 0$, $\delta_\Delta(t)$ becomes narrower and higher, as it maintains

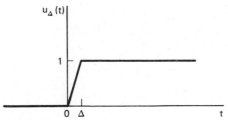

Figure 2.19 Continuous approximation to the unit step.

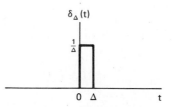

Figure 2.20 Derivative of $u_\Delta(t)$.

its unit area. Its limiting form,

$$\delta(t) = \lim_{\Delta \to 0} \delta_\Delta(t) \tag{2.22}$$

will be depicted as shown in Figure 2.21. More generally, a scaled impulse $k\delta(t)$ will have an area k and thus

$$\int_{-\infty}^{t} k\delta(\tau)\,d\tau = ku(t)$$

A scaled impulse is shown in Figure 2.22. Although the "value" at $t = 0$ is infinite, the height of the arrow used to depict the scaled impulse will be chosen to be representative of its *area*.

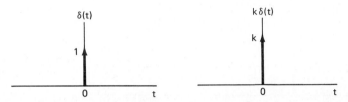

Figure 2.21 Unit impulse.　　　　**Figure 2.22** Scaled impulse.

The graphical interpretation of the running integral of eq. (2.19) is illustrated in Figure 2.23. Since the area of the continuous-time unit impulse $\delta(\tau)$ is concentrated at $\tau = 0$, we see that the running integral is 0 for $t < 0$ and 1 for $t > 0$. Also, we note that the relationship in eq. (2.19) between the continuous-time unit step and impulse can be rewritten in a different form by changing the variable of integration from τ to $\sigma = t - \tau$,

$$u(t) = \int_{-\infty}^{t} \delta(\tau)\,d\tau = \int_{\infty}^{0} \delta(t - \sigma)(-d\sigma)$$

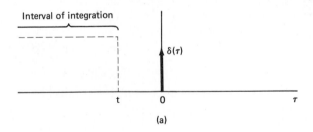

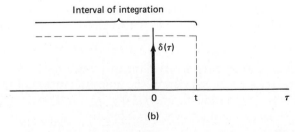

Figure 2.23 Running integral given in eq. (2.19): (a) $t < 0$; (b) $t > 0$.

or, equivalently,

$$u(t) = \int_0^\infty \delta(t - \sigma)\, d\sigma \qquad (2.23)$$

The interpretation of this form of the relationship between $u(t)$ and $\delta(t)$ is given in Figure 2.24. Since in this case the area of $\delta(t - \sigma)$ is concentrated at the point $\sigma = t$, we again see that the integral in eq. (2.23) is 0 for $t < 0$ and 1 for $t > 0$. This type of graphical interpretation of the behavior of the unit impulse under integration will be extremely useful in Chapter 3.

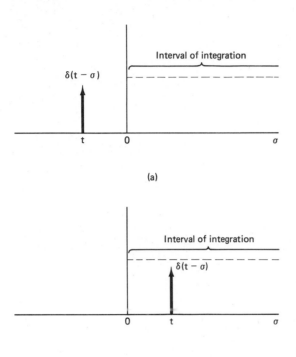

Figure 2.24 Relationship given in eq. (2.23): (a) $t < 0$; (b) $t > 0$.

Although the preceding discussion of the unit impulse is somewhat informal, it is adequate for our present purposes and does provide us with some important intuition into the behavior of this signal. For example, it will be important on occasion to consider the product of an impulse and a more well-behaved continuous-time function. The interpretation of this quantity is most readily developed using the definition of $\delta(t)$ according to eq. (2.22). Thus, let us consider $x_1(t)$ given by

$$x_1(t) = x(t)\, \delta_\Delta(t)$$

In Figure 2.25(a) we have depicted the two time functions $x(t)$ and $\delta_\Delta(t)$, and in Figure 2.25(b) we see an enlarged view of the nonzero portion of their product. By construction, $x_1(t)$ is zero outside the interval $0 \le t \le \Delta$. For Δ sufficiently small so that $x(t)$ is approximately constant over this interval,

$$x(t)\, \delta_\Delta(t) \simeq x(0)\, \delta_\Delta(t)$$

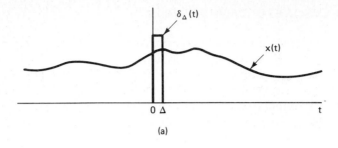

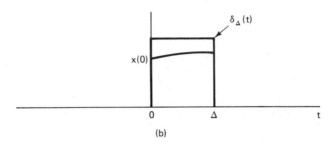

Figure 2.25 The product $x(t)\delta_\Delta(t)$: (a) graphs of both functions; (b) enlarged view of the nonzero portion of their product.

Since $\delta(t)$ is the limit as $\Delta \longrightarrow 0$ of $\delta_\Delta(t)$, it follows that

$$x(t)\,\delta(t) = x(0)\,\delta(t) \tag{2.24}$$

By the same argument we have an analogous expression for an impulse concentrated at an arbitrary point, say, t_0. That is,

$$x(t)\delta(t - t_0) = x(t_0)\delta(t - t_0)$$

In Chapter 3 we provide another interpretation of the unit impulse using some of the concepts that we will develop for our study of systems. The interpretation of $\delta(t)$ that we have given in the present section, combined with this later discussion, will provide us with the insight that we require in order to use the impulse in our study of signals and systems.†

---

†The unit impulse and other related functions (which are often collectively referred to as *singularity functions*) have been thoroughly studied in the field of mathematics under the alternative names of *generalized functions* and the *theory of distributions*. For a discussion of this subject see the book *Distribution Theory and Transform Analysis*, by A. H. Zemanian (New York: McGraw-Hill Book Company, 1965) or the more advanced text *Fourier Analysis and Generalized Functions*, by M. J. Lighthill (New York: Cambridge University Press, 1958). For brief introductions to the subject, see *The Fourier Integral and Its Applications*, by A. Papoulis (New York: McGraw-Hill Book Company, 1962), or *Linear Systems Analysis*, by C. L. Liu and J. W. S. Liu (New York: McGraw-Hill Book Company, 1975). Our discussion of singularity functions in Section 3.7 is closely related in spirit to the mathematical theory described in these texts and thus provides an informal introduction to concepts that underlie this topic in mathematics as well as a discussion of the basic properties of these functions that we will use in our treatment of signals and systems.

2.4 BASIC DISCRETE-TIME SIGNALS

For the discrete-time case, there are also a number of basic signals that play an important role in the analysis of signals and systems. These signals are direct counterparts of the continuous-time signals described in Section 2.3, and, as we will see, many of the characteristics of basic discrete-time signals are directly analogous to properties of basic continuous-time signals. There are, however, several important differences in discrete time, and we will point these out as we examine the properties of these signals.

2.4.1 The Discrete-Time Unit Step and Unit Impulse Sequences

The counterpart of the continuous-time step function is the discrete-time *unit step*, denoted by $u[n]$ and defined by

$$u[n] = \begin{cases} 0, & n < 0 \\ 1, & n \geq 0 \end{cases} \tag{2.25}$$

The unit step sequence is shown in Figure 2.26. As we discussed in Section 2.3, a second

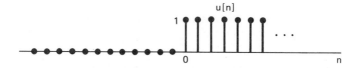

Figure 2.26 Unit step sequence.

very important continuous-time signal is the unit impulse. In discrete time we define the *unit impulse* (or *unit sample*) as

$$\delta[n] = \begin{cases} 0, & n \neq 0 \\ 1, & n = 0 \end{cases} \tag{2.26}$$

which is shown in Figure 2.27. Throughout the book we will refer to $\delta[n]$ interchangeably as the unit sample or unit impulse. Note that unlike its continuous-time counterpart, there are no analytical difficulties in defining $\delta[n]$.

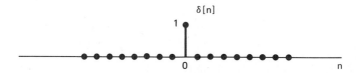

Figure 2.27 Unit sample (impulse).

The discrete-time unit sample possesses many properties that closely parallel the characteristics of the continuous-time unit impulse. For example, since $\delta[n]$ is nonzero (and equal to 1) only for $n = 0$, it is immediately seen that

$$x[n]\,\delta[n] = x[0]\,\delta[n] \tag{2.27}$$

which is the discrete-time counterpart of eq. (2.24). In addition, while the continuous-time impulse is formally the first derivative of the continuous-time unit step, the discrete-time unit impulse is the *first difference* of the discrete-time step

$$\delta[n] = u[n] - u[n - 1] \tag{2.28}$$

Similarly, while the continuous-time unit step is the running integral of $\delta(t)$, the discrete-time unit step is the *running sum* of the unit sample. That is,

$$u[n] = \sum_{m=-\infty}^{n} \delta[m] \tag{2.29}$$

which is illustrated in Figure 2.28. Since the only nonzero value of the unit sample is

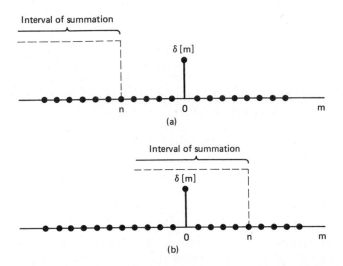

Figure 2.28 Running sum of eq. (2.29): (a) $n < 0$; (b) $n > 0$.

at the point at which its argument is zero, we see from the figure that the running sum in eq. (2.29) is 0 for $n < 0$ and 1 for $n \geq 0$. Also, in analogy with the alternative form of eq. (2.23) for the relationship between the continuous-time unit step and impulse, the discrete-time unit step can also be written in terms of the unit sample as

$$u[n] = \sum_{k=0}^{\infty} \delta[n - k] \tag{2.30}$$

which can be obtained from eq. (2.29) by changing the variable of summation from m to $k = n - m$. Equation (2.30) is illustrated in Figure 2.29, which is the discrete-time counterpart of Figure 2.24.

2.4.2 Discrete-Time Complex Exponential and Sinusoidal Signals

As in continuous time, an important signal in discrete time is the *complex exponential signal* or *sequence*, defined by

$$x[n] = C\alpha^n \tag{2.31}$$

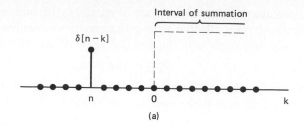

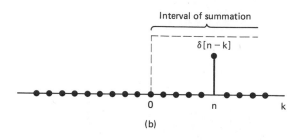

Figure 2.29 Relationship given in eq. (2.30): (a) $n < 0$; (b) $n > 0$.

where C and α are in general complex numbers. This could alternatively be expressed in the form

$$x[n] = Ce^{\beta n} \qquad (2.32)$$

where

$$\alpha = e^{\beta}$$

Although the discrete-time complex exponential sequence in the form of eq. (2.32) is more analogous to the form of the continuous-time complex exponential, it is often more convenient to express the discrete-time complex exponential sequence in the form of eq. (2.31).

If C and α are real, we can have one of several types of behavior, as illustrated in Figure 2.30. Basically if $|\alpha| > 1$, the signal grows exponentially with n, while if $|\alpha| < 1$, we have a decaying exponential. Furthermore, if α is positive, all the values of $C\alpha^n$ are of the same sign, but if α is negative, then the sign of $x[n]$ alternates. Note also that if $\alpha = 1$, then $x[n]$ is a constant, whereas if $\alpha = -1$, $x[n]$ alternates in value between $+C$ and $-C$. Real discrete-time exponentials are often used to describe population growth as a function of generation and return on investment as a function of day, month, or quarter.

Another important complex exponential is obtained by using the form given in eq. (2.32) and by constraining β to be purely imaginary. Specifically, consider

$$x[n] = e^{j\Omega_0 n} \qquad (2.33)$$

As in the continuous-time case, this signal is closely related to the sinusoidal signal

$$x[n] = A \cos(\Omega_0 n + \phi) \qquad (2.34)$$

If we take n to be dimensionless, then both Ω_0 and ϕ have units of radians. Three examples of sinusoidal sequences are shown in Figure 2.31. As before, Euler's relation allows us to relate complex exponentials and sinusoids:

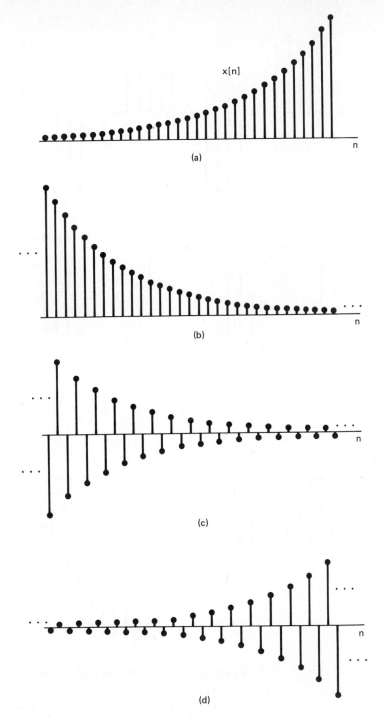

Figure 2.30 $x[n] = C\alpha^n$: (a) $\alpha > 1$; (b) $0 < \alpha < 1$; (c) $-1 < \alpha < 0$; (d) $\alpha < -1$.

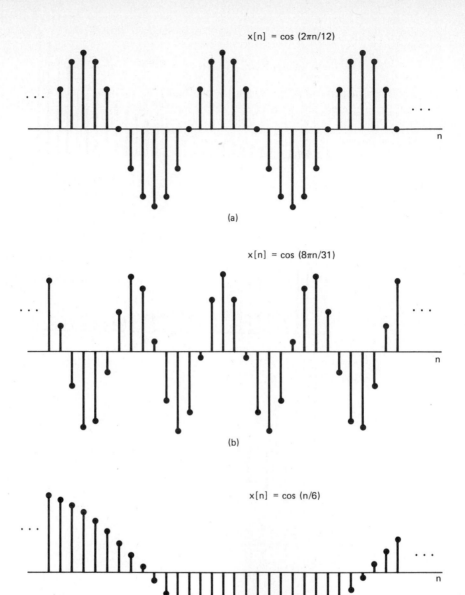

Figure 2.31 Discrete-time sinusoidal signals.

$$e^{j\Omega_0 n} = \cos \Omega_0 n + j \sin \Omega_0 n \qquad (2.35)$$

and

$$A \cos (\Omega_0 n + \phi) = \frac{A}{2} e^{j\phi} e^{j\Omega_0 n} + \frac{A}{2} e^{-j\phi} e^{-j\Omega_0 n} \qquad (2.36)$$

Similarly, a general complex exponential can be written and interpreted in terms of real exponentials and sinusoidal signals. Specifically, if we write C and α in polar form

$$C = |C| e^{j\theta}$$

$$\alpha = |\alpha| e^{j\Omega_0}$$

then

$$C\alpha^n = |C| |\alpha|^n \cos (\Omega_0 n + \theta) + j|C| |\alpha|^n \sin (\Omega_0 n + \theta) \qquad (2.37)$$

Thus for $|\alpha| = 1$, the real and imaginary parts of a complex exponential sequence are sinusoidal. For $|\alpha| < 1$, they correspond to sinusoidal sequences multiplied by a decaying exponential, and for $|\alpha| > 1$, they correspond to sinusoidal sequences multiplied by a growing exponential. Examples of these signals are depicted in Figure 2.32.

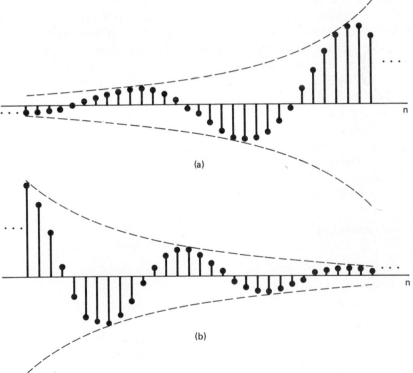

(a)

(b)

Figure 2.32 (a) Growing discrete-time sinusoidal signal; (b) decaying discrete-time sinusoid.

2.4.3 Periodicity Properties of Discrete-Time Complex Exponentials

Let us now continue our examination of the signal $e^{j\Omega_0 n}$. Recall first the following two properties of its continuous-time counterpart $e^{j\omega_0 t}$: (1) the larger the magnitude of ω_0, the higher the rate of oscillation in the signal; and (2) $e^{j\omega_0 t}$ is periodic for any value of ω_0. In this section we describe the discrete-time versions of both of these properties, and as we will see, there are definite differences between each of these and its continuous-time counterpart.

The fact that the discrete-time version of the first property is different from the continuous-time property is a direct consequence of another extremely important distinction between discrete-time and continuous-time complex exponentials. To see what this difference is, consider the complex exponential with frequency $(\Omega_0 + 2\pi)$:

$$e^{j(\Omega_0 + 2\pi)n} = e^{j2\pi n}e^{j\Omega_0 n} = e^{j\Omega_0 n} \tag{2.38}$$

From eq. (2.38) we see that the exponential at frequency $(\Omega_0 + 2\pi)$ is the *same* as that at frequency Ω_0. Thus, we have a very different situation from the continuous-time case, in which the signals $e^{j\omega_0 t}$ are all distinct for distinct values of ω_0. In discrete time, these signals are not distinct, as the signal with frequency Ω_0 is identical to the signals with frequencies $(\Omega_0 \pm 2\pi)$, $(\Omega_0 \pm 4\pi)$, and so on. Therefore, in considering discrete-time exponentials, we need only consider an interval of length 2π in which to choose Ω_0. Although, according to eq. (2.38), any 2π interval will do, on most occasions we will use the interval $0 \leq \Omega_0 < 2\pi$ or the interval $-\pi \leq \Omega_0 < \pi$.

Because of the periodicity implied by eq. (2.38), the signal $e^{j\Omega_0 n}$ does *not* have a continually increasing rate of oscillation as Ω_0 is increased in magnitude. Rather, as we increase Ω_0 from 0, we obtain signals with increasing rates of oscillation until we reach $\Omega_0 = \pi$. Then, however, as we continue to increase Ω_0, we *decrease* the rate of oscillation until we reach $\Omega_0 = 2\pi$, which is the same as $\Omega_0 = 0$. We have illustrated this point in Figure 2.33. Therefore, the low-frequency (that is, slowly varying) discrete-time exponentials have values of Ω_0 near 0, 2π, or any other even multiple of π, while the high frequencies (corresponding to rapid variations) are located near $\Omega_0 = \pm\pi$ and other odd multiples of π.

The second property we wish to consider concerns the periodicity of the discrete-time complex exponential. In order for the signal $e^{j\Omega_0 n}$ to be periodic with period $N > 0$ we must have that

$$e^{j\Omega_0(n+N)} = e^{j\Omega_0 n} \tag{2.39}$$

or, equivalently,

$$e^{j\Omega_0 N} = 1 \tag{2.40}$$

For eq. (2.40) to hold, $\Omega_0 N$ must be a multiple of 2π. That is, there must be an integer m so that

$$\Omega_0 N = 2\pi m \tag{2.41}$$

or, equivalently,

$$\frac{\Omega_0}{2\pi} = \frac{m}{N} \tag{2.42}$$

According to eq. (2.42), the signal $e^{j\Omega_0 n}$ is *not* periodic for arbitrary values of Ω_0. It is periodic only if $\Omega_0/2\pi$ is a rational number, as in eq. (2.42). Clearly, these same observations also hold for discrete-time sinusoidal signals. For example, the sequence in Figure 2.31(a) is periodic with period 12, the signal in Figure 2.31(b) is periodic with period 31, and the signal in Figure 2.31(c) is not periodic.

Using the calculations that we have just made, we can now examine the fundamental period and frequency of discrete-time complex exponentials, where we define the fundamental frequency of a discrete-time periodic signal as we did in continuous

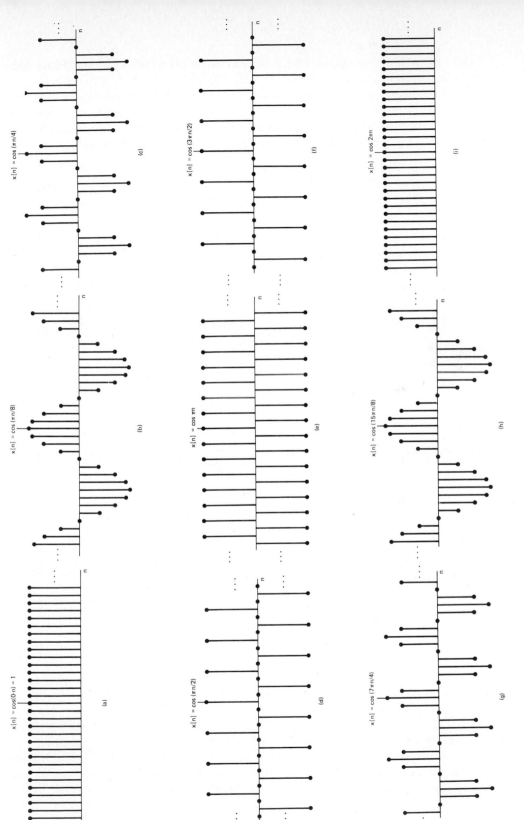

Figure 2.33 Discrete-time sinusoidal sequences for several different frequencies.

time. That is, if $x[n]$ is periodic with fundamental period N, its fundamental frequency is $2\pi/N$. Consider, then, a periodic complex exponential $x[n] = e^{j\Omega_0 n}$ with $\Omega_0 \neq 0$. As we have just seen, Ω_0 must satisfy eq. (2.42) for some pair of integers m and N, with $N > 0$. In Problem 2.17 it is shown that if $\Omega_0 \neq 0$, and if N and m have no factors in common, then the fundamental period of $x[n]$ is N. Assuming that this is the case and using eq. (2.42), we find that the fundamental frequency of the periodic signal $e^{j\Omega_0 n}$ is

$$\frac{2\pi}{N} = \frac{\Omega_0}{m} \tag{2.43}$$

Note that the fundamental period can also be written as

$$N = m\left(\frac{2\pi}{\Omega_0}\right) \tag{2.44}$$

These last two expressions again differ from their continuous-time counterparts as can be seen in Table 2.1 in which we have summarized some of the differences between the continuous-time signal $e^{j\omega_0 t}$ and the discrete-time signal $e^{j\Omega_0 n}$. Note that as in the continuous-time case, the constant discrete-time signal resulting from setting $\Omega_0 = 0$ has a fundamental frequency of 0 and its fundamental period is undefined. For more discussion of the properties of periodic discrete-time exponentials, see Problems 2.17 and 2.18.

TABLE 2.1 DIFFERENCES BETWEEN THE SIGNALS $e^{j\omega_0 t}$ AND $e^{j\Omega_0 n}$.

$e^{j\omega_0 t}$	$e^{j\Omega_0 n}$
Distinct signals for distinct values of ω_0	Identical signals for exponentials at frequencies separated by 2π
Periodic for any choice of ω_0	Periodic only if $$\Omega_0 = \frac{2\pi m}{N}$$ for some integers $N > 0$ and m.
Fundamental frequency ω_0	Fundamental frequency† $$\frac{\Omega_0}{m}$$
Fundamental period $\omega_0 = 0$: undefined $\omega_0 \neq 0$: $\dfrac{2\pi}{\omega_0}$	Fundamental period† $\Omega_0 = 0$: undefined $\Omega_0 \neq 0$: $m\left(\dfrac{2\pi}{\Omega_0}\right)$

†These statements assume that m and N do not have any factors in common.

As in continuous time, we will find it useful on occasion to consider sets of harmonically related periodic exponentials, that is, periodic exponentials that are all periodic with period N. From eq. (2.42) we know that these are precisely the signals that are at frequencies that are multiples of $2\pi/N$. That is,

$$\phi_k[n] = e^{jk(2\pi/N)n}, \qquad k = 0, \pm 1, \ldots \qquad (2.45)$$

In the continuous-time case all of the harmonically related complex exponentials, $e^{jk(2\pi/T)t}$, $k = 0, \pm 1, \pm 2, \ldots$ are distinct. However, because of eq. (2.38), this is *not* the case in discrete time. Specifically,

$$\phi_{k+N}[n] = e^{j(k+N)(2\pi/N)n}$$
$$= e^{j2\pi n}e^{jk(2\pi/N)n} = \phi_k[n] \qquad (2.46)$$

This implies that there are only N distinct periodic exponentials in the set given in eq. (2.45). For example, $\phi_0[n], \phi_1[n], \ldots, \phi_{N-1}[n]$ are all distinct, and any other $\phi_k[n]$ is identical to one of these (e.g., $\phi_N[n] = \phi_0[n]$ and $\phi_{-1}[n] = \phi_{N-1}[n]$).

Finally, in order to gain some additional insight into the issue of periodicity for discrete-time complex exponentials, consider a discrete-time sequence obtained by taking samples of a continuous-time exponential, $e^{j\omega_0 t}$ at equally spaced points in time:

$$x[n] = e^{j\omega_0 nT} = e^{j(\omega_0 T)n} \qquad (2.47)$$

From eq. (2.47) we see that $x[n]$ is itself a discrete-time exponential with $\Omega_0 = \omega_0 T$. Therefore, according to our preceding analysis, $x[n]$ will be periodic only if $\omega_0 T/2\pi$ is a rational number. Identical statements can be made for discrete-time sequences obtained by taking equally spaced samples of continuous-time periodic sinusoidal signals. For example, if

$$x(t) = \cos 2\pi t \qquad (2.48)$$

then the three discrete-time signals in Figure 2.31 can be thought of as being defined by

$$x[n] = x(nT) = \cos 2\pi nT \qquad (2.49)$$

for different choices of T. Specifically, $T = \frac{1}{12}$ for Figure 2.31(a), $T = \frac{4}{31}$ for Figure 2.31(b), and $T = \frac{1}{12}\pi$ for Figure 2.31(c). If we think of discrete-time sinusoidal sequences as being obtained as in eq. (2.47), then we see that although the sequence $x[n]$ may not be periodic, its envelope $x(t)$ *is* periodic. This can be directly seen in Figure 2.31(c), where the eye provides the visual interpolation between the discrete sequence values to produce the continuous-time periodic envelope. The use of the concept of sampling to gain insight into the periodicity of discrete-time sinusoidal sequences is explored further in Problem 2.18.

2.5 SYSTEMS

A *system* can be viewed as any process that results in the transformation of signals. Thus, a system has an input signal and an output signal which is related to the input through the system transformation. For example, a high-fidelity system takes a recorded audio signal and generates a reproduction of that signal. If the hi-fi system has tone controls, we can change the characteristics of the system, that is, the tonal quality of the reproduced signal, by adjusting the controls. An automobile can also be viewed as a system in which the input is the depression of the accelerator pedal and the output is the motion of the vehicle. An image-enhancement system transforms an input image into an output image which has some desired properties, such as improved contrast.

As we have stated earlier, we will be interested in both continuous-time and discrete-time systems. A *continuous-time system* is one in which continuous-time input signals are transformed into continuous-time output signals. Such a system will be represented pictorially as in Figure 2.34(a), where $x(t)$ is the input and $y(t)$ is the output. Alternatively, we will represent the input–output relation of a continuous-time system by the notation

$$x(t) \longrightarrow y(t) \tag{2.50}$$

Similarly, a *discrete-time system*, that is, one that transforms discrete-time inputs into discrete-time outputs, will be depicted as in Figure 2.34(b) and will be represented symbolically as

$$x[n] \longrightarrow y[n] \tag{2.51}$$

In most of this book we will treat discrete-time systems and continuous-time systems separately but in parallel. As we have already mentioned, this will allow us to use insights gained in one setting to aid in our understanding of the other. In Chapter 8 we will bring continuous-time and discrete-time systems together through the concept of sampling and will develop some insights into the use of discrete-time systems to process continuous-time signals that have been sampled. In the remainder of this section and continuing through the following section, we develop some of the basic concepts for both continuous-time and discrete-time systems.

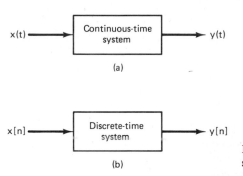

Figure 2.34 (a) Continuous-time system; (b) discrete-time system.

One extremely important idea that we will use throughout this book is that of an interconnection of systems. A *series* or *cascade interconnection* of two systems is illustrated in Figure 2.35(a). We will refer to diagrams such as this as *block diagrams*. Here the output of System 1 is the input to System 2, and the overall system transforms an input by processing it first by System 1 and then by System 2. Similarly, one can define a series interconnection of three or more systems. A *parallel interconnection* of two systems is illustrated in Figure 2.35(b). Here the same input signal is applied to Systems 1 and 2. The symbol "$\oplus$" in the figure denotes addition, so that the output of the parallel interconnection is the sum of the outputs of Systems 1 and 2. We can also define parallel interconnections of more than two systems, and we can combine both cascade and parallel interconnections to obtain more complicated interconnections. An example of such an interconnection is given in Figure 2.35(c).†

Interconnections such as these can be used to construct new systems out of

---

†On occasion we will also use the symbol $\otimes$ in our pictorial representation of systems to denote the operation of multiplying two signals (see, for example, Figure P2.24-1 in Problem 2.24).

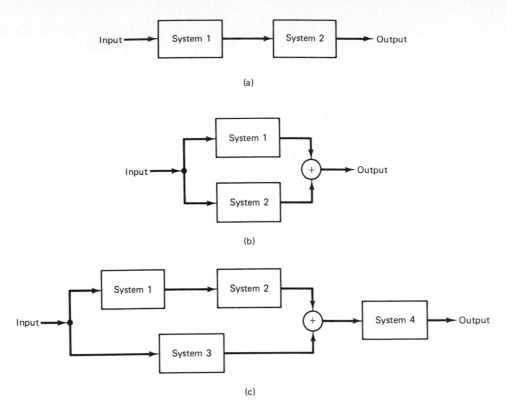

Figure 2.35 Interconnections of systems: (a) series (cascade) interconnection; (b) parallel interconnection; (c) series/parallel interconnection.

existing ones. For example, we can design systems to compute complicated arithmetic expressions by interconnecting basic arithmetic building blocks, as illustrated in Figure 2.36 for the calculation of

$$y[n] = (2x[n] - x[n]^2)^2 \qquad (2.52)$$

In this figure the "+" and "−" signs next to the "⊕" symbol indicate that the signal $x[n]^2$ is to be subtracted from the signal $2x[n]$. By convention, if no "+" or "−" signs are present next to a "⊕" symbol, we will assume that the corresponding signals are to be added.

In addition to providing a mechanism that allows us to build new systems, interconnections also allow us to view an existing system as an interconnection of its component parts. For example, electrical circuits involve interconnections of basic

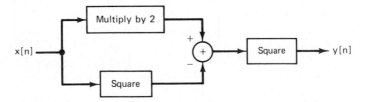

Figure 2.36 System for the calculation of $y[n] = (2x[n] - x[n]^2)^2$.

circuit elements (resistors, capacitors, inductors). Similarly, the operation of an automobile can be broken down into the interconnected operation of the carburetor, pistons, crankshaft, and so on. Viewing a complex system in this manner is often useful in facilitating the analysis of the properties of the system. For example, the response characteristics of an *RLC* circuit can be directly determined from the characteristics of its components and the specification of how they are interconnected.

Another important type of system interconnection is a *feedback interconnection*, an example of which is illustrated in Figure 2.37. Here the output of System 1 is the input to System 2, while the output of System 2 is fed back and added to the external input to produce the actual input to System 1. Feedback systems arise in a wide variety of applications. For example, a speed governor on an automobile senses vehicle velocity and adjusts the input from the driver in order to keep the speed at a safe level. Also, electrical circuits are often usefully viewed as containing feedback interconnections. As an example, consider the circuit depicted in Figure 2.38(a). As indicated in

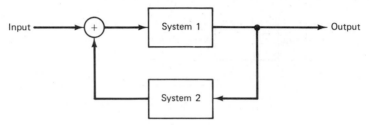

Figure 2.37 Feedback interconnection.

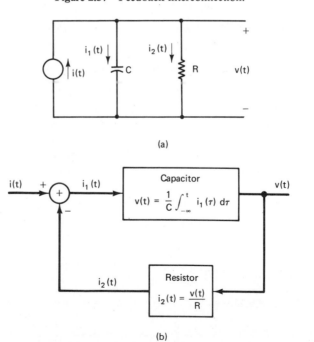

(a)

(b)

Figure 2.38 (a) Simple electrical circuit; (b) block diagram in which the circuit is depicted as the feedback interconnection of the two circuit elements.

Figure 2.38(b), this system can be viewed as the feedback interconnection of the two circuit elements. In Section 3.6 we use feedback interconnections in our description of the structure of a particularly important class of systems, and Chapter 11 is devoted to a detailed analysis of the properties of feedback systems.

2.6 PROPERTIES OF SYSTEMS

In this section we introduce and discuss a number of basic properties of continuous-time and discrete-time systems. These properties have both physical and mathematical interpretations, and thus by examining them we will also develop some insights into and facility with the mathematical representation that we have described for signals and systems.

2.6.1 Systems with and without Memory

A system is said to be *memoryless* if its output for each value of the independent variable is dependent only on the input at that same time. For example, the system in eq. (2.52) and illustrated in Figure 2.36 is memoryless, as the value of $y[n]$ at any particular time n_0 depends only on the value of $x[n]$ at that time. Similarly, a resistor is a memoryless system; with the input $x(t)$ taken as the current and with the voltage taken as the output $y(t)$, the input–output relationship of a resistor is

$$y(t) = Rx(t) \tag{2.53}$$

where R is the resistance. One particularly simple memoryless system is the *identity system*, whose output is identical to its input. That is,

$$y(t) = x(t)$$

is the input–output relationship for the continuous-time identity system, and

$$y[n] = x[n]$$

is the corresponding relationship in discrete time.

An example of a *system with memory* is

$$y[n] = \sum_{k=-\infty}^{n} x[k] \tag{2.54}$$

and a second example is

$$y(t) = x(t - 1) \tag{2.55}$$

A capacitor is another example of a system with memory, since if the input is taken to be the current and voltage is the output, then

$$y(t) = \frac{1}{C} \int_{-\infty}^{t} x(\tau)\, d\tau \tag{2.56}$$

where C is the capacitance.

2.6.2 Invertibility and Inverse Systems

A system is said to be *invertible* if distinct inputs lead to distinct outputs. Said another way, a system is invertible if by observing its output, we can determine its input. That is, as illustrated in Figure 2.39(a) for the discrete-time case, we can construct an *inverse system* which when cascaded with the original system yields an output $z[n]$ equal to the input $x[n]$ to the first system. Thus, the series interconnection in Figure 2.39(a) has an overall input–output relationship that is the same as that for the identity system.

An example of an invertible continuous-time system is

$$y(t) = 2x(t) \tag{2.57}$$

for which the inverse system is

$$z(t) = \tfrac{1}{2}y(t) \tag{2.58}$$

This example is illustrated in Figure 2.39(b). Another example of an invertible system is that defined by eq. (2.54). For this system the difference between two successive values of the output is precisely the last input value. Therefore, in this case the inverse system is

$$z[n] = y[n] - y[n - 1] \tag{2.59}$$

as illustrated in Figure 2.39(c). Examples of noninvertible systems are

$$y[n] = 0 \tag{2.60}$$

that is, the system that produces the zero output sequence for any input sequence, and

$$y(t) = x^2(t) \tag{2.61}$$

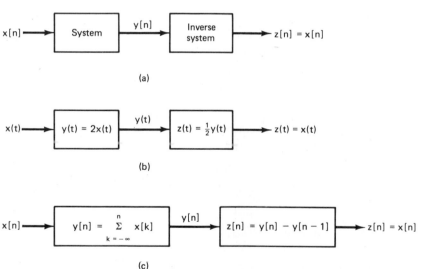

Figure 2.39 Concept of an inverse system for: (a) a general invertible system; (b) the invertible system described by eq. (2.57); (c) the invertible system defined in eq. (2.54).

2.6.3 Causality

A system is *causal* if the output at any time depends only on values of the input at the present time and in the past. Such a system is often referred to as being *nonanticipative,* as the system output does not anticipate future values of the input. Consequently, if two inputs to a causal system are identical up to some time t_0 or n_0, the corresponding outputs must also be equal up to this same time. The motion of an automobile is causal since it does not anticipate future actions of the driver. Similarly, the systems described by eqs. (2.55) and (2.56) are causal, but the systems defined by

$$y[n] = x[n] - x[n+1] \tag{2.62}$$

and

$$y(t) = x(t+1) \tag{2.63}$$

are not. Note also that all memoryless systems are causal.

Although causal systems are of great importance, they do not by any means constitute the only systems that are of practical significance. For example, causality is not of fundamental importance in applications, such as image processing, in which the independent variable is not time. Furthermore, in processing data for which time is the independent variable but which have already been recorded, as often happens with speech, geophysical, or meteorological signals, to name a few, we are by no means constrained to process those data causally. As another example, in many applications, including stock market analysis and demographic studies, we may be interested in determining a slowly varying trend in data that also contain high-frequency fluctuations about this trend. In this case, a possible approach is to average data over an interval in order to smooth out the fluctuations and keep only the trend. An example of a noncausal averaging system is

$$y[n] = \frac{1}{2M+1} \sum_{k=-M}^{+M} x[n-k] \tag{2.64}$$

2.6.4 Stability

Stability is another important system property. Intuitively, a stable system is one in which small inputs lead to responses that do not diverge. Suppose that we consider the situation depicted in Figure 2.40. Here we have a ball resting on a surface. In Figure 2.40(a) that surface is a hill with the ball at the crest, while in Figure 2.40(b) the surface is a valley, with the ball at the base. If we imagine a system whose input is a horizontal acceleration applied to the ball and whose output is the ball's vertical position, then the system depicted in Figure 2.40(a) is unstable, because an arbitrarily small perturbation in the horizontal position of the ball leads to the ball rolling down the hill. On the other hand, the system of Figure 2.40(b) is stable, because small horizontal accelerations lead to small perturbations in vertical position. Similarly, any of the phenomena mentioned in preceding sections, such as chain reactions and population growth, that can be represented by growing exponentials are examples of the responses of unstable systems, while phenomena, such as the response of a passive *RC* circuit, that lead to decaying exponentials are examples of the responses of stable systems.

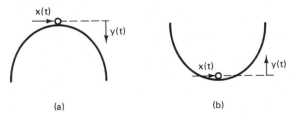

(a)   (b)

Figure 2.40 Examples of (a) an unstable system; and (b) a stable system. Here, the input is a horizontal acceleration applied to the ball, and the output is its vertical position.

The preceding paragraph provides us with an intuitive understanding of the concept of stability. Basically, if the input to a stable system is bounded (i.e., if its magnitude does not grow without bound), then the output must also be bounded and therefore cannot diverge. This is the definition of stability that we will use throughout this book. To illustrate the use of this definition, consider the system defined by eq. (2.64). Suppose that the input $x[n]$ is bounded in magnitude by some number, say, B, for all values of n. Then it is easy to see that the largest possible magnitude for $y[n]$ is also B, because $y[n]$ is the average of a finite set of values of the input. Therefore, $y[n]$ is bounded and the system is stable. On the other hand, consider the system described by eq. (2.54). Unlike the system in eq. (2.64), this system sums *all* of the past values of the input rather than just a finite set of values, and the system is unstable, as this sum can grow continually even if $x[n]$ is bounded. For example, suppose that $x[n] = u[n]$, the unit step, which is obviously a bounded input since its largest value is 1. In this case the output of the system of eq. (2.54) is

$$y[n] = \sum_{k=-\infty}^{n} u[k] = (n+1)u[n] \tag{2.65}$$

That is, $y[0] = 1$, $y[1] = 2$, $y[2] = 3$, and so on, and $y[n]$ grows without bound.

The properties and concepts that we have examined so far in this section are of great importance, and we examine some of these in far greater detail later in the book. There remain, however, two additional properties—time invariance and linearity— that play a central role in the subsequent chapters of this book, and in the remainder of this section we introduce and provide initial discussions of these two very important concepts.

2.6.5 Time Invariance

A system is *time-invariant* if a time shift in the input signal causes a time shift in the output signal. Specifically, if $y[n]$ is the output of a discrete-time, time-invariant system when $x[n]$ is the input, then $y[n - n_0]$ is the output when $x[n - n_0]$ is applied. In continuous time with $y(t)$ the output corresponding to the input $x(t)$, a time-invariant system will have $y(t - t_0)$ as the output when $x(t - t_0)$ is the input.

To illustrate the procedure for checking whether a system is time-invariant or not and at the same time to gain some insight into this property, let us consider the continuous-time system defined by

$$y(t) = \sin [x(t)] \tag{2.66}$$

To check if this system is time-invariant or time-varying, we proceed as follows. Let $x_1(t)$ be any input to this system, and let

$$y_1(t) = \sin [x_1(t)] \tag{2.67}$$

be the corresponding output. Then consider a second input obtained by shifting $x_1(t)$:

$$x_2(t) = x_1(t - t_0) \tag{2.68}$$

The output corresponding to this input is

$$y_2(t) = \sin [x_2(t)] = \sin [x_1(t - t_0)] \tag{2.69}$$

Similarly, from eq. (2.67),

$$y_1(t - t_0) = \sin [x_1(t - t_0)] \tag{2.70}$$

Comparing eqs. (2.69) and (2.70), we see that $y_2(t) = y_1(t - t_0)$, and therefore this system is time-invariant.

As a second example, consider the discrete-time system

$$y[n] = nx[n] \tag{2.71}$$

and consider the responses to two inputs $x_1[n]$ and $x_2[n]$, where $x_2[n] = x_1[n - n_0]$:

$$y_1[n] = nx_1[n] \tag{2.72}$$
$$y_2[n] = nx_2[n] = nx_1[n - n_0] \tag{2.73}$$

However, if we shift the output $y_1[n]$, we obtain

$$y_1[n - n_0] = (n - n_0)x_1[n - n_0] \neq y_2[n] \tag{2.74}$$

Thus we conclude that this system is time-varying. Equation (2.71) represents a system with a time-varying gain. Therefore, shifting the input will result in different values of the gain multiplying values of the shifted input. Note that if the gain is constant, as in eq. (2.57), then the system is time-invariant. Other examples of time-invariant systems are given by eqs. (2.53)–(2.64).

2.6.6 Linearity

A *linear system*, in continuous time or discrete time, is one that possesses the important property of superposition: If an input consists of the weighted sum of several signals, then the output is simply the superposition, that is, the weighted sum, of the responses of the system to each of those signals. Mathematically, let $y_1(t)$ be the response of a continuous-time system to $x_1(t)$ and let $y_2(t)$ be the output corresponding to the input $x_2(t)$. Then the system is linear if:

  1. The response to $x_1(t) + x_2(t)$ is $y_1(t) + y_2(t)$.
  2. The response to $ax_1(t)$ is $ay_1(t)$, where a is any complex constant.

The first of these two properties is referred to as the *additivity* property of a linear system; the second is referred to as the *scaling* or *homogeneity* property. Although we have written this definition using continuous-time signals, the same definition holds in discrete time. The systems specified by eqs. (2.53)–(2.60), (2.62)–(2.64), and (2.71) are linear, while those defined by eqs. (2.61) and (2.66) are nonlinear. Note that a system can be linear without being time-invariant, as in eq. (2.71), and it can be time-invariant without being linear, as in eqs. (2.61) and (2.66).†

The two properties defining a linear system can be combined into a single statement which is written below for the discrete-time case:

$$ax_1[n] + bx_2[n] \longrightarrow ay_1[n] + by_2[n] \tag{2.75}$$

---

†It is also possible for a system to be additive but not homogeneous or homogeneous but not additive. In either case the system is nonlinear, as it violates one of the two properties of linearity. We will not be particularly concerned with such systems, but we have included several examples in Problem 2.27.

where a and b are any complex constants. Furthermore, it is straightforward to show from the definition of linearity that if $x_k[n]$, $k = 1, 2, 3, \ldots$, are a set of inputs to a discrete-time linear system with corresponding outputs $y_k[n]$, $k = 1, 2, 3, \ldots$, then the response to a linear combination of these inputs given by

$$x[n] = \sum_k a_k x_k[n] = a_1 x_1[n] + a_2 x_2[n] + a_3 x_3[n] + \ldots \tag{2.76}$$

is

$$y[n] = \sum_k a_k y_k[n] = a_1 y_1[n] + a_2 y_2[n] + a_3 y_3[n] + \ldots \tag{2.77}$$

This very important fact is known as the *superposition property*, which holds for linear systems in both continuous time and discrete time.

Linear systems possess another important property, which is that zero input yields zero output. For example, if $x[n] \longrightarrow y[n]$, then the scaling property tells us that

$$0 = 0 \cdot x[n] \longrightarrow 0 \cdot y[n] = 0 \tag{2.78}$$

Consider then the system

$$y[n] = 2x[n] + 3 \tag{2.79}$$

From eq. (2.78) we see that this system is not linear, since $y[n] = 3$ if $x[n] = 0$. This may seem surprising, since eq. (2.79) is a linear equation, but this system does violate the zero-in/zero-out property of linear systems. On the other hand, this system falls into the class of *incrementally linear systems* described in the next paragraph.

An incrementally linear system in continuous or discrete time is one that responds linearly to *changes* in the input. That is, the *difference* in the responses to any two inputs to an incrementally linear system is a linear (i.e., additive and homogeneous) function of the *difference* between the two inputs. For example, if $x_1[n]$ and $x_2[n]$ are two inputs to the system specified by eq. (2.79), and if $y_1[n]$ and $y_2[n]$ are the corresponding outputs, then

$$y_1[n] - y_2[n] = 2x_1[n] + 3 - \{2x_2[n] + 3\} = 2\{x_1[n] - x_2[n]\} \tag{2.80}$$

It is straightforward to verify (Problem 2.33) that any incrementally linear system can be visualized as shown in Figure 2.41 for the continuous-time case. That is, the

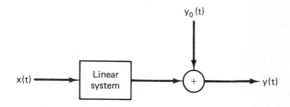

Figure 2.41 Structure of an incrementally linear system.

response of such a system equals the sum of the response of a linear system and of another signal that is unaffected by the input. Since the output of the linear system is zero if the input is zero, we see that this added signal is precisely the zero-input response of the overall system. For example, for the system specified by eq. (2.79) the output consists of the sum of the response of the linear system

$$x[n] \longrightarrow 2x[n]$$

and the zero-input response

$$y_0[n] = 3$$

Because of the structure of incrementally linear systems suggested by Figure 2.41 many of the characteristics of such systems can be analyzed using the techniques we will develop for linear systems. In this book we analyze one particularly important class of incrementally linear systems which we introduce in Section 3.5.

2.7 SUMMARY

In this chapter we have developed a number of basic concepts related to continuous- and discrete-time signals and systems. In particular we introduced a graphical representation of signals and used this representation in performing transformations of the independent variable. We also defined and examined several basic signals both in continuous time and in discrete time, and we investigated the concept of periodicity for continuous- and discrete-time signals.

In developing some of the elementary ideas related to systems, we introduced block diagrams to facilitate our discussions concerning the interconnection of systems, and we defined a number of important properties of systems, including causality, stability, time invariance, and linearity. The primary focus in this book will be on systems possessing the last two of these properties, that is, on the class of linear, time-invariant (LTI) systems, both in continuous time and in discrete time. These systems play a particularly important role in system analysis and design, in part due to the fact that many systems encountered in nature can be successfully modeled as linear and time-invariant. Furthermore, as we shall see, the properties of linearity and time invariance allow us to analyze in detail the characteristics of LTI systems. In Chapter 3 we develop a fundamental representation for this class of systems that will be of great use in developing many of the important tools of signal and system analysis.

PROBLEMS

The first seven problems for this chapter serve as a review of the topic of complex numbers, their representation, and several of their basic properties. As we will use complex numbers extensively in this book, it is important that readers familiarize themselves with the fundamental ideas considered and used in these problems.

The complex number z can be expressed in several ways. The *Cartesian* or *rectangular* form for z is given by

$$z = x + jy$$

where $j = \sqrt{-1}$ and x and y are real numbers referred to respectively as the *real part* and the *imaginary part* of z. As we indicated in the chapter, we will often use the notation

$$x = \mathcal{R}e\,\{z\}, \qquad y = \mathcal{I}m\,\{z\}$$

The complex number z can also be represented in *polar form* as

$$z = re^{j\theta}$$

where $r > 0$ is the *magnitude* of z and θ is the *angle* or *phase* of z. These quantities will often be written as

$$r = |z|, \qquad \theta = \sphericalangle z$$

The relationship between these two representations of complex numbers can be determined either from *Euler's relation*

$$e^{j\theta} = \cos\theta + j\sin\theta \qquad \text{(P2.0-1)}$$

or by plotting z in the complex plane, as shown in Figure P2.0. Here the coordinate axes are $\mathcal{Re}\{z\}$ along the horizontal axis and $\mathcal{Im}\{z\}$ along the vertical axis. With respect to this graphical representation, x and y are the Cartesian coordinates of z, and r and θ are its polar coordinates.

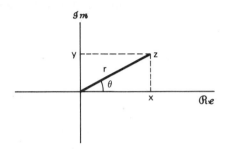

Figure P2.0

2.1. (a) Using Euler's relation or Figure P2.0, determine expressions for x and y in terms of r and θ.

(b) Determine expressions for r and θ in terms of x and y.

(c) If we are given only r and $\tan\theta$, can we uniquely determine x and y? Explain your answer.

2.2. Using Euler's relation, derive the following relationships.

(a) $\cos\theta = \frac{1}{2}(e^{j\theta} + e^{-j\theta})$

(b) $\sin\theta = \frac{1}{2j}(e^{j\theta} - e^{-j\theta})$

(c) $\cos^2\theta = \frac{1}{2}(1 + \cos 2\theta)$

(d) $(\sin\theta)(\sin\phi) = \frac{1}{2}\cos(\theta - \phi) - \frac{1}{2}\cos(\theta + \phi)$

(e) $\sin(\theta + \phi) = \sin\theta\cos\phi + \cos\theta\sin\phi$

2.3. Let z_0 be a complex number with polar coordinates r_0, θ_0 and Cartesian coordinates x_0, y_0. Determine expressions for the Cartesian coordinates of the following complex numbers in terms of x_0 and y_0. Plot the points $z_0, z_1, z_2, z_3, z_4,$ and z_5 in the complex plane when $r_0 = 2$, $\theta_0 = \pi/4$ and when $r_0 = 2$, $\theta_0 = \pi/2$. Indicate on your plots the real and imaginary parts of each point.

(a) $z_1 = r_0 e^{-j\theta_0}$ **(b)** $z_2 = r_0$ **(c)** $z_3 = r_0 e^{j(\theta_0 + \pi)}$

(d) $z_4 = r_0 e^{j(-\theta_0 + \pi)}$ **(e)** $z_5 = r_0 e^{j(\theta_0 + 2\pi)}$

2.4. Let z denote a complex variable

$$z = x + jy = re^{j\theta}$$

The *complex conjugate* of z is denoted by z^* and is given by

$$z^* = x - jy = re^{-j\theta}$$

Derive each of the following relations, where z, z_1, and z_2 are arbitrary complex numbers.

(a) $zz^* = r^2$

(b) $\dfrac{z}{z^*} = e^{j2\theta}$

(c) $z + z^* = 2\mathcal{Re}\{z\}$

(d) $z - z^* = 2j \, \mathcal{I}m \, \{z\}$

(e) $(z_1 + z_2)^* = z_1^* + z_2^*$

(f) $(az_1z_2)^* = az_1^*z_2^*$, where a is any real number

(g) $\left(\dfrac{z_1}{z_2}\right)^* = \dfrac{z_1^*}{z_2^*}$

(h) $\mathcal{R}e\left\{\dfrac{z_1}{z_2}\right\} = \dfrac{1}{2}\left[\dfrac{z_1z_2^* + z_1^*z_2}{z_2z_2^*}\right]$

2.5. Express each of the following complex numbers in Cartesian form and plot them in the complex plane, indicating the real and imaginary parts of each number.

(a) $\dfrac{3 + 4j}{1 - 2j}$

(b) $\dfrac{j(2 + j)}{(1 + j)(2 - j)}$

(c) $2j\dfrac{(1 + j)^2}{(3 - j)}$

(d) $4e^{j(\pi/6)}$

(e) $\sqrt{2} \, e^{j(25\pi/4)}$

(f) $je^{j(11\pi/4)}$

(g) $3e^{j4\pi} + 2e^{j7\pi}$

(h) The complex number z whose magnitude is $|z| = \sqrt{2}$ and whose angle is $\sphericalangle z = -\pi/4$

(i) $(1 - j)^9$

(j) $\dfrac{6e^{-j\pi/3}}{1 - j}$

2.6. Express each of the following complex numbers in polar form and plot them in the complex plane, indicating the magnitude and angle of each number.

(a) $1 + j\sqrt{3}$ **(b)** -5 **(c)** $-5 - 5j$

(d) $3 + 4j$ **(e)** $(1 - j\sqrt{3})^3$ **(f)** $(1 + j)^5$

(g) $(\sqrt{3} + j^3)(1 - j)$ **(h)** $\dfrac{2 - j(6/\sqrt{3})}{2 + j(6/\sqrt{3})}$ **(i)** $\dfrac{1 + j\sqrt{3}}{\sqrt{3} + j}$

(j) $j(1 + j)e^{j\pi/6}$ **(k)** $(\sqrt{3} + j)2\sqrt{2} \, e^{-j\pi/4}$ **(l)** $\dfrac{e^{j\pi/3} - 1}{1 + j\sqrt{3}}$

2.7. Derive the following relations, where z, z_1, and z_2 are arbitrary complex numbers.

(a) $(e^z)^* = e^{z^*}$

(b) $z_1z_2^* + z_1^*z_2 = 2 \, \mathcal{R}e \, \{z_1z_2^*\} = 2 \, \mathcal{R}e \, \{z_1^*z_2\}$

(c) $|z| = |z^*|$

(d) $|z_1z_2| = |z_1||z_2|$

(e) $\mathcal{R}e \, \{z\} \leq |z|, \quad \mathcal{I}m \, \{z\} \leq |z|$

(f) $|z_1z_2^* + z_1^*z_2| \leq 2|z_1z_2|$

(g) $(|z_1| - |z_2|)^2 \leq |z_1 + z_2|^2 \leq (|z_1| + |z_2|)^2$

2.8. The relations considered in this problem are used on many occasions throughout this book.

(a) Prove the validity of the following expression:

$$\sum_{n=0}^{N-1} \alpha^n = \begin{cases} N, & \alpha = 1 \\ \dfrac{1 - \alpha^N}{1 - \alpha}, & \text{for any complex number } \alpha \neq 1 \end{cases}$$

(b) Show that if $|\alpha| < 1$, then

$$\sum_{n=0}^{\infty} \alpha^n = \frac{1}{1 - \alpha}$$

(c) Show also if $|\alpha| < 1$, then

$$\sum_{n=0}^{\infty} n\alpha^n = \frac{\alpha}{(1-\alpha)^2}$$

(d) Evaluate

$$\sum_{n=k}^{\infty} \alpha^n$$

assuming that $|\alpha| < 1$.

2.9. (a) A continuous-time signal $x(t)$ is shown in Figure P2.9(a). Sketch and label carefully each of the following signals.

  (i) $x(t-2)$
  (ii) $x(1-t)$
  (iii) $x(2t+2)$
  (iv) $x(2-t/3)$
  (v) $[x(t) + x(2-t)]u(1-t)$
  (vi) $x(t)[\delta(t+\frac{3}{2}) - \delta(t-\frac{3}{2})]$

(b) For the signal $h(t)$ depicted in Figure P2.9(b), sketch and label carefully each of the following signals.

  (i) $h(t+3)$
  (ii) $h(t/2-2)$
  (iii) $h(1-2t)$
  (iv) $4h(t/4)$
  (v) $\frac{1}{2}h(t)u(t) + h(-t)u(t)$
  (vi) $h(t/2)\,\delta(t+1)$
  (vii) $h(t)[u(t+1) - u(t-1)]$

(c) Consider again the signals $x(t)$ and $h(t)$ shown in Figure P2.9(a) and (b), respectively. Sketch and label carefully each of the following signals.

  (i) $x(t)h(t+1)$
  (ii) $x(t)h(-t)$

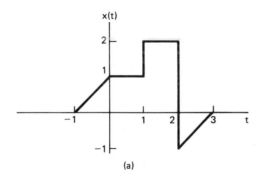

(a)

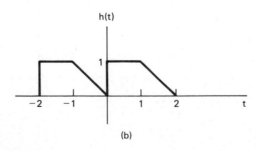

(b)

Figure P2.9

Signals and Systems   Chap. 2

(iii) $x(t - 1)h(1 - t)$

(iv) $x(1 - t)h(t - 1)$

(v) $x(2 - t/2)h(t + 4)$

2.10. (a) A discrete-time signal $x[n]$ is shown in Figure P2.10(a). Sketch and label carefully each of the following signals.

(i) $x[n - 2]$

(ii) $x[4 - n]$

(iii) $x[2n]$

(iv) $x[2n + 1]$

(v) $x[n]u[2 - n]$

(vi) $x[n - 1]\delta[n - 3]$

(vii) $\frac{1}{2}x[n] + \frac{1}{2}(-1)^n x[n]$

(viii) $x[n^2]$

(b) For the signal $h[n]$ depicted in Figure P2.10(b), sketch and label carefully each of the following signals.

(i) $h[2 - n]$

(ii) $h[n + 2]$

(iii) $h[-n]u[n] + h[n]$

(iv) $h[n + 2] + h[-1 - n]$

(v) $h[3n]\,\delta[n - 1]$

(vi) $h[n + 1]\{u[n + 3] - u[-n]\}$

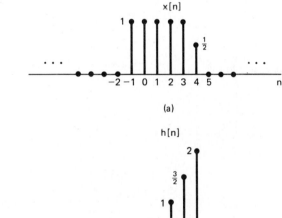

(a)

(b)

Figure P2.10

(c) Consider the signals $x[n]$ and $h[n]$ used in parts (a) and (b). Sketch and label carefully each of the following signals.

   (i) $h[n]x[-n]$ (ii) $x[n+2]h[1-2n]$

  (iii) $x[1-n]h[n+4]$ (iv) $x[n-1]h[n-3]$

2.11. Although, as mentioned in the text, we will focus our attention on signals with one independent variable, on occasion it will be instructive to consider signals with two independent variables in order to illustrate particular concepts involving signals and systems. A two-dimensional signal $d(x, y)$ can often be usefully visualized as a picture where the brightness of the picture at any point is used to represent the value of $d(x, y)$ at that point. For example, in Figure P.2.11(a) we have depicted a picture representing the signal $d(x, y)$ which takes on the value 1 in the shaded portion of the (x, y)-plane and zero elsewhere.

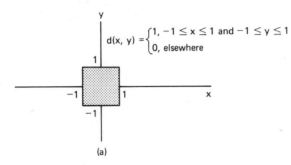

(a)

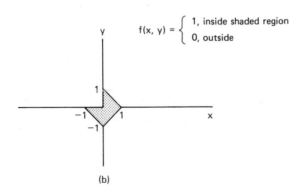

(b)

Figure P2.11

(a) Consider the signal $d(x, y)$ depicted in Figure P2.11(a). Sketch each of the following.

   (i) $d(x+1, y-2)$ (ii) $d(x/2, 2y)$

  (iii) $d(y, 3x)$ (iv) $d(x-y, x+y)$

   (v) $d(1/x, 1/y)$

(b) For the signal $f(x, y)$ illustrated in Figure P2.11(b), sketch each of the following.

   (i) $f(x-3, y+2)$

   (ii) $f(x, -y)$

  (iii) $f(-\frac{1}{2}y, 2x)$

  (iv) $f(2-x, -1-y)$

   (v) $f(2y-1, x/3+2)$

  (vi) $f(x\cos\theta - y\sin\theta, x\sin\theta + y\cos\theta)$, $\theta = \pi/4$

 (vii) $f(x, y)u(\frac{1}{2} - y)$

2.12. Determine and sketch the even and odd parts of the signals depicted in Figure P2.12. Label your sketches carefully.

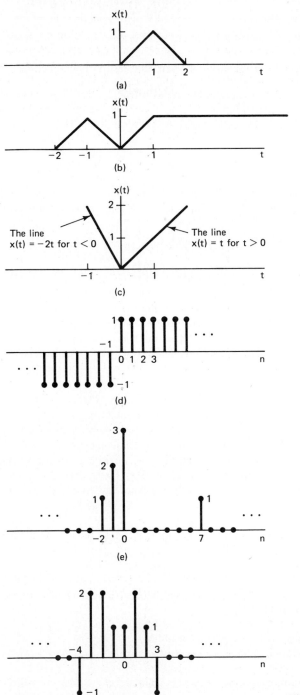

Figure P2.12

2.13. In this problem we explore several of the properties of even and odd signals.

(a) Show that if $x[n]$ is an odd signal, then

$$\sum_{n=-\infty}^{+\infty} x[n] = 0$$

(b) Show that if $x_1[n]$ is an odd signal and $x_2[n]$ is an even signal, then $x_1[n]x_2[n]$ is an odd signal.

(c) Let $x[n]$ be an arbitrary signal with even and odd parts denoted by

$$x_e[n] = \mathcal{E}v\,\{x[n]\}$$

$$x_o[n] = \mathcal{O}d\,\{x[n]\}$$

Show that

$$\sum_{n=-\infty}^{+\infty} x^2[n] = \sum_{n=-\infty}^{+\infty} x_e^2[n] + \sum_{n=-\infty}^{+\infty} x_o^2[n]$$

(d) Although parts (a)–(c) have been stated in terms of discrete-time signals, the analogous properties are also valid in continuous time. To demonstrate this, show that

$$\int_{-\infty}^{+\infty} x^2(t)\,dt = \int_{-\infty}^{+\infty} x_e^2(t)\,dt + \int_{-\infty}^{+\infty} x_o^2(t)\,dt$$

where $x_e(t)$ and $x_o(t)$ are, respectively, the even and odd parts of $x(t)$.

2.14. (a) Let $x_e[n]$ shown in Figure P2.14(a) be the even part of a signal $x[n]$. Given that $x[n] = 0$ for $n < 0$, determine and carefully sketch $x[n]$ for all n.

(b) Let $x_o[n]$ shown in Figure P2.14(b) be the odd part of a signal $x[n]$. Given that $x[n] = 0$ for $n < 0$ and $x[0] = 1$, determine and carefully sketch $x[n]$.

(c) Let $x_e(t)$ shown in Figure P2.14(c) be the even part of a signal $x(t)$. Also, in Figure P2.14(d) we have depicted the signal $x(t + 1)u(-t - 1)$. Determine and carefully sketch the odd part of $x(t)$.

2.15. If $x(t)$ is a continuous-time signal, we have seen that $x(2t)$ is a "speeded-up" version of $x(t)$, in the sense that the duration of the signal is cut in half. Similarly, $x(t/2)$ represents a "slowed-down" version of $x(t)$, with the time scale of the signal spread out to twice its original scale. The concepts of "slowing down" or "speeding up" a signal are somewhat different in discrete time, as we will see in this problem.

To begin, consider a discrete-time signal $x[n]$, and define two related signals, which in some sense represent, respectively, "speeded-up" and "slowed-down" versions of $x[n]$:

$$y_1[n] = x[2n]$$

$$y_2[n] = \begin{cases} x[n/2], & n \text{ even} \\ 0, & n \text{ odd} \end{cases}$$

(a) For the signal $x[n]$ depicted in Figure P2.15, plot $y_1[n]$ and $y_2[n]$ as defined above.

(b) Let $x(t)$ be a continuous-time signal, and let $y_1(t) = x(2t)$, $y_2(t) = x(t/2)$. Consider the following statements:

(1) If $x(t)$ is periodic, then $y_1(t)$ is periodic.

(2) If $y_1(t)$ is periodic, then $x(t)$ is periodic.

(3) If $x(t)$ is periodic, then $y_2(t)$ is periodic.

(4) If $y_2(t)$ is periodic, then $x(t)$ is periodic.

Determine if each of these statements is true, and if so, determine the relationship between the fundamental periods of the two signals considered in the statement. If

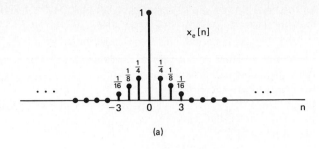

(a)

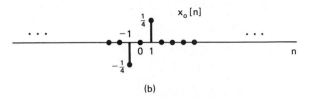

(b)

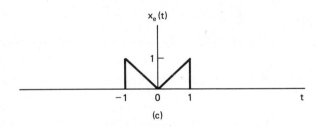

(c)

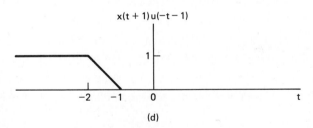

(d)

Figure P2.14

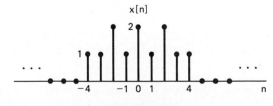

Figure P2.15

the statement is not true, produce a counterexample. Do the same for the following statements.

(i) If $x[n]$ is periodic, then $y_1[n]$ is periodic.

(ii) If $y_1[n]$ is periodic, then $x[n]$ is periodic.

(iii) If $x[n]$ is periodic, then $y_2[n]$ is periodic.

(iv) If $y_2[n]$ is periodic, then $x[n]$ is periodic.

2.16. Determine whether or not each of the following signals is periodic. If a signal is periodic, determine its fundamental period.

(a) $x(t) = 2 \cos (3t + \pi/4)$

(b) $x(t) = e^{j(\pi t - 1)}$

(c) $x[n] = \cos (8\pi n/7 + 2)$

(d) $x[n] = e^{j(n/8 - \pi)}$

(e) $x(t) = [\sin (t - \pi/6)]^2$

(f) $x[n] = \cos (\pi n^2/8)$

(g) $x[n] = \sum_{m=-\infty}^{+\infty} \{\delta[n - 3m] - \delta[n - 1 - 3m]\}$

(h) $x(t) = [\cos 2\pi t] u(t)$

(i) $x(t) = \mathcal{E}v \{[\cos (2\pi t)] u(t)\}$

(j) $x(t) = \mathcal{E}v \{[\cos (2\pi t + \pi/4)] u(t)\}$

(k) $x[n] = \cos (n/4) \cos (\pi n/4)$

(l) $x[n] = 2 \cos (\pi n/4) + \sin (\pi n/8) - 2 \cos (\pi n/2 + \pi/6)$

(m) $x(t) = \sum_{n=-\infty}^{+\infty} e^{-(t - 3n)^2}$

2.17. (a) Consider the periodic discrete-time exponential signal

$$x[n] = e^{jm(2\pi/N)n}$$

Show that the fundamental period N_0 of this signal is given by

$$N_0 = N/\gcd (m, N) \qquad \text{(P2.17-1)}$$

where $\gcd (m, N)$ is the *greatest common divisor* of m and N, that is, the largest integer that divides both m and N an integral number of times. For example,

$$\gcd (2, 3) = 1, \qquad \gcd (2, 4) = 2, \qquad \gcd (8, 12) = 4$$

Note that $N_0 = N$ if m and N have no factors in common.

(b) Consider the following set of harmonically related periodic exponential signals

$$\phi_k[n] = e^{jk(2\pi/7)n}$$

Find the fundamental period and/or frequency for these signals for all integer values of k.

(c) Repeat part (b) for

$$\phi_k[n] = e^{jk(2\pi/8)n}$$

2.18. Let $x(t)$ be the continuous-time complex exponential signal

$$x(t) = e^{j\omega_0 t}$$

with fundamental frequency ω_0 and fundamental period $T_0 = 2\pi/\omega_0$. Consider the discrete-time signal obtained by taking equally spaced samples of $x(t)$. That is,

$$x[n] = x(nT) = e^{j\omega_0 nT}$$

(a) Show that $x[n]$ is periodic if and only if T/T_0 is a rational number, that is, if and only if some multiple of the sampling interval *exactly equals* a multiple of the period of $x(t)$.

(b) Suppose that $x[n]$ is periodic, that is, that

$$\frac{T}{T_0} = \frac{p}{q} \qquad \text{(P2.18-1)}$$

54

where p and q are integers. What are the fundamental period and fundamental frequency of $x[n]$? Express the fundamental frequency as a fraction of $\omega_0 T$.

(c) Again assuming that T/T_0 satisfies eq. (P2.18-1), determine precisely how many periods of $x(t)$ are needed to obtain the samples that form a single period of $x[n]$.

2.19. (a) Let $x(t)$ and $y(t)$ be periodic signals with fundamental periods T_1 and T_2, respectively. Under what conditions is the sum

$$x(t) + y(t)$$

periodic, and what is the fundamental period of this signal if it is periodic?

(b) Let $x[n]$ and $y[n]$ be periodic signals with fundamental periods N_1 and N_2, respectively. Under what conditions is the sum

$$x[n] + y[n]$$

periodic, and what is the fundamental period of this signal if it is periodic?

(c) Consider the signals

$$x(t) = \cos \frac{2\pi t}{3} + 2 \sin \frac{16\pi t}{3}$$

$$y(t) = \sin \pi t$$

Show that

$$z(t) = x(t)y(t)$$

is periodic, and write $z(t)$ as a linear combination of harmonically related complex exponentials. That is, find a number T and complex numbers c_k so that

$$z(t) = \sum_k c_k e^{jk(2\pi/T)t}$$

2.20. (a) Consider a system with input $x(t)$ and with output $y(t)$ given by

$$y(t) = \sum_{n=-\infty}^{+\infty} x(t)\, \delta(t - nT)$$

(i) Is this system linear?

(ii) Is this system time-invariant?

For each part, if your answer is yes, show why this is so. If your answer is no, produce a counterexample.

(b) Suppose that the input to this system is

$$x(t) = \cos 2\pi t$$

Sketch and label carefully the output $y(t)$ for each of the following values of T.

$$T = 1, \tfrac{1}{2}, \tfrac{1}{4}, \tfrac{1}{8}, \tfrac{1}{12}$$

All of your sketches should have the same horizontal and vertical scales.

(c) Repeat part (b) for

$$x(t) = e^t \cos 2\pi t$$

2.21. In this problem we examine a few of the properties of the unit impulse function.

(a) Show that

$$\delta(2t) = \tfrac{1}{2}\, \delta(t)$$

Hint: Examine $\delta_\Delta(2t)$ (see Figure 2.20).

(b) What is $\delta[2n]$?

(c) In Section 2.3 we defined the continuous-time unit impulse as the limit of the signal $\delta_\Delta(t)$. More precisely, we defined several of the *properties* of $\delta(t)$ by examining the corresponding properties of $\delta_\Delta(t)$. For example, since the signal $u_\Delta(t)$ defined by

$$u_\Delta(t) = \int_{-\infty}^{t} \delta_\Delta(\tau)\, d\tau$$

converges to the unit step

$$u(t) = \lim_{\Delta \to 0} u_\Delta(t) \qquad\qquad \text{(P2.21-1)}$$

we could then interpret $\delta(t)$ through the equation

$$u(t) = \int_{-\infty}^{t} \delta(\tau)\, d\tau$$

or by viewing $\delta(t)$ as the formal derivative of $u(t)$.

    This type of discussion is important, as we are in effect trying to define $\delta(t)$ through its properties rather than by specifying its value for each t, which is not possible. In Chapter 3 we provide a very simple characterization of the behavior of the unit impulse that is extremely useful in the study of linear, time-invariant systems. For the present, however, we concentrate on demonstrating that the important concept in using the unit impulse is to understand *how* it behaves. To do this, consider the six signals depicted in Figure P2.21. Show that each "behaves like an

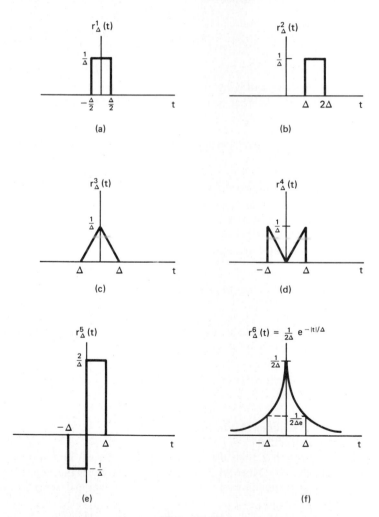

Figure P2.21

Signals and Systems   Chap. 2

impulse" as $\Delta \longrightarrow 0$ in that, if we let

$$u_\Delta^i(t) = \int_{-\infty}^t r_\Delta^i(\tau)\, d\tau$$

then

$$\lim_{\Delta \to 0} u_\Delta^i(t) = u(t)$$

In each case sketch and label carefully the signal $u_\Delta^i(t)$. Note that

$$r_\Delta^2(0) = r_\Delta^4(0) = 0 \qquad \text{for all } \Delta$$

Therefore, it is not enough to define or to think of $\delta(t)$ as being zero for $t \neq 0$ and infinite for $t = 0$. Rather, it is properties such as eq. (P2.21-1) that define the impulse. In Section 3.7 we will define a whole class of signals known as *singularity functions* which are related to the unit impulse and which are also defined in terms of their properties rather than their values.

(d) The role played by $u(t)$, $\delta(t)$, and other singularity functions in the study of linear, time-invariant systems is that of *idealizations* of physical phenomena, and, as we will see, the use of these idealizations allows us to obtain an exceedingly important and very simple representation of such systems. In using singularity functions we need, however, to be careful. In particular we must remember that they are idealizations, and thus whenever we perform a calculation using them we are implicitly assuming that this calculation represents an accurate description of the behavior of the signals that they are intended to idealize. To illustrate this consider the equation

$$x(t)\, \delta(t) - x(0)\, \delta(t) \tag{P2.21-2}$$

This equation is based on the observation that

$$x(t)\, \delta_\Delta(t) \simeq x(0)\, \delta_\Delta(t) \tag{P2.21-3}$$

Taking the limit of this relationship then yields the idealized one given by eq. (P2.21-2). However, a more careful examination of our derivation of eq. (P2.21-3) shows that the approximate equality (P2.21-3) really only makes sense if $x(t)$ is continuous at $t = 0$. If it is not, then we will not have $x(t) \simeq x(0)$ for t small.

To make this point clearer, consider the unit step signal $u(t)$. Recall from eq. (2.18) that $u(t) = 0$ for $t < 0$ and $u(t) = 1$ for $t > 0$, but that its value at $t = 0$ is not defined [note, for example, that $u_\Delta(0) = 0$ for all Δ while $u_\Delta^1(0) = \frac{1}{2}$ (from part (c))]. The fact that $u(0)$ is not defined is not particularly bothersome, as long as the calculations we perform using $u(t)$ do not rely on a specific choice for $u(0)$. For example, if $f(t)$ is a signal that is continuous at $t = 0$, then the value of

$$\int_{-\infty}^{+\infty} f(\sigma)u(\sigma)\, d\sigma$$

does not depend upon a choice for $u(0)$. On the other hand, the fact that $u(0)$ is undefined is significant in that it means that certain calculations involving singularity functions are undefined. Consider trying to define a value for the product $u(t)\, \delta(t)$. To see that this *cannot* be defined, show that

$$\lim_{\Delta \to 0} [u_\Delta(t)\, \delta(t)] = 0$$

but

$$\lim_{\Delta \to 0} [u_\Delta(t)\, \delta_\Delta(t)] = \tfrac{1}{2}\delta(t)$$

In general, we can define the product of two signals without any difficulty as long as the signals do not contain singularities (discontinuities, impulses or the other singularities introduced in Section 3.7) whose locations coincide. When the locations do coincide, the product is undefined. As an example, show that the signal

$$g(t) = \int_{-\infty}^{+\infty} u(\tau)\, \delta(t - \tau)\, d\tau$$

is identical to $u(t)$; that is, it is 0 for $t < 0$, it equals 1 for $t > 0$, and it is undefined for $t = 0$.

2.22. In this chapter we introduced a number of general properties of systems. In particular, a system may or may not be

    (1) Memoryless
    (2) Time-invariant
    (3) Linear
    (4) Causal
    (5) Stable

Determine which of these properties hold and which do not hold for each of the following systems. Justify your answers. In each example $y(t)$ or $y[n]$ denotes the system output, and $x(t)$ or $x[n]$ is the system input.

  (a) $y(t) = e^{x(t)}$
  (b) $y[n] = x[n]x[n - 1]$
  (c) $y(t) = \dfrac{dx(t)}{dt}$
  (d) $y[n] = x[-n]$
  (e) $y[n] = x[n - 2] - 2x[n - 17]$
  (f) $y(t) = x(t - 1) - x(1 - t)$
  (g) $y(t) = [\sin(6t)]x(t)$
  (h) $y[n] = \displaystyle\sum_{k=n-2}^{n+4} x[k]$
  (i) $y[n] = nx[n]$
  (j) $y(t) = \displaystyle\int_{-\infty}^{3t} x(\tau)\, d\tau$
  (k) $y[n] = \mathcal{E}v\,\{x[n]\}$
  (l) $y(t) = \begin{cases} 0, & t < 0 \\ x(t) + x(t - 100), & t \geq 0 \end{cases}$
  (m) $y(t) = \begin{cases} 0, & x(t) < 0 \\ x(t) + x(t - 100), & x(t) \geq 0 \end{cases}$
  (n) $y[n] = \begin{cases} x[n], & n \geq 1 \\ 0, & n = 0 \\ x[n + 1], & n \leq -1 \end{cases}$
  (o) $y[n] = \begin{cases} x[n], & n \geq 1 \\ 0, & n = 0 \\ x[n], & n \leq -1 \end{cases}$
  (p) $y(t) = x(t/2)$
  (q) $y[n] = x[2n]$

2.23. An important concept in many communications applications is the *correlation* between two signals. In the problems at the end of Chapter 3 we will have more to say about this topic and will provide some indication of how it is used in practice. For now we content ourselves with a brief introduction to correlation functions and some of their properties.

       Let $x(t)$ and $y(t)$ be two signals; then the correlation function $\phi_{xy}(t)$ is defined as

$$\phi_{xy}(t) = \int_{-\infty}^{+\infty} x(t + \tau)y(\tau)\, d\tau$$

The function $\phi_{xx}(t)$ is usually referred to as the *autocorrelation function* of the signal $x(t)$, while $\phi_{xy}(t)$ is often called a *cross-correlation function*.

(a) What is the relationship between $\phi_{xy}(t)$ and $\phi_{yx}(t)$?

(b) Compute the odd part of $\phi_{xx}(t)$.

(c) Suppose that $y(t) = x(t + T)$. Express $\phi_{xy}(t)$ and $\phi_{yy}(t)$ in terms of $\phi_{xx}(t)$.

(d) It is often important in practice to compute the correlation function $\phi_{hx}(t)$, where $h(t)$ is a fixed given signal but where $x(t)$ may be any of a wide variety of signals. In this case what is done is to design a system with input $x(t)$ and output $\phi_{hx}(t)$. Is this system linear? Is it time-invariant? Is it causal? Explain your answers.

(e) Do any of your answers to part (d) change if we take as the output $\phi_{xh}(t)$ rather than $\phi_{hx}(t)$?

2.24. Consider the system shown in Figure P2.24-1. Here the square root operation produces the positive square root.

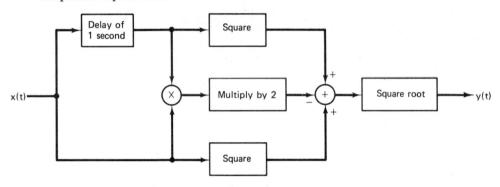

Figure P2.24-1

(a) Find an explicit relationship between $y(t)$ and $x(t)$.

(b) Is this system linear?

(c) Is it time-invariant?

(d) What is the response $y(t)$ when the input is as shown in Figure P2.24-2?

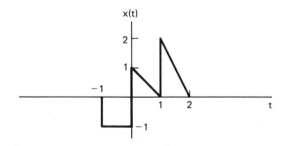

Figure P2.24-2

2.25. (a) Is the following statement true or false?
The series interconnection of two linear, time-invariant systems is itself a linear, time-invariant system.
Justify your answer.

(b) Is the following statement true or false?

The series connection of two nonlinear systems is itself nonlinear. Justify your answer.

(c) Consider three systems with the following input–output relationships:

$$\text{System 1} \quad y[n] = \begin{cases} x[n/2], & n \text{ even} \\ 0, & n \text{ odd} \end{cases}$$

$$\text{System 2} \quad y[n] = x[n] + \tfrac{1}{2}x[n-1] + \tfrac{1}{4}x[n-2]$$

$$\text{System 3} \quad y[n] = x[2n]$$

Suppose that these systems are connected in series as depicted in Figure P2.25. Find the input–output relationship for the overall interconnected system. Is this system linear? Is it time-invariant?

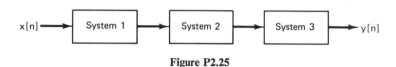

Figure P2.25

(d) Consider a second series interconnection of the form of Figure P2.25, where in this case the three systems are specified by the following equations:

$$\text{System 1} \quad y[n] = x[\tau n]$$

$$\text{System 2} \quad y[n] = ax[n-1] + bx[n] + cx[n+1]$$

$$\text{System 3} \quad y[n] = x[-n]$$

Here a, b, and c are real numbers. Find the input–output relationship for the overall interconnected system. Under what conditions on the numbers a, b, and c does the overall system have each of the following properties?

(i) The overall system is linear and time-invariant.

(ii) The input–output relationship of the overall system is identical to that of System 2.

(iii) The overall system is causal.

2.26. Determine if each of the following systems is invertible. If it is, construct the inverse system. If it is not, find two input signals to the system that have the same output.

(a) $y(t) = x(t-4)$

(b) $y(t) = \cos[x(t)]$

(c) $y[n] = nx[n]$

(d) $y(t) = \int_{-\infty}^{t} x(\tau)\,d\tau$

(e) $y[n] = \begin{cases} x[n-1], & n \geq 1 \\ 0, & n = 0 \\ x[n], & n \leq -1 \end{cases}$

(f) $y[n] = x[n]x[n-1]$

(g) $y[n] = x[1-n]$

(h) $y(t) = \int_{-\infty}^{t} e^{-(t-\tau)}x(\tau)\,d\tau$

(i) $y[n] = \sum_{k=-\infty}^{n} (\tfrac{1}{2})^{n-k}x[k]$

(j) $y(t) = \dfrac{dx(t)}{dt}$

(k) $y[n] = \begin{cases} x[n+1], & n \geq 0 \\ x[n], & n \leq -1 \end{cases}$

(l) $y(t) = x(2t)$

(m) $y[n] = x[2n]$

(n) $y[n] = \begin{cases} x[n/2], & n \text{ even} \\ 0, & n \text{ odd} \end{cases}$

2.27. In the text we discussed the fact that the property of linearity for a system is equivalent to the system possessing both the additivity property and the homogeneity property. For convenience we repeat these two properties here:

1. Let $x_1(t)$ and $x_2(t)$ be any two inputs to a system with corresponding outputs $y_1(t)$ and $y_2(t)$. Then the system is *additive* if

$$x_1(t) + x_2(t) \longrightarrow y_1(t) + y_2(t)$$

2. Let $x(t)$ be any input to a system with corresponding output $y(t)$. Then the system is *homogeneous* if

$$cx(t) \longrightarrow cy(t) \tag{P2.27-1}$$

where c is an arbitrary complex constant.

The analogous definitions can be stated for discrete-time systems.
   (a) Determine if each of the systems defined in parts (i)–(iv) is additive and/or homogeneous. Justify your answers by providing a proof if one of these two properties holds, or a counterexample if it does not hold.

   (i) $y[n] = \mathcal{Re}\,\{x[n]\}$

   (ii) $y(t) = \dfrac{1}{x(t)}\left[\dfrac{dx(t)}{dt}\right]^2$

   (iii) $y[n] = \begin{cases} \dfrac{x[n]x[n-2]}{x[n-1]}, & x[n-1] \neq 0 \\ 0, & x[n-1] = 0 \end{cases}$

   (iv) The continuous-time system whose output $y(t)$ is zero for all times at which the input $x(t)$ is *not* zero. At each point at which $x(t) = 0$ the output is an impulse of area equal to the derivative of $x(t)$ at that instant. Assume that all inputs permitted for this system have continuous derivatives.

   (b) A system is called *real-linear* if it is additive and if equation (P2.27-1) holds for c an arbitrary *real* number. One of the systems considered in part (a) is not linear but is real-linear. Which one is it?

   (c) Show that if a system is *either* additive or homogeneous, it has the property that if the input is identically zero, then the output is also identically zero.

   (d) Determine a system (either in continuous or in discrete time) that is *neither* additive *nor* homogeneous but which has a zero output if the input is identically zero.

   (e) From part (c) can you conclude that if the input to a linear system is zero between times t_1 and t_2 in continuous time or between times n_1 and n_2 in discrete time, then its output must also be zero between these same times? Explain your answer.

2.28. Consider the discrete-time system that performs the following operation. At each time n, it computes

$$r_+[n] = |x[n] - x[n-1]|$$
$$r_0[n] = |x[n+1] - x[n-1]|$$
$$r_-[n] = |x[n+1] - x[n]|$$

It then determines the largest of these. Then the system output $y[n]$ is given by

$$y[n] = x[n+1] \text{ if } r_+[n] = \max\,(r_+[n], r_0[n], r_-[n])$$
$$y[n] = x[n] \qquad \text{if } r_0[n] = \max\,(r_+[n], r_0[n], r_-[n])$$
$$y[n] = x[n-1] \text{ if } r_-[n] = \max\,(r_+[n], r_0[n], r_-[n])$$

(a) Show that if $x[n]$ is real-valued, this system simply chooses the middle value of the three numbers $x[n-1]$, $x[n]$, and $x[n+1]$. This operation is referred to as *median filtering*, and it is sometimes used if the signal $x[n]$ contains occasional large, spurious values that represent distortion of the information carried in the signal. For example, in deep-space communication, the signal received on earth may contain short bursts of erratic behavior, resulting from some type of interference. Median filtering represents one method for partially removing such bursts. There are many alternative methods for attempting to achieve this same goal, and the relative merits of these depend upon the characteristics of the original, undisturbed signal and of the interfering signal.

(b) Show that the system described above is homogeneous but not additive.

(c) In Chapter 6 we consider in greater detail the problem of filtering, that is, of using a system to process a received signal in order to remove interference that may be present. As we will see, there are trade-offs that must be considered in the design of such a system. In the remainder of this problem we provide a brief look at filtering by examining three candidate systems:

System 1:   A median filter

System 2:   An averaging system:

$$y[n] = \tfrac{1}{3}\{x[n-1] + x[n] + x[n+1]\}$$

System 3:   An averager that places more weight on the signal value at the present time:

$$y[n] = \tfrac{1}{4}x[n-1] + \tfrac{1}{2}x[n] + \tfrac{1}{4}x[n+1]$$

We suppose that the input signal $x[n]$ consists of the sum of a desired signal $x_d[n]$ and an interfering signal $x_i[n]$:

$$x[n] = x_d[n] + x_i[n]$$

(i) Assume first that $x_i[n] = 0$, and compute the outputs from each of the three systems with $x_d[n]$ taken first as in Figure P2.28(a), and then as in Figure P2.28(b). As can be seen, each of these systems introduces *distortion*; that is, the output is a distorted version of the input. For median filtering, peaks get clipped, while for the two averagers the peaks are attenuated and spread out. Which type of distortion is least troublesome depends, of course, on the application.

(ii) Consider next the case in which $x[n]$ is the sum of $x_d[n]$ as given in Figure P2.28(a) and $x_i[n]$ as depicted in Figure P2.28(c). Determine the outputs of each of the three systems. As can be seen, the median filter does a better job of removing the sporadic burst of noise, and this is for exactly the same reason that this filter distorted the signal of Figure P2.28(b). That is, median filtering suppresses sporadic peaks, and whether one chooses to use such a filter depends on the character of both the desired *and* interfering signals: if the desired signal is expected to have isolated peaks, distortion can be expected, but if the desired signal varies only mildly, and the interference consists only of occasional bursts, then the peak-suppressing character of median filtering will help in removing the interference.

(iii) Consider the case in which $x_d[n]$ is again given by Figure P2.28(a), but where $x_i[n] = 5(-1)^n$. Compute the outputs of the three systems in this case. Note that System 3 is perfectly suited to the rejection of this noise.

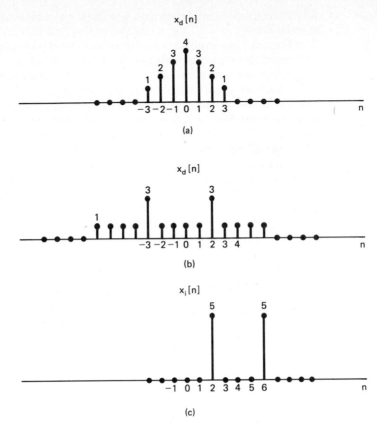

$x_d[n]$

(a)

$x_d[n]$

(b)

$x_i[n]$

(c)

Figure P2.28

(iv) Note that Systems 2 and 3 are linear, and therefore a scaling of the amplitude of the interfering signal simply scales that portion of the output due to the interference. On the other hand, System 1 is *not* linear, and thus one might expect a very different-looking response to the inputs $x_d[n] + x_i[n]$ and $x_d[n] + kx_i[n]$. To illustrate this point, calculate the outputs of Systems 1 and 3 when $x_d[n]$ is as in Figure P2.28(a) and $x_i[n] = \frac{1}{2}(-1)^n$. Compare these responses to those you determined in part (iii).

2.29. (a) Consider a time-invariant system with input $x(t)$ and output $y(t)$. Show that if $x(t)$ is periodic with period T, then so is $y(t)$. Show that the analogous result also holds in discrete time.

(b) Consider the system

$$y[n] = x^2[n]$$

Find a periodic input so that the fundamental period of the output is smaller than the fundamental period of $x[n]$. Find a second periodic input where the fundamental periods of $x[n]$ and $y[n]$ are the same.

(c) Provide an example (either in continuous time or in discrete time) of a linear *time-varying* system and a periodic input to this system for which the corresponding output is *not* periodic.

2.30. (a) Show that causality for a continuous-time linear system is equivalent to the following statement:

> For any time t_o and any input $x(t)$ so that $x(t) = 0$ for $t < t_0$, the corresponding output $y(t)$ must also be zero for $t < t_0$.

The analogous statement can be made for discrete-time linear systems.

(b) Find a nonlinear system that satisfies this condition but is not causal.

(c) Find a nonlinear system that is causal but does not satisfy this condition.

(d) Show that invertibility for a discrete-time linear system is equivalent to the following statement:

> The only input that produces the output $y[n] = 0$ for all n is $x[n] = 0$ for all n.

The analogous statement is also true for continuous-time linear systems.

(e) Find a nonlinear system that satisfies the condition of part (d) but is not invertible.

2.31. In this problem we illustrate one of the most important consequences of the properties of linearity and time invariance. Specifically, once we know the response of a linear system or of a linear time-invariant (LTI) system to a single input or the responses to several inputs, we can directly compute the responses to many other input signals. Much of the remainder of this book deals with a thorough exploitation of this fact in order to develop results and techniques for analyzing and synthesizing LTI systems.

(a) Consider an LTI system whose response to the signal $x_1(t)$ in Figure P2.31-1(a) is

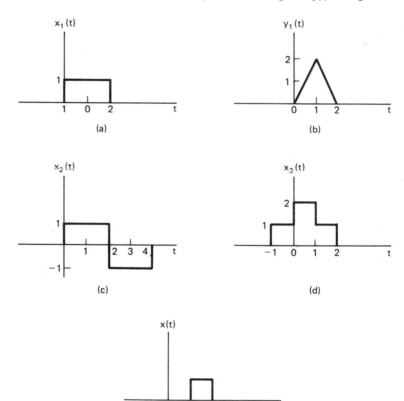

Figure P2.31-1

the signal $y_1(t)$ illustrated in Figure P2.31-1(b). Determine and sketch carefully the response of the system to the input $x_2(t)$ depicted in Figure P2.31-1(c).

(b) Determine and sketch the response of the system considered in part (a) to the input $x_3(t)$ shown in Figure P2.31-1(d).

(c) Suppose that a second LTI system has the following output $y(t)$ when the input is the unit step $x(t) = u(t)$:

$$y(t) = e^{-t}u(t) + u(-1 - t)$$

Determine and sketch the response of this system to the input $x(t)$ shown in Figure P2.31-1(e).

(d) Suppose that a particular discrete-time linear (but possibly not time-invariant) system has the responses $y_1[n]$, $y_2[n]$, and $y_3[n]$ to the input signals $x_1[n]$, $x_2[n]$, and $x_3[n]$, respectively, as illustrated in Figure P2.31-2(a). If the input to this system is $x[n]$ as illustrated in Figure P2.31-2(b), what is the output $y[n]$?

(e) If an LTI system has the response $y_1[n]$ to the input $x_1[n]$ as in Figure P2.31-2(a), what would its responses be to $x_2[n]$ and $x_3[n]$?

(f) A particular linear system has the property that the response to t^k is $\cos kt$. What is the response of this system to the input

$$x_1(t) = \pi + 6t^2 - 47t^5 + \sqrt{e}\,t^6?$$

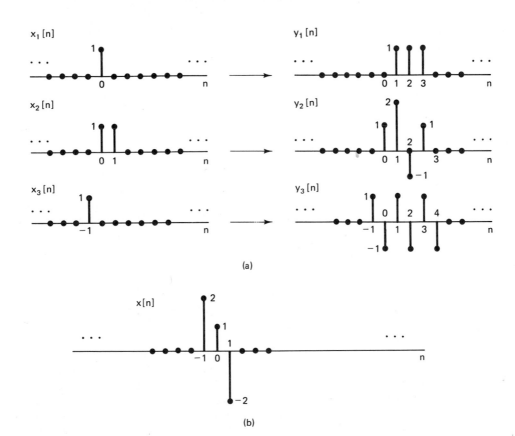

(a)

(b)

Figure P2.31-2

What is the response to the input

$$x_2(t) = \frac{1 + t^{10}}{1 + t^2} ?$$

Hint: See part (a) of Problem 2.8.

2.32. Consider the feedback system of Figure P2.32. Assume that $y[n] = 0$ for $n < 0$.
  (a) Sketch the output when $x[n] = \delta[n]$.
  (b) Sketch the output when $x[n] = u[n]$.

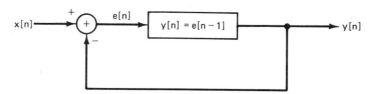

Figure P2.32

2.33. (a) Let S denote an incrementally linear system, and let $x_1[n]$ be an arbitrary input signal to S with corresponding output $y_1[n]$. Consider the system illustrated in Figure P2.33-1. Show that this system is linear and that, in fact, the overall input-output relationship between $x[n]$ and $y[n]$ does not depend on the particular choice of $x_1[n]$.

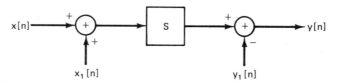

Figure P2.33-1

  (b) Use the result of part (a) to show that S can be represented in the form shown in Figure 2.41.
  (c) Which of the following systems are incrementally linear? Justify your answers, and if a system is incrementally linear, identify the linear system L and the zero-input response $y_0[n]$ or $y_0(t)$ for the representation of the system as shown in Figure 2.41.
  (i) $y[n] = n + x[n] + 2x[n + 4]$
  (ii) $y[n] = \begin{cases} n/2, & n \text{ even} \\ (n - 1)/2 + \displaystyle\sum_{k=-\infty}^{(n-1)/2} x[k], & n \text{ odd} \end{cases}$
  (iii) $y[n] = \begin{cases} x[n] - x[n - 1] + 3 & \text{if } x[0] \geq 0 \\ x[n] - x[n - 1] - 3 & \text{if } x[0] < 0 \end{cases}$
  (iv) The system depicted in Figure P2.33-2(a).
  (v) The system depicted in Figure P2.33-2(b).
  (d) Suppose that a particular incrementally linear system has a representation as in Figure 2.41, with L denoting the linear system and $y_0[n]$ the zero-input response. Show that S is time-invariant if and only if L is a time-invariant system *and* $y_0[n]$ is constant.

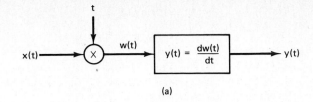

(a)

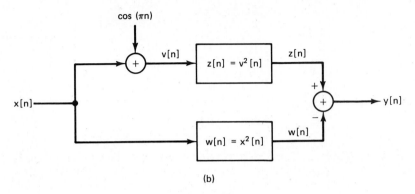

(b)

Figure P2.33-2

In Chapter 2 we introduced and discussed a number of basic system properties. Two of these, linearity and time invariance, play a fundamental role in signal and system analysis because of the many physical processes that can be modeled by linear time-invariant (LTI) systems and because such systems can be analyzed in great detail. The basic objectives of this book are to develop an understanding of the properties and tools for analyzing signals and LTI systems and to provide an introduction to several of the very important applications of these tools. In this chapter we begin this development by deriving and examining a fundamental and extremely useful representation for LTI systems and by introducing an important class of such systems.

One of the primary reasons for the amenability to analysis of LTI systems is that any such system possesses the superposition property described in eqs. (2.75) and (2.76) for discrete-time LTI systems. Similarly, if the input to a continuous-time LTI system consists of a linear combination of signals

$$x(t) = a_1 x_1(t) + a_2 x_2(t) + a_3 x_3(t) + \ldots \tag{3.1}$$

then by superposition the output is given by

$$y(t) = a_1 y_1(t) + a_2 y_2(t) + a_3 y_3(t) + \ldots \tag{3.2}$$

where $y_k(t)$ is the response to $x_k(t)$, $k = 1, 2, 3, \ldots$. Therefore, if we can represent the input to an LTI system in terms of a set of basic signals, we can then use superposition to compute the output of the system in terms of its responses to these basic signals.

Linear Time-Invariant Systems

As we will see in the next section, one of the important characteristics of the unit impulse, both in discrete and in continuous time, is that it can be used as a building block to represent very general signals. This fact, together with the properties of superposition and time invariance, will allow us to develop a complete characterization of any LTI system in terms of its response to a unit impulse. This representation, referred to as the convolution sum in the discrete-time case and the convolution integral in continuous time, provides considerable analytical convenience in dealing with LTI systems. Following our development of the convolution sum in Section 3.2 and the convolution integral in Section 3.3, we use these characterizations in Section 3.4 to examine some of the other properties of LTI systems. In Sections 3.5 and 3.6 we introduce the class of continuous-time systems described by linear constant-coefficient differential equations and its discrete-time counterpart, the class of systems described by linear constant-coefficient difference equations. We will return to examine these two very important classes of systems on numerous occasions in subsequent chapters as we continue our development of the techniques for analyzing signals and LTI systems.

3.1 THE REPRESENTATION OF SIGNALS IN TERMS OF IMPULSES

As mentioned in Section 3.0, the continuous-time unit impulse and the discrete-time unit impulse can each be used as the basic signal from which we can construct an extremely broad class of signals. To see how this construction is developed, let us first examine the discrete-time case. Consider the signal $x[n]$ depicted in Figure 3.1(a). In the remaining parts of this figure we have depicted five time-shifted, scaled unit impulse sequences, where the scaling on each impulse equals the value of $x[n]$ at the particular time instant at which the unit sample is located. For example,

$$x[-1]\delta[n+1] = \begin{cases} x[-1], & n = -1 \\ 0, & n \neq -1 \end{cases}$$

$$x[0]\delta[n] = \begin{cases} x[0], & n = 0 \\ 0, & n \neq 0 \end{cases}$$

$$x[1]\delta[n-1] = \begin{cases} x[1], & n = 1 \\ 0, & n \neq 1 \end{cases}$$

Therefore, the sum of the five sequences, in the figure, that is

$$x[-2]\,\delta[n+2] + x[-1]\,\delta[n+1] + x[0]\,\delta[n] + x[1]\,\delta[n-1] + x[2]\,\delta[n-2] \qquad (3.3)$$

equals $x[n]$ for $-2 \leq n \leq 2$. More generally, by including additional shifted, scaled impulses, we can write that

$$x[n] = \ldots + x[-3]\,\delta[n+3] + x[-2]\,\delta[n+2] + x[-1]\,\delta[n+1] + x[0]\,\delta[n]$$
$$+ x[1]\,\delta[n-1] + x[2]\,\delta[n-2] + x[3]\,\delta[n-3] + \ldots \qquad (3.4)$$

For any value of n only one of the terms on the right-hand side of eq. (3.4) is nonzero, and the scaling on that term is precisely $x[n]$. Writing this summation in a more

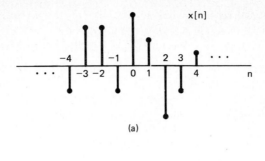

(a)

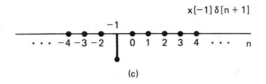

(b)

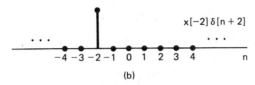

(c)

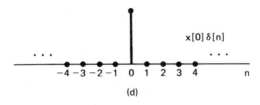

(d)

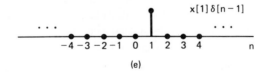

(e)

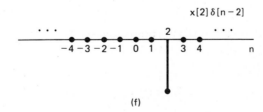

(f)

Figure 3.1 Decomposition of a discrete-time signal into a weighted sum of shifted impulses.

compact form, we have

$$x[n] = \sum_{k=-\infty}^{+\infty} x[k]\, \delta[n-k] \tag{3.5}$$

This corresponds to the representation of an arbitrary sequence as a linear combination of shifted unit impulses $\delta[n-k]$, where the weights in this linear combination are $x[k]$. As an example, consider $x[n] = u[n]$, the unit step. In this case since $u[k] = 0$ for $k < 0$ and $u[k] = 1$ for $k \geq 0$, eq. (3.5) becomes

$$u[n] = \sum_{k=0}^{+\infty} \delta[n-k]$$

which is identical to the expression we derived in Section 2.4 [see eq. (2.30)]. Equation (3.5) is called the *sifting* property of the discrete-time unit impulse.

For the continuous-time case a corresponding representation can be developed again in terms of the unit impulse. Specifically, consider the pulse or "staircase" approximation, $\hat{x}(t)$, to a continuous-time signal $x(t)$, as illustrated in Figure 3.2(a). In a manner similar to the discrete-time case, this approximation can be expressed as a linear combination of delayed pulses as illustrated in Figure 3.2(a)–(e). If we define

$$\delta_\Delta(t) = \begin{cases} \dfrac{1}{\Delta}, & 0 < t < \Delta \\ 0, & \text{otherwise} \end{cases} \tag{3.6}$$

then, since $\Delta\delta_\Delta(t)$ has unit amplitude, we have the expression

$$\hat{x}(t) = \sum_{k=-\infty}^{\infty} x(k\Delta)\, \delta_\Delta(t - k\Delta)\Delta \tag{3.7}$$

From Figure 3.2 we see that, as in the discrete-time case, for any value of t, exactly one term in the summation on the right-hand side of eq. (3.7) is nonzero.

As we let Δ approach 0, the approximation $\hat{x}(t)$ becomes better and better, and in the limit equals $x(t)$. Therefore,

$$x(t) = \lim_{\Delta \to 0} \sum_{k=-\infty}^{+\infty} x(k\Delta)\, \delta_\Delta(t - k\Delta)\Delta \tag{3.8}$$

Also, as $\Delta \to 0$, the summation in eq. (3.8) approaches an integral. This can be seen most easily by considering the graphical interpretation of eq. (3.8) illustrated in Figure 3.3. Here we have illustrated the signals $x(\tau)$, $\delta_\Delta(t - \tau)$, and their product. We have also indicated a shaded region whose area approaches the area under $x(\tau)\delta_\Delta(t - \tau)$ as $\Delta \to 0$. Note that the shaded region has an area equal to $x(m\Delta)$ where $t - \Delta < m\Delta < t$. Furthermore, for this value of t only the term with $k = m$ is nonzero in the summation in eq. (3.8), and thus the right-hand side of this equation also equals $x(m\Delta)$. Consequently, we have from eq. (3.8) and from the preceding argument that $x(t)$ equals the limit as $\Delta \to 0$ of the area under $x(\tau)\delta_\Delta(t - \tau)$. Moreover, from eq. (2.22), we know that the limit as $\Delta \to 0$ of $\delta_\Delta(t)$ is the unit impulse function $\delta(t)$. Consequently,

$$x(t) = \int_{-\infty}^{+\infty} x(\tau)\, \delta(t - \tau)\, d\tau \tag{3.9}$$

As in discrete time we refer to eq. (3.9) as the *sifting* property of the continuous-time impulse. We note that for the specific example of $x(t) = u(t)$, eq. (3.9) becomes

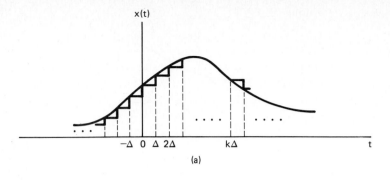

(a)

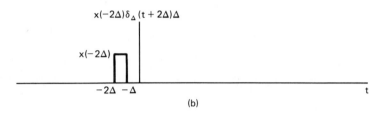

(b)

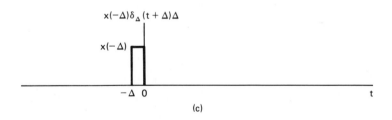

(c)

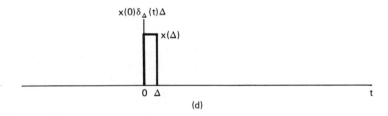

(d)

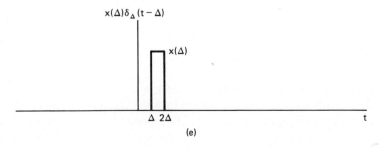

(e)

Figure 3.2 Staircase approximation to a continuous-time signal.

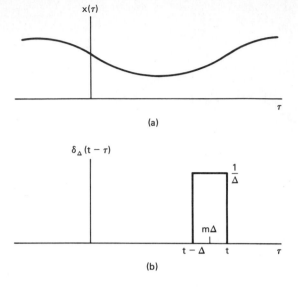

(a)

(b)

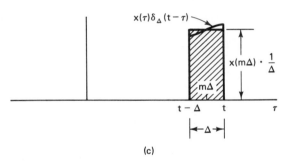

(c)

$x(\tau)\delta_\Delta(t-\tau)$

$x(m\Delta)\cdot\dfrac{1}{\Delta}$

Figure 3.3 Graphical interpretation of eq. (3.8).

$$u(t) = \int_{-\infty}^{+\infty} u(\tau)\,\delta(t-\tau)\,d\tau = \int_{0}^{\infty} \delta(t-\tau)\,d\tau \qquad (3.10)$$

since $u(\tau) = 0$ for $\tau < 0$ and $u(\tau) = 1$ for $\tau > 0$. Equation (3.10) is identical to the expression derived in Section 2.3 [see eq. (2.23)].

Note that we could have derived eq. (3.9) directly using properties of the unit impulse discussed in Section 2.3.2. Specifically, as illustrated in Figure 3.4(b) the signal $\delta(t-\tau)$ (for t fixed) is a unit impulse located at $\tau = t$. Thus, as shown in Figure 3.4(c), the signal $x(\tau)\,\delta(t-\tau)$ equals $x(t)\,\delta(t-\tau)$ [i.e., it is a scaled impulse at $\tau = t$ with area equal to the value of $x(t)$]. Consequently, the integral of this signal from $\tau = -\infty$ to $\tau = +\infty$ equals $x(t)$, that is,

$$\int_{-\infty}^{+\infty} x(\tau)\,\delta(t-\tau)\,d\tau = \int_{-\infty}^{+\infty} x(t)\,\delta(t-\tau)\,d\tau = x(t)\int_{-\infty}^{+\infty} \delta(t-\tau)\,d\tau = x(t)$$

Although this derivation follows directly from Section 2.3.2, we have included the derivation given in eqs. (3.6)–(3.9) to stress the similarities with the discrete-time case and in particular to emphasize the interpretation of eq. (3.9) as representing the

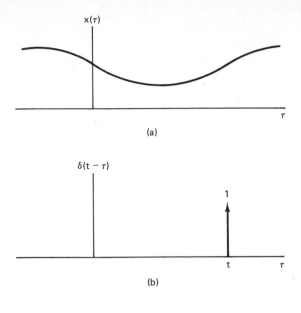

(a)

(b)

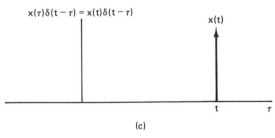

(c)

Figure 3.4 (a) Arbitrary signal $x(\tau)$; (b) impulse $\delta(t - \tau)$ as a function of τ with t fixed; (c) product of these two signals.

signal $x(t)$ as a "sum" (i.e., an integral) of weighted, shifted impulses. In the next two sections we use this representation of signals in discrete and continuous time to help us obtain an important characterization of the response of LTI systems.

3.2 DISCRETE-TIME LTI SYSTEMS: THE CONVOLUTION SUM

Consider a discrete-time linear system and an arbitrary input $x[n]$ to that system. As we saw in Section 3.1, we can express $x[n]$ as a linear combination of shifted unit samples, in the form of eq. (3.5), which we repeat here for convenience.

[eq. (3.5)]
$$x[n] = \sum_{k=-\infty}^{+\infty} x[k]\, \delta[n - k]$$

Using the superposition property of linear systems [eqs. (2.75) and (2.76)], the output $y[n]$ can be expressed as a linear combination of the responses of the system to shifted unit samples. Specifically, if we let $h_k[n]$ denote the response of a linear system to the shifted unit sample $\delta[n - k]$, then the response of the system to an arbitrary input

$x[n]$ can be expressed as

$$y[n] = \sum_{k=-\infty}^{+\infty} x[k]\, h_k[n] \tag{3.11}$$

According to eq. (3.11), if we know the response of a linear system to the set of displaced unit samples, we can construct the response to an arbitrary input. An interpretation of eq. (3.11) is illustrated in Figure 3.5. In Figure 3.5(a) we have illustrated a particular signal $x[n]$ that is nonzero only for $n = -1, 0$, and 1. This signal is applied as the input to a linear system whose responses to the signals $\delta[n + 1]$, $\delta[n]$, and $\delta[n - 1]$ are depicted in Figure 3.5(b). Since $x[n]$ can be written as a linear combination of $\delta[n + 1]$, $\delta[n]$, and $\delta[n - 1]$, superposition allows us to write the response to $x[n]$ as a linear combination of the responses to the individual shifted impulses. The individual shifted and scaled impulses that comprise $x[n]$ are illustrated on the left-hand side of Figure 3.5(c), while the responses to these component signals are pictured on the right-hand side. Finally, in Figure 3.5(d) we have depicted the actual input $x[n]$, which is the sum of its components in Figure 3.5(c) and the actual output $y[n]$, which by superposition is the sum of *its* components in Figure 3.5(c). Thus, the response at time n of a linear system is simply the superposition of the responses due to each successive input value.

In general, of course, the responses $h_k[n]$ need not be related to each other for different values of k. However, if the linear system is also time-invariant, then

$$h_k[n] = h_0[n - k] \tag{3.12}$$

Specifically, since $\delta[n - k]$ is a time-shifted version of $\delta[n]$, the response $h_k[n]$ is a

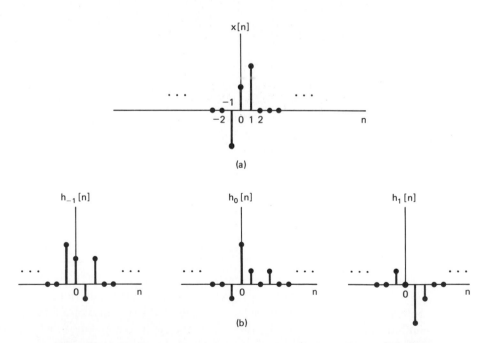

Figure 3.5 Graphical interpretation of the response of a discrete-time linear system as expressed in eq. (3.11).

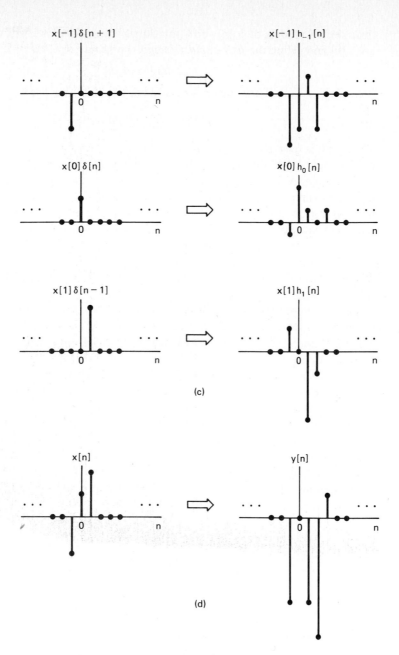

Figure 3.5 (cont.)

time-shifted version of $h_0[n]$. For notational convenience we will drop the subscript on $h_0[n]$ and define the *unit impulse (sample) response*, $h[n]$, as

$$h[n] = h_0[n] \qquad (3.13)$$

(i.e., $\delta[n] \rightarrow h[n]$). Then for an LTI system, eq. (3.11) becomes

$$y[n] = \sum_{k=-\infty}^{+\infty} x[k]h[n-k] \qquad (3.14)$$

This result is referred to as the *convolution sum* or *superposition sum*, and the operation on the right-hand side of eq. (3.14) is known as the *convolution* of the sequences $x[n]$ and $h[n]$ which we will represent symbolically as $y[n] = x[n] * h[n]$. Note that eq. (3.14) expresses the response of an LTI system to an arbitrary input in terms of its response to the unit impulse. From this fact we immediately see that an LTI system is completely characterized by its impulse response. We will develop a number of the implications of this observation in this and in the following chapters.

The interpretation of eq. (3.14) is similar to that given previously for eq. (3.11), where in this case the response due to the input $x[k]$ applied at time k is $x[k] h[n-k]$, which is simply a shifted and scaled version of $h[n]$. As before, the actual output is the superposition of all these responses. Thus at any fixed time n, the output $y[n]$ consists of the sum over all values of k of the numbers $x[k] h[n-k]$. As illustrated in Figure 3.6, this interpretation of eq. (3.14) leads directly to a very useful way in which to visualize the calculation of $y[n]$ using the convolution sum. Specifically, consider the evaluation of the output for some specific value of n. In Figure 3.6(a) we have depicted $h[k]$, and in Figure 3.6(b) we have shown $h[n-k]$ as a function of k with n fixed. Note that $h[n-k]$ is obtained from $h[k]$ by reflection about the origin, followed by a shift to the right by n if n is positive and to the left by $|n|$ if n is negative. Finally, in Figure 3.6(c) we have illustrated $x[k]$. The output $y[n]$ for this specific value of n is then calculated by weighting each value of $x[k]$ by the corresponding value of $h[n-k]$, that is, by multiplying the corresponding points in Figure 3.6(b) and (c), and then summing these products. To illustrate this procedure, we now consider two examples.

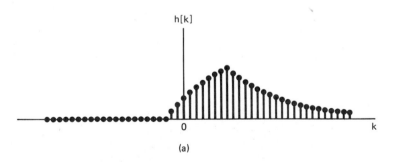

(a)

Figure 3.6 Interpretation of eq. (3.14). The signal $h[n-k]$ (as a function of k with n fixed) is obtained by reflection and shifting from the unit sample response $h[k]$. The response $y[n]$ is obtained by multiplying the signals $x[k]$ and $h[n-k]$ in (b) and (c) and then by summing the products over all values of k.

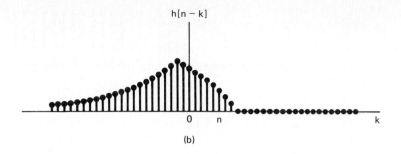

(b)

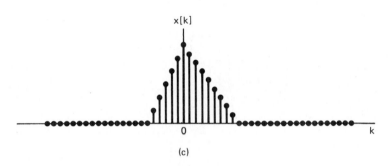

(c)

Figure 3.6 cont.

Example 3.1

Let us consider an input $x[n]$ and unit sample response $h[n]$ given by

$$x[n] = \alpha^n u[n]$$

$$h[n] = u[n]$$

with $0 < \alpha < 1$. In Figure 3.7 we have shown $h[k]$, $h[-k]$, $h[-1-k]$, and $h[1-k]$, that is, $h[n-k]$ for $n = 0$, -1, and $+1$, and $h[n-k]$ for an arbitrary positive value of n and an arbitrary negative value of n. Finally, $x[k]$ is illustrated in Figure 3.7(g). From this figure we note that for $n < 0$ there is no overlap between the nonzero points in $x[k]$ and $h[n-k]$. Thus, for $n < 0$, $x[k]h[n-k] = 0$ for all values of k, and hence $y[n] = 0$, $n < 0$. For $n \geq 0$, $x[k]h[n-k]$ is given by

$$x[k]h[n-k] = \begin{cases} \alpha^k, & 0 \leq k \leq n \\ 0, & \text{otherwise} \end{cases}$$

Thus, for $n \geq 0$,

$$y[n] = \sum_{k=0}^{n} \alpha^k$$

and using the result of Problem 2.8, we can write this as

$$y[n] = \frac{1 - \alpha^{n+1}}{1 - \alpha} \qquad \text{for } n \geq 0$$

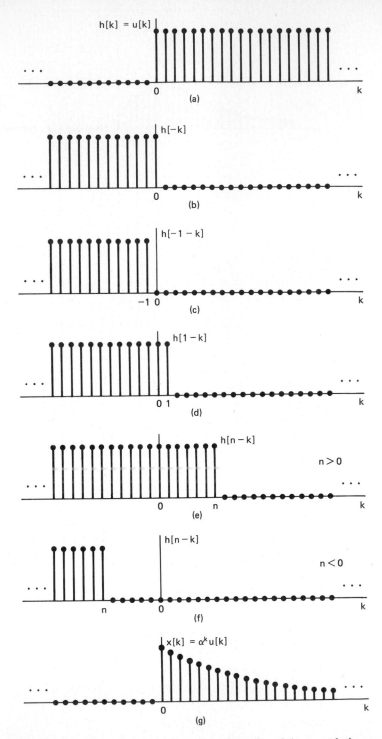

Figure 3.7 Graphical interpretation of the calculation of the convolution sum for Example 3.1.

Thus, for all n, $y[n]$ is given by

$$y[n] = \left(\frac{1 - \alpha^{n+1}}{1 - \alpha}\right) u[n]$$

and is sketched in Figure 3.8.

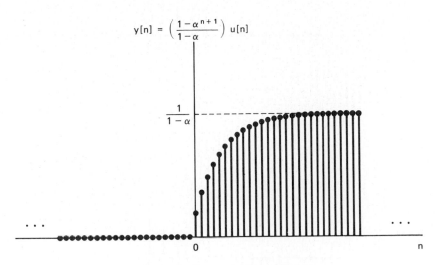

Figure 3.8 Output for Example 3.1.

Example 3.2

As a second example, consider the two sequences $x[n]$ and $h[n]$ given by

$$x[n] = \begin{cases} 1, & 0 \le n \le 4 \\ 0, & \text{otherwise} \end{cases}$$

$$h[n] = \begin{cases} \alpha^n, & 0 \le n \le 6 \\ 0, & \text{otherwise} \end{cases}$$

These signals are depicted in Figure 3.9. In order to calculate the convolution of these two signals, it is convenient to consider five separate intervals for n. This is illustrated in Figure 3.10.

Interval 1. For $n < 0$ there is no overlap between the nonzero portions of $x[k]$ and $h[n - k]$, and consequently $y[n] = 0$.

Interval 2. For $0 \le n \le 4$ the product $x[k]h[n - k]$ is given by

$$x[k]h[n - k] = \begin{cases} \alpha^{n-k}, & 0 \le k \le n \\ 0, & \text{otherwise} \end{cases}$$

Thus, in this interval,

$$y[n] = \sum_{k=0}^{n} \alpha^{n-k}$$

Changing the variable of summation from k to $r = n - k$ we obtain

$$y[n] = \sum_{r=0}^{n} \alpha^r = \frac{1 - \alpha^{n+1}}{1 - \alpha}$$

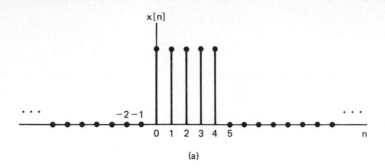

(a)

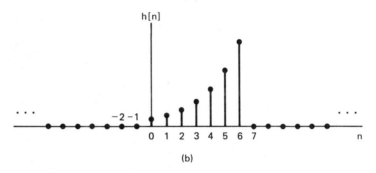

(b)

Figure 3.9 The signals to be convolved in Example 3.2.

Interval 3. For $n > 4$ but $n - 6 \leq 0$ (i.e., $4 < n \leq 6$), $x[k]h[n - k]$ is given by

$$x[k]h[n - k] = \begin{cases} \alpha^{n-k}, & 0 \leq k \leq 4 \\ 0, & \text{otherwise} \end{cases}$$

Thus, in this interval

$$y[n] = \sum_{k=0}^{4} \alpha^{n-k} = \alpha^n \sum_{k=0}^{4} (\alpha^{-1})^k = \alpha^n \frac{1 - \alpha^{-5}}{1 - \alpha^{-1}} = \frac{\alpha^{n-4} - \alpha^{n+1}}{1 - \alpha}$$

Interval 4. For $n > 6$ but $n - 6 \leq 4$ (i.e., for $6 < n \leq 10$),

$$x[k]h[n - k] = \begin{cases} \alpha^{n-k}, & (n - 6) \leq k \leq 4 \\ 0, & \text{otherwise} \end{cases}$$

so that

$$y[n] = \sum_{k=n-6}^{4} \alpha^{n-k}$$

Letting $r = k - n + 6$ we obtain

$$y[n] = \sum_{r=0}^{10-n} \alpha^{6-r} = \alpha^6 \sum_{r=0}^{10-n} (\alpha^{-1})^r = \alpha^6 \frac{1 - \alpha^{n-11}}{1 - \alpha^{-1}} = \frac{\alpha^{n-4} - \alpha^7}{1 - \alpha}$$

Interval 5. For $(n - 6) < 4$, or equivalently $n > 10$, there is no overlap between the nonzero portions of $x[k]$ and $h[n - k]$, and hence

$$y[n] = 0$$

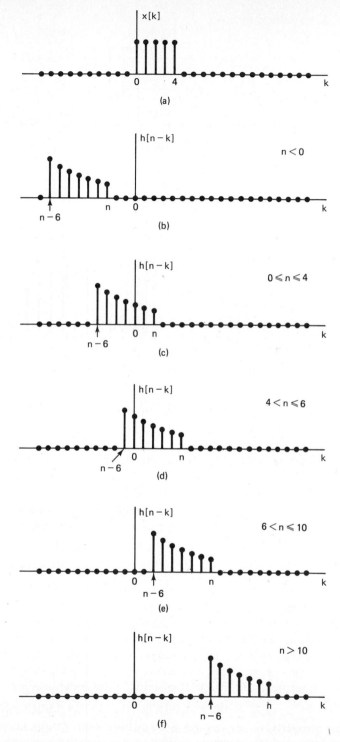

Figure 3.10 Graphical interpretation of the convolution performed in Example 3.2.

Summarizing, then,

$$y[n] = \begin{cases} 0, & n < 0 \\ \dfrac{1 - \alpha^{n+1}}{1 - \alpha}, & 0 \le n \le 4 \\ \dfrac{\alpha^{n-4} - \alpha^{n+1}}{1 - \alpha}, & 4 < n \le 6 \\ \dfrac{\alpha^{n-4} - \alpha^{7}}{1 - \alpha}, & 6 < n \le 10 \\ 0, & 10 < n \end{cases}$$

which is pictured in Figure 3.11.

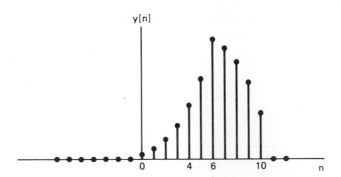

Figure 3.11 Result of performing the convolution in Example 3.2.

These two examples illustrate the usefulness of interpreting the calculation of the convolution sum graphically. In the remainder of this section we examine a number of important properties of convolution that we will find useful on many occasions.

The first basic property of convolution is that it is a *commutative* operation, that is,

$$x[n] * h[n] = h[n] * x[n] \tag{3.15}$$

This is proved in a straightforward manner by means of a substitution of variables in eq. (3.14). In particular, letting $r = n - k$ or, equivalently, $k = n - r$, eq. (3.14) becomes

$$x[n] * h[n] = \sum_{k=-\infty}^{+\infty} x[k]\, h[n - k] = \sum_{r=-\infty}^{+\infty} x[n - r]\, h[r] = h[n] * x[n] \tag{3.16}$$

Thus, we see that with this substitution of variables, the roles of $x[n]$ and $h[n]$ are interchanged. According to eq. (3.16), the output of an LTI system with input $x[n]$ and unit sample response $h[n]$ is identical to the output of an LTI system with input $h[n]$ and unit sample response $x[n]$. For example, we could have calculated the convolution in Example 3.2 by first reflecting and shifting $x[k]$, then by multiplying the signals $x[n - k]$ and $h[k]$, and finally by summing the products for all values of k.

A second useful property of convolution is that it is *associative*, that is,

$$x[n] * (h_1[n] * h_2[n]) = (x[n] * h_1[n]) * h_2[n] \tag{3.17}$$

This property is proven by a straightforward manipulation of the summations involved, and an example verifying this property is given in Problem 3.5. The interpretation of the associative property is indicated in Figure 3.12(a) and (b). The systems shown in these block diagrams are LTI systems with the indicated unit sample responses. This pictorial representation is a particularly convenient way in which to denote LTI systems in block diagrams, and it also reemphasizes the fact that the impulse response of an LTI system completely characterizes its behavior.

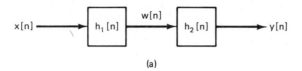

(a)

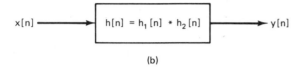

(b)

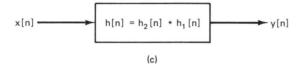

(c)

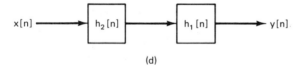

(d)

Figure 3.12 Associative property of convolution and the implication of this and the commutative property for the series interconnection of LTI systems.

In Figure 3.12(a),

$$y[n] = w[n] * h_2[n]$$
$$= (x[n] * h_1[n]) * h_2[n]$$

In Figure 3.12(b),

$$y[n] = x[n] * h[n]$$
$$= x[n] * (h_1[n] * h_2[n])$$

According to the associative property, the series interconnection of the two systems in Figure 3.12(a) is equivalent to the single system in Figure 3.12(b). This can be generalized to an arbitrary number of LTI systems in cascade. Also, as a consequence of the associative property in conjunction with the commutative property of convolution, the overall unit sample response of a cascade of LTI systems is independent of the order in which the systems are cascaded. This can be seen for the case of two systems, again by reference to Figure 3.12. From the commutative property, the system of Figure 3.12(b) is equivalent to the system of Figure 3.12(c). Then, from the associative property, this is in turn equivalent to the system of Figure 3.12(d), which we note is a cascade combination of two systems as in Figure 3.12(a) but with the order of the cascade interchanged. We note also that because of the associative property of convolution, the expression

$$y[n] = x[n] * h_1[n] * h_2[n] \qquad (3.18)$$

is unambiguous. That is, according to eq. (3.17), it does not matter in which order we convolve these signals.

A third property of convolution is the *distributive* property, that is, that convolution distributes over addition so that

$$x[n] * (h_1[n] + h_2[n]) = x[n] * h_1[n] + x[n] * h_2[n] \qquad (3.19)$$

This can also be verified in a straightforward manner. Again, the distributive property has a useful interpretation. Consider two LTI systems in parallel, as indicated in Figure 3.13(a). The two systems $h_1[n]$ and $h_2[n]$ have identical inputs and their outputs are added.

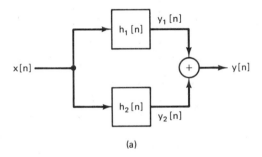

(a)

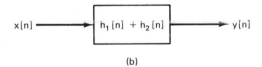

(b)

Figure 3.13 Interpretation of the distributive property of convolution for a parallel interconnection of LTI systems.

Since

$$y_1[n] = x[n] * h_1[n]$$

and

$$y_2[n] = x[n] * h_2[n]$$

the system of Figure 3.13(a) has output

$$y[n] = x[n] * h_1[n] + x[n] * h_2[n] \qquad (3.20)$$

corresponding to the right-hand side of eq. (3.19). The system of Figure 3.13(b) has output

$$y[n] = x[n] * (h_1[n] + h_2[n]) \qquad (3.21)$$

corresponding to the left-hand side of eq. (3.19). Thus, by virtue of the distributive property of convolution, a parallel combination of LTI systems can be replaced by a single LTI system whose unit sample response is the sum of the individual unit sample responses in the parallel combination.

In this section we have derived several important results for discrete-time LTI systems. Specifically, we have derived the convolution sum formula for the output of an LTI system, and from this have seen that the unit sample response completely characterizes the behavior of the system. We have also examined several important properties of convolution (commutativity, associativity, and distributivity) and from these have deduced several properties concerning the interconnection of LTI systems. It is important to emphasize that the results in this section hold in general *only* for LTI systems. In particular, the unit impulse response of a nonlinear system does *not* completely characterize the behavior of the system. For example, consider a system with unit impulse response

$$h[n] = \begin{cases} 1, & n = 0, 1 \\ 0, & \text{otherwise} \end{cases} \tag{3.22}$$

There is exactly one LTI system with this as its unit impulse response, and we can find it by substituting eq. (3.22) into the convolution sum, eq. (3.14), to obtain

$$y[n] = x[n] + x[n-1] \tag{3.23}$$

However, there are *many* nonlinear systems with this response to the input $\delta[n]$. For example, both of the following systems have this property:

$$y[n] = (x[n] + x[n-1])^2$$

$$y[n] = \max (x[n], x[n-1])$$

In addition, it is not true in general that the order in which nonlinear systems are cascaded can be changed without changing the overall response. For example, if we have two memoryless systems, one being multiplication by 2 and the other squaring the input, then if we multiply first and square second, we obtain

$$y[n] = 4x[n]^2$$

However, if we multiply by 2 after squaring, we have

$$y[n] = 2x[n]^2$$

Thus, being able to interchange the order of systems in a cascade is a characteristic particular to LTI systems. In fact, as shown in Problem 3.20, we need both linearity and time invariance in order for this property to be true in general.

As the preceding discussion indicates, discrete-time LTI systems have a number of properties not possessed by other classes of systems. As we will see, these properties help to facilitate the analysis of LTI systems and allow us to gain a detailed understanding of their behavior. The basis for our initial analysis in this section into the properties of LTI systems was the convolution sum, which we derived using the properties of superposition and time-invariance together with the representation of input signals as weighted sums of shifted impulses. In the next section we derive an analogous representation for the response of a continuous-time LTI system, again using superposition, time invariance, and the sifting property of the unit impulse. In Section 3.4 we use the representations in this and the next section to obtain very explicit characterizations for LTI systems of the properties introduced in Section 2.6.

3.3 CONTINUOUS-TIME LTI SYSTEMS: THE CONVOLUTION INTEGRAL

In analogy with the results derived and discussed in the preceding section, the goal of this section is to obtain a complete characterization of a continuous-time LTI system in terms of its unit impulse response. Specifically, consider a linear system with input $x(t)$ and output $y(t)$. In Section 3.1, eqs. (3.8) and (3.9), we saw how an arbitrary continuous-time signal could be expressed as the limiting form of a linear combination of shifted pulses:

$$x(t) = \lim_{\Delta \to 0} \sum_{k=-\infty}^{+\infty} x(k\Delta)\, \delta_\Delta(t - k\Delta)\Delta \qquad (3.24)$$

where $\delta_\Delta(t)$ is given by eq. (3.6). Proceeding in an analogous manner to Section 3.2, let us define $\hat{h}_{k\Delta}(t)$ as the response of an LTI system to the input $\delta_\Delta(t - k\Delta)$. Then, from eq. (3.24) and the superposition property of linear systems,

$$y(t) = \lim_{\Delta \to 0} \sum_{k=-\infty}^{+\infty} x(k\Delta)\, \hat{h}_{k\Delta}(t)\Delta \qquad (3.25)$$

The interpretation of eq. (3.25) is similar to that for eq. (3.11) in discrete time. In particular, consider Figure 3.14, which is the continuous-time counterpart of Figure 3.5. In Figure 3.14(a) we have depicted the input $x(t)$ and its approximation $\hat{x}(t)$, while in Figure 3.14(b)–(d), we have shown the responses of the system to three of the weighted pulses in the expression for $\hat{x}(t)$. Then the output $\hat{y}(t)$ corresponding to $\hat{x}(t)$ is the superposition of all of these responses [Figure 3.14(e)]. What remains is to take the limit of $\hat{y}(t)$ as $\Delta \to 0$, as indicated in eq. (3.25) and as illustrated in Figure 3.14(f), to obtain the response $y(t)$ to the actual input $x(t)$. Note that since the pulse $\delta_\Delta(t - k\Delta)$ corresponds to a shifted unit impulse as $\Delta \to 0$, the response $\hat{h}_{k\Delta}(t)$ becomes the response to such an impulse in the limit. Therefore, if we let $h_\tau(t)$ denote the response at time t to the unit impulse $\delta(t - \tau)$ located at time τ, then

$$y(t) = \lim_{\Delta \to 0} \sum_{k=-\infty}^{+\infty} x(k\Delta)\, h_{k\Delta}(t)\Delta \qquad (3.26)$$

As $\Delta \to 0$, the summation on the right-hand side becomes an integral, as can be seen graphically in Figure 3.15. Therefore,

$$y(t) = \int_{-\infty}^{+\infty} x(\tau)\, h_\tau(t)\, d\tau \qquad (3.27)$$

The interpretation of eq. (3.27) is exactly analogous to the one for eq. (3.25). As we showed in Section 3.1, any input $x(t)$ can be represented as

$$x(t) = \int_{-\infty}^{+\infty} x(\tau)\, \delta(t - \tau)\, d\tau$$

That is, we can intuitively think of $x(t)$ as a "sum" of weighted shifted impulses, where the weight on the impulse $\delta(t - \tau)$ is $x(\tau)\, d\tau$. With this interpretation, eq. (3.27) simply represents the superposition of the responses to each of these inputs, and, by linearity, the weight on the response $h_\tau(t)$ to the shifted impulse $\delta(t - \tau)$ is also $x(\tau)\, d\tau$.

Equation (3.27) represents the general form of the response of a linear system in continuous time. If in addition to being linear the system is also time-invariant,

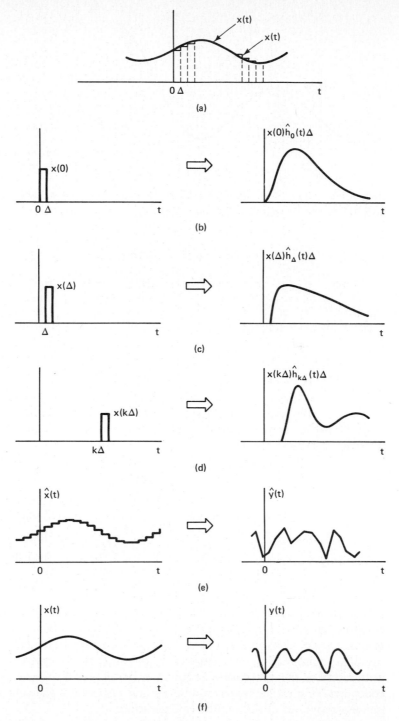

Figure 3.14 Graphical interpretation of the response of a continuous-time linear system as expressed in eq. (3.25).

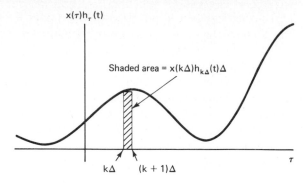

Figure 3.15 Graphical illustration of eqs. (3.26) and (3.27).

then $h_\tau(t) = h_0(t - \tau)$. Again for notational convenience we will drop the subscript and define the *unit impulse response* $h(t)$ as

$$h(t) = h_0(t) \qquad (3.28)$$

[i.e., $\delta(t) \longrightarrow h(t)$]. In this case, eq. (3.27) becomes

$$y(t) = \int_{-\infty}^{+\infty} x(\tau)h(t - \tau) \, d\tau \qquad (3.29)$$

Equation (3.29), referred to as the *convolution integral* or the *superposition integral*, is the continuous-time counterpart of the convolution sum of eq. (3.14) and corresponds to a representation of a continuous-time LTI system in terms of its response to a unit impulse. The convolution of two signals $x(t)$ and $h(t)$ will be represented symbolically as

$$y(t) = x(t) * h(t) \qquad (3.30)$$

While we have chosen to use the same symbol * to denote both discrete-time and continuous-time convolution, the context will generally be sufficient to distinguish the two cases.

Continuous-time convolution satisfies the same properties discussed for discrete-time convolution. In particular, continuous-time convolution is *commutative, associative*, and *distributive*. That is,

$$x(t) * h(t) = h(t) * x(t) \qquad \text{(commutativity)} \qquad (3.31)$$

$$x(t) * [h_1(t) * h_2(t)] = [x(t) * h_1(t)] * h_2(t) \qquad \text{(associativity)} \qquad (3.32)$$

$$x(t) * [h_1(t) + h_2(t)] = [x(t) * h_1(t)] + [x(t) * h_2(t)] \qquad \text{(distributivity)} \qquad (3.33)$$

These properties all have the same implications as those discussed for discrete-time convolution. As a consequence of the commutative property, the roles of input signal and impulse response are interchangeable. From the associative property a cascade combination of LTI systems can be condensed into a single system whose impulse response is the convolution of the individual impulse responses. Furthermore, the overall impulse response is unaffected by the order of the cascaded systems. Finally, as a result of the distributive property, a parallel combination of LTI systems is equivalent to a single system whose impulse response is the sum of the individual impulse responses in the parallel configuration.

It is important to emphasize again that these properties are particular to LTI systems. Just as in discrete time, a nonlinear continuous-time system is not completely characterized by its response to a unit impulse. Also, the overall impulse response of a cascade of two nonlinear systems (or even linear but time-varying systems) does depend upon the order in which the systems are cascaded.

The procedure for evaluating the convolution integral is quite similar to that for its discrete-time counterpart, the convolution sum. Specifically, in eq. (3.29) we see that for any value of t the output $y(t)$ is a weighted integral of the input, where the weight on $x(\tau)$ is $h(t - \tau)$. To evaluate this integral for a specific value of t we first obtain the signal $h(t - \tau)$ (regarded as a function of τ with t fixed) from $h(\tau)$ by a reflection about the origin plus a shift to the right by t if $t > 0$ or a shift to the left by $|t|$ for $t < 0$. We next multiply together the signals $x(\tau)$ and $h(t - \tau)$, and $y(t)$ is obtained by integrating the resulting product from $\tau = -\infty$ to $\tau = +\infty$. To illustrate the evaluation of the convolution integral, let us consider two examples.

Example 3.3

Let $x(t)$ be the input to an LTI system with unit impulse response $h(t)$, where

$$x(t) = e^{-at}u(t)$$

$$h(t) = u(t)$$

where $a > 0$. In Figure 3.16 we have depicted the functions $h(\tau)$, $x(\tau)$, and $h(t - \tau)$ for a negative value of t and for a positive value of t. From this figure we see that for $t < 0$ the product of $x(\tau)$ and $h(t - \tau)$ is zero and consequently $y(t)$ is zero. For $t > 0$,

$$x(\tau)h(t - \tau) = \begin{cases} e^{-a\tau}, & 0 < \tau < t \\ 0, & \text{otherwise} \end{cases}$$

From this expression we can compute $y(t)$ for $t > 0$ as

$$y(t) = \int_0^t e^{-a\tau}\, dt = -\frac{1}{a}e^{-a\tau}\Big|_0^t$$

$$= \frac{1}{a}(1 - e^{-at})$$

Thus, for all t, $y(t)$ is

$$y(t) = \frac{1}{a}(1 - e^{-at})u(t)$$

which is pictured in Figure 3.17.

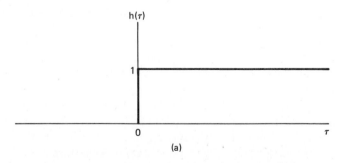

Figure 3.16 Calculation of the convolution integral for Example 3.3.

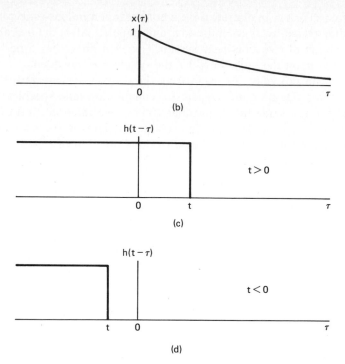

(b)

(c)

(d)

Figure 3.16 cont.

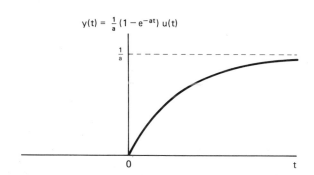

$$y(t) = \frac{1}{a}(1 - e^{-at})u(t)$$

Figure 3.17 Response of the system in Example 3.3 with impulse response $h(t) = u(t)$ to the input $x(t) = e^{-at}u(t)$.

Example 3.4

Consider the convolution of the following two signals:

$$x(t) = \begin{cases} 1, & 0 < t < T \\ 0, & \text{otherwise} \end{cases}$$

$$h(t) = \begin{cases} t, & 0 < t < 2T \\ 0, & \text{otherwise} \end{cases}$$

As in Example 3.2 for discrete-time convolution, it is convenient to consider the evaluation of $y(t)$ in separate intervals. In Figure 3.18 we have sketched $x(\tau)$ and have

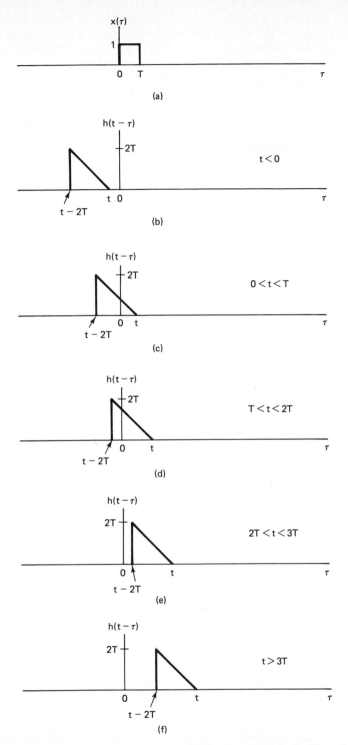

Figure 3.18 Signals $x(\tau)$ and $h(t - \tau)$ for different values of t for Example 3.4.

illustrated $h(t - \tau)$ in each of the intervals of interest. For $t < 0$ and for $3T < t$, $x(\tau)h(t - \tau) = 0$ and consequently $y(t) = 0$. For the other intervals, the product $x(\tau)h(t - \tau)$ is as indicated in Figure 3.19. Thus, for these three intervals, the integra-

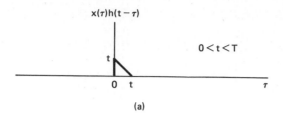

x(τ)h(t − τ)

0 < t < T

(a)

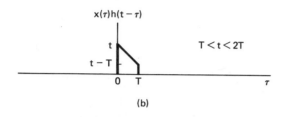

x(τ)h(t − τ)

T < t < 2T

(b)

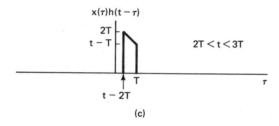

x(τ)h(t − τ)

2T < t < 3T

(c)

Figure 3.19 Product $x(\tau)h(t - \tau)$ for Example 3.4 for the three ranges of values of t for which this product is not identically zero. (See Figure 3.18.)

tion can be carried out graphically, with the result that

$$y(t) = \begin{cases} 0, & t < 0 \\ \tfrac{1}{2}t^2, & 0 < t < T \\ Tt - \tfrac{1}{2}T^2, & T < t < 2T \\ -\tfrac{1}{2}t^2 + Tt + \tfrac{3}{2}T^2, & 2T < t < 3T \\ 0, & 3T < t \end{cases}$$

which is depicted in Figure 3.20.

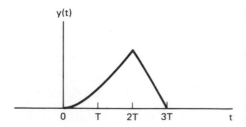

y(t)

Figure 3.20 Signal $y(t) = x(t) * h(t)$ for Example 3.4.

Linear Time-Invariant Systems   Chap. 3

From these examples we see, as we did for discrete-time convolution, that it is generally useful to consider the graphical evaluation of the convolution integral.

3.4 PROPERTIES OF LINEAR TIME-INVARIANT SYSTEMS

In the preceding two sections we developed the extremely important representations of continuous-time and discrete-time LTI systems in terms of their unit impulse responses. In discrete time this representation takes the form of the convolution sum, while its continuous-time counterpart is the convolution integral, both of which we repeat here for convenience.

$$y[n] = \sum_{k=-\infty}^{+\infty} x[k]h[n-k] = \sum_{k=-\infty}^{+\infty} h[k]x[n-k] = x[n] * h[n] \tag{3.34}$$

$$y(t) = \int_{-\infty}^{+\infty} x(\tau)h(t-\tau)\,d\tau = \int_{-\infty}^{+\infty} h(\tau)x(t-\tau)\,d\tau = x(t) * h(t) \tag{3.35}$$

As we have emphasized, one consequence of these representations is that the characteristics of an LTI system are completely determined by its impulse response. We have already deduced properties concerning series and parallel interconnections of LTI systems from this fact, and in this section we use the characterization of LTI systems in terms of their impulse responses in examining several other important system properties.

3.4.1 LTI Systems with and without Memory

Recall from Section 2.6.1 that a system is memoryless if its output at any time depends only on the value of the input at that same time. From eq. (3.34) we see that the only way that this can be true for a discrete-time LTI system is if $h[n] = 0$ for $n \neq 0$. In this case the impulse response has the form

$$h[n] = K\,\delta[n] \tag{3.36}$$

where $K = h[0]$ is a constant, and the system is then specified by the relation

$$y[n] = Kx[n] \tag{3.37}$$

If a discrete-time LTI system has an impulse response $h[n]$ which is not identically zero for $n \neq 0$, then the system has memory. An example of an LTI system with memory is the system given by eq. (3.23). The impulse response for this system, given in eq. (3.22), is nonzero for $n = 1$.

From eq. (3.35) we can deduce similar properties of continuous-time LTI systems with and without memory. In particular, a continuous-time LTI system is memoryless if $h(\tau) = 0$ for $\tau \neq 0$, and such a memoryless system has the form

$$y(t) = Kx(t) \tag{3.38}$$

for some constant K. This system has the impulse response

$$h(t) = K\,\delta(t) \tag{3.39}$$

Note that if $K = 1$ in eqs. (3.36) and (3.39), then these systems become *identity systems*, with input equal to the output and with unit impulse response equal to the unit impulse. In this case, the convolution sum and integral formulas imply that

$$x[n] = x[n] * \delta[n]$$
$$x(t) = x(t) * \delta(t)$$

which are nothing more than the familiar sifting properties

$$x[n] = \sum_{k=-\infty}^{+\infty} x[k]\, \delta[n - k]$$

$$x(t) = \int_{-\infty}^{+\infty} x(\tau)\, \delta(t - \tau)\, d\tau$$

3.4.2 Invertibility of LTI Systems

A second system property that we discussed in Section 2.6.2 was that of invertibility. Consider a continuous-time LTI system with impulse response $h(t)$. Recall that this system is invertible only if we can design an inverse system that, when connected in series with the original system, produces an output equal to the input to the first system. Furthermore, if an LTI system is invertible, then it has an LTI inverse (see Problem 3.22). Therefore, we have the picture depicted in Figure 3.21. We are given a system with impulse response $h(t)$ and wish to design a system with impulse response $h_1(t)$ so that $z(t) = x(t)$, that is, so that the series interconnection in Figure 3.21(a) is identical to the identity system in Figure 3.21(b). Since the overall impulse response in Figure 3.21(a) is $h(t) * h_1(t)$, we have the condition that $h_1(t)$ must satisfy for it to be the impulse response of the inverse system:

$$h(t) * h_1(t) = \delta(t) \tag{3.40}$$

Similarly, in discrete time the unit sample response of the inverse system for an LTI system with unit sample response $h[n]$ must satisfy

$$h[n] * h_1[n] = \delta[n] \tag{3.41}$$

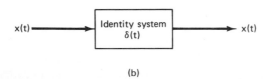

(a)

(b)

Figure 3.21 Concept of an inverse system for continuous-time LTI systems. The system with impulse response $h_1(t)$ is the inverse of the system with impulse response $h(t)$ if $h(t) * h_1(t) = \delta(t)$.

Example 3.5

Consider the LTI system consisting of a pure time shift

$$y(t) = x(t - t_0) \tag{3.42}$$

Such a system is called a *delay* if $t_0 > 0$ and an *advance* if $t_0 < 0$. The impulse response for this system can be obtained immediately from eq. (3.42) by taking the input equal to $\delta(t)$,

$$h(t) = \delta(t - t_0) \tag{3.43}$$

Therefore,

$$x(t - t_0) = x(t) * \delta(t - t_0) \tag{3.44}$$

That is, the convolution of a signal with a shifted impulse simply shifts the signal.

The inverse for this system can be obtained very easily. All we must do to recover the input is to shift the output back. That is, if we take

$$h_1(t) = \delta(t + t_0)$$

then

$$h(t) * h_1(t) = \delta(t - t_0) * \delta(t + t_0) = \delta(t)$$

Similarly, a pure time shift in discrete time has the unit sample response $\delta[n - n_0]$ and its inverse has sample response $\delta[n + n_0]$.

Example 3.6

Consider an LTI system with unit impulse response

$$h[n] = u[n] \tag{3.45}$$

Then using the convolution sum, we can calculate the response of this system to an arbitrary input:

$$y[n] = \sum_{k=-\infty}^{\infty} x[k]u[n - k] \tag{3.46}$$

Since $u[n - k]$ equals 0 for $(n - k) < 0$ and 1 for $(n - k) \geq 0$, eq. (3.46) becomes

$$y[n] = \sum_{k=-\infty}^{n} x[k] \tag{3.47}$$

That is, this system, which we first encountered in Section 2.6 [see eq. (2.54)], is a *summer* or *accumulator* that computes the running sum of all the values of the input up to the present time. As we saw in Section 2.6.2, this system is invertible, and its inverse is given by

$$y[n] = x[n] - x[n - 1] \tag{3.48}$$

which is simply a *first difference* operation. If we let $x[n] = \delta[n]$, we find that the impulse response of the inverse system is

$$h_1[n] = \delta[n] - \delta[n - 1] \tag{3.49}$$

That this is the unit sample response of the inverse to the LTI system specified by eq. (3.45), or equivalently by eq. (3.47), can be verified by direct calculation:

$$
\begin{aligned}
h[n] * h_1[n] &= u[n] * \{\delta[n] - \delta[n - 1]\} \\
&= u[n] * \delta[n] - u[n] * \delta[n - 1] \\
&= u[n] - u[n - 1] \\
&= \delta[n]
\end{aligned}
\tag{3.50}
$$

3.4.3 Causality for LTI Systems

In Section 2.6 we introduced the concept of causality—the output of a causal system depends only on the present and past values of the input. By using the convolution sum and integral given in eqs. (3.34) and (3.35), we can relate this property to a corresponding property of the impulse response of an LTI system. Specifically, in order for a discrete-time LTI system to be causal, $y[n]$ must not depend on $x[k]$ for $k > n$. From eq. (3.34) we see that this will be the case if

$$h[n] = 0 \qquad \text{for } n < 0 \tag{3.51}$$

In this case, eq. (3.34) becomes

$$y[n] = \sum_{k=-\infty}^{n} x[k]\, h[n-k] = \sum_{k=0}^{\infty} h[k]\, x[n-k] \tag{3.52}$$

Similarly, a continous-time LTI system is causal if

$$h(t) = 0 \qquad \text{for } t < 0 \tag{3.53}$$

and in this case the convolution integral is given by

$$y(t) = \int_{-\infty}^{t} x(\tau)\, h(t-\tau)\, d\tau = \int_{0}^{\infty} h(\tau)\, x(t-\tau)\, d\tau \tag{3.54}$$

Both the summer ($h[n] = u[n]$) and its inverse ($h[n] = \delta[n] - \delta[n-1]$), described in Example 3.6, satisfy eq. (3.51) and therefore are causal. The pure time shift with impulse response $h(t) = \delta(t - t_0)$ is causal for $t_0 \geq 0$ (when the time shift is a delay), but is noncausal for $t_0 < 0$ (in which case the time shift is an advance).

3.4.4 Stability for LTI Systems

Recall from Section 2.6.4 that a system is *stable* if every bounded input produces a bounded output. In order to determine conditions under which LTI systems are stable, consider an input $x[n]$ that is bounded in magnitude:

$$|x[n]| < B \qquad \text{for all } n \tag{3.55}$$

Suppose that we apply this input to an LTI system with unit sample response $h[n]$. Then, using the convolution sum, we obtain an expression for the magnitude of the output

$$|y[n]| = \left| \sum_{k=-\infty}^{+\infty} h[k]\, x[n-k] \right| \tag{3.56}$$

Since the magnitude of a sum of a set of numbers is no larger than the sum of the magnitudes of the numbers, we can deduce from eq. (3.56) that

$$|y[n]| \leq \sum_{k=-\infty}^{+\infty} |h[k]|\,|x[n-k]| \tag{3.57}$$

From eq. (3.55) we have that $|x[n-k]| < B$ for all values of k and n. This together with eq. (3.57) implies that

$$|y[n]| \leq B \sum_{k=-\infty}^{+\infty} |h[k]| \qquad \text{for all } n \tag{3.58}$$

From this we can conclude that if the unit impulse response is *absolutely summable*, that is, if

$$\sum_{k=-\infty}^{+\infty} |h[k]| < \infty \tag{3.59}$$

then eq. (3.58) implies that $y[n]$ is bounded in magnitude and hence that the system is stable. Therefore, eq. (3.59) is a sufficient condition to guarantee the stability of a discrete-time LTI system. In fact, this condition is also a necessary condition, since, as shown in Problem 3.21, if eq. (3.59) is not satisfied, there are bounded inputs that result in unbounded outputs. Thus, the stability of a discrete-time LTI system is completely equivalent to eq. (3.59).

In continuous time we obtain an analogous characterization of stability in terms of the impulse response of an LTI system. Specifically, if $|x(t)| < B$ for all t, then, in analogy with eqs. (3.56)–(3.58), we find that

$$|y(t)| = \left| \int_{-\infty}^{+\infty} h(\tau)x(t-\tau)\, d\tau \right|$$

$$\leq \int_{-\infty}^{+\infty} |h(\tau)|\,|x(t-\tau)|\, d\tau$$

$$\leq B \int_{-\infty}^{+\infty} |h(\tau)|\, d\tau$$

Therefore, the system is stable if the impulse response is *absolutely integrable*,

$$\int_{-\infty}^{+\infty} |h(\tau)|\, d\tau < \infty \tag{3.60}$$

and, as in discrete time, if eq. (3.60) is not satisfied, there are bounded inputs that produce unbounded outputs. Therefore, the stability of a continuous-time LTI system is equivalent to eq. (3.60).

As an example, consider a system that is a pure time shift in either continuous or discrete time. In this case

$$\sum_{n=-\infty}^{+\infty} |h[n]| = \sum_{n=-\infty}^{+\infty} |\delta[n-n_0]| = 1$$

$$\int_{-\infty}^{+\infty} |h(\tau)|\, d\tau = \int_{-\infty}^{+\infty} |\delta(\tau-t_0)|\, d\tau = 1$$

and we conclude that both these systems are stable. This should not be surprising, since if a signal is bounded in magnitude, so is any time-shifted version of that signal.

As a second example, consider the accumulator described in Example 3.6. As we discussed in Section 2.6.4, this is an unstable system since if we apply a constant input to an accumulator, the output grows without bound. That this system is unstable can also be seen from the fact that its impulse response $u[n]$ is not absolutely summable:

$$\sum_{n=-\infty}^{\infty} |u[n]| = \sum_{n=0}^{\infty} u[n] = \infty$$

Similarly, consider an *integrator*, the continuous-time counterpart of the accumulator:

$$y(t) = \int_{-\infty}^{t} x(\tau)\, d\tau \tag{3.61}$$

This is an unstable system for precisely the same reason as that given for the accumulator (i.e., a constant input gives rise to an output that grows without bound). The impulse response for this system can be found by letting $x(t) = \delta(t)$, in which case we find that

$$h(t) = \int_{-\infty}^{t} \delta(\tau)\, d\tau = u(t)$$

and

$$\int_{-\infty}^{+\infty} |u(\tau)|\, d\tau = \int_0^\infty d\tau = \infty$$

which corroborates our statement that eq. (3.61) represents an unstable system.

3.4.5 The Unit Step Response of an LTI System

In the preceding parts of this section we have seen that the representation of an LTI system in terms of its unit impulse response allows us to obtain very explicit characterizations of system properties. Specifically, since $h[n]$ or $h(t)$ completely determines the behavior of an LTI system, we have been able to relate properties such as stability and causality to properties of those signals, to deduce that if $h[n] = \delta[n - n_0]$, then the system must be a time shift, to determine that $h(t) = u(t)$ corresponds to an integrator, and so on.

There is another signal that is also used quite often in describing the behavior of LTI systems. This is the *unit step response*, $s[n]$ or $s(t)$, that is, the output when $x[n] = u[n]$ or $x(t) = u(t)$. We will find it useful on occasion to refer to the step response, and therefore it is worthwhile relating it to the impulse response. From the convolution-sum representation, we know that the step response of a discrete-time LTI system is the convolution of the unit step with the unit sample response

$$s[n] = u[n] * h[n]$$

However, by the commutative property of convolution, we know that $s[n] = h[n] * u[n]$, and therefore $s[n]$ can be viewed as the response of a discrete-time LTI system with unit sample response $u[n]$ to the input $h[n]$. As we have seen, $u[n]$ is the unit impulse response of an accumulator. Therefore,

$$s[n] = \sum_{k=-\infty}^{n} h[k] \tag{3.62}$$

From this equation and from Example 3.6 we see that $h[n]$ can be recovered from $s[n]$ using the relation

$$h[n] = s[n] - s[n - 1] \tag{3.63}$$

Similarly, in continuous time $s(t)$ equals the response of an integrator [with impulse response $u(t)$] to the input $h(t)$. That is, the unit step response is the integral of the unit impulse response:

$$s(t) = \int_{-\infty}^{t} h(\tau)\, d\tau \tag{3.64}$$

and, from eq. (3.64),†

$$h(t) = \frac{ds(t)}{dt} = s'(t) \tag{3.65}$$

Therefore, in both continuous and discrete time, the unit step response can also be

---

†Throughout this book we will use both of the notations indicated in eq. (3.65) to denote first derivatives. Analogous notation will also be used for higher derivatives.

used to characterize an LTI system, since we can calculate the unit impulse response from it. In Problem 3.14 expressions analogous to the convolution sum and convolution integral are derived for the representation of an LTI system in terms of its unit step response.

3.5 SYSTEMS DESCRIBED BY DIFFERENTIAL AND DIFFERENCE EQUATIONS

An extremely important class of continuous-time systems is that for which the input and output are related through a *linear constant-coefficient differential equation*. Equations of this type arise in the description of a wide variety of systems and physical phenomena. The response of an *RLC* circuit can be described in terms of a differential equation, as can the response of a mechanical system containing restoring and damping forces. Also, the kinetics of a chemical reaction and the kinematics of the motion of an object or vehicle are usually described by differential equations.

Correspondingly, an important class of discrete-time systems is that for which the input and output are related through a *linear constant-coefficient difference equation*. Equations of this type are used to describe the sequential behavior of many different processes, including return on investment as a function of time and a sampled speech signal as the response of the human vocal tract to excitation from the vocal cords. Such equations also arise quite frequently in the specification of discrete-time systems designed to perform particular desired operations on the input signal. For example, the system discussed in Section 2.6 [see eq. (2.64)] that computes the average value of the input over an interval is described by a difference equation.

Throughout this book there will be many occasions in which we will consider and examine systems described by linear constant-coefficient differential and difference equations. In this section we introduce and describe these classes of systems, first in continuous time and then in discrete time. In subsequent chapters we develop additional tools for the analysis of signals and systems that will provide us with useful methods for solving linear constant-coefficient differential and difference equations and for analyzing the properties of systems described by such equations.

3.5.1 Linear Constant-Coefficient Differential Equations

To bring out the important ideas concerning systems specified by linear constant-coefficient differential equations, we begin by examining an example. Specifically, consider a continuous-time system whose input and output are related by the equation

$$\frac{dy(t)}{dt} + 2y(t) = x(t) \tag{3.66}$$

Equation (3.66) describes the system response implicitly, and we must solve the differential equation to obtain an explicit expression for the system output as a function of the input. To review the usual method for the solution of linear constant-coefficient differential equations,† let us consider this example with a particular input signal:

$$x(t) = K[\cos \omega_0 t]u(t) \tag{3.67}$$

where K is a real number.

The complete solution to eq. (3.66) consists of the sum of a *particular solution*, $y_p(t)$, and a *homogeneous solution*, $y_h(t)$:

$$y(t) = y_p(t) + y_h(t) \tag{3.68}$$

where the particular solution satisfies eq. (3.66) and $y_h(t)$ is a solution of the homogeneous differential equation

$$\frac{dy(t)}{dt} + 2y(t) = 0 \tag{3.69}$$

To find a particular solution for the input given in eq. (3.67), we observe that for $t > 0$, $x(t)$ may be written as

$$x(t) = \mathcal{R}e\{Ke^{j\omega_0 t}\} \tag{3.70}$$

We then hypothesize a solution of the form

$$y_p(t) = \mathcal{R}e\{Ye^{j\omega_0 t}\} \tag{3.71}$$

where Y is a complex number. Substituting these two expressions into eq. (3.66) yields

$$\mathcal{R}e\{j\omega_0 Ye^{j\omega_0 t} + 2Ye^{j\omega_0 t}\} = \mathcal{R}e\{Ke^{j\omega_0 t}\} \tag{3.72}$$

Since eq. (3.70) must be true for all $t > 0$, the complex amplitudes of the exponentials on both sides of the equation must be equal. That is,

$$j\omega_0 Y + 2Y = K$$

or

$$Y = \frac{K}{j\omega_0 + 2} = \frac{K}{\sqrt{4 + \omega_0^2}} e^{-j\theta} \tag{3.73}$$

where

$$\theta = \tan^{-1}\left(\frac{\omega_0}{2}\right) \tag{3.74}$$

Therefore,

$$
\begin{aligned}
y_p(t) &= \mathcal{R}e\{Ye^{j\omega_0 t}\} \\
&= \frac{K}{\sqrt{4 + \omega_0^2}} \cos(\omega_0 t - \theta), \qquad t > 0
\end{aligned}
\tag{3.75}
$$

†Our discussion of the solution of linear constant-coefficient differential equations is brief, since we assume that the reader has some familiarity with this material. For review, we recommend a text on the solution of ordinary differential equations such as *Ordinary Differential Equations* (2nd ed.) by G. Birkhoff and G. C. Rota (Waltham, Mass.: Blaisdell Publishing Co., 1969) or *An Introduction to Ordinary Differential Equations* by E. A. Coddington (Englewood Cliffs, N.J.: Prentice-Hall, Inc., 1961). There are also numerous texts that discuss differential equations in the context of circuit theory. See, for example, *Basic Circuit Theory* by C. A. Desoer and E. S. Kuh (New York: McGraw-Hill Book Company, 1969). As mentioned in the text, in the following chapters we present another very useful method for solving linear differential equations that will be sufficient for our purposes. In addition a number of exercises involving the solution of differential equations are included in the problems at the end of the chapter.

In order to determine $y_h(t)$, we hypothesize a solution of the form

$$y_h(t) = Ae^{st} \tag{3.76}$$

Substituting this into eq. (3.69) gives

$$Ase^{st} + 2Ae^{st} = Ae^{st}(s + 2) = 0 \tag{3.77}$$

From this equation we see that we must take $s = -2$ and that Ae^{-2t} is a solution to eq. (3.69) for *any* choice of A. Combining this with eqs. (3.68) and (3.75), we find that for $t > 0$, the solution of the differential equation is

$$y(t) = Ae^{-2t} + \frac{K}{\sqrt{4 + \omega_0^2}} \cos{(\omega_0 t - \theta)}, \qquad t > 0 \tag{3.78}$$

Note from eq. (3.78) that the differential equation (3.66) does not completely specify the output $y(t)$ in terms of the input $x(t)$. Specifically, in eq. (3.78) the constant A has not as yet been determined. This is a basic property of a system described by a differential equation: In order to have the output completely determined by the input, we need to specify *auxiliary conditions* on the differential equation. In the example, we need to specify the value of $y(t)$ at some given instant of time. This will determine A and consequently will determine $y(t)$ for all time. For example, if we specify

$$y(0) = y_0 \tag{3.79}$$

then, from eq. (3.78),

$$A = y_0 - \frac{K \cos{\theta}}{\sqrt{4 + \omega_0^2}} \tag{3.80}$$

Therefore, for $t > 0$,

$$y(t) = y_0 e^{-2t} + \frac{K}{\sqrt{4 + \omega_0^2}} [\cos{(\omega_0 t - \theta)} - e^{-2t} \cos{\theta}], \qquad t > 0$$

For $t < 0$, $x(t) = 0$ and therefore $y(t)$ satisfies the homogeneous differential equation (3.69). As we have seen, solutions to this equation are of the form Be^{-2t}, and using the auxiliary condition given in eq. (3.79), we find that

$$y(t) = y_0 e^{-2t}, \qquad t < 0$$

Combining the solutions for $t > 0$ and $t < 0$, we then have that

$$y(t) = y_0 e^{-2t} + \frac{K}{\sqrt{4 + \omega_0^2}} [\cos{(\omega_0 t - \theta)} - e^{-2t} \cos{\theta}]u(t) \tag{3.81}$$

Therefore, by specifying an auxiliary condition for the differential equation (3.66), we obtain an explicit expression for the output $y(t)$ in terms of the input $x(t)$, and we can then investigate the properties of the system specified in this fashion. For example, consider the system specified by eq. (3.66) with auxiliary condition given by eq. (3.79). Let us first determine if this system is linear. Recall that as shown in Section 2.6.6, a linear system has the property that zero input produces zero output. However, if we let $K = 0$ in our example, $x(t) = 0$ [see eq. (3.67)], but from eq. (3.81) we see that

$$y(t) = y_0 e^{-2t}$$

Therefore, this system is definitely *not* linear if $y_0 \neq 0$. It *is* linear, however, if the auxiliary condition is zero. To see this, let $x_1(t)$ and $x_2(t)$ be two input signals, and let $y_1(t)$ and $y_2(t)$ be the corresponding responses. That is,

$$\frac{dy_1(t)}{dt} + 2y_1(t) = x_1(t) \tag{3.82}$$

$$\frac{dy_2(t)}{dt} + 2y_2(t) = x_2(t) \tag{3.83}$$

and also $y_1(t)$ and $y_2(t)$ must satisfy the auxiliary condition

$$y_1(0) = y_2(0) = 0 \tag{3.84}$$

Consider next the input $x_3(t) = \alpha x_1(t) + \beta x_2(t)$, where α and β are any complex numbers. Then, using eqs. (3.82) and (3.83), it is not difficult to see that $y_3(t) = \alpha y_1(t) + \beta y_2(t)$ satisfies the differential equation

$$\frac{dy_3(t)}{dt} + 2y_3(t) = x_3(t) \tag{3.85}$$

and also, from eq. (3.84),

$$y_3(0) = \alpha y_1(0) + \beta y_2(0) = 0 \tag{3.86}$$

Therefore, $y_3(t)$ is the response corresponding to $x_3(t)$, and thus the system is linear.

Although the system specified by eq. (3.66) with the auxiliary condition of eq. (3.79) is not linear for a nonzero auxiliary condition, it is incrementally linear. Specifically, in eq. (3.81) we see that the solution consists of two terms. The first is the response due to the nonzero auxiliary condition alone, while the second term is the response if $y_0 = 0$, that is, the *linear* response of the system assuming that the auxiliary condition is zero. This fact generalizes to all systems characterized by linear constant-coefficient differential equations. That is, any such system is incrementally linear and therefore can be thought of conceptually as having the form depicted in Figure 3.22. Thus, if the auxiliary conditions are zero for a system specified by a

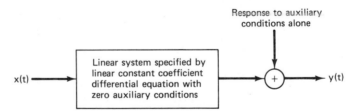

Figure 3.22 Incrementally linear structure of a system specified by a linear constant-coefficient differential equation.

linear constant-coefficient differential equation, the system is linear, and the overall response of a system with nonzero auxiliary conditions is simply the sum of the response of the linear system with zero auxiliary conditions and the response to the auxiliary conditions alone. In most of this book we will be concerned with linear systems, and therefore in the remainder of our discussion in this section we will restrict our attention to the linear case (i.e., to the case of zero auxiliary conditions). Because of the decomposition shown in Figure 3.22, analysis of the linear case yields considerable insights into the properties of systems with nonzero auxiliary conditions. In Section 9.8 we adapt one of the tools we will develop for the analysis of linear systems to allow us to analyze systems specified by linear constant-coefficient differential equations with nonzero auxiliary conditions.

In addition to linearity, a second question about the properties of linear systems specified by differential equations concerns their causality. Specifically, in order to have causality for a linear system specified by a linear constant-coefficient differential equation, we must make a particular choice for the auxiliary conditions for this system. This is the condition of *initial rest* which specifies that if the input $x(t)$ is applied to the system, and $x(t) = 0$ for $t \leq t_0$, then $y(t)$ also is zero for $t \leq t_0$. To gain some understanding of this condition, consider the following example.

Example 3.7

Let us first consider a linear system described by eq. (3.66) and an auxiliary condition specified at a fixed point in time. Specifically,

$$y(0) = 0 \qquad (3.87)$$

As we have just seen, these two equations together specify a linear system. Consider the following two inputs:

$$x_1(t) = 0 \qquad \text{for all } t \qquad (3.88)$$

$$x_2(t) = \begin{cases} 0, & t < -1 \\ 1, & t > -1 \end{cases} \qquad (3.89)$$

Since the system is linear, the response $y_1(t)$ to the input $x_1(t)$ is

$$y_1(t) = 0 \qquad \text{for all } t \qquad (3.90)$$

Now consider the solution to the differential equation for $x(t) = x_2(t)$. For $t > -1$, $x_2(t) = 1$. Therefore, if we seek a particular solution that is constant,

$$y_p(t) = Y, \qquad t > -1$$

we find, upon substitution into eq. (3.66) that

$$2Y = 1$$

Including the homogeneous solution, we obtain

$$y_2(t) = Ae^{-2t} + \tfrac{1}{2}, \qquad t > -1 \qquad (3.91)$$

and, to satisfy eq. (3.87) we must take $A = -\tfrac{1}{2}$, which yields

$$y_2(t) = \tfrac{1}{2} - \tfrac{1}{2}e^{-2t}, \qquad t > -1 \qquad (3.92)$$

To find $y_2(t)$ for $t < -1$, we first note that $x_2(t) = 0$ for $t < -1$. Thus the particular solution is zero for $t < -1$, and consequently

$$y_2(t) = Be^{-2t}, \qquad t < -1 \qquad (3.93)$$

Since the two pieces of the solution in eq. (3.92) and (3.93) must match at $t = -1$, we can determine B from the equation

$$\tfrac{1}{2} - \tfrac{1}{2}e^2 = Be^2$$

which yields

$$y_2(t) = (\tfrac{1}{2} - \tfrac{1}{2}e^2)e^{-2(t+1)}, \qquad t < -1 \qquad (3.94)$$

Note that since $x_1(t) = x_2(t)$ for $t < -1$, it must be true that $y_1(t) = y_2(t)$ for $t < -1$ if this system is causal. However, comparing eqs. (3.90) and (3.94) we see that this is not the case and conclude that the system is not causal.

Consider now a second linear system specified by eq. (3.66) and the assumption of initial rest. Then the response $y_1(t)$ to the input $x_1(t)$ is still given by eq. (3.90), but the response to $x_2(t)$ given in eq. (3.89) is different from that given in eqs. (3.92) and (3.94). Specifically, since $x_2(t) = 0$ for $t < -1$, initial rest implies that $y_2(t) = 0$ for $t < -1$. To find the response for $t > -1$, we must solve eq. (3.66) with the condition specified by initial rest, that is,

$$y_2(-1) = 0 \tag{3.95}$$

As before, the general form of the solution for $y_2(t)$ for $t > -1$ is given by eq. (3.91), but in this case A is chosen to satisfy eq. (3.95). This yields the solution

$$y_2(t) = \tfrac{1}{2} - \tfrac{1}{2}e^{-2(t+1)}, \qquad t > -1 \tag{3.96}$$

and by the assumption of initial rest we can write the solution for all time as

$$y_2(t) = [\tfrac{1}{2} - \tfrac{1}{2}e^{-2(t+1)}]u(t+1) \tag{3.97}$$

From this we see that we do satisfy the condition for causality. That is, $x_1(t) = x_2(t) = 0$ for $t < -1$, and also $y_1(t) = y_2(t) = 0$ for $t < -1$.

Intuitively, specifying the auxiliary condition at a *fixed* point in time as we did in eq. (3.87), leads to a noncausal system, as the response to inputs that are nonzero *before* this fixed time, such as $x_2(t)$ in eq. (3.89), must in some sense anticipate this future condition. On the other hand, the condition of initial rest does *not* specify the auxiliary condition at a fixed point in time but rather adjusts this point in time so that the response is zero *until* the input becomes nonzero, and consequently the phenomenon of the response anticipating the future does not occur. In fact, Problem 2.30 demonstrates that a linear system is causal if and only if whenever $x(t) = 0$ for $t \leq t_0$, then $y(t) = 0$ for $t \leq t_0$. This is nothing more than the condition of initial rest.

Note that in the example we made use of a basic consequence of initial rest. Specifically, if we make the initial rest assumption, and if $x(t) = 0$ for $t \leq t_0$, then we need only solve for $y(t)$ for $t > t_0$, and this solution can be obtained from the differential equation and the condition $y(t_0) = 0$, which in this case is called an *initial condition*.

In addition to guaranteeing linearity and causality, initial rest also implies time invariance. To see this consider the system described by eq. (3.66), which is initially at rest, and let $y_1(t)$ be the response to an input $x_1(t)$, which is zero for $t \leq t_0$. That is,

$$\frac{dy_1(t)}{dt} + 2y_1(t) = x_1(t) \tag{3.98}$$

$$y_1(t_0) = 0 \tag{3.99}$$

Now consider the input

$$x_2(t) = x_1(t - T) \tag{3.100}$$

From Figure 3.23 it can be seen that $x_2(t)$ is zero for $t \leq t_0 + T$. Therefore, the response $y_2(t)$ to this input must satisfy the differential equation

$$\frac{dy_2(t)}{dt} + 2y_2(t) = x_2(t) \tag{3.101}$$

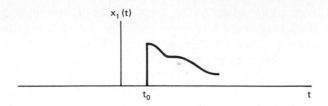

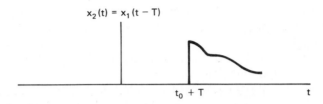

Figure 3.23 Illustration of the fact that if $x_1(t) = 0$ for $t < t_0$ and $x_2(t) = x_1(t - T)$, then $x_2(t) = 0$ for $t < t_0 + T$.

with the initial condition

$$y_2(t_0 + T) = 0 \tag{3.102}$$

Using eqs. (3.98) and (3.99), it is straightforward to show that $y_1(t - T)$ satisfies eqs. (3.101) and (3.102), and thus that

$$y_2(t) = y_1(t - T) \tag{3.103}$$

A general Nth-order linear constant-coefficient differential equation is given by

$$\sum_{k=0}^{N} a_k \frac{d^k y(t)}{dt^k} = \sum_{k=0}^{M} b_k \frac{d^k x(t)}{dt^k} \tag{3.104}$$

The order refers to the highest derivative of the output $y(t)$ appearing in the equation. In the case when $N = 0$, eq. (3.101) reduces to

$$y(t) = \frac{1}{a_0} \sum_{k=0}^{M} b_k \frac{d^k x(t)}{dt^k} \tag{3.105}$$

That is, $y(t)$ is an explicit function of the input $x(t)$ and its derivatives. For $N \geq 1$, eq. (3.104) specifies the output implicitly in terms of the input. In this case the analysis of eq. (3.104) proceeds just as in our example. The solution $y(t)$ consists of two parts, a homogeneous solution and a particular solution. Also, as in the example, the differential equation (3.104) does not completely specify the output in terms of the input. In the general case we need a *set* of auxiliary conditions corresponding to the values of

$$y(t), \quad \frac{dy(t)}{dt}, \quad \ldots, \quad \frac{d^{N-1}y(t)}{dt^{N-1}}$$

at some point in time. Furthermore, the system specified by eq. (3.104) and these auxiliary conditions will be linear only if *all* of these auxiliary conditions are zero. Otherwise, the system is incrementally linear with the response due to the auxiliary conditions alone added to the response due to the input assuming zero auxiliary conditions (Figure 3.22). Furthermore, for the system to be linear and causal, we must assume initial rest. That is, if $x(t) = 0$ for $t \leq t_0$, we assume that $y(t) = 0$ for $t \leq t_0$, and therefore the response for $t > t_0$ can be calculated from the differential equation (3.104) with the initial conditions.

$$y(t_0) = \frac{dy(t_0)}{dt} = \cdots = \frac{d^{N-1}y(t_0)}{dt^{N-1}} = 0$$

In this case, not only is the system linear and causal, it is also time-invariant. In the remainder of this book we will be focusing our attention primarily on LTI systems, and therefore when we consider systems described by differential equations, we will make the assumption of initial rest. In this case the output $y(t)$ can, of course, be computed by solving the differential equation in the manner outlined in this section and illustrated in more detail in several problems at the end of the chapter. However, in Chapter 4 we will develop some tools for the analysis, of continuous-time LTI systems that will greatly facilitate the solution of differential equations and in particular will provide us with a convenient method for calculating the impulse response for LTI systems specified by linear constant-coefficient differential equations that are initially at rest.

3.5.2 Linear Constant-Coefficient Difference Equations

The discrete-time counterpart of eq. (3.104) is the Nth-order linear constant-coefficient difference equation

$$\sum_{k=0}^{N} a_k y[n-k] = \sum_{k=0}^{M} b_k x[n-k] \tag{3.106}$$

An equation of this type can be solved in a manner exactly analogous to that for differential equations (see Problem 3.31).† Specifically, the solution $y[n]$ can be written as the sum of a particular solution to eq. (3.106) and a solution to the homogeneous equation

$$\sum_{k=0}^{N} a_k y[n-k] = 0 \tag{3.107}$$

As in the continuous-time case, eq. (3.106) does not completely specify the output in terms of the input, and to do this, we must also specify some auxiliary conditions. Furthermore, the system described by eq. (3.106) and the auxiliary conditions is incrementally linear, consisting of the sum of the response to the auxiliary conditions alone and the linear response to the input assuming zero auxiliary conditions. Thus, a system described by eq. (3.106) is linear if the auxiliary conditions are zero.

Although all these properties can be developed following an approach that

---

†For a detailed treatment of the methods for solving linear constant-coefficient difference equations, we refer the reader to *Finite Difference Equations* by H. Levy and F. Lessman (New York: Macmillan, Inc., 1961) or *The Calculus of Finite Differences* by L. M. Milne-Thomson (New York: Macmillan, Inc., 1933). In Chapter 5 we present another method for solving difference equations that greatly facilitates the analysis of LTI systems that are so-described. In addition, we refer the reader to the problems at the end of this chapter that deal with the solution of difference equations.

directly parallels our discussion for differential equations, the discrete-time case offers an alternative path. This stems from the observation that eq. (3.106) can be rearranged in the form

$$y[n] = \frac{1}{a_0} \left\{ \sum_{k=0}^{M} b_k x[n-k] - \sum_{k=1}^{N} a_k y[n-k] \right\} \qquad (3.108)$$

This equation directly expresses the output at time n in terms of previous values of the input and output. From this we can immediately see the need for auxiliary conditions. In order to calculate $y[n]$, we need to know $y[n-1], \ldots, y[n-N]$. Therefore, if we are given the input for all n and a set of auxiliary conditions such as $y[-N], y[-N+1], \ldots, y[-1]$, eq. (3.108) can then be solved for successive values of $y[n]$.

An equation of the form of eq. (3.106) or (3.108) is called a *recursive equation*, since it specifies a recursive procedure for determining the output in terms of the input and previous outputs. In the special case when $N = 0$, eq. (3.108) reduces to

$$y[n] = \sum_{k=0}^{M} \left(\frac{b_k}{a_0} \right) x[n-k] \qquad (3.109)$$

This is the discrete-time counterpart of the continuous-time system given in eq. (3.105). In this case $y[n]$ is an explicit function of the present and previous values of the input. For this reason eq. (3.109) is called a *nonrecursive equation*, since we do not recursively use previously computed values of the output to compute the present value of the output. Therefore, just as in the case of the system given in eq. (3.105), we do not need auxiliary conditions in order to determine $y[n]$. Furthermore, eq. (3.109) describes an LTI system, and by direct computation the impulse response of this system is found to be

$$h[n] = \begin{cases} \dfrac{b_n}{a_0}, & 0 \le n \le M \\ 0, & \text{otherwise} \end{cases} \qquad (3.110)$$

That is, eq. (3.109) is nothing more than the convolution sum. Note that the impulse response for this system has finite duration; that is, it is nonzero only over a finite time interval. Because of this property the system specified by eq. (3.109) is often called a *finite impulse response (FIR) system*.

Although we do not require auxiliary conditions for the case of $N = 0$, such conditions are needed for the recursive case when $N \ge 1$. To gain some insight into the behavior and properties of recursive difference equations, consider the first-order example

$$y[n] - \tfrac{1}{2}y[n-1] = x[n] \qquad (3.111)$$

which can be expressed in the form

$$y[n] = x[n] + \tfrac{1}{2}y[n-1] \qquad (3.112)$$

Suppose it is known that $y[-1] = a$ and that the input is

$$x[n] = K\,\delta[n] \qquad (3.113)$$

where K is an arbitrary complex number. We can solve for succesive values of $y[n]$ for $n \ge 0$ as follows:

$$y[0] = x[0] + \tfrac{1}{2}y[-1] = K + \tfrac{1}{2}a$$
$$y[1] = x[1] + \tfrac{1}{2}y[0] = (\tfrac{1}{2})(K + \tfrac{1}{2}a)$$
$$y[2] = x[2] + \tfrac{1}{2}y[1] = (\tfrac{1}{2})^2(K + \tfrac{1}{2}a) \qquad\qquad (3.114)$$

.
.
.

$$y[n] = x[n] + \tfrac{1}{2}y[n-1] = (\tfrac{1}{2})^n(K + \tfrac{1}{2}a), \qquad n \geq 0$$
$$= (\tfrac{1}{2})^n K + (\tfrac{1}{2})^{n+1}a, \qquad n \geq 0$$

By rearranging eq. (3.111) in the form

$$y[n-1] = 2\{y[n] - x[n]\} \qquad\qquad (3.115)$$

and again using the known value of $y[-1] = a$, we can also determine $y[n]$ for $n < 0$. Specifically,

$$y[-2] = 2\{y[-1] - x[-1]\} = 2a$$
$$y[-3] = 2\{y[-2] - x[-2]\} = 2^2 a$$
$$y[-4] = 2\{y[-3] - x[-3]\} = 2^3 a \qquad\qquad (3.116)$$

.
.
.

$$y[-n] = 2\{y[-n+1] - x[-n+1]\} = 2^{n-1}a = (\tfrac{1}{2})^{-n+1}a$$

Combining eqs. (3.114) and (3.116) we have that for all values of n,

$$y[n] = (\tfrac{1}{2})^{n+1} a + K(\tfrac{1}{2})^n u[n] \qquad\qquad (3.117)$$

From eq. (3.117) we again see the need for auxiliary conditions. Since this equation is a valid solution of eq. (3.111) for any value of a, this value must be specified in order for $y[n]$ to be determined as a function of the input. Also, if $K = 0$, the input is zero, and we see that the output will be zero only if the auxiliary condition is zero. If this is the case, then the system described by eq. (3.111) is linear. This can be verified directly, much as we did in the continuous-time case (see Problem 3.31).

Furthermore, as in continuous time, to ensure the linearity *and* causality of the system described by eq. (3.111), we must make the assumption of initial rest. That is, we assume that if $x[n] = 0$ for $n \leq n_0$, then $y[n] = 0$ for $n \leq n_0$. In this case we need only solve the difference equation forward in time for $n > n_0$ starting with the initial condition $y[n_0] = 0$. Under this condition the system described by eq. (3.111) is not only linear and causal, but it is also time-invariant. Linearity follows from the zero initial condition, while time invariance can be verified in exactly the same fashion as was used in continuous time (Problem 3.31). From eq. (3.117) we then have that under the assumption of initial rest, the unit impulse response for this LTI system is

$$h[n] = (\tfrac{1}{2})^n u[n]$$

Note that this system has an impulse response that has infinite duration.

As indicated at the beginning of this section, all the observations we have made for this simple example carry over to the general case of a system described by eq. (3.106). For the system to be linear, the auxiliary conditions must be zero, and if the initial-rest assumption is made, the system is causal, linear, and time-invariant. As we will be concentrating our attention on LTI systems, we will usually make the assumption of initial rest when considering systems described by difference equations. In Chapter 5, we will develop tools for the analysis of discrete-time LTI systems that will provide us with a very useful and efficient method for solving linear constant-coefficient difference equations and for analyzing the properties of the LTI systems that they describe. As in our example, we will see that the LTI system specified by the general recursive difference equation (3.108) has an impulse response of infinite duration as long as it truly is recursive, that is, as long as at least one of the a_k, $k = 1$, ..., N, is nonzero. Because of this characteristic, a system specified by a recursive difference equation is often called an *infinite impulse response* (*IIR*) system. We refer the reader to the problems at the end of the chapter for more detailed illustrations of the ideas introduced in this section concerning the solution of difference equations and the analysis of the systems they describe.

3.6 BLOCK-DIAGRAM REPRESENTATIONS OF LTI SYSTEMS DESCRIBED BY DIFFERENTIAL AND DIFFERENCE EQUATIONS

One of the important uses of the tools of LTI system analysis is in the design of systems with specified characteristics. Often in practice, the type of system chosen for a design is described by a differential equation in continuous time or a difference equation in discrete time. The reason for these choices is that many of the physical components and systems that are commonly used to implement LTI system designs are themselves described by differential and difference equations. For example, *RLC* circuits and circuits containing operational amplifiers are described by linear differential equations. Also, systems described by linear difference equations are readily implemented as computer algorithms on a general-purpose digital computer or with special-purpose hardware. In this section we introduce block-diagram representations for systems described by differential and difference equations. We will do this first in discrete time and then in continuous time. The representations described here and in Chapters 4 and 5 allow us to develop some understanding about the issues associated with the implementation of such systems.

3.6.1 Representations for LTI Systems Described by Difference Equations

As expressed in eq. (3.108) a linear constant-coefficient difference equation can be viewed as an algorithm for computing successive values of $y[n]$ as a linear combination of its past values and the present and past values of the input. In implementing such an LTI system on a digital computer or in special-purpose hardware we would explicitly implement this algorithm. While eq. (3.108) suggests one way of organizing the computations, there are, in fact, a variety of alternatives, each of which represents a different *structure* or *realization* of the LTI system described by eq. (3.108). In this section we discuss two of these structures, and other realizations are described in Chapter 5.

It is generally convenient to develop and describe each of these alternative implementations by representing and manipulating eq. (3.108) pictorially. To develop such a pictorial representation we note that the evaluation of eq. (3.108) requires three basic operations: addition, multiplication by a coefficient, and delay. Thus, let us define three basic network elements, as indicated in Figure 3.24. To see how these

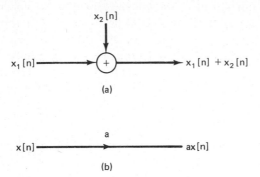

(a)

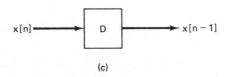

(b)

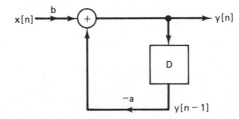

(c)

Figure 3.24 Basic elements for the block-diagram representation of discrete-time LTI systems described by linear constant-coefficient difference equations: (a) an adder; (b) multiplication by a coefficient; (c) a unit delay.

basic elements can be used, we consider several examples. First consider the LTI system initially at rest and described by the first-order equation

$$y[n] + ay[n-1] = bx[n] \qquad (3.118)$$

which can be rewritten in a form that directly suggests a recursive algorithm

$$y[n] = -ay[n-1] + bx[n] \qquad (3.119)$$

This algorithm is represented pictorially in Figure 3.25. Note that the delay element

Figure 3.25 Block-diagram representation for the LTI system described by eq. (3.118).

requires memory: at any point in time n, we need to store $y[n-1]$ so that it can be used in the computation of $y[n]$. Also note that Figure 3.25 is an example of a feedback system, since the output is fed back through a delay and a coefficient multiplication and is then added to $bx[n]$. The presence of feedback is a direct consequence of the recursive nature of eq. (3.119).

Consider next the nonrecursive LTI system

$$y[n] = b_0x[n] + b_1x[n-1] \tag{3.120}$$

The algorithm suggested by eq. (3.120) is illustrated in Figure 3.26. Note that this system also requires one delay element. Also, there is no feedback in this block diagram, since previous values of the output are *not* used in the calculation of the present value.

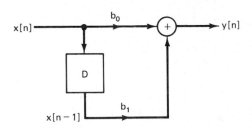

Figure 3.26 Block-diagram representation for the LTI system described by eq. (3.120).

As a third example, consider the LTI system initially at rest and described by the equation

$$y[n] + ay[n-1] = b_0x[n] + b_1x[n-1] \tag{3.121}$$

We can again interpret eq. (3.121) as specifying an algorithm for calculating $y[n]$ recursively:

$$y[n] = -ay[n-1] + b_0x[n] + b_1x[n-1] \tag{3.122}$$

This algorithm is represented graphically in Figure 3.27. Note that this algorithm can be viewed as the cascade of the two LTI systems depicted in Figures 3.25 and 3.26 (with $b = 1$ in Figure 3.25). That is, we calculate

$$w[n] = b_0x[n] + b_1x[n-1] \tag{3.123}$$

and

$$y[n] = -ay[n-1] + w[n] \tag{3.124}$$

However, since the overall response of the cascade of two LTI systems does not depend upon the order in which the two systems are cascaded, we can reverse the order of the two systems in Figure 3.27 to obtain an alternative algorithm for computing the response of the LTI system specified by eq. (3.121). This system is illustrated in Figure 3.28. From this figure we see that

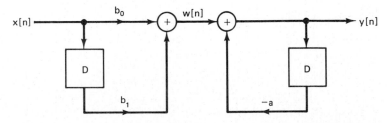

Figure 3.27 Block-diagram representation for the LTI system specified by eq. (3.121).

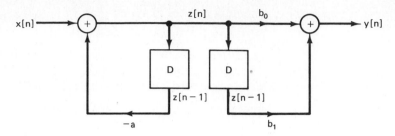

Figure 3.28 Alternative block-diagram representation for the LTI system described by eq. (3.121); (compare to Figure 3.27).

$$z[n] = -az[n-1] + x[n] \qquad (3.125)$$

$$y[n] = b_0 z[n] + b_1 z[n-1] \qquad (3.126)$$

It can be directly verified that $y[n]$ as defined in eqs. (3.125) and (3.126) does satisfy the difference equation (3.121), but we already know that this must be true because of the commutativity of convolution and the implication this has for interchanging the order of LTI systems in cascade.

With Figure 3.28 drawn in the form shown, there is no obvious advantage to this configuration over that in Figure 3.27. However, upon examining Figure 3.28, we see that the two delays have the same input (i.e., they require the storage of the same quantity) and consequently can be collapsed into a single delay, as indicated in Figure 3.29. Since each delay element requires memory, the configuration in Figure 3.29 is more efficient than the one in Figure 3.27, since it requires the storage of only one number, while the other requires that two values be stored at each point in time.

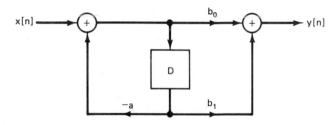

Figure 3.29 Block-diagram representation requiring a single delay element for the LTI system described by eq. (3.121); (compare to Figure 3.28).

This same basic idea can be applied to the general recursive equation (3.108). For convenience we repeat that equation here with $M = N$. If $M \neq N$, then the appropriate coefficients a_k or b_k can be set to zero:

$$y[n] = \frac{1}{a_0} \left\{ \sum_{k=0}^{N} b_k x[n-k] - \sum_{k=1}^{N} a_k y[n-k] \right\} \qquad (3.127)$$

The algorithm implied by this equation is illustrated in Figure 3.30. This algorithm for realizing the difference equation (3.127) is called the *direct form I realization*, and from the figure we see that we can interpret this algorithm as the cascade of a non-

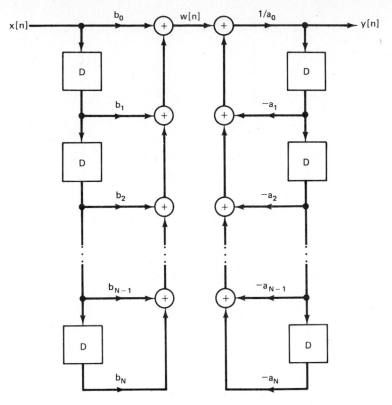

Figure 3.30 Direct form I realization for the LTI system described by eq. (3.127).

recursive system

$$w[n] = \sum_{k=0}^{N} b_k x[n-k] \tag{3.128}$$

and a recursive system

$$y[n] = \frac{1}{a_0}\left\{-\sum_{k=1}^{N} a_k y[n-k] + w[n]\right\} \tag{3.129}$$

By reversing the order of these two systems, we obtain the configuration depicted in Figure 3.31, which represents an alternative structure for the realization of eq. (3.127). The corresponding difference equations are

$$z[n] = \frac{1}{a_0}\left\{-\sum_{k=1}^{N} a_k z[n-k] + x[n]\right\} \tag{3.130}$$

$$y[n] = \sum_{k=0}^{N} b_k z[n-k] \tag{3.131}$$

As before, we note that the two chains of delays in Figure 3.31 have the same input and therefore can be collapsed into a single chain, resulting in the *direct form II realization* depicted in Figure 3.32. In this configuration, implementation of the difference equation requires only N delay elements instead of the $2N$ delay elements required in Figure 3.30. The direct form II realization is sometimes referred to as a

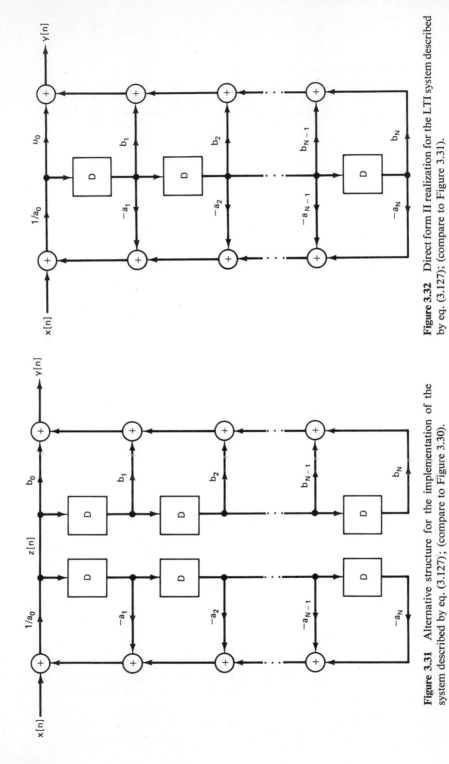

Figure 3.31 Alternative structure for the implementation of the system described by eq. (3.127); (compare to Figure 3.30).

Figure 3.32 Direct form II realization for the LTI system described by eq. (3.127); (compare to Figure 3.31).

116

canonic realization, since it requires the minimum number of delay elements (i.e., it requires the minimum amount of storage) needed in order to implement eq. (3.127).

3.6.2 Representations for LTI Systems Described by Differential Equations

In analyzing continuous-time systems specified by differential equations, we can proceed in an exactly analogous fashion. Consider the general linear constant-coefficient differential equation (3.104). For convenience, we assume that $M = N$ (where if this is not true we simply set the appropriate a_k or b_k to zero), and in this case we can rewrite eq. (3.104) in the form

$$y(t) = \frac{1}{a_0}\left\{\sum_{k=0}^{N} b_k \frac{d^k x(t)}{dt^k} - \sum_{k=1}^{N} a_k \frac{d^k y(t)}{dt^k}\right\} \tag{3.132}$$

The right-hand side of this equation involves three basic operations: addition, multiplication by a coefficient, and differentiation. Therefore, if we define the three basic network elements indicated in Figure 3.33, we can consider implementing

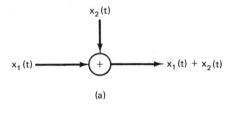

(a)

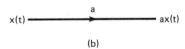

(b)

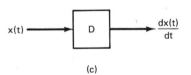

(c)

Figure 3.33 One possible set of basic elements for the block-diagram representation of continuous-time LTI systems described by linear constant-coefficient differential equations: (a) an adder; (b) multiplication by a coefficient; (c) a differentiator.

eq. (3.132) as an interconnection of these basic elements in a manner exactly analogous to that carried out for the implementation of difference equations. In fact, comparing eqs. (3.127) and (3.132), it is straightforward to verify that Figure 3.30 represents the direct form I realization of eq. (3.132), and Figure 3.32 is its direct form II realization, where D is interpreted here as a differentiator and $x[n]$ and $y[n]$ are replaced by $x(t)$ and $y(t)$.

A practical difficulty with the hardware implementation of a linear constant-coefficient differential equation using the elements of Figure 3.33 is that a differentiation element is often difficult to realize. An alternative, which is particularly well suited to the use of operational amplifiers, is suggested by converting the Nth-order differential equation (3.104) into an integral equation. Specifically, define the successive integrals of $y(t)$:

$$y_{(0)}(t) = y(t) \tag{3.133}$$

$$y_{(1)}(t) = y(t) * u(t) = \int_{-\infty}^{t} y(\tau)\, dt \tag{3.134}$$

$$y_{(2)}(t) = y(t) * u(t) * u(t)$$
$$= y_{(1)}(t) * u(t) = \int_{-\infty}^{t} \left(\int_{-\infty}^{\tau} y(\sigma)\, d\sigma \right) d\tau \tag{3.135}$$

and, more generally, the kth integral of $y(t)$ is given by

$$y_{(k)}(t) = y_{(k-1)}(t) * u(t) = \int_{-\infty}^{t} y_{(k-1)}(\tau)\, d\tau \tag{3.136}$$

Similarly, we can define the successive integrals $x_{(k)}(t)$ of $x(t)$.

Consider again the differential eq. (3.104), which we repeat here for convenience, again assuming that $M = N$:

$$\sum_{k=0}^{N} a_k \frac{d^k y(t)}{dt^k} = \sum_{k=0}^{N} b_k \frac{d^k x(t)}{dt^k} \tag{3.137}$$

Note that if we assume initial rest, then the Nth integral of $d^k y(t)/dt^k$ is precisely $y_{(N-k)}(t)$ since the initial conditions for the integration are zero. Similarly the Nth integral of $d^k x(t)/dt^k$ is $x_{(N-k)}(t)$. Therefore, taking the Nth integral of eq. (3.137), we obtain the equation

$$\sum_{k=0}^{N} a_k y_{(N-k)}(t) = \sum_{k=0}^{N} b_k x_{(N-k)}(t) \tag{3.138}$$

Since $y_{(0)}(t) = y(t)$, eq. (3.138) can be reexpressed as

$$y(t) = \frac{1}{a_N} \left\{ \sum_{k=0}^{N} b_k x_{(N-k)}(t) - \sum_{k=0}^{N-1} a_k y_{(N-k)}(t) \right\} \tag{3.139}$$

Implementation of eq. (3.139) utilizes the adder and coefficient multiplier as were indicated in Figure 3.33. In place of the differentiator, we use an integrator as defined in Figure 3.34. This element has $u(t)$ as its impulse response, and it can be

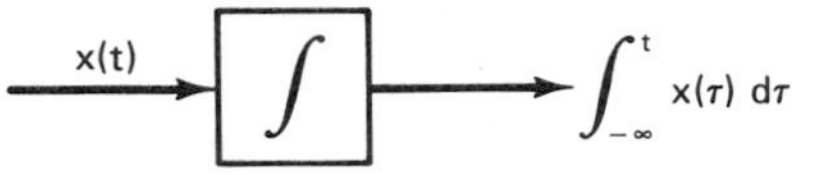

Figure 3.34 Pictorial representation of an integrator.

implemented using an operational amplifier. The development of the direct form I and direct form II realization of eq. (3.139) exactly parallels that for the implementation of difference equations. The resulting realizations are illustrated in Figures 3.35 and 3.36. Note that the direct form II realization requires only N integrators while direct form I uses $2N$ integrators. As before, the direct form II configuration is sometimes called canonic, as it requires the minimum number of integrators needed in any realization of the LTI system described by eq. (3.139) or, equivalently, by eq. (3.132). In Chapter 4 we introduce two other canonic structures for the realization of LTI systems described by such equations.

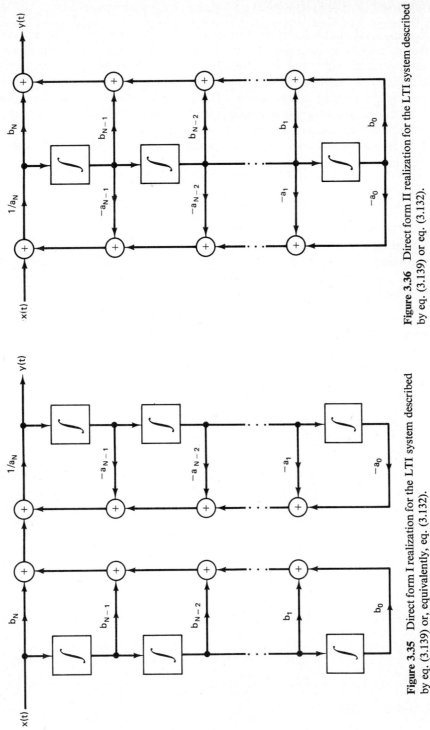

Figure 3.35 Direct form I realization for the LTI system described by eq. (3.139) or, equivalently, eq. (3.132).

Figure 3.36 Direct form II realization for the LTI system described by eq. (3.139) or eq. (3.132).

3.7 SINGULARITY FUNCTIONS

In our initial discussion of the continuous-time unit impulse function, we described the unit impulse as the limiting form of a rectangular pulse which became progressively narrower and higher but maintained unity area. Although this interpretation does provide some valuable intuition into the nature of $\delta(t)$, there are limitations to viewing the impulse in this fashion. In this section we would like to reconsider the unit impulse and other similar signals and to interpret them in the context of convolution and continuous-time LTI systems.

In Section 3.4 we saw that $\delta(t)$ is the impulse response of the identity system. That is,

$$x(t) = x(t) * \delta(t) \tag{3.140}$$

for any signal $x(t)$. Therefore, if we take $x(t) = \delta(t)$, we have that

$$\delta(t) = \delta(t) * \delta(t) \tag{3.141}$$

Equation (3.140) is a basic property of the unit impulse, and it also has a significant implication for our previous definition of $\delta(t)$ as the limiting form of a rectangular pulse. Specifically, let $\delta_\Delta(t)$ correspond to the rectangular pulse as defined in Figure 2.20, and let

$$r_\Delta(t) = \delta_\Delta(t) * \delta_\Delta(t) \tag{3.142}$$

Then $r_\Delta(t)$ is as sketched in Figure 3.37. If we wish to interpret $\delta(t)$ as the limit as $\Delta \longrightarrow 0$ of $\delta_\Delta(t)$, then, by virtue of eq. (3.141), the limit as $\Delta \longrightarrow 0$ for $r_\Delta(t)$ must also be a unit impulse. In a similar manner, we can argue that the limit as $\Delta \longrightarrow 0$ of $r_\Delta(t) * r_\Delta(t)$ or $r_\Delta(t) * \delta_\Delta(t)$ must be a unit impulse, and so on. Thus, we see that for consistency, if we define the unit impulse as the limiting form of some signal, then in fact there are an infinite number of very dissimilar-looking signals all of which behave like an impulse in the limit.

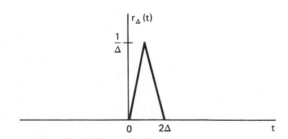

Figure 3.37 The signal $r_\Delta(t)$ defined in eq. (3.142).

The key words in the preceding paragraph are "behave like an impulse." Most frequently, a function or signal is defined by what it is at each value of the independent variable, and it is this perspective that leads us to choose a definition for the impulse as the limiting form of a signal such as a rectangular pulse. However, the primary importance of the unit impulse is not what it *is* at each value of t, but rather what it *does* under convolution. Thus, from the point of view of linear systems analysis, we may alternatively *define* the unit impulse as that signal for which

$$x(t) = x(t) * \delta(t) \tag{3.143}$$

for any $x(t)$. In this sense, all the signals referred to in the preceding paragraph behave like a unit impulse in the limit because if we replace $\delta(t)$ by any of these signals, then eq. (3.143) is satisfied in the limit.

All the properties of the unit impulse that we need can be obtained from the *operational definition* given by eq. (3.143). For example, if we let $x(t) = 1$ for all t, then

$$1 = x(t) = x(t) * \delta(t) = \delta(t) * x(t) = \int_{-\infty}^{+\infty} \delta(\tau)x(t - \tau)\, d\tau$$

$$= \int_{-\infty}^{+\infty} \delta(\tau)\, d\tau$$

so that the unit impulse has unit area. Furthermore, if we take an arbitrary signal $g(t)$, reverse it to obtain $g(-t)$, and then convolve this with $\delta(t)$, we obtain

$$g(-t) = g(-t) * \delta(t) = \int_{-\infty}^{+\infty} g(\tau - t)\, \delta(\tau)\, d\tau$$

which for $t = 0$ yields

$$g(0) = \int_{-\infty}^{+\infty} g(\tau)\, \delta(\tau)\, d\tau \tag{3.144}$$

Therefore, the operational definition of $\delta(t)$ given by eq. (3.143) implies eq. (3.144). On the other hand, eq. (3.144) implies eq. (3.143). To see this, let $x(t)$ be a given signal, fix a time t, and define

$$g(\tau) = x(t - \tau)$$

Then, using eq. (3.144), we have that

$$x(t) = g(0) = \int_{-\infty}^{+\infty} g(\tau)\, \delta(\tau)\, d\tau = \int_{-\infty}^{+\infty} x(t - \tau)\, \delta(\tau)\, d\tau$$

which is precisely eq. (3.143). Therefore eq. (3.144) is an equivalent operational definition of the unit impulse. That is, the unit impulse is the signal which when multiplied by an arbitrary signal $g(t)$ and then integrated from $-\infty$ to $+\infty$ produces the value $g(0)$. In a similar fashion we can define *any* signal operationally by its behavior when multiplied by any other signal $g(t)$ and then integrated from $-\infty$ to $+\infty$.

Since we will be concerned principally with LTI systems and thus with convolution, the characterization of $\delta(t)$ given in eq. (3.143) will be the one to which we will refer most. On the other hand, eq. (3.144) is useful in determining some of the other properties of the unit impulse. For example, consider the signal $f(t)\, \delta(t)$, where $f(t)$ is an arbitrary signal. Then, from eq. (3.144),

$$\int_{-\infty}^{+\infty} g(\tau)f(\tau)\, \delta(\tau)\, d\tau = g(0)f(0) \tag{3.145}$$

On the other hand, if we consider the signal $f(0)\, \delta(t)$, we see that

$$\int_{-\infty}^{+\infty} g(\tau)f(0)\, \delta(\tau)\, d\tau = g(0)f(0) \tag{3.146}$$

Comparing eqs. (3.145) and (3.146), we see that the two signals $f(t)\, \delta(t)$ and $f(0)\, \delta(t)$ behave identically when multiplied by $g(t)$ and then integrated from $-\infty$ to $+\infty$. Consequently, using this form of the operational definition of signals, we conclude that

$$f(t)\,\delta(t) = f(0)\,\delta(t) \tag{3.147}$$

which is a property that we derived by alternative means in Section 2.3 [see eq. (2.24)].

The unit impulse is one of a class of signals known as *singularity functions*, each of which can be defined operationally in terms of its behavior under convolution. Consider the LTI system for which the output is the derivative of the input

$$y(t) = \frac{dx(t)}{dt} \tag{3.148}$$

The unit impulse response of this system is the derivative of the unit impulse which is called the *unit doublet*, $u_1(t)$. From the convolution representation for LTI systems, we then have that

$$\frac{dx(t)}{dt} = x(t) * u_1(t) \tag{3.149}$$

for any signal $x(t)$. Just as eq. (3.143) serves as the operational definition of $\delta(t)$, we will take eq. (3.149) as the operational definition of $u_1(t)$. Similarly, we can define $u_2(t)$, the second derivative of $\delta(t)$, as the impulse response of an LTI system which takes the second derivative of the input

$$\frac{d^2x(t)}{dt^2} = x(t) * u_2(t) \tag{3.150}$$

From eq. (3.149) we see that

$$\frac{d^2x(t)}{dt^2} = \frac{d}{dt}\left(\frac{dx(t)}{dt}\right) = x(t) * u_1(t) * u_1(t) \tag{3.151}$$

and therefore

$$u_2(t) = u_1(t) * u_1(t) \tag{3.152}$$

In general, $u_k(t)$, $k > 0$, is the kth derivative of $\delta(t)$ and thus is the impulse response of a system that takes the kth derivative of the input. Since this system can be obtained as the cascade of k differentiators, we have that

$$u_k(t) = \underbrace{u_1(t) * \ldots * u_1(t)}_{k \text{ times}} \tag{3.153}$$

Each of these singularity functions has properties which, as with the unit impulse, can be derived from its operational definition. For example, if we consider the constant signal $x(t) = 1$, then

$$0 = \frac{dx(t)}{dt} = x(t) * u_1(t) = \int_{-\infty}^{+\infty} u_1(\tau)x(t - \tau)\,d\tau$$

$$= \int_{-\infty}^{+\infty} u_1(\tau)\,d\tau$$

so that the unit doublet has zero area. Moreover, if we convolve the signal $g(-t)$ with $u_1(t)$, we obtain

$$\int_{-\infty}^{+\infty} g(\tau - t)u_1(\tau)\,d\tau = g(-t) * u_1(t) = \frac{dg(-t)}{dt} = -\frac{dg(-t)}{dt}$$

which for $t = 0$ yields

$$-g'(0) = \int_{-\infty}^{+\infty} g(\tau)u_1(\tau)\,d\tau \tag{3.154}$$

In an analogous manner we can derive related properties of $u_1(t)$ and higher-order singularity functions, and several of these properties are considered in Problem 3.39.

In addition to singularity functions which are derivatives of different orders of the unit impulse, we can also define signals that represent successive integrals of the unit impulse function. As we saw in Section 3.4, the unit step is the impulse response of an integrator

$$y(t) = \int_{-\infty}^{t} x(\tau)\,d\tau$$

Therefore,

$$u(t) = \int_{-\infty}^{t} \delta(\tau)\,d\tau \tag{3.155}$$

and we also have an operational definition of $u(t)$:

$$x(t) * u(t) = \int_{-\infty}^{t} x(\tau)\,d\tau \tag{3.156}$$

Similarly, we can define the system that consists of a cascade of two integrators. Its impulse response is denoted by $u_{-2}(t)$, which is simply the convolution of $u(t)$, the impulse response of one integrator, with itself:

$$u_{-2}(t) = u(t) * u(t) = \int_{-\infty}^{t} u(\tau)\,d\tau \tag{3.157}$$

Since $u(t)$ equals 0 for $t < 0$ and equals 1 for $t > 0$, we see that

$$u_{-2}(t) = tu(t) \tag{3.158}$$

This signal, which is referred to as the *unit ramp function*, is shown in Figure 3.38.

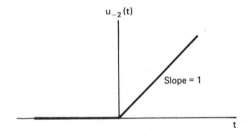

Slope = 1

Figure 3.38 Unit ramp function.

Also, we can obtain an operational definition for the behavior of $u_{-2}(t)$ under convolution from eqs. (3.156) and (3.157):

$$x(t) * u_{-2}(t) = x(t) * u(t) * u(t)$$
$$= \left(\int_{-\infty}^{t} x(\sigma)\,d\sigma \right) * u(t) = \int_{-\infty}^{t} \left(\int_{-\infty}^{\tau} x(\sigma)\,d\sigma \right) d\tau \tag{3.159}$$

In an analogous fashion we can define higher-order integrals of $\delta(t)$ as the impulse responses of cascades of integrators:

$$u_{-k}(t) = \underbrace{u(t) * \ldots * u(t)}_{k \text{ times}} = \int_{-\infty}^{t} u_{-(k-1)}(\tau)\,d\tau \tag{3.160}$$

The convolution of $x(t)$ with $u_{-3}(t)$, $u_{-4}(t)$, ..., generate correspondingly higher-order integrals of $x(t)$. Also, note that the integrals in eq. (3.160) can be evaluated directly (Problem 3.40), as was done in eq. (3.158), to obtain

$$u_{-k}(t) = \frac{t^{k-1}}{(k-1)!}u(t) \tag{3.161}$$

Thus, unlike the derivatives of $\delta(t)$, the successive integrals of the unit impulse are functions that can be defined for each value of t [eq. (3.161)] as well as by their behavior under convolution.

At times it will be useful to use an alternative notation for $\delta(t)$ and $u(t)$, specifically

$$\delta(t) = u_0(t) \tag{3.162}$$

$$u(t) = u_{-1}(t) \tag{3.163}$$

With this notation $u_k(t)$ for $k > 0$ denotes the impulse response of a cascade of k differentiators, $u_0(t)$ is the impulse response of the identity system, and for $k < 0$, $u_k(t)$ is the impulse response of a cascade of $|k|$ integrators. Furthermore, since a differentiator is the inverse system to an integrator,

$$u(t) * u_1(t) = \delta(t)$$

or, in our alternative notation,

$$u_{-1}(t) * u_1(t) = u_0(t) \tag{3.164}$$

More generally, from eqs. (3.153), (3.160), and (3.164) we see that for k and r any integers,

$$u_k(t) * u_r(t) = u_{k+r}(t) \tag{3.165}$$

If k and r are both positive, eq. (3.165) states that a cascade of k differentiators followed by r more differentiators yields an output that is the $(k + r)$th derivative of the input. Similarly, if k is negative and r is negative, we have a cascade of $|k|$ integrators followed by another $|r|$ integrators. Also, if k is negative and r is positive, we have a cascade of k integrators followed by r differentiators, and eq. (3.165) tells us that the overall system is equivalent to a cascade of $|k + r|$ integrators if $(k + r) < 0$, a cascade of $(k + r)$ differentiators if $(k + r) > 0$, or the identity system if $(k + r) = 0$. Therefore, by defining singularity functions in terms of their behavior under convolution, we obtain a characterization for them that allows us to manipulate them with relative ease and to interpret them directly in terms of their significance for LTI systems. Since this is our primary concern in this book, the operational definition for singularity functions that we have given in this section will suffice for our purposes.†

---

†As mentioned in the footnote on p. 25, singularity functions have been heavily studied in the field of mathematics under the alternative names of generalized functions and distribution theory. The approach we have taken in this section is actually closely allied in spirit with the rigorous approach taken in the references given on p. 25.

3.8 SUMMARY

In this chapter we have developed very important representations for LTI systems, both in discrete time and continuous time. In discrete time we derived a representation of signals as weighted sums of shifted unit impulses, and we then used this to derive the convolution-sum representation for the response of a discrete-time LTI system. In continuous time we derived an analogous representation of continuous-time signals as weighted integrals of shifted unit impulses, and we used this to derive the convolution integral representation for continuous-time LTI systems. These representations are extremely important, as they allow us to compute the response of an LTI system to an arbitrary input in terms of its response to a unit impulse. Moreover, the convolution sum and integral provided us with the means in Section 3.4 to analyze the properties of LTI systems and in particular to relate many LTI system properties, including causality and stability, to corresponding properties of the unit impulse response. Also in Section 3.7 we developed an interpretation of the unit impulse and other related singularity functions in terms of their behavior under convolution. This interpretation is particularly useful in the analysis of LTI systems.

An important class of continuous-time systems are those described by linear constant-coefficient differential equations, and their discrete-time counterpart, linear constant-coefficient difference equations, play an equally important role in discrete time. In Section 3.5 we reviewed methods for solving equations of these types, and discussed the properties of causality, linearity, and time invariance for systems described by such equations. We also introduced pictorial representations for these systems which are useful in describing structures for their implementation.

PROBLEMS

3.1. Compute the convolution $y[n] = x[n] * h[n]$ of the following pairs of signals.

(a) $x[n] = \alpha^n u[n]$, $\alpha \neq \beta$
     $h[n] = \beta^n u[n]$,

(b) $x[n] = h[n] = \alpha^n u[n]$

(c) $x[n] = 2^n u[-n]$
     $h[n] = u[n]$

(d) $x[n] = (-1)^n \{u[-n] - u[-n - 8]\}$
     $h[n] = u[n] - u[n - 8]$

(e) $x[n]$ and $h[n]$ as in Figure P3.1(a).

(f) $x[n]$ and $h[n]$ as in Figure P3.1(b).

(g) $x[n]$ and $h[n]$ as in Figure P3.1(c).

(h) $x[n] = 1$ for all n, $h[n] = \begin{cases} (\frac{1}{2})^n, & n \geq 0 \\ 4^n, & n < 0 \end{cases}$

(i) $x[n] = u[n] - u[-n]$, $h[n] = \begin{cases} (\frac{1}{2})^n, & n \geq 0 \\ 4^n, & n < 0 \end{cases}$

(j) $x[n] = (-\frac{1}{2})^n u[n - 4]$
     $h[n] = 4^n u[2 - n]$

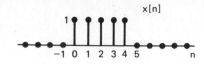

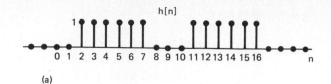

(a)

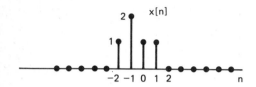

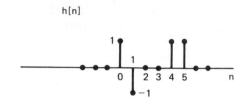

(b)

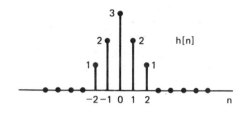

(c)

Figure P3.1

3.2. For each of the following pairs of waveforms, use the convolution integral to find the response $y(t)$ of the LTI system with impulse response $h(t)$ to the input $x(t)$. Sketch your results.

(a) $x(t) = e^{-\alpha t}u(t)$
$h(t) = e^{-\beta t}u(t)$ (Do this both when $\alpha \neq \beta$ and when $\alpha = \beta$.)

(b) $x(t) = u(t) - 2u(t-2) + u(t-5)$
$h(t) = e^{2t}u(1-t)$

(c) $x(t) = e^{-3t}u(t)$
$h(t) = u(t-1)$

(d) $x(t) = e^{-2t}u(t+2) + e^{3t}u(-t+2)$
$h(t) = e^{t}u(t-1)$

(e) $x(t) = \begin{cases} e^{t}, & t < 0 \\ e^{5t} - 2e^{-t}, & t > 0 \end{cases}$
$h(t)$ as in Figure P3.2(a)

(f) $x(t)$ and $h(t)$ as in Figure P3.2(b).

(g) $x(t)$ as in Figure P3.2(c).
$h(t) = u(-2-t)$

(h) $x(t) = \delta(t) - 2\delta(t-1) + \delta(t-2)$, and $h(t)$ as in Figure P3.2(d).

(i) $x(t)$ and $h(t)$ as in Figure P3.2(e).

(j) $x(t)$ and $h(t)$ as in Figure P3.2(f).

(k) $x(t)$ and $h(t)$ as in Figure P3.2(g).

(l) $x(t)$ as in Figure P3.2(h).
$h(t) = e^{-t}[u(t - 1) - u(t - 2)]$
(m) $x(t)$ and $h(t)$ as in Figure P3.2(i).

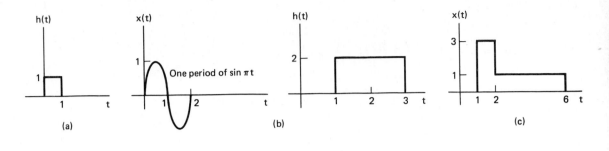

(a)

(b)

(c)

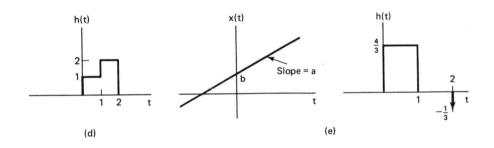

(d)

(e)

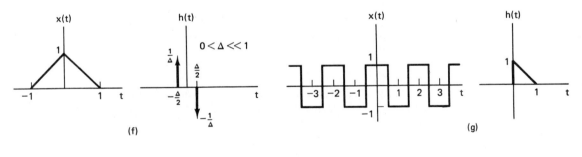

(f)

(g)

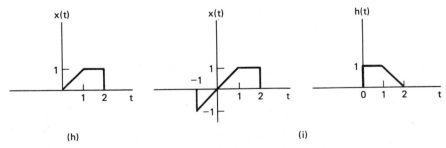

(h)

(i)

Figure P3.2

3.3. (a) As we have seen in Section 3.1, a discrete-time linear (and possibly time-varying) system is characterized by its responses $h_k[n]$ to delayed unit samples $\delta[n - k]$. For each of the following choices for $h_k[n]$, determine an explicit relationship between the input $x[n]$ and output $y[n]$ of the linear system so specified.

(i) $h_k[n] = \delta[n - k]$

(ii) $h_k[n] = \begin{cases} \delta[n - k], & k \text{ even} \\ 0, & k \text{ odd} \end{cases}$

(iii) $h_k[n] = \delta[2n - k]$

(iv) $h_k[n] = ku[n - k]$

(v) $h_k[n] = k\delta[n - 2k] + 3k\delta[n - k]$

(vi) $h_k[n] = \begin{cases} \delta[n - k + 1], & k \text{ odd} \\ 5u[n - k], & k \text{ even} \end{cases}$

(b) Which of the systems in part (a) are causal and which are not? Justify your answers.

(c) Determine and sketch the response of each of the systems of part (a) to the input $x[n] = u[n]$.

3.4. Consider a linear system with the following response to $\delta(t - \tau)$:

$$h_\tau(t) = u(t - \tau) - u(t - 2\tau)$$

(a) Is this system time-invariant?

(b) Is it causal?

(c) Determine the response of this system to each of the following two inputs:

(i) $x_1(t) = u(t - 1) - u(t - 3)$

(ii) $x_2(t) = e^{-t}u(t)$

3.5. One of the important properties of convolution, in both continuous and discrete time, is the associativity property. In this problem we will check and illustrate this property.

(a) Prove the equality

$$[x(t) * h(t)] * g(t) = x(t) * [h(t) * g(t)] \tag{P3.5-1}$$

by showing that both sides of eq. (P3.5-1) equal

$$\int_{-\infty}^{+\infty} \int_{-\infty}^{+\infty} x(\tau)h(\sigma)g(t - \tau - \sigma) \, d\tau \, d\sigma$$

(b) Consider two LTI systems with unit sample responses $h_1[n]$ and $h_2[n]$ shown in Figure P3.5-1. These two systems are cascaded as shown in Figure P3.5-2. Let $x[n] = u[n]$.

(i) Compute $y[n]$ by first computing $w[n] = x[n] * h_1[n]$ and then $y[n] = w[n] * h_2[n]$, that is, $y[n] = \{x[n] * h_1[n]\} * h_2[n]$.

(ii) Now find $y[n]$ by first convolving $h_1[n]$ and $h_2[n]$ to obtain $g[n] = h_1[n] * h_2[n]$, and then convolving $x[n]$ with $g[n]$ to obtain $y[n] = x[n] * \{h_1[n] * h_2[n]\}$.

The answer to (i) and (ii) should be identical, illustrating the associative property of discrete-time convolution.

(c) Consider the cascade of two LTI systems as in Figure P3.5-2, where in this case

$$h_1[n] = \sin 8n$$
$$h_2[n] = a^n u[n], \qquad |a| < 1$$

and where the input is

$$x[n] = \delta[n] - a\,\delta[n - 1]$$

Determine the output $y[n]$. (*Hint:* The use of the associative and commutative properties of convolution should greatly facilitate the solution in this case.)

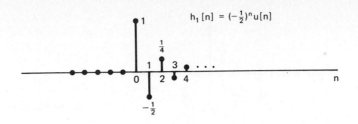

$h_1[n] = (-\frac{1}{2})^n u[n]$

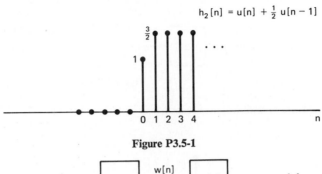

$h_2[n] = u[n] + \frac{1}{2} u[n-1]$

Figure P3.5-1

$$x[n] \longrightarrow \boxed{h_1[n]} \xrightarrow{w[n]} \boxed{h_2[n]} \longrightarrow y[n]$$

Figure P3.5-2

3.6. (a) Consider the interconnection of LTI systems depicted in Figure P3.6-1. Express the overall impulse response $h[n]$ in terms of $h_1[n]$, $h_2[n]$, $h_3[n]$, $h_4[n]$, and $h_5[n]$.

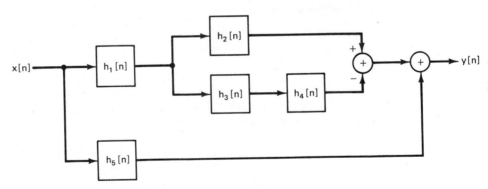

Figure P3.6-1

(b) Determine $h[n]$ when

$$h_1[n] = 4(\tfrac{1}{2})^n\{u[n] - u[n-3]\}$$
$$h_2[n] = h_3[n] = (n+1)u[n]$$
$$h_4[n] = \delta[n-1]$$
$$h_5[n] = \delta[n] - 4\delta[n-3]$$

(c) Sketch the response of the system of part (b) if $x[n]$ is as in Figure P3.6-2.

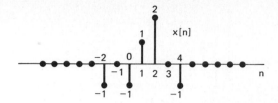

Figure P3.6-2

3.7. Consider the cascade interconnection of three causal LTI systems illustrated in Figure P3.7-1. The impulse response $h_2[n]$ is given by

$$h_2[n] = u[n] - u[n - 2]$$

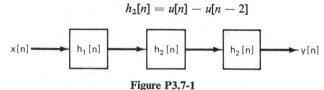

Figure P3.7-1

and the overall impulse response is as shown in Figure P3.7-2.

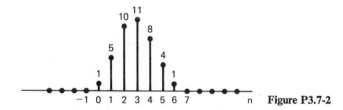

Figure P3.7-2

(a) Find the impulse response $h_1[n]$.
(b) Find the response of the overall system to the input

$$x[n] = \delta[n] - \delta[n - 1]$$

3.8. (a) Consider an LTI system with input and output related through the equation

$$y(t) = \int_{-\infty}^{t} e^{-(t-\tau)} x(\tau - 2) \, d\tau$$

What is the impulse response $h(t)$ for this system?

(b) Determine the response of this system when the input $x(t)$ is as shown in Figure P3.8-1.

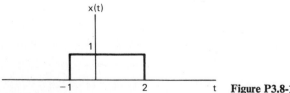

Figure P3.8-1

(c) Consider the interconnection of LTI systems depicted in Figure P3.8-2. Here $h(t)$ is as in part (a). Determine the output $y(t)$ when the input $x(t)$ is again given by Figure P3.8-1. Perform this calculation in two ways:

Linear Time-Invariant Systems   Chap. 3

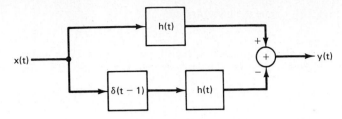

Figure P3.8-2

   (i) Compute the overall impulse response of the interconnected system and then use the convolution integral to evaluate $y(t)$.

   (ii) Use the result of part (b), together with the properties of convolution, to determine $y(t)$ without evaluating a convolution integral.

3.9. **(a)** Let $h(t)$ be the triangular pulse shown in Figure P3.9-1(a), and let $x(t)$ be the

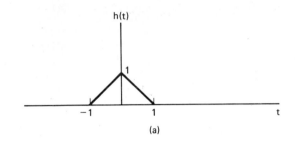

(a)

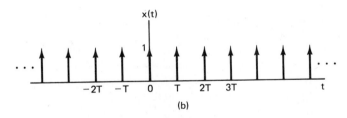

(b)

Figure P3.9-1

impulse train depicted in Figure P3.9-1(b). That is,

$$x(t) = \sum_{k=-\infty}^{+\infty} \delta(t - kT) \qquad \text{(P3.9-1)}$$

Determine and sketch $y(t) = x(t) * h(t)$ for the following values of T:

   (i) $T = 4$
   (ii) $T = 2$
   (iii) $T = 3/2$
   (iv) $T = 1$

 (b) Consider an LTI system with impulse response $h(t) = e^{-t}u(t)$. Determine and sketch the output $y(t)$ when the input $x(t)$ is the impulse train of eq. (P3.9-1) with $T = 1$.

(c) Let $x(t)$ be the impulse train

$$x(t) = \sum_{k=0}^{\infty} (-1)^k \, \delta(t - k) \qquad \text{(P3.9-2)}$$

Determine and sketch the output of the LTI system with impulse response $h(t)$ depicted in Figure P3.9-2 when $x(t)$ in eq. (P3.9-2) is the input.

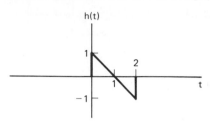

Figure P3.9-2

3.10. Determine if each of the following statements or equations is true in general. Provide proofs for those you believe to be true and counterexamples for those that you think are false.

(a) $x[n] * \{h[n]g[n]\} = \{x[n] * h[n]\}g[n]$

(b) $\alpha^n x[n] * \alpha^n h[n] = \alpha^n \{x[n] * h[n]\}$

(c) If $y(t) = x(t) * h(t)$, then $y(2t) = 2x(2t) * h(2t)$.

(d) If $y[n] = x[n] * h[n]$, then $y[2n] = 2x[2n] * h[2n]$.

(e) If $x(t)$ and $h(t)$ are odd, then $y(t) = x(t) * h(t)$ is even.

(f) If $y(t) = x(t) * h(t)$, then $\mathcal{E}v\,\{y(t)\} = x(t) * \mathcal{E}v\,\{h(t)\} + \mathcal{E}v\,\{x(t)\} * h(t)$.

3.11. (a) If

$$x(t) = 0, \qquad |t| > T_1$$

and

$$h(t) = 0, \qquad |t| > T_2$$

then

$$x(t) * h(t) = 0, \qquad |t| > T_3$$

for some positive number T_3. Express T_3 in terms of T_1 and T_2.

(b) A discrete-time LTI system has input $x[n]$, impulse response $h[n]$, and output $y[n]$. If $h[n]$ is known to be zero everywhere outside the interval $N_0 \le n \le N_1$ and $x[n]$ is known to be zero everywhere outside the interval $N_2 \le n \le N_3$, then the output $y[n]$ is constrained to be zero everywhere except on some interval $N_4 \le n \le N_5$.

(i) Determine N_4 and N_5 in terms of N_0, N_1, N_2, and N_3.

(ii) If the interval $N_0 \le n \le N_1$, is of length M_h, $N_2 \le n \le N_3$ is of length M_x, and $N_4 \le n \le N_5$ is of length M_y, express M_y in terms of M_h and M_x.

(c) Consider a discrete-time LTI system, with the property that if the input $x[n] = 0$ for all $n \ge 10$, then the output $y[n] = 0$ for all $n \ge 15$. What condition must $h[n]$, the impulse response of the system, satisfy for this to be true?

(d) Consider an LTI system with impulse response in Figure P3.11. Over what interval must we know $x(t)$ in order to determine $y(0)$?

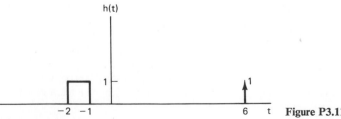

Figure P3.11

3.12. Let $x_1(t)$ and $x_2(t)$ be two periodic signals, with a common period of T_0. It is not too difficult to check that the convolution of $x_1(t)$ and $x_2(t)$ does not converge. However, it is sometimes useful to consider a form of convolution for such signals that is referred to as *periodic convolution*. Specifically, we define the periodic convolution of $x_1(t)$ and $x_2(t)$ as

$$y(t) = \int_0^{T_0} x_1(\tau)x_2(t - \tau)\, d\tau = x_1(t) \circledast x_2(t) \qquad \text{(P3.12-1)}$$

Note that we are integrating over exactly one period.

(a) Show that $y(t)$ is periodic with period T_0.

(b) Consider the signal

$$y_a(t) = \int_a^{a+T_0} x_1(\tau)x_2(t - \tau)\, d\tau$$

where a is an arbitrary real number. Show that

$$y(t) = y_a(t)$$

Hint: Write $a = kT_0 + b$, where $0 \le b < T_0$.

(c) Compute the periodic convolution of the signals depicted in Figure P3.12-1, where we take $T_0 = 1$.

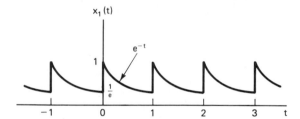

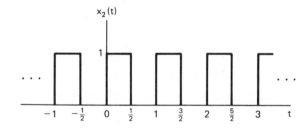

Figure P3.12-1

(d) Show that periodic convolution is commutative and associative; that is,

$$x_1(t) \circledast x_2(t) = x_2(t) \circledast x_1(t)$$

and

$$x_1(t) \circledast [x_2(t) \circledast x_3(t)] = [x_1(t) \circledast x_2(t)] \circledast x_3(t)$$

In a similar fashion we can define the periodic convolution $y[n]$ of two discrete-time periodic signals $x_1[n]$ and $x_2[n]$ with common period N_0. This is given by the expression

$$y[n] = \sum_{k=0}^{N_0-1} x_1[k]x_2[n - k]$$

Again we can check that $y[n]$ is periodic with period N_0 and that we can actually calculate it by summing over *any* N_0 consecutive values of k, that is,

$$y[n] = \sum_{k=m}^{N_0+m-1} x_1[k]x_2[n-k]$$

for any integer m.

(e) Consider the signals $x_1[n]$ and $x_2[n]$ depicted in Figure P3.12-2. These signals are periodic with period 6. Compute and sketch their periodic convolution using $N_0 = 6$.

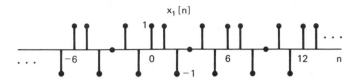

$x_1[n]$

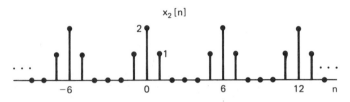

$x_2[n]$

Figure P3.12-2

(f) Since these signals are periodic with period 6, they are also periodic with period 12. Compute the periodic convolution of $x_1[n]$ and $x_2[n]$ using $N_0 = 12$.

(g) In general, if $x_1[n]$ and $x_2[n]$ are periodic with period N, what is the relationship between their periodic convolution computed using $N_0 = N$ and that obtained using $N_0 = kN$ for k a positive integer?

3.13. We define the area under a continuous-time signal $v(t)$ as

$$A_v = \int_{-\infty}^{+\infty} v(t)\, dt$$

Show that if $y(t) = x(t) * h(t)$, then

$$A_y = A_x A_h$$

3.14. (a) Show that if the response of an LTI system to $x(t)$ is the output $y(t)$, then the response of the system to

$$x'(t) = \frac{dx(t)}{dt}$$

is $y'(t)$. Do this problem in three different ways:

(i) Directly from the properties of linearity and time invariance and the fact that

$$x'(t) = \lim_{h \to 0} \frac{x(t) - x(t-h)}{h}$$

(ii) By differentiating the convolution integral.

(iii) By examining the system in Figure P3.14.

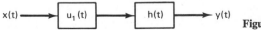

$x(t) \longrightarrow \boxed{u_1(t)} \longrightarrow \boxed{h(t)} \longrightarrow y(t)$

Figure P3.14

(b) Demonstrate the validity of the following relationships.

(i) $y'(t) = x(t) * h'(t)$

(ii) $y(t) = \left(\int_{-\infty}^{t} x(\tau) \, d\tau \right) * h'(t) = \int_{-\infty}^{t} [x'(\tau) * h(\tau)] \, d\tau = x'(t) * \left(\int_{-\infty}^{t} h(\tau) \, d\tau \right)$

Hint: These are easily done using block diagrams as in (iii) of part (a) and the fact that $u_1(t) * u_{-1}(t) = \delta(t)$.

(c) An LTI system has the response $y(t) = \sin \omega_0 t$ to the input $x(t) = e^{-5t}u(t)$. Use the result of part (a) to aid you in determining the impulse response of this system.

(d) A second LTI system has the response $y(t) = (e^t - 1)u(t)$ to the input $x(t) = (\sin t)u(t)$. What is the impulse response for this system?

(e) Compute the convolution of the two signals in Problem 3.2(b) using the fact that

$$y(t) = \int_{-\infty}^{t} [x'(\tau) * h(\tau)] \, d\tau$$

(f) Let $s(t)$ be the unit step response of a continuous-time LTI system. Use part (b) to deduce that the response $y(t)$ to the input $x(t)$ is given by

$$y(t) = \int_{-\infty}^{+\infty} x'(\tau) s(t - \tau) \, d\tau \qquad \text{(P3.14-1)}$$

Show also that

$$x(t) = \int_{-\infty}^{+\infty} x'(\tau) u(t - \tau) \, d\tau \qquad \text{(P3.14-2)}$$

(g) Use eq. (P3.14-1) to determine the response of an LTI system with step response

$$s(t) = (e^{-3t} - 2e^{-2t} + 1)u(t)$$

to the input $x(t) = e^t u(t)$.

(h) Let $s[n]$ be the unit step response of a discrete-time LTI system. What are the discrete-time counterparts of eqs. (P3.14-1) and (P3.14-2)?

(i) Use the discrete-time version of equation (P3.14-1) to determine the response of an LTI system with step response

$$s[n] = (\tfrac{1}{2})^n u[n + 1]$$

to the input

$$x[n] = (-\tfrac{1}{2})^n u[n]$$

3.15. We are given a certain linear time-invariant system with impulse response $h_0(t)$. We are told that when the input is $x_0(t)$, the output is $y_0(t)$, which is sketched in Figure P3.15. We are then given the following set of inputs to linear time-invariant systems with the indicated impulse responses.

Input $x(t)$	*Impulse Response $h(t)$*
(a) $x(t) = 2x_0(t)$	$h(t) = h_0(t)$
(b) $x(t) = x_0(t) - x_0(t - 2)$	$h(t) = h_0(t)$
(c) $x(t) = x_0(t - 2)$	$h(t) = h_0(t + 1)$
(d) $x(t) = x_0(-t)$	$h(t) = h_0(t)$
(e) $x(t) = x_0(-t)$	$h(t) = h_0(-t)$
(f) $x(t) = x_0'(t)$	$h(t) = h_0'(t)$

[Here $x_0'(t)$, $h_0'(t)$ denote the first derivatives of $x_0(t)$ and $h_0(t)$.]

In each of these cases, determine whether or not we have given enough information to determine the output $y(t)$ when the input is $x(t)$ and the system has impulse

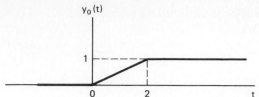

$y_0(t)$

1

0   2   t   **Figure P3.15**

response $h(t)$. If it is possible to determine $y(t)$, provide an accurate sketch of it with numerical values clearly indicated on the graph.

3.16. An important class of continuous-time LTI systems, are those whose responses to the unit impulse consist of *trains* of equally spaced impulses. That is, if the input to such a system is $x(t) = \delta(t)$, then output is

$$h(t) = \sum_{n=-\infty}^{+\infty} h_n\delta(t - nT) \tag{P3.16-1}$$

where $T > 0$ is the impulse spacing and the h_k are specified constants.

Systems of this type can be implemented in the form of *tapped delay lines*. An example of a tapped delay line is illustrated in Figure P3.16-1 for a system with impulse response

$$\sum_{n=0}^{3} h_n\delta(t - nT)$$

The system labeled "Delay T" in the figure is an ideal delay, whose output $y(t)$ is a delayed version of its input $x(t)$, i.e., $y(t) = x(t - T)$. After each such delay (and before the first one) the signal is "tapped" off, multiplied by a constant gain (h_0, h_1, h_2, etc.), and the output is the sum of these weighted, delayed versions of the input. As we will see at the end of this problem, tapped delay lines are of great practical value for the design and implementation of LTI systems with specified response characteristics.

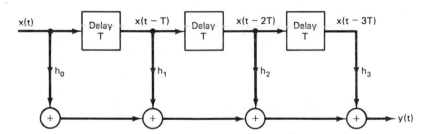

Figure P3.16-1

(a) Consider the LTI system with impulse response given by eq. (P3.16-1). Suppose that the input to this system is itself an impulse train

$$x(t) = \sum_{n=-\infty}^{+\infty} x_n\delta(t - nT)$$

Show that the output $y(t)$ is also an impulse train

$$y(t) = \sum_{n=-\infty}^{+\infty} y_n\delta(t - nT)$$

where the coefficients $\{y_n\}$ are determined as the discrete-time convolution of the sequences $\{x_n\}$ and $\{h_n\}$:

$$y_n = \sum_{k=-\infty}^{+\infty} x_k h_{n-k} \qquad (P3.16\text{-}2)$$

(b) What is the impulse response of a system consisting of the cascade of three identical tapped delay lines, each as in Figure P3.16-1, with $h_0 = 1, h_1 = -1, h_2 = h_3 = 0$?

(c) Consider the tapped delay line illustrated in Figure P3.16-2, where the gain at each tap is equal to the delay time T. Suppose that the input to this system is

$$x(t) = e^{-t}u(t)$$

Carefully sketch the output $y(t)$. What does $y(t)$ look like in the limit as $T \to 0$, i.e., as we have shorter and shorter delays between the taps and as the tap gains get smaller and smaller. If we call this limiting form of the output $y_0(t)$, show that for any given value of $T > 0$

$$\frac{T}{1 - e^{-T}} y_0(t + T)$$

forms an envelope for $y(t)$, i.e., that this signal is a smooth curve connecting the successive peaks of $y(t)$.

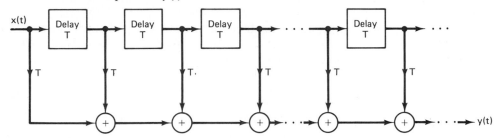

Figure P3.16-2

(d) Consider a system with an impulse response that is piecewise constant. That is,

$$h(t) = h_n \qquad \text{for } n < t < n + 1 \qquad (P3.16\text{-}3)$$

The signal $h(t)$ is illustrated in Figure P3.16-3. Show that this system can be implemented as the cascade of a tapped-delay line with impulse response as in eq. (P3.16-1) and the LTI system depicted in Figure P3.16-4.

(e) Show that we can also implement the system specified by eq. (P3.16-3) as the cascade of a tapped-delay line and an integrator. Do this by determining what the tap coefficients must be so that $h(t)$ has the desired form.

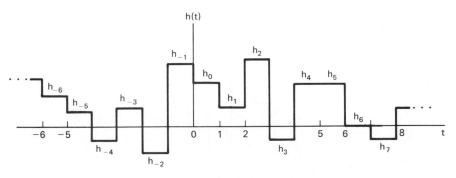

Figure P3.16-3

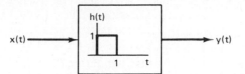

Figure P3.16-4

Note that by using either of the two methods described above we can obtain a system with an impulse response that is a staircase approximation of any desired impulse response. In this problem we have chosen the time interval T between taps to be unity, but any value of T could be used, and the result is that we can obtain arbitrarily accurate staircase approximations using tapped delay lines. This fact, together with the development of inexpensive components for the implementation of tapped-delay lines, has made the use of systems of this type quite attractive in many applications.

(f) As a final observation it is worth noting that it is relatively easy to compute the convolution of $h(t)$ in eq. (P3.16-3) with another signal of the same kind

$$x(t) = x_n \quad \text{for} \quad n < t < n + 1 \qquad \text{(P3.16-4)}$$

Specifically, use the result of part (a) and (d), together with associative and commutative properties of convolution to show that the convolution $x(t) * h(t)$ can be determined as follows:

(i) Compute the discrete convolution in eq. (P3.16-2) where h_n and x_n are given in eqs. (P3.16-3) and (P3.16-4) respectively.

(ii) Then $x(t) * h(t) = y_n$ for $t = n + 1$, and its values for $n + 1 < t < n + 2$ lie on the straight line connecting y_n and y_{n+1} (see Figure P3.16-5)

(g) Apply the technique of part (f) to convolve the signals depicted in Figure P3.16-6.

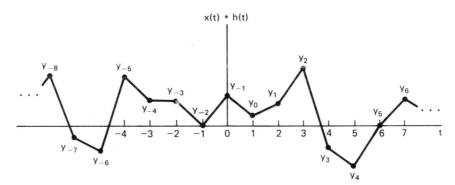

Figure P3.16-5

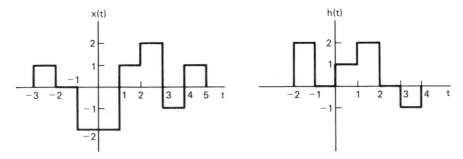

Figure P3.16-6

3.17. Our development of the convolution sum representation for discrete-time LTI systems was based on using the unit sample function as a building block for the representation of arbitrary input signals. This representation, together with knowledge of the response to $\delta[n]$ and the property of superposition, allowed us to represent the system response to an arbitrary input in terms of a convolution. In this problem we consider the use of other signals as building blocks for the construction of arbitrary input signals.

Consider the following set of signals:

$$\phi[n] = (\tfrac{1}{2})^n u[n]$$
$$\phi_k[n] = \phi[n - k], \qquad k = 0, \pm1, \pm2, \pm3, \dots$$

(a) Show that an arbitrary signal can be represented in the form

$$x[n] = \sum_{k=-\infty}^{+\infty} a_k \phi[n - k]$$

by determining an explicit expression for the coefficient a_k in terms of the values of the signal $x[n]$. (*Hint:* What is the representation for $\delta[n]$?)

(b) Let $r[n]$ be the response of an LTI system to the input $x[n] = \phi[n]$. Find an expression for the response $y[n]$ to an arbitrary input $x[n]$ in terms of $r[n]$ and $x[n]$.

(c) Show that $y[n]$ can written as

$$y[n] = \psi[n] * x[n] * r[n]$$

by finding the signal $\psi[n]$

(d) Use the result of part (c) to express the impulse response of the system in terms of $r[n]$. Also, show that

$$\psi[n] * \phi[n] = \delta[n]$$

3.18. Just as we saw for discrete-time signals and systems in Problem 3.17, it is possible to consider representing continuous-time LTI systems using basic input signals other than shifted unit impulses. For example, in this problem we consider an LTI system that has the response $q(t)$ depicted in Figure P3.18(b) to the input $p(t)$ shown in Figure P3.18(a).

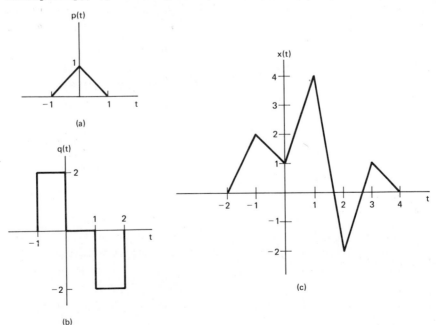

Figure P3.18

(a) Show that the input signal $x(t)$ of Figure P3.18(c) can be represented by

$$x(t) = \sum_{n=-\infty}^{+\infty} a_n p(t-n)$$

and find the values of a_n.

(b) Write down an expression for the response $y(t)$ to the input $x(t)$ shown in Figure P3.18(c) in terms of the responses to the building block inputs $p(t-n)$. Sketch $y(t)$.

(c) Find the system response to the unit ramp input $u_{-2}(t)$.

(d) Find the step and impulse responses of this system. [*Hint:* Part (c) and Problem 3.14 should be of use here.]

(e) Find a block diagram representation of the system in terms of the following elements: integrators, differentiators, ideal unit delay elements (i.e., the output of such a delay equals the input delayed by 1 second), adders, and elements for the multiplication of a signal by a constant coefficient.

3.19. Consider a discrete-time LTI system with unit sample response

$$h[n] = (n+1)\alpha^n u[n]$$

where $|\alpha| < 1$.

Show that the step response of this system is given by

$$s[n] = \left[\frac{1}{(\alpha-1)^2} - \frac{\alpha}{(\alpha-1)^2}\alpha^n + \frac{\alpha}{(\alpha-1)}(n+1)\alpha^n \right]u[n]$$

Hint: Note that

$$\sum_{k=0}^{N}(k+1)\alpha^k = \frac{d}{d\alpha}\sum_{k=0}^{N+1}\alpha^k$$

3.20. In the text we saw that the overall input–output relationship of the cascade of two LTI systems does not depend on the order in which they are cascaded. This fact, known as the commutativity property, depends on both the linearity and the time invariance of both systems. In this problem we provide several examples to illustrate this point.

(a) Consider two discrete-time systems A and B, where system A is an LTI system with unit sample response $h[n] = (\frac{1}{2})^n u[n]$. System B, on the other hand, is linear but time-varying. Specifically, if the input to system B is $w[n]$, its output $z[n]$ is given by

$$z[n] = nw[n]$$

Show that the commutativity property does not hold for these two systems by computing the response of each of the systems depicted in Figure P3.20-1 to the input $x[n] = \delta[n]$.

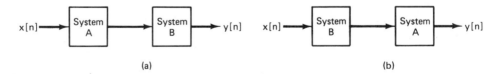

(a)   (b)

Figure P3.20-1

(b) Suppose that we replace system B in each of the interconnected systems of Figure P3.20-1, by the system with the following relationship between its input $w[n]$ and output $z[n]$:

$$z[n] = w[n] + 2$$

Repeat the calculations of part (a) in this case.

(c) What is the overall input–output relationship for the system of Figure P3.20-2?

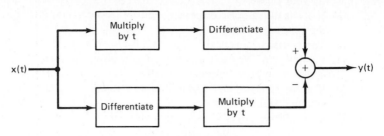

Figure P3.20-2

3.21. In the text we showed that if $h[n]$ is absolutely summable,

$$\sum_{k=-\infty}^{+\infty} |h[k]| < \infty$$

then the LTI system is stable. This means that absolute summability is a *sufficient* condition for stability. In this problem we shall show that it is also a *necessary* condition. Consider an LTI system with impulse response $h[n]$, which we assume is not absolutely summable, that is,

$$\sum_{k=-\infty}^{+\infty} |h[k]| = \infty$$

(a) Suppose that the input to this system is

$$x[n] = \begin{cases} 0 & \text{if } h[-n] = 0 \\ \dfrac{h[-n]}{|h[-n]|} & \text{if } h[-n] \neq 0 \end{cases}$$

Does this input signal represent a bounded input? If so, what is the smallest number B such that

$$|x[n]| \leq B \qquad \text{for all } n?$$

(b) Calculate the output at $n = 0$ for this particular choice of input. Does this result prove the contention that absolute summability is a necessary condition for stability?

(c) In a similar fashion, show that a continuous-time LTI system is stable if and only if its impulse response is absolutely integrable.

3.22. Consider the cascade of two systems shown in Figure P3.22. The first system, A, is known to be LTI. The second system, B, is known to be the inverse of system A. Let $y_1(t)$ denote the response of system A to $x_1(t)$, and let $y_2(t)$ be the response of system A to $x_2(t)$.

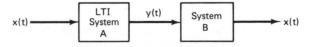

Figure P3.22

(a) What is the response of system B to the input $ay_1(t) + by_2(t)$, where a and b are constants?

(b) What is the response of system B to the input $y_1(t - \tau)$?

3.23. Determine if each of the following statements concerning LTI systems is true or false. Justify your answers.

(a) If $h(t)$ is the impulse response of an LTI system, and $h(t)$ is periodic and nonzero, the system is unstable.

(b) The inverse of a causal LTI system is always causal.

(c) If $|h[n]| \leq K$ for each n, where K is a given number, then the LTI system with $h[n]$ as its impulse response is stable.

(d) If a discrete-time LTI system has an impulse response $h[n]$ of finite duration, the system is stable.

(e) If an LTI system is causal, it is stable.

(f) The cascade of a noncausal LTI system with a causal one is necessarily noncausal.

(g) A continuous-time LTI system is stable if and only if its step response $s(t)$ is absolutely integrable, that is,

$$\int_{-\infty}^{+\infty} |s(t)| \, dt < \infty$$

(h) A discrete-time LTI system is causal if and only if its step response $s[n]$ is zero for $n < 0$.

3.24. The following are the impulse responses of LTI systems, in either continuous or discrete time. Determine whether each system is stable and/or causal. Justify your answers.

(a) $h[n] = (\frac{1}{2})^n u[n]$

(b) $h[n] = (0.99)^n u[n + 3]$

(c) $h[n] = (0.99)^n u[-n]$

(d) $h[n] = (4)^n u[2 - n]$

(e) $h[n] = (-\frac{1}{2})^n u[n] + (1.01)^n u[n - 1]$

(f) $h[n] = (-\frac{1}{2})^n u[n] + (1.01)^n u[1 - n]$

(g) $h[n] = n(\frac{1}{2})^n u[n]$

(h) $h(t) = e^{-3t} u(t - 1)$

(i) $h(t) = e^{-3t} u(1 - t)$

(j) $h(t) = e^{-t} u(t + 100)$

(k) $h(t) = e^t u(-1 - t)$

(l) $h(t) = e^{-4|t|}$

(m) $h(t) = te^{-t} u(t)$

(n) $h(t) = (2e^{-t} - e^{(t-100)/100}) u(t)$

3.25. One important use of inverse systems is in situations in which one wishes to remove distortions of some type. A good example of this is the problem of removing echoes from acoustic signals. For example, if an auditorium has a perceptible echo, then an initial acoustic impulse will be followed by attenuated versions of the sound at regularly spaced intervals. Consequently, an often used model for this phenomenon is an LTI system with an impulse response consisting of a train of impulses.

$$h(t) = \sum_{k=0}^{\infty} h_k \delta(t - kT) \qquad (3.25\text{-}1)$$

Here the echoes occur T seconds apart, and h_k represents the gain factor on the kth echo resulting from an initial acoustic impulse.

(a) Suppose that $x(t)$ represents the original acoustic signal (the music produced by an orchestra, for example) and that $y(t) = x(t) * h(t)$ is the actual signal that is heard if no processing is done to remove the echoes. In order to remove the distortion introduced by the echoes, assume that a microphone is used to sense $y(t)$ and that the resulting signal is transduced into an electrical signal. We will also use

$y(t)$ to denote this signal, as it represents the electrical equivalent of the acoustic signal, and we can go from one to the other via acoustic–electrical conversion systems.

The important point to note is that the system with impulse response given by eq. (P3.25-1) is invertible. Therefore, we can find an LTI system with impulse response $g(t)$ so that

$$y(t) * g(t) = x(t)$$

and thus, by processing the electrical signal $y(t)$ in this fashion and then converting back to an acoustic signal, we can remove the troublesome echoes.

The required impulse response $g(t)$ is also an impulse train,

$$g(t) = \sum_{k=0}^{\infty} g_k \, \delta(t - kT)$$

Determine the algebraic equations that the successive g_k must satisfy, and solve for $g_0, g_1,$ and g_2 in terms of the h_k. (*Hint:* you may find the result of part (a) of Problem 3.16 to be of value)

(b) Suppose that $h_0 = 1$, $h_1 = \frac{1}{2}$, and $h_i = 0$ for all $i \geq 2$. What is $g(t)$ in this case?

(c) A good model for the generation of echos is illustrated in Figure P3.25. Hence each successive echo represents a fed-back version of $y(t)$, delayed by T seconds and scaled by α. Typically $0 < \alpha < 1$, as successive echos are attenuated.

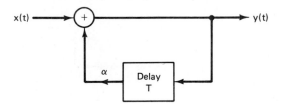

Figure P3.25

(i) What is the impulse response of this system (assume initial rest, i.e., $y(t) = 0$ for $t < 0$ if $x(t) = 0$ for $t < 0$)?

(ii) Show that the system is stable if $0 < \alpha < 1$ and unstable if $\alpha > 1$.

(iii) What is $g(t)$ in this case? Construct a realization of this inverse system using adders, coefficient multipliers, and T-second delay elements.

Although we have phrased this discussion in terms of continuous-time systems because of the application we have been considering, the same general ideas hold in discrete time. That is, the LTI system with impulse response

$$h[n] = \sum_{k=0}^{\infty} \delta[n - kN]$$

is invertible and has as its inverse an LTI system with impulse response

$$g[n] = \sum_{k=0}^{\infty} g_k[n - kN]$$

It is not difficult to check that the g_i satisfy the same algebraic equations as in part (a).

(d) Consider now the discrete-time LTI system with impulse response

$$h[n] = \sum_{k=-\infty}^{\infty} \delta[n - kN]$$

This system is *not* invertible. Find two inputs that produce the same output.

3.26. In Problem 2.23 we introduced and examined some of the basic properties of correlation functions for continuous-time signals. The discrete-time counterpart of the correlation function has essentially the same properties as those in continous time, and both are extremely important in numerous applications (as is discussed in the following two problems). In this problem we introduce the discrete-time correlation function and examine several more of its properties.

Let $x[n]$ and $y[n]$ be two real-valued discrete-time signals. The *autocorrelation functions* $\phi_{xx}[n]$ and $\phi_{yy}[n]$ of $x[n]$ and $y[n]$, respectively, are defined by the expressions

$$\phi_{xx}[n] = \sum_{m=-\infty}^{+\infty} x[m+n]x[m]$$

$$\phi_{yy}[n] = \sum_{m=-\infty}^{+\infty} y[m+n]y[m]$$

and the *cross-correlation functions* are given by

$$\phi_{xy}[n] = \sum_{m=-\infty}^{+\infty} x[m+n]y[m]$$

$$\phi_{yx}[n] = \sum_{m=-\infty}^{+\infty} y[m+n]x[m]$$

As in continuous time, these functions possess certain symmetry properties. Specifically, $\phi_{xx}[n]$ and $\phi_{yy}[n]$ are even functions, while $\phi_{xy}[n] = \phi_{yx}[-n]$.

(a) Compute the autocorrelation sequences for the signals $x_1[n]$, $x_2[n]$, $x_3[n]$, and $x_4[n]$ depicted in Figure P3.26.

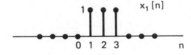
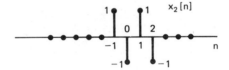

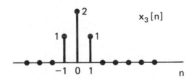
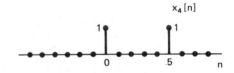

Figure P3.26

(b) Compute the cross-correlation sequences

$$\phi_{x_i x_j}[n], \qquad i \neq j, \quad i, j = 1, 2, 3, 4$$

for $x_i[n]$, $i = 1, 2, 3, 4$, as shown in Figure P3.26.

(c) Let $x[n]$ be the input to an LTI system with unit sample response $h[n]$, and let the corresponding output be $y[n]$. Find expressions for $\phi_{xy}[n]$ and $\phi_{yy}[n]$ in terms of $\phi_{xx}[n]$ and $h[n]$. Show how $\phi_{xy}[n]$ and $\phi_{yy}[n]$ can each be viewed as the output of LTI systems with $\phi_{xx}[n]$ as the input (do this by explicitly specifying the impulse response of each of the two systems).

(d) Let $h[n] = x_1[n]$ in Figure P3.26, and let $y[n]$ be the output of the LTI system with impulse response $h[n]$ when the input $x[n]$ also equals $x_1[n]$. Calculate $\phi_{xy}[n]$ and $\phi_{yy}[n]$ using the results of part (c).

3.27. Let $h_1(t)$, $h_2(t)$, and $h_3(t)$, as sketched in Figure P3.27, be the impulse responses of three LTI systems. These three signals are known as *Walsh functions* and are of considerable practical importance because they can be easily generated by digital logic circuitry and because multiplication by these functions can be simply implemented by a polarity-reversing switch.

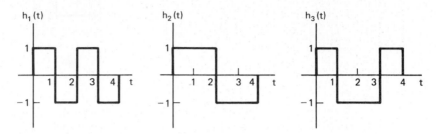

Figure P3.27

(a) Determine and sketch a choice for $x_1(t)$, a continuous-time signal, that has the following properties.
  (i) $x_1(t)$ is real.
  (ii) $x_1(t) = 0$ for $t < 0$.
  (iii) $|x_1(t)| \le 1$ for all $t \ge 0$.
  (iv) $y_1(t) = x_1(t) * h_1(t)$ is as large as possible at $t = 4$.
(b) Repeat part (a) for $x_2(t)$ and $x_3(t)$, by making $y_2(t) = x_2(t) * h_2(t)$ and $y_3(t) = x_3(t) * h_3(t)$ each as large as possible at $t = 4$.
(c) What is the value of

$$y_{ij}(t) = x_i(t) * h_j(t), \qquad i \ne j$$

at time $t = 4$ for $i, j = 1, 2, 3$?
(d) Show how to realize an LTI system for each impulse response $h_i(t)$ as the cascade of a tapped-delay line and an integrator (see Problem 3.16).

   The system with impulse response $h_i(t)$ is known as the *matched filter* for the signal $x_i(t)$ because the impulse response is tuned to $x_i(t)$ in order to produce the maximum output signal. In the next problem we relate the concept of a matched filter to that of the correlation function for continuous-time signals and provide some insight into the applications of these ideas.

3.28. The *cross-correlation function* between two continuous-time real signals $x(t)$ and $y(t)$ is given by

$$\phi_{xy}(t) = \int_{-\infty}^{+\infty} x(t + \tau) y(\tau) \, d\tau \qquad \text{(P3.28-1)}$$

The *autocorrelation function* of a signal $x(t)$ is obtained by setting $y(t) = x(t)$ in eq. (P3.28-1).

$$\phi_{xx}(t) = \int_{-\infty}^{+\infty} x(t + \tau) x(\tau) \, d\tau$$

(a) Compute the autocorrelation function for each of the two signals $x_1(t)$ and $x_2(t)$ depicted in Figure P3.28-1.
(b) Let $x(t)$ be a given signal and assume that $x(t)$ is of finite duration, i.e., that $x(t) = 0$ for $t < 0$ and $t > T$. Find the impulse response of an LTI system so that $\phi_{xx}(t - T)$ is the output if $x(t)$ is the input.

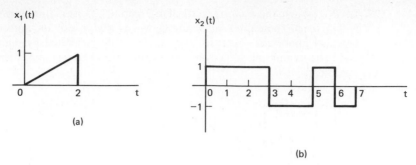

Figure P3.28-1

The system determined in part (b) is a *matched filter* for the signal $x(t)$. That this definition of matched filter is identical to the one introduced in Problem 3.27 can be seen from the following.

(c) Let $x(t)$ be as in part (b), and let $y(t)$ denote the response to this signal of an LTI system with real impulse response $h(t)$. Assume that $h(t) = 0$ for $t < 0$ and for $t > T$. Show that the choice for $h(t)$ that maximizes $y(T)$ subject to the constraint that

$$\int_0^T h^2(t)\, dt = M \qquad \text{a fixed positive number} \qquad \text{(P3.28-2)}$$

is a scalar multiple of the impulse response determined in part (b). [*Hint:* Schwartz's inequality states that

$$\int_b^a u(t)v(t)\, dt \le \left[\int_a^b u^2(t)\, dt \right]^{1/2} \left[\int_a^b v^2(t)\, dt \right]^{1/2}$$

for any two signals $u(t)$ and $v(t)$. Use this to obtain a bound on $y(T)$.]

The constraint given by eq. (P3.28-2) simply provides a scaling to the impulse response, as increasing M merely changes the scalar multiplier mentioned in part (c). Thus, we see that the particular choice for $h(t)$ in parts (b) and (c) is matched to the signal $x(t)$ to produce maximum output. This is an extremely important property in a number of applications, as we will now indicate.

In communication problems one often wishes to transmit one of a small number of possible pieces of information. For example, if a complex message is encoded into a sequence of binary digits, we can imagine a system that transmits the information bit by bit. Each bit can then be transmitted by sending one signal, say $x_0(t)$, if the bit is a zero, or a different signal $x_1(t)$, if a 1 is to be communicated. In this case the receiving system for these signals must be capable of recognizing if $x_0(t)$ or $x_1(t)$ has been received. Intuitively, what makes sense is to have two systems in the receiver, one "tuned" to $x_0(t)$ and one to $x_1(t)$, where by "tuned" we mean that the system gives a large output after the signal to which it is tuned is received. The property of producing a large output when a particular signal is received is exactly what the matched filter possesses.

In practice there is always distortion and interference in the transmission and reception processes. Consequently we want to maximize the difference between the response of a matched filter to the input to which it is matched and the response of the filter to one of the other possible signals that can be transmitted. To illustrate this point, consider the following.

(d) Consider the two signals $x_0(t)$ and $x_1(t)$ depicted in Figure P3.28-2. Let L_0 denote the matched filter for $x_0(t)$, and let L_1 denote the matched filter for $x_1(t)$.

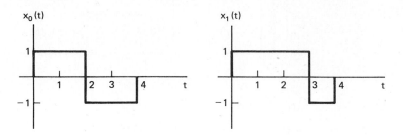

Figure P3.28-2

(i) Sketch the responses of L_0 to $x_0(t)$ and $x_1(t)$. Do the same for L_1.

(ii) Compare the values of these responses at $t = 4$. How might you modify $x_0(t)$ so that the receiver would have even an easier job of distinguishing between $x_0(t)$ and $x_1(t)$ in that the response of L_0 to $x_1(t)$ and L_1 to $x_0(t)$ would both be zero at $t = 4$?

(e) Another application in which matched filters and correlation functions play an important role is in radar systems. The underlying principle of radar is that an electromagnetic pulse transmitted at a target will be reflected by the target and will subsequently return to the sender with a delay proportional to the distance to the target. Ideally, the received signal will simply be a shifted and possibly scaled version of the original transmitted signal.

(i) Let $p(t)$ be the original pulse that is sent out. Show that

$$\phi_{pp}(0) = \max_t \phi_{pp}(t)$$

Use this to deduce that if the waveform received back by the sender is

$$x(t) = \alpha p(t - t_0)$$

where α is a positive constant, then

$$\phi_{xp}(t_0) = \max_t \phi_{xp}(t)$$

Hint: Use Schwartz's inequality.

Thus, the way in which simple radar ranging systems work is based on using a matched filter for the transmitted waveform $p(t)$ and noting the time at which the output of this system reaches a maximum.

(ii) For $p(t)$ and $x(t)$ as in Figure P3.28-3, sketch $\phi_{xp}(t)$. Assuming a propagation velocity of $c = 3 \times 10^8$ meters/sec, determine the distance from the transmitter to the target.

(iii) Since this technique for estimating travel time looks at the peak of a correlation function, it is useful to use pulse shapes $p(t)$ that have sharply peaked correlation functions. This is important due to the inevitable presence of distortion and interference in the received waveform. Which of the two pulses in Figure P3.28-4 would you prefer to use?

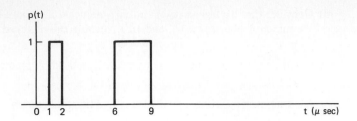

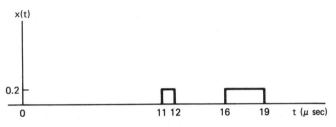

Figure P3.28-3

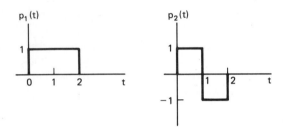

Figure P3.28-4

3.29. A discrete-time LTI system has its input–output relation characterized by the difference equation

$$y[n] = x[n] - 2x[n-2] + x[n-3] - 3x[n-4]$$

Determine and sketch the unit sample response for this system.

3.30. Consider the LTI system initially at rest and described by the difference equation

$$y[n] + 2y[n-1] = x[n] + 2x[n-2]$$

Find the response of this system to the input depicted in Figure P3.30 by solving the difference equation recursively.

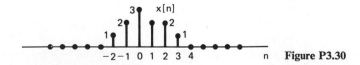

Figure P3.30

3.31. In this problem we parallel our discussion of Section 3.5.1 in order to present one standard method for solving linear constant-coefficient difference equations. Specifically, consider the difference equation

$$y[n] - \tfrac{1}{2}y[n-1] = x[n] \qquad \text{(P3.31-1)}$$

and suppose that

$$x[n] = K(\cos \Omega_0 n)u[n] \qquad \text{(P3.31-2)}$$

Assume that the solution $y[n]$ consists of the sum of a particular solution $y_p[n]$ to eq. (P3.31-1) for $n \geq 0$ and a homogeneous solution $y_h[n]$ satisfying the equation

$$y_h[n] - \tfrac{1}{2}y_h[n-1] = 0$$

(a) If we assume that $y_h[n] = Az_0^n$, what value must be chosen for z_0?
(b) If we assume that for $n \geq 0$

$$y_p[n] = B \cos (\Omega_0 n + \theta)$$

what are the values of B and θ? (*Hint:* In analogy with continuous time, it is convenient to view

$$x[n] = \Re e \, \{Ke^{j\Omega_0 n}u[n]\} \qquad \text{and} \qquad y[n] = \Re e \, \{Ye^{j\Omega_0 n}u[n]\}$$

where Y is a complex number to be determined.)
(c) Suppose that we provide the auxiliary condition

$$y[0] = 0 \qquad \text{(P3.31-3)}$$

Verify that eq. (P3.31-1) together with eq. (P3.31-3) specify a linear system. Show that this system is not causal by comparing the responses of the system to the following two inputs

$$x_1[n] = 0 \qquad \text{for all } n$$

$$x_2[n] = \begin{cases} 0, & n < -2 \\ 1, & n \geq -2 \end{cases}$$

(d) Consider the system described by eq. (P3.31-1) and the assumption of initial rest. Show that this is a causal LTI system.
(e) Suppose that the LTI system described by eq. (P3.31-1) and initially at rest has as its input the signal specified by eq. (P3.31-2). Since $x[n] = 0$ for $n < 0$, we have that $y[n] = 0$ for $n < 0$. Also, from parts (a) and (b) we have that $y[n]$ has the form

$$y[n] = Az_0^n + B \cos (\Omega_0 n + \theta)$$

for $n \geq 0$. In order to solve for the unknown constant A, we must specify a value for $y[n]$ for some $n \geq 0$. Use the condition of initial rest and eqs. (P3.31-1) and (P3.31-2) to determine $y[0]$. From this value determine the constant A. The result of this calculation yields the solution to the difference equation (P3.31-1) under the condition of initial rest, when the input is given by eq. (P3.31-2).

3.32. A \$100,000 mortgage is to be retired by *equal* monthly payments of D dollars. Interest, compounded monthly, is charged at a rate of 12% per annum on the unpaid balance; for example, after the first month, the total debt equals

$$\$100,000 + \left(\frac{0.12}{12}\right)\$100,000 = \$101,000$$

The problem is to determine D such that after a specified time the mortgage is paid in full, leaving a net balance of zero.
(a) To set up the problem, let $y[n]$ denote the unpaid balance just after the nth monthly payment. Assume that the principal is borrowed in month 0 and monthly payments begin in month 1. Show that $y[n]$ satisfies the difference equation

$$y[n] - \gamma y[n-1] = -D \qquad n \geq 1 \qquad \text{(P3.32-1)}$$

with initial condition

$$y[0] = \$100{,}000$$

where γ is a constant that you should determine.

(b) Solve the difference equation of part (a) to determine

$$y[n] \qquad \text{for } n \geq 0$$

Hint: The particular solution of eq. (P3.32-1) is a constant Y. Find the value of Y and express $y[n]$ for $n \geq 1$ as the sum of the particular and homogeneous solutions. Determine the unknown constant in the homogeneous solution by directly calculating $y[1]$ from eq. (P3.32-1) and comparing it to your solution.

(c) If the mortgage is to be retired in 30 years, after 360 monthly payments of D dollars, determine the appropriate value of D.

(d) What is the total payment to the bank over the 30-year period?

(e) Why do banks make loans?

3.33. **(a)** Consider the homogeneous differential equation

$$\sum_{k=0}^{N} a_k \frac{d^k y(t)}{dt^k} = 0 \qquad\qquad \text{(P3.33-1)}$$

Show that if s_0 is a solution of the equation

$$p(s) = \sum_{k=0}^{N} a_k s^k = 0 \qquad\qquad \text{(P3.33-2)}$$

then $Ae^{s_0 t}$ is a solution of eq. (P3.33-1), where A is an arbitrary complex constant.

(b) The polynomial $p(s)$ in eq. (P3.33-2) can be factored in terms of its roots, $s_1, \ldots, s_r$,

$$p(s) = a_N (s - s_1)^{\sigma_1} (s - s_2)^{\sigma_2} \ldots (s - s_r)^{\sigma_r}$$

where the s_i are the distinct solutions of eq. (P3.33-2) and the σ_i are their *multiplicities*. Note that

$$\sigma_1 + \sigma_2 + \ldots + \sigma_r = N$$

In general, if $\sigma_i > 1$, then not only is $Ae^{s_i t}$ a solution of eq. (P3.33-1), but so is $At^j e^{s_i t}$ as long as j is an integer greater than or equal to zero and less than or equal to $(\sigma_i - 1)$. To illustrate this, show that if $\sigma_i = 2$, then $Ate^{s_i t}$ is a solution of eq. (P3.33-1). [*Hint:* Show that if s is an arbitrary complex number, then

$$\sum_{k=0}^{N} a_k \frac{d^k (Ate^{st})}{dt^k} = Ap(s)te^{st} + A\frac{dp(s)}{ds} e^{st}.]$$

Thus, the most general solution of eq. (P3.33-1) is

$$\sum_{i=1}^{r} \sum_{j=0}^{\sigma_i - 1} A_{ij} t^j e^{s_i t}$$

where the A_{ij} are arbitrary complex constants.

(c) Solve the following homogeneous differential equations with the specified auxiliary conditions.

(i) $\dfrac{d^2 y(t)}{dt^2} + 3\dfrac{dy(t)}{dt} + 2y(t) = 0;$ $\qquad\qquad y(0) = 0, \quad y'(0) = 2$

(ii) $\dfrac{d^2 y(t)}{dt^2} + 3\dfrac{dy(t)}{dt} + 2y(t) = 0;$ $\qquad\qquad y(0) = 1, \quad y'(0) = -1$

(iii) $\dfrac{d^2 y(t)}{dt^2} + 3\dfrac{dy(t)}{dt} + 2y(t) = 0;$ $\qquad\qquad y(0) = 0, \quad y'(0) = 0$

(iv) $\dfrac{d^2 y(t)}{dt^2} + 2\dfrac{dy(t)}{dt} + y(t) = 0;$ $\qquad\qquad y(0) = 1, \quad y'(0) = 1$

(v) $\dfrac{d^3y(t)}{dt^3} + \dfrac{d^2y(t)}{dt^2} - \dfrac{dy(t)}{dt} - y(t) = 0;$ $\quad y(0) = 1,$ $\quad y'(0) = 1,$ $\quad y''(0) = -2$

(vi) $\dfrac{d^2y(t)}{dt^2} + 2\dfrac{dy(t)}{dt} + 5y(t) = 0;$ $\qquad\qquad y(0) = 1,$ $\quad y'(0) = 1$

(d) Consider the homogeneous difference equation

$$\sum_{k=0}^{N} a_k y[n-k] = 0 \qquad\qquad\text{(P3.33-3)}$$

Show that if z_0 is a solution of the equation

$$\sum_{k=0}^{N} a_k z^{-k} = 0 \qquad\qquad\text{(P3.33-4)}$$

then Az_0^n is a solution of eq. (P3.33-3), where A is an arbitrary constant.

As it is more convenient for the moment to work with polynomials that have only nonnegative powers of z, consider the equation obtained by multiplying both sides of eq. (P3.33-4) by z^N:

$$p(z) = \sum_{k=0}^{N} a_k z^{N-k} = 0 \qquad\qquad\text{(P3.33-5)}$$

The polynomial $p(z)$ can be factored as

$$p(z) = a_0(z - z_1)^{\sigma_1} \dots (z - z_r)^{\sigma_r}$$

where $z_1, \dots, z_r$ are the distinct roots of $p(z)$.

(e) Show that if $y[n] = nz^{n-1}$, then

$$\sum_{k=0}^{N} a_k y[n-k] = \frac{dp(z)}{dz} z^{n-N} + (n-N)p(z)z^{n-N-1}$$

Use this fact to show that if $\sigma_i = 2$, then both Az_i^n and Bnz_i^{n-1} are solutions of eq. (P3.33-3), where A and B are arbitrary complex constants. More generally, one can use this same procedure to show that if $\sigma_i > 1$, then

$$A\frac{n!}{r!(n-r)!}z^{n-r}$$

is a solution of eq. (P3.33-3), for $r = 0, 1, \dots, \sigma_i - 1.$[†]

(f) Solve the following homogeneous difference equations with the specified auxiliary conditions:

(i) $y[n] + \frac{3}{4}y[n-1] + \frac{1}{8}y[n-2] = 0;$ $\quad y[0] = 1,$ $\quad y[-1] = -6$
(ii) $y[n] - 2y[n-1] + y[n-2] = 0;$ $\quad y[0] = 1,$ $\quad y[1] = 0$
(iii) $y[n] - 2y[n-1] + y[n-2] = 0;$ $\quad y[0] = 1,$ $\quad y[10] = 21$
(iv) $y[n] - \dfrac{\sqrt{2}}{2}y[n-1] + \frac{1}{4}y[n-2] = 0;$ $\quad y[0] = 0,$ $\quad y[-1] = 1$

3.34. In the text we described one method for solving linear constant-coefficient difference equations, and another method for doing this was illustrated in Problem 3.31. If the assumption of initial rest is made so that the system described by the differential equation is LTI and causal, then in principle we can determine the unit impulse response $h[n]$ using either of these procedures. In Chapter 5 we describe another method that allows us to determine $h[n]$ in a much neater way. In this problem we describe yet another approach, which basically shows that $h[n]$ can be determined by solving the homogeneous equation with appropriate initial conditions.

[†]Here we are using the factorial notation. That is, $k! = k(k-1)(k-2) \cdots 2 \cdot 1$, where 0! is defined to be 1.

(a) Consider the system initially at rest and described by the equation

$$y[n] - \tfrac{1}{2}y[n-1] = x[n] \qquad\qquad \text{(P3.34-1)}$$

Assuming that $x[n] = \delta[n]$, what is $y[0]$? What equation does $h[n]$ satisfy for $n \geq 1$, and with what auxiliary condition? Solve this equation to obtain a closed-form expression for $h[n]$.

(b) Consider next the LTI system initially at rest and described by the difference equation

$$y[n] - \tfrac{1}{2}y[n-1] = x[n] + 2x[n-1] \qquad\qquad \text{(P3.34-2)}$$

This system is depicted in Figure P3.34(a) as a cascade of two LTI systems that are initially at rest. Because of the properties of LTI systems we can reverse the order of the systems in the cascade to obtain an alternative representation of the same overall system, as illustrated in Figure P3.34(b). From this fact, use the result of part (a) to determine the impulse response for the system described by eq. (P3.34-2).

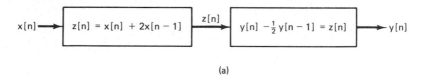

(a)

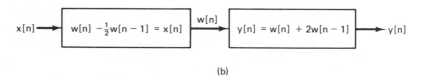

(b)

Figure P3.34

(c) Consider again the system of part (a) with $h[n]$ denoting its impulse response. Show that the response $y[n]$ to an arbitrary input $x[n]$ is in fact given by the convolution sum

$$y[n] = \sum_{m=-\infty}^{+\infty} h[n-m]x[m] \qquad\qquad \text{(P3.34-3)}$$

Do this by verifying that eq. (P3.34-3) satisfies the difference equation (P3.34-1).

(d) Consider the LTI system initially at rest and described by the difference equation

$$\sum_{k=0}^{N} a_k y[n-k] = x[n] \qquad\qquad \text{(P3.34-4)}$$

Assuming that $a_0 \neq 0$, what is $y[0]$ if $x[n] = \delta[n]$? Using this, specify the homogeneous equation and initial conditions that the impulse response of this system must satisfy.

Consider next the causal LTI system described by the difference equation

$$\sum_{k=0}^{N} a_k y[n-k] = \sum_{k=0}^{M} b_k x[n-k] \qquad\qquad \text{(P3.34-5)}$$

Express the impulse response of this system in terms of that for the LTI system described by eq. (P3.34-4).

(e) There is an alternative method for determining the impulse response of the LTI system described by eq. (P3.34-5). Specifically, given the condition of initial rest, i.e., in this case $y[-N] = y[-N+1] = \ldots = y[-1] = 0$, solve eq. (P3.34-5) recursively when $x[n] = \delta[n]$ in order to determine $y[0], \ldots, y[M]$. What equation does $h[n]$ satisfy for $n \geq M$? What are the appropriate initial conditions for this equation?

(f) Find the impulse responses of the causal LTI systems described by the following equations.

(i) $y[n] - y[n-2] = x[n]$

(ii) $y[n] - y[n-2] = x[n] + 2x[n-1]$

(iii) $y[n] - y[n-2] = 2x[n] - 3x[n-4]$

(iv) $y[n] - (\sqrt{3}/2)y[n-1] + \frac{1}{4}y[n-2] = x[n]$

Use either of the methods outlined in parts (d) and (e).

3.35. In this problem we consider a procedure that is the continuous-time counterpart of the technique developed in Problem 3.34. Again we will see that the problem of determining the impulse response $h(t)$ for $t > 0$ for an LTI system initially at rest and described by a linear constant-coefficient differential equation reduces to the problem of solving the homogeneous equation with appropriate initial conditions.

(a) Consider the LTI system initially at rest and described by the differential equation

$$\frac{dy(t)}{dt} + 2y(t) = x(t) \qquad (\text{P3.35-1})$$

Suppose that $x(t) = \delta(t)$. In order to determine the value of $y(t)$ *immediately after* the application of the unit impulse, consider integrating eq. (P3.35-1) from $t = 0^-$ to $t = 0^+$ (i.e., from "just before" to "just after" the application of the impulse). This yields

$$y(0^+) - y(0^-) + 2\int_{0^-}^{0^+} y(\tau)\, d\tau = \int_{0^-}^{0^+} \delta(\tau)\, d\tau = 1 \qquad (\text{P3.35-2})$$

Since the system is initially at rest and $x(t) = 0$ for $t < 0$, $y(0^-) = 0$. Therefore, we see that $y(0^+) = 1$, as if this is the case $y(t)$ contains only a step at $t = 0$, and consequently its integral from $t = 0^-$ to $t = 0^+$ is 0. Since $x(t) = 0$ for $t > 0$, we see that the impulse response of our system is the solution of the homogeneous equation

$$\frac{dy(t)}{dt} + 2y(t) = 0$$

with initial condition

$$y(0^+) = 1$$

Solve this differential equation to obtain the impulse response $h(t)$ for this system. Check your result by showing that

$$y(t) = \int_{-\infty}^{+\infty} h(t-\tau)x(\tau)\, d\tau$$

satisfies eq. (P3.35-1) for any input $x(t)$.

(b) To generalize this argument, consider an LTI system initially at rest and described by the differential equation

$$\sum_{k=0}^{N} a_k \frac{d^k y(t)}{dt^k} = x(t) \qquad (\text{P3.35-3})$$

with $x(t) = \delta(t)$. Assume the condition of initial rest which, since $x(t) = 0$ for

$t < 0$, implies that

$$y(0^-) = \frac{dy}{dt}(0^-) = \ldots = \frac{d^{N-1}y}{dt^{N-1}}(0^-) = 0 \qquad \text{(P3.35-4)}$$

Integrate both sides of eq. (P3.35-3) once from $t = 0^-$ to $t = 0^+$, and use eq. (P3.35-4) and an argument similar to that used in part (a) to show that the resulting equation is satisfied with

$$y(0^+) = \frac{dy}{dt}(0^+) = \ldots \frac{d^{N-2}y}{dt}(0^+) = 0 \qquad \text{(P3.35-5a)}$$

and

$$\frac{d^{N-1}y}{dt}(0^+) = \frac{1}{a_N} \qquad \text{(P3.35-5b)}$$

Consequently the system's impulse response for $t > 0$ can be obtained by solving the homogeneous equation

$$\sum_{k=0}^{N} a_k \frac{d^k y(t)}{dt^k} = 0$$

with initial conditions given by eq. (P3.35-5).

(c) Consider now the causal LTI system described by the differential equation

$$\sum_{k=0}^{N} a_k \frac{d^k y(t)}{dt^k} = \sum_{k=0}^{M} b_k \frac{d^k x(t)}{dt^k} \qquad \text{(P3.35-6)}$$

Express the impulse response of this system in terms of that for the system of part (b). (*Hint:* examine Figure P3.35).

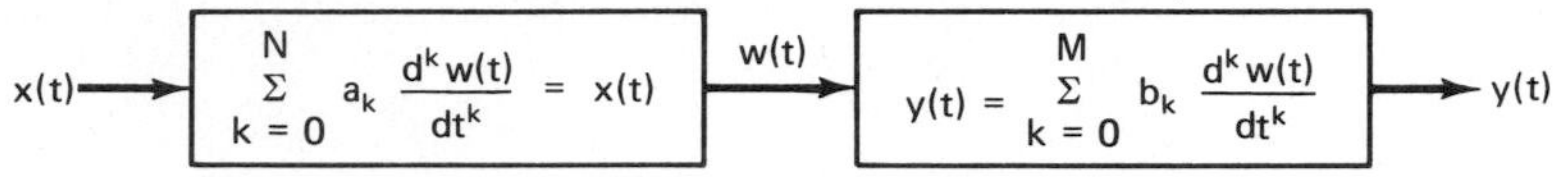

Figure P3.35

(d) Apply the procedures outlined in parts (b) and (c) to find the impulse responses for the LTI systems initially at rest and described by the following differential equations.

(i) $\dfrac{d^2 y(t)}{dt^2} + 3\dfrac{dy(t)}{dt} + 2y(t) = x(t)$

(ii) $\dfrac{d^2 y(t)}{dt} + 2\dfrac{dy(t)}{dt} + 2y(t) = x(t)$

(e) Use the result of parts (b) and (c) to deduce that if $M \geq N$ in eq. (P3.35-6), then the impulse response $h(t)$ will contain singularity terms concentrated at $t = 0$. In particular, $h(t)$ will contain a term of the form

$$\sum_{r=0}^{M-N} \alpha_r u_r(t)$$

where the α_r are constants and the $u_r(t)$ are the singularity functions defined in Section 3.7.

(f) Find the impulse responses for the causal LTI systems described by the following differential equations.

(i) $\dfrac{dy(t)}{dt} + 2y(t) = 3\dfrac{dx(t)}{dt} + x(t)$

(ii) $\dfrac{d^2 y(t)}{dt^2} + 5\dfrac{dy(t)}{dt} + 6y(t) = \dfrac{d^3 x(t)}{dt^3} + 2\dfrac{d^2 x(t)}{dt^2} + 4\dfrac{dx(t)}{dt} + 3x(t)$

3.36. Consider the LTI system initially at rest and described by the difference equation

$$y[n] - \tfrac{1}{2}y[n-1] + y[n-2] = 6x[n] - 7x[n-1] + 5x[n-2]$$

(a) Determine a closed-form expression for the unit step response of this system in two ways:
   (i) By solving the difference equation directly by (1) assuming a particular solution of the form $y[n] = Y$ for $n > 0$ and solving for Y; (2) finding the homogeneous solution, which will have two unspecified constants; and (3) directly calculating $y[0]$ and $y[1]$ and using these values to determine the constants by equating $y[n]$ and $y_p[n] + y_h[n]$ for $n = 0$ and 1.
   (ii) By first solving for the unit impulse response (see Problem 3.34) and then using the convolution sum.

(b) Find the output of the system when the input is

$$x[n] = (-\tfrac{1}{2})^n$$

Use either of the two methods given above, where in the first case you should assume a particular solution of the form

$$y[n] = Y(-\tfrac{1}{2})^n \qquad \text{for } n > 0$$

(c) What is the response of the system if

$$x[n] = (-\tfrac{1}{2})^n u[n-2] + 3u[n-4]$$

(d) Construct the direct form II realization for this system.

3.37. Consider the continuous-time LTI system initially at rest and described by the differential equation

$$\frac{d^2y(t)}{dt^2} + \frac{dy(t)}{dt} - 2y(t) = x(t) \qquad \text{(P3.37-1)}$$

(a) Determine the step response of this system in two ways:
   (i) By solving the differential equation directly, that is, by (1) assuming a particular solution $y(t) = Y$ for $t > 0$ and solving for Y; (2) finding the homogeneous solution, which will have two unspecified constants; and (3) by using the initial rest conditions $y(0) = y'(0) = 0$ to determine the constants by equating $y(t)$ and $y'(t)$ with $y_p(t) + y_h(t)$ and $y'_p(t) + y'_h(t)$, respectively, at $t = 0$.
   (ii) By first solving for the unit impulse response (see Problem 3.35) and then using the convolution integral.

(b) Repeat (i) of part (a) for the input

$$x(t) = e^{-2t}(\cos 3t)u(t)$$

In this case the particular solution for $t > 0$ is of the form

$$y(t) = \mathcal{R}e\{Ye^{-2t}e^{j3t}\}$$

(c) Construct the direct form II realizations for the LTI system described by eq. (P3.37-1).

3.38. Determine the direct form II realization for each of the following LTI systems, all of which are assumed to be at rest initially.

(a) $2y[n] - y[n-1] + y[n-3] = x[n] - 5x[n-4]$

(b) $4\dfrac{d^2y(t)}{dt^2} + 2\dfrac{dy(t)}{dt} = x(t) - 3\dfrac{d^2x(t)}{dt^2}$

(c) $y[n] = x[n] - x[n-1] + 2x[n-3] - 3x[n-4]$

(d) $\dfrac{d^4y(t)}{dt^4} = x(t) - 2\dfrac{dx(t)}{dt}$

(e) $\dfrac{d^2 y(t)}{dt^2} + 2\dfrac{dy(t)}{dt} - 2y(t) = x(t) + \dfrac{dx(t)}{dt} + 3\displaystyle\int_{-\infty}^{t} x(\tau)\, d\tau$

3.39. In Section 3.7 we characterized the unit doublet through the equation

$$x(t) * u_1(t) = \int_{-\infty}^{+\infty} x(t - \tau)u_1(\tau)\, d\tau = x'(t) \qquad \text{(P3.39-1)}$$

for any signal $x(t)$. From this equation we derived the fact that

$$\int_{-\infty}^{+\infty} g(\tau)u_1(\tau)\, d\tau = -g'(0) \qquad \text{(P3.39-2)}$$

(a) Show that eq. (P3.39-2) is an equivalent characterization of $u_1(t)$ by showing that eq. (P3.39-2) implies eq. (P3.39-1). [*Hint:* Fix t and define the signal $g(\tau) = x(t - \tau)$.]

   Thus we have seen that characterizing the unit impulse or unit doublet by how it behaves under convolution is equivalent to characterizing how it behaves under integration when multiplied by an arbitrary signal $g(t)$. In fact, as indicated in Section 3.7, the equivalence of these operational definitions holds for all signals and in particular for all singularity functions.

(b) Let $f(t)$ be a given signal. Show that

$$f(t)u_1(t) = f(0)u_1(t) - f'(0)\, \delta(t)$$

by showing that both have the same operational definitions.

(c) What is the value of

$$\int_{-\infty}^{\infty} x(\tau)u_2(\tau)\, d\tau ?$$

(d) Find an expression for $f(t)u_2(t)$ analogous to that considered in part (b) for $f(t)u_1(t)$.

3.40. Show by induction that

$$u_{-k}(t) = \frac{t^{k-1}}{(k-1)!}u(t) \qquad \text{for } k = 1, 2, 3, \ldots$$

3.41. In analogy with the continuous-time singularity functions, we can define a set of discrete-time signals. Specifically, let

$$u_{-1}[n] = u[n]$$
$$u_0[n] = \delta[n]$$
$$u_1[n] = \delta[n] - \delta[n - 1]$$

and define

$$u_k[n] = \underbrace{u_1[n] * u_1[n] * \ldots * u_1[n]}_{k \text{ times}}, \qquad k > 0$$

$$u_k[n] = \underbrace{u_{-1}[n] * u_{-1}[n] * \ldots * u_{-1}[n]}_{|k| \text{ times}}, \qquad k < 0$$

Note that

$$x[n] * \delta[n] = x[n]$$

$$x[n] * u[n] = \sum_{m=-\infty}^{n} x[m]$$

$$x[n] * u_1[n] = x[n] - x[n - 1]$$

which are the counterparts of the operational definitions of $\delta(t)$, $u(t)$, and $u_1(t)$ given in the text in eqs. (3.143), (3.156), and (3.149), respectively.

(a) What is

$$\sum_{m=-\infty}^{+\infty} x[m]u_1[m]?$$

(b) Show that

$$x[n]u_1[n] = x[0]u_1[n] - \{x[1] - x[0]\}\,\delta[n-1]$$

$$= x[1]u_1[n] - \{x[1] - x[0]\}\,\delta[n]$$

(c) Sketch the signals $u_2[n]$ and $u_3[n]$.

(d) Sketch $u_{-2}[n]$ and $u_{-3}[n]$.

(e) Show that in general for $k > 0$,

$$u_k[n] = \frac{(-1)^n k!}{n!(k-n)!}\{u[n] - u[n-k-1]\} \qquad \text{(P3.41-1)}$$

Hint: Use induction. From part (c) you should have that $u_k[n]$ satisfies eq. (P3.41-1) for $k = 2$ and 3. Then, assuming it is true for $u_k[n]$, write $u_{k+1}[n]$ in terms of $u_k[n]$ and show it is also true for $u_{k+1}[n]$.

(f) Show that in general for $k > 0$,

$$u_{-k}[n] = \frac{(n+k-1)!}{n!(k-1)!}u[n] \qquad \text{(P3.41-2)}$$

Hint: Again use induction. Note that

$$u_{-(k+1)}[n] - u_{-(k+1)}[n-1] = u_{-k}[n] \qquad \text{(P3.41-3)}$$

Assuming that eq. (P3.41-2) is valid for $u_{-k}[n]$, use eq. (P3.41-3) to show that it is valid for $u_{-(k+1)}[n]$ as well.

3.42. In this chapter we have used several properties and ideas that greatly facilitate the analysis of LTI systems. Among these are two that we wish to examine a bit more closely in this problem. As we will see, in certain very special cases one must be careful in using these properties, which otherwise hold without qualification.

(a) One of the basic and most important properties of convolution (in both continuous and discrete time) is that of associativity. That is, if $x(t)$, $h(t)$, and $g(t)$ are three signals, then

$$x(t) * [g(t) * h(t)] = [x(t) * g(t)] * h(t) = [x(t) * h(t)] * g(t) \qquad \text{(P3.42-1)}$$

This fact is true, in general, as long as all three expressions in eq. (P3.42-1) are well defined and finite. As this is usually the case in practice, we will in general use the associativity property without comment or assumption. However, there are some cases in which this is *not* the case. For example, consider the system depicted in Figure P3.42 with $h(t) = u_1(t)$ and $g(t) = u(t)$. Compute the response of this system to the input

$$x(t) = 1 \qquad \text{for all } t$$

Do this in the three different ways suggested by eq. (P3.42-1) and by the figure:
   (i)   By first convolving the two impulse responses and then convolving the result with $x(t)$.
   (ii) By convolving $x(t)$ first with $u_1(t)$ and then convolving the result with $u(t)$.
   (iii) By convolving $x(t)$ first with $u(t)$ and then convolving the result with $u_1(t)$.

(b) Repeat part (a) for

$$x(t) = e^{-t}$$

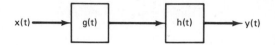

Figure P3.42

and

$$h(t) = e^{-t}u(t)$$
$$g(t) = u_1(t) + \delta(t)$$

(c) Do the same for

$$x[n] = (\tfrac{1}{2})^n$$
$$h[n] = (\tfrac{1}{2})^n u[n]$$
$$g[n] = \delta[n] - \tfrac{1}{2}\,\delta[n-1]$$

Thus, in general, we have that the associativity property of convolution holds if and only if the three expressions in eq. (P3.42-1) make sense (i.e., their interpretations in terms of LTI systems are meaningful). For example, in part (a) differentiating a constant and then integrating makes sense, but the process of integrating the constant from $t = -\infty$ and *then* differentiating does not, and it is only in such cases that associativity breaks down.

Closely related to the discussion above is an issue involving inverse systems. Specifically, consider the LTI system with impulse response $h(t) = u(t)$. As we saw in part (a), there are inputs, specifically $x(t) = $ nonzero constant, for which the output of this system is infinite, and thus it is meaningless to consider the question of inverting such outputs to recover the input. However, if we limit ourselves to inputs that do yield finite outputs, that is, that satisfy

$$\left| \int_{-\infty}^{t} x(\tau)\, d\tau \right| < \infty \qquad \text{(P3.42-2)}$$

then the system *is* invertible, and the LTI system with impulse response $u_1(t)$ is its inverse.

(d) Show that the LTI system with impulse response $u_1(t)$ is *not* invertible. (*Hint:* Find two different inputs that both yield zero output for all time.) However, show that it is invertible if we limit ourselves to inputs that satisfy eq. (P3.42-2). [*Hint:* In Problem 2.30 we showed that an LTI system is invertible if no input other than $x(t) = 0$ yields an output that is zero for all time; are there two inputs $x(t)$ that satisfy eq. (P3.42-2) and that yield identically zero responses when convolved with $u_1(t)$?]

What we have illustrated in this problem is the following:

1. If $x(t)$, $h(t)$, and $g(t)$ are three signals and if $x(t) * g(t)$, $x(t) * h(t)$, and $h(t) * g(t)$ are *all* well defined and finite, then the associativity property, eq. (P3.42-1), holds.
2. Let $h(t)$ be the impulse response of an LTI system, and suppose that the

impulse response $g(t)$ of a second system has the property that

$$h(t) * g(t) = \delta(t) \qquad \text{(P3.42-3)}$$

Then from (1) we know that *for all inputs $x(t)$ for which $x(t) * h(t)$ and $x(t) * g(t)$ are both well defined and finite*, the two cascades of systems depicted in Figure P3.42 both act as the identity system, and thus the two LTI systems can be regarded as inverses of one another. For example, if $h(t) = u(t)$ and $g(t) = u_1(t)$, then as long as we restrict ourselves to inputs satisfying eq. (P3.42-2), we can regard these two systems as inverses.

Therefore, we see that the associativity property of eq. (P3.42-1) and the definition of LTI inverses as given in eq. (P3.42-3) are valid as long as all the convolutions that are involved are finite. As this is certainly the case in any realistic problem, we will in general use these properties without comment or qualification. Note that although we have phrased most of our discussion in terms of continuous-time signals and systems, the same points can also be made in discrete time [as should be evident from part (c)].

The starting point for our development of the convolution sum in Chapter 3 was the representation of the input to a discrete-time LTI system as a weighted sum of shifted unit impulses. Similarly, by representing the input to a continuous-time LTI system as a weighted integral of shifted impulses, we were able to derive the convolution integral. These representations for discrete- and continuous-time LTI systems indicate how the response of such a system to an arbitrary input is constructed from the system's responses to elementary inputs, namely its responses to shifted unit impulses. Thus, the convolution sum and integral not only provide us with a convenient way in which to calculate the response of an LTI system assuming that we know its response to a unit impulse, but they also indicate that the characteristics of an LTI system are completely determined by its unit impulse response. Because of this fact we were able to analyze many of the properties of LTI systems in some detail and to relate these properties to equivalent characteristics of the impulse responses of such systems.

In this and the following chapter we develop an alternative representation for signals and LTI systems. In this chapter we focus on the continuous-time case, and the discrete-time version is described in Chapter 5. As in Chapter 3, the starting point for our discussion is the development of a representation of signals as weighted sums and integrals of a set of basic signals. In Chapter 3 we used shifted unit impulses as basic signals, whereas in this and the next chapter we use complex exponentials. The resulting representations are known as the continuous- and discrete-time Fourier series and transform. As we will see, these representations can be used to construct broad and useful classes of signals.

Fourier Analysis for Continuous–Time Signals and Systems

Once we have these representations we can again proceed as we did in Chapter 3. That is, because of the superposition property, the response of an LTI system to any input consisting of a linear combination of basic signals is the linear combination of the individual responses to each of these basic signals. In Chapter 3 these responses were all shifted versions of the unit impulse response, and it was this fact that gave us the representations for LTI systems that we found so useful. As we will find in this and the next chapter, the response of an LTI system to a complex exponential also has a particularly simple form, and it is this fact that will provide us with another convenient representation for LTI systems and consequently with another way in which to analyze these systems and gain insight into their properties. The framework of Fourier analysis that we build in this chapter and in Chapter 5 plays an extremely important role in the study of signals and systems. We will use this framework extensively throughout the remainder of the book.

The development of the techniques of Fourier analysis has a long history involving a great many individuals and the investigation of many different physical phenomena.† The concept of using "trigonometric sums," that is, sums of harmonically related sines and cosines or periodic complex exponentials, to describe periodic phenomena goes back at least as far as the Babylonians, who used ideas of this type in order to predict astronomical events.‡ The modern history of this subject begins in 1748 with L. Euler, who examined the motion of a vibrating string. In Figure 4.1 we have indicated the first few normal modes of such a string. If we consider the vertical deflection $f_t(x)$ of the string at time t and at a distance x along the string, then for any fixed instant of time t, the normal modes are harmonically related sinusoidal functions of x. What Euler noted was that if the configuration of a vibrating string at some point in time is a linear combination of these normal modes, so is its configuration at any subsequent time. Furthermore, Euler showed that one could calculate the coefficients for the linear combination at the later time in a very straightforward manner from the coefficients at the earlier time. In doing this Euler had performed the same type of calculation as we will in the next section in deriving one of the properties of trigonometric sums that make them so useful for the analysis of LTI systems. Specifically, we will see that if the input to an LTI system is expressed as a linear combination of periodic complex exponentials or sinusoids, the output can also be expressed in this form, with coefficients that are conveniently expressed

†The historical material in this chapter was taken from several references: I. Gratton-Guiness, *Joseph Fourier 1768–1830* (Cambridge, Mass.: The MIT Press, 1972); G. F. Simmons, *Differential Equations: With Applications and Historical Notes* (New York: McGraw-Hill Book Company, 1972); C. Lanczos, *Discourse on Fourier Series* (London: Oliver and Boyd, 1966); R. E. Edwards, *Fourier Series: A Modern Introduction* (New York: Holt, Rinehart and Winston, Inc., 1967); and A. D. Aleksandrov, A. N. Kolmogorov, and M. A. Lavrent'ev, *Mathematics: Its Content, Methods, and Meaning*, trans. S. H. Gould, Vol. II; trans. K. Hirsch, Vol. III (Cambridge, Mass.: The MIT Press, 1963). In particular, a far more complete account of Fourier's life and contributions can be found in the book of Gratton-Guiness. Other specific references are cited in several places in the chapter.

‡H. Dym and H. P. McKean, *Fourier Series and Integrals* (New York: Academic Press, 1972). This text and the book of Simmons referenced above also contain discussions of the vibrating-string problem and its role in the development of Fourier analysis.

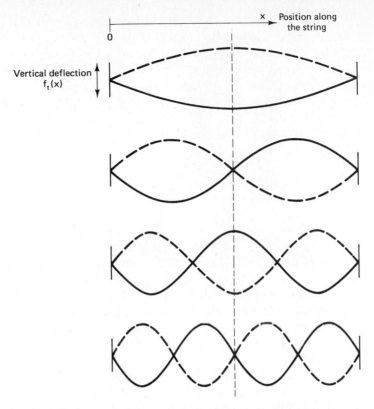

Figure 4.1 Normal modes of a vibrating string (solid lines indicate the configuration of each of these modes at some fixed time instant t).

in terms of those of the input. As we will see, this greatly facilitates the analysis of LTI systems.

The property described in the preceding paragraph would not be particularly useful unless it were true that a large class of interesting functions could be represented by linear combinations of complex exponentials. In the middle of the eighteenth century this point was the subject of heated debate. In 1753, D. Bernoulli argued on physical grounds that all physical motions of a string could be represented by linear combinations of normal modes, but he did not pursue this mathematically and his ideas were not widely accepted. In fact, Euler himself discarded trigonometric series, and in 1759 J. L. Lagrange strongly criticized the use of trigonometric series in the examination of vibrating strings. His criticism was based on his own belief that it was impossible to represent signals with corners (i.e., with discontinuous slopes), using trigonometric series. Since such a configuration arises from the plucking of a string (i.e., pulling it taut and then releasing it), Lagrange argued that trigonometric series were of very limited use.

It was in this somewhat hostile and skeptical environment that Jean Baptiste Joseph Fourier (Figure 4.2) presented his ideas half a century later. Fourier was

Figure 4.2 Jean Baptiste Joseph Fourier [picture from J.B.J. Fourier, *Oeuvres de Fourier*, Vol. II (Paris: Gauthier-Villars et Fils, 1890)].

born on March 21, 1768, in Auxerre, France, and by the time of his entrance into the controversy concerning trigonometric series he had already had a lifetime of experiences. His many contributions, and in particular those concerned with the series and transform that carry his name, are made even more impressive by the circumstances under which he worked. His revolutionary discoveries, although not completely appreciated during his own lifetime, have had a major impact on the development of mathematics and have been and still are of great importance in an extremely wide range of scientific and engineering disciplines.

In addition to his studies in mathematics, Fourier led an active political life. In fact, during the years that followed the French Revolution his activities almost led to his downfall, as he narrowly avoided the guillotine on two separate occasions. Subsequently, Fourier became an associate of Napoleon Bonaparte, accompanied him on his expeditions to Egypt (during which Fourier collected the information he would use later as the basis for his treatises on egyptology), and in 1802 was appointed by Bonaparte to the position of prefect of a region of France centered in Grenoble. It was there, while serving as prefect, that Fourier developed his ideas on trigonometric series.

The physical motivation for Fourier's work was the phenomenon of heat propagation and diffusion. This in itself was a significant step in that most previous research in mathematical physics had dealt with rational and celestial mechanics. By 1807, Fourier had completed a substantial portion of his work on heat diffusion, and on December 21, 1807, he presented these results to the Institut de France. In his work Fourier had found series of harmonically related sinusoids to be useful in representing the temperature distribution through a body. In addition, he claimed that "any" periodic signal could be represented by such a series. While his treatment of this topic was significant, many of the basic ideas behind it had been discovered by others. Also, Fourier's mathematical arguments were still imprecise, and it remained for P. L. Dirichlet in 1829 to provide precise conditions under which a periodic signal could be represented by a Fourier series.† Thus, Fourier did not actually contribute to the mathematical theory of Fourier series. However, he did have the clear insight to see the potential for this series representation, and it was to a great extent his work and his claims that spurred much of the subsequent work on Fourier series. In addition, Fourier took this type of representation one very large step further than any of his predecessors. Specifically, he obtained a representation for *aperiodic* signals—not as weighted *sums* of harmonically related sinusoids—but as weighted *integrals* of sinusoids that are *not* all harmonically related. As with the Fourier series, the Fourier integral or transform remains one of the most powerful tools for the analysis of LTI systems.

Four distinguished mathematicians and scientists were appointed to examine the 1807 paper of Fourier. Three of the four, S. F. Lacroix, G. Monge, and P. S. Laplace, were in favor of publication of the paper, but the fourth, J. L. Lagrange, remained adamant in the rejection of trigonometric series that he had put forth 50 years earlier. Because of Lagrange's vehement objections, Fourier's paper never appeared. After several other attempts to have his work accepted and published by the Institut de France, Fourier undertook the writing of another version of his work, which appeared as the text *Théorie analytique de la chaleur*.‡ This book was published in 1822, 15 years after Fourier had first presented his results to the Institut de France.

Toward the end of his life Fourier received some of the recognition he deserved, but the most significant tribute to him has been the enormous impact of his work on so many disciplines within the fields of mathematics, science, and engineering. The theory of integration, point-set topology, and eigenfunction expansions are just a few examples of topics in mathematics that have their roots in the analysis of Fourier series and integrals.§ Much as with the original studies of vibration and heat diffusion, there are numerous problems in science and engineering in which sinusoidal signals, and therefore Fourier series and transforms, play an important role. For example, sinusoidal signals arise naturally in describing the periodic behavior of the earth's climate. Alternating-current sources generate sinusoidal voltages and currents, and, as we will see, the tools of Fourier analysis enable us to analyze the response of an LTI system, such as a circuit, to such sinusoidal inputs. Also, as illustrated in Figure 4.3, waves in the ocean consist of the linear combination of sinusoidal waves with different spatial periods or *wavelengths*. Signals transmitted by radio and television stations are sinusoidal in nature as well, and as a quick perusal of any text on Fourier analysis will show, the range of applications in which sinusoidal signals arise and in which the tools of Fourier analysis are useful extends far beyond these few examples.

†Both S. D. Poisson and A. L. Cauchy had obtained results about convergence of Fourier series before 1829, but Dirichlet's work represented such a significant extension of their results that he is usually credited with being the first to consider Fourier series convergence in a rigorous fashion.

‡See the English translation: J. B. J. Fourier, *The Analytical Theory of Heat*, trans. A. Freeman (Cambridge, 1878).

§For more on the impact of Fourier's work on mathematics, see W. A. Coppel, "J. B. Fourier—On the Occasion of His Two Hundredth Birthday," *American Mathematical Monthly*, 76 (1969), 468–83.

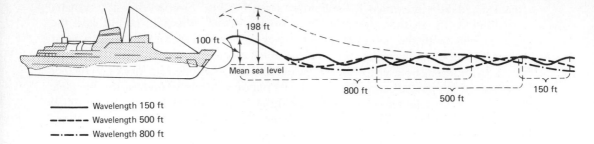

_____ Wavelength 150 ft
- - - - - Wavelength 500 ft
—·—·— Wavelength 800 ft

Figure 4.3 Ship encountering the superposition of three wave trains, each with a different spatial period. When these waves reinforce one another, a very large wave can result. In more severe sea conditions, a giant wave indicated by the dotted line could result. Whether such a reinforcement occurs at any location depends upon the relative phases of the components that are superposed (see Section 4.10). [Adapted from an illustration by P. Mion in "Nightmare waves are all too real to deepwater sailors," by P. Britton, *Smithsonian* 8 (February 1978), pp. 64–65].

In this chapter we develop some of the basic tools of continuous-time Fourier analysis in much the same manner as that used by Euler, Bernoulli, Lagrange, Fourier, Dirichlet, and those that followed after them. As we introduce and describe these tools we will begin to establish some insight into the utility of Fourier analysis and the frequency-domain (i.e., Fourier series or Fourier transform) representation of signals and systems. Although the original work of Fourier and his contemporaries was concerned solely with phenomena in continuous time, the essential ideas of Fourier analysis carry over to discrete time and provide us with extremely powerful techniques for the analysis of discrete-time LTI systems. The development of discrete-time Fourier analysis is the subject of Chapter 5. With this understanding of frequency-domain methods in hand in both continuous and discrete time, we will then be in a position to apply these techniques to several topics of great practical importance. In Chapter 6 we take up the subject of filtering (i.e., the design of systems with particular response characteristics to sinusoids at different frequencies). The topic of modulation is addressed in Chapter 7, and, in particular, in that chapter we develop the basic ideas behind the operation of amplitude modulation (AM) communication systems. In Chapter 8 we use continuous- and discrete-time Fourier analysis techniques together as we investigate the sampling of continuous-time signals and the processing of sampled signals using discrete-time systems.

4.1 THE RESPONSE OF CONTINUOUS-TIME LTI SYSTEMS TO COMPLEX EXPONENTIALS

As we indicated in the preceding section, it is advantageous in the study of LTI systems to represent signals as linear combinations of basic signals that possess the following two properties:

1. The set of basic signals can be used to construct a broad and useful class of signals.

2. The response of an LTI system to each basic signal should be simple enough in structure to provide us with a convenient representation for the response of the system to any signal constructed as a linear combination of these basic signals.

For continuous-time LTI systems both of these advantages are provided by the set of complex exponentials of the form e^{st}, where s is a general complex number. In Sections 4.2–4.5 we consider in some detail the first of these properties. In this section we consider the second property and in this way provide motivation for the use of Fourier series and transforms in the analysis of LTI systems.

The importance of complex exponentials in the study of LTI systems stems from the fact shown below that the response of an LTI system to a complex exponential input is the same complex exponential with only a change in amplitude, that is,

$$e^{st} \longrightarrow H(s)e^{st} \tag{4.1}$$

where the complex amplitude factor $H(s)$ will in general be a function of the complex variable s. A signal for which the system output is just a (possibly complex) constant times the input is referred to as an *eigenfunction* of the system, and the amplitude factor is referred to as the *eigenvalue*.

To show that complex exponentials are indeed eigenfunctions of LTI systems, let us consider an LTI system with impulse response $h(t)$. For an input $x(t)$ we can determine the output through the use of the convolution integral, so that with $x(t)$ of the form $x(t) = e^{st}$ we have from eq. (3.29) that

$$y(t) = \int_{-\infty}^{+\infty} h(\tau)x(t - \tau)\, d\tau$$

$$= \int_{-\infty}^{+\infty} h(\tau)e^{s(t-\tau)}\, d\tau \tag{4.2}$$

Expressing $e^{s(t-\tau)}$ as $e^{st}e^{-s\tau}$ and noting that e^{st} can be moved outside the integral, eq. (4.2) becomes

$$y(t) = e^{st} \int_{-\infty}^{+\infty} h(\tau)e^{-s\tau}\, d\tau \tag{4.3}$$

Thus, the response to e^{st} is of the form

$$y(t) = H(s)e^{st} \tag{4.4}$$

where $H(s)$ is a complex constant whose value depends on s and which is related to the system impulse response by

$$H(s) = \int_{-\infty}^{+\infty} h(\tau)e^{-s\tau}\, d\tau \tag{4.5}$$

Thus, we have shown that any complex exponential is an eigenfunction of an LTI system. The constant $H(s)$ for a specified value of s is then the eigenvalue associated with the eigenfunction e^{st}.

The usefulness for the analysis of LTI systems of decomposing more general signals in terms of eigenfunctions can be seen from an example. Let $x(t)$ correspond to a linear combination of three complex exponentials, that is,

$$x(t) = a_1 e^{s_1 t} + a_2 e^{s_2 t} + a_3 e^{s_3 t} \tag{4.6}$$

The response to each separately is just

$$a_1 e^{s_1 t} \longrightarrow a_1 H(s_1) e^{s_1 t}$$

$$a_2 e^{s_2 t} \longrightarrow a_2 H(s_2) e^{s_2 t}$$

$$a_3 e^{s_3 t} \longrightarrow a_3 H(s_3) e^{s_3 t}$$

and from the superposition property given in eqs. (3.1) and (3.2), the response to the sum is the sum of the responses, so that

$$y(t) = a_1 H(s_1) e^{s_1 t} + a_2 H(s_2) e^{s_2 t} + a_3 H(s_3) e^{s_3 t} \tag{4.7}$$

More generally,

$$\sum_k a_k e^{s_k t} \longrightarrow \sum_k a_k H(s_k) e^{s_k t} \tag{4.8}$$

Thus, for an LTI system, if we know the eigenvalues $H(s_k)$, then the response to a linear combination of complex exponentials can be constructed in a straightforward manner.

It was precisely this fact that Euler discovered for the problem of the vibrating string, and this observation provided Fourier with the motivation to consider the question of how broad a class of signals could be represented as a linear combination of complex exponentials. In Sections 4.2 and 4.3 we use an approach very similar to that used by Fourier as we consider the representation of periodic signals in terms of complex exponentials and the properties of such representations. In Section 4.4 we extend these results in precisely the manner originally used by Fourier to more general aperiodic signals. Although in general the variable s is complex and of the form $\sigma + j\omega$, throughout this chapter we restrict it to be purely imaginary so that $s = j\omega$, and thus we consider only complex exponentials of the form $e^{j\omega t}$. In Chapter 9 we consider the more general case of s complex as we develop the transform that carries the name of one of Fourier's 1807 examiners, P. S. Laplace.

4.2 REPRESENTATION OF PERIODIC SIGNALS: THE CONTINUOUS-TIME FOURIER SERIES

4.2.1 Linear Combinations of Harmonically Related Complex Exponentials

Recall from Chapter 2 that a signal is periodic if for some positive, nonzero value of T,

$$x(t) = x(t + T) \qquad \text{for all } t \tag{4.9}$$

The fundamental period T_0 of $x(t)$ is the minimum positive, nonzero value of T for which eq. (4.9) is satisfied, and the value $2\pi/T_0$ is referred to as the fundamental frequency.

In Chapter 2 we also introduced two basic periodic signals, the sinusoid

$$x(t) = \cos \omega_0 t \tag{4.10}$$

and the periodic complex exponential

$$x(t) = e^{j\omega_0 t} \tag{4.11}$$

Both of these signals are periodic with fundamental frequency ω_0 and fundamental period $T_0 = 2\pi/\omega_0$. Associated with the signal in eq. (4.11) is the set of *harmonically related* complex exponentials

$$\phi_k(t) = e^{jk\omega_0 t}, \qquad k = 0, \pm 1, \pm 2, \ldots \tag{4.12}$$

As we discussed in Section 2.3, each of these signals has a fundamental frequency that is a multiple of ω_0, and therefore each is periodic with period T_0 (although for $|k| \geq 2$ the fundamental period of $\phi_k(t)$ is a fraction of T_0). Thus, a linear combination of harmonically related complex exponentials of the form

$$x(t) = \sum_{k=-\infty}^{+\infty} a_k e^{jk\omega_0 t} \tag{4.13}$$

is also periodic with period T_0. In eq. (4.13), the term for $k = 0$ is a dc or constant term. The two terms for $k = +1$ and $k = -1$ both have fundamental period equal to T_0 and are collectively referred to as the *fundamental components*, or as the *first harmonic components*. The two terms for $k = +2$ and $k = -2$ are periodic with half the period (or equivalently twice the frequency) of the fundamental components and are referred to as the *second harmonic components*. More generally, the components for $k = +N$ and $k = -N$ are referred to as the Nth harmonic components.

The representation of a periodic signal in the form of eq. (4.13) is referred to as the *Fourier series* representation. Before developing the properties of this representation, let us consider an example.

Example 4.1

Consider a periodic signal $x(t)$, with fundamental frequency 2π, which is expressed in the form of eq. (4.13) as

$$x(t) = \sum_{k=-3}^{+3} a_k e^{jk2\pi t} \tag{4.14}$$

where

$$a_0 = 1$$
$$a_1 = a_{-1} = \tfrac{1}{4}$$
$$a_2 = a_{-2} = \tfrac{1}{2}$$
$$a_3 = a_{-3} = \tfrac{1}{3}$$

Rewriting eq. (4.14) collecting together each pair of harmonic components, we obtain

$$x(t) = 1 + \tfrac{1}{4}(e^{j2\pi t} + e^{-j2\pi t}) + \tfrac{1}{2}(e^{j4\pi t} + e^{-j4\pi t})$$
$$+ \tfrac{1}{3}(e^{j6\pi t} + e^{-j6\pi t}) \tag{4.15}$$

Equivalently, using Euler's relation we can write $x(t)$ in the form

$$x(t) = 1 + \tfrac{1}{2}\cos 2\pi t + \cos 4\pi t + \tfrac{2}{3}\cos 6\pi t \tag{4.16}$$

In Figure 4.4 we illustrate graphically for this example how the signal $x(t)$ is built up from its harmonic components.

Equation (4.16) is an example of an alternate form for the Fourier series of real periodic signals. Specifically, suppose that $x(t)$ is real and can be represented in the form of eq. (4.13). Then since $x^*(t) = x(t)$, we obtain

$$x(t) = \sum_{k=-\infty}^{+\infty} a_k^* e^{-jk\omega_0 t}$$

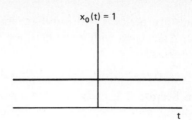

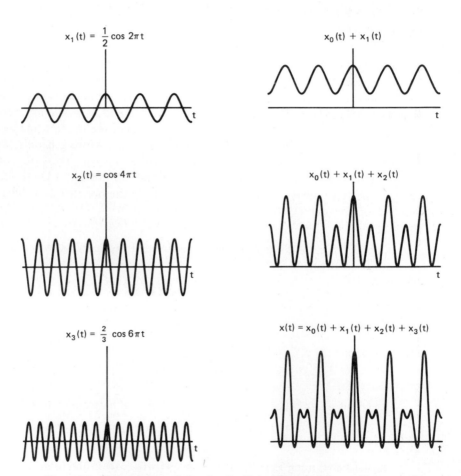

Figure 4.4 Construction of the signal $x(t)$ in Example 4.1 as a linear combination of harmonically related sinusoidal signals.

Replacing k by $-k$ in the summation, we have equivalently

$$x(t) = \sum_{k=-\infty}^{+\infty} a^*_{-k} e^{jk\omega_0 t}$$

which, by comparison with eq. (4.13), requires that $a_k = a^*_{-k}$ or equivalently that

$$a^*_k = a_{-k} \tag{4.17}$$

Note that this is the case in Example 4.1, where the a_k are in fact real and $a_k = a_{-k}$.

To derive the alternative forms of the Fourier series, we first rearrange the summation in eq. (4.13) as

$$x(t) = a_0 + \sum_{k=1}^{\infty} [a_k e^{jk\omega_0 t} + a_{-k} e^{-jk\omega_0 t}]$$

Using (4.17), this becomes

$$x(t) = a_0 + \sum_{k=1}^{\infty} [a_k e^{jk\omega_0 t} + a_k^* e^{-jk\omega_0 t}]$$

Since the two terms inside the summation are complex conjugates of each other, this can be expressed as

$$x(t) = a_0 + \sum_{k=1}^{\infty} 2\Re e\{a_k e^{jk\omega_0 t}\} \tag{4.18}$$

If a_k is expressed in polar form as†

$$a_k = A_k e^{j\theta_k}$$

then eq. (4.18) becomes

$$x(t) = a_0 + \sum_{k=1}^{\infty} 2\Re e\{A_k e^{j(k\omega_0 t + \theta_k)}\}$$

That is,

$$x(t) = a_0 + 2 \sum_{k=1}^{\infty} A_k \cos(k\omega_0 t + \theta_k) \tag{4.19}$$

Equation (4.19) is one commonly encountered form for the Fourier series of real periodic signals in continuous time. Another form is obtained by writing a_k in rectangular form as

$$a_k = B_k + jC_k$$

where B_k and C_k are both real. With this expression for a_k, eq. (4.18) takes the form

$$x(t) = a_0 + 2 \sum_{k=1}^{\infty} [B_k \cos k\omega_0 t - C_k \sin k\omega_0 t] \tag{4.20}$$

In Example 4.1 the a_k are all real and therefore both representations, eqs. (4.19) and (4.20), reduce to the same form, eq. (4.16).

Thus, we see that for real periodic functions the Fourier series in terms of complex exponentials as given in eq. (4.13) is mathematically equivalent to either of the two forms in eqs. (4.19) and (4.20) using trigonometric functions. Although the latter two are common forms for Fourier series,‡ the complex exponential form of eq. (4.13) is particularly convenient for our purposes and thus we will use that form almost exclusively.

Equation (4.17) is an example of one of many properties possessed by Fourier series and transforms. These properties are often quite useful in gaining insight and for computational purposes. We defer a discussion of the majority of these properties

---

†If z is a complex number, then its polar form is $z = re^{j\theta}$, where r and θ are real numbers. The number r is positive and is called the *magnitude* of z, which we will usually write as $|z|$. Also, the number θ is called the *angle* or *phase* of z and will be denoted by $\sphericalangle z$. For more on the manipulation of polar and rectangular forms for complex numbers, see the problems at the end of Chapter 2.

‡In fact, in his original work Fourier used the sine–cosine form of the Fourier series given in eq. (4.20).

until Section 4.6 in which we develop them within the broader context of the Fourier transform. One of the most important of these properties, however, we have essentially developed already. Let $x(t)$ be a periodic signal with a Fourier series representation given by eq. (4.13), and suppose that we apply this signal as the input to an LTI system with impulse response $h(t)$. In general, if the input to an LTI system is periodic with period T, then the output is also periodic with the same period. We can verify this fact directly by calculating the Fourier series coefficients of the output in terms of those of the input. Using the fact that each of the complex exponentials in eq. (4.13) is an eigenfunction of the system, it follows from eq. (4.8) that the output $y(t)$ is given by

$$y(t) = \sum_{k=-\infty}^{+\infty} a_k H(k\omega_0)e^{jk\omega_0 t} \tag{4.21}$$

where, from eq. (4.5), the eigenvalues $H(k\omega_0)$ are†

$$H(k\omega_0) = \int_{-\infty}^{+\infty} h(\tau)e^{-jk\omega_0\tau}\, d\tau \tag{4.22}$$

That is, if $\{a_k\}$ is the set of Fourier series coefficients for the input $x(t)$, then $\{a_k H(k\omega_0)\}$ is the set of coefficients for the output $y(t)$.

Example 4.2

Consider the periodic signal $x(t)$ discussed in Example 4.1, and let this signal be the input to an LTI system with impulse response

$$h(t) = e^{-t}u(t)$$

To calculate the output $y(t)$, we first compute $H(k\omega_0)$:

$$H(k\omega_0) = \int_0^\infty e^{-\tau}e^{-jk\omega_0\tau}\, d\tau$$

$$= -\frac{1}{1+jk\omega_0}e^{-\tau}e^{-jk\omega_0\tau}\Big|_0^\infty$$

$$= \frac{1}{1+jk\omega_0} \tag{4.23}$$

Therefore, using eqs. (4.14), (4.21), and (4.23), together with the fact that $\omega_0 = 2\pi$ in this example, we obtain

$$y(t) = \sum_{k=-3}^{+3} b_k e^{jk2\pi t} \tag{4.24}$$

with $b_k = a_k H(k2\pi)$:

$$b_0 = 1$$

$$b_1 = \frac{1}{4}\left(\frac{1}{1+j2\pi}\right), \qquad b_{-1} = \frac{1}{4}\left(\frac{1}{1-j2\pi}\right)$$

$$b_2 = \frac{1}{2}\left(\frac{1}{1+j4\pi}\right), \qquad b_{-2} = \frac{1}{2}\left(\frac{1}{1-j4\pi}\right) \tag{4.25}$$

$$b_3 = \frac{1}{3}\left(\frac{1}{1+j6\pi}\right), \qquad b_{-3} = \frac{1}{3}\left(\frac{1}{1-j6\pi}\right)$$

†Equation (4.22) is just the special case of eq. (4.5) when $s = jk\omega_0$. Note that in terms of the notation of eq. (4.5), we should write $H(jk\omega_0)$ in eq. (4.22). However, in this chapter we will be concerned only with values of $s = j\omega$. Therefore, to simplify the notation, we will suppress j and write $H(\omega)$ instead of $H(j\omega)$.

Note that $y(t)$ must be a real signal, since it is the convolution of $x(t)$ and $h(t)$, which are both real. This can be checked by examining eq. (4.25) and observing that $b_k^* = b_{-k}$. Therefore, $y(t)$ can also be expressed in either of the forms given by eqs. (4.19) and (4.20):

$$y(t) = 1 + 2 \sum_{k=1}^{3} D_k \cos(2\pi kt + \theta_k) \tag{4.26}$$

$$y(t) = 1 + 2 \sum_{k=1}^{3} [E_k \cos 2\pi kt - F_k \sin 2\pi kt] \tag{4.27}$$

where

$$b_k = D_k e^{j\theta_k} = E_k + jF_k, \qquad k = 1, 2, 3 \tag{4.28}$$

These coefficients can be evaluated directly from eq. (4.25). For example,

$$D_1 = |b_1| = \frac{1}{4\sqrt{1 + 4\pi^2}}, \qquad \theta_1 = \sphericalangle b_1 = -\tan^{-1}(2\pi)$$

$$E_1 = \mathcal{R}e\{b_1\} = \frac{1}{4(1 + 4\pi^2)}, \qquad F_1 = \mathcal{I}m\{b_1\} = -\frac{\pi}{2(1 + 4\pi^2)}$$

In Figure 4.5 we have sketched $y(t)$ and each of the terms in its Fourier series representation given by eq. (4.26). Comparing this with Figure 4.4, we see that each term is a sinusoid at the same frequency as the corresponding term of $x(t)$ but with a different magnitude and phase.

4.2.2 Determination of the Fourier Series Representation of a Periodic Signal

Assuming that a given periodic signal can be represented with the series of eq. (4.13), it is necessary to know how to determine the coefficients a_k. Multiplying both sides of eq. (4.13) by $e^{-jn\omega_0 t}$, we obtain

$$x(t)e^{-jn\omega_0 t} = \sum_{k=-\infty}^{+\infty} a_k e^{jk\omega_0 t} e^{-jn\omega_0 t} \tag{4.29}$$

Integrating both sides from 0 to $T_0 = 2\pi/\omega_0$, we have

$$\int_0^{T_0} x(t)e^{-jn\omega_0 t} \, dt = \int_0^{T_0} \sum_{k=-\infty}^{+\infty} a_k e^{jk\omega_0 t} e^{-jn\omega_0 t} \, dt$$

Here T_0 is the fundamental period of $x(t)$, and consequently we are integrating over one period. Interchanging the order of integration and summation yields

$$\int_0^{T_0} x(t)e^{-jn\omega_0 t} \, dt = \sum_{k=-\infty}^{+\infty} a_k \left[\int_0^{T_0} e^{j(k-n)\omega_0 t} \, dt \right] \tag{4.30}$$

The evaluation of the bracketed integral is straightforward. Rewriting this integral using Euler's formula, we obtain

$$\int_0^{T_0} e^{j(k-n)\omega_0 t} \, dt = \int_0^{T_0} \cos(k-n)\omega_0 t \, dt + j \int_0^{T_0} \sin(k-n)\omega_0 t \, dt \tag{4.31}$$

For $k \neq n$, $\cos(k-n)\omega_0 t$ and $\sin(k-n)\omega_0 t$ are periodic sinusoids with fundamental period $(T_0/|k-n|)$. Therefore, in eq. (4.31) we are integrating over an interval (of length T_0) that is an integral number of periods of these signals. Since the integral may be viewed as measuring the total area under these functions over the interval,

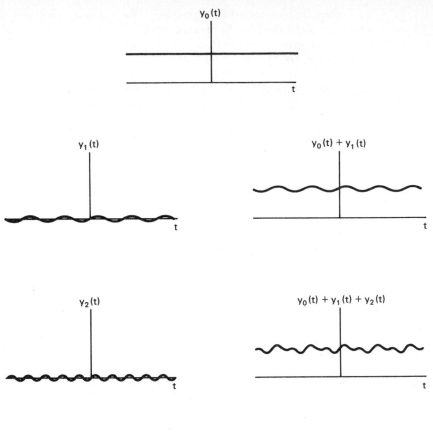

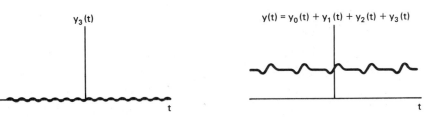

Figure 4.5 The signal $y(t)$ in Example 4.2, obtained by applying $x(t)$ in Example 4.1 (Figure 4.4) as the input to the LTI system with impulse response $h(t) = e^{-t}u(t)$. Referring to Figure 4.4, $y_0(t)$ is the response to $x_0(t)$, $y_1(t)$ is the response to $x_1(t)$, and so on.

we see that for $k \neq n$, both of the integrals on the right-hand side of eq. (4.31) are zero. For $k = n$, the integrand on the left-hand side of eq. (4.31) equals 1, and thus the integral equals T_0. In summary we then have that

$$\int_0^{T_0} e^{j(k-n)\omega_0 t}\, dt = \begin{cases} T_0, & k = n \\ 0, & k \neq n \end{cases}$$

and consequently the right-hand side of eq. (4.30) reduces to $T_0 a_n$. Therefore,

$$a_n = \frac{1}{T_0} \int_0^{T_0} x(t)e^{-jn\omega_0 t}\, dt \qquad (4.32)$$

which is the equation we have sought for determining the coefficients. Furthermore, note that in evaluating eq. (4.31) the only fact that we used concerning the interval of integration was that we were integrating over an interval of length T_0, which is an integral number of periods of $\cos(k - n)\omega_0 t$ and $\sin(k - n)\omega_0 t$. Therefore, if we let $\int_{T_0}$ denote integration over *any* interval of length T_0, we have that

$$\int_{T_0} e^{j(k-n)\omega_0 t}\, dt = \begin{cases} T_0, & k = n \\ 0, & k \neq n \end{cases}$$

Consequently, if we integrate eq. (4.29) over any interval of length T_0, we can perform exactly the same steps as we did in going from eq. (4.30) to eq. (4.32). The result is the expression

$$a_n = \frac{1}{T_0} \int_{T_0} x(t)e^{-jn\omega_0 t}\, dt \qquad (4.33)$$

To summarize, if $x(t)$ has a Fourier series representation [i.e., if it can be expressed as a linear combination of harmonically related complex exponentials in the form of eq. (4.13)], then the coefficients are given by eq. (4.33). This pair of equations, rewritten below, defines the Fourier series of a periodic signal.

$$x(t) = \sum_{k=-\infty}^{+\infty} a_k e^{jk\omega_0 t} \qquad (4.34)$$

$$a_k = \frac{1}{T_0} \int_{T_0} x(t)e^{-jk\omega_0 t}\, dt \qquad (4.35)$$

Equation (4.34) is often referred to as the *synthesis* equation and eq. (4.35) as the *analysis* equation. The coefficients $\{a_k\}$ are often called the *Fourier series coefficients* or the *spectral coefficients* of $x(t)$. These complex coefficients measure the portion of the signal $x(t)$ that is at each harmonic of the fundamental component. The coefficient a_0 is the dc or constant component of $x(t)$ and is given by eq. (4.35) with $k = 0$. That is,

$$a_0 = \frac{1}{T_0} \int_{T_0} x(t)\, dt \qquad (4.36)$$

which is simply the average value of $x(t)$ over one period. We note that the term "spectral coefficient" is derived from problems such as the spectroscopic decomposition of light into spectral lines (i.e., into its elementary components at different frequencies). The intensity of any line in such a decomposition is a direct measure of the fraction of the total light energy at the frequency corresponding to the line.

Equations (4.34) and (4.35) were known to both Euler and Lagrange in the middle of the eighteenth century. However, they discarded this line of analysis without having examined the question of how large a class of periodic signals could, in fact, be represented in this fashion. Before we turn to this question in the next section, let us illustrate the Fourier series by means of a few examples.

Example 4.3

Consider the signal

$$x(t) = \sin \omega_0 t$$

One approach to determining the Fourier series coefficients for this example would be to apply eq. (4.35). For this simple case, however, it is easier simply to expand the sinusoidal function as a linear combination of complex exponentials and identify by inspection the Fourier series coefficients. Specifically, we can express $\sin \omega_0 t$ as

$$\sin \omega_0 t = \frac{1}{2j} e^{j\omega_0 t} - \frac{1}{2j} e^{-j\omega_0 t}$$

Thus,

$$a_1 = \frac{1}{2j}, \qquad a_{-1} = -\frac{1}{2j}$$

$$a_k = 0, \qquad k \neq +1 \text{ or } -1$$

Example 4.4

Let

$$x(t) = 1 + \sin \omega_0 t + 2 \cos \omega_0 t + \cos \left(2\omega_0 t + \frac{\pi}{4}\right)$$

As with Example 4.3, we can again expand $x(t)$ directly in terms of complex exponentials so that

$$x(t) = 1 + \frac{1}{2j}[e^{j\omega_0 t} - e^{-j\omega_0 t}] + [e^{j\omega_0 t} + e^{-j\omega_0 t}] + \frac{1}{2}[e^{j(2\omega_0 t + \pi/4)} + e^{-j(2\omega_0 t + \pi/4)}]$$

Collecting terms, we obtain

$$x(t) = 1 + \left(1 + \frac{1}{2j}\right)e^{j\omega_0 t} + \left(1 - \frac{1}{2j}\right)e^{-j\omega_0 t} + \left(\frac{1}{2}e^{j(\pi/4)}\right)e^{j2\omega_0 t} + \left(\frac{1}{2}e^{-j(\pi/4)}\right)e^{-j2\omega_0 t}$$

Thus, the Fourier series coefficients for this example are

$$a_0 = 1$$

$$a_1 = \left(1 + \frac{1}{2j}\right) = 1 - \frac{1}{2}j$$

$$a_{-1} = \left(1 - \frac{1}{2j}\right) = 1 + \frac{1}{2}j$$

$$a_2 = \frac{1}{2}e^{j(\pi/4)} = \frac{\sqrt{2}}{4}(1 + j)$$

$$a_{-2} = \frac{1}{2}e^{-j(\pi/4)} = \frac{\sqrt{2}}{4}(1 - j)$$

$$a_k = 0, \qquad |k| > 2$$

In Figure 4.6 we have plotted the magnitude and phase of a_k on bar graphs in which each line represents either the magnitude or the phase of the corresponding harmonic component of $x(t)$.

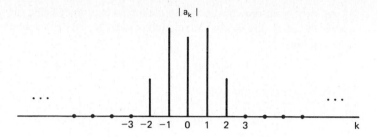

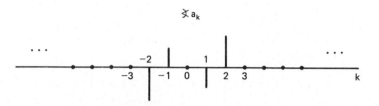

Figure 4.6 Plots of the magnitude and phase of the Fourier coefficients of the signal considered in Example 4.4.

Example 4.5

The periodic square wave, sketched in Figure 4.7 and defined over one period as

$$x(t) = \begin{cases} 1, & |t| < T_1 \\ 0, & T_1 < |t| < \dfrac{T_0}{2} \end{cases} \tag{4.37}$$

is a signal that we will encounter on several occasions. This signal is periodic with fundamental period T_0 and fundamental frequency $\omega_0 = 2\pi/T_0$.

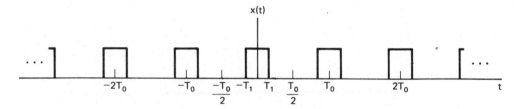

Figure 4.7 Periodic square wave.

To determine the Fourier series coefficients for $x(t)$, we will use eq. (4.35). Because of the symmetry of $x(t)$ about $t = 0$ it is most convenient to choose the interval over which the integration is performed as $-(T_0/2) \leq t < T_0/2$, although any interval of length T_0 is equally valid and thus will lead to the same result. Using these limits of integration and substituting from eq. (4.37), we have first for $k = 0$ that

$$a_0 = \frac{1}{T_0} \int_{-T_1}^{T_1} dt = \frac{2T_1}{T_0} \tag{4.38}$$

As mentioned previously, a_0 has the interpretation of being the average value of $x(t)$, which in this case equals the fraction of each period during which $x(t) = 1$. For $k \neq 0$ we obtain

$$a_k = \frac{1}{T_0} \int_{-T_1}^{T_1} e^{-jk\omega_0 t} \, dt = -\frac{1}{jk\omega_0 T_0} e^{-jk\omega_0 t} \Big|_{-T_1}^{T_1}$$

We may rewrite this as

$$a_k = \frac{2}{k\omega_0 T_0} \left[\frac{e^{jk\omega_0 T_1} - e^{-jk\omega_0 T_1}}{2j} \right] \tag{4.39}$$

Noting that the term in brackets is $\sin k\omega_0 T_1$, the coefficients a_k can be expressed as

$$a_k = \frac{2 \sin k\omega_0 T_1}{k\omega_0 T_0} = \frac{\sin k\omega_0 T_1}{k\pi}, \qquad k \neq 0 \tag{4.40}$$

Here we have used the fact that $\omega_0 T_0 = 2\pi$.

Graphical representations of the Fourier series coefficients for this example for a fixed value of T_1 and several values of T_0 are shown in Figure 4.8. For $T_0 = 4T_1$, $x(t)$ is a *symmetric square wave* (i.e., one that is unity for half the period and zero for half the period). In this case $\omega_0 T_1 = \pi/2$, and from eq. (4.40),

$$a_k = \frac{\sin (\pi k/2)}{k\pi}, \qquad k \neq 0 \tag{4.41}$$

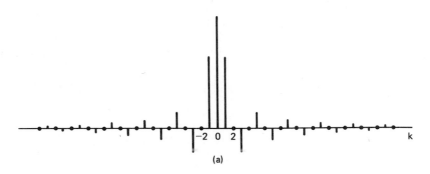

(a)

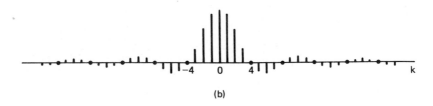

(b)

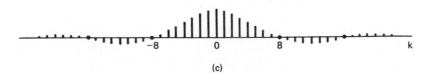

(c)

Figure 4.8 Fourier series coefficients for the periodic square wave: (a) $T_0 = 4T_1$; (b) $T_0 = 8T_1$; (c) $T_0 = 16T_1$.

Fourier Analysis for Continuous-Time Signals and Systems   Chap. 4

while from eq. (4.38),

$$a_0 = \tfrac{1}{2}$$

From eq. (4.41) it is clear that $a_k = 0$ for k even. Also, $\sin(\pi k/2)$ alternates between ± 1 for successive odd values of k. Therefore,

$$a_1 = a_{-1} = \frac{1}{\pi}$$

$$a_3 = a_{-3} = -\frac{1}{3\pi}$$

$$a_5 = a_{-5} = \frac{1}{5\pi}$$

$$\cdot$$
$$\cdot$$
$$\cdot$$

For this particular example, the Fourier coefficients are real, and consequently they can be depicted graphically with only a single graph. More generally, of course, the Fourier coefficients are complex, and consequently two graphs corresponding to real and imaginary parts or magnitude and phase of each coefficient would be required.

4.3 APPROXIMATION OF PERIODIC SIGNALS USING FOURIER SERIES AND THE CONVERGENCE OF FOURIER SERIES

Although Euler and Lagrange would have been happy with the results of Examples 4.3 and 4.4, they would have objected to Example 4.5, since $x(t)$ is discontinuous while each of the harmonic components is continuous. Fourier, on the other hand, considered the same example and maintained that the Fourier series representation of the square wave is valid. In fact, Fourier maintained that *any* periodic signal could be represented by a Fourier series. Although this is not quite true, it *is* true that Fourier series can be used to represent an extremely large class of periodic signals, which includes the square wave and all other signals with which we will be concerned in this book.

To gain an understanding of the square-wave example and more generally of the question of the validity of Fourier series representations, let us first look at the problem of approximating a given periodic signal $x(t)$ by a linear combination of a finite number of harmonically related complex exponentials, that is, by a finite series of the form

$$x_N(t) = \sum_{k=-N}^{N} a_k e^{jk\omega_0 t} \tag{4.42}$$

Let $e_N(t)$ denote the approximation error, which is given by

$$e_N(t) = x(t) - x_N(t) = x(t) - \sum_{k=-N}^{+N} a_k e^{jk\omega_0 t} \tag{4.43}$$

In order to determine how good any particular approximation is, we need to specify a quantitative measure of the size of the approximation error. The criterion that we will use is the total squared-error magnitude over one period:

$$E_N = \int_{T_0} |e_N(t)|^2 \, dt = \int_{T_0} e_N(t) e_N^*(t) \, dt \tag{4.44}$$

In general, for any signal $z(t)$ the quantity

$$E = \int_a^b |z(t)|^2 \, dt$$

is often referred to as the *energy* in $z(t)$ over the time interval $a \le t \le b$. This terminology is motivated by the fact that if $z(t)$ corresponds to the current flowing into a 1-Ω resistor, then E is the total energy dissipated in the resistor over the time interval $a \le t \le b$. With respect to eq. (4.44), E_N then represents the energy in the approximation error over one period.

As shown in Problem 4.8, the particular choice for the coefficients a_k in eq. (4.42) that minimize the energy in the error is given by

$$a_k = \frac{1}{T_0} \int_{T_0} x(t) e^{-jk\omega_0 t} \, dt \qquad (4.45)$$

Comparing eqs. (4.45) and (4.35), we see that this is identical to the expression used to determine the Fourier series coefficients. Thus, if $x(t)$ has a Fourier series representation, the best approximation using only a finite number of harmonically related complex exponentials is obtained by truncating the Fourier series to the desired number of terms. As N increases, new terms are added but the previous ones remain unchanged and E_N decreases. If, in fact, $x(t)$ has a Fourier series representation, then the limit of E_N as $N \longrightarrow \infty$ is zero.

Let us turn now to the question of the validity of the Fourier series representation for periodic signals. For any such signal we can attempt to obtain a set of Fourier coefficients through the use of eq. (4.35). However, in some cases the integral in eq. (4.35) may diverge; that is, the value obtained for some of the a_k may be infinite. Moreover, even if all of the coefficients obtained from eq. (4.35) are finite, when these coefficients are substituted into the synthesis equation (4.34), the resulting infinite series may not converge to the original signal $x(t)$.

It happens, however, that there are no convergence difficulties if $x(t)$ is continuous. That is, every continuous periodic signal has a Fourier series representation so that the energy E_N in the approximation error approaches 0 as N goes to ∞. This is also true for many discontinuous signals. Since we will find it very useful to use discontinuous signals, such as the square wave of Example 4.5, it is worthwhile to investigate the issue of convergence in a bit more detail. There are two somewhat different conditions that a periodic signal can satisfy to guarantee that it can be represented by a Fourier series. In discussing these we will not attempt to provide a complete mathematical justification. More rigorous treatments can be found in many texts on Fourier analysis.[†]

One class of periodic signals that are representable through the Fourier series is that comprising signals which are square-integrable over a period. That is, any signal $x(t)$ in this class has finite energy over a single period:

[†]See, for example, R. V. Churchill, *Fourier Series and Boundary Value Problems*, 2nd ed. (New York: McGraw-Hill Book Company, 1963); W. Kaplan, *Operational Methods for Linear Systems* (Reading, Mass.: Addison-Wesley Publishing Company, 1962); and the book by Dym and McKean referenced on p. 162.

$$\int_{T_0} |x(t)|^2 < \infty \qquad (4.46)$$

When this condition is satisfied, we are guaranteed that the coefficients a_k obtained from eq. (4.35) are finite. Furthermore, let $x_N(t)$ be the approximation to $x(t)$ obtained by using these coefficients for $|k| \leq N$:

$$x_N(t) = \sum_{k=-N}^{+N} a_k e^{jk\omega_0 t} \qquad (4.47)$$

Then, we are guaranteed that $\lim_{N\to\infty} E_N = 0$, where E_N is defined in eq. (4.44). That is, if we define

$$e(t) = x(t) - \sum_{k=-\infty}^{+\infty} a_k e^{jk\omega_0 t} \qquad (4.48)$$

then

$$\int_{T_0} |e(t)|^2\, dt = 0 \qquad (4.49)$$

As we will see in an example at the end of this section, eq. (4.49) does not imply that the signal $x(t)$ and its Fourier series representation

$$\sum_{k=-\infty}^{+\infty} a_k e^{jk\omega_0 t} \qquad (4.50)$$

are equal at every value of t. What it does say is that there is no energy in their difference.

The type of convergence guaranteed when $x(t)$ is square-integrable is often useful. In fact, most of the periodic signals that we consider do have finite energy over a single period and consequently do have Fourier series representations. However, an alternate set of conditions, developed by P. L. Dirichlet, and also satisfied by essentially all of the signals with which we will be concerned guarantees that $x(t)$ will in fact be *equal* to $x(t)$ except at isolated values of t for which $x(t)$ is discontinuous. At these values of t the infinite series of eq. (4.50) converges to the "average" value of the discontinuity (i.e., halfway between the values on either side of the discontinuity).

The Dirichlet conditions are as follows:

Condition 1. Over any period $x(t)$ must be *absolutely integrable*, that is,

$$\int_{T_0} |x(t)|\, dt < \infty \qquad (4.51)$$

As with square integrability, this guarantees that each coefficient a_k will be finite, since

$$|a_k| \leq \frac{1}{T_0} \int_{T_0} |x(t) e^{-jk\omega_0 t}|\, dt = \frac{1}{T_0} \int_{T_0} |x(t)|\, dt$$

So if

$$\int_{T_0} |x(t)|\, dt < \infty$$

then

$$|a_k| < \infty$$

A periodic signal that violates the first Dirichlet condition is

$$x(t) = \frac{1}{t}, \qquad 0 < t \le 1$$

where $x(t)$ is periodic with period 1. This signal is illustrated in Figure 4.9(a).

Condition 2. In any finite interval of time, $x(t)$ is of bounded variation, that is, there are no more than a finite number of maxima and minima during any single period of the signal.

An example of a time function that meets condition 1 but not condition 2 is

$$x(t) = \sin\left(\frac{2\pi}{t}\right), \qquad 0 < t \le 1 \tag{4.52}$$

as illustrated in Figure 4.9(b). For this function (periodic with $T_0 = 1$)

$$\int_0^1 |x(t)| \, dt < 1$$

It has, however, an infinite number of maxima and minima in the interval.

Condition 3. In any finite interval of time there are only a finite number of discontinuities. Furthermore, each of these discontinuities must be finite.

An example of a time function that violates condition 3 is illustrated in Figure 4.9(c). The signal $x(t)$ (of period $T = 8$) is composed of an infinite number of sections each of which is half the height and half the width of the previous section. Thus, the area under one period of the function is clearly less than 8. However, there are an infinite number of finite discontinuities, thereby violating condition 3.

As can be seen from the examples given in Figure 4.9, the signals that do not satisfy the Dirichlet conditions are generally pathological in nature and thus are not particularly important in the study of signals and systems. For this reason the question of convergence of Fourier series will not play a particularly significant role in the remainder of the book. For a periodic signal that varies continuously, we know that the Fourier series representation converges and equals the original signal at every value of t. For a periodic signal with discontinuities the Fourier series representation equals the signal everywhere except at the isolated points of discontinuity, at which the series converges to the average value of the signal on either side of the discontinuity. In this case the difference between the original signal and its Fourier series representation contains no energy, and consequently the two signals can be thought of as being the same for all practical purposes. Specifically, since the signals differ only at isolated points, the integrals of both signals over any interval are identical. For this reason, the two signals behave identically under convolution and consequently *are* identical from the standpoint of the analysis of LTI systems.

Thus, the Fourier series of a periodic signal with several discontinuities does

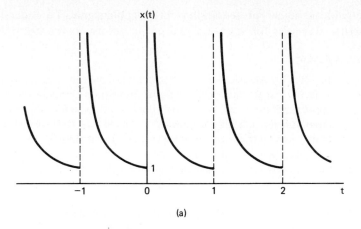

(a)

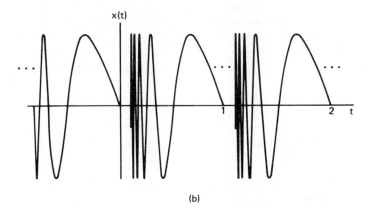

(b)

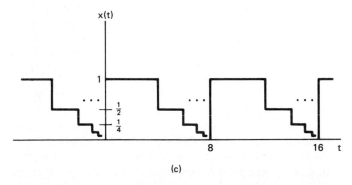

(c)

Figure 4.9 Signals that violate the Dirichlet conditions: (a) the signal $x(t)$, periodic with period 1, with $x(t) = 1/t$ for $0 < t \leq 1$ (this signal violates the first Dirichlet condition); (b) the periodic signal of eq. (4.52) which violates the second Dirichlet condition; (c) a signal, periodic with period 8, that violates the third Dirichlet condition [for $0 \leq t < 8$ the value of $x(t)$ decreases by a factor of 2 whenever the distance from t to 8 decreases by a factor of 2; that is $x(t) = 1$, $0 \leq t < 4$, $x(t) = 1/2$, $4 \leq t < 6$, $x(t) = 1/4$, $6 \leq t < 7$, $x(t) = 1/8$, $7 \leq t < 7.5$, etc.].

provide a useful representation of the original signal. To understand *how* the Fourier series converges for such a signal, let us return to the example of a square wave.

Example 4.6†

In 1898, an American physicist, Albert Michelson, constructed a harmonic analyzer, a device that for any periodic signal $x(t)$ would compute the truncated Fourier series approximation of eq. (4.47) for values of N up to 80. Michelson tested his device on many functions, with the expected result that $x_N(t)$ looked very much like $x(t)$. However, when he tried the square wave, he obtained an important and to him, very surprising result. Michelson was concerned about the behavior he observed and thought that his device might have had a defect. He wrote about this problem to the famous mathematical physicist Josiah Gibbs, who investigated it and reported his explanation in 1899.

What Michelson had observed is illustrated in Figure 4.10, where we have shown $x_N(t)$ for several values of N for $x(t)$ a symmetric square wave ($T_0 = 4T_1$). In each case the partial sum is superimposed on the original square wave. Since the square wave satisfies the Dirichlet conditions, the limit as $N \rightarrow \infty$ of $x_N(t)$ at the discontinuities should be the average value of the discontinuity. We see from Figure 4.10 that this is in fact the case, since for any N, $x_N(t)$ has exactly that value at the discontinuities. Furthermore, for any other value of t, say $t = t_1$, we are guaranteed that

$$\lim_{N \to \infty} x_N(t_1) = x(t_1)$$

Therefore, we also have that the squared error in the Fourier series representation of the square wave has zero area, as in eq. (4.49).

For this example, the interesting effect that Michelson observed is that the behavior of the partial sum in the vicinity of the discontinuity exhibits ripples and that the peak amplitude of these ripples does not seem to decrease with increasing N. Gibbs showed that this is in fact the case. For a discontinuity of unity height the partial sum exhibits a maximum value of 1.09 (i.e., an overshoot of 9% of the height of the discontinuity) no matter how large N becomes. One must be careful to interpret this correctly. As stated before, for any *fixed* value of t, say $t = t_1$, the partial sums will converge to the correct value, and at the discontinuity will converge to one-half the sum of the values of the signal on either side of the discontinuity. However, the closer t_1 is chosen to the point of discontinuity, the larger N must be in order to reduce the error below a specified amount. Thus, as N increases, the ripples in the partial sums become compressed toward the discontinuity, but for *any* finite value of N, the peak amplitude of the ripples remains constant. This behavior has come to be known as the *Gibbs phenomenon*. The implication of this phenomenon is that the truncated Fourier series approximation $x_N(t)$ of a discontinuous signal $x(t)$ will in general exhibit high-frequency ripples and overshoot near the discontinuities, and if such an approximation is used in practice, a large enough value of N should be chosen so as to guarantee that the total energy in these ripples is insignificant. In the limit, of course, we know that the energy in the approximation error vanishes and that the Fourier series representation of a discontinuous signal such as the square wave converges.

†The historical information used in this example is taken from the book by Lanczos referenced on p. 162.

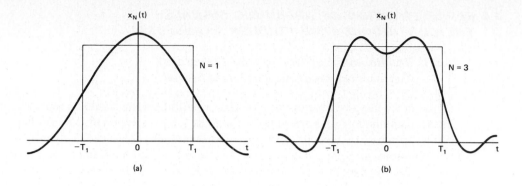

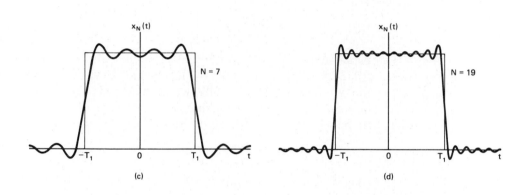

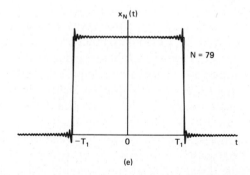

Figure 4.10 Convergence of the Fourier series representation of a square wave: an illustration of the Gibbs phenomenon. Here we have depicted the finite series approximation $x_N(t) = \sum_{k=-N}^{N} a_k e^{jk\omega_0 t}$ for several values of N.

4.4 REPRESENTATION OF APERIODIC SIGNALS: THE CONTINUOUS-TIME FOURIER TRANSFORM

4.4.1 Development of the Fourier Transform Representation of an Aperiodic Signal

In the preceding two sections we saw how a periodic signal could be represented as a linear combination of harmonically related complex exponentials. In fact, these results can be extended to develop a representation of *aperiodic* signals as linear combinations of complex exponentials. The introduction of this representation is one of Fourier's most important contributions, and our development of the Fourier transform follows very closely the approach he used in his original work.

Consider again the periodic square wave discussed in Example 4.5. The Fourier series coefficients are given by eq. (4.40) as

$$a_k = \frac{2 \sin k\omega_0 T_1}{k\omega_0 T_0} \tag{4.53}$$

where T_0 is the period and $\omega_0 = 2\pi/T_0$. In Figure 4.8 we plotted these coefficients for a fixed value of T_1 and several choices for T_0. In Figure 4.11 we have repeated this figure with several modifications. Specifically, we have plotted $T_0 a_k$ rather than a_k, and we have also modified the horizontal spacing in each plot. The significance of these changes can be seen by examining eq. (4.53). Multiplying a_k by T_0, we obtain

$$T_0 a_k = \frac{2 \sin k\omega_0 T_1}{k\omega_0} = \left. \frac{2 \sin \omega T_1}{\omega} \right|_{\omega = k\omega_0} \tag{4.54}$$

Thus, with ω thought of as a continuous variable, the function $(2 \sin \omega T_1)/\omega$ represents the envelope of $T_0 a_k$, and these coefficients are simply equally spaced samples of this envelope. Also, for fixed T_1, the envelope of $T_0 a_k$ is *independent of* T_0. However, from Figure 4.11 we see that as T_0 increases (or, equivalently, ω_0 decreases), the envelope is sampled with a closer and closer spacing. As T_0 becomes arbitrarily large, the original periodic square wave approaches a rectangular pulse (i.e., all that remains in the time domain is an aperiodic signal corresponding to one period of the square wave). Also, the Fourier series coefficients, multiplied by T_0, become more and more closely spaced samples of the envelope, so in some sense (which we will specify shortly) the set of Fourier series coefficients approaches the envelope function as $T_0 \longrightarrow \infty$.

This example illustrates the basic idea behind Fourier's development of a representation for aperiodic signals. Specifically, we think of an aperiodic signal as the limit of a periodic signal as the period becomes arbitrarily large, and we examine the limiting behavior of the Fourier series representation for this signal. Consider a general aperiodic signal $x(t)$ that is of finite duration. That is, for some number T_1, $x(t) = 0$ if $|t| > T_1$. Such a signal is illustrated in Figure 4.12(a). From this aperiodic signal we can construct a periodic signal $\tilde{x}(t)$ for which $x(t)$ is one period, as indicated in Figure 4.12(b). As we choose the period T_0 to be larger, $\tilde{x}(t)$ is identical to $x(t)$ over a longer interval, and as $T_0 \longrightarrow \infty$, $\tilde{x}(t)$ is equal to $x(t)$ for any finite value of t.

Let us now examine the effect of this on the Fourier series representation of $\tilde{x}(t)$. Rewriting eqs. (4.34) and (4.35) here for convenience, with the integral in eq. (4.35) carried out over the interval $-T_0/2 \leq t \leq T_0/2$, we have

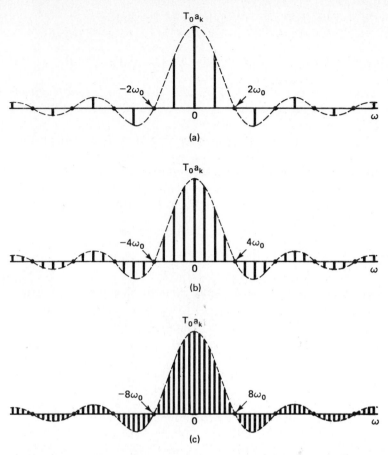

Figure 4.11 Fourier coefficients and their envelope for the periodic square wave: (a) $T_0 = 4T_1$; (b) $T_0 = 8T_1$; (c) $T_0 = 16T_1$.

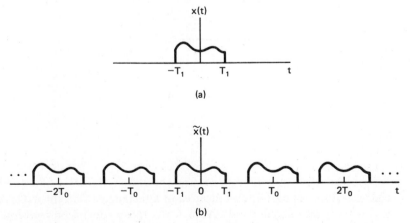

Figure 4.12 (a) Aperiodic signal $x(t)$; (b) periodic signal $\tilde{x}(t)$, constructed to be equal to $x(t)$ over one period.

$$\tilde{x}(t) = \sum_{k=-\infty}^{+\infty} a_k e^{jk\omega_0 t} \tag{4.55}$$

$$a_k = \frac{1}{T_0} \int_{-T_0/2}^{T_0/2} \tilde{x}(t) e^{-jk\omega_0 t}\, dt \tag{4.56}$$

Since $\tilde{x}(t) = x(t)$ for $|t| < T_0/2$ and also since $x(t) = 0$ outside this interval, eq. (4.56) can be rewritten as

$$a_k = \frac{1}{T_0} \int_{-T_0/2}^{T_0/2} x(t) e^{-jk\omega_0 t}\, dt = \frac{1}{T_0} \int_{-\infty}^{+\infty} x(t) e^{-jk\omega_0 t}\, dt$$

Therefore, defining the envelope $X(\omega)$ of $T_0 a_k$ as

$$X(\omega) = \int_{-\infty}^{+\infty} x(t) e^{-j\omega t}\, dt \tag{4.57}$$

we have that the coefficients a_k can be expressed as

$$a_k = \frac{1}{T_0} X(k\omega_0) \tag{4.58}$$

Combining eqs. (4.58) and (4.55), $\tilde{x}(t)$ can be expressed in terms of $X(\omega)$ as

$$\tilde{x}(t) = \sum_{k=-\infty}^{+\infty} \frac{1}{T_0} X(k\omega_0) e^{jk\omega_0 t}$$

or equivalently, since $2\pi/T_0 = \omega_0$,

$$\tilde{x}(t) = \frac{1}{2\pi} \sum_{k=-\infty}^{+\infty} X(k\omega_0) e^{jk\omega_0 t} \omega_0 \tag{4.59}$$

As $T_0 \to \infty$, $\tilde{x}(t)$ approaches $x(t)$, and consequently eq. (4.59) becomes a representation of $x(t)$. Furthermore, $\omega_0 \to 0$ as $T_0 \to \infty$, and the right-hand side of eq. (4.59) passes to an integral. This can be seen by considering the graphical interpretation of eq. (4.59), illustrated in Figure 4.13. Each term in the summation on the

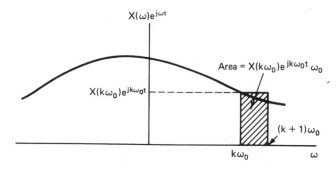

Figure 4.13 Graphical interpretation of eq. (4.59).

right-hand side of eq. (4.59) is the area of a rectangle of height $X(k\omega_0)e^{jk\omega_0 t}$ and width ω_0 (here t is regarded as fixed). As $\omega_0 \to 0$, this by definition converges to the integral of $X(\omega)e^{j\omega t}$. Therefore, using the fact that $\tilde{x}(t) \rightarrow x(t)$ as $T_0 \rightarrow \infty$, eqs. (4.59) and (4.57) become

$$x(t) = \frac{1}{2\pi} \int_{-\infty}^{+\infty} X(\omega)e^{j\omega t}\, d\omega \qquad (4.60)$$

$$X(\omega) = \int_{-\infty}^{+\infty} x(t)e^{-j\omega t}\, dt \qquad (4.61)$$

Equations (4.60) and (4.61) are referred to as the *Fourier transform pair* with the function $X(\omega)$ as given by eq. (4.61) referred to as the *Fourier transform* or *Fourier integral* of $x(t)$ and eq. (4.60) as the *inverse Fourier transform* equation. The *synthesis* equation (4.60) plays a role for aperiodic signals similar to that of eq. (4.34) for periodic signals, since both correspond to a decomposition of a signal into a linear combination of complex exponentials. For periodic signals these complex exponentials have amplitudes $\{a_k\}$ as given by eq. (4.35) and occur at a discrete set of harmonically related frequencies $k\omega_0$, $k = 0, \pm 1, \pm 2, \ldots$. For aperiodic signals these complex exponentials occur at a continuum of frequencies and, according to the synthesis equation (4.60), have "amplitude" $X(\omega)(d\omega/2\pi)$. In analogy with the terminology used for the Fourier series coefficients of a periodic signal, the transform $X(\omega)$ of an aperiodic signal $x(t)$ is commonly referred to as the *spectrum* of $x(t)$, as it provides us with the information concerning how $x(t)$ is composed of sinusoidal signals at different frequencies.

4.4.2 Convergence of Fourier Transforms

Although the argument we used in deriving the Fourier transform pair assumed that $x(t)$ was of arbitrary but finite duration, eqs. (4.60) and (4.61) remain valid for an extremely broad class of signals of infinite duration. In fact, our derivation of the Fourier transform suggests that a set of conditions like those required for the convergence of Fourier series should also apply here, and indeed that can be shown to be the case.† Specifically, consider $X(\omega)$ evaluated according to eq. (4.61), and let $\hat{x}(t)$ denote the signal obtained by using $X(\omega)$ in the right-hand side of eq. (4.60). That is,

$$\hat{x}(t) = \frac{1}{2\pi} \int_{-\infty}^{+\infty} X(\omega)e^{j\omega t}\, d\omega$$

What we would like to know is when eq. (4.60) is valid [i.e., when $\hat{x}(t)$ is a valid representation of the original signal $x(t)$]. If $x(t)$ is square-integrable so that

†For a mathematically rigorous discussion of the Fourier transform and of its properties and applications, see R. Bracewell, *The Fourier Transform and Its Applications* (New York: McGraw-Hill Book Company, 1965); A. Papoulis, *The Fourier Integral and Its Applications* (New York: McGraw-Hill Book Company, 1962); E. C. Titchmarsh, *Introduction to the Theory of Fourier Integrals* (Oxford: Clarendon Press, 1948); and the book by Dym and McKean referenced on p. 162.

$$\int_{-\infty}^{+\infty} |x(t)|^2 \, dt < \infty \qquad (4.62)$$

then we are guaranteed that $X(\omega)$ is finite [i.e., eq. (4.61) converges] and that, with $e(t)$ denoting the error between $\hat{x}(t)$ and $x(t)$ [i.e., $e(t) = \hat{x}(t) - x(t)$],

$$\int_{-\infty}^{+\infty} |e(t)|^2 \, dt = 0 \qquad (4.63)$$

Equations (4.62) and (4.63) are the aperiodic counterparts of eqs. (4.46) and (4.49) for periodic signals. Thus, as with periodic signals, if $x(t)$ is square-integrable, then although $x(t)$ and its Fourier representation $\hat{x}(t)$ may differ significantly at individual values of t, there is no energy in their difference.

Just as with periodic signals, there is an alternative set of conditions which are sufficient to ensure that $\hat{x}(t)$ is equal to $x(t)$ for any t except at a discontinuity, where it is equal to the average value of the discontinuity. These conditions, again referred to as the Dirichlet conditions, require that:

1. $x(t)$ be absolutely integrable, that is,

$$\int_{-\infty}^{+\infty} |x(t)| \, dt < \infty \qquad (4.64)$$

2. $x(t)$ have a finite number of maxima and minima within any finite interval.
3. $x(t)$ have a finite number of discontinuities within any finite interval. Furthermore, each of these discontinuities must be finite.

Therefore, absolutely integrable signals that are continuous or have several discontinuities have Fourier transforms.

Although the two alternative sets of conditions that we have given are sufficient to guarantee that a signal has a Fourier transform, we will see in the next section that periodic signals, which are neither absolutely integrable nor square-integrable over an *infinite* interval, can be considered to have Fourier transforms if impulse functions are permitted in the transform. This has the advantage that the Fourier series and Fourier transform can be incorporated in a common framework, and we will find this to be very convenient in subsequent chapters. Before examining this point further in the next section, let us first consider several examples of the Fourier transform.

4.4.3 Examples of Continuous-Time Fourier Transforms

Example 4.7

Consider the signal

$$x(t) = e^{-at}u(t)$$

If $a < 0$, then $x(t)$ is not absolutely integrable and hence $X(\omega)$ does not exist. For $a > 0$, $X(\omega)$ is obtained from eq. (4.61) as

$$X(\omega) = \int_{0}^{\infty} e^{-at}e^{-j\omega t} \, dt = -\frac{1}{a+j\omega}e^{-(a+j\omega)t}\Big|_0^{\infty}$$

That is,

$$X(\omega) = \frac{1}{a + j\omega}, \qquad a > 0$$

Since this particular Fourier transform has both real and imaginary parts, to plot it as a function of ω we express $X(\omega)$ in terms of its magnitude and phase:

$$|X(\omega)| = \frac{1}{\sqrt{a^2 + \omega^2}}, \qquad \sphericalangle X(\omega) = -\tan^{-1}\left(\frac{\omega}{a}\right)$$

Each of these components is sketched in Figure 4.14. Note that if a is complex rather than real, then $x(t)$ is absolutely integrable as long as $\mathcal{R}e\{a\} > 0$, and in this case the preceding calculation yields the same form for $X(\omega)$. That is,

$$X(\omega) = \frac{1}{a + j\omega}, \qquad \mathcal{R}e\{a\} > 0$$

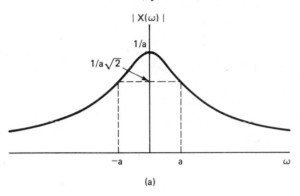

(a)

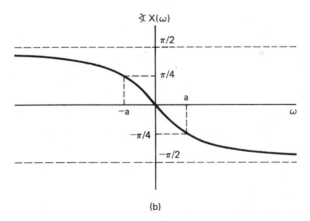

(b)

Figure 4.14 Fourier transform of the signal $x(t) = e^{-at}u(t)$, $a > 0$, considered in Example 4.7.

Example 4.8

Let

$$x(t) = e^{-a|t|}$$

where $a > 0$. This signal is sketched in Figure 4.15. The spectrum of this signal is

$$X(\omega) = \int_{-\infty}^{+\infty} e^{-a|t|}e^{-j\omega t}\, dt = \int_{-\infty}^{0} e^{at}e^{-j\omega t}\, dt + \int_{0}^{\infty} e^{-at}e^{-j\omega t}\, dt$$

$$= \frac{1}{a - j\omega} + \frac{1}{a + j\omega}$$

$$= \frac{2a}{a^2 + \omega^2}$$

In this case $X(\omega)$ is real, and it is illustrated in Figure 4.16.

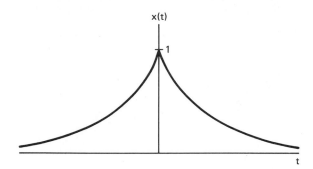

Figure 4.15 Signal $x(t) = e^{-a|t|}$ of Example 4.8.

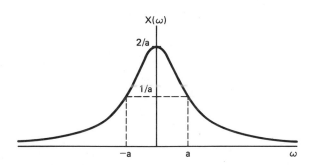

Figure 4.16 Fourier transform of the signal considered in Example 4.8 and depicted in Figure 4.15.

Example 4.9

Now let us determine the spectrum of the unit impulse

$$x(t) = \delta(t) \tag{4.65}$$

Substituting into eq. (4.61) we see that

$$X(\omega) = \int_{-\infty}^{+\infty} \delta(t)e^{-j\omega t}\, dt = 1 \tag{4.66}$$

That is, the unit impulse has a Fourier transform representation consisting of equal contributions at *all* frequencies.

Example 4.10

Consider the rectangular pulse signal

$$x(t) = \begin{cases} 1, & |t| < T_1 \\ 0, & |t| > T_1 \end{cases} \tag{4.67}$$

as shown in Figure 4.17(a). Applying eq. (4.61), we find that the Fourier transform of this signal is

$$X(\omega) = \int_{-T_1}^{T_1} e^{-j\omega t}\, dt = 2\frac{\sin \omega T_1}{\omega} \tag{4.68}$$

as sketched in Figure 4.17(b).

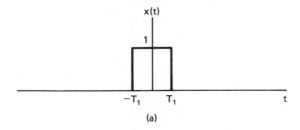

(a)

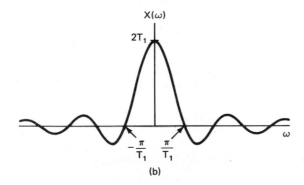

(b)

Figure 4.17 The rectangular pulse signal of Example 4.9 and its Fourier transform.

As we discussed at the beginning of this section, the signal given by eq. (4.67) can be thought of as the limiting form of a periodic square wave as the period becomes arbitrarily large. Therefore, we might expect that the convergence of the synthesis equation for this signal would behave in a manner similar to that observed for the square wave (see Example 4.6). This is, in fact, the case. Specifically, consider the inverse Fourier transform representation for the rectangular pulse signal:

$$\hat{x}(t) = \frac{1}{2\pi} \int_{-\infty}^{+\infty} 2\frac{\sin \omega T_1}{\omega} e^{j\omega t}\, d\omega$$

Then, since $x(t)$ is square-integrable, we know that

$$\int_{-\infty}^{+\infty} |x(t) - \hat{x}(t)|^2\, dt = 0$$

Furthermore, because $x(t)$ satisfies the Dirichlet conditions, we know that $\hat{x}(t) = x(t)$ except at the points of discontinuity, $t = \pm T_1$, where it converges to $\frac{1}{2}$, which is the average of the values of $x(t)$ on both sides of the discontinuity. In addition, the convergence of $\hat{x}(t)$ to $x(t)$ exhibits the Gibbs phenomenon, much as was illustrated for the periodic square wave in Figure 4.10. Specifically, in analogy with the finite Fourier series approximation of eq. (4.42), consider the following integral over a finite length interval of frequencies:

$$\frac{1}{2\pi} \int_{-W}^{W} 2\frac{\sin \omega T_1}{\omega} e^{j\omega t} d\omega$$

Then as $W \longrightarrow \infty$, this signal converges to $x(t)$ everywhere except at the discontinuities. Moreover, this signal exhibits ripples near the discontinuities. The peak amplitude of these ripples does not decrease as W increases, although the ripples do become compressed toward the discontinuity, and the energy in the ripples converges to zero.

Example 4.11

Consider the signal $x(t)$ whose Fourier transform is given by

$$X(\omega) = \begin{cases} 1, & |\omega| < W \\ 0, & |\omega| > W \end{cases} \tag{4.69}$$

This transform is illustrated in Figure 4.18(b). Using the synthesis equation (4.60), we can determine $x(t)$:

$$x(t) = \frac{1}{2\pi} \int_{-W}^{W} e^{j\omega t} d\omega = \frac{\sin Wt}{\pi t} \tag{4.70}$$

which is depicted in Figure 4.18(a).

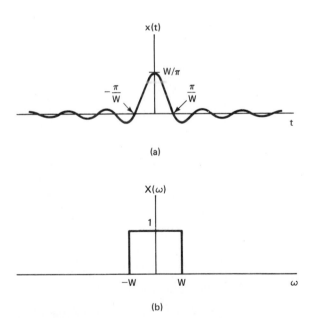

(a)

(b)

Figure 4.18 Fourier transform pair of Example 4.11.

Comparing Figures 4.17 and 4.18 or, equivalently, eqs. (4.67) and (4.68) with eqs. (4.69) and (4.70), we see an interesting relationship. In each case the Fourier transform pair consists of a $(\sin x)/x$ function and a rectangular pulse. However in Example 4.10 it is the *signal* $x(t)$ that is a pulse, while in Example 4.11, it is the *transform* $X(\omega)$. The special relationship that is apparent here is a direct consequence of the *duality property* for Fourier transforms, which we discuss in detail in Section 4.6.6.

The functions given in eqs. (4.68) and (4.70) and others of this same general form play a very important role in Fourier analysis and in the study of LTI systems. For this reason one such function has been given a special name, the *sinc function*, which is defined as

$$\text{sinc}(x) = \frac{\sin \pi x}{\pi x} \tag{4.71}$$

and which is plotted in Figure 4.19. Both of the signals in eqs. (4.68) and (4.70) can be expressed in terms of the sinc function:

$$\frac{2 \sin \omega T_1}{\omega} = 2T_1 \, \text{sinc}\left(\frac{\omega T_1}{\pi}\right)$$

$$\frac{\sin Wt}{\pi t} = \frac{W}{\pi} \, \text{sinc}\left(\frac{Wt}{\pi}\right)$$

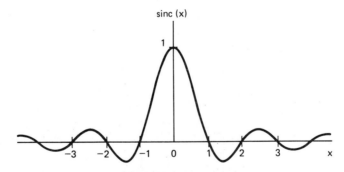

Figure 4.19 The sinc function.

As one last comment we note that we can gain some insight into one other property of the Fourier transform by examining Figure 4.18, which we have redrawn in Figure 4.20 for several different values of W. From this figure we see that as W increases, $X(\omega)$ becomes broader while the main peak of $x(t)$ at $t = 0$ becomes higher and the width of the first lobe of this signal (i.e., the part of the signal for $|t| < \pi/W$) becomes narrower. In fact, in the limit as $W \to \infty$, $X(\omega) = 1$ for all W, and consequently from Example 4.9, we see that $x(t)$ in eq. (4.70) converges to an impulse as $W \to \infty$. The behavior depicted in Figure 4.20 is an example of the inverse relationship that exists between the time and frequency domains, and we can see a similar effect in Figure 4.17, where an increase in T_1 broadens $x(t)$ but makes $X(\omega)$ narrower. In Section 4.6.5 we provide an explanation for this behavior in the context of the scaling property of the Fourier transform.

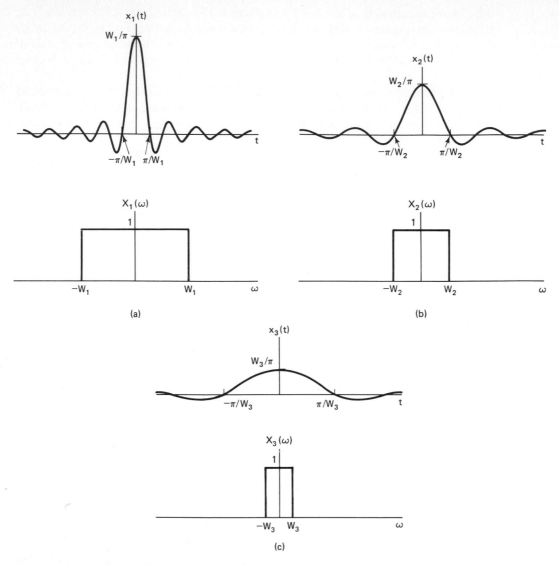

Figure 4.20 Fourier transform pair of Figure 4.18 for several different values of W.

4.5 PERIODIC SIGNALS AND THE CONTINUOUS-TIME FOURIER TRANSFORM

In the preceding section we developed the Fourier transform for aperiodic signals by considering the behavior of the Fourier series of periodic signals as the period is made arbitrarily long. As this result indicates, the Fourier series and Fourier transform representations are closely related, and in this section we investigate this relationship further and also develop a Fourier transform representation for periodic signals.

4.5.1 Fourier Series Coefficients as Samples of the Fourier Transform of One Period

As a first step, recall from our derivation of the Fourier transform that the important observation that we made was that the Fourier coefficients of a periodic signal $\tilde{x}(t)$ could be obtained from samples of an envelope that we found to be equal to the Fourier transform of an aperiodic signal $x(t)$ which is equal to one period of $\tilde{x}(t)$. Specifically, let $\tilde{x}(t)$ have fundamental period T_0, as illustrated in Figure 4.21. As we saw in the preceding section, if $x(t)$ is taken as

$$x(t) = \begin{cases} \tilde{x}(t), & -\dfrac{T_0}{2} \leq t \leq \dfrac{T_0}{2} \\ 0, & t < -\dfrac{T_0}{2} \quad \text{or} \quad t > \dfrac{T_0}{2} \end{cases} \tag{4.72}$$

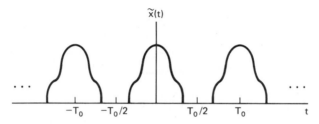

Figure 4.21 Periodic signal.

then the Fourier coefficients a_k of $\tilde{x}(t)$ can be expressed in terms of samples of the Fourier transform $X(\omega)$ of $x(t)$:

$$a_k = \frac{1}{T_0} \int_{-T_0/2}^{T_0/2} \tilde{x}(t) e^{-jk\omega_0 t} \, dt = \frac{1}{T_0} \int_{-T_0/2}^{T_0/2} x(t) e^{-jk\omega_0 t} \, dt$$

$$= \frac{1}{T_0} \int_{-\infty}^{+\infty} x(t) e^{-jk\omega_0 t} \, dt = \frac{1}{T_0} X(k\omega_0) \tag{4.73}$$

However, since the Fourier coefficients a_k can be obtained by integrating over *any* interval of length T_0 [see eq. (4.35)], we can actually obtain a more general statement than that given in eq. (4.73). Specifically, let s be an arbitrary point in time, and define the signal $x(t)$ to be equal to $\tilde{x}(t)$ over the interval $s \leq t \leq s + T_0$ and zero elsewhere. That is,

$$x(t) = \begin{cases} \tilde{x}(t), & s \leq t \leq s + T_0 \\ 0, & t < s \quad \text{or} \quad t > s + T_0 \end{cases} \tag{4.74}$$

Then the Fourier series coefficients of $\tilde{x}(t)$ are given by

$$a_k = \frac{1}{T_0} X(k\omega_0) \tag{4.75}$$

where $X(\omega)$ is the Fourier transform of $x(t)$ as defined in eq. (4.74). Note that eq. (4.75) is valid for *any* choice of s, and not just the choice of $s = -T_0/2$ used in eqs. (4.72) and (4.73). This does *not*, however, mean that the transform $X(\omega)$ is the same for all values of s, but it does imply that the set of *samples* $X(k\omega_0)$ is independent of s.

Rather than provide a demonstration of the validity of eq. (4.75) in general, we illustrate it by means of the following example.

Example 4.12

Let $\tilde{x}(t)$ be the periodic square wave with period T_0 illustrated in Figure 4.22(a), and let $x_1(t)$ and $x_2(t)$ be as depicted in Figure 4.22(b) and (c). These signals are each equal

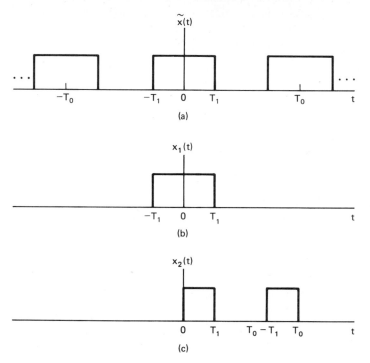

(a)

(b)

(c)

Figure 4.22 (a) Periodic square wave $\tilde{x}(t)$; (b, c) two aperiodic signals each of which equals $\tilde{x}(t)$ over a different interval of length T_0.

to $\tilde{x}(t)$ over different intervals of length T_0. As we saw in Example 4.10, the Fourier transform of $x_1(t)$ is given by

$$X_1(\omega) = \frac{2 \sin \omega T_1}{\omega} \tag{4.76}$$

The Fourier transform of $x_2(t)$ can be calculated from eq. (4.61):

$$X_2(\omega) = \int_{-\infty}^{\infty} x_2(t)e^{-j\omega t}\, dt = \int_{0}^{T_1} e^{-j\omega t}\, dt + \int_{T_0-T_1}^{T_0} e^{-j\omega t}\, dt$$

$$= \frac{1}{j\omega}[1 - e^{-j\omega T_1}] + \frac{1}{j\omega}e^{-j\omega T_0}[e^{j\omega T_1} - 1]$$

$$= \frac{1}{j\omega}e^{-j\omega T_1/2}[e^{j\omega T_1/2} - e^{-j\omega T_1/2}] + \frac{1}{j\omega}e^{-j\omega(T_0-T_1/2)}[e^{j\omega T_1/2} - e^{-j\omega T_1/2}]$$

$$= \frac{2}{\omega}\sin\left(\frac{\omega T_1}{2}\right)[e^{-j\omega T_1/2} + e^{-j\omega(T_0-T_1/2)}] \tag{4.77}$$

The transforms $X_1(\omega)$ and $X_2(\omega)$ are definitely not equal. In fact, $X_1(\omega)$ is real for all values of ω, whereas $X_2(\omega)$ is not. However, for $\omega = k\omega_0$, eq. (4.77) becomes

Fourier Analysis for Continuous-Time Signals and Systems   Chap. 4

$$X_2(k\omega_0) = \frac{2}{k\omega_0} \sin\left(\frac{k\omega_0 T_1}{2}\right)[e^{-jk\omega_0 T_1/2} + e^{-jk\omega_0 T_0}e^{jk\omega_0 T_1/2}]$$

Since $\omega_0 T_0 = 2\pi$, this reduces to

$$X_2(k\omega_0) = \frac{2}{k\omega_0} \sin\left(\frac{k\omega_0 T_1}{2}\right)[e^{-jk\omega_0 T_1/2} + e^{jk\omega_0 T_1/2}]$$

$$= \frac{4}{k\omega_0} \sin\left(\frac{k\omega_0 T_1}{2}\right) \cos\left(\frac{k\omega_0 T_1}{2}\right)$$

Then using the trigonometric identity $\sin 2x = 2 (\sin x)(\cos x)$, we find that

$$X_2(k\omega_0) = \frac{2 \sin(k\omega_0 T_1)}{k\omega_0} = X_1(k\omega_0)$$

which substantiates the result stated in eq. (4.75) that the Fourier coefficients of a periodic signal can be obtained from samples of the Fourier transform of an aperiodic signal that equals the original periodic signal over *any arbitrary* interval of length T_0 and that is zero outside this interval.

4.5.2 The Fourier Transform for Periodic Signals

We now wish to consider the Fourier transform of a periodic signal. As we will see, we can construct the Fourier transform of such a signal directly from its Fourier series representation. The resulting Fourier transform for a periodic signal consists of a train of impulses in frequency, with the areas of the impulses proportional to the Fourier series coefficients. This will turn out to be a very important representation, as it will facilitate our treatment of the application of Fourier analysis techniques to problems of modulation and sampling.

To suggest the general result, let us consider a signal $x(t)$ with Fourier transform $X(\omega)$ which is a single impulse of area 2π at $\omega = \omega_0$, that is,

$$X(\omega) = 2\pi\delta(\omega - \omega_0) \tag{4.78}$$

To determine the signal $x(t)$ for which this is the Fourier transform we can apply the inverse transform relation (4.60) to obtain

$$x(t) = \frac{1}{2\pi} \int_{-\infty}^{+\infty} 2\pi\delta(\omega - \omega_0)e^{j\omega t}\, d\omega$$

$$= e^{j\omega_0 t}$$

More generally, if $X(\omega)$ is of the form of a linear combination of impulses equally spaced in frequency, that is,

$$X(\omega) = \sum_{k=-\infty}^{+\infty} 2\pi a_k \delta(\omega - k\omega_0) \tag{4.79}$$

then the application of eq. (4.60) yields

$$x(t) = \sum_{k=-\infty}^{+\infty} a_k e^{jk\omega_0 t} \tag{4.80}$$

We see that eq. (4.80) corresponds exactly to the Fourier *series* representation of a periodic signal, as specified by eq. (4.34). Thus, the Fourier transform of a periodic signal with Fourier series coefficients $\{a_k\}$ can be interpreted as a train of impulses occurring at the harmonically related frequencies and for which the area of the impulse at the kth harmonic frequency $k\omega_0$ is 2π times the kth Fourier series coefficient a_k.

Example 4.13

Consider again the square wave illustrated in Figure 4.22(a). The Fourier series coefficients for this signal are

$$a_k = \frac{\sin k\omega_0 T_1}{\pi k}$$

and its Fourier transform is

$$X(\omega) = \sum_{k=-\infty}^{+\infty} \frac{2 \sin k\omega_0 T_1}{k} \delta(\omega - k\omega_0)$$

which is sketched in Figure 4.23 for $T_0 = 4T_1$. In comparison with Figure 4.8, the only differences are a proportionality factor of 2π and the use of impulses rather than a bar graph.

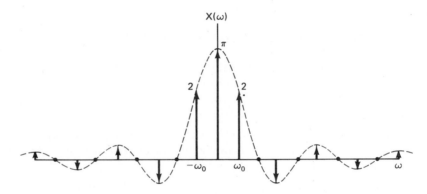

Figure 4.23   Fourier transform of a symmetric periodic square wave.

Example 4.14

Let

$$x(t) = \sin \omega_0 t$$

The Fourier series coefficients for this example are

$$a_1 = \frac{1}{2j}$$

$$a_{-1} = -\frac{1}{2j}$$

$$a_k = 0, \qquad k \neq 1 \text{ or } -1$$

Thus, the Fourier transform is as shown in Figure 4.24(a). Similarly, for

$$x(t) = \cos \omega_0 t$$

the Fourier series coefficients are

$$a_1 = a_{-1} = \tfrac{1}{2}$$

$$a_k = 0, \qquad k \neq 1 \text{ or } -1$$

The Fourier transform of this signal is depicted in Figure 4.24(b). These two transforms will be of great importance when we analyze modulation systems in Chapter 7.

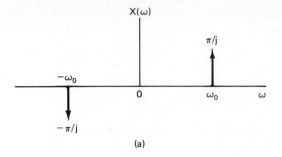

(a)

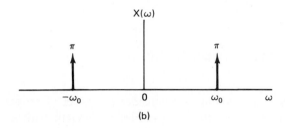

(b)

Figure 4.24 Fourier transforms of (a) $x(t) = \sin \omega_0 t$; (b) $x(t) = \cos \omega_0 t$.

Example 4.15

A signal that we will find extremely useful in our analysis of sampling systems in Chapter 8 is the periodic impulse train given by

$$x(t) = \sum_{k=-\infty}^{+\infty} \delta(t - kT)$$

as drawn in Figure 4.25(a). This signal is periodic with fundamental period T.

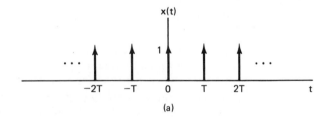

(a)

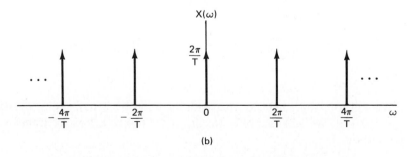

(b)

Figure 4.25 (a) Periodic impulse train; (b) its Fourier transform.

To determine the Fourier transform of this signal, we first compute its Fourier series coefficients:

$$a_k = \frac{1}{T} \int_{-T/2}^{+T/2} \delta(t) e^{-jk\omega_0 t} \, dt = \frac{1}{T}$$

Inserting this into eq. (4.79) gives

$$X(\omega) = \frac{2\pi}{T} \sum_{k=-\infty}^{+\infty} \delta\left(\omega - \frac{2\pi k}{T}\right)$$

The transform of an impulse train in time is thus itself an impulse train in frequency, as sketched in Figure 4.25(b). Here again we see an illustration of the relationship between the time and the frequency domains. As the spacing between the impulses in time (i.e., the period) gets longer, the spacing between impulses in frequency (the fundamental frequency) gets smaller.

4.6 PROPERTIES OF THE CONTINUOUS-TIME FOURIER TRANSFORM

In this and in the following two sections, we consider a number of properties of the Fourier transform. As we shall see, these properties provide us with a significant amount of insight into the transform and into the relationship between the time-domain and frequency-domain descriptions of a signal. In addition, many of these properties are often useful in reducing the complexity of the evaluation of Fourier transforms or inverse transforms. Furthermore, as described in the preceding section, there is a close relationship between the Fourier series and Fourier transform representations of a periodic signal, and using this relationship it is possible to translate many of the properties we develop for Fourier transforms into corresponding properties for Fourier series. For this reason we have omitted the derivation of these Fourier series properties. Several of the omitted derivations are considered in Problem 4.9. In Section 4.9 we summarize all the properties for both the series and transform and in addition provide a table of the series and transform representations of some of the basic signals that we have already encountered or will find of use in the remainder of this chapter and in subsequent chapters.

Throughout this discussion we will be referring frequently to time functions and their Fourier transforms, and we will find it convenient to use a shorthand notation to indicate the relationship between a signal and its transform. Recall that a signal $x(t)$ and its Fourier transform $X(\omega)$ are related by the Fourier transform synthesis and analysis equations

[eq. (4.60)] $$x(t) = \frac{1}{2\pi} \int_{-\infty}^{+\infty} X(\omega) e^{j\omega t} \, d\omega \qquad (4.81)$$

[eq. (4.61)] $$X(\omega) = \int_{-\infty}^{+\infty} x(t) e^{-j\omega t} \, dt \qquad (4.82)$$

We will sometimes refer to $X(\omega)$ with the notation $\mathcal{F}\{x(t)\}$ and to $x(t)$ with the notation $\mathcal{F}^{-1}\{X(\omega)\}$. We will also refer to $x(t)$ and $X(\omega)$ as a Fourier transform pair with the notation

$$x(t) \overset{\mathcal{F}}{\longleftrightarrow} X(\omega)$$

Thus, with reference to Example 4.7,

$$\frac{1}{a + j\omega} = \mathcal{F}\{e^{-at}u(t)\}$$

$$e^{-at}u(t) = \mathcal{F}^{-1}\left\{\frac{1}{a + j\omega}\right\}$$

and

$$e^{-at}u(t) \quad \overset{\mathcal{F}}{\longleftrightarrow} \quad \frac{1}{a + j\omega}$$

4.6.1 Linearity of the Fourier Transform

If

$$x_1(t) \quad \overset{\mathcal{F}}{\longleftrightarrow} \quad X_1(\omega)$$

and

$$x_2(t) \quad \overset{\mathcal{F}}{\longleftrightarrow} \quad X_2(\omega)$$

then

$$\boxed{ax_1(t) + bx_2(t) \quad \overset{\mathcal{F}}{\longleftrightarrow} \quad aX_1(\omega) + bX_2(\omega)} \tag{4.83}$$

In words, the Fourier transform of a linear combination of two signals is the same linear combination of the transforms of the individual components. The linearity property is easily extended to a linear combination of an arbitrary number of components. The proof of eq. (4.83) follows directly by application of eq. (4.82).

4.6.2 Symmetry Properties of the Fourier Transform

If $x(t)$ is a real-valued time function, then

$$\boxed{X(-\omega) = X^*(\omega) \qquad [x(t) \text{ real}]} \tag{4.84}$$

where * denotes the complex conjugate. This is referred to as *conjugate symmetry*. The conjugate symmetry of the Fourier transform follows by evaluating the complex conjugate of eq. (4.82):

$$X^*(\omega) = \left[\int_{-\infty}^{+\infty} x(t)e^{-j\omega t}\, dt\right]^*$$

$$= \int_{-\infty}^{+\infty} x^*(t)e^{j\omega t}\, dt$$

Using the fact that $x(t)$ is real so that $x^*(t) = x(t)$, we have

$$X^*(\omega) = \int_{-\infty}^{+\infty} x(t)e^{j\omega t}\, dt = X(-\omega)$$

where the second equality follows from eq. (4.82) evaluated at $-\omega$.

Referring to Example 4.7, with $x(t) = e^{-at}u(t)$,

$$X(\omega) = \frac{1}{a + j\omega}$$

and

$$X(-\omega) = \frac{1}{a - j\omega} = X^*(\omega)$$

Also, in Section 4.2 we discussed the Fourier series analog of this property. Specifically, if $x(t)$ is periodic and real, then from eq. (4.17),

$$a_k = a^*_{-k}$$

An illustration of this property was given in Example 4.2.

As one consequence of eq. (4.84), if we express $X(\omega)$ in rectangular form as

$$X(\omega) = \mathcal{R}e\{X(\omega)\} + j\mathcal{I}m\{X(\omega)\}$$

then if $x(t)$ is real,

$$\mathcal{R}e\{X(\omega)\} = \mathcal{R}e\{X(-\omega)\}$$
$$\mathcal{I}m\{X(\omega)\} = -\mathcal{I}m\{X(-\omega)\}$$

That is, the real part is an *even* function of frequency and the imaginary part is an *odd* function of frequency. Similarly, if we express $X(\omega)$ in polar form as

$$X(\omega) = |X(\omega)| e^{j\theta(\omega)}$$

then it follows from eq. (4.84) that $|X(\omega)|$ is an even function of ω and $\theta(\omega)$ is an odd function of ω. Thus, when computing or displaying the Fourier transform of a real-valued function of time, the real and imaginary parts or magnitude and phase of the transform need only be generated or displayed for positive frequencies, as the values for negative frequencies can be determined directly from the values for $\omega > 0$ using the relationships just derived.

As a further consequence of eq. (4.84), if $x(t)$ is both real and even, then $X(\omega)$ will also be real and even. To see this, we write

$$X(-\omega) = \int_{-\infty}^{+\infty} x(t)e^{j\omega t}\, dt$$

or with the substitution of variables $\tau = -t$,

$$X(-\omega) = \int_{-\infty}^{+\infty} x(-\tau)e^{-j\omega \tau}\, d\tau$$

Since $x(-\tau) = x(\tau)$, we have

$$X(-\omega) = \int_{-\infty}^{+\infty} x(\tau)e^{-j\omega \tau}\, d\tau$$
$$= X(\omega)$$

Thus, $X(\omega)$ is an even function. This, together with eq. (4.84), also requires that $X^*(\omega) = X(\omega)$ [i.e., that $X(\omega)$ is real]. Example 4.8 illustrates this property for the real, even signal $e^{-a|t|}$. In a similar manner it can be shown that if $x(t)$ is an odd time function so that $x(t) = -x(-t)$, then $X(\omega)$ is pure imaginary and odd.

The analogous property for Fourier series is that a periodic, real, even signal $x(t)$ has real and even Fourier coefficients (i.e., $a_k = a_{-k}$), whereas if $x(t)$ is odd, the coefficients are odd and pure imaginary. The first of these two cases is illustrated in Example 4.5 for the periodic square wave.

Finally, as was discussed in Chapter 2, a real function $x(t)$ can always be expressed in terms of the sum of an even function $x_e(t) = \mathcal{E}v\{x(t)\}$ and an odd function $x_o(t) = \mathcal{O}d\{x(t)\}$: that is,

$$x(t) = x_e(t) + x_o(t)$$

From the linearity of the Fourier transform,

$$\mathcal{F}\{x(t)\} = \mathcal{F}\{x_e(t)\} + \mathcal{F}\{x_o(t)\}$$

and from the discussion above, $\mathcal{F}\{x_e(t)\}$ is a real function and $\mathcal{F}\{x_o(t)\}$ is pure imaginary. Thus, we can conclude that with $x(t)$ real,

$$x(t) \quad \overset{\mathcal{F}}{\longleftrightarrow} \quad X(\omega)$$

$$\mathcal{E}v\{x(t)\} \quad \overset{\mathcal{F}}{\longleftrightarrow} \quad \mathcal{R}e\{X(\omega)\}$$

$$\mathcal{O}d\{x(t)\} \quad \overset{\mathcal{F}}{\longleftrightarrow} \quad j\mathcal{I}m\{X(\omega)\}$$

4.6.3 Time Shifting

If

$$x(t) \quad \overset{\mathcal{F}}{\longleftrightarrow} \quad X(\omega)$$

then

$$x(t - t_0) \quad \overset{\mathcal{F}}{\longleftrightarrow} \quad e^{-j\omega t_0} X(\omega) \qquad (4.85)$$

To establish this property, consider

$$\mathcal{F}\{x(t - t_0)\} = \int_{-\infty}^{+\infty} x(t - t_0) e^{-j\omega t}\, dt \qquad (4.86)$$

Letting $\sigma = t - t_0$ in (4.86), we have

$$\mathcal{F}\{x(t - t_0)\} = \int_{-\infty}^{+\infty} x(\sigma) e^{-j\omega(\sigma + t_0)}\, d\sigma = e^{-j\omega t_0} X(\omega)$$

One consequence of this property is that a signal which is shifted in time does not have the *magnitude* of its Fourier transform altered. That is, if we express $X(\omega)$ in polar form as

$$\mathcal{F}\{x(t)\} = X(\omega) = |X(\omega)| e^{j\theta(\omega)}$$

then

$$\mathcal{F}\{x(t - t_0)\} = e^{-j\omega t_0} X(\omega) = |X(\omega)| e^{j[\theta(\omega) - \omega t_0]}$$

Thus, the effect of a time shift on a signal is to introduce a phase shift in its transform which is a linear function of ω.

4.6.4 Differentiation and Integration

Let $x(t)$ be a signal with Fourier transform $X(\omega)$. Then, by differentiating both sides of the Fourier transform synthesis equation (4.81), we obtain

$$\frac{dx(t)}{dt} = \frac{1}{2\pi} \int_{-\infty}^{+\infty} j\omega X(\omega) e^{j\omega t}\, d\omega$$

Therefore,

$$\frac{dx(t)}{dt} \overset{\mathcal{F}}{\longleftrightarrow} j\omega X(\omega) \qquad\qquad (4.87)$$

This is a particularly important property as it replaces the operation of differentiation in the time domain with that of multiplication by $j\omega$ in the frequency domain. We will find this to be extremely useful in our discussion in Section 4.11 on the use of Fourier transforms for the analysis of LTI systems described by differential equations.

Since differentiation in time corresponds to multiplication by $j\omega$ in the frequency domain, one might conclude that integration should involve division by $j\omega$ in the frequency domain. This is indeed the case, but it is only one part of the picture. The precise relationship is

$$\int_{-\infty}^{t} x(\tau)\, d\tau \overset{\mathcal{F}}{\longleftrightarrow} \frac{1}{j\omega} X(\omega) + \pi X(0)\, \delta(\omega) \qquad\qquad (4.88)$$

The impulse term on the right-hand side of eq. (4.88) reflects the dc or average value that can result from the integration.

To gain some understanding of this property, consider the unit step signal $u(t)$. In Figure 4.26 we have illustrated the even–odd decomposition of $u(t)$, which can

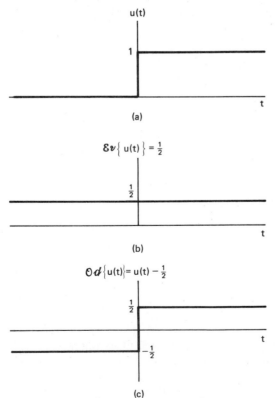

Figure 4.26 Even-odd decomposition of the continuous-time unit step.

be written as

$$u(t) = \tfrac{1}{2} + [u(t) - \tfrac{1}{2}] \tag{4.89}$$

Consider first the odd part $v(t) = u(t) - \tfrac{1}{2}$. Since $v'(t) = u'(t) = \delta(t)$, we have from the differentiation property that

$$\mathcal{F}\{\delta(t)\} = \mathcal{F}\left\{\frac{dv(t)}{dt}\right\} = j\omega V(\omega) \tag{4.90}$$

and since the Fourier transform of the unit impulse is 1, we conclude that

$$V(\omega) = \frac{1}{j\omega} \tag{4.91}$$

Note that since $v(t)$ is real and odd, we must have that $V(\omega)$ is pure imaginary and odd, which is readily verified from eq. (4.91).

Consider next the even part of $u(t)$, which is the constant signal $\tfrac{1}{2}$. That is, $\mathcal{E}v\{u(t)\}$ is a periodic signal at zero frequency, and hence its Fourier transform is an impulse at $\omega = 0$. Specifically, using eq. (4.79), we obtain

$$\mathcal{F}\{\tfrac{1}{2}\} = \pi\delta(\omega) \tag{4.92}$$

and combining the transforms of the even and odd parts of $u(t)$, we find that

$$\mathcal{F}\{u(t)\} = \frac{1}{j\omega} + \pi\delta(\omega) \tag{4.93}$$

This result agrees with the integration property, eq. (4.88). That is, with $x(t) = \delta(t)$, we have that $X(\omega) = 1$ and the integral of $x(t)$ is $u(t)$. With these substitutions, eq. (4.88) yields eq. (4.93). Furthermore, we have seen that the impulse term in eq. (4.93) comes directly from the nonzero dc value of $u(t)$.

Finally, note that we can apply the differentiation property (4.87) to recover the transform of the impulse, that is,

$$\delta(t) = \frac{du(t)}{dt} \quad \overset{\mathcal{F}}{\longleftrightarrow} \quad j\omega\left[\frac{1}{j\omega} + \pi\delta(\omega)\right]$$

The right-hand side of this expression reduces to 1 since $\omega\delta(\omega) = 0$, and thus

$$\delta(t) \quad \overset{\mathcal{F}}{\longleftrightarrow} \quad 1$$

4.6.5 Time and Frequency Scaling

If

$$x(t) \quad \overset{\mathcal{F}}{\longleftrightarrow} \quad X(\omega)$$

then

$$\boxed{x(at) \quad \overset{\mathcal{F}}{\longleftrightarrow} \quad \frac{1}{|a|}X\left(\frac{\omega}{a}\right)} \tag{4.94}$$

where a is a real constant. This property follows directly from the definition of the Fourier transform. Specifically,

$$\mathcal{F}\{x(at)\} = \int_{-\infty}^{+\infty} x(at)e^{-j\omega t}\, dt$$

Using the substitution of variables $\tau = at$, we obtain

$$\mathcal{F}\{x(at)\} = \begin{cases} \dfrac{1}{a} \displaystyle\int_{-\infty}^{+\infty} x(\tau)e^{-j(\omega/a)\tau}\, d\tau, & a > 0 \\[2em] -\dfrac{1}{a} \displaystyle\int_{-\infty}^{+\infty} x(\tau)e^{-j(\omega/a)\tau}\, d\tau, & a < 0 \end{cases}$$

which corresponds to relation (4.94). Thus, aside from the amplitude factor of $1/|a|$, linear scaling in time by a factor of a corresponds to a linear scaling in frequency by a factor of $1/a$, and vice versa. A common illustration of relation (4.94) is the effect on frequency content that results when an audio tape is recorded at one speed and played back at a different speed. If the playback speed is higher than the recording speed, corresponding to compression in time (i.e., $a > 1$), then the spectrum is expanded in frequency (i.e., the audible effect is that the playback frequencies are higher). Conversely, the played-back signal will contain lower frequencies if the playback speed is slower than the recording speed ($a < 1$). For example, if a recording of the sound of a small bell ringing is played back at a reduced speed, the result will sound like the chiming of a larger and deeper-sounding bell.

The scaling property is another example of the inverse relationship between time and frequency that we have already encountered on several occasions. For example, we have seen that as we increase the period of a sinusoidal signal, we decrease its frequency. Also, as we saw in Example 4.11 (see Figure 4.20), if we consider the transform

$$X(\omega) = \begin{cases} 1, & |\omega| < W \\ 0, & |\omega| > W \end{cases}$$

then as we increase W, the inverse transform of $X(\omega)$ becomes narrower and higher and approaches an impulse as $W \rightarrow \infty$. Finally, in Example 4.15 we saw that the spacing in frequency between impulses in the Fourier transform of a periodic impulse train is inversely proportional to the spacing in time.

On several occasions in the remainder of this book we will encounter the consequences of the inverse relationship between the time and frequency domains. In addition, the reader may very well come across the implications of this property in a wide variety of other topics in science and engineering. One example is the uncertainty principle in physics. Another such implication is illustrated in Problem 4.35.

4.6.6 Duality

By comparing the transform and inverse transform relations (4.81) and (4.82), we observe that there is a definite symmetry (i.e., that these equations are similar but not quite identical in form). In fact, this symmetry leads to a property of the Fourier transform known as *duality*. In Example 4.11 we alluded to this property when we noted the striking relationship that exists between the Fourier transform pairs of Examples 4.10 and 4.11. In the first of these examples we derived the Fourier transform pair

$$x_1(t) = \begin{cases} 1, & |t| < T_1 \\ 0, & |t| > T_1 \end{cases} \quad \overset{\mathcal{F}}{\longleftrightarrow} \quad X_1(\omega) = \frac{2\sin \omega T_1}{\omega} = 2T_1 \operatorname{sinc}\left(\frac{\omega T_1}{\pi}\right) \qquad (4.95)$$

while in Example 4.11 we considered the pair

$$x_2(t) = \frac{\sin Wt}{\pi t} = \frac{W}{\pi}\operatorname{sinc}\left(\frac{Wt}{\pi}\right) \overset{\mathcal{F}}{\longleftrightarrow} X_2(\omega) = \begin{cases} 1, & |\omega| < W \\ 0, & |\omega| > W \end{cases} \tag{4.96}$$

These two Fourier transform pairs and the relationship between them are depicted in Figure 4.27.

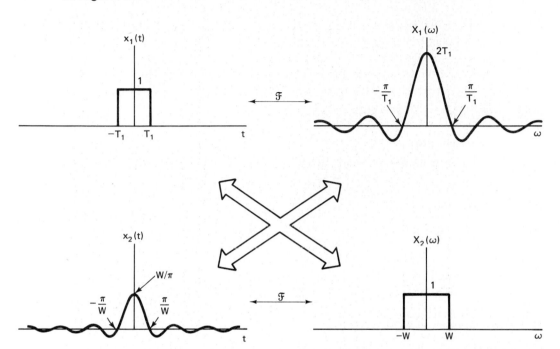

Figure 4.27 Relationship between the Fourier transform pair of eqs. (4.95) and (4.96).

The symmetry exhibited by these two examples extends to Fourier transforms in general. Specifically, consider two functions related through the integral expression

$$f(u) = \int_{-\infty}^{+\infty} g(v)e^{-juv}\,dv \tag{4.97}$$

By comparing eq. (4.97) with the Fourier synthesis and analysis equations (4.81) and (4.82), we see that with $u = \omega$ and $v = t$,

$$f(\omega) = \mathcal{F}\{g(t)\} \tag{4.98}$$

while with $u = t$ and $v = \omega$,

$$g(-\omega) = \frac{1}{2\pi}\mathcal{F}\{f(t)\} \tag{4.99}$$

That is, if we are given the Fourier transform pair for the time function $g(t)$,

$$g(t) \overset{\mathcal{F}}{\longleftrightarrow} f(\omega) \tag{4.100}$$

and then consider the function of *time* $f(t)$, its Fourier transform pair is

$$f(t) \overset{\mathcal{F}}{\longleftrightarrow} 2\pi g(-\omega) \tag{4.101}$$

The implications of these last two equations are significant. For example, suppose that

$$g(v) = \begin{cases} 1, & |v| < M \\ 0, & |v| > M \end{cases} \tag{4.102}$$

Then, from eq. (4.97),

$$f(u) = \frac{2 \sin uM}{u} = 2M \text{ sinc} \left(\frac{uM}{\pi} \right) \tag{4.103}$$

This result, together with eq. (4.98) or, equivalently, eq. (4.100), yields the transform pair in eq. (4.95) for $M = T_1$, while if we use eq. (4.99) or (4.101) we obtain the pair in eq. (4.96) with $M = W$. Therefore, the property of duality allows us to obtain both of these dual transform pairs from one evaluation of eq. (4.97). This can often be useful in reducing the complexity of the calculations involved in determining transforms and inverse transforms. To illustrate the use of duality, we consider the following example.

Example 4.16

Suppose that we would like to calculate the Fourier transform of the signal

$$x(t) = \frac{2}{t^2 + 1} \tag{4.104}$$

If we let

$$f(u) = \frac{2}{u^2 + 1}$$

then from eq. (4.100) we have the Fourier transform pair

$$g(t) \quad \overset{\mathcal{F}}{\longleftrightarrow} \quad f(\omega) = \frac{2}{\omega^2 + 1}$$

From Example 4.8 we then see that

$$g(t) = e^{-|t|}$$

Furthermore, using the transform pair given by eq. (4.101), we conclude that since $f(t) = x(t)$, then

$$X(\omega) = \mathcal{F}\{f(t)\} = 2\pi g(-\omega) = 2\pi e^{-|\omega|} \tag{4.105}$$

The duality property can also be used to determine or to suggest other properties of Fourier transforms. Specifically, if there are characteristics of a time function that have implications with regard to the Fourier transform, then the same characteristics associated with a frequency function will have *dual* implications in the time domain. For example, we know that a periodic time function has a Fourier transform that is a train of weighted, equally spaced impulses. Because of duality, a *time* function that is a train of weighted, equally spaced impulses will have a Fourier transform that is periodic in frequency. This is a consequence of eqs. (4.98) and (4.99) and can be verified directly from eqs. (4.81) and (4.82). Similarly, the properties of the Fourier transform considered in Sections 4.6.2–4.6.5 also imply dual properties. For example, in Section 4.6.4 we saw that differentiation in the time domain corresponds to multiplication by $j\omega$ in the frequency domain. From the preceding discussion we might then suspect that multiplication by jt in the time domain corresponds roughly to differentiation in the frequency domain. To determine the precise form of this dual property, we can proceed in a fashion exactly analogous to that used in Section

4.6.4. Specifically, if we differentiate the analysis equation (4.82) with respect to ω, we obtain

$$\frac{dX(\omega)}{d\omega} = \int_{-\infty}^{+\infty} -jtx(t)e^{-j\omega t}\, dt \tag{4.106}$$

That is,

$$-jtx(t) \quad \overset{\mathcal{F}}{\longleftrightarrow} \quad \frac{dX(\omega)}{d\omega} \tag{4.107}$$

Similarly, we can derive the dual properties to eqs. (4.85) and (4.88). These are

$$e^{j\omega_0 t}x(t) \quad \overset{\mathcal{F}}{\longleftrightarrow} \quad X(\omega - \omega_0) \tag{4.108}$$

and

$$-\frac{1}{jt}x(t) + \pi x(0)\,\delta(t) \quad \overset{\mathcal{F}}{\longleftrightarrow} \quad \int_{-\infty}^{\omega} X(\eta)\, d\eta \tag{4.109}$$

4.6.7 Parseval's Relation

If $x(t)$ and $X(\omega)$ are a Fourier transform pair, then

$$\int_{-\infty}^{+\infty} |x(t)|^2\, dt = \frac{1}{2\pi} \int_{-\infty}^{+\infty} |X(\omega)|^2\, d\omega \tag{4.110}$$

This expression, referred to as Parseval's relation, follows from direct application of the Fourier transform. Specifically,

$$\int_{-\infty}^{+\infty} |x(t)|^2\, dt = \int_{-\infty}^{+\infty} x(t)x^*(t)\, dt = \int_{-\infty}^{+\infty} x(t) \left[\frac{1}{2\pi} \int_{-\infty}^{+\infty} X^*(\omega)e^{-j\omega t}\, d\omega \right] dt$$

Reversing the order of integration gives

$$\int_{-\infty}^{+\infty} |x(t)|^2\, dt = \frac{1}{2\pi} \int_{-\infty}^{+\infty} X^*(\omega) \left[\int_{-\infty}^{+\infty} x(t)e^{-j\omega t}\, dt \right] d\omega$$

But the bracketed term is simply the Fourier transform of $x(t)$; thus,

$$\int_{-\infty}^{+\infty} |x(t)|^2\, dt = \frac{1}{2\pi} \int_{-\infty}^{+\infty} |X(\omega)|^2\, d\omega$$

The quantity on the left-hand side of eq. (4.110) is the total energy in the signal $x(t)$. Parseval's relation, eq. (4.110), says that this total energy may be determined either by computing the energy per unit time ($|x(t)|^2$) and integrating over all time, or by computing the energy per unit frequency ($|X(\omega)|^2/2\pi$) and integrating over all frequencies. For this reason $|X(\omega)|^2$ is often referred to as the *energy-density spectrum* of the signal $x(t)$ (see also Problem 6.6).

The energy in a periodic signal is infinite and consequently eq. (4.110) is not useful for that class of signals. However, as considered in Problem 4.14, for periodic signals there is an analogous relationship. Specifically,

$$\boxed{\frac{1}{T_0} \int_{T_0} |x(t)|^2 \, dt = \sum_{k=-\infty}^{+\infty} |a_k|^2} \tag{4.111}$$

where the a_k are the Fourier series coefficients of $x(t)$ and T_0 is its period. Thus, for the periodic case Parseval's relation relates the energy in one period of the time function to the energy in the Fourier series coefficients. Furthermore, the quantity $|a_k|^2$ has the interpretation as that part of the energy per period contributed by the kth harmonic.

There are many other properties of the Fourier transform pair in addition to those we have already discussed. In the next two sections we present two specific properties that play particularly central roles in the study of LTI systems and their applications. The first of these, discussed in Section 4.7, is referred to as the *convolution property* and it forms the basis for our discussion of filtering in Chapter 6. The second, discussed in Section 4.8, is referred to as the *modulation property* and it provides the foundation for our discussion of modulation in Chapter 7 and sampling in Chapter 8. In Section 4.9 we summarize the properties of the Fourier transform.

4.7 THE CONVOLUTION PROPERTY

One of the most important properties of the Fourier transform with regard to its use in dealing with LTI systems is its effect on the convolution operation. To derive the relation, consider an LTI system with impulse response $h(t)$, output $y(t)$, and input $x(t)$, so that

$$y(t) = \int_{-\infty}^{+\infty} x(\tau)h(t-\tau) \, d\tau \tag{4.112}$$

We desire $Y(\omega)$, which is

$$Y(\omega) = \mathcal{F}\{y(t)\} = \int_{-\infty}^{+\infty} \int_{-\infty}^{+\infty} [x(\tau)h(t-\tau) \, d\tau]e^{-j\omega t} \, dt \tag{4.113}$$

Interchanging the order of integration and noting that $x(\tau)$ does not depend on t, we have

$$Y(\omega) = \int_{-\infty}^{+\infty} x(\tau) \left[\int_{-\infty}^{+\infty} h(t-\tau)e^{-j\omega t} \, dt \right] d\tau \tag{4.114}$$

By the shifting property (4.85), the bracketed term is simply $e^{-j\omega\tau}H(\omega)$. Substituting this into eq. (4.114) yields

$$Y(\omega) = \int_{-\infty}^{+\infty} x(\tau)e^{-j\omega\tau}H(\omega) \, d\tau = H(\omega) \int_{-\infty}^{+\infty} x(\tau)e^{-j\omega\tau} \, d\tau \tag{4.115}$$

The integral is $\mathcal{F}\{x(t)\}$, and hence

$$Y(\omega) = H(\omega)X(\omega)$$

That is,

$$\boxed{y(t) = h(t) * x(t) \quad \overset{\mathscr{F}}{\longleftrightarrow} \quad Y(\omega) = H(\omega)X(\omega)} \tag{4.116}$$

This property is essentially a consequence of the fact that complex exponentials are eigenfunctions of LTI systems, and can alternatively be derived by recalling our interpretation of the Fourier transform synthesis equation as an expression for $x(t)$ as a linear combination of complex exponentials. Specifically, referring back to eq. (4.59), we expressed $x(t)$ as the limit of a sum, that is,

$$x(t) = \frac{1}{2\pi} \int_{-\infty}^{+\infty} X(\omega)e^{j\omega t}\, d\omega = \lim_{\omega_0 \to 0} \frac{1}{2\pi} \sum_{k=-\infty}^{+\infty} X(k\omega_0)e^{jk\omega_0 t}\omega_0 \tag{4.117}$$

The response of a linear system with impulse response $h(t)$ to a complex exponential $e^{jk\omega_0 t}$ is, from eq. (4.4), $H(k\omega_0)e^{jk\omega_0 t}$, where

$$H(k\omega_0) = \int_{-\infty}^{+\infty} h(t)e^{-jk\omega_0 t}\, dt \tag{4.118}$$

From superposition [see eq. (4.8)] we then have that

$$\frac{1}{2\pi} \sum_{k=-\infty}^{+\infty} X(k\omega_0)e^{jk\omega_0 t}\omega_0 \longrightarrow \frac{1}{2\pi} \sum_{k=-\infty}^{+\infty} X(k\omega_0)H(k\omega_0)e^{jk\omega_0 t}\omega_0$$

and thus from eq. (4.117) we see that the response of the linear system to $x(t)$ is

$$\begin{aligned} y(t) &= \lim_{\omega_0 \to 0} \frac{1}{2\pi} \sum_{k=-\infty}^{+\infty} X(k\omega_0)H(k\omega_0)e^{jk\omega_0 t}\omega_0 \\ &= \frac{1}{2\pi} \int_{-\infty}^{+\infty} X(\omega)H(\omega)e^{j\omega t}\, d\omega \end{aligned} \tag{4.119}$$

Since $y(t)$ and its Fourier transform $Y(\omega)$ are related by

$$y(t) = \frac{1}{2\pi} \int_{-\infty}^{+\infty} Y(\omega)e^{j\omega t}\, d\omega \tag{4.120}$$

we can identify $Y(\omega)$ from eq. (4.119), yielding

$$Y(\omega) = X(\omega)H(\omega) \tag{4.121}$$

as we had derived previously.

This alternative derivation of this very important property of Fourier transforms emphasizes again that $H(\omega)$, the Fourier transform of the system impulse response, is simply the change in complex amplitude experienced by a complex exponential of frequency ω, as it passes through a linear time-invariant system. The function $H(\omega)$ is generally referred to as the *frequency response* of the system, and it plays as important a role in the analysis of LTI systems as does its inverse transform, the unit impulse response. For one thing, since $h(t)$ completely characterizes an LTI system, then so must $H(\omega)$. In addition, many of the properties of LTI systems can be conveniently interpreted in terms of $H(\omega)$. For example, in Section 3.3 we saw that the impulse response of the cascade of two LTI systems is the convolution of the impulse responses of the individual systems and that the overall response does not depend on the order in which the systems are cascaded. Using eq. (4.116), we can rephrase this in terms of frequency responses. As illustrated in Figure 4.28, the overall frequency response of the cascade of two systems is simply the product of the individual frequency responses, and from this it is clear that the overall response does not depend on the order of cascade.

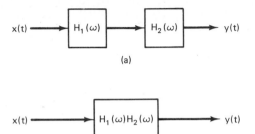

(a)

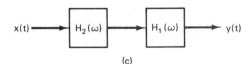

(b)

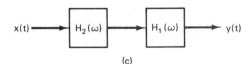

(c)

Figure 4.28 Three equivalent LTI systems. Here each block represents an LTI system with the indicated frequency response.

It is also important to note that the frequency response cannot be defined for every LTI system. If, however, an LTI system is stable, then, as we saw in Section 3.4 and Problem 3.21, its impulse response is absolutely integrable. That is,

$$\int_{-\infty}^{+\infty} |h(t)| \, dt < \infty \qquad (4.122)$$

Equation (4.122) is one of the three Dirichlet conditions which together guarantee the existence of the Fourier transform $H(\omega)$ of $h(t)$. Thus, assuming that $h(t)$ satisfies the other two conditions, as essentially all signals of physical or practical significance do, we see that a stable LTI system has a frequency response $H(\omega)$. If, however, an LTI system is unstable, that is, if

$$\int_{-\infty}^{+\infty} |h(t)| \, dt = \infty$$

then the Fourier transform may not exist, and in this case the response of the system to a sinusoidal input is infinite. (See Problem 5.27 for an illustration and explanation of this point in the context of discrete-time LTI systems.)

Therefore, in using Fourier analysis to study LTI systems, we will be restricting ourselves to systems with impulse responses that possess Fourier transforms. In order to use transform techniques to examine those unstable LTI systems that do not have finite-valued frequency responses, we will have to consider a generalization of the continuous-time Fourier transform, the Laplace transform. We defer this discussion to Chapter 9, and until then we will consider the many problems and practical applications that we can analyze using Fourier transforms.

To illustrate the convolution property further, let us consider several examples.

Example 4.17

Consider a continuous-time LTI system with impulse response

$$h(t) = \delta(t - t_0) \tag{4.123}$$

The frequency response of this system is given by

$$H(\omega) = e^{-j\omega t_0} \tag{4.124}$$

Thus, for any input $x(t)$ with Fourier transform $X(\omega)$, the Fourier transform of the output is

$$Y(\omega) = H(\omega)X(\omega)$$
$$= e^{-j\omega t_0} X(\omega) \tag{4.125}$$

This result, in fact, is consistent with the shifting property of Section 4.6.3. Specifically, as was discussed in Chapter 3 (see Example 3.5), a system for which the impulse response is $\delta(t - t_0)$ applies a time shift of t_0 to the input, that is,

$$y(t) = x(t - t_0)$$

Thus, the shifting property (4.85) also yields eq. (4.125). Note that either from our discussion in Section 4.6.3 or directly from eq. (4.124), we see that the frequency response of a system which is a pure time shift has unity magnitude at all frequencies and has a phase characteristic that is a linear function of ω.

Example 4.18

As a second example, let us examine a differentiator, that is, an **LTI** system for which the input $x(t)$ and the output $y(t)$ are related by

$$y(t) = \frac{dx(t)}{dt}$$

From the differentiation property of Section 4.6.4,

$$Y(\omega) = j\omega X(\omega) \tag{4.126}$$

Consequently, from eq. (4.116) it follows that the frequency response $H(\omega)$ of a differentiator is

$$H(\omega) = j\omega \tag{4.127}$$

Example 4.19

Suppose that we now have an integrator, that is, an **LTI** system specified by the equation

$$y(t) = \int_{-\infty}^{t} x(\tau) \, d\tau$$

The impulse response for this system is the unit step $u(t)$, and, therefore from eq. (4.93) the frequency response of this system is

$$H(\omega) = \frac{1}{j\omega} + \pi\delta(\omega)$$

Then, using eq. (4.116), we have that

$$Y(\omega) = H(\omega)X(\omega) = \frac{1}{j\omega} X(\omega) + \pi X(\omega)\,\delta(\omega) = \frac{1}{j\omega} X(\omega) + \pi X(0)\,\delta(\omega)$$

which is precisely the integration property of eq. (4.88).

Example 4.20

Consider the response of an LTI system with impulse response

$$h(t) = e^{-at}u(t), \qquad a > 0$$

to the input signal

$$x(t) = e^{-bt}u(t), \qquad b > 0$$

From Example 4.7 we have that the Fourier transforms of $x(t)$ and $h(t)$, are

$$X(\omega) = \frac{1}{b + j\omega}$$

$$H(\omega) = \frac{1}{a + j\omega}$$

Therefore,

$$Y(\omega) = \frac{1}{(a + j\omega)(b + j\omega)} \tag{4.128}$$

To determine the output $y(t)$ we wish to obtain the inverse transform of $Y(\omega)$. This is most simply done by expanding $Y(\omega)$ in a *partial fraction expansion*. As we shall see, such expansions are extremely useful in evaluating many inverse Fourier transforms, including those that arise in calculating the response of an LTI system described by a differential equation. We discuss this at more length in Section 4.11, and in the Appendix we develop the general method for performing a partial fraction expansion. We will illustrate the basic idea behind the partial fraction expansion for this example.

Suppose first that $b \neq a$. In this case we can express $Y(\omega)$ in the form

$$Y(\omega) = \frac{A}{a + j\omega} + \frac{B}{b + j\omega} \tag{4.129}$$

where A and B are constants to be determined. Setting the right-hand sides of eqs. (4.128) and (4.129) equal, we obtain

$$\frac{A}{a + j\omega} + \frac{B}{b + j\omega} = \frac{1}{(a + j\omega)(b + j\omega)}$$

or

$$A(b + j\omega) + B(a + j\omega) = 1 \tag{4.130}$$

Since this must hold for all values of ω, the coefficient of ω in the left-hand side of eq. (4.130) must be zero, and the remaining terms must sum to 1. That is,

$$A + B = 0$$

and

$$Ab + Ba = 1$$

Solving these equations, we find that

$$A = \frac{1}{b - a} = -B$$

and therefore

$$Y(\omega) = \frac{1}{b - a}\left[\frac{1}{a + j\omega} - \frac{1}{b + j\omega}\right] \tag{4.131}$$

In the Appendix we present a more efficient but equivalent method for performing partial fraction expansions in general.

The inverse transform of each of the two terms in eq. (4.131) can be recognized by inspection, and, using the linearity property of Section 4.6.1, we have

$$y(t) = \frac{1}{b-a}[e^{-at}u(t) - e^{-bt}u(t)]$$

The partial fraction expansion of eq. (4.131) is not valid if $b = a$. However, for this case,

$$Y(\omega) = \frac{1}{(a+j\omega)^2}$$

Recognizing this as

$$\frac{1}{(a+j\omega)^2} = j\frac{d}{d\omega}\left[\frac{1}{a+j\omega}\right]$$

we can use the dual of the differentiation property as given in eq. (4.107). Thus,

$$e^{-at}u(t) \quad \overset{\mathcal{F}}{\longleftrightarrow} \quad \frac{1}{a+j\omega}$$

$$te^{-at}u(t) \quad \overset{\mathcal{F}}{\longleftrightarrow} \quad j\frac{d}{d\omega}\left[\frac{1}{a+j\omega}\right] = \frac{1}{(a+j\omega)^2}$$

and consequently,

$$y(t) = te^{-at}u(t)$$

Example 4.21

Suppose that we now consider the LTI system with impulse response

$$h(t) = e^{-t}u(t)$$

and consider the input

$$x(t) = \sum_{k=-3}^{3} a_k e^{jk2\pi t}$$

where

$$a_0 = 1, \qquad a_1 = a_{-1} = \tfrac{1}{4}$$
$$a_2 = a_{-2} = \tfrac{1}{2}, \qquad a_3 = a_{-3} = \tfrac{1}{3}$$

Then

$$H(\omega) = \frac{1}{j\omega + 1}$$

$$X(\omega) = \sum_{k=-3}^{3} 2\pi a_k\, \delta(\omega - 2\pi k)$$

and thus

$$Y(\omega) = H(\omega)X(\omega) = \sum_{k=-3}^{3} 2\pi a_k H(2\pi k)\, \delta(\omega - 2\pi k)$$

$$= \sum_{k=-3}^{3} \left(\frac{2\pi a_k}{1+j2\pi k}\right) \delta(\omega - 2\pi k)$$

Converting this back to a Fourier series representation, we obtain

$$y(t) = \sum_{k=-3}^{3} \left(\frac{a_k}{1+j2\pi k}\right) e^{j2\pi kt}$$

which is identical to the response obtained in Example 4.2 [see eqs. (4.24) and (4.25)].

4.7.1 Periodic Convolution

In Example 4.21 the convolution property was applied to the convolution of a periodic signal with an aperiodic signal. If both signals are periodic, then the convolution integral does not converge. This reflects the fact that an LTI system with periodic impulse response is unstable and does not have a finite-valued frequency response. However, it is sometimes useful to consider a form of convolution for periodic signals with equal periods, referred to as *periodic convolution*. The periodic convolution of two signals $\tilde{x}_1(t)$ and $\tilde{x}_2(t)$ with common period T_0 is defined as

$$\tilde{y}(t) = \int_{T_0} \tilde{x}_1(\tau)\tilde{x}_2(t - \tau)\,d\tau \tag{4.132}$$

This operation is similar to usual convolution, which is sometimes called *aperiodic convolution*. As illustrated in Figure 4.29, we see that, just as with aperiodic convolu-

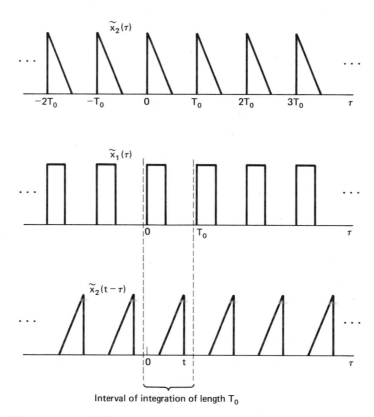

Figure 4.29 Periodic convolution of two continuous-time periodic signals.

tion, periodic convolution also involves multiplying $\tilde{x}_1(\tau)$ by a reversed and shifted version of $\tilde{x}_2(\tau)$, but in this case the product is integrated over a single period. As t changes, one period of $\tilde{x}_2(t - \tau)$ slides out of the interval of integration and the next one slides in. If t is changed by T_0, then the periodic signal $\tilde{x}_2(t - \tau)$ will have been shifted through a full period and therefore will look exactly as it did before the shift. From this we can deduce that the result of periodic convolution is a periodic signal $\tilde{y}(t)$. Furthermore, as shown in Problem 3.12, the result of periodic convolution does not depend on which interval of length T_0 is chosen for the integration in eq. (4.132), and also from Problem 4.16 we have that if $\{a_k\}$, $\{b_k\}$, and $\{c_k\}$ denote the Fourier series coefficients of $\tilde{x}_1(t)$, $\tilde{x}_2(t)$, and $\tilde{y}(t)$, respectively, then

$$c_k = T_0 a_k b_k \tag{4.133}$$

which is the counterpart of the convolution property for periodic convolution.

4.8 THE MODULATION PROPERTY

The convolution property states that convolution in the *time* domain corresponds to multiplication in the *frequency* domain. Because of the duality between the time and frequency domains, we would expect a dual property to also hold (i.e., that multiplication in the time domain corresponds to convolution in the frequency domain). Specifically,

$$\boxed{r(t) = s(t)p(t) \quad \xleftrightarrow{\mathcal{F}} \quad R(\omega) = \frac{1}{2\pi}[S(\omega) * P(\omega)]} \tag{4.134}$$

This can be shown by using the duality relations of Section 4.6.6 together with the convolution property, or by directly using the Fourier transform relations in a manner analogous to the procedure used in deriving the convolution property.

Multiplication of one signal by another can be thought of as using one signal to scale or *modulate* the amplitude of the other, and consequently the multiplication of two signals is often referred to as *amplitude modulation*. For this reason eq. (4.134) is called the modulation property. As we shall see in Chapters 7 and 8 this property has several very important applications. To illustrate eq. (4.134) and to suggest several of the applications that we will discuss in these subsequent chapters, let us consider several examples.

Example 4.22

Let $s(t)$ be a signal whose spectrum $S(\omega)$ is depicted in Figure 4.30(a). Also consider the signal $p(t)$ defined by

$$p(t) = \cos \omega_0 t$$

Then

$$P(\omega) = \pi\delta(\omega - \omega_0) + \pi\delta(\omega + \omega_0)$$

as sketched in Figure 4.30(b), and the spectrum $R(\omega)$ of $r(t) = s(t)p(t)$ is obtained by an application of eq. (4.134), yielding

$$R(\omega) = \frac{1}{2\pi}S(\omega) * P(\omega) = \tfrac{1}{2}S(\omega - \omega_0) + \tfrac{1}{2}S(\omega + \omega_0) \tag{4.135}$$

which is sketched in Figure 4.30(c). Here we have assumed that $\omega_0 > \omega_1$, so that the two nonzero portions of $R(\omega)$ do not overlap. Thus, we see that the spectrum of $r(t)$ consists of the sum of two shifted and scaled versions of $S(\omega)$.

From eq. (4.135) and from Figure 4.30 it is intuitively clear that all of the information in the signal $s(t)$ is preserved when we multiply this signal by a sinusoidal signal, although the information has been shifted to higher frequencies. This fact forms the basis for sinusoidal amplitude modulation systems, and in the next example we provide a glimpse of how we can recover the original signal $s(t)$ from the amplitude modulated signal $r(t)$.

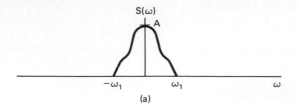

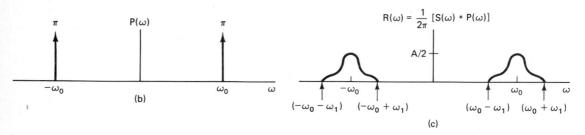

Figure 4.30   Use of the modulation property in Example 4.22.

Example 4.23

Let us now consider $r(t)$ as obtained in Example 4.22, and let

$$g(t) = r(t)p(t)$$

where, again, $p(t) = \cos \omega_0 t$. Then, $R(\omega)$, $P(\omega)$, and $G(\omega)$ are as shown in Figure 4.31.

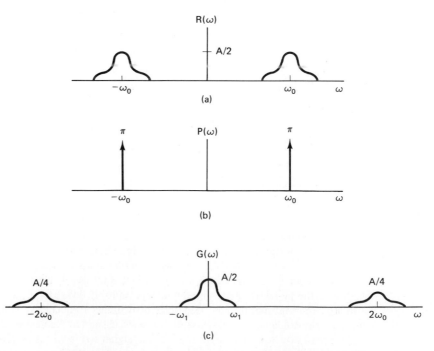

Figure 4.31   Spectra of the signals considered in Example 4.23.

From Figure 4.31(c) and the linearity of the Fourier transform, we see that $g(t)$ is the sum of $\frac{1}{2}s(t)$ and a signal with a spectrum that is nonzero only at higher frequencies (centered around $\pm 2\omega_0$). Suppose then that we apply the signal $g(t)$ as the input to an LTI system with frequency response $H(\omega)$ that is constant at low frequencies (say, for $|\omega| < \omega_1$) and zero at high frequencies (for $|\omega| > \omega_0$). Then the output of this system will have as its spectrum $H(\omega)G(\omega)$, which, because of the particular choice of $H(\omega)$, will be a scaled replica of $S(\omega)$. Therefore, the output itself will be a scaled version of $s(t)$. In Chapter 7 we expand significantly on this idea as we develop in detail the fundamentals of amplitude modulation.

Example 4.24

Again consider a signal $r(t)$ defined as the product of two signals

$$r(t) = s(t)p(t)$$

where $p(t)$ is now taken to be a periodic impulse train

$$p(t) = \sum_{k=-\infty}^{+\infty} \delta(t - kT) \tag{4.136}$$

As is illustrated in Figure 4.32, we see that $r(t)$ is an impulse train with the amplitudes of the impulses equal to *samples* of $s(t)$ spaced at time intervals of length T apart. That is,

$$r(t) = s(t)p(t) = \sum_{k=-\infty}^{+\infty} s(t)\,\delta(t - kT) = \sum_{k=-\infty}^{+\infty} s(kT)\,\delta(t - kT)$$

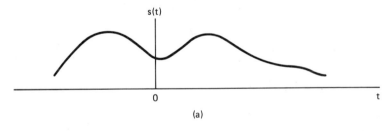

(a)

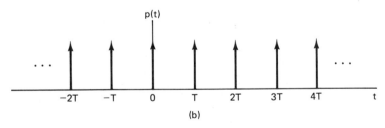

(b)

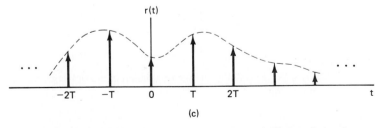

(c)

Figure 4.32 Product of a signal $s(t)$ and a periodic impulse train.

Recall from Example 4.15 that the spectrum of the periodic impulse train $p(t)$ is itself a periodic impulse train. Specifically,

$$P(\omega) = \frac{2\pi}{T} \sum_{k=-\infty}^{+\infty} \delta\left(\omega - \frac{2\pi k}{T}\right)$$

and consequently from the modulation property

$$R(\omega) = \frac{1}{2\pi}[S(\omega) * P(\omega)] = \frac{1}{T} \sum_{k=-\infty}^{+\infty} S(\omega) * \delta\left(\omega - \frac{2\pi k}{T}\right)$$

$$= \frac{1}{T} \sum_{k=-\infty}^{+\infty} S\left(\omega - \frac{2\pi k}{T}\right)$$

That is, $R(\omega)$ consists of periodically repeated replicas of $S(\omega)$ as illustrated in Figure 4.33(c) for $2\pi/T \gg \omega_1$, so that the nonzero portions of $R(\omega)$ do not overlap. From this figure we see that, just as in the preceding example, if $r(t)$ is applied as the input to an LTI system with frequency response that is constant for $|\omega| < \omega_1$ and zero for higher frequencies, then the output of this system will be proportional to $s(t)$. This is the basic idea behind sampling, and in Chapter 8 we explore the implications of this example in some depth.

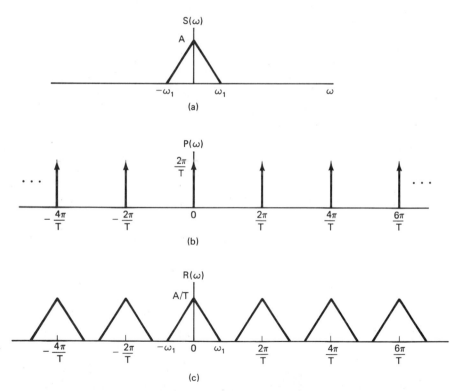

Figure 4.33 Effect in the frequency domain of multiplying a signal by a periodic impulse train.

4.9 TABLES OF FOURIER PROPERTIES AND OF BASIC FOURIER TRANSFORM AND FOURIER SERIES PAIRS

In the preceding three sections and in the problems, we have considered some of the important properties of the Fourier transform. These are summarized in Table 4.1. As we discussed at the beginning of Section 4.6, many of these properties have counterparts for Fourier series, and these are summarized in Table 4.2. In most

TABLE 4.1 PROPERTIES OF THE FOURIER TRANSFORM

Aperiodic signal	Fourier transform
$x(t)$	$X(\omega)$
$y(t)$	$Y(\omega)$
$ax(t) + by(t)$	$aX(\omega) + bY(\omega)$
$x(t - t_0)$	$e^{-j\omega t_0} X(\omega)$
$e^{j\omega_0 t} x(t)$	$X(\omega - \omega_0)$
$x^*(t)$	$X^*(-\omega)$
$x(-t)$	$X(-\omega)$
$x(at)$	$\dfrac{1}{\|a\|} X\left(\dfrac{\omega}{a}\right)$
$x(t) * y(t)$	$X(\omega) Y(\omega)$
$x(t)y(t)$	$\dfrac{1}{2\pi} X(\omega) * Y(\omega)$
$\dfrac{d}{dt} x(t)$	$j\omega X(\omega)$
$\displaystyle\int_{-\infty}^{t} x(t)\, dt$	$\dfrac{1}{j\omega} X(\omega) + \pi X(0)\, \delta(\omega)$
$tx(t)$	$j\dfrac{d}{d\omega} X(\omega)$
$x(t)$ real	$\begin{cases} X(\omega) = X^*(-\omega) \\ \mathcal{Re}\{X(\omega)\} = \mathcal{Re}\{X(-\omega)\} \\ \mathcal{Im}\{X(\omega)\} = -\mathcal{Im}\{X(-\omega)\} \\ \|X(\omega)\| = \|X(-\omega)\| \\ \sphericalangle X(\omega) = -\sphericalangle X(-\omega) \end{cases}$
$x_e(t) = \mathcal{Ev}\{x(t)\}$ [$x(t)$ real]	$\mathcal{Re}\{X(\omega)\}$
$x_o(t) = \mathcal{Od}\{x(t)\}$ [$x(t)$ real]	$j\mathcal{Im}\{X(\omega)\}$

Duality

$$f(u) = \int_{-\infty}^{+\infty} g(v)e^{-juv}\, dv$$

$$g(t) \xleftrightarrow{\;\mathcal{F}\;} f(\omega)$$

$$f(t) \xleftrightarrow{\;\mathcal{F}\;} 2\pi g(-\omega)$$

Parseval's Relation for Aperiodic Signals

$$\int_{-\infty}^{+\infty} |x(t)|^2\, dt = \frac{1}{2\pi} \int_{-\infty}^{+\infty} |X(\omega)|^2\, d\omega$$

cases these properties can be deduced directly from the corresponding property in Table 4.1 using the discussion of Section 4.5, in which the Fourier series was incor-

TABLE 4.2 PROPERTIES OF FOURIER SERIES

Periodic signal	Fourier series coefficients
$x(t)$ ⎱ periodic with $y(t)$ ⎰ period T_0	a_k b_k
$Ax(t) + By(t)$	$Aa_k + Bb_k$
$x(t - t_0)$	$a_k e^{-jk(2\pi/T_0)t_0}$
$e^{jM(2\pi/T_0)t}x(t)$	a_{k-M}
$x^*(t)$	a^*_{-k}
$x(-t)$	a_{-k}
$x(\alpha t),\ \alpha > 0\ \left(\text{periodic with period } \dfrac{T_0}{\alpha}\right)$	a_k
$\displaystyle\int_{T_0} x(\tau)y(t - \tau)\,d\tau$	$T_0 a_k b_k$
$x(t)y(t)$	$\displaystyle\sum_{l=-\infty}^{+\infty} a_l b_{k-l}$
$\dfrac{dx(t)}{dt}$	$jk\dfrac{2\pi}{T_0}a_k$
$\displaystyle\int_{-\infty}^{t} x(t)\,dt$ (finite-valued and periodic only if $a_0 = 0$)	$\left(\dfrac{1}{jk(2\pi/T_0)}\right)a_k$
$x(t)$ real	$\begin{cases} a_k = a^*_{-k} \\ \mathcal{R}e\{a_k\} = \mathcal{R}e\{a_{-k}\} \\ \mathcal{I}m\{a_k\} = -\mathcal{I}m\{a_{-k}\} \\ |a_k| = |a_{-k}| \\ \sphericalangle a_k = -\sphericalangle a_{-k} \end{cases}$
$x_e(t) = \mathcal{E}v\{x(t)\}\quad [x(t)\text{ real}]$	$\mathcal{R}e\{a_k\}$
$x_o(t) = \mathcal{O}d\{x(t)\}\quad [x(t)\text{ real}]$	$j\mathcal{I}m\{a_k\}$

Parseval's Relation for Periodic Signals

$$\frac{1}{T_0}\int_{T_0} |x(t)|^2\,dt = \sum_{k=-\infty}^{+\infty} |a_k|^2$$

porated into the framework of the Fourier transform. In some cases, however, as with the convolution property or Parseval's relation, the transform property as it applies to aperiodic signals and the Fourier transform is not useful for periodic signals and Fourier series, although a modified version is.

Finally, in Table 4.3 we have assembled a list of many of the basic and important Fourier transform pairs. We will encounter many of these repeatedly as we apply the tools of Fourier analysis in our examination of signals and systems. All of these transform pairs, except for the last one in the table, have been considered as examples in the preceding sections. The last one is considered in Problem 4.19.

TABLE 4.3 BASIC FOURIER TRANSFORM PAIRS

Signal	Fourier transform	Fourier series coefficients (if periodic)
$\displaystyle\sum_{k=-\infty}^{+\infty} a_k e^{jk\omega_0 t}$	$\displaystyle 2\pi \sum_{k=-\infty}^{+\infty} a_k \delta(\omega - k\omega_0)$	a_k
$e^{j\omega_0 t}$	$2\pi\delta(\omega - \omega_0)$	$a_1 = 1$ $a_k = 0,$ otherwise
$\cos\omega_0 t$	$\pi[\delta(\omega - \omega_0) + \delta(\omega + \omega_0)]$	$a_1 = a_{-1} = \dfrac{1}{2}$ $a_k = 0,$ otherwise
$\sin\omega_0 t$	$\dfrac{\pi}{j}[\delta(\omega - \omega_0) - \delta(\omega + \omega_0)]$	$a_1 = -a_{-1} = \dfrac{1}{2j}$ $a_k = 0,$ otherwise
$x(t) = 1$	$2\pi\delta(\omega)$	$a_0 = 1, \quad a_k = 0, k \neq 0$ $\left(\begin{array}{l}\text{has this Fourier series represen-}\\ \text{tation for any choice of } T_0 > 0\end{array}\right)$
Periodic square wave $x(t) = \begin{cases} 1, & \|t\| < T_1 \\ 0, & T_1 < \|t\| \le \dfrac{T_0}{2} \end{cases}$ and $x(t + T_0) = x(t)$	$\displaystyle\sum_{k=-\infty}^{+\infty} \dfrac{2\sin k\omega_0 T_1}{k}\,\delta(\omega - k\omega_0)$	$\dfrac{\omega_0 T_1}{\pi}\,\mathrm{sinc}\left(\dfrac{k\omega_0 T_1}{\pi}\right) = \dfrac{\sin k\omega_0 T_1}{k\pi}$
$\displaystyle\sum_{n=-\infty}^{+\infty} \delta(t - nT)$	$\dfrac{2\pi}{T} \displaystyle\sum_{k=-\infty}^{+\infty} \delta\left(\omega - \dfrac{2\pi k}{T}\right)$	$a_k = \dfrac{1}{T}$ for all k
$x(t) = \begin{cases} 1, & \|t\| < T_1 \\ 0, & \|t\| > T_1 \end{cases}$	$2T_1\,\mathrm{sinc}\left(\dfrac{\omega T_1}{\pi}\right) = \dfrac{2\sin\omega T_1}{\omega}$	—
$\dfrac{W}{\pi}\,\mathrm{sinc}\left(\dfrac{Wt}{\pi}\right) = \dfrac{\sin Wt}{\pi t}$	$X(\omega) = \begin{cases} 1, & \|\omega\| < W \\ 0, & \|\omega\| > W \end{cases}$	—
$\delta(t)$	1	—
$u(t)$	$\dfrac{1}{j\omega} + \pi\delta(\omega)$	—
$\delta(t - t_0)$	$e^{-j\omega t_0}$	—
$e^{-at}u(t),\ \mathcal{R}e\{a\} > 0$	$\dfrac{1}{a + j\omega}$	—
$te^{-at}u(t),\ \mathcal{R}e\{a\} > 0$	$\dfrac{1}{(a + j\omega)^2}$	—
$\dfrac{t^{n-1}}{(n-1)!}e^{-at}u(t),$ $\mathcal{R}e\{a\} > 0$	$\dfrac{1}{(a + j\omega)^n}$	—

4.10 THE POLAR REPRESENTATION OF CONTINUOUS-TIME FOURIER TRANSFORMS

4.10.1 The Magnitude and Phase of Fourier Transforms

In the preceding sections we have developed some insights into the properties of Fourier transforms and have seen some of the ways in which the tools of Fourier analysis can be used effectively in the study of signals and systems. From the synthesis equation (4.60) we know that we can recover a signal $x(t)$ from its Fourier transform $X(\omega)$, and therefore we can conclude that $X(\omega)$ must contain all the information in $x(t)$. We have already gained some understanding of how this information is embedded in $X(\omega)$ by examining how various possible characteristics of $x(t)$ (real, even, time-shifted, etc.) manifest themselves as corresponding properties of $X(\omega)$. In this section we discuss the polar representation of the Fourier transform and by doing so we will gain more insight into its characteristics.

The polar or magnitude-phase representation of $X(\omega)$ is

$$X(\omega) = |X(\omega)| e^{j \sphericalangle X(\omega)} \qquad (4.137)$$

From the synthesis equation (4.60) we can think of $X(\omega)$ as providing us with a decomposition of the signal $x(t)$ into a "sum" of periodic complex exponentials at different frequencies. In fact, as mentioned earlier $|X(\omega)|^2$ has the interpretation as the energy-density spectrum of $x(t)$. That is, $|X(\omega)|^2 d\omega/2\pi$ can be thought of as the amount of the energy in the signal $x(t)$ that lies in the frequency band between ω and $\omega + d\omega$.

While $|X(\omega)|$ provides us with the information about the relative magnitudes of the complex exponentials that make up $x(t)$, $\sphericalangle X(\omega)$ provides us with information concerning the relative phases of these exponentials. Depending upon what this phase function is, we can obtain very different looking signals, even if the magnitude function remains unchanged. For example, consider again the example illustrated in Figure 4.3. In this case, a ship encounters the superposition of three wave trains, each of which can be modeled as a sinusoidal signal. Depending upon the relative phases of these three sinusoids (and, of course, on their magnitudes), the amplitude of their sum may be quite small or very large. The implications of phase for the ship, therefore, are quite significant. To see the effect of phase in more detail, consider the signal

$$x(t) = 1 + \tfrac{1}{2} \cos\left(2\pi t + \phi_1\right) + \cos\left(4\pi t + \phi_2\right) + \tfrac{2}{3} \cos\left(6\pi t + \phi_3\right) \qquad (4.138)$$

In Figure 4.4 we depicted $x(t)$ in the case when $\phi_1 = \phi_2 = \phi_3 = 0$. In Figure 4.34 we illustrate $x(t)$ for this and for several other choices for the ϕ_i. As this figure demonstrates, the resulting signals can differ significantly, depending upon the values of the ϕ_i.

Therefore, we see that changes in the phase function of $X(\omega)$ lead to changes in the time-domain characteristics of the signal $x(t)$. In some instances phase distortion may be important, whereas in others it is not. For example, consider the human auditory system. If $X(\omega)$ is the Fourier transform of a signal corresponding to an individual spoken sound (such as a vowel), then a human being would be able to recognize this sound even if the signal were distorted by a change in the phase of $X(\omega)$.

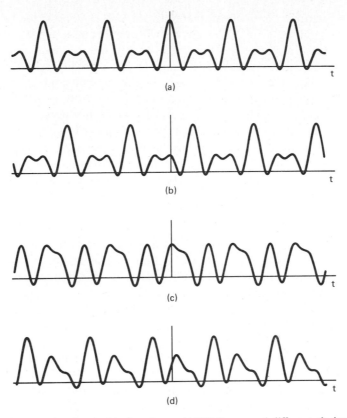

Figure 4.34 The signal $x(t)$ given in eq. (4.138) for several different choices of the phase angles ϕ_1, ϕ_2, ϕ_3: (a) $\phi_1 = \phi_2 = \phi_3 = 0$; (b) $\phi_1 = 4$ rad., $\phi_2 = 8$ rad., $\phi_3 = 12$ rad.; (c) $\phi_1 = 6$ rad., $\phi_2 = -2.7$ rad., $\phi_3 = 0.93$ rad.; (d) $\phi_1 = 1.2$ rad., $\phi_2 = 4.1$ rad., $\phi_3 = -7.02$ rad.

For example, the sound would be recognizable if the acoustic signal produced had a Fourier transform equal to $|X(\omega)|$ (i.e., a signal with zero phase and the same magnitude function). On the other hand, although mild phase distortions such as those affecting individual sounds do not lead to a loss of intelligibility, more severe phase distortions of speech certainly do. For example, if $x(t)$ is a tape recording of a sentence, then the signal $x(-t)$ represents the sentence being played backward. From Table 4.1, we know that

$$\mathcal{F}\{x(-t)\} = X(-\omega)$$

where $X(\omega)$ is the Fourier transform of $x(t)$. Furthermore, since $x(t)$ is real, $|X(-\omega)| = |X(\omega)|$ and $\sphericalangle X(-\omega) = -\sphericalangle X(\omega)$. That is, the spectrum of a sentence played in reverse has the same magnitude function as the spectrum of the original sentence and differs only in phase, where there is a sign reversal. Clearly, this phase change has a significant impact on the recognizability of the recording.

A second illustrative example of the effect and importance of phase is found in examining images. A black-and-white picture can be thought of as a signal $x(t_1, t_2)$ with two independent variables. Here t_1 denotes the horizontal coordinate of a point on a picture, t_2 is the vertical coordinate, and $x(t_1, t_2)$ is the brightness of the image at the point (t_1, t_2). Although we will not examine signals with two independent variables at any length in this book, we will use examples of them (i.e., pictures) on several occasions because such examples are extremely useful in visualizing some of the concepts that we will develop. In the present context, in order to discuss the effect of phase for images, we need to introduce Fourier analysis for signals with two independent variables. For our purposes we do not need to develop this topic in any detail but can simply observe that the techniques of two-dimensional Fourier analysis are quite similar to those that we have developed for signals with one independent variable. Specifically, in two dimensions we decompose a signal $x(t_1, t_2)$ into a sum (integral) of products of complex exponentials that oscillate at possibly different rates in each of the two directions, that is, signals of the form

$$e^{j\omega_1 t_1} e^{j\omega_2 t_2}$$

The result of this decomposition is a two-dimensional Fourier transform $X(\omega_1, \omega_2)$ which contains the information about how the signal $x(t_1, t_2)$ is constructed from these basic signals. Several elementary aspects of two-dimensional Fourier analysis are addressed in Problem 4.26.

Returning to the question of phase in pictures, we note that in viewing a picture, some of the most important information for the eyes is contained in the edges and regions of high contrast. Intuitively, regions of maximum and minimum intensity in a picture are places at which the complex exponentials at different frequencies are in phase. Therefore, it seems plausible to expect the phase of the Fourier transform of a picture to contain much of the information in the picture, and in particular the phase should capture the information about the edges. To substantiate this expectation, in Figure 4.35(a) we have repeated the picture of Fourier shown in Figure 4.2. In Figure 4.35(b) we have depicted the magnitude of Fourier's transform, where in this image the horizontal axis is ω_1, the vertical is ω_2, and the brightness of the image at the point (ω_1, ω_2) is proportional to the magnitude of the transform $X(\omega_1, \omega_2)$ of the image in Figure 4.35(a). Similarly, the phase of this transform is depicted in Figure 4.35(c). Figure 4.35(d) is the result of setting the phase [Figure 4.35(c)] of $X(\omega_1, \omega_2)$ to zero (without changing its magnitude) and inverse transforming. In Figure 4.35(e) the magnitude of $X(\omega_1, \omega_2)$ was set equal to 1, but the phase was kept unchanged. Finally, in Figure 4.35(f) we have depicted the image obtained by inverse transforming the function obtained by using the phase in Figure 4.35(c) and the magnitude of the transform of a *completely different* image, specifically the picture shown in Figure 2.2! These figures clearly illustrate the importance of phase in representing images.

Returning to signals with a single independent variable, there is one particular type of phase distortion that is quite easy to visualize. This is the case of *linear phase* in which the phase shift at frequency ω is a linear function of ω. Specifically, if we modify $\sphericalangle X(\omega)$ by adding to it $\alpha\omega$, then, from eq. (4.137), the resulting Fourier transform is $X(\omega)e^{j\alpha\omega}$, and from the time-shifting property (4.85), the resulting signal is $x(t + \alpha)$. That is, it is simply a time-shifted version of the original signal. In this case, the phases of the complex exponentials at different frequencies are shifted so that the relative phases of these signals at time $t + \alpha$ are identical to the relative phases of the original signal at time t. Therefore, when these exponentials are superposed, we obtain a shifted version of $x(t)$. If the phase shift is a nonlinear function of ω, then each complex exponential will be shifted in a manner that results in a change in the relative phases. When these exponentials are superposed, we obtain a signal that may look considerably different than $x(t)$. This is precisely what is illustrated in Figure 4.34. Figure 4.34(b) is an example of linear phase, while Figure 4.34(c) and (d) depict two examples of nonlinear phase.

4.10.2 Bode Plots

In Section 4.7 we saw that the Fourier transforms of the input and output of an LTI system with frequency response $H(\omega)$ are related by

$$Y(\omega) = H(\omega)X(\omega)$$

or, equivalently, in terms of the polar representation, we have that

$$|Y(\omega)| = |H(\omega)||X(\omega)| \tag{4.139}$$

$$\sphericalangle Y(\omega) = \sphericalangle H(\omega) + \sphericalangle X(\omega) \tag{4.140}$$

Because of the multiplicative form of eq. (4.139) the magnitude of the frequency response of an LTI system is sometimes referred to as the *gain* of the system.

As we will find in the remainder of this chapter and in subsequent chapters, it is often convenient to represent Fourier transforms graphically when using frequency-domain techniques to examine LTI systems. Earlier in this chapter we used a graphical representation for $X(\omega)$ consisting of separate plots of $|X(\omega)|$ and $\sphericalangle X(\omega)$ as functions of ω. Although this representation is useful and in fact will be used extensively throughout this book, eqs. (4.139) and (4.140) suggest a modification to this representation that is also of great value in LTI system analysis. Specifically, note that if we have plotted $\sphericalangle H(\omega)$ and $\sphericalangle X(\omega)$ as a function of ω, then $\sphericalangle Y(\omega)$ can be obtained by adding the corresponding points on these two graphs. Similarly, if we plot $\log|H(\omega)|$ and $\log|X(\omega)|$, we can add these to obtain $\log|Y(\omega)|$. Such a representation using the logarithm of the magnitude function often facilitates the graphical manipulations that are performed in analyzing LTI systems. For example, since the frequency response of the cascade of LTI systems is the product of the individual frequency responses, we can obtain plots of the log magnitude and phase of the overall frequency response simply by adding the corresponding plots for each of the component systems.

The most widely known graphical representation of the type just described is the *Bode plot*. In this representation the quantities $\sphericalangle H(\omega)$ and $20\log_{10}|H(\omega)|$ are plotted versus frequency. The latter of these quantities, which is proportional to the log-magnitude, is referred to as the magnitude expressed in *decibels* (abbreviated dB). Thus, 0 dB corresponds to a value of $|H(\omega)|$ equal to 1; 20 dB is equivalent to $|H(\omega)| = 10$; -20 dB corresponds to $|H(\omega)| = 0.1$; 40 dB is the same as $|H(\omega)| = 100$; and so on. Also it is useful to note that 1 dB is approximately equivalent to $|H(\omega)| = 1.12$, and 6 dB approximately corresponds to $|H(\omega)| = 2$. Typical Bode plots are illustrated in Figure 4.36. In these plots a logarithmic scale is usually used for ω. Not only does this allow us to obtain adequate resolution when the frequency range of interest is large, but also the shape of a particular response curve does not change if frequency is scaled (see Problem 4.46). Also, as we shall see in the next two sections, the use of a logarithmic frequency scale greatly facilitates the plotting

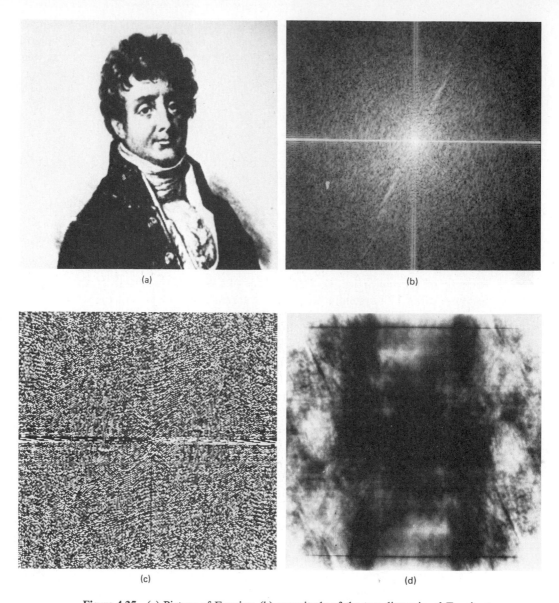

Figure 4.35 (a) Picture of Fourier; (b) magnitude of the two-dimensional Fourier transform of (a); (c) phase of the Fourier transform of (a); (d) picture whose Fourier transform has magnitude as in (b) and phase equal to zero; (e) picture whose Fourier transform has magnitude equal to 1 and phase as in (c); (f) picture whose Fourier transform has phase as in (c) and magnitude equal to that of the transform of the picture in Figure 2.2.

of these curves for LTI systems described by differential equations. Note also that in Figure 4.36 we have plotted the magnitude and phase curves for positive ω only. As we discussed in Section 4.6, if $h(t)$ is real, then $|H(\omega)|$ is an even function of ω and $\not\prec H(\omega)$ is an odd function. Because of this, the plots for negative ω are superfluous and can be obtained immediately from the plots for positive ω.

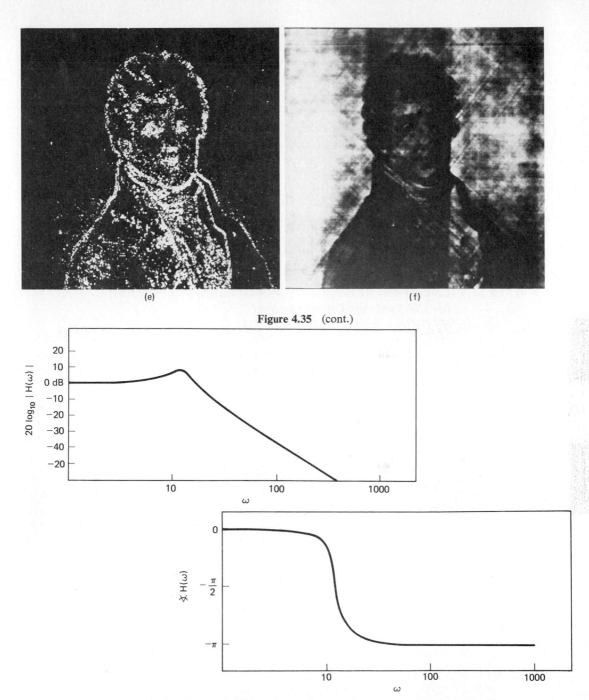

Figure 4.35   (cont.)

Figure 4.36   Typical Bode plots (note that ω is plotted using a logarithmic scale).

As we continue our development of the tools of signal and system analysis, we will encounter numerous situations in which it is more convenient to plot $|X(\omega)|$ on a linear scale, and we will also find many places in which plotting $\log|X(\omega)|$ is to be preferred. For example, in Chapters 6 to 8 we will encounter LTI systems whose magnitude functions are zero over different ranges of frequency and 1 over others. Since log (0) = −∞, the Bode plot is not particularly useful for these systems, and the use of a linear scale for $|X(\omega)|$ is far more convenient. On the other hand, in the next two sections we will consider a class of systems for which Bode plots are of great value. In addition, because the logarithm expands the scale for small values of $|X(\omega)|$, Bode plots are often useful in displaying the fine detail of frequency responses near zero. For these reasons we have introduced both the linear and logarithmic graphical representations for the magnitude of Fourier transforms, and we will use each as is appropriate.

4.11 THE FREQUENCY RESPONSE OF SYSTEMS CHARACTERIZED BY LINEAR CONSTANT-COEFFICIENT DIFFERENTIAL EQUATIONS

4.11.1 Calculation of Frequency and Impulse Responses for LTI Systems Characterized by Differential Equations

As we discussed in Sections 3.5 and 3.6, a particularly important and useful class of continuous-time LTI systems are those for which the input and output satisfy a linear constant-coefficient differential equation of the form

$$\sum_{k=0}^{N} a_k \frac{d^k y(t)}{dt^k} = \sum_{k=0}^{M} b_k \frac{d^k x(t)}{dt^k} \tag{4.141}$$

In this section we consider the question of determining the frequency response of such an LTI system. In principle, this could be done using the technique reviewed in Chapter 3 for solving equations such as eq. (4.141). Specifically, we could use that procedure to determine the impulse response of the LTI system, and then by taking the Fourier transform of the impulse response, we would obtain the frequency response. However, because of the properties of the Fourier transform discussed in preceding sections, there is a much simpler and much more direct procedure which we will describe for obtaining the frequency response. Once the frequency response is so obtained, the impulse response can then be determined using the inverse transform, and as we will see, the technique of partial fraction expansion makes this procedure an extremely useful method for calculating the impulse response and thus for characterizing and computing responses of LTI systems described by linear constant-coefficient differential equations.

To outline the procedure alluded to in the preceding paragraph, consider an LTI system characterized by eq. (4.141). We know from the convolution property that

$$Y(\omega) = H(\omega)X(\omega)$$

or, equivalently,

$$H(\omega) = \frac{Y(\omega)}{X(\omega)} \tag{4.142}$$

where $X(\omega)$, $Y(\omega)$, and $H(\omega)$ are the Fourier transforms of the input $x(t)$, output $y(t)$, and impulse response $h(t)$, respectively. Here, of course, we are assuming implicitly that these three Fourier transforms all exist. Next, consider applying the Fourier transform to both sides of eq. (4.141) to obtain

$$\mathcal{F}\left\{\sum_{k=0}^{N} a_k \frac{d^k y(t)}{dt^k}\right\} = \mathcal{F}\left\{\sum_{k=0}^{M} b_k \frac{d^k x(t)}{dt^k}\right\} \tag{4.143}$$

From the linearity property (4.83) this becomes

$$\sum_{k=0}^{N} a_k \mathcal{F}\left\{\frac{d^k y(t)}{dt^k}\right\} = \sum_{k=0}^{M} b_k \mathcal{F}\left\{\frac{d^k x(t)}{dt^k}\right\} \tag{4.144}$$

and from the differentiation property (4.87),

$$\sum_{k=0}^{N} a_k (j\omega)^k Y(\omega) = \sum_{k=0}^{M} b_k (j\omega)^k X(\omega)$$

or, equivalently,

$$Y(\omega)\left[\sum_{k=0}^{N} a_k (j\omega)^k\right] = X(\omega)\left[\sum_{k=0}^{M} b_k (j\omega)^k\right]$$

Thus, from eq. (4.142),

$$H(\omega) = \frac{Y(\omega)}{X(\omega)} = \frac{\displaystyle\sum_{k=0}^{M} b_k (j\omega)^k}{\displaystyle\sum_{k=0}^{N} a_k (j\omega)^k} \tag{4.145}$$

From eq. (4.145) we observe that $H(\omega)$ is a *rational function*, that is, it is a ratio of polynomials in $(j\omega)$. The coefficients of the numerator polynomial are the same coefficients as those that appear on the right-hand side of eq. (4.141), and the coefficients of the denominator polynomial are the same coefficients as appear on the left side of eq. (4.141). Thus, we see that the frequency response given in eq. (4.145) for the LTI system characterized by eq. (4.141) can be written down directly by inspection.

Example 4.25

Consider the LTI system that is initially at rest and that is characterized by

$$\frac{dy(t)}{dt} + ay(t) = x(t) \tag{4.146}$$

with $a > 0$. From eq. (4.145), the frequency response is

$$H(\omega) = \frac{1}{j\omega + a} \tag{4.147}$$

Comparing this with Example 4.7, we see that eq. (4.147) is the Fourier transform of $e^{-at}u(t)$. Thus, the impulse response of the system is recognized as

$$h(t) = e^{-at}u(t)$$

Example 4.26

Consider an LTI system initially at rest that is characterized by the differential equation

$$\frac{d^2 y(t)}{dt^2} + 4\frac{dy(t)}{dt} + 3y(t) = \frac{dx(t)}{dt} + 2x(t)$$

From eq. (4.145), the frequency response is

$$H(\omega) = \frac{(j\omega) + 2}{(j\omega)^2 + 4(j\omega) + 3} \tag{4.148}$$

To determine the corresponding impulse response we require the inverse Fourier transform of $H(\omega)$. This can be done using the technique of partial fraction expansion that was used in Example 4.20 and is discussed in detail in the Appendix (in particular see Example A.1, in which the details of the calculations for this example are worked out). As a first step, we factor the denominator of the right-hand side of eq. (4.148) into a product of lower-order terms as

$$H(\omega) = \frac{j\omega + 2}{(j\omega + 1)(j\omega + 3)} \tag{4.149}$$

Next, using the method of partial fraction expansion developed in the Appendix, we obtain the following expression for $H(\omega)$:

$$H(\omega) = \frac{\frac{1}{2}}{j\omega + 1} + \frac{\frac{1}{2}}{j\omega + 3}$$

The inverse transform of each term can be recognized by inspection with the result that

$$h(t) = \tfrac{1}{2}e^{-t}u(t) + \tfrac{1}{2}e^{-3t}u(t)$$

The procedure used in Example 4.26 to obtain the inverse Fourier transform is generally useful in inverting transforms that are ratios of polynomials in $(j\omega)$. In particular, we can use eq. (4.145) to determine the frequency response of any LTI system described by a linear constant-coefficient differential equation and then can calculate the impulse response by performing a partial fraction expansion that puts the frequency response into a form in which the inverse transform of each term can be recognized by inspection. In addition, if the Fourier transform $X(\omega)$ of the input to such a system is also a ratio of polynomials in $j\omega$, then so is $Y(\omega) = H(\omega)X(\omega)$. In this case we can use the same technique to solve the differential equation, that is, to find the response $y(t)$ to the input $x(t)$. This is illustrated in the next example.

Example 4.27

Consider the system of Example 4.26 and suppose that the input is

$$x(t) = e^{-t}u(t)$$

Then, using eq. (4.149), we have that

$$Y(\omega) = H(\omega)X(\omega) = \left[\frac{j\omega + 2}{(j\omega + 1)(j\omega + 3)}\right]\left[\frac{1}{j\omega + 1}\right]$$

$$= \frac{j\omega + 2}{(j\omega + 1)^2(j\omega + 3)} \tag{4.150}$$

As developed in the Appendix, in this case we seek a partial fraction expansion of the form

$$Y(\omega) = \frac{A_{11}}{j\omega + 1} + \frac{A_{12}}{(j\omega + 1)^2} + \frac{A_{21}}{j\omega + 3} \tag{4.151}$$

where A_{11}, A_{12}, and A_{21} are constants to be determined. In Example A.2 in the Appendix, the technique of partial fraction expansion is applied to this example in order to determine these constants. The values obtained are

$$A_{11} = \tfrac{1}{4}, \qquad A_{12} = \tfrac{1}{2}, \qquad A_{21} = -\tfrac{1}{4}$$

so that

$$Y(\omega) = \frac{\frac{1}{4}}{j\omega + 1} + \frac{\frac{1}{2}}{(j\omega + 1)^2} - \frac{\frac{1}{4}}{j\omega + 3} \tag{4.152}$$

Again, the inverse Fourier transform for each term in eq. (4.152) can be obtained by inspection. The first and third terms are of the same type that we have encountered in the preceding two examples, while the inverse transform of the second term can be obtained from Table 4.2 or, as was done in Example 4.20, by applying the dual of the differentiation property, as given in eq. (4.107), to $1/(j\omega + 1)$. The inverse transform of eq. (4.152) is then found to be

$$y(t) = [\tfrac{1}{4}e^{-t} + \tfrac{1}{2}te^{-t} - \tfrac{1}{4}e^{-3t}]u(t)$$

From this and the preceding examples we see how the techniques of Fourier analysis allow us to reduce questions concerning LTI systems characterized by differential equations to elementary problems in algebra. This important fact is illustrated further in a number of the problems at the end of the chapter.

4.11.2 Cascade and Parallel-Form Structures

In Section 3.6 we described several realizations for the implementation of LTI systems characterized by differential equations. These realizations consisted of interconnections of three basic network elements: adders, coefficient multipliers, and integrators, and could be directly implemented using operational amplifiers. In addition to the realization structures discussed in Section 3.6, there are two other very important and widely used structures that we can now describe with the aid of Fourier analysis.

Consider the frequency response of eq. (4.145) for an LTI system characterized by a linear constant-coefficient differential equation. A structure referred to as the *cascade structure* is then obtained by first factoring the numerator and denominator into products of first-order terms:

$$H(\omega) = \frac{b_M \displaystyle\prod_{k=1}^{M} (\lambda_k + j\omega)}{a_N \displaystyle\prod_{k=1}^{N} (\nu_k + j\omega)} \tag{4.153}$$

Some of the λ_k and ν_k may be complex; however, if they are, then they appear in complex-conjugate pairs. By multiplying together the two first-order terms involving complex conjugate λ_k's or ν_k's, we obtain second-order terms with *real* coefficients. For example,

$$(\lambda + j\omega)(\lambda^* + j\omega) = |\lambda|^2 + 2\mathcal{Re}\{\lambda\}j\omega + (j\omega)^2$$

Therefore, assuming that there are P conjugate pairs in the numerator and Q in the denominator, $H(\omega)$ can be written as a ratio of products of first- and second-order polynomials with real coefficients:[†]

$$H(\omega) = \frac{b_M}{a_N} \frac{\displaystyle\prod_{k=1}^{P} [\beta_{0k} + \beta_{1k}(j\omega) + (j\omega)^2] \prod_{k=1}^{M-2P} (\lambda_k + j\omega)}{\displaystyle\prod_{k=1}^{Q} [\alpha_{0k} + \alpha_{1k}(j\omega) + (j\omega)^2] \prod_{k=1}^{N-2Q} (\nu_k + j\omega)} \tag{4.154}$$

What eq. (4.154) states is that the frequency response of *any* LTI system described

---

†In eq. (4.154) we have renumbered the remaining real λ_k and ν_k so that the products run from 1 to $M - 2P$ and $N - 2Q$, respectively.

by a linear constant-coefficient differential equation can be written as the product of first- and second-order terms. This implies that such an LTI system can be realized as the cascade of first- and second-order systems. For this reason first- and second-order systems play an extremely important role in the analysis and synthesis of linear systems, and we will discuss their properties in more detail in the next section.

To illustrate the form of a cascade structure, consider for convenience the case when N is even, $M = N$, and $H(\omega)$ is represented as the product of second-order terms alone:

$$H(\omega) = \frac{b_N}{a_N} \prod_{k=1}^{N/2} \frac{\beta_{0k} + \beta_{1k}(j\omega) + (j\omega)^2}{\alpha_{0k} + \alpha_{1k}(j\omega) + (j\omega)^2} \tag{4.155}$$

This corresponds to multiplying together pairs of the remaining first-order terms in both the numerator and denominator of eq. (4.154).

Having $H(\omega)$ in the form of eq. (4.155), we can then realize the LTI system with this frequency response as a cascade of $N/2$ second-order systems, each of which is described by a differential equation of the form

$$\frac{d^2y(t)}{dt^2} + \alpha_{1k}\frac{dy(t)}{dt} + \alpha_{0k}y(t) = \beta_{0k}x(t) + \beta_{1k}\frac{dx(t)}{dt} + \frac{d^2x(t)}{dt^2} \tag{4.156}$$

In Section 3.6 we saw how to realize such a differential equation using adders, coefficient multipliers, and integrators, and in Figure 4.37 we have illustrated the cascade realization of a sixth-order system using the direct form II realization (Figure 3.36) for each subsystem of the form of eq. (4.156). Note that the cascade structure for a given $H(\omega)$ is by no means unique. For example, we have arbitrarily paired second-order numerator polynomials with second-order denominator polynomials and have also chosen an arbitrary pairing of the first-order terms in eq. (4.154) that are to be multiplied together. Problems 4.50 and 4.51 contain examples that illustrate the cascade structure and also indicate the flexibility in the choice of a cascade structure for an LTI system with a given rational frequency response.

The second realization that we can now describe is the *parallel-form* structure which is obtained by performing a partial fraction expansion of $H(\omega)$ in eq. (4.145) or, equivalently, in eq. (4.153). For simplicity, let us assume that all of the v_k in eq. (4.153) are distinct and that $M = N$. In this case, a partial fraction expansion yields

$$H(\omega) = \left(\frac{b_N}{a_N}\right) + \sum_{k=1}^{N} \frac{A_k}{v_k + j\omega} \tag{4.157}$$

Again, in order to obtain an implementation involving only real coefficients, we can add together the pairs involving complex conjugate v_k's to obtain

$$H(\omega) = \left(\frac{b_N}{a_N}\right) + \sum_{k=1}^{Q} \frac{\gamma_{0k} + \gamma_{1k}(j\omega)}{\alpha_{0k} + \alpha_{1k}(j\omega) + (j\omega)^2} + \sum_{l=1}^{N-2Q} \frac{A_l}{v_l + j\omega} \tag{4.158}$$

Thus, using eq. (4.158), we can realize the LTI system with frequency response $H(\omega)$ as the parallel interconnection of LTI systems with frequency responses corresponding to each term in eq. (4.158). To illustrate the parallel-form structure, consider the case in which N is even and $H(\omega)$ is represented as a sum of only second-order terms:

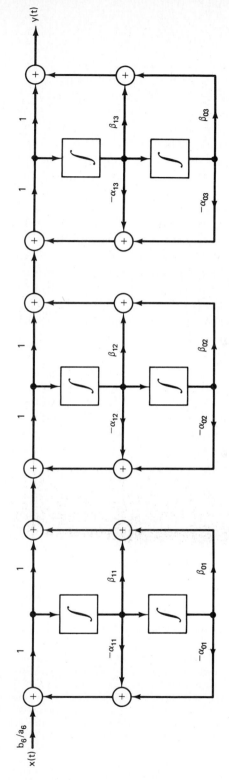

Figure 4.37 Cascade structure with a direct form II realization of each second-order subsystem.

$$H(\omega) = \left(\frac{b_N}{a_N}\right) + \sum_{k=1}^{N/2} \frac{\gamma_{0k} + \gamma_{1k}(j\omega)}{\alpha_{0k} + \alpha_{1k}(j\omega) + (j\omega)^2} \qquad (4.159)$$

This corresponds to adding together pairs of the remaining first-order terms in eq. (4.158). In Figure 4.38 we have illustrated the parallel-form realization for a sixth-order system where we have used the direct form II realization of Figure 3.36 for each term in eq. (4.159). Other examples of parallel-form structures are considered in Problems 4.50 and 4.51.

From the discussions in this section and in Section 3.6 the reader can see that there are a wide variety of possible structures that can be chosen for the implementation of LTI systems described by linear constant-coefficient differential equations. While all of these are equivalent in the sense that they ideally provide realizations of the same specified frequency response, there is the unavoidable fact that in practice implementations corresponding to different structures do *not* perform identically. For example, in any real system it is impossible to set the coefficients in a structure to the precise values desired and in fact these coefficient values may drift over time due, for example, to temperature variations. This raises the issue of the sensitivity of a realization to changes in its parameters, and in general different structures have different sensitivity properties. While we will not discuss this topic in this book, the methods of analysis we have developed provide the foundation for the examination of this and related questions which are of great importance in the choice of a structure for the implementation of an LTI system.†

In this section we have seen that the properties of the Fourier transform greatly facilitate the analysis of LTI systems characterized by linear constant-coefficient differential equations and in fact reduce many of the necessary calculations to straightforward algebraic manipulations. It is important to note, however, that not all LTI systems described by linear constant-coefficient differential equations have frequency responses. For example, if we had considered the case of $a < 0$ in Example 4.25, then the impulse response of the system specified by eq. (4.146) and the assumption of initial rest is still given by $e^{-at}u(t)$, but in this case $h(t)$ is not absolutely integrable and $H(\omega)$ does not exist. Thus, the expression for the frequency response in eq. (4.147) or more generally in eq. (4.145) yields the frequency response of an LTI system only when the system *has* a frequency response (i.e., when its impulse response is absolutely integrable or equivalently when the system is stable). Therefore, whenever we consider the use of the tools of Fourier analysis for LTI systems described by differential equations we will be assuming implicitly that the system has a frequency response, which can, of course, be checked by computing the impulse response and seeing if it is absolutely integrable. In Chapter 9 we develop techniques very similar to those described here that can be used for stable *and* unstable systems.

†We refer the interested reader to S. J. Mason and H. J. Zimmermann, *Electronic Circuits, Signals, and Systems* (New York: John Wiley and Sons, Inc., 1960) for a brief, general introduction to the subject of sensitivity and to A. V. Oppenheim and R. W. Schafer, *Digital Signal Processing* (Englewood Cliffs, N.J.: Prentice-Hall, Inc., 1975) for a discussion of sensitivity and a number of other issues that arise in the choice of a structure for implementation. While the discussion in *Digital Signal Processing* focuses on discrete-time systems, the general concepts introduced and discussed therein are also relevant for the implementation of continuous-time systems.

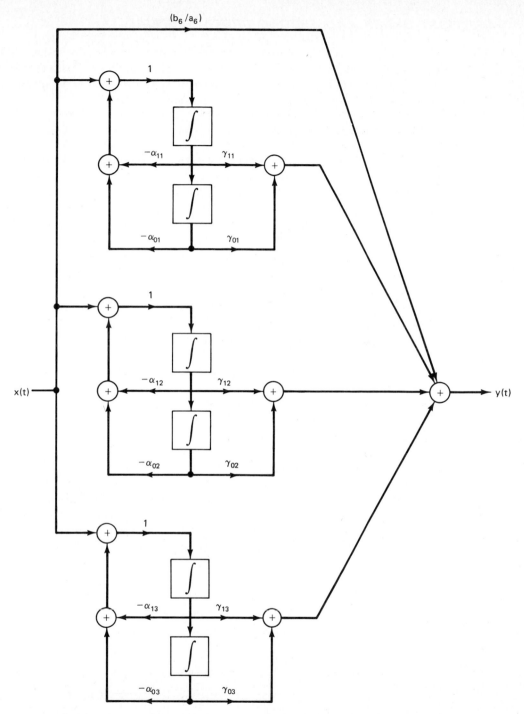

Figure 4.38 Parallel-form realization with a direct form II realization for each second-order subsystem.

4.12 FIRST-ORDER AND SECOND-ORDER SYSTEMS

As we have just seen in the preceding section, first- and second-order systems represent basic building blocks out of which we can construct parallel or cascade realizations of systems with higher-order frequency responses. In this section we investigate the properties of these basic systems in somewhat greater detail.

4.12.1 First-Order Systems

The differential equation for a first-order system is often expressed in the form

$$\tau \frac{dy(t)}{dt} + y(t) = x(t) \tag{4.160}$$

where τ is a coefficient whose significance will be made clear shortly. The corresponding frequency response for the first-order system is

$$H(\omega) = \frac{1}{j\omega\tau + 1} \tag{4.161}$$

and its impulse response is

$$h(t) = \frac{1}{\tau} e^{-t/\tau} u(t) \tag{4.162}$$

which is sketched in Figure 4.39(a). In addition, the step response of the system is given by

$$s(t) = h(t) * u(t) = [1 - e^{-t/\tau}]u(t) \tag{4.163}$$

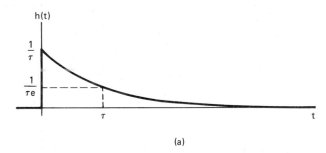

(a)

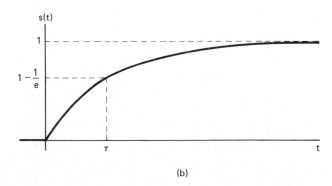

(b)

Figure 4.39   (a) Impulse response and (b) step response of a first-order system.

This is sketched in Figure 4.39(b). The parameter τ is called the *time constant* of the system, and it controls the rate at which the first-order system responds. For example, as illustrated in Figure 4.39, at $t = \tau$ the impulse response has reached $(1/e)$ times its value at $t = 0$, and the step response is within $1/e$ of its final value. Therefore, as τ is decreased, the impulse response decays more sharply, and the rise time of the step response becomes shorter. That is, the step response rises more sharply toward its final value.

Figure 4.40 depicts the Bode plot of the frequency response of eq. (4.161), where we have plotted the log magnitude and angle of $H(\omega)$ versus ω, using a logarithmic scale for ω. In this figure we illustrate another advantage in using a logarithmic frequency scale, as we can, without too much difficulty, obtain a useful approximate Bode plot for a first-order system. To see this, let us first examine the plot of the log magnitude of the frequency response. Specifically, from eq. (4.161) we obtain

$$20 \log_{10} |H(\omega)| = -10 \log_{10} [(\omega\tau)^2 + 1] \tag{4.164}$$

From this we see that for $\omega\tau \ll 1$, the log magnitude is approximately zero, while for $\omega\tau \gg 1$, the log magnitude is approximately a *linear* function of $\log_{10}(\omega)$. That is,

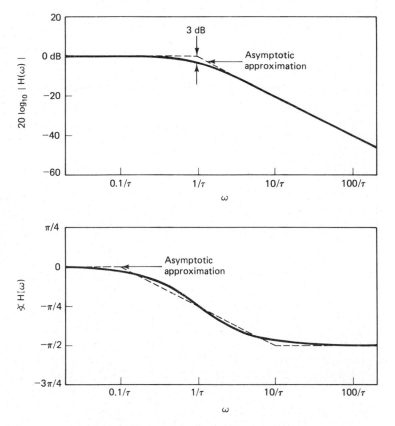

Figure 4.40   Bode plot for a first-order system.

$$20 \log_{10} |H(\omega)| \simeq 0 \qquad \text{for } \omega \ll 1/\tau \tag{4.165a}$$

$$20 \log_{10} |H(\omega)| \simeq -20 \log_{10} (\omega\tau) = -20 \log_{10} (\omega) - 20 \log_{10} (\tau)$$
$$\text{for } \omega \gg 1/\tau \tag{4.165b}$$

In other words, for a first-order system, the low and high frequency asymptotes of the log magnitude are straight lines. The low frequency asymptote [given by eq. (4.165a)] is just the 0 dB line, while the high frequency asymptote [specified by eq. (4.165b)] corresponds to a decrease of 20 dB in $|H(\omega)|$ for every decade, i.e., factor of 10, in ω. This is sometimes referred to as the "20 dB per decade" asymptote.

Note that the two asymptotic approximations given in eq. (4.165) are equal at the point $\log_{10} (\omega) = -\log_{10} (\tau)$, or equivalently $\omega = 1/\tau$. Interpreted graphically, this means that the two straight-line asymptotes meet at $\omega = 1/\tau$, and this suggests a straight-line approximation to the magnitude plot. That is, our approximation to $20 \log_{10} |H(\omega)|$ equals 0 for $\omega \le 1/\tau$ and is given by eq. (4.165b) for $\omega \ge 1/\tau$. This approximation is also sketched (as a dashed line) in Figure 4.40. The point at which the slope of the approximation changes is precisely $\omega = 1/\tau$ which, for this reason, is often referred to as the *break frequency*. Also, note that at $\omega = 1/\tau$ the two terms $[(\omega\tau)^2$ and $1]$ in the logarithm in eq. (4.164) are equal. Thus at this point the actual value of the magnitude is

$$20 \log_{10} \left| H\left(\frac{1}{\tau}\right) \right| = -10 \log_{10} (2) \simeq -3 \text{ dB} \tag{4.166}$$

For this reason the point $\omega = 1/\tau$ is sometimes called the *3dB point*. From Figure 4.40 we see that only near the break frequency is there any significant error in the straight-line approximate Bode plot. Thus, if we wish to obtain a more accurate sketch of the Bode plot, we need only modify the approximation near the break frequency.

It is also possible to obtain a useful straight-line approximation to $\measuredangle H(\omega)$. Specifically,

$$\measuredangle H(\omega) = \tan^{-1}(\omega\tau) \simeq \begin{cases} 0 & \omega \le \dfrac{0.1}{\tau} \\[2mm] -\dfrac{\pi}{4}[\log_{10}(\omega\tau) + 1] & \dfrac{0.1}{\tau} \le \omega \le \dfrac{10}{\tau} \\[2mm] -\dfrac{\pi}{2} & \omega \ge \dfrac{10}{\tau} \end{cases} \tag{4.167}$$

Note that this approximation decreases linearly (from 0 to $-\pi/2$) as a function of ω in the range

$$\frac{0.1}{\tau} \le \omega \le \frac{10}{\tau}$$

Also, zero is the correct asymptotic value of $\measuredangle H(\omega)$ for $\omega \ll 1/\tau$, and $-\pi/2$ is the correct asymptotic value of $\measuredangle H(\omega)$ for $\omega \gg 1/\tau$. Furthermore, the approximation agrees with the actual value of $\measuredangle H(\omega)$ at the break frequency $\omega = 1/\tau$, at which point

$$\measuredangle H\left(\frac{1}{\tau}\right) = -\frac{\pi}{4} \tag{4.168}$$

This asymptotic approximation is also plotted in Figure 4.40, and from this we can see how, if desired, we can modify the straight-line approximation to obtain a more accurate sketch of $\angle H(\omega)$.

From this first-order system we can again see the inverse relationship between time and frequency. As we make τ smaller, we speed up the time response of the system [i.e., $h(t)$ becomes more compressed toward the origin] and we simultaneously make the break frequency large [i.e., $H(\omega)$ becomes broader since $|H(\omega)| \simeq 1$ for a larger range of frequencies]. This can also be seen by multiplying the impulse response by τ and observing the relationship between $\tau h(t)$ and $H(\omega)$:

$$\tau h(t) = e^{-t/\tau} u(t), \qquad H(\omega) = \frac{1}{j\omega\tau + 1}$$

Thus, $\tau h(t)$ is a function of t/τ and $H(\omega)$ is a function of $\omega\tau$, and from this we see that changing τ is essentially equivalent to a scaling in time and frequency.

4.12.2 Second-Order Systems

The linear constant-coefficient differential equation for a second-order system is

$$\frac{d^2 y(t)}{dt^2} + 2\zeta\omega_n \frac{dy(t)}{dt} + \omega_n^2 y(t) = \omega_n^2 x(t) \tag{4.169}$$

Equations of this type arise in many physical systems, including *RLC* circuits and mechanical systems, such as the one illustrated in Figure 4.41, composed of a spring,

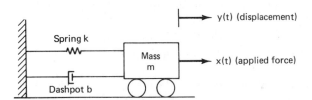

Figure 4.41 Second-order system consisting of a spring and dashpot attached to a movable mass and a fixed support.

a mass, and a viscous damper or dashpot. The input is the applied force $x(t)$, and the output is the displacement of the mass $y(t)$ from some equilibrium position at which the spring exerts no restoring force. The equation of motion for this system is

$$m \frac{d^2 y(t)}{dt^2} = x(t) - k y(t) - b \frac{dy(t)}{dt}$$

or

$$\frac{d^2 y(t)}{dt^2} + \left(\frac{b}{m}\right) \frac{dy(t)}{dt} + \left(\frac{k}{m}\right) y(t) = \frac{1}{m} x(t)$$

Comparing this to eq. (4.169), we see that if we identify

$$\omega_n = \sqrt{\frac{k}{m}}$$

$$\zeta = \frac{b}{2\sqrt{km}} \tag{4.170}$$

then [except for a scale factor of k on $x(t)$] the equation of motion for the system of Figure 4.41 reduces to eq. (4.169).

The frequency response for the second-order system of eq. (4.169) is

$$H(\omega) = \frac{\omega_n^2}{(j\omega)^2 + 2\zeta\omega_n(j\omega) + \omega_n^2} \qquad (4.171)$$

The denominator of $H(\omega)$ can be factored to yield

$$H(\omega) = \frac{\omega_n^2}{(j\omega - c_1)(j\omega - c_2)}$$

where

$$c_1 = -\zeta\omega_n + \omega_n\sqrt{\zeta^2 - 1}$$
$$c_2 = -\zeta\omega_n - \omega_n\sqrt{\zeta^2 - 1} \qquad (4.172)$$

For $\zeta \neq 1$, c_1 and c_2 are different, and we can perform a partial fraction expansion to obtain

$$H(\omega) = \frac{M}{j\omega - c_1} - \frac{M}{j\omega - c_2} \qquad (4.173)$$

where

$$M = \frac{\omega_n}{2\sqrt{\zeta^2 - 1}} \qquad (4.174)$$

In this case we can immediately obtain the impulse response for the system as

$$h(t) = M[e^{c_1 t} - e^{c_2 t}]u(t) \qquad (4.175)$$

If $\zeta = 1$, then $c_1 = c_2 = -\omega_n$, and

$$H(\omega) = \frac{\omega_n^2}{(j\omega + \omega_n)^2} \qquad (4.176)$$

From Table 4.1 we find that in this case the impulse response is

$$h(t) = \omega_n^2 t e^{-\omega_n t} u(t) \qquad (4.177)$$

Note that from eqs. (4.172), (4.174), (4.175), and (4.177), we can see that $h(t)/\omega_n$ is a function of $\omega_n t$. Furthermore, eq. (4.171) can be rewritten as

$$H(\omega) = \frac{1}{\left(j\dfrac{\omega}{\omega_n}\right)^2 + 2\zeta\left(j\dfrac{\omega}{\omega_n}\right) + 1}$$

from which we see that the frequency response is a function of ω/ω_n. Thus, changing ω_n is essentially identical to a time and frequency scaling.

The parameter ζ is referred to as the *damping ratio* and the parameter ω_n as the *undamped natural frequency*. The motivation for this terminology becomes clear when we take a more detailed look at the impulse response and the step response of a second-order system. First, from eq. (4.172) we see that for $0 < \zeta < 1$, c_1 and c_2 are complex, and we can rewrite the impulse response in eq. (4.175) in the form

$$h(t) = \frac{\omega_n e^{-\zeta\omega_n t}}{2j\sqrt{1 - \zeta^2}}\{\exp[j(\omega_n\sqrt{1 - \zeta^2})t] - \exp[-j(\omega_n\sqrt{1 - \zeta^2})t]\}u(t)$$

$$= \frac{\omega_n e^{-\zeta\omega_n t}}{\sqrt{1 - \zeta^2}}[\sin(\omega_n\sqrt{1 - \zeta^2})t]u(t) \qquad (4.178)$$

Thus, for $0 < \zeta < 1$ the second-order system has an impulse response that has a damped oscillatory behavior, and in this case the system is referred to as being *underdamped*. If $\zeta > 1$, both c_1 and c_2 are real and the impulse response is the difference between two decaying exponentials. In this case the system is *overdamped*. The case of $\zeta = 1$, when $c_1 = c_2$, is called the *critically damped* case. The impulse responses (multiplied by $1/\omega_n$) for second-order systems with different values of ζ are plotted versus t in Figure 4.42(a).

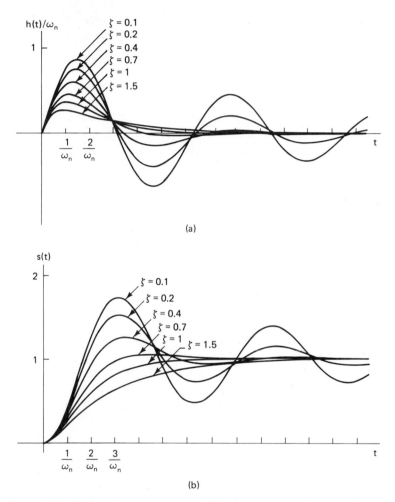

Figure 4.42 (a) Impulse responses and (b) step responses for second-order systems with different values of the damping ratio ζ.

The step response of a second-order system can be calculated from eq. (4.175) for $\zeta \neq 1$. This yields the expression

$$s(t) = h(t) * u(t) = \left\{ 1 + M\left[\frac{e^{c_1 t}}{c_1} - \frac{e^{c_2 t}}{c_2} \right] \right\} u(t) \tag{4.179}$$

For $\zeta = 1$, we can use eq. (4.177) to obtain

$$s(t) = [1 - e^{-\omega_n t} - \omega_n t e^{-\omega_n t}] u(t) \tag{4.180}$$

The step response of a second-order system is plotted versus t in Figure 4.42(b) for several values of ζ. From this figure we see that in the underdamped case, the step response exhibits both *overshoot* (i.e., the step response exceeds its final value) and *ringing* (i.e., oscillatory behavior). For $\zeta = 1$, the step response has the fastest response (i.e., the shortest rise time) that is possible without overshoot. As ζ increases beyond 1, the response becomes slower. This can be seen from eqs. (4.172) and (4.179). As ζ increases, c_1 becomes smaller in magnitude, while c_2 increases in magnitude. Therefore, although the time constant $(1/|c_2|)$ of $e^{c_2 t}$ decreases, the time constant $(1/|c_1|)$ of $e^{c_1 t}$ increases, and it is this fact that leads to the slow response for large values of ζ. In terms of our spring–dashpot example, as we increase the magnitude of the damping coefficient b beyond the critical value at which ζ in eq. (4.170) equals 1, the motion of the mass becomes more and more sluggish. Finally, note that, as we have said, the value of ω_n essentially controls the time scale of the responses $h(t)$ and $s(t)$. For example, in the underdamped case we have that the larger ω_n is, the more compressed is the impulse response as a function of t and the higher the frequency of the oscillations in both $h(t)$ and $s(t)$. In fact, from eq. (4.178) we see that the frequency of the oscillations in $h(t)$ and $s(t)$ is $\omega_n \sqrt{1 - \zeta^2}$, which does increase with increasing ω_n. Note, however, that this frequency depends explicitly on the damping ratio and does not equal (and is in fact smaller than) ω_n except in the *undamped* case, $\zeta = 0$. For our spring–dashpot example, we therefore conclude that the rate of oscillation of the mass equals ω_n when no dashpot is present, and the oscillation frequency decreases when we include the dashpot.

In Figure 4.43 we have depicted the Bode plot of the frequency response given in eq. (4.171) for several values of ζ. In this case we have plotted $20 \log_{10} |H(\omega)|$ and $\sphericalangle H(\omega)$ versus ω, using a logarithmic frequency scale. As in the first-order case, the logarithmic frequency scale leads to linear high- and low-frequency asymptotes for the log magnitude. Specifically, from eq. (4.171) we have that

$$20 \log_{10} |H(\omega)| = -10 \log_{10} \left\{ \left[1 - \left(\frac{\omega}{\omega_n} \right)^2 \right]^2 + 4\zeta^2 \left(\frac{\omega}{\omega_n} \right)^2 \right\} \qquad (4.181)$$

From this expression we can deduce that

$$20 \log_{10} |H(\omega)| \simeq 0 \qquad \qquad \text{for } \omega \ll \omega_n \qquad (4.182a)$$

$$20 \log_{10} |H(\omega)| \simeq -40 \log_{10} \left(\frac{\omega}{\omega_n} \right) = -40 \log_{10} (\omega) + 40 \log_{10} (\omega_n)$$

$$\text{for } \omega \gg \omega_n \qquad (4.182b)$$

Therefore the low frequency asymptote of the log magnitude is the 0 dB line, while the high frequency asymptote (given by eq. (4.182b)) has a slope of -40 dB per decade, i.e. $|H(\omega)|$ decreases by 40 dB for every increase in ω of a factor of 10. Also, note that the two straight line asymptotes meet at the point $\omega = \omega_n$. Thus, we obtain a straight-line approximation to the log magnitude by using the approximation given in eq. (4.182a) for $\omega \leq \omega_n$ and using the straight line of eq. (4.182b) for $\omega \geq \omega_n$. For this reason, ω_n is known as the break frequency of the second-order system. This approximation is also plotted (as a dashed line) in Figure 4.43.

We can also obtain a straight-line approximation to $\sphericalangle H(\omega)$ whose exact expression can be obtained from eq. (4.171) as

$$\sphericalangle H(\omega) = -\tan^{-1} \left(\frac{2\zeta(\omega/\omega_n)}{1 - (\omega/\omega_n)^2} \right) \qquad (4.183)$$

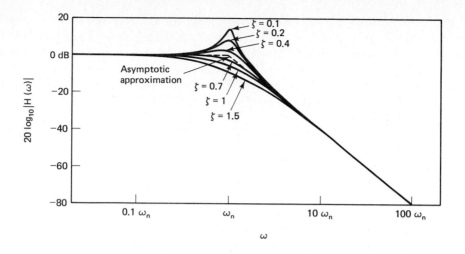

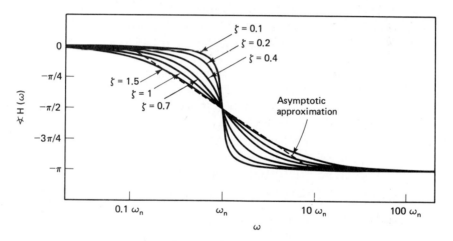

Figure 4.43 Bode plots for second-order systems with several different values of damping ratio ζ.

Our approximation to $\sphericalangle\, H(\omega)$ is

$$\sphericalangle\, H(\omega) \simeq \begin{cases} 0 & \omega \leq 0.1\omega_n \\ -\dfrac{\pi}{2}\left[\log_{10}\left(\dfrac{\omega}{\omega_n}\right) + 1\right] & 0.1\omega_n \leq \omega \leq 10\omega_n \\ -\pi & \omega \geq 10\omega_n \end{cases} \qquad (4.184)$$

which is also plotted in Figure 4.43. Note that the approximation and the actual value again are equal at the break frequency $\omega = \omega_n$, where

$$\sphericalangle\, H(\omega_n) = -\frac{\pi}{2}$$

It is important to observe that the asymptotic approximations we have obtained for a second-order system do not depend on ζ, while the actual plots of $|H(\omega)|$ and

$\angle H(\omega)$ certainly do, and thus to obtain an accurate sketch, especially near the break frequency $\omega = \omega_n$, one must take this discrepancy into account by modifying the approximations to conform more closely to the actual plots. This discrepancy is most pronounced for small values of ζ. In particular, note that in this case the actual log-magnitude has a peak around $\omega = \omega_n$. In fact, straightforward calculations using eq. (4.181) show that for $\zeta < \sqrt{2}/2 \simeq 0.707$, $|H(\omega)|$ has a maximum value at

$$\omega_{max} = \omega_n\sqrt{1 - 2\zeta^2} \qquad (4.185)$$

and the value at this maximum point is

$$|H(\omega_{max})| = \frac{1}{2\zeta\sqrt{1 - \zeta^2}} \qquad (4.186)$$

For $\zeta > 0.707$, however, $H(\omega)$ decreases monotonically as ω increases from zero. The fact that $H(\omega)$ can have a peak is extremely important in the design of *RLC* circuits. In some applications one may want to design such a circuit so that it has a sharp peak in the magnitude of its frequency response at some specified frequency, thereby providing large amplifications for sinusoids at frequencies within a narrow band. The *quality Q* of such a circuit is defined to be a measure of the sharpness of the peak. For a second-order circuit described by an equation of the form of eq. (4.169), Q is usually taken as

$$Q = \frac{1}{2\zeta}$$

and from Figure 4.43 and eq. (4.186) we see that this definition has the proper behavior: the less damping there is in the system, the sharper the peak in $|H(\omega)|$.

In the preceding section and again at the start of this section, we indicated that first- and second-order systems can be used as basic building blocks for more complex LTI systems with rational frequency responses. One consequence of this is that the Bode plots given in this section essentially provide us with all of the information we need to construct Bode plots for arbitrary rational frequency responses. Specifically, in this section we have described the Bode plots for the frequency responses given by eqs. (4.161) and (4.171). In addition, we can readily obtain the Bode plots for

$$H(\omega) = 1 + j\omega\tau$$

and

$$H(\omega) = 1 + 2\zeta\left(\frac{j\omega}{\omega_n}\right) + \left(\frac{j\omega}{\omega_n}\right)^2$$

from Figures 4.40 and 4.43 by using the fact that

$$20\log_{10}|H(\omega)| = -20\log_{10}\left|\frac{1}{H(\omega)}\right|$$

and

$$\angle(H(\omega)) = -\angle\left(\frac{1}{H(\omega)}\right)$$

Furthermore, since a rational frequency response can be factored into the product of first- and second-order terms, its Bode plot can be obtained by summing the plots for each of the terms. For example, consider

$$H(\omega) = \frac{1}{(1 + j\omega)(1 + j\omega/100)}$$

The Bode plots for the two first-order factors in $H(\omega)$ are shown in Figure 4.44(a)

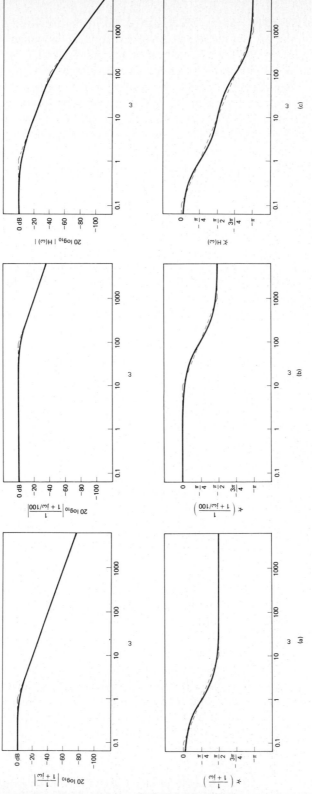

Figure 4.44 Construction of the Bode plot of a higher-order frequency response from those of its first-order factors: (a) Bode plot for $1/(1+j\omega)$; (b) Bode plot for $1/(1+j\omega/100)$; (c) Bode plot for $H(\omega) = 1/[(1+j\omega)(1+j\omega/100)]$.

and (b), while the Bode plot for $H(\omega)$ is depicted in Figure 4.44(c). Here the actual Bode plot and the asymptotic approximation are each obtained by summing the corresponding curves for the two first-order factors. Several other examples of the construction of Bode plots in this manner are considered in Problem 4.45.

Finally, note that in our discussion of first-order systems we restricted our attention to values of $\tau > 0$, and for second-order systems we examined only the case when $\zeta > 0$ and $\omega_n^2 > 0$. It is not difficult to check that if any of these parameters are negative, then the corresponding impulse response is not absolutely integrable. Thus, in this section we have restricted attention to those first- and second-order systems which are stable and consequently for which we can define frequency responses.

4.13 SUMMARY

In this chapter we have developed the tools of Fourier analysis for continuous-time signals and systems. As we discussed in Sections 4.1 and 4.2, one of the primary motivations for the use of Fourier analysis is the fact that complex exponential signals are eigenfunctions of continuous-time LTI systems. That is, if the input to an LTI system is a complex exponential, then the output is simply that same complex exponential scaled by a complex constant. The other important property of complex exponentials is that a wide variety of signals can be represented as weighted sums or integrals of these basic signals. In this chapter we have focused our attention on the set of periodic complex exponentials. Using these signals we considered the Fourier series representation of periodic signals and the Fourier transform representation of aperiodic signals, and we also described in detail the relationship between these representations.

The Fourier transform possesses a wide variety of important properties that describe how different characteristics of signals are reflected in their transforms. Among these properties are two that have particular significance for our study of signals and systems. The first of these is the convolution property. Because of this property one is led to the description of an LTI system in terms of its frequency response. This description plays a fundamental role in the frequency-domain approach to the analysis of LTI systems which we will explore at greater length in subsequent chapters. A second property of the Fourier transform that has extremely important implications is the modulation property. This property provides the basis for the frequency-domain analysis of modulation and sampling systems which we examine further in Chapters 7 and 8.

Finally, in this chapter we have seen that the tools of Fourier analysis are particularly well suited to the examination of LTI systems characterized by linear constant-coefficient differential equations. Specifically, we have found that the frequency response for such a system could be determined by inspection and that the technique of partial fraction expansion could then be used to facilitate the calculation of the impulse response of the system. The form of the frequency response for LTI systems specified by differential equations also led us directly to the development of the cascade and parallel-form structures for the implementation of such LTI systems. These structures point out the important role played by first- and second-order systems. We discussed the properties of these basic systems at some length and in the process utilized a convenient graphical representation, the Bode plot, for displaying the magnitude and phase of the frequency response of an LTI system.

The purpose of this chapter has been to introduce and to develop some facility with the tools of Fourier analysis and some appreciation for the value of the frequency domain in analyzing and understanding the properties of continuous-time signals and systems. In Chapter 5 we develop an analogous set of tools for the discrete-time case, and in Chapters 6 through 8 we will use the techniques of continuous- and discrete-time Fourier analysis as we examine the topics of filtering, modulation, and sampling.

PROBLEMS

4.1. Determine the Fourier series representations for each of the following signals.

    (a) e^{j200t}

    (b) $\cos{[\pi(t-1)/4]}$

    (c) $\cos 4t + \sin 8t$

    (d) $\cos 4t + \sin 6t$

    (e) $x(t)$ is periodic with period 2, and

$$x(t) = e^{-t} \quad \text{for } -1 < t < 1$$

    (f) $x(t)$ as illustrated in Figure P4.1(a)

    (g) $x(t) = [1 + \cos 2\pi t][\cos (10\pi t + \pi/4)]$

    (h) $x(t)$ is periodic with period 2, and

$$x(t) = \begin{cases} (1-t) + \sin 2\pi t, & 0 < t < 1 \\ 1 + \sin 2\pi t, & 1 < t < 2 \end{cases}$$

    (i) $x(t)$ as depicted in Figure P4.1(b)

    (j) $x(t)$ as depicted in Figure P4.1(c)

    (k) $x(t)$ as depicted in Figure P4.1(d).

    (l) $x(t)$ is periodic with period 4 and

$$x(t) = \begin{cases} \sin \pi t & 0 \le t \le 2 \\ 0 & 2 \le t \le 4 \end{cases}$$

    (m) $x(t)$ as depicted in Figure P4.1(e)

    (n) $x(t)$ as depicted in Figure P4.1(f).

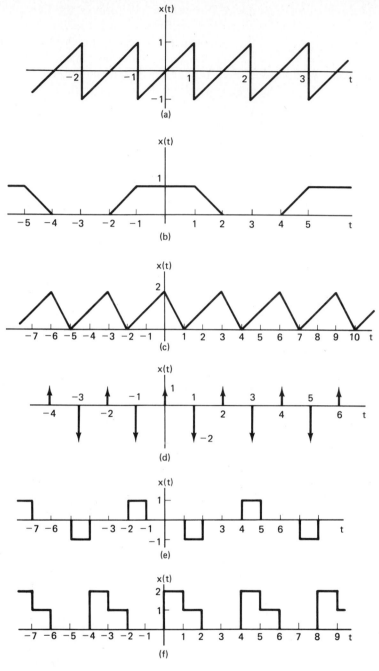

Figure P4.1

4.2. One technique for building a dc power supply is to take an ac signal and full-wave-rectify it. That is, we put the ac signal $x(t)$ through a system which produces $y(t) = |x(t)|$ as its output.

(a) Sketch the input and output waveforms if $x(t) = \cos t$. What are the fundamental periods of the input and output?

(b) If $x(t) = \cos t$, determine the coefficients of the Fourier series for the output $y(t)$.

(c) What is the amplitude of the dc component of the input signal? What is the amplitude of the dc component of the output signal?

4.3. As we have seen in this chapter, the concept of an eigenfunction is an extremely important tool in the study of LTI systems. It is also true that the same can be said of linear but time-varying systems. Specifically, consider such a system with input $x(t)$ and output $y(t)$. We say that a signal $\phi(t)$ is an *eigenfunction* of the system if

$$\phi(t) \longrightarrow \lambda\phi(t)$$

That is, if $x(t) = \phi(t)$, then $y(t) = \lambda\phi(t)$, where the complex constant λ is called the *eigenvalue associated with* $\phi(t)$.

(a) Suppose that we can represent the input $x(t)$ to our system as a linear combination of eigenfunctions $\phi_k(t)$, each of which has a corresponding eigenvalue λ_k.

$$x(t) = \sum_{k=-\infty}^{+\infty} c_k \phi_k(t)$$

Express the output $y(t)$ of the system in terms of $\{c_k\}$, $\{\phi_k(t)\}$, and $\{\lambda_k\}$.

(b) Consider the system characterized by the differential equation

$$y(t) = t^2 \frac{d^2 x(t)}{dt^2} + t \frac{dx(t)}{dt}$$

Is this system linear? Is this system time-invariant?

(c) Show that the set of functions

$$\phi_k(t) = t^k$$

are eigenfunctions of the system in part (b). For each $\phi_k(t)$, determine the corresponding eigenvalue λ_k.

(d) Determine the output of this system if

$$x(t) = 10t^{-10} + 3t + \tfrac{1}{2}t^4 + \pi$$

4.4. (a) Consider an LTI system with impulse response

$$h(t) = e^{-4t}u(t)$$

Find the Fourier series representation of the output $y(t)$ for each of the following inputs.

(i) $x(t) = \cos 2\pi t$

(ii) $x(t) = \sin 4\pi t + \cos(6\pi t + \pi/4)$

(iii) $x(t) = \sum_{n=-\infty}^{+\infty} \delta(t-n)$

(iv) $x(t) = \sum_{n=-\infty}^{+\infty} (-1)^n \delta(t-n)$

(v) $x(t)$ is the periodic square wave depicted in Figure P4.4.

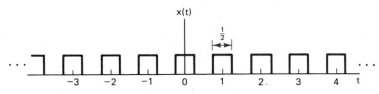

Figure P4.4

(b) Repeat part (a) for

$$h(t) = \begin{cases} \sin 2\pi t + \cos 4\pi t, & 0 \le t < 1 \\ 0, & \text{otherwise} \end{cases}$$

(c) Repeat part (a) for

$$h(t) = e^{-4|t|}$$

4.5. As we have seen, the techniques of Fourier analysis are of value in examining continuous-time LTI systems because periodic complex exponentials are eigenfunctions for LTI systems. In this problem we wish to substantiate the following statement: Although some LTI systems may have additional eigenfunctions, the complex exponentials are the *only* signals that are eigenfunctions for *every* LTI system.

(a) What are the eigenfunctions of the LTI system with unit impulse response $h(t) = \delta(t)$? What are the associated eigenvalues?

(b) Consider the LTI system with unit impulse response $h(t) = \delta(t - T)$. Find a signal that is not of the form e^{st} but that is an eigenfunction with eigenvalue 1. Similarly, find eigenfunctions with eigenvalues $\frac{1}{2}$ and 2 which are not complex exponentials. (*Hint:* You can find impulse trains that meet these requirements.)

(c) Consider a stable LTI system with impulse response $h(t)$ that is real and even. Show that $\cos \omega t$ and $\sin \omega t$ are eigenfunctions of this system.

(d) Consider the LTI system with impulse response $h(t) = u(t)$. Suppose that $\phi(t)$ is an eigenfunction of this system with eigenvalue λ. Find the differential equation that $\phi(t)$ must satisfy and solve this differential equation. This result together with those of the preceding parts should prove the validity of the statement made at the beginning of the problem.

4.6. Consider the signal

$$x(t) = \cos 2\pi t$$

Since $x(t)$ is periodic with a fundamental period of 1, it is also periodic with a period of N, where N is any positive integer. What are the Fourier series coefficients of $x(t)$ if we regard it as a periodic signal with period 3?

4.7. Two time functions $u(t)$ and $v(t)$ are said to be *orthogonal over the interval* (a, b) if

$$\int_a^b u(t) v^*(t) \, dt = 0 \qquad \text{(P4.7-1)}$$

If, in addition,

$$\int_a^b |u(t)|^2 \, dt = 1 = \int_a^b |v(t)|^2 \, dt$$

the functions are said to be *normalized* and hence are called *orthonormal*. A set of functions $\{\phi_k(t)\}$ is called an *orthogonal (orthonormal) set* if each pair of functions in this set is orthogonal (orthonormal).

(a) Consider the pairs of signals $u(t)$ and $v(t)$ depicted in Figure P4.7. Determine if each pair is orthogonal over the interval $(0, 4)$.

(b) Are the functions $\sin m\omega_0 t$ and $\sin n\omega_0 t$ orthogonal over the interval $(0, T)$, where $T = 2\pi/\omega_0$? Are they orthonormal?

(c) Repeat part (b) for the functions $\phi_m(t)$ and $\phi_n(t)$, where

$$\phi_k(t) = \frac{1}{\sqrt{T}}[\cos k\omega_0 t + \sin k\omega_0 t]$$

(d) Show that the set of functions $\phi_k(t) = e^{jk\omega_0 t}$ are orthogonal over *any* interval of length $T = 2\pi/\omega_0$. Are they also orthonormal?

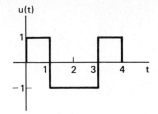

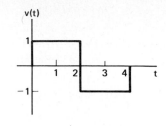

(a)

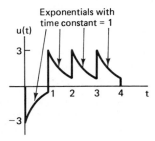

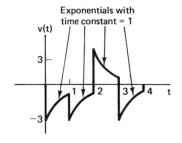

(b)

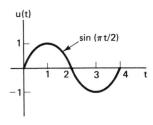

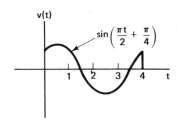

(c)

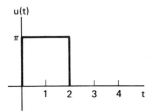

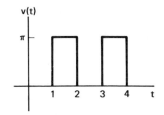

(d)

Figure P4.7

(e) Let $x(t)$ be an arbitrary signal and let $x_o(t)$ and $x_e(t)$ be, respectively, the odd and even parts of $x(t)$. Show that $x_o(t)$ and $x_e(t)$ are orthogonal over the interval $(-T, T)$ for any T.

(f) Show that if $\{\phi_k(t)\}$ is a set of orthogonal signals on the interval (a, b), then the set $\{(1/\sqrt{A_k})\phi_k(t)\}$ is orthonormal, where

$$A_k = \int_a^b |\phi_k(t)|^2 \, dt$$

(g) Let $\{\phi_i(t)\}$ be a set of orthonormal signals on the interval (a, b), and consider a signal of the form

$$x(t) = \sum_i a_i \phi_i(t)$$

where the a_i are complex constants. Show that

$$\int_a^b |x(t)|^2 \, dt = \sum_i |a_i|^2$$

(h) Suppose that $\phi_1(t), \ldots, \phi_N(t)$ are nonzero only over the time interval $0 \le t \le T$ and that they are orthonormal over this time interval. Let L_i denote the LTI system with impulse response

$$h_i(t) = \phi_i(T - t) \qquad \text{(P4.7-2)}$$

Show that if $\phi_j(t)$ is applied to this system, then the output at time T is 1 if $i = j$ and 0 if $i \ne j$. The system with impulse response given by eq. (P4.7-2) was referred to in Problems 3.27 and 3.28 as the *matched filter* for the signal $\phi_i(t)$.

4.8. The purpose of this problem is to show that the representation of an arbitrary periodic signal by a Fourier series, or more generally as a linear combination of any set of orthogonal functions, is computationally efficient and in fact is very useful for obtaining good approximations of signals.†

Specifically, let $\{\phi_i(t)\}$, $i = 0, \pm 1, \pm 2, \ldots$, be a set of orthonormal functions on the interval $a \le t \le b$, and let $x(t)$ be a given signal. Consider the following approximation of $x(t)$ over the interval $a \le t \le b$:

$$\hat{x}_N(t) = \sum_{i=-N}^{+N} a_i \phi_i(t) \qquad \text{(P4.8-1)}$$

where the a_i are constants (in general, complex). To measure the deviation between $x(t)$ and the series approximation $\hat{x}_N(t)$, we consider the error $e_N(t)$ defined as

$$e_N(t) = x(t) - \hat{x}_N(t) \qquad \text{(P4.8-2)}$$

A reasonable and widely used criterion for measuring the quality of the approximation is the energy in the error signal over the interval of interest, that is, the integral of the squared-error magnitude over the interval $a \le t \le b$:

$$E = \int_a^b |e_N(t)|^2 \, dt \qquad \text{(P4.8-3)}$$

(a) Show that E is minimized by choosing

$$a_i = \int_a^b x(t) \phi_i^*(t) \, dt \qquad \text{(P4.8-4)}$$

Hint: Use eqs. (P4.8-1)–(P4.8-3) to express E in terms of a_i, $\phi_i(t)$, and $x(t)$. Then, express a_i in rectangular coordinates as $a_i = b_i + jc_i$, and show that the equations

$$\frac{\partial E}{\partial b_i} = 0 \quad \text{and} \quad \frac{\partial E}{\partial c_i} = 0, \qquad i = 0, \pm 1, \pm 2, \ldots, \pm N$$

are satisfied by the a_i as given in eq. (P4.8-4).

(b) How does the result of part (a) change if the $\{\phi_i(t)\}$ are orthogonal but not orthonormal, with

$$A_i = \int_a^b |\phi_i(t)|^2 \, dt$$

†See Problem 4.7 for the definitions of orthogonal and orthonormal functions.

(c) Let $\phi_n(t) = e^{jn\omega_0 t}$ and choose any interval of length $T_0 = 2\pi/\omega_0$. Show that the a_i that minimize E are as given in eq. (4.45).

(d) The set of *Walsh functions* are an often-used set of orthonormal functions (see Problem 3.27). The first five Walsh functions, $\phi_0(t)$, $\phi_1(t)$, $\ldots$, $\phi_4(t)$, are illustrated in Figure P4.8, where we have scaled time so that the $\phi_i(t)$ are nonzero over the interval $0 \le t \le 1$ and are orthonormal on this interval. Let $x(t) = \sin \pi t$. Find the approximation of $x(t)$ of the form

$$\hat{x}(t) = \sum_{i=0}^{4} a_i \phi_i(t)$$

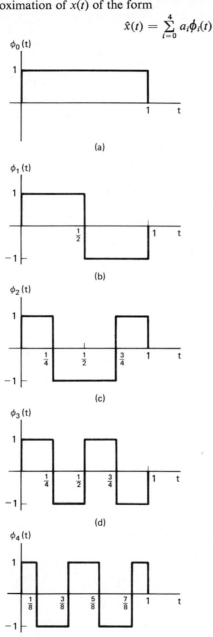

(a)

(b)

(c)

(d)

(e)

Figure P4.8

such that

$$\int_0^1 |x(t) - \hat{x}(t)|^2 \, dt$$

is minimized.

(e) Show that $\hat{x}_N(t)$ in eq. (P4.8-1) and $e_N(t)$ in eq. (P4.8-2) are orthogonal if the a_i are chosen as in eq. (P4.8-4).

The results of parts (a) and (b) are extremely important in that they show that each coefficient a_i is *independent* of all other a_j's, $i \neq j$. Thus, if we add more terms to the approximation [e.g., if we compute the approximation $\hat{x}_{N+1}(t)$], the coefficients of $\phi_i(t)$, $i = 1, \ldots, N$, previously determined will not change. In contrast to this, consider another type of series expansion, the polynomial Taylor series. The *infinite* Taylor series for e^t is given by $e^t = 1 + t + t^2/2! + \ldots$ but as we shall show, when we consider a *finite* polynomial series and the error criterion of eq. (P4.8-3), we get a very different result.

Specifically, let $\phi_0(t) = 1$, $\phi_1(t) = t$, $\phi_2(t) = t^2$, and so on.

(f) Are the $\phi_i(t)$ orthogonal over the interval $0 \leq t \leq 1$?

(g) Consider an approximation of $x(t) = e^t$ over the interval $0 \leq t \leq 1$ of the form

$$\hat{x}_0(t) = a_0 \phi_0(t)$$

Find the value of a_0 that minimizes the energy in the error signal over the interval.

(h) We now wish to approximate e^t by a Taylor series using two terms $\hat{x}_1(t) = a_0 + a_1 t$. Find the optimum values for a_0 and a_1. [*Hint:* Compute E in terms of a_0 and a_1, and then solve the simultaneous equations

$$\frac{\partial E}{\partial a_0} = 0 \quad \text{and} \quad \frac{\partial E}{\partial a_1} = 0$$

Note that your answer for a_0 has changed from its value in part (g), where there was only one term in the series. Further, as you increase the number of terms in the series, that coefficient and all others will continue to change. We can thus see the advantage to be gained in expanding a function using orthogonal terms.]

4.9. Let $x(t)$ be a periodic signal, with fundamental period T_0 and Fourier series coefficients a_k. Consider each of the following signals. The Fourier series coefficients for each can be expressed in terms of the a_k as is done in Table 4.2. Show that the expression in Table 4.2 is correct for each of these signals.

(a) $x(t - t_0)$

(b) $x(-t)$

(c) $x^*(t)$

(d) $\int_{-\infty}^t x(\tau) \, d\tau$ (for this part, assume that $a_0 = 0$)

(e) $\dfrac{dx(t)}{dt}$

(f) $x(\alpha t)$, $\alpha > 0$ (for this part, determine the period of the signal)

4.10.† As we discussed in the text, the origins of Fourier analysis can be found in problems of mathematical physics. In particular, the work of Fourier was motivated by his investigation of heat diffusion. In this problem we illustrate how Fourier series enter into this investigation.

Consider the problem of determining the temperature at a given depth beneath

†This problem has been adapted from A. Sommerfeld, *Partial Differential Equations in Physics* (New York: Academic Press, 1949), pp. 68–71.

the surface of the earth as a function of time, where we assume that the temperature at the surface is a given function of time $T(t)$, periodic with period 1 (where the unit of time is one year). Let $T(x, t)$ denote the temperature at a depth x below the surface at time t. This function obeys the heat diffusion equation

$$\frac{\partial T(x, t)}{\partial t} = \frac{1}{2} k^2 \frac{\partial^2 T(x, t)}{\partial x^2} \tag{P4.10-1}$$

with auxiliary condition

$$T(0, t) = T(t) \tag{P4.10-2}$$

Here k is the heat diffusion constant for the earth ($k > 0$). Suppose that we expand $T(t)$ in a Fourier series

$$T(t) = \sum_{n=-\infty}^{+\infty} a_n e^{jn2\pi t} \tag{P4.10-3}$$

Similarly, let us expand $T(x, t)$ at any given depth x in a Fourier series in t:

$$T(x, t) = \sum_{n=-\infty}^{+\infty} b_n(x) e^{jn2\pi t} \tag{P4.10-4}$$

where the Fourier coefficients $b_n(x)$ depend upon depth x.

(a) Use eqs. (P4.10-1)–(P4.10-4) to show that $b_n(x)$ satisfies the differential equation

$$\frac{d^2 b_n(x)}{dx^2} = \frac{4\pi jn}{k^2} b_n(x) \tag{P4.10-5a}$$

with auxiliary condition

$$b_n(0) = a_n \tag{P4.10-5b}$$

Since eq. (P4.10-5a) is a second-order equation, we need a second auxiliary condition. On physical grounds we can argue that far below the earth's surface the variations in temperature due to surface fluctuations should disappear. That is,

$$\lim_{x \to \infty} T(x, t) = \text{a constant} \tag{P4.10-5c}$$

(b) Show that the solution to eq. (P4.10-5) is

$$b_n(x) = \begin{cases} a_n \exp\left[-\sqrt{2\pi|n|}\,(1+j)x/k\right], & n \geq 0 \\ a_n \exp\left[-\sqrt{2\pi|n|}\,(1-j)x/k\right], & n \leq 0 \end{cases}$$

(c) Thus, the temperature oscillations at depth x are damped and phase-shifted versions of the temperature oscillations at the surface. To see this more clearly, let

$$T(t) = a_0 + a_1 \sin 2\pi t$$

(so that a_0 represents the mean yearly temperature). Sketch $T(t)$ and $T(x, t)$ over a one-year period for

$$x = k\sqrt{\frac{\pi}{2}}$$

$a_0 = 2$, and $a_1 = 1$. Note that at this depth the temperature oscillations are not only significantly damped, but the phase shift is such that it is warmest in winter and coldest in summer. This is exactly why vegetable cellars are constructed!

4.11. Consider the closed contour, shown in Figure P4.11. As illustrated, we can view this curve as being traced out by the tip of a rotating vector of varying length. Let $r(\theta)$ denote the length of the vector as a function of the angle θ. Then $r(\theta)$ is periodic in θ with period 2π, and thus has a Fourier series representation. Let $\{a_k\}$ denote the Fourier coefficients of $r(\theta)$.

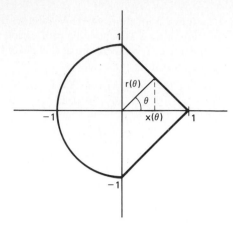

Figure P4.11

(a) Consider now the projection $x(\theta)$ of the vector $r(\theta)$ onto the x axis, as indicated in the figure. Determine the Fourier coefficients for $x(\theta)$ in terms of the a_k's.

(b) Consider the sequence of coefficients

$$b_k = a_k e^{jk\pi/4}$$

Sketch the figure in the plane that corresponds to this set of coefficients.

(c) Repeat part (b) if

$$b_k = a_k \delta[k]$$

(d) Sketch figures in the plane such that $r(\theta)$ is *not* constant but does have each of the following properties:

(i) $r(\theta)$ is even.

(ii) The fundamental period of $r(\theta)$ is π.

(iii) The fundamental period of $r(\theta)$ is $\pi/2$.

4.12. **(a)** A continuous-time periodic signal $x(t)$ with period T is said to be *odd-harmonic* if in its Fourier series representation

$$x(t) = \sum_{k=-\infty}^{+\infty} a_k e^{jk(2\pi/T)t} \qquad \text{(P4.12-1)}$$

$a_k = 0$ for every even integer k.

(i) Show that if $x(t)$ is odd-harmonic, then

$$x(t) = -x\left(t + \frac{T}{2}\right) \qquad \text{(P4.12-2)}$$

(ii) Show that if $x(t)$ satisfies eq. (P4.12-2), then it is odd-harmonic.

(b) Suppose that $x(t)$ is an odd-harmonic periodic signal with period 2, such that

$$x(t) = t \qquad \text{for } 0 < t < 1$$

Sketch $x(t)$ and find its Fourier series coefficients.

(c) Analogously, we could define an even-harmonic function as one for which $a_k = 0$ for k odd in the representation in eq. (P4.12-1). Could T be the fundamental period for such a signal? Explain your answer.

(d) More generally, show that T is the fundamental period of $x(t)$ in eq. (P4.12-1) if one of two things happens:

    1. Either a_1 or a_{-1} is nonzero.

or

    2. There are two integers k and l that have no common factors and are such that both a_k and a_l are nonzero.

4.13. Let $x(t)$ be a real, periodic signal with Fourier series representation given in the sine–cosine form of eq. (4.20):

$$x(t) = a_0 + 2 \sum_{k=1}^{\infty} [B_k \cos k\omega_0 t - C_k \sin k\omega_0 t] \qquad \text{(P4.13-1)}$$

(a) Find the exponential Fourier series representation of the even and odd parts of $x(t)$; that is, find the coefficients α_k and β_k in terms of the coefficients in eq. (P4.13-1) so that

$$\mathcal{E}v\{x(t)\} = \sum_{k=-\infty}^{+\infty} \alpha_k e^{jk\omega_0 t}$$

$$\mathcal{O}d\{x(t)\} = \sum_{k=-\infty}^{+\infty} \beta_k e^{jk\omega_0 t}$$

(b) What is the relationship between α_k and α_{-k} of part (a)? What is the relationship between β_k and β_{-k}?

(c) Suppose that the signals $x(t)$ and $z(t)$ shown in Figure P4.13 have the sine–cosine series representations

$$x(t) = a_0 + 2 \sum_{k=1}^{\infty} \left[B_k \cos \left(\frac{2\pi k t}{3} \right) - C_k \sin \left(\frac{2\pi k t}{3} \right) \right]$$

$$z(t) = d_0 + 2 \sum_{k=1}^{\infty} \left[E_k \cos \left(\frac{2\pi k t}{3} \right) - F_k \sin \left(\frac{2\pi k t}{3} \right) \right]$$

Sketch the signal

$$y(t) = 4(a_0 + d_0) + 2 \sum_{k=1}^{\infty} \left\{ \left[B_k + \frac{1}{2} E_k \right] \cos \left(\frac{2\pi k t}{3} \right) + F_k \sin \left(\frac{2\pi k t}{3} \right) \right\}$$

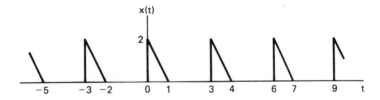

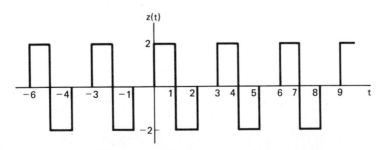

Figure P4.13

4.14. In this problem we derive the Fourier series counterparts of two important properties of the continuous-time Fourier transform—the modulation property and Parseval's theorem. Let $x(t)$ and $y(t)$ both be continuous-time periodic signals having period T_0

with Fourier series representations given by

$$x(t) = \sum_{k=-\infty}^{+\infty} a_k e^{jk\omega_0 t}, \qquad y(t) = \sum_{k=-\infty}^{+\infty} b_k e^{jk\omega_0 t} \qquad \text{(P4.14-1)}$$

(a) Show that the Fourier series coefficients of the signal

$$z(t) = x(t)y(t) = \sum_{k=-\infty}^{+\infty} c_k e^{jk\omega_0 t}$$

are given by the discrete convolution

$$c_k = \sum_{n=-\infty}^{+\infty} a_n b_{k-n}$$

(b) Use the result of part (a) to compute the Fourier series coefficients of the signals $x_1(t)$, $x_2(t)$, and $x_3(t)$ depicted in Figure P4.14.

(c) Suppose that $y(t)$ in eq. (P4.14-1) equals $x^*(t)$. Express the b_k in eq. (P4.14-1) in terms of the a_k and use the result of part (a) to prove Parseval's theorem for peri-

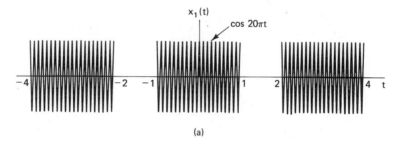

(a)

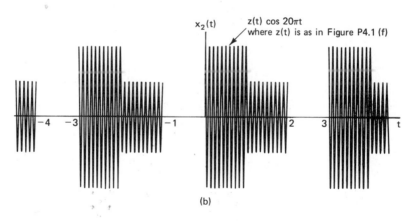

(b)

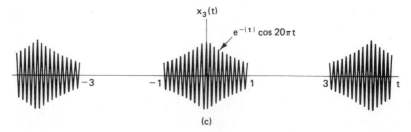

(c)

Figure P4.14

Fourier Analysis for Continuous-Time Signals and Systems   Chap. 4

odic signals, that is, to show that

$$\frac{1}{T_0} \int_0^{T_0} |x(t)|^2 \, dt = \sum_{k=-\infty}^{+\infty} |a_k|^2$$

4.15. Suppose that a periodic continuous-time signal is the input to an LTI system. The signal $x(t)$ has a Fourier series representation

$$x(t) = \sum_{k=-\infty}^{+\infty} \alpha^{|k|} e^{jk(\pi/4)t}$$

where α is a real number between 0 and 1, and the frequency response of the system is given by

$$H(\omega) = \begin{cases} 1 & |\omega| < W \\ 0 & |\omega| > W \end{cases}$$

How large must W be in order for the output of the system to have at least 90% of the average energy per period of $x(t)$?

4.16. In the text and in Problem 3.12 we defined the periodic convolution of two periodic signals $\tilde{x}_1(t)$ and $\tilde{x}_2(t)$ that have the same period T_0. Specifically, the periodic convolution of these signals is defined as

$$\tilde{y}(t) = \tilde{x}_1(t) \circledast \tilde{x}_2(t) = \int_{T_0} \tilde{x}_1(\tau) \tilde{x}_2(t - \tau) \, d\tau \tag{P4.16-1}$$

As shown in Problem 3.12, any interval of length T_0 can be used in the integral in eq. (P4.16-1), and $\tilde{y}(t)$ is also periodic with period T_0.

(a) If $\tilde{x}_1(t)$, $\tilde{x}_2(t)$, and $\tilde{y}(t)$ have Fourier series representations

$$\tilde{x}_1(t) = \sum_{k=-\infty}^{+\infty} a_k e^{jk(2\pi/T_0)t}, \qquad \tilde{x}_2(t) = \sum_{k=-\infty}^{+\infty} b_k e^{jk(2\pi/T_0)t},$$

$$\tilde{y}(t) = \sum_{k=-\infty}^{+\infty} c_k e^{jk(2\pi/T_0)t}$$

show that

$$c_k = T_0 a_k b_k$$

(b) Consider the periodic signal $\tilde{x}(t)$ depicted in Figure P4.16-1. This signal is the result of the periodic convolution of another periodic signal $\tilde{z}(t)$ with itself. Find $\tilde{z}(t)$ and then use part (a) to determine the Fourier series representation for $\tilde{x}(t)$.

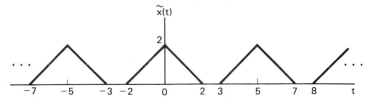

Figure P4.16-1

(c) Suppose now that $x_1(t)$ and $x_2(t)$ are the finite-duration signals illustrated in Figure P4.16-2(a) and (b). Consider forming the periodic signals $\tilde{x}_1(t)$ and $\tilde{x}_2(t)$, which consist of periodically repeated versions of $x_1(t)$ and $x_2(t)$ as illustrated for $\tilde{x}_1(t)$ in Figure P4.16-2(c). Let $y(t)$ be the usual, aperiodic convolution of $x_1(t)$ and $x_2(t)$,

$$y(t) = x_1(t) * x_2(t)$$

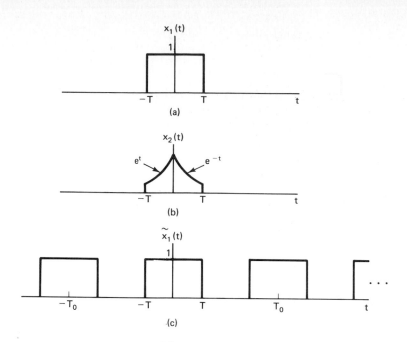

Figure P4.16-2

and let $\bar{y}(t)$ be the periodic convolution of $\tilde{x}_1(t)$ and $\tilde{x}_2(t)$,

$$\bar{y}(t) = \tilde{x}_1(t) \circledast \tilde{x}_2(t)$$

Show that if T_0 is large enough, we can recover $y(t)$ completely from one period of $\bar{y}(t)$, that is,

$$y(t) = \begin{cases} \bar{y}(t), & |t| \leq T_0/2 \\ 0, & |t| > T_0/2 \end{cases}$$

(d) Let $T = 1$ and let $x_1(t)$ and $x_2(t)$ be as in Figure P4.16-2(a) and (b). Use the results of parts (a) and (c) to find a representation of $y(t) = x_1(t) * x_2(t)$ of the form

$$y(t) = \begin{cases} \displaystyle\sum_{k=-\infty}^{+\infty} c_k e^{jk(2\pi/T_0)t}, & |t| \leq T_0/2 \\ 0, & |t| > T_0/2 \end{cases}$$

4.17. Compute the Fourier transform of each of the following signals.

(a) $[e^{-\alpha t} \cos \omega_0 t] u(t)$, $\alpha > 0$

(b) $e^{2+t} u(-t + 1)$

(c) $e^{-3|t|} \sin 2t$

(d) $e^{-3t}[u(t + 2) - u(t - 3)]$

(e) $x(t)$ as in Figure P4.17(a)

(f) $u_1(t) + 2\delta(3 - 2t)$

(g) $x(t) = \begin{cases} 1 + \cos \pi t, & |t| \leq 1 \\ 0, & |t| > 1 \end{cases}$

(h) $\displaystyle\sum_{k=0}^{\infty} \alpha^k \delta(t - kT)$, $|\alpha| < 1$

(i) $[te^{-2t} \sin 4t] u(t)$

(j) $\sin t + \cos (2\pi t + \pi/4)$

(k) $\left[\dfrac{\sin \pi t}{\pi t}\right]\left[\dfrac{\sin 2\pi(t - 1)}{\pi(t - 1)}\right]$

(l) $x(t)$ as in Figure P4.17(b)

(m) $x(t)$ as in Figure P4.17(c)

(n) $x(t)$ as in Figure P4.17(d)

(o) $x(t) = \begin{cases} 1 - t^2, & 0 < t < 1 \\ 0, & \text{otherwise} \end{cases}$

(p) $\displaystyle\sum_{n=-\infty}^{+\infty} e^{-|t-2n|}$

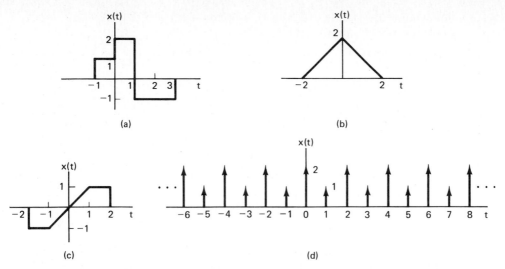

(a)

(b)

(c)

(d)

Figure P4.17

4.18. The following are the Fourier transforms of continuous-time signals. Determine the continuous-time signal corresponding to each transform.

(a) $X(\omega) = \dfrac{2 \sin [3(\omega - 2\pi)]}{(\omega - 2\pi)}$

(b) $X(\omega) = \cos (4\omega + \pi/3)$

(c) $X(\omega)$ as given by the magnitude and phase plots of Figure P4.18(a)

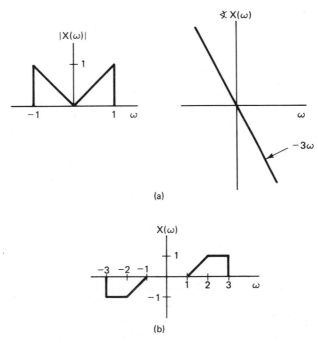

(a)

(b)

Figure P4.18

(d) $X(\omega) = 2[\delta(\omega - 1) - \delta(\omega + 1)] + 3[\delta(\omega - 2\pi) + \delta(\omega + 2\pi)]$

(e) $X(\omega)$ as in Figure P4.18(b)

4.19. Use properties of the Fourier transform to show by induction that the Fourier transform of

$$x(t) = \frac{t^{n-1}}{(n-1)!}e^{-at}u(t), \qquad a > 0$$

is

$$\frac{1}{(a+j\omega)^n}$$

4.20. Often one would like to determine the Fourier transform of a signal which is defined experimentally, by a set of measured values or by a trace on an oscilloscope, for which no closed-form analytic expression for the signal is known. Alternatively, one might wish to evaluate the Fourier transform of a signal that is defined precisely in closed form but is so complicated that the evaluation of its transform is virtually impossible. In both cases, however, one can use numerical methods to obtain an approximation to the Fourier transform to any desired accuracy. One such method is outlined in part (a) and is based on a piece-wise polynomial approximation of the signal to be transformed.

(a) If the function $x(t)$ to be transformed is sufficiently smooth, $x(t)$ may be approximated by a small number of polynomial pieces. In the following discussion, only first-order polynomials will be considered; the extension to higher-order polynomials follows in a straightforward manner. Figure P4.20-1 shows a function $x(t)$

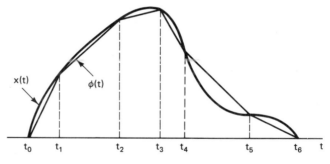

Figure P4.20-1

and a function $\phi(t)$ comprising straight-line segments (the points t_i indicate the beginning and end points of the linear segments). Since $\phi(t) \simeq x(t)$, then $\Phi(\omega) \simeq X(\omega)$. By evaluating the transform $\Phi(\omega)$, show that

$$\Phi(\omega) = \frac{1}{\omega^2} \sum_i k_i e^{-j\omega t_i}$$

Determine the k_i in terms of the time instants $t_0, t_1, t_2, \ldots$, and the values of $x(t)$ at these times.

(b) Suppose that $x(t)$ is the trapezoid shown in Figure P4.20-2. Since $x(t)$ already consists of linear segments no approximation is necessary. Determine $X(\omega)$ in this case.

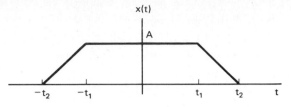

Figure P4.20-2

(c) Suppose that we want to evaluate numerically the Fourier transform of a signal $x(t)$ which is known to be time-limited, that is,

$$x(t) = 0 \qquad \text{for } |t| > T$$

Furthermore, suppose that we choose the points t_i sufficiently close together so that the first-order polynomial approximation $\phi(t)$ is sufficiently accurate to ensure that the absolute error, $|x(t) - \phi(t)|$, is bounded by some constant ϵ for $|t| < T$, that is,

$$E(t) = |x(t) - \phi(t)| \le \epsilon \qquad \text{for } |t| < T$$

[*Note:* Since $x(t)$ is time-limited, $E(t) = 0$ for $|t| > T$.] Show that the energy of the error in the Fourier transform approximation, $\Phi(\omega)$, to $X(\omega)$ is less than $4\pi T\epsilon^2$, that is

$$\int_{-\infty}^{\infty} |X(\omega) - \Phi(\omega)|^2 \, d\omega \le 4\pi T\epsilon^2$$

4.21. (a) Let $x(t)$ be a real, odd signal. Show that $X(\omega) = \mathfrak{F}\{x(t)\}$ is pure imaginary and odd.
(b) What property does the Fourier transform of a signal $x(t)$ have if $x(-t) = x^*(t)$?
(c) Consider a system with input $x(t)$ and output

$$y(t) = \mathfrak{Re}\{x(t)\}$$

Express the Fourier transform of $y(t)$ in terms of that for $x(t)$.
(d) Show that if $x(t)$ and $y(t)$ are two arbitrary signals with Fourier transforms $X(\omega)$ and $Y(\omega)$, respectively, then

$$\int_{-\infty}^{+\infty} x(t)y^*(t) \, dt = \frac{1}{2\pi} \int_{-\infty}^{+\infty} X(\omega) Y^*(\omega) \, d\omega$$

This result is a generalization of Parseval's theorem.
(e) Let $x(t)$ be a given signal with Fourier transform $X(\omega)$. Define the signal

$$f(t) = \frac{d^2 x(t)}{dt^2}$$

(i) Suppose that

$$X(\omega) = \begin{cases} 1 & |\omega| < 1 \\ 0 & |\omega| > 1 \end{cases}$$

What is the value of

$$\int_{-\infty}^{+\infty} |f(t)|^2 \, dt ?$$

(ii) What is the inverse Fourier transform of $f(\omega/4)$?
(f) Derive the modulation property

$$x(t)y(t) \quad \overset{\mathfrak{F}}{\longleftrightarrow} \quad \frac{1}{2\pi}[X(\omega) * Y(\omega)]$$

(g) Show that if the impulse response $h(t)$ of a continuous-time LTI system is real, then the response of the system to $x(t) = \cos \omega_0 t$ is given by

$$y(t) = \mathcal{R}e\{H(\omega_0)e^{j\omega_0 t}\}$$

where $H(\omega)$ is the system frequency response.

4.22. Consider the signal

$$x_0(t) = \begin{cases} e^{-t}, & 0 \le t \le 1 \\ 0, & \text{elsewhere} \end{cases}$$

Determine the Fourier transform of each of the signals shown in Figure P4.22. You should be able to do this by explicitly evaluating *only* the transform of $x_0(t)$ and then using properties of the Fourier transform.

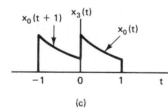

(a)

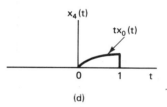

(b)

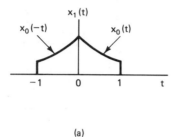

(c)

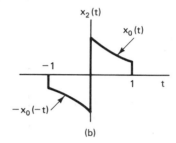

(d)

Figure P4.22

4.23. A real, continuous-time function $x(t)$ has a Fourier transform $X(\omega)$ whose magnitude obeys the relation

$$\ln |X(\omega)| = -|\omega|$$

Find $x(t)$ if $x(t)$ is known to be:
(a) an even function of time
(b) an odd function of time

4.24. (a) Determine which of the real signals depicted in Figure P4.24, if any, have Fourier transforms that satisfy each of the following:
(i) $\mathcal{R}e\{X(\omega)\} = 0$
(ii) $\mathcal{I}m\{X(\omega)\} = 0$
(iii) There exists a real α such that $e^{j\alpha\omega}X(\omega)$ is real.
(iv) $\displaystyle\int_{-\infty}^{\infty} X(\omega)\,d\omega = 0$
(v) $\displaystyle\int_{-\infty}^{\infty} \omega X(\omega)\,d\omega = 0$
(vi) $X(\omega)$ periodic
(b) Construct a signal that has properties (i), (iv), and (v) and does *not* have the others.

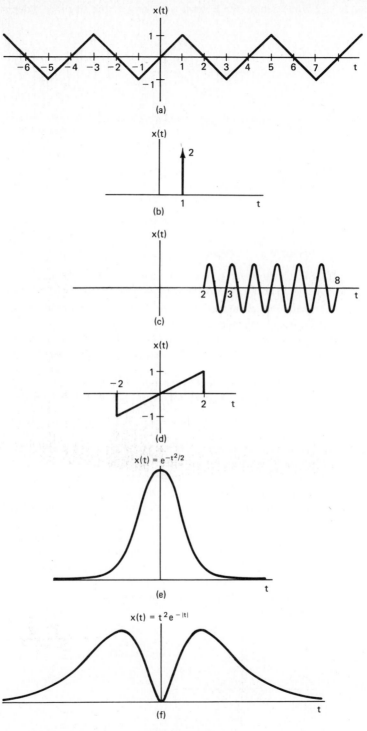

Figure P4.24

4.25. Let $X(\omega)$ denote the Fourier transform of the signal $x(t)$ depicted in Figure P4.25.

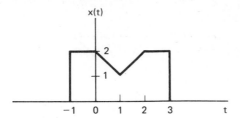

Figure P4.25

(a) Find $\sphericalangle X(\omega)$.

(b) Find $X(0)$.

(c) Find $\displaystyle\int_{-\infty}^{+\infty} X(\omega)\, d\omega$.

(d) Evaluate $\displaystyle\int_{-\infty}^{+\infty} X(\omega)\frac{2\sin\omega}{\omega}e^{j2\omega}\, d\omega$.

(e) Evaluate $\displaystyle\int_{-\infty}^{+\infty} |X(\omega)|^2\, d\omega$.

(f) Sketch the inverse Fourier transform of $\mathcal{R}e\{X(\omega)\}$.

Note: you should perform all of these calculations without explicitly evaluating $X(\omega)$.

4.26. As mentioned in the text, the techniques of Fourier analysis can be extended to signals having two independent variables. These techniques play as important a role in some applications, such as image processing, as their one-dimensional counterparts do in others. In this problem we introduce some of the elementary ideas of two-dimensional Fourier analysis.

Let $x(t_1, t_2)$ be a signal depending upon the two-independent variables t_1 and t_2. The *two-dimensional Fourier transform* of $x(t_1, t_2)$ is defined as

$$X(\omega_1, \omega_2) = \int_{-\infty}^{+\infty} \int_{-\infty}^{+\infty} x(t_1, t_2)e^{-j(\omega_1 t_1 + \omega_2 t_2)}\, dt_1\, dt_2$$

(a) Show that this double integral can be performed as two successive one-dimensional Fourier transforms, first in t_1, with t_2 regarded as fixed, and then in t_2.

(b) Use the result of part (a) to determine the inverse transform, that is, an expression for $x(t_1, t_2)$ in terms of $X(\omega_1, \omega_2)$.

(c) Determine the two-dimensional Fourier transforms of the following signals:

(i) $x(t_1, t_2) = e^{-t_1 + 2t_2}u(t_1 - 1)u(2 - t_2)$

(ii) $x(t_1, t_2) = \begin{cases} e^{-|t_1| - |t_2|} & \text{if } -1 < t_1 \le 1 \text{ and } -1 \le t_2 \le 1 \\ 0 & \text{otherwise} \end{cases}$

(iii) $x(t_1, t_2) = \begin{cases} e^{-|t_1| - |t_2|} & \text{if } 0 \le t_1 \le 1 \text{ or } 0 \le t_2 \le 1 \text{ (or both)} \\ 0 & \text{otherwise} \end{cases}$

(iv) $x(t_1, t_2)$ as depicted in Figure P4.26-1

(v) $e^{-|t_1 + t_2| - |t_1 - t_2|}$

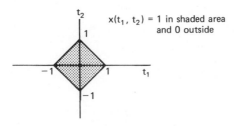

$x(t_1, t_2) = 1$ in shaded area and 0 outside

Figure P4.26-1

(d) Determine the signal $x(t_1, t_2)$ whose two-dimensional Fourier transform is

$$X(\omega_1, \omega_2) = \frac{2\pi}{4 + j\omega_1} \delta(\omega_2 - 2\omega_1)$$

(e) Let $x(t_1, t_2)$ and $h(t_1, t_2)$ be two signals with two-dimensional Fourier transforms $X(\omega_1, \omega_2)$ and $H(\omega_1, \omega_2)$. Determine the transforms of the following signals in terms of $X(\omega_1, \omega_2)$ and $H(\omega_1, \omega_2)$.

(i) $x(t_1 - T_1, t_2 - T_2)$
(ii) $x(at_1, bt_2)$
(iii) $y(t_1, t_2) = \displaystyle\int_{-\infty}^{+\infty} \int_{-\infty}^{+\infty} x(\tau_1, \tau_2) h(t_1 - \tau_1, t_2 - \tau_2)\, d\tau_1\, d\tau_2$

(f) Just as one can define two-dimensional Fourier transforms, one can also consider two-dimensional Fourier series for periodic signals with two independent variables. Specifically, consider a signal $x(t_1, t_2)$ which satisfies the equation

$$x(t_1, t_2) = x(t_1 + T_1, t_2) = x(t_1, t_2 + T_2) \qquad \text{for all } t_1, t_2$$

This signal is periodic with period T_1 in the t_1 direction and with period T_2 in the t_2 direction. Such a signal has a series representation of the form

$$x(t_1, t_2) = \sum_{n=-\infty}^{+\infty} \sum_{m=-\infty}^{+\infty} a_{mn} e^{j(m\omega_1 t_1 + n\omega_2 t_2)}$$

where

$$\omega_1 = 2\pi/T_1, \qquad \omega_2 = 2\pi/T_2$$

Find an expression for a_{mn} in terms of $x(t_1, t_2)$.

(g) Determine the Fourier series coefficients a_{mn} for the following signals:
(1) $\cos(2\pi t_1 + 2t_2)$
(2) The signal illustrated in Figure P4.26-2.

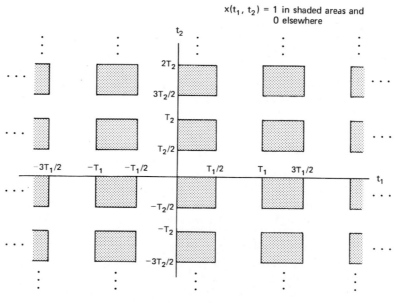

Figure P4.26-2

4.27. (a) Compute the convolution of each of the following pairs of signals $x(t)$ and $h(t)$ by calculating $X(\omega)$ and $H(\omega)$, using the convolution property, and inverse transforming.

    (i) $x(t) = te^{-2t}u(t)$, $h(t) = e^{-4t}u(t)$

    (ii) $x(t) = te^{-2t}u(t)$, $h(t) = te^{-4t}u(t)$

    (iii) $x(t) = e^{-t}u(t)$, $h(t) = e^{t}u(-t)$

(b) Suppose that $x(t) = e^{-(t-2)}u(t-2)$ and $h(t)$ is as depicted in Figure P4.27. Verify

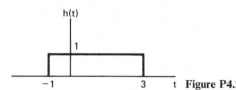

the convolution property for this pair of signals by showing that the Fourier transform of $y(t) = x(t) * h(t)$ equals $H(\omega)X(\omega)$.

4.28. As we pointed out in Chapter 3, the convolution integral representation for continuous-time LTI systems emphasizes the fact that an LTI system is completely specified by its response to $\delta(t)$. It is also true, as illustrated in Problems 3.17 and 3.18, that continuous-time or discrete-time LTI systems are also completely specified by their responses to other specific inputs. On the other hand, there are some inputs for which this is not the case; that is, many different LTI systems may have the identical response to one of these inputs.

(a) To illustrate this point, show that the three LTI systems with impulse responses

$$h_1(t) = u(t)$$
$$h_2(t) = -2\delta(t) + 5e^{-2t}u(t)$$
$$h_3(t) = 2te^{-t}u(t)$$

all have the same response to $x(t) = \cos t$.

(b) Find the impulse response of another LTI system with the same response to $\cos t$.

4.29. In Problems 2.23 and 3.28 we defined and examined several of the properties and uses of correlation functions. In this problem we examine the properties of such functions in the frequency domain. Let $x(t)$ and $y(t)$ be two real signals. The cross-correlation function $\phi_{xy}(t)$ is defined as

$$\phi_{xy}(t) = \int_{-\infty}^{+\infty} x(t + \tau)y(\tau)\, d\tau$$

Similarly, we can define $\phi_{yx}(t)$, $\phi_{xx}(t)$, and $\phi_{yy}(t)$ [the last two of these are called the autocorrelation functions of the signals $x(t)$ and $y(t)$, respectively]. Let $\Phi_{xy}(\omega)$, $\Phi_{yx}(\omega)$, $\Phi_{xx}(\omega)$, and $\Phi_{yy}(\omega)$ denote the Fourier transforms of $\phi_{xy}(t)$, $\phi_{yx}(t)$, $\phi_{xx}(t)$, and $\phi_{yy}(t)$, respectively.

(a) What is the relationship between $\Phi_{xy}(\omega)$ and $\Phi_{yx}(\omega)$?

(b) Find an expression for $\Phi_{xy}(\omega)$ in terms of $X(\omega)$ and $Y(\omega)$.

(c) Show that $\Phi_{xx}(\omega)$ is real and nonnegative for every ω.

(d) Suppose now that $x(t)$ is the input to an LTI system with a real-valued impulse response and with frequency response $H(\omega)$, and that $y(t)$ is the output. Find expressions for $\Phi_{xy}(\omega)$ and $\Phi_{yy}(\omega)$ in terms of $\Phi_{xx}(\omega)$ and $H(\omega)$.

(e) Let $x(t)$ be as is illustrated in Figure P4.29, and let the LTI system impulse response be $h(t) = e^{-at}u(t)$, $a > 0$. Compute $\Phi_{xx}(\omega)$, $\Phi_{xy}(\omega)$, and $\Phi_{yy}(\omega)$ using the results of the preceding parts of this problem.

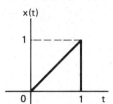

Figure P4.29

(f) Suppose that we are given the following Fourier transform of a function $\phi(t)$:

$$\Phi(\omega) = \frac{\omega^2 + 100}{\omega^2 + 25}$$

Find the impulse responses of *two* causal, stable LTI systems that both have autocorrelation functions equal to $\phi(t)$. Which one of these has a stable, causal inverse?

4.30. As we pointed out in this chapter, LTI systems with impulse responses that have the general form of the sinc function play an important role in LTI system analysis. The reason for this importance will become clear in Chapters 6 through 8 when we discuss the topics of filtering, modulation, and sampling. In this problem and several that follow we give some indication of the special properties of systems of this type.

(a) Consider the signal

$$x(t) = \cos 2\pi t + \sin 6\pi t$$

Suppose that this signal is the input to each of the LTI systems with impulse responses given below. Determine the output in each case.

(i) $\quad h(t) = \dfrac{\sin 4\pi t}{\pi t}$

(ii) $\quad h(t) = \dfrac{[\sin 4\pi t][\sin 8\pi t]}{\pi t^2}$

(iii) $\quad h(t) = \dfrac{[\sin 4\pi t][\cos 8\pi t]}{\pi t}$

(b) Consider an LTI system with impulse response

$$h(t) = \frac{\sin 2\pi t}{\pi t}$$

Determine the output $y_i(t)$ for each of the following input waveforms $x_i(t)$.

(i) $\quad x_1(t) = $ the symmetric square wave depicted in Figure P4.30(a)

(ii) $\quad x_2(t) = $ the symmetric square wave depicted in Figure P4.30(b)

(iii) $x_3(t) = x_1(t) \cos 5\pi t$

(iv) $x_4(t) = \displaystyle\sum_{k=-\infty}^{+\infty} \delta\left(t - \frac{10k}{3}\right)$

(v) $\quad x_5(t)$ is a real signal whose frequency response for positive frequencies has a constant phase angle of $\pi/2$ and whose magnitude for $\omega > 0$ is sketched in Figure P4.30(c).

(vi) $x_6(t) = \dfrac{1}{1 + t^2}$

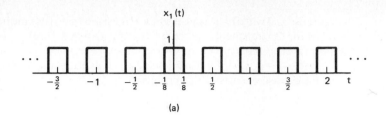

(a)

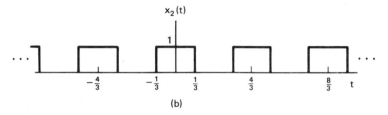

(b)

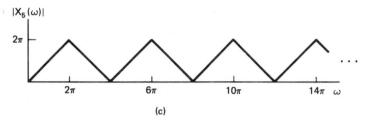

(c)

Figure P4.30

4.31. Consider the interconnection of four LTI systems depicted in Figure P4.31, where

$$h_1(t) = \frac{d}{dt}\left[\frac{\sin \omega_c t}{2\pi t}\right]$$

$$H_2(\omega) = e^{-j2\pi\omega/\omega_c}$$

$$h_3(t) = \frac{\sin 3\omega_c t}{\pi t}$$

$$h_4(t) = u(t)$$

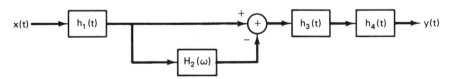

Figure P4.31

(a) Determine and sketch $H_1(\omega)$.

(b) What is the impulse response $h(t)$ of the entire system?

(c) What is the output $y(t)$ when the input is

$$x(t) = \sin 2\omega_c t + \cos (\omega_c t/2)?$$

4.32. An important property of the frequency response $H(\omega)$ of a continuous-time LTI system with a real, causal impulse response $h(t)$ is that $H(\omega)$ is completely specified by its real part, $\mathcal{R}e\{H(\omega)\}$. This problem is concerned with deriving and examining some of the implications of this property, which is generally referred to as *real-part sufficiency*.

(a) Prove the property of real-part sufficiency by examining the signal $h_e(t)$, which is the even part of $h(t)$. What is the Fourier transform of $h_e(t)$? Indicate how $h(t)$ can be recovered from $h_e(t)$.

(b) If the real part of the frequency response of a causal system is

$$\mathcal{R}e\{H(\omega)\} = \cos \omega$$

what is $h(t)$?

(c) Show that $h(t)$ can be recovered from $h_o(t)$, the odd part of $h(t)$, for every value of t except $t = 0$. Note that if $h(t)$ does not contain any singularities $[\delta(t), u_1(t), u_2(t),$ etc.] at $t = 0$, then the frequency response

$$H(\omega) = \int_{-\infty}^{+\infty} h(t) e^{-j\omega t} \, dt$$

will not change if the value of $h(t)$ is set to some arbitrary finite value at the single point $t = 0$. Thus, in this case, show that $H(\omega)$ is also completely specified by its imaginary part.

(d) Assume now that $h(t)$ does not have any singularities at $t = 0$. Then, in parts (a) and (b) we have seen that either the real or imaginary part of $H(\omega)$ completely determines $H(\omega)$. In this part we will derive an explicit relationship between $H_R(\omega)$ and $H_I(\omega)$, the real and imaginary parts of $H(\omega)$, under these conditions. To begin, note that since $h(t)$ is causal,

$$h(t) = h(t)u(t) \tag{P4.32-1}$$

except perhaps at $t = 0$. Since $h(t)$ contains no singularities at $t = 0$, the Fourier transforms of both sides of eq. (P4.32-1) must be identical. Use this fact, together with the modulation property, to show that

$$H(\omega) = \frac{1}{j\pi} \int_{-\infty}^{+\infty} \frac{H(\eta)}{\omega - \eta} \, d\eta \tag{P4.32-2}$$

Use eq. (P4.32-2) to determine an expression for $H_R(\omega)$ in terms of $H_I(\omega)$ and one for $H_I(\omega)$ in terms of $H_R(\omega)$.

(e) The operation

$$y(t) = \frac{1}{\pi} \int_{-\infty}^{+\infty} \frac{x(\tau)}{t - \tau} \, d\tau \tag{P4.32-3}$$

is called the *Hilbert transform*. We have just seen that for a real, causal impulse response $h(t)$, the real and imaginary parts of its transform can be determined from one another using the Hilbert transform.

Now consider eq. (P4.32-3) and regard $y(t)$ as the output of an LTI system with input $x(t)$. Show that the frequency response of this system is

$$H(\omega) = \begin{cases} -j, & \omega > 0 \\ j, & \omega < 0 \end{cases}$$

(f) What is the Hilbert transform of the signal $x(t) = \cos 3t$?

4.33. (a) Let $x(t)$ have Fourier transform $X(\omega)$, and let $p(t)$ be periodic with fundamental frequency ω_0 and Fourier series representation

$$p(t) = \sum_{n=-\infty}^{+\infty} a_n e^{jn\omega_0 t}$$

What is the Fourier transform of

$$y(t) = x(t)p(t) \qquad \text{(P4.33-1)}$$

(b) Suppose that $X(\omega)$ is as depicted in Figure P4.33-1. Sketch the spectrum of $y(t)$ in eq. (P4.33-1) for each of the following choices for $p(t)$.

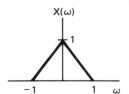

Figure P4.33-1

(i) $p(t) = \cos(t/2)$

(ii) $p(t) = \cos t$

(iii) $p(t) = \cos 2t$

(iv) $p(t) = (\sin t)(\sin 2t)$

(v) $p(t) = \cos 2t - \cos t$

(vi) $p(t) = \displaystyle\sum_{n=-\infty}^{+\infty} \delta(t - \pi n)$

(vii) $p(t) = \displaystyle\sum_{n=-\infty}^{+\infty} \delta(t - 2\pi n)$

(viii) $p(t) = \displaystyle\sum_{n=-\infty}^{+\infty} \delta(t - 4\pi n)$

(ix) $p(t) = \displaystyle\sum_{n=-\infty}^{+\infty} \delta(t - 2\pi n) - \frac{1}{2}\sum_{n=-\infty}^{+\infty} \delta(t - \pi n)$

(x) $p(t) =$ the periodic square wave shown in Figure P4.33-2

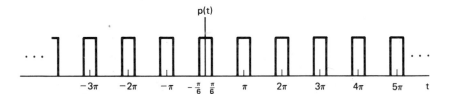

Figure P4.33-2

4.34. (a) Consider the system depicted in Figure P4.34, where

$$x(t) = \frac{\sin(3Wt/2)}{\pi t}$$

$$p(t) = \cos 2Wt + 4\cos 8Wt$$

$$h(t) = \sum_{k=-\infty}^{+\infty} c_k e^{jkWt}$$

Find the Fourier series representation for $y(t)$.

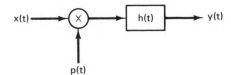

Figure P4.34

(b) Suppose now that $x(t)$ is real and is given by

$$x(t) = \sum_{k=-\infty}^{+\infty} a_k e^{jkWt}$$

$$p(t) = \cos Wt$$

$$h(t) = \frac{\sin (Wt/2)}{\pi t}$$

What is $y(t)$?

(c) What is $y(t)$ if $x(t)$ and $h(t)$ are as in part (b), but

$$p(t) = \sin Wt$$

(d) Again consider $x(t)$ and $h(t)$ as in part (b). How would you choose $p(t)$ in the system of Figure P4.34 if you wished to determine the real part of any particular Fourier coefficient of $x(t)$? How would you choose $p(t)$ to determine the imaginary part of the coefficient?

4.35. In the text we pointed out on several occasions that there is an inverse relationship between time and frequency. In this problem we examine in more detail one particular example of the consequences of this inverse relationship and in so doing introduce some qualitative concepts that are of great importance in signal and system analysis.

Let $H(\omega)$ be the frequency response of a continuous-time LTI system, and suppose that $H(\omega)$ is real, even, and positive. Also, assume that

$$\max_{\omega} \{H(\omega)\} = H(0)$$

(a) Show that:

(i) The impulse response, $h(t)$, is real.

(ii) $\max \{|h(t)|\} = h(0)$.

Hint: If $f(t, \omega)$ is a complex function of two variables, then

$$\left| \int_{-\infty}^{+\infty} f(t, \omega) \, d\omega \right| \le \int_{-\infty}^{+\infty} |f(t, \omega)| \, d\omega$$

One important concept in system analysis is the *bandwidth* of an LTI system. There are many different mathematical ways in which to define bandwidth, but they all are related to the qualitative and intuitive idea that a system with frequency response $G(\omega)$ essentially "stops" signals of the form $e^{j\omega t}$ for values of ω where $G(\omega)$ vanishes or is small and "passes" those complex exponentials in the band of frequency where $G(\omega)$ is not small. The width of this band is the bandwidth. These ideas will be made much clearer in our discussion of filtering in Chapter 6, but for now we will consider a special definition of bandwidth for those systems with frequency responses that have the properties specified previously for $H(\omega)$. Specifically, one definition of the bandwidth B_w of such a system is the width of a rectangle of height $H(0)$, which has an area equal to the area under $H(\omega)$. This is illustrated in Figure P4.35-1. Note that since $H(0) = \max_{\omega} H(\omega)$, the frequencies within the band indicated in the figure are those for which $H(\omega)$ is largest. The exact choice of the width in Figure P4.35-1 is of course a bit arbitrary, but we have chosen one definition that allows us to compare different systems and to make precise a very important time–frequency relationship.

(b) What is the bandwidth of the system with frequency response

$$H(\omega) = \begin{cases} 1, & |\omega| < W \\ 0, & |\omega| > W \end{cases}$$

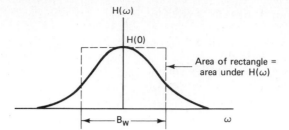

$H(\omega)$

$H(0)$

Area of rectangle =
area under $H(\omega)$

B_w

ω

Figure P4.35-1

(c) Find an expression for the bandwidth B_w in terms of $H(\omega)$.

Let $s(t)$ denote the step response of our system. An important measure of the speed of response of a system is the *rise time*, which again has a qualitative definition, leading to many possible mathematical definitions, one of which we will use. Intuitively, the rise time of a system is a measure of how fast the step response rises from zero to its final value,

$$s(\infty) = \lim_{t \to \infty} s(t)$$

Thus, the smaller the rise time, the faster the response of the system. For our system we will define the rise time t_r to be

$$t_r = \frac{s(\infty)}{h(0)}$$

Since

$$s'(t) = h(t)$$

and also because of the property that $h(0) = \max_t h(t)$, we see that t_r has the interpretation as the time it would take to go from zero to $s(\infty)$ while maintaining the maximum rate of change of $s(t)$. This is illustrated in Figure P4.35-2.

(d) Find an expression for t_r in terms of $H(\omega)$.

(e) Combine the results of parts (c) and (d) to show that

$$B_w t_r = 2\pi \qquad \text{(P4.35-1)}$$

Thus, we *cannot* independently specify both the rise time and bandwidth of our system. For example, if we want a fast system (t_r small), then eq. (P4.35-1) implies

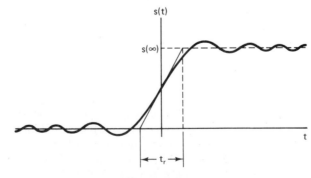

$s(t)$

$s(\infty)$

t

t_r

Figure P4.35-2

that the system must have a large bandwidth. This is a fundamental trade-off that is of central importance in many problems of system design.

4.36. (a) Consider two LTI systems with impulse responses $h(t)$ and $g(t)$, respectively, and suppose that these systems are inverses of one another. Suppose also that both systems have frequency responses, denoted by $H(\omega)$ and $G(\omega)$, respectively. What is the relationship between $H(\omega)$ and $G(\omega)$?

(b) Consider the continuous-time LTI system with frequency response

$$H(\omega) = \begin{cases} 1, & 2 < |\omega| < 3 \\ 0, & \text{otherwise} \end{cases}$$

(i) Is it possible to find an input $x(t)$ to this system so that the output is as depicted in Figure P4.29? If so, find such an $x(t)$. If not, explain why not.

(ii) Is this system invertible? Explain your answer.

(c) Consider an auditorium with an echo problem. As discussed in Problem 3.25, we can model the acoustics of the auditorium as an LTI system with an impulse response consisting of an impulse train, with the kth impulse in the train corresponding to the kth echo. Suppose that in this particular case the impulse response is given by

$$h(t) = \sum_{k=0}^{\infty} e^{-kT}\delta(t - kT)$$

where the factor e^{-kT} represents the attenuation of the kth echo.

In order to make a high-quality recording from the stage, the effect of the echoes must be removed by performing some processing of the sounds sensed by the recording equipment. In Problem 3.25 we used convolutional techniques to consider one example of the design of such a processor (for a different echo model). In this problem we will use frequency-domain techniques. Specifically, let $G(\omega)$ denote the frequency response of the LTI system to be used to process the sensed acoustic signal. Choose $G(\omega)$ so that the echoes are completely removed and the resulting signal is a faithful reproduction of the original stage sounds.

(d) Find a differential equation for the inverse of the system with impulse response

$$h(t) = 2\delta(t) + u_1(t)$$

(e) Consider the LTI system initially at rest and described by the differential equation

$$\frac{d^2y(t)}{dt^2} + 6\frac{dy(t)}{dt} + 9y(t) = \frac{d^2x(t)}{dt^2} + 3\frac{dx(t)}{dt} + 2x(t)$$

The inverse of this system is also initially at rest and described by a differential equation. Find the differential equation describing the inverse. Also, find the impulse responses $h(t)$ and $g(t)$ of the original system and its inverse.

4.37. A real-valued continuous-time function $x(t)$ has a Fourier transform $X(\omega)$ whose magnitude and phase are illustrated in Figure P4.37-1.

The functions $x_a(t)$, $x_b(t)$, $x_c(t)$, and $x_d(t)$ have Fourier transforms whose magnitudes are identical to $X(\omega)$ but whose phase functions differ, as shown in Figure P4.37-2. The phase functions $\sphericalangle X_a(\omega)$ and $\sphericalangle X_b(\omega)$ are formed by adding linear phase to $\sphericalangle X(\omega)$. The function $\sphericalangle X_c(\omega)$ is formed by reflecting $\sphericalangle X(\omega)$ about $\omega = 0$, and $\sphericalangle X_d(\omega)$ is obtained by a combination of reflection and addition of linear phase. Using the properties of Fourier transforms, determine expressions for $x_a(t)$, $x_b(t)$, $x_c(t)$, and $x_d(t)$ in terms of $x(t)$.

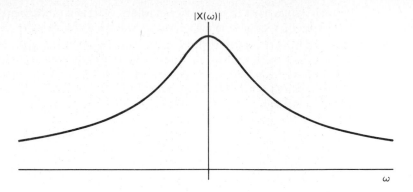

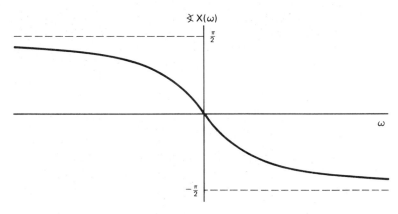

Figure P4.37-1

4.38. In this problem we provide additional examples of the effects of nonlinear changes in phase.

(a) Consider a continuous-time LTI system with frequency response

$$H(\omega) = \frac{a - j\omega}{a + j\omega}$$

where $a > 0$. What is the magnitude of $H(\omega)$? What is $\measuredangle H(\omega)$? What is the impulse response of this system?

(b) Let the input to the system of part (a) be

$$x(t) = e^{-bt}u(t), \qquad b > 0$$

What is the output $y(t)$ when $b \neq a$? What is the output when $b = a$?

(c) Let $b = 2$. Sketch $y(t)$ in part (b) when a takes on the following values:
  (i) $a = 1$
  (ii) $a = 2$
  (iii) $a = 4$

(d) Determine the output of the system of part (a) with $a = 1$, when the input is

$$\cos (t/\sqrt{3}) + \cos t + \cos \sqrt{3}\, t$$

Roughly sketch both the input and the output.

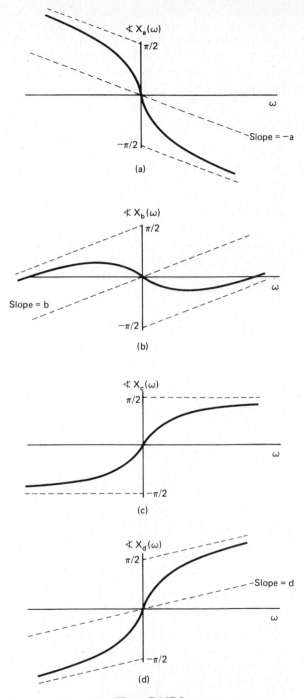

Figure P4.37-2

4.39. Inverse systems often find application in problems involving imperfect measuring devices. For example, consider a device for measuring the temperature of a liquid. It is often reasonable to model such a device as an LTI system, which, because of the response characteristics of the measuring element (e.g., the mercury in a thermometer), does not respond instantaneously to temperature changes. In particular, assume that the response of this device to a unit step in temperature is

$$s(t) = (1 - e^{-t/2})u(t) \tag{P4.39-1}$$

(a) Design a compensating system which, when provided with the output of the measuring device, produces an output equal to the instantaneous temperature of the liquid.

(b) One of the problems that often arises in using inverse systems as compensators for measuring devices is that gross inaccuracies in indicated temperature may occur if the actual output of the measuring device contains errors, due to small, erratic phenomena in the device. Since there always are such error sources in real systems, one must take them into account. To illustrate this, consider a measuring device whose overall output can be modeled as the sum of the response of the measuring device characterized by eq. (P4.39-1) and an interfering "noise" signal $n(t)$. This is depicted in Figure P4.39(a), where we have also included the inverse system of part (a), which now has as its input the *overall* output of the measuring device. Suppose that $n(t) = \sin \omega t$. What is the contribution of $n(t)$ to the output of the inverse system, and how does this output change as ω is increased?

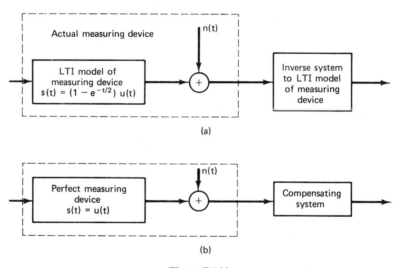

(a)

(b)

Figure P4.39

(c) The issue raised in part (b) is an important one in many applications of LTI system analysis. Specifically, we are confronted with the fundamental trade-off between the speed of response of a system and its abilities to attenuate high-frequency interference.

In part (b) we saw that this trade-off implied that by attempting to speed up the response of a measuring device (by means of an inverse system), we produced a system that would also amplify corrupting sinusoidal signals. To illustrate this concept further, consider a measuring device that responds instantaneously to

Fourier Analysis for Continuous-Time Signals and Systems   Chap. 4

changes in temperature but which also is corrupted by noise. The response of such a system can be modeled, as depicted in Figure P4.39(b), as the sum of the response of a perfect measuring device and a corrupting signal $n(t)$. Suppose that as shown in Figure P4.39(b), we wish to design a compensating system that will *slow down* the response to actual temperature variations but also will attenuate the noise $n(t)$. Let the impulse response of the compensating system be

$$h(t) = ae^{-at}u(t)$$

Choose a so that the overall system of Figure P4.39(b) responds as quickly as possible to a step change in temperature subject to the constraint that the amplitude of the portion of the output due to the noise $n(t) = \sin 6t$ is no larger than $\frac{1}{4}$.

4.40. Consider an LTI system whose response to the input

$$x(t) = [e^{-t} + e^{-3t}]u(t)$$

is

$$y(t) = [2e^{-t} - 2e^{-4t}]u(t)$$

(a) Find the frequency response of this system.
(b) Determine the system impulse response.
(c) Find the differential equation relating the input and output and construct a realization of this system using integrators, adders, and coefficient multipliers.

4.41. The output $y(t)$ of a causal LTI system is related to the input $x(t)$ by the equation

$$\frac{dy(t)}{dt} + 10y(t) = \int_{-\infty}^{+\infty} x(\tau)z(t - \tau)\, d\tau - x(t)$$

where $z(t) = e^{-t}u(t) + 3\delta(t)$.
(a) Find the frequency response of this system $H(\omega) = Y(\omega)/X(\omega)$ and sketch the Bode plot of its magnitude and phase.
(b) Determine the impulse response of this system.

4.42. The output $y(t)$ of a causal LTI system is related to the input $x(t)$ by the differential equation

$$\frac{dy(t)}{dt} + 2y(t) = x(t)$$

(a) Determine the frequency response

$$H(\omega) = \frac{Y(\omega)}{X(\omega)}$$

of the system and sketch its Bode plot.
(b) If $x(t) = e^{-t}u(t)$, determine $Y(\omega)$, the Fourier transform of the output.
(c) Using the technique of partial fraction expansion, determine the output $y(t)$ for the input $x(t)$ in part (b).
(d) Repeat parts (b) and (c) first if the input has as its Fourier transform

(i) $X(\omega) = \dfrac{1 + j\omega}{2 + j\omega}$

then if

(ii) $X(\omega) = \dfrac{2 + j\omega}{1 + j\omega}$

and finally if

(iii) $X(\omega) = \dfrac{1}{(2 + j\omega)(1 + j\omega)}$

4.43. The input and the output of a causal LTI system are related by the differential equation

$$\frac{d^2y(t)}{dt^2} + 6\frac{dy(t)}{dt} + 8y(t) = 2x(t)$$

(a) Find the impulse response of this system.

(b) What is the response of this system if $x(t) = te^{-2t}u(t)$?

(c) Repeat part (a) for the causal LTI system described by the equation

$$\frac{d^2y(t)}{dt^2} + \sqrt{2}\frac{dy(t)}{dt} + y(t) = 2\frac{d^2x(t)}{dt^2} - 2x(t)$$

4.44. In Section 4.12 we introduced the idea of a time constant for a first-order system. As we saw, the time constant provides a measure of how fast a first-order system responds to inputs. The idea of measuring the speed of response of a system is also important for higher-order systems, and in this problem we investigate the extension to such systems.

(a) Recall that the time constant of the first-order system with impulse response

$$h(t) = ae^{-at}u(t), \qquad a > 0$$

is $1/a$, which is the amount of time from $t = 0$ that it takes the system step response $s(t)$ to settle to within $(1/e)$ of its final value [i.e., $s(\infty) = \lim_{t\to\infty} s(t)$]. Using this same quantitative definition, find the equation that must be solved in order to determine the time constant of the causal LTI system described by the differential equation

$$\frac{d^2y(t)}{dt^2} + 11\frac{dy(t)}{dt} + 10y(t) = 9x(t) \qquad \text{(P4.44-1)}$$

(b) As can be seen from part (a), if we use the precise definition of time constant given in part (a), we obtain a simple expression for the time constant of a first-order system, but the calculations are decidedly more complex for the system of eq. (P4.44-1). However, show that this system can be viewed as the parallel interconnection of two first-order systems. Thus, we usually think of the system of eq. (P4.44-1) as having *two* time constants, corresponding to the two first-order factors. What are the two time constants for this example?

(c) The discussion given in part (b) can be directly generalized to all systems with impulse responses that are linear combinations of decaying exponentials. In any system of this type, one can identify the *dominant* time constants of the system, which are simply the largest of the time constants. These represent the slowest parts of the system response and consequently they have the dominant effect on how fast the system as a whole can respond. What is the dominant time constant of the system of eq. (P4.44-1)? Substitute this time constant into the equation determined in part (a). Although this number will not satisfy the equation exactly, you should see that it nearly does, which is an indication that it is very close to the time constant defined as in part (a). Thus the approach we have outlined in parts (b) and (c) is of value in providing insight into the speed of response of LTI systems without requiring excessive calculation.

(d) One important use of the concept of dominant time constants is in the reduction of the order of LTI systems. This is of great practical significance in problems involving the analysis of complex systems having a few dominant time constants and other very small time constants. In order to reduce the complexity of the model of the system to be analyzed, one often can simplify the fast parts of the system. That is, suppose that we regard a complex system as a parallel interconnection of first- and second-order systems. Suppose that one of these subsystems, with impulse response $h(t)$ and step response $s(t)$, is fast, that is, that $s(t)$ settles to its final value $s(\infty)$ very quickly. Then in this case we can approximate this subsystem by the subsystem that settles to this same final value *instantaneously* That is, if $\hat{s}(t)$ is the step response of our approximation, then

$$\hat{s}(t) = s(\infty)u(t)$$

This is illustrated in Figure P4.44. Note that the impulse response of our approximate system is then

$$\hat{h}(t) = s(\infty)\delta(t)$$

which indicates that our approximate system is *memoryless.*

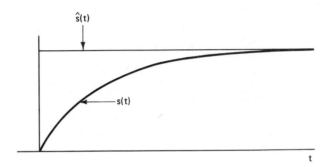

Figure P4.44

Consider again the causal LTI system described by eq. (P4.44-1) and in particular the representation of it as a parallel interconnection of two first-order systems determined in part (b). Use the method outlined above to replace the faster of the two subsystems by a memoryless system. What is the differential equation that then describes the resulting overall system? What is the frequency response of this system? Sketch the magnitudes of $|H(\omega)|$ (*not* log $H(\omega)|$) and $\angle H(\omega)$ for both the original and approximate systems. Over what range of frequencies are these frequency responses nearly equal? Sketch the step responses for both systems. Over what range of time are the step responses nearly equal? From these plots you will see some of the similarities *and* differences between the original system and its approximation. The utility of an approximation such as this depends upon the specific application. In particular, one must take into account both how widely separated the different time constants are and also the nature of the inputs to be considered. As you will see from your answers to this part of the problem, the frequency response of the approximate system is essentially the same as the frequency response of the original system at low frequencies. That is, when the fast parts of the system are sufficiently fast compared to the rate of fluctuation of the input, the approximation becomes a useful one.

4.45. **(a)** Sketch the Bode plots for the following frequency responses.

(i) $1 + (j\omega/10)$

(ii) $1 - (j\omega/10)$

(iii) $\dfrac{16}{(j\omega + 2)^4}$

(iv) $\dfrac{1 - (j\omega/10)}{1 + j\omega}$

(v) $\dfrac{(j\omega/10) - 1}{1 + j\omega}$

(vi) $\dfrac{1 + (j\omega/10)}{1 + j\omega}$

(vii) $\dfrac{1 - (j\omega/10)}{(j\omega)^2 + (j\omega) + 1}$

(viii) $\dfrac{10 + 5j\omega + 10(j\omega)^2}{1 + (j\omega/10)}$

(ix) $1 + j\omega + (j\omega)^2$

(x) $1 - j\omega + (j\omega)^2$

(xi) $\dfrac{(j\omega + 10)(10j\omega + 1)}{[(j\omega/100) + 1][(j\omega)^2 + j\omega + 1)]}$

(b) Determine and sketch the impulse response and the step response for the system with frequency response given by (iv). Do the same for the system with frequency response given by (vi).

The system given in (iv) is often referred to as a nonminimum phase system, while the system specified in (vi) is referred to as being minimum phase. The corresponding impulse responses of (iv) and (vi) are referred to as a nonminimum phase signal and a minimum phase signal, respectively. By comparing the Bode plots of these two frequency responses we can see that they have identical magnitudes; however, the magnitude of the *phase* of the system of (iv) is larger than that of the system of (vi).

We can also note differences in the time-domain behavior of these two systems. For example, the impulse response of the minimum phase system has more of its energy concentrated near $t = 0$ than does the impulse response of the nonminimum phase system. In addition, the step response for (iv) initially has the opposite sign from its asymptotic value as $t \longrightarrow \infty$, while this is not the case for the system of (vi).

The important concept of minimum and nonminimum phase systems can be extended to more general LTI systems than the simple first-order systems we have treated here, and the distinguishing characteristics of these systems can be described far more thoroughly than we have done. We will come across such systems again in the problems in subsequent chapters.

4.46. Let $x(t)$ have the Bode plot depicted in Figure 4.44(c). Sketch the Bode plots for $10x(10t)$.

4.47. Recall that an integrator has as its frequency response

$$H(\omega) = \frac{1}{j\omega} + \pi\,\delta(\omega)$$

where the impulse at $\omega = 0$ is a result of the fact that the integration of a constant input from $t = -\infty$ results in an infinite output. Thus, if we avoid inputs that are constant, or equivalently only examine $H(\omega)$ for $\omega > 0$, we see that

$$20\log|H(\omega)| = -20\log(\omega)$$
$$\sphericalangle H(\omega) = -\pi/2 \qquad \text{for } \omega > 0$$

In other words, the Bode plot for an integrator, as illustrated in Figure P4.47, consists of two straight-line plots. These plots reflect the principal characteristics of an integrator: 90° of negative phase shift at all positive values of frequency and the amplification of low frequencies.

(a) A useful, simple model of an electric motor is as an LTI system, with input equal to the applied voltage and output given by the motor shaft angle. This system can be visualized as the cascade of a stable LTI system (with the voltage as input and shaft

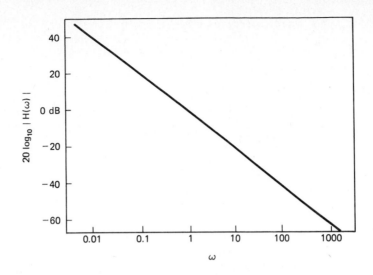

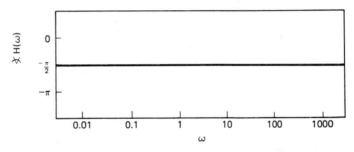

Figure P4.47

angular velocity as output) and an integrator (representing the integration of the angular velocity). Often a first-order system model is used for the first part of the cascade. Assuming, for example, that this first-order system has a time constant of 0.1 second, we obtain an overall motor frequency response of the form

$$H(\omega) = \frac{1}{j\omega(1 + j\omega/10)} + \pi\delta(\omega)$$

Sketch the Bode plot for this system for $\omega > 0.001$.

(b) Sketch the Bode plot for a differentiator.

(c) Do the same for the systems with the following frequency responses:

(i) $H(\omega) = \dfrac{j\omega}{1 + j\omega/100}$

(ii) $\dfrac{j\omega}{(1 + (j\omega)/10 + (j\omega)^2/100)}$

4.48. An LTI system is said to have *phase lead* at a particular frequency $\omega = \omega_0$ if $\measuredangle\, H(\omega_0) > 0$. The terminology stems from the fact that if $e^{j\omega_0 t}$ is the input to this system, then the phase of the output will exceed or lead the phase of the input. Similarly, if $\measuredangle\, H(\omega_0) < 0$, the system is said to have *phase lag* at this frequency. Note that the

system with frequency response

$$\frac{1}{1 + j\omega\tau}$$

has phase lag for all $\omega > 0$, while

$$1 + j\omega\tau$$

has phase lead for all $\omega > 0$.

(a) Construct the Bode plots for the following two systems. Which has phase lead and which phase lag? Also, which one amplifies signals at certain frequencies?

(i) $\dfrac{1 + (j\omega/10)}{1 + 10j\omega}$

(ii) $\dfrac{1 + 10j\omega}{1 + (j\omega/10)}$

(b) Repeat part (a) for the following three frequency responses:

(i) $\dfrac{(1 + j\omega/10)^2}{(1 + 10j\omega)^3}$

(ii) $\dfrac{1 + j\omega/10}{100(j\omega)^2 + 10j\omega + 1}$

(iii) $\dfrac{1 + 10j\omega}{0.01(j\omega)^2 + 0.2j\omega + 1}$

4.49. Consider the system depicted in Figure P4.49. The "compensator" box is a continuous-time LTI system.

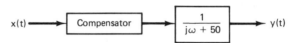

Figure P4.49

(a) Suppose that it is desired to choose the frequency response of the compensator so that the overall frequency response $H(\omega)$ of the cascade satisfies the following two conditions:

(1) The log magnitude of $H(\omega)$ has a slope of -40 dB/decade beyond $\omega = 1000$.
(2) For all $0 < \omega < 1000$, the log magnitude of $H(\omega)$ should be between ± 10 dB.

Design a suitable compensator, that is, determine a frequency response for a compensator that meets the requirements, and draw the Bode plot for the resulting $H(\omega)$.

(b) Repeat part (a) if the specifications on the log magnitude of $H(\omega)$ are:

(1) It should have a slope of $+20$ dB/decade for $0 < \omega < 10$.
(2) It should be between $+10$ and $+30$ dB for $10 < \omega < 100$.
(3) It should have a slope of -20 dB/decade for $100 < \omega < 1000$.
(4) It should have a slope of -40 dB/decade for $\omega > 1000$.

4.50. Consider the causal LTI system with frequency response

$$H(\omega) = \frac{5(j\omega) + 7}{(j\omega + 4)\,[(j\omega)^2 + (j\omega) + 1]}$$

(a) Determine the impulse response of this system.
(b) Determine a cascade structure (using adders, integrators, and coefficient multipliers) for this system consisting of the cascade of a first-order system and a second-order system.
(c) Determine a parallel-form structure consisting of the parallel interconnection of a first-order system and a second-order system.

4.51. (a) Consider the LTI system with frequency response

$$H(\omega) = \frac{1}{(j\omega + 2)^3}$$

Construct a realization of this system consisting of the cascade of three first-order systems. Does this system have a parallel-form realization consisting of the parallel interconnection of three first-order systems? Explain your answer.

(b) Repeat part (a) for the frequency response

$$H(\omega) = \frac{j\omega + 3}{(j\omega + 2)^2(j\omega + 1)}$$

Also find a realization consisting of the parallel interconnection of a second-order system and a first-order system.

(c) Consider the frequency response

$$H(\omega) = \frac{1}{(j\omega + 2)^3(j\omega + 1)}$$

Find a realization consisting of the parallel interconnection of a third-order system and a first-order system. Can you construct a parallel realization consisting of two second-order systems? What about one consisting of one second-order system and two first-order systems? Explain your answers.

(d) Construct 4 different cascade realizations of the frequency response of part (b). Each such realization should consist of the cascade of a different second-order system and first-order system.

time-series analysis techniques to a wide range of problems, including economic forecasting, the analysis of demographic data, and the use of observational data to draw inferences concerning the nature of particular physical phenomena.

In the 1940s and 1950s major steps forward were taken in the development of discrete-time techniques in general and in the use of the tools of discrete-time Fourier analysis in particular. The reasons for this push forward were the increasing use and capabilities of digital computers and the development of design methods for sampled-data systems, that is, discrete-time systems for the processing of sampled continuous-time signals. As more and more use was being made of the computer, the overlap in the realms of applications of discrete- and continuous-time techniques grew, and this provided a natural connection between the two methodologies that heretofore had developed essentially independently. In addition, the great flexibility of the digital computer spurred experimentation with the design of increasingly sophisticated discrete-time systems that had no apparent practical implementation in analog equipment. Many of these discrete-time systems require the calculation of numerous Fourier transforms, which at the time seemed to be a prohibitive computational burden. Nevertheless, the possibilities that were opened up by the digital computer were sufficiently tempting that active work began on the investigation of digital voice encoders, digital spectrum analyzers, and other all-digital systems, with the hope that eventually such systems would be practical. In the mid-1960s an algorithm, now known as the fast Fourier transform or FFT, was developed. This algorithm proved to be perfectly suited for efficient digital implementation, and it reduced the computation time for transforms by orders of magnitude. With this tool many interesting but previously impractical ideas suddenly became practical, and the development of discrete-time signal and system analysis techniques moved forward at an accelerated pace.

What has emerged out of this long history is a cohesive framework for the analysis of discrete-time signals and systems and a very large class of existing and potential applications. In this chapter we describe one of the most important and fundamental parts of this framework and in subsequent chapters we provide an introduction to several of the important applications. As we will see, there are many similarities between the techniques of discrete-time Fourier analysis and their continuous-time counterparts. For example, the basic reasons for the utility of representing signals in terms of complex exponentials are the same in discrete time as they are in continuous time. In particular, if the input and output of a discrete-time LTI system are expressed as linear combinations of complex exponentials, then the coefficients in the representation of the output can be expressed in an extremely convenient form in terms of the coefficients of the linear combination representing the input. Furthermore, as in continuous time, a broad and useful class of signals can be represented by such linear combinations.

On the other hand, although much of discrete-time Fourier analysis is very similar to the continuous-time case, there are some important differences. For example in Section 5.2 we will see that the Fourier series representation of a discrete-time periodic signal is a *finite* series, as opposed to the infinite series representation required for continuous-time periodic signals. In fact, the FFT relies on this finiteness in an intrinsic way and consequently is inherently a discrete-time concept. As we proceed with our development in this chapter we will use the many similarities in the continuous- and discrete-time cases to enhance our understanding of the basic concepts of Fourier analysis that are important to both, and we will also use the several differences to deepen our understanding of the distinct characteristics of each.

5.1 THE RESPONSE OF DISCRETE-TIME LTI SYSTEMS TO COMPLEX EXPONENTIALS

The motivation for developing a Fourier representation in the discrete-time case (i.e., for expressing general sequences as linear combinations of complex exponentials) is identical to that in the continuous-time case. Specifically, complex exponential sequences are eigenfunctions of discrete-time LTI systems. That is, suppose that an LTI system with impulse response $h[n]$ has as its input the signal

$$x[n] = z^n \tag{5.1}$$

where z is a complex number. The output of the system can be determined from the convolution sum as

$$y[n] = h[n] * x[n] = \sum_{k=-\infty}^{+\infty} h[k]x[n-k]$$

$$= \sum_{k=-\infty}^{+\infty} h[k]z^{n-k} = z^n \sum_{k=-\infty}^{+\infty} h[k]z^{-k}$$

From this expression we see that if the input $x[n]$ is the complex exponential given by eq. (5.1), then the output is the same complex exponential multiplied by a constant that depends on the value of z, that is,

$$y[n] = H(z)z^n \tag{5.2}$$

where

$$H(z) = \sum_{k=-\infty}^{+\infty} h[k]z^{-k} \tag{5.3}$$

Here $H(z)$ is the eigenvalue associated with the eigenfunction z^n.

More generally, we see that eq. (5.2), together with the superposition property, implies that the representation of signals in terms of complex exponentials leads to a convenient expression for the response of an LTI system. Specifically, if the input to a discrete-time LTI system is represented as a linear combination of complex exponentials, that is, if

$$x[n] = \sum_k a_k z_k^n \tag{5.4}$$

then the output $y[n]$ will be

$$y[n] = \sum_k a_k H(z_k)z_k^n \tag{5.5}$$

In other words, the output can also be represented as a linear combination of the same complex exponential signals, and each coefficient in this representation of the output is obtained as the product of the corresponding coefficient a_k of the input and the system's eigenvalue $H(z_k)$ associated with the eigenfunction z_k^n.

Analogous to our discussion in Chapter 4, we will restrict ourselves in this chapter to complex exponentials of the form $e^{j\Omega n}$ (i.e., z^n with $|z| = 1$). In Chapter 10 we consider the more general case. In the next section we consider the Fourier series representation of periodic signals, while in Section 5.3 we develop the discrete-time Fourier transform as an extension of the Fourier series in a manner that exactly parallels the development in Section 4.4. As mentioned previously, we will find many similarities and several important differences with the continuous-time case.

5.2 REPRESENTATION OF PERIODIC SIGNALS: THE DISCRETE-TIME FOURIER SERIES

5.2.1 Linear Combinations of Harmonically Related Complex Exponentials

As discussed in Chapter 2, a discrete-time signal $x[n]$ is periodic if for some positive value of N,

$$x[n] = x[n + N] \tag{5.6}$$

For example, as we saw in Section 2.4.3, the complex exponential $e^{j(2\pi/N)n}$ is periodic with period N. Furthermore, the set of all discrete-time complex exponential signals that are periodic with period N is given by

$$\phi_k[n] = e^{jk(2\pi/N)n}, \qquad k = 0, \pm 1, \pm 2, \ldots \tag{5.7}$$

All of these signals have frequencies that are multiples of the same fundamental frequency, $2\pi/N$, and thus are harmonically related.

An important distinction between the set of harmonically related signals in discrete and continuous time is that while all of the continuous-time signals in eq. (4.12) are distinct, there are only N different signals in the set given in eq. (5.7). As discussed in Section 2.4.3, the reason for this is that discrete-time complex exponentials which differ in frequency by a multiple of 2π are identical. That is,

$$e^{j(\Omega+2\pi r)n} = e^{j\Omega n}e^{j2\pi rn} = e^{j\Omega n}$$

A direct consequence of this fact is that $\phi_0[n] = \phi_N[n]$, $\phi_1[n] = \phi_{N+1}[n]$, and more generally

$$\phi_k[n] = \phi_{k+rN}[n] \tag{5.8}$$

That is, when k is changed by any integer multiple of N, we generate the identical sequence.

We now wish to consider the representation of more general periodic sequences in terms of linear combinations of the sequences $\phi_k[n]$ in eq. (5.7). Such a linear combination has the form

$$x[n] = \sum_k a_k\phi_k[n] = \sum_k a_k e^{jk(2\pi/N)n} \tag{5.9}$$

Since the sequences $\phi_k[n]$ are distinct only over a range of N successive values of k, the summation in eq. (5.9) need only include terms over this range. Thus, the summation in eq. (5.9) is on k, as k varies over a range of N successive integers, and for convenience we indicate this by expressing the limits of the summation as $k = \langle N \rangle$. That is,

$$x[n] = \sum_{k=\langle N \rangle} a_k\phi_k[n] = \sum_{k=\langle N \rangle} a_k e^{jk(2\pi/N)n} \tag{5.10}$$

For example, k could take on the values $k = 0, 1, \ldots, N-1$ or $k = 3, 4, \ldots,$

$N + 2$. In either case, by virtue of eq. (5.8), exactly the same set of complex exponential sequences appears in the summation on the right-hand side of eq. (5.10). Equation (5.10) will be referred to as the *discrete-time Fourier series* and the coefficients a_k as the *Fourier series coefficients*. As mentioned in the introduction, in discrete time this is a finite series, and as we have just seen, this is a direct consequence of eq. (5.8).

5.2.2 Determination of the Fourier Series Representation of a Periodic Signal

Suppose now that we are given a sequence $x[n]$ which is periodic with period N. We would like to determine if a representation of $x[n]$ in the form of eq. (5.10) exists, and if so, what the values of the coefficients a_k are. This question can be phrased in terms of finding a solution to a set of linear equations. Specifically, if we evaluate eq. (5.10) for successive values of n, we obtain

$$x[0] = \sum_{k=\langle N \rangle} a_k$$

$$x[1] = \sum_{k=\langle N \rangle} a_k e^{j2\pi k/N}$$

$$\cdot$$
$$\cdot$$
$$\cdot$$

$$x[N-1] = \sum_{k=\langle N \rangle} a_k e^{j2\pi k(N-1)/N} \qquad (5.11)$$

Since both sides of eq. (5.10) are periodic with period N, the equation for $x[N]$ is identical to that for $x[0]$ given in eq. (5.11). Thus, eq. (5.11) represents a set of N linear equations for the N unknown coefficients a_k as k ranges over a set of N successive integers. One can show that this set of N equations is linearly independent and consequently can be solved to obtain the coefficients a_k in terms of the given values of $x[n]$.

In Problem 5.2 we consider an example in which the Fourier series coefficients are obtained by explicitly solving the set of N equations given in eq. (5.11). However, by following steps parallel to those used in continuous time, it is possible to obtain a closed-form expression for the coefficients a_k in terms of the sequence values $x[n]$. To determine this relationship it is helpful first to show that

$$\sum_{n=0}^{N-1} e^{jk(2\pi/N)n} = \begin{cases} N, & k = 0, \pm N, \pm 2N, \ldots \\ 0, & \text{otherwise} \end{cases} \qquad (5.12)$$

What eq. (5.12) says is that the sum over one period of the values of a periodic complex exponential is zero unless that complex exponential is constant. That is, this equation is simply the discrete-time counterpart of the equation

$$\int_0^T e^{jk(2\pi/T)t} \, dt = \begin{cases} T, & k = 0 \\ 0, & \text{otherwise} \end{cases}$$

To derive eq. (5.12), note first that the left-hand side of eq. (5.12) is the sum of a finite number of terms in a geometric series. That is, it is of the form

$$\sum_{n=0}^{N-1} \alpha^n$$

with $\alpha = e^{jk(2\pi/N)}$. As considered in Problem 2.8, this sum can be expressed in closed form as

$$\sum_{n=0}^{N-1} \alpha^n = \begin{cases} N, & \alpha = 1 \\ \dfrac{1 - \alpha^N}{1 - \alpha}, & \alpha \neq 1 \end{cases} \qquad (5.13)$$

We also know that $e^{jk(2\pi/N)} = 1$ only when k is a multiple of N (i.e., when $k = 0, \pm N, \pm 2N, \ldots$). Therefore, applying eq. (5.13) with $\alpha = e^{jk(2\pi/N)}$ we obtain

$$\sum_{n=0}^{N-1} e^{jk(2\pi/N)n} = \begin{cases} N, & k = 0, \pm N, \pm 2N, \ldots \\ \dfrac{1 - e^{jk(2\pi/N)N}}{1 - e^{jk(2\pi/N)}}, & \text{otherwise} \end{cases}$$

which reduces to eq. (5.12) since $e^{jk(2\pi/N)N} = e^{jk2\pi} = 1$.

Since each of the complex exponentials in the summation in eq. (5.12) is periodic with period N, the equality in eq. (5.12) will remain valid with the summation carried out over any interval of length N, that is,

$$\sum_{n=\langle N \rangle} e^{jk(2\pi/N)n} = \begin{cases} N, & k = 0, \pm N, \pm 2N, \ldots \\ 0, & \text{otherwise} \end{cases} \qquad (5.14)$$

A graphical interpretation of eq. (5.14) is depicted in Figure 5.1 for the case of $N = 6$. Here complex numbers are represented as vectors in the complex plane with the abscissa as the real axis and the ordinate as the imaginary axis. Since each value of a periodic complex exponential sequence is a complex number with unit magnitude, all of the vectors in Figure 5.1 are of unit length. Furthermore, from the symmetry of each of these figures we can deduce that the sum of the vectors $e^{jk(2\pi/6)n}$ over one period will be equal to zero unless $k = 0, 6, 12$, and so on.

Now let us consider the Fourier series representation of eq. (5.10). Multiplying both sides by $e^{-jr(2\pi/N)n}$ and summing over N terms, we obtain

$$\sum_{n=\langle N \rangle} x[n] e^{-jr(2\pi/N)n} = \sum_{n=\langle N \rangle} \sum_{k=\langle N \rangle} a_k e^{j(k-r)(2\pi/N)n} \qquad (5.15)$$

Interchanging the order of summation on the right-hand side, we have

$$\sum_{n=\langle N \rangle} x[n] e^{-jr(2\pi/N)n} = \sum_{k=\langle N \rangle} a_k \sum_{n=\langle N \rangle} e^{j(k-r)(2\pi/N)n} \qquad (5.16)$$

From the identity in eq. (5.14) the innermost sum on n on the right-hand side of eq. (5.16) is zero unless $k - r$ is zero or an integer multiple of N. Therefore, if we choose values for r over the same range as that over which k varies in the outer summation, the innermost sum on the right-hand side of eq. (5.16) equals N if $k = r$ and 0 if $k \neq r$. Thus the right-hand side of eq. (5.16) reduces to Na_r, and we have

$$a_r = \frac{1}{N} \sum_{n=\langle N \rangle} x[n] e^{-jr(2\pi/N)n} \qquad (5.17)$$

This then provides a closed-form expression for obtaining the Fourier series coefficients, and we have the *discrete-time Fourier series pair* which we summarize in eqs. (5.18) and (5.19).

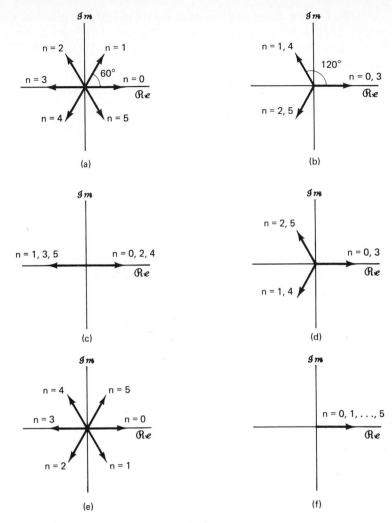

Figure 5.1 The complex exponential sequence $\phi_k[n] = e^{jk(2\pi/6)n}$ over one period ($n = 0, 1, \ldots, 5$) for different values of k: (a) $k = 1$; (b) $k = 2$; (c) $k = 3$; (d) $k = 4$; (e) $k = 5$; (f) $k = 6$.

$$x[n] = \sum_{k=\langle N \rangle} a_k e^{jk(2\pi/N)n} \tag{5.18}$$

$$a_k = \frac{1}{N} \sum_{n=\langle N \rangle} x[n] e^{-jk(2\pi/N)n} \tag{5.19}$$

These equations play the role for discrete-time periodic signals that eqs. (4.34) and (4.35) play for continuous-time periodic signals, with eq. (5.18) representing the *synthesis* equation and eq. (5.19) the *analysis* equation.

The Fourier series coefficients a_k are often referred to as the *spectral coefficients* of $x[n]$. These coefficients specify a decomposition of $x[n]$ into a sum of N harmonically related complex exponentials. Referring to eq. (5.10), we see that if we take k in the range from 0 to $N - 1$, we have

$$x[n] = a_0\phi_0[n] + a_1\phi_1[n] + \ldots + a_{N-1}\phi_{N-1}[n] \tag{5.20}$$

Similarly, if we take k to range from 1 to N, we obtain

$$x[n] = a_1\phi_1[n] + a_2\phi_2[n] + \ldots + a_N\phi_N[n] \tag{5.21}$$

From eq. (5.8) we know that $\phi_0[n] = \phi_N[n]$, and therefore upon comparing eqs. (5.20) and (5.21), we must conclude that $a_0 = a_N$. Similarly, by letting k range over any set of N consecutive integers and using eq. (5.8), we can conclude that

$$a_k = a_{k+N} \tag{5.22}$$

That is, if we consider more than N sequential values of k, the values a_k will repeat periodically with period N. It is important that this fact be interpreted carefully. In particular, since there are only N distinct complex exponentials that are periodic with period N, the discrete-time Fourier series representation is a finite series with N terms. Therefore, if we fix the N consecutive values of k over which we define the Fourier series in eq. (5.18), we will obtain a set of exactly N Fourier coefficients from eq. (5.19). On the other hand, at times it will be convenient to use different sets of N values of k, and consequently it is useful to regard eq. (5.18) as a sum over any *arbitrary* set of N successive values of k. For this reason it is sometimes convenient to think of a_k as a sequence defined for all values of k, where only N successive elements in this sequence will be used in the Fourier series representation. Furthermore, since the $\phi_k[n]$ repeat periodically with period N as we vary k [eq. (5.8)], so must the a_k [eq. (5.22)].

Example 5.1

Consider the signal

$$x[n] = \sin \Omega_0 n \tag{5.23}$$

which is the discrete-time counterpart of Example 4.3. There are three different situations, depending upon whether $2\pi/\Omega_0$ is an integer, a ratio of integers, or an irrational number. From Chapter 2 we know that $x[n]$ is periodic in the first two cases but not in the third. Consequently, the Fourier series representation of this signal applies only in the first two cases. For the case when $2\pi/\Omega_0$ is an integer N, that is, when

$$\Omega_0 = \frac{2\pi}{N}$$

$x[n]$ is periodic with fundamental period N, and we obtain a result that is exactly analogous to the continuous-time case. Expanding the signal as a sum of two complex exponentials, we obtain

$$x[n] = \frac{1}{2j}e^{j(2\pi/N)n} - \frac{1}{2j}e^{-j(2\pi/N)n} \tag{5.24}$$

Comparing eq. (5.24) with eq. (5.18), we see by inspection that

$$a_1 = \frac{1}{2j}, \qquad a_{-1} = -\frac{1}{2j} \tag{5.25}$$

and the remaining coefficients over the interval of summation are zero. As described previously, these coefficients repeat with period N and thus for example a_{N+1} is also equal to $1/2j$ and a_{N-1} equals $-1/2j$. The Fourier series coefficients for this example with $N = 5$ are illustrated in Figure 5.2. The fact that they periodically repeat is indicated. However, only one period is utilized in the synthesis equation (5.18).

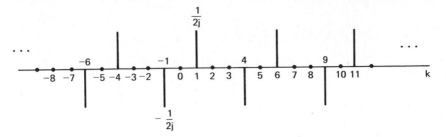

Figure 5.2 Fourier coefficients for $x[n] = \sin(2\pi/5)n$.

Consider now the case when $2\pi/\Omega_0$ is a ratio of integers, that is, when

$$\Omega_0 = \frac{2\pi m}{N}$$

Assuming that m and N do not have any common factors, we know from Chapter 2 that $x[n]$ has a fundamental period of N. Again expanding $x[n]$ as a sum of two complex exponentials, we have

$$x[n] = \frac{1}{2j}e^{jm(2\pi/N)n} - \frac{1}{2j}e^{-jm(2\pi/N)n}$$

from which we can determine by inspection that $a_m = (1/2j)$, $a_{-m} = (-1/2j)$, and the remaining coefficients over one period of length N are zero. The Fourier coefficients for this example with $m = 3$ and $N = 5$ are depicted in Figure 5.3. Again we have

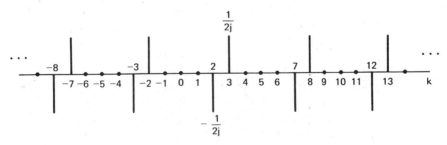

Figure 5.3 Fourier coefficients for $x[n] = \sin 3(2\pi/5)n$.

indicated the periodicity of the coefficients. For example, for $N = 5$, $a_2 = a_{-3}$, which in our example equals $(-1/2j)$. Note, however, that over any period of length 5 there are only two nonzero Fourier coefficients, and therefore there are only two nonzero terms in the synthesis equation.

Example 5.2

Suppose that we now consider the signal

$$x[n] = 1 + \sin\left(\frac{2\pi}{N}\right)n + 3\cos\left(\frac{2\pi}{N}\right)n + \cos\left(\frac{4\pi}{N}n + \frac{\pi}{2}\right)$$

This signal is periodic with period N, and, as in Example 5.1, we can expand $x[n]$ directly in terms of complex exponentials to obtain

$$x[n] = 1 + \frac{1}{2j}[e^{j(2\pi/N)n} - e^{-j(2\pi/N)n}]$$

$$+ \frac{3}{2}[e^{j(2\pi/N)n} + e^{-j(2\pi/N)n}]$$

$$+ \frac{1}{2}[e^{j(4\pi n/N + \pi/2)} + e^{-j(4\pi n/N + \pi/2)}]$$

Collecting terms, we find that

$$x[n] = 1 + \left(\frac{3}{2} + \frac{1}{2j}\right)e^{j(2\pi/N)n} + \left(\frac{3}{2} - \frac{1}{2j}\right)e^{-j(2\pi/N)n}$$

$$+ \left(\frac{1}{2}e^{j\pi/2}\right)e^{j2(2\pi/N)n} + \left(\frac{1}{2}e^{-j\pi/2}\right)e^{-j2(2\pi/N)n}$$

Thus, the Fourier series coefficients for this example are

$$a_0 = 1$$

$$a_1 = \frac{3}{2} + \frac{1}{2j} = \frac{3}{2} - \frac{1}{2}j$$

$$a_{-1} = \frac{3}{2} - \frac{1}{2j} = \frac{3}{2} + \frac{1}{2}j$$

$$a_2 = \frac{1}{2}j$$

$$a_{-2} = -\frac{1}{2}j$$

with $a_k = 0$ for other values of k in the interval of summation in the synthesis equation (5.18). Again, the Fourier coefficients are periodic with period N, so, for example, $a_N = 1$, $a_{3N-1} = \frac{3}{2} + \frac{1}{2}j$, and $a_{2-N} = \frac{1}{2}j$. In Figure 5.4(a) we have plotted the real and

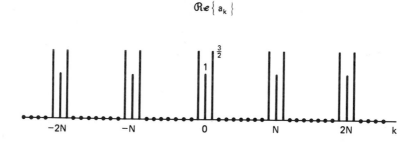

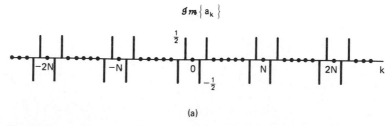

(a)

Figure 5.4   (a) Real and imaginary parts of the Fourier series coefficients in Example 5.2; (b) magnitude and phase of the same coefficients.

imaginary parts of these coefficients for $N = 10$, while the magnitude and phase of the coefficients are depicted in Figure 5.4(b).

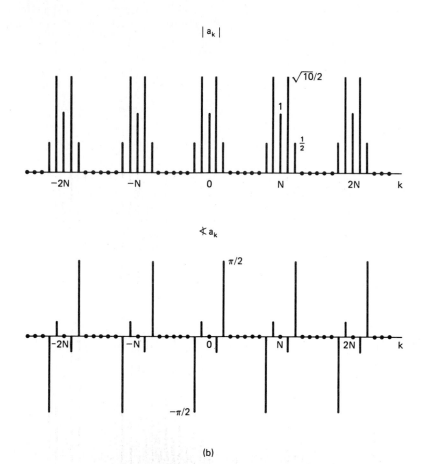

(b)

Figure 5.4 (cont.)

Note that in this example $a_{-k} = a_k^*$ for all values of k. In fact, this equality holds whenever $x[n]$ is real. This property is identical to one that we discussed in Section 4.2 for continuous-time periodic signals, and as in continuous time, one implication of this fact is that there are two alternative forms for the discrete-time Fourier series of real periodic sequences. These forms are analogous to the continuous-time Fourier series representations given in eqs. (4.19) and (4.20) and are examined in Problem 5.7. For our purposes the exponential form of the Fourier series as given in eq. (5.15) is particularly convenient, and we will use it exclusively.

The discrete-time Fourier series possesses a number of other useful properties, many of which are direct counterparts of properties of continuous-time Fourier series. We defer the discussion of these properties until Section 5.5, where they are developed in the more general context of the discrete-time Fourier transform.

Example 5.3

In this example we consider the discrete-time periodic square wave shown in Figure 5.5. We can evaluate the Fourier series using eq. (5.19). Because of the symmetry of this sequence around $n = 0$, it is particularly convenient to choose a symmetric interval over which the summation in eq. (5.19) is carried out. Thus, we express eq. (5.19) as

$$a_k = \frac{1}{N} \sum_{n=-N_1}^{N_1} e^{-jk(2\pi/N)n} \tag{5.26}$$

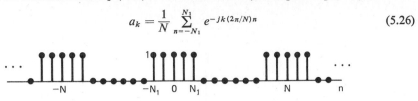

Figure 5.5 Discrete-time periodic square wave.

Letting $m = n + N_1$, eq. (5.26) becomes

$$a_k = \frac{1}{N} \sum_{m=0}^{2N_1} e^{-jk(2\pi/N)(m-N_1)}$$

$$= \frac{1}{N} e^{jk(2\pi/N)N_1} \sum_{m=0}^{2N_1} e^{-jk(2\pi/N)m} \tag{5.27}$$

The summation in eq. (5.27) consists of the sum of the first $(2N_1 + 1)$ terms in a geometric series, which can be evaluated using eq. (5.13). This yields

$$a_k = \frac{1}{N} e^{jk(2\pi/N)N_1} \left(\frac{1 - e^{-jk2\pi(2N_1+1)/N}}{1 - e^{-jk(2\pi/N)}} \right)$$

$$= \frac{1}{N} \frac{e^{-jk(2\pi/2N)}[e^{jk2\pi(N_1+1/2)/N} - e^{-jk2\pi(N_1+1/2)/N}]}{e^{-jk(2\pi/2N)}[e^{jk(2\pi/2N)} - e^{-jk(2\pi/2N)}]} \tag{5.28a}$$

$$= \frac{1}{N} \frac{\sin[2\pi k(N_1 + 1/2)/N]}{\sin(2\pi k/2N)}, \qquad k \neq 0, \pm N, \pm 2N, \ldots$$

and

$$a_k = \frac{2N_1 + 1}{N}, \qquad k = 0, \pm N, \pm 2N, \ldots \tag{5.28b}$$

This expression for the Fourier series coefficients can be written more compactly if we express the coefficients as samples of an envelope:

$$Na_k = \left. \frac{\sin[(2N_1 + 1)\Omega/2]}{\sin(\Omega/2)} \right|_{\Omega = 2\pi k/N} \tag{5.29}$$

In Figure 5.6 the coefficients Na_k are sketched for $2N_1 + 1 = 5$ and with $N = 10, 20,$ and 40 in Figure 5.6(a), (b), and (c), respectively. As N increases but N_1 remains fixed, the envelope of Na_k stays the same but the sample spacing to obtain the Fourier coefficients decreases.

It is useful to compare this example with the continuous-time square wave of Example 4.5. In that case, referring to eq. (4.54) we see that the functional form of the envelope is that of a sinc function. The same form could not result in the discrete-time case since the discrete-time Fourier series coefficients (and hence also their envelope) must be periodic. The discrete-time counterpart of the sinc function is of the form $(\sin \beta x/\sin x)$, as is the case in eq. (5.29).

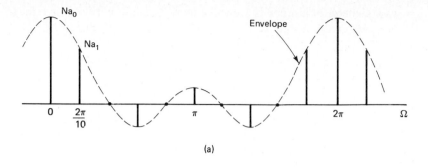

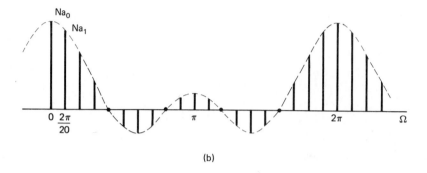

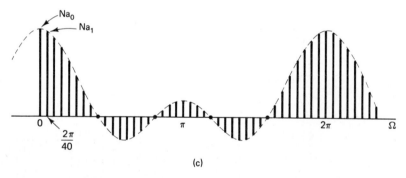

Figure 5.6 Fourier series coefficients for the periodic square wave of Example 5.3; plots of Na_k for $2N_1 + 1 = 5$ and (a) $N = 10$; (b) $N = 20$; and (c) $N = 40$.

In discussing the convergence of the continuous-time Fourier series in Section 4.3, we considered the specific example of a symmetric square wave and observed how the finite sum in eq. (4.47) converged to the square wave as the number of terms approached infinity. In particular, we observed the Gibbs phenomenon at the discontinuity, whereby as the number of terms increased, the ripples in the partial sum (Figure 4.10) became compressed toward the discontinuity with the peak amplitude of the ripples remaining constant independent of the number of terms in the partial sum. Let us consider the analogous sequence of partial sums for the discrete-time square wave, where for simplicity we will assume that the period N is odd. In Figure 5.7 we have depicted the signals

$$\hat{x}[n] = \sum_{k=-M}^{M} a_k e^{jk(2\pi/N)n} \qquad (5.30a)$$

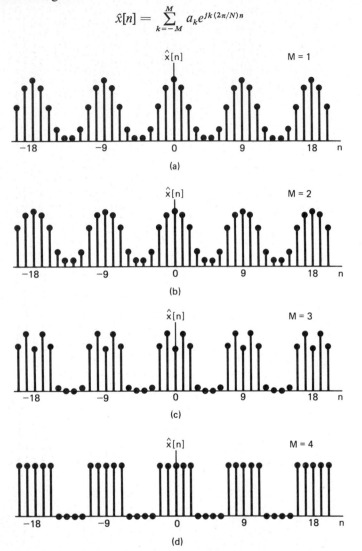

(a)

(b)

(c)

(d)

Figure 5.7 Partial sums of eq. (5.30) for the periodic square wave of Figure 5.5 with $N = 9$ and $2N_1 + 1 = 5$: (a) $M = 1$; (b) $M = 2$; (c) $M = 3$; (d) $M = 4$.

for the example of Figure 5.5 with $N = 9$, $2N_1 + 1 = 5$, and for several values of M. For $M = 4$, the partial sum exactly equals $x[n]$. That is, in contrast to the continuous-time case, there are no convergence issues and no Gibbs phenomenon. In fact, there are no convergence issues with discrete-time Fourier series in general. The reason for this stems from the fact that any discrete-time periodic sequence $x[n]$ is completely specified by a *finite* number, N, of parameters, namely the values of the sequence over one period. The Fourier series analysis equation (5.19) simply transforms this set of N

parameters into an equivalent set, the values of the N Fourier coefficients, and the synthesis equation (5.18) tells us how to recover the original sequence values in terms of a *finite* series. Thus if we take $M = (N-1)/2$ in eq. (5.30a), this sum includes exactly N terms and consequently from the synthesis equations we have that $\hat{x}[n] = x[n]$. Similarly, if N is even and we let

$$\hat{x}[n] = \sum_{k=-M+1}^{M} a_k e^{jk(2\pi/N)n} \tag{5.30b}$$

then with $M = N/2$ this sum consists of N terms and again we can conclude from eq. (5.18) that $\hat{x}[n] = x[n]$. In contrast, a continuous-time periodic signal takes on a continuum of values over a single period, and an infinite number of Fourier coefficients are required to represent it. Thus, in general, *none* of the finite partial sums in eq. (4.47) yield the exact values of $x(t)$, and convergence issues arise as we consider the problem of evaluating the limit as the number of terms approaches infinity.

In our discussion to this point of the discrete-time Fourier series, we have seen that there are many similarities and several differences between the discrete- and continuous-time cases. One of the most important of these similarities arises from the fact that complex exponentials are the eigenfunctions of LTI systems in both continuous and discrete time. Because of this, if the input $x[n]$ to a discrete-time LTI system is periodic with period N, we obtain a convenient expression for the Fourier coefficients of the output $y[n]$ in terms of those for the input. Specifically, if

$$x[n] = \sum_{k=\langle N \rangle} a_k e^{jk(2\pi/N)n}$$

and if $h[n]$ is the impulse response of the system, then from eq. (5.5),

$$y[n] = \sum_{k=\langle N \rangle} a_k H\!\left(\frac{2\pi k}{N}\right) e^{jk(2\pi/N)n} \tag{5.31}$$

where, from eq. (5.3),†

$$H\!\left(\frac{2\pi k}{N}\right) = \sum_{n=-\infty}^{+\infty} h[n] e^{-jk(2\pi/N)n} \tag{5.32}$$

The use of this property is illustrated in the following example.

Example 5.4

Suppose that an LTI system with impulse response $h[n] = \alpha^n u[n]$, $-1 < \alpha < 1$, has as its input the signal

$$x[n] = \cos\!\left(\frac{2\pi n}{N}\right) \tag{5.33}$$

As in Example 5.1, $x[n]$ can be written in Fourier series form as

$$x[n] = \tfrac{1}{2} e^{j(2\pi/N)n} + \tfrac{1}{2} e^{-j(2\pi/N)n}$$

Also, from eq. (5.32),

$$H\!\left(\frac{2\pi k}{N}\right) = \sum_{n=0}^{\infty} \alpha^n e^{-jk(2\pi/N)n} = \sum_{n=0}^{\infty} (\alpha e^{-j2\pi k/N})^n$$

†Equation (5.32) represents a change of notation. Using the notation of eq. (5.3), we see that the quantity on the right-hand side of eq. (5.32) is $H(e^{j2\pi k/N})$. However, since in this chapter we will be evaluating $H(z)$ only for $|z| = 1$ (i.e., for $z = e^{j\Omega}$), we will use the simpler notation of $H(\Omega)$ rather than write $H(e^{j\Omega})$. This is essentially the same type of notation simplification that was made in Chapter 4 (see the footnote on p. 172).

This geometric series can be evaluated using the result of Problem 2.8, yielding

$$H\left(\frac{2\pi k}{N}\right) = \frac{1}{1 - \alpha e^{-j2\pi k/N}}$$

Using eq. (5.31), we then obtain the Fourier series for the output

$$
\begin{aligned}
y[n] &= \frac{1}{2}H\left(\frac{2\pi}{N}\right)e^{j(2\pi/N)n} + \frac{1}{2}H\left(-\frac{2\pi}{N}\right)e^{-j(2\pi/N)n} \\
&= \frac{1}{2}\left(\frac{1}{1 - \alpha e^{-j2\pi/N}}\right)e^{j(2\pi/N)n} + \frac{1}{2}\left(\frac{1}{1 - \alpha e^{j2\pi/N}}\right)e^{-j(2\pi/N)n}
\end{aligned}
\tag{5.34}
$$

If we write

$$\frac{1}{1 - \alpha e^{-j2\pi/N}} = re^{j\theta}$$

then eq. (5.34) reduces to

$$y[n] = r\cos\left(\frac{2\pi}{N}n + \theta\right) \tag{5.35}$$

For example, if $N = 4$,

$$\frac{1}{1 - \alpha e^{-j2\pi/4}} = \frac{1}{1 + \alpha j}$$

and thus

$$y[n] = \frac{1}{\sqrt{1 + \alpha^2}}\cos\left(\frac{\pi n}{2} - \tan^{-1}(\alpha)\right)$$

5.3 REPRESENTATION OF APERIODIC SIGNALS: THE DISCRETE-TIME FOURIER TRANSFORM

In Example 5.3 [eq. (5.29) and Figure 5.6] we saw that the Fourier series coefficients for a periodic square wave could be viewed as samples of an envelope function, and that as the period of the sequence increases, these samples become more and more finely spaced. A similar observation in Section 4.4 for the continuous-time case suggested a representation of an aperiodic signal $x(t)$ by first constructing a periodic signal $\tilde{x}(t)$, which equaled $x(t)$ over one period. As this period then approached infinity, $\tilde{x}(t)$ was equal to $x(t)$ over larger and larger intervals of time, and the Fourier series representation for $\tilde{x}(t)$ approached the Fourier transform representation for $x(t)$. In this section we apply an exactly analogous procedure to discrete-time signals in order to develop the Fourier transform representation for aperiodic sequences.

Consider a general aperiodic sequence $x[n]$ which is of finite duration. That is, for some integer N_1, $x[n] = 0$ if $|n| > N_1$. A signal of this type is illustrated in Figure 5.8(a). From this aperiodic signal we can construct a periodic sequence $\tilde{x}[n]$ for which $\tilde{x}[n]$ is one period, as illustrated in Figure 5.8(b). As we choose the period N to be larger, $x[n]$ is identical to $x[n]$ over a longer interval, and as $N \longrightarrow \infty$, $\tilde{x}[n] = x[n]$ for any finite value of n.

Let us now examine the Fourier series representation of $\tilde{x}[n]$. Specifically, rewriting eqs. (5.18) and (5.19), we have

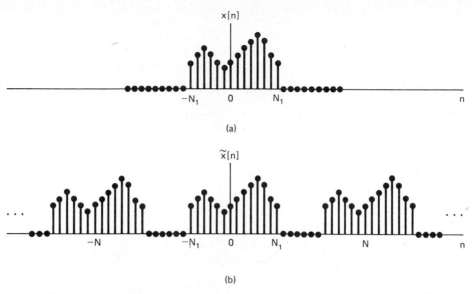

Figure 5.8 (a) Finite duration signal $x[n]$; (b) periodic signal $\tilde{x}[n]$ constructed to be equal to $x[n]$ over one period.

$$\tilde{x}[n] = \sum_{k=\langle N \rangle} a_k e^{jk(2\pi/N)n} \tag{5.36}$$

$$a_k = \frac{1}{N} \sum_{n=\langle N \rangle} \tilde{x}[n] e^{-jk(2\pi/N)n} \tag{5.37}$$

Since $x[n] = \tilde{x}[n]$ over a period that includes the interval $|n| \leq N_1$, it is convenient to choose the interval of summation in eq. (5.37) to be this period, so that $\tilde{x}[n]$ can be replaced by $x[n]$ in the summation. Therefore,

$$a_k = \frac{1}{N} \sum_{n=-N_1}^{N_1} x[n] e^{-jk(2\pi/N)n} = \frac{1}{N} \sum_{n=-\infty}^{+\infty} x[n] e^{-jk(2\pi/N)n} \tag{5.38}$$

where we have used the fact that $x[n]$ is zero outside the interval $|n| \leq N_1$. Defining the envelope $X(\Omega)$ of $N a_k$ as

$$X(\Omega) = \sum_{n=-\infty}^{+\infty} x[n] e^{-j\Omega n} \tag{5.39}$$

we have that the coefficients a_k are given by

$$a_k = \frac{1}{N} X(k\Omega_0) \tag{5.40}$$

where Ω_0 is used to denote the sample spacing $2\pi/N$. Thus, the coefficients a_k are proportional to equally spaced samples of this envelope function. Combining eqs. (5.36) and (5.40) yields

$$\tilde{x}[n] = \sum_{k=\langle N \rangle} \frac{1}{N} X(k\Omega_0) e^{jk\Omega_0 n} \tag{5.41}$$

Since $\Omega_0 = 2\pi/N$, or equivalently $1/N = \Omega_0/2\pi$, eq. (5.41) can be rewritten as

$$\tilde{x}[n] = \frac{1}{2\pi} \sum_{k=\langle N \rangle} X(k\Omega_0)e^{jk\Omega_0 n}\Omega_0 \tag{5.42}$$

As with eq. (4.59), as $N \rightarrow \infty$, $\tilde{x}[n]$ equals $x[n]$ for any finite value of n and $\Omega_0 \rightarrow 0$. Thus, eq. (5.42) passes to an integral. To see this more clearly, consider $X(\Omega)e^{j\Omega n}$ as sketched in Figure 5.9. From eq. (5.39), $X(\Omega)$ is seen to be periodic in Ω with period

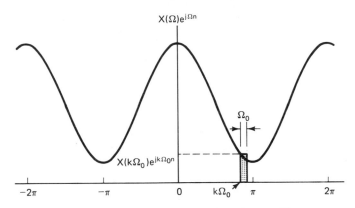

Figure 5.9 Graphical interpretation of eq. (5.42).

2π, and so is $e^{j\Omega n}$. Thus, the product $X(\Omega)e^{j\Omega n}$ will also be periodic. As depicted in the figure, each term in the summation in eq. (5.42) represents the area of a rectangle of height $X(k\Omega_0)e^{jk\Omega_0 n}$ and width Ω_0. As $\Omega_0 \rightarrow 0$ this becomes an integral. Furthermore, since the summation in eq. (5.42) is carried out over N consecutive intervals of width $\Omega_0 = 2\pi/N$, the total interval of integration will always have a width of 2π. Therefore, as $N \rightarrow \infty$, $\tilde{x}[n] = x[n]$ and eq. (5.42) becomes

$$x[n] = \frac{1}{2\pi} \int_{2\pi} X(\Omega)e^{j\Omega n} \, d\Omega$$

where, since $X(\Omega)e^{j\Omega n}$ is periodic with period 2π, the interval of integration can be taken as *any* interval of length 2π. Thus, we have the pair of equations

$$x[n] = \frac{1}{2\pi} \int_{2\pi} X(\Omega)e^{j\Omega n} \, d\Omega \tag{5.43}$$

$$X(\Omega) = \sum_{n=-\infty}^{+\infty} x[n]e^{-j\Omega n} \tag{5.44}$$

Equations (5.43) and (5.44) are the discrete-time counterparts of eqs. (4.60) and (4.61). The function $X(\Omega)$ is referred to as the *discrete-time Fourier transform* and the pair of equations as the *Fourier transform pair*. Equation (5.43) is the *synthesis equation* and eq. (5.44), the *analysis equation*. Our derivation of these equations indicates heuristically how an aperiodic sequence can be thought of as a linear combination of complex exponentials. In particular, the synthesis equation is in effect a representation of $x[n]$ as a linear combination of complex exponentials infinitesimally close in frequency and with amplitudes $X(\Omega)(d\Omega/2\pi)$. For this reason, as in continuous time, the transform $X(\Omega)$ will often be referred to as the *spectrum* of $x[n]$, as it provides us with the information on how $x[n]$ is composed of complex exponentials at different frequencies.

Although the argument that we have used was constructed assuming that $x[n]$ was of arbitrary but finite duration, eqs. (5.43) and (5.44) remain valid for an extremely broad class of signals with infinite duration. In this case, we again must consider the question of convergence of the infinite summation in eq. (5.44), and the conditions on $x[n]$ that guarantee the convergence of this sum are direct counterparts of the convergence conditions for the continuous-time Fourier transform.† Specifically, eq. (5.44) will converge either if $x[n]$ is absolutely summable, that is,

$$\sum_{n=-\infty}^{+\infty} |x[n]| < \infty \tag{5.45}$$

or if the sequence has finite energy, that is,

$$\sum_{n=-\infty}^{+\infty} |x[n]|^2 < \infty \tag{5.46}$$

Therefore, we see that the discrete-time Fourier transform possesses many similarities with the continuous-time case. The major differences between the two cases are the periodicity of the discrete-time transform $X(\Omega)$ and the finite interval of integration in the synthesis equation. Both of these stem from a fact that we have noted several time before: Discrete-time complex exponentials that differ in frequency by a multiple of 2π are identical. In Section 5.2 we saw that for periodic discrete-time signals, the implications of this statement are that the Fourier coefficients are periodic and that the Fourier series representation is a finite sum. For aperiodic signals the analogous implications are the periodicity of $X(\Omega)$ and the fact that the synthesis equation involves an integration only over a frequency interval that produces distinct complex exponentials (i.e., any interval of length 2π). In Chapter 2 we noted one further consequence of the periodicity of $e^{j\Omega n}$ as a function of Ω. Specifically, since $\Omega = 0$ and $\Omega = 2\pi$ yield the same signal, the frequencies near these values or any other even multiple of π correspond to low frequencies. Similarly, the high frequencies in discrete time are the values near odd multiples of π. Thus, the signal $x_1[n]$ shown in Figure 5.10(a) with Fourier transform depicted in Figure 5.10(b) varies more slowly than the signal $x_2[n]$ in Figure 5.10(c), whose transform is shown in Figure 5.10(d).

To illustrate the Fourier transform, let us consider several examples.

---

†For discussions of the convergence issues for the discrete-time Fourier transform, see A. V. Oppenheim and R. W. Schafer, *Digital Signal Processing* (Englewood Cliffs, N.J.: Prentice-Hall, Inc., 1975), and J. A. Cadzow and H. R. Martens, *Discrete-Time and Computer Control Systems* (Englewood Cliffs, N.J.: Prentice-Hall, Inc., 1970).

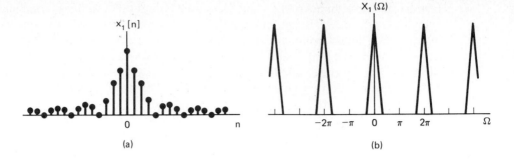

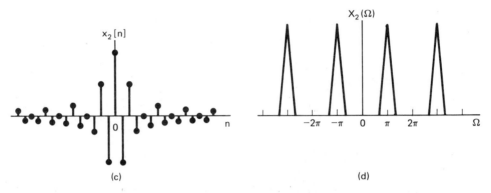

Figure 5.10 (a) Discrete-time signal $x_1[n]$ with (b) Fourier transform $X_1(\Omega)$ concentrated near $\Omega = 0, \pm 2\pi, \pm 4\pi, \ldots$; (c) discrete-time signal $x_2[n]$ with (d) Fourier transform $X_2(\Omega)$ concentrated near $\Omega = \pm \pi, \pm 3\pi, \ldots$.

Example 5.5

Consider the signal

$$x[n] = a^n u[n], \qquad |a| < 1$$

In this case

$$X(\Omega) = \sum_{n=-\infty}^{+\infty} a^n u[n] e^{-j\Omega n}$$

$$= \sum_{n=0}^{\infty} (ae^{-j\Omega})^n = \frac{1}{1 - ae^{-j\Omega}}$$

The amplitude and phase of $X(\Omega)$ for this example are drawn in Figure 5.11(a) for $a > 0$ and in Figure 5.11(b) for $a < 0$. Note that all of these functions are periodic with period 2π.

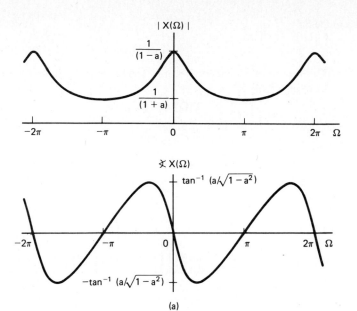

(a)

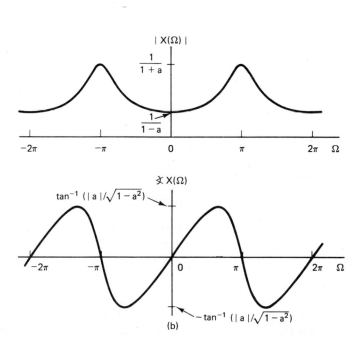

(b)

Figure 5.11 Magnitude and phase of the Fourier transform of Example 5.5 for (a) $a > 0$; and (b) $a < 0$.

Example 5.6

Let
$$x[n] = a^{|n|}, \qquad |a| < 1$$

This signal is sketched for $0 < a < 1$ in Figure 5.12(a). Its Fourier transform is obtained from eq. (5.44) as

$$X(\Omega) = \sum_{n=-\infty}^{+\infty} a^{|n|} e^{-j\Omega n}$$

$$= \sum_{n=0}^{\infty} a^n e^{-j\Omega n} + \sum_{n=-\infty}^{-1} a^{-n} e^{-j\Omega n}$$

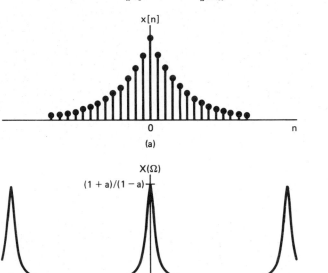

Figure 5.12 Signal $x[n] = a^{|n|}$ of Example 5.6 and its Fourier transform $(0 < a < 1)$.

Making the substitution of variables $m = -n$ in the second summation, we obtain

$$X(\Omega) = \sum_{n=0}^{\infty} (ae^{-j\Omega})^n + \sum_{m=1}^{\infty} (ae^{j\Omega})^m$$

The first summation is recognized as an infinite geometric series; the second is also a geometric series but is missing the first term. Therefore, we see that

$$X(\Omega) = \frac{1}{1 - ae^{-j\Omega}} + \frac{1}{1 - ae^{j\Omega}} - 1$$

$$= \frac{1 - a^2}{1 - 2a \cos \Omega + a^2}$$

In this case $X(\Omega)$ is real and is illustrated in Figure 5.12(b), again for $0 < a < 1$.

Example 5.7

Consider the rectangular pulse

$$x[n] = \begin{cases} 1, & |n| \leq N_1 \\ 0, & |n| > N_1 \end{cases} \tag{5.47}$$

which is illustrated in Figure 5.13(a) for $N_1 = 2$. In this case

$$X(\Omega) = \sum_{n=-N_1}^{N_1} e^{-j\Omega n} \qquad (5.48)$$

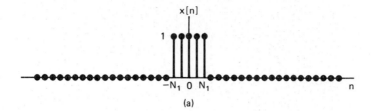

(a)

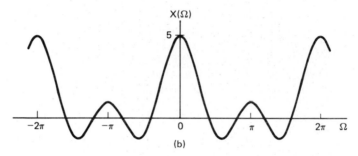

(b)

Figure 5.13 (a) Rectangular pulse signal of Example 5.7 for $N_1 = 2$, and (b) its Fourier transform.

Using the same type of calculations as we used to obtain eq. (5.28) in Example 5.3, we find that $X(\Omega)$ can be written as

$$X(\Omega) = \frac{\sin \Omega(N_1 + \frac{1}{2})}{\sin (\Omega/2)} \qquad (5.49)$$

This Fourier transform is sketched in Figure 5.13(b) for $N_1 = 2$. Note that as mentioned in the preceding section, the function in eq. (5.49) is the discrete-time counterpart of the sinc function, which appears in the Fourier transform of the continuous-time rectangular pulse (see Example 4.10). The most important difference between these two functions is that the function in eq. (5.49) is periodic with period 2π, as are all discrete-time Fourier transforms.

In the preceding section we saw that there are no convergence difficulties with the discrete-time Fourier series representation of a periodic sequence because the synthesis equation (5.18) is a finite sum. Similarly, since the *interval of integration* is finite for the discrete-time Fourier transform synthesis equation (5.43), there are no convergence problems in this case either. That is, if we approximate an aperiodic signal $x[n]$ by an integral of complex exponentials with frequencies taken over the interval $|\Omega| \leq W$,

$$\hat{x}[n] = \frac{1}{2\pi} \int_{-W}^{W} X(\Omega)e^{j\Omega n} \qquad (5.50)$$

then $\hat{x}[n] = x[n]$ for $W = \pi$. Thus, much as in Figure 5.7, we would expect not to see anything like the Gibbs phenomenon. This is illustrated in the following example.

Example 5.8

Let $x[n]$ be the unit sample

$$x[n] = \delta[n]$$

In this case the analysis equation (5.44) is easily evaluated, yielding

$$X(\Omega) = 1$$

That is, just as in continuous time, the unit impulse has a Fourier transform representation consisting of equal contributions at all frequencies. If we then apply eq. (5.50) to this example, we obtain

$$\hat{x}[n] = \frac{1}{2\pi} \int_{-W}^{W} e^{j\Omega n} \, d\Omega = \frac{\sin Wn}{\pi n} \tag{5.51}$$

This is plotted in Figure 5.14 for several values of W. As can be seen, the frequency of the oscillations in the approximation increases as W is increased, which is similar to what we observed in the continuous-time case. On the other hand, in contrast to the continuous-time case, the amplitude of these oscillations decreases relative to the magnitude of $\hat{x}[0]$ as W is increased, and these oscillations disappear entirely for $W = \pi$.

5.4 PERIODIC SIGNALS AND THE DISCRETE-TIME FOURIER TRANSFORM

As in the continuous-time case, there are several very important relationships between the Fourier series representation for periodic signals and the Fourier transform for aperiodic signals. In the preceding section we paralleled the development in Section 4.4 by deriving the Fourier transform from an examination of the behavior of the Fourier series for periodic signals as the period is made arbitrarily long. In this section we proceed in an identical fashion to our discussion in Section 4.5 in order to develop further the relationship between the two representations. Specifically, we first discuss how the Fourier series coefficients of a periodic sequence can be obtained from the Fourier transform of one period of that sequence. We then show how the Fourier series representation of periodic signals can be incorporated within the framework of the Fourier transform by interpreting the transform of a periodic signal as an impulse train in frequency.

5.4.1 Fourier Series Coefficients as Samples of the Fourier Transform of One Period

Let $\tilde{x}[n]$ be a periodic signal with period N, and let $x[n]$ represent one period of $\tilde{x}[n]$, that is,

$$x[n] = \begin{cases} \tilde{x}[n], & M \le n \le M + N - 1 \\ 0, & \text{otherwise} \end{cases} \tag{5.52}$$

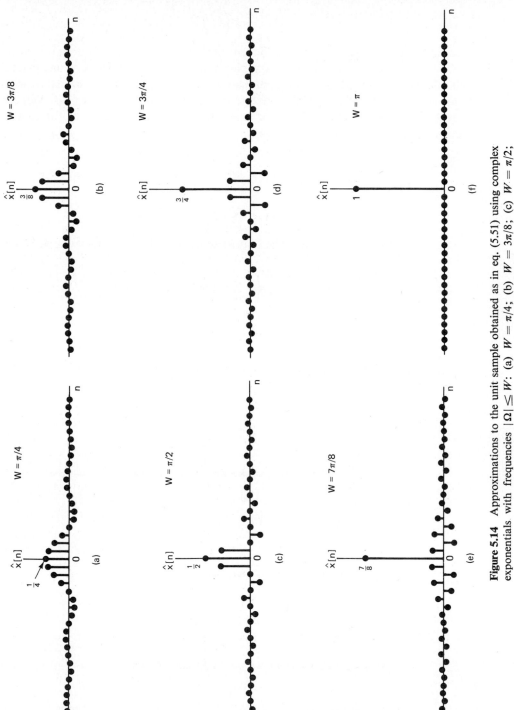

Figure 5.14 Approximations to the unit sample obtained as in eq. (5.51) using complex exponentials with frequencies $|\Omega| \leq W$: (a) $W = \pi/4$; (b) $W = 3\pi/8$; (c) $W = \pi/2$; (d) $W = 3\pi/4$; (e) $W = 7\pi/8$; (f) $W = \pi$. Note that for $W = \pi$, $\hat{x}[n] = \delta[n]$.

315

where M is arbitrary. Then it can be shown by direct calculation that

$$Na_k = X\left(k\frac{2\pi}{N}\right) \tag{5.53}$$

where the a_k are the Fourier series coefficients of $\tilde{x}[n]$ and $X(\Omega)$ is the Fourier transform of $x[n]$. Thus, the Na_k correspond to samples of the Fourier transform of one period. It should be noted that as in the continuous-time case, eq. (5.53) holds no matter what choice we make for M in eq. (5.52). That is, although $x[n]$ and therefore also $X(\Omega)$ can be expected to change markedly as M is varied, the values of $X(\Omega)$ at the sample frequencies $2\pi k/N$ do not depend on M.

Example 5.9

The discrete-time counterpart of the periodic impulse train of Example 4.15 is the sequence

$$x[n] = \sum_{k=-\infty}^{+\infty} \delta[n - kN] \tag{5.54}$$

as sketched in Figure 5.15(a). The Fourier series coefficients for this example can be calculated directly from eq. (5.19):

[eq. (5.19)] $\qquad\qquad a_k = \frac{1}{N} \sum_{n=\langle N\rangle} \tilde{x}[n]e^{-jk(2\pi/N)n}$

Choosing the interval of summation as $0 \le n \le N - 1$, we have

$$a_k = \frac{1}{N} \tag{5.55}$$

If we then define $x_1[n]$ as in eq. (5.52) with $M = 0$, we have that

$$x_1[n] = \delta[n]$$

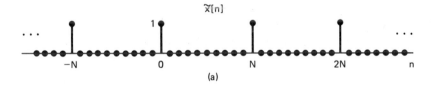

(a)

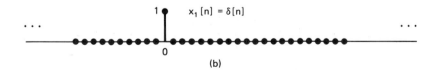

(b)

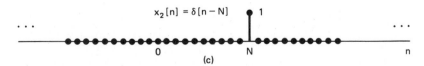

(c)

Figure 5.15 (a) Discrete-time periodic impulse train $\tilde{x}[n]$; (b, c) two aperiodic sequences each of which equals $\tilde{x}[n]$ over a single period.

which is shown in Figure 5.15(b). The Fourier transform of this signal is

$$X_1(\Omega) = 1 \tag{5.56}$$

Using eqs. (5.55) and (5.56) we see that eq. (5.53) holds. If instead we choose M differently in eq. (5.52), say $0 < M < N$, then we obtain a second signal,

$$x_2[n] = \delta[n - N]$$

which is depicted in Figure 5.15(c) and whose transform is

$$X(\Omega) = e^{-j\Omega N} \tag{5.57}$$

which clearly differs from eq. (5.56). However, at the set of sample frequencies $\Omega = 2\pi k/N$, eqs. (5.57) and (5.56) are identical, consistent with our prior observation that eq. (5.53) is valid independent of the choice of M.

5.4.2 The Fourier Transform for Periodic Signals

We now turn to the calculation of the Fourier transform representation for periodic signals. To derive the form of this representation, consider the signal

$$x[n] = e^{j\Omega_0 n} \tag{5.58}$$

In continuous time we saw that the Fourier transform of $e^{j\omega_0 t}$ was an impulse at $\omega = \omega_0$. Therefore, we might expect the same type of transform to result for the discrete-time signal of eq. (5.58). Recall, however, that the discrete-time Fourier transform is periodic in Ω with period 2π, and this is a direct consequence of the fact that

$$e^{j\Omega_0 n} = e^{j(\Omega_0 + 2\pi r)n} \tag{5.59}$$

for any integer r. This observation then suggests that the Fourier transform of $x[n]$ in eq. (5.58) should have impulses at $\Omega_0, \Omega_0 \pm 2\pi, \Omega_0 \pm 4\pi$, and so on. In fact, the Fourier transform of $x[n]$ is the impulse train

$$X(\Omega) = \sum_{l=-\infty}^{+\infty} 2\pi\delta(\Omega - \Omega_0 - 2\pi l) \tag{5.60}$$

which is illustrated in Figure 5.16. In order to check the validity of this expression,

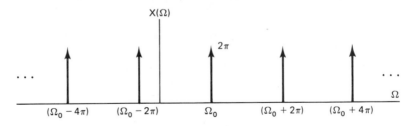

Figure 5.16   Fourier transform of $x[n] = e^{ji\Omega_0 n}$.

we must evaluate the inverse transform of eq. (5.60). Substituting eq. (5.60) into the synthesis equation (5.43), we find that

$$\frac{1}{2\pi}\int_{2\pi} X(\Omega)e^{j\Omega n}\,d\Omega = \frac{1}{2\pi}\int_{2\pi} \sum_{l=-\infty}^{+\infty} 2\pi\delta(\Omega - \Omega_0 - 2\pi l)e^{j\Omega n}\,d\Omega$$

Note that any interval of length 2π includes exactly one impulse in the summation given in eq. (5.60). Therefore, if the interval of integration chosen includes the impulse located at $\Omega_0 + 2\pi r$, we find that

$$\frac{1}{2\pi} \int_{2\pi} X(\Omega) e^{j\Omega n} \, d\Omega = e^{j(\Omega_0 + 2\pi r)n} = e^{j\Omega_0 n}$$

More generally, if $x[n]$ is the sum of an arbitrary set of complex exponentials

$$x[n] = b_1 e^{j\Omega_1 n} + b_2 e^{j\Omega_2 n} + \ldots + b_M e^{j\Omega_M n} \tag{5.61}$$

then its Fourier transform is

$$X(\Omega) = b_1 \sum_{l=-\infty}^{+\infty} 2\pi\delta(\Omega - \Omega_1 - 2\pi l) + b_2 \sum_{l=-\infty}^{+\infty} 2\pi\delta(\Omega - \Omega_2 - 2\pi l)$$

$$+ \ldots + b_M \sum_{l=-\infty}^{+\infty} 2\pi\delta(\Omega - \Omega_M - 2\pi l) \tag{5.62}$$

That is, $X(\Omega)$ is a periodic impulse train, with impulses located at the frequencies $\Omega_1, \ldots, \Omega_M$ of each of the complex exponentials and at all points that are multiples of 2π from these frequencies. Consequently, any 2π interval includes exactly one impulse from each of the summations on the right-hand side of eq. (5.62).

Note that eq. (5.60) is the Fourier transform of the signal given in eq. (5.58), whether this signal is periodic or not, that is, whether Ω_0 is or is not of the form $2\pi m/N$ for some integers m and N. Similarly, $x[n]$ in eq. (5.61) is periodic with period N only when *all* of the Ω_i have this form (with possibly different values of m for each). In particular, suppose that $x[n]$ is periodic with period N and that it has the Fourier series representation

$$x[n] = \sum_{k=\langle N \rangle} a_k e^{jk(2\pi/N)n} \tag{5.63}$$

If we choose the interval of summation in eq. (5.63) as $k = 0, 1, \ldots, N - 1$, then

$$x[n] = a_0 + a_1 e^{j(2\pi/N)n} + a_2 e^{j2(2\pi/N)n} + \ldots + a_{N-1} e^{j(N-1)(2\pi/N)n} \tag{5.64}$$

That is, $x[n]$ is of the form given in eq. (5.61) with

$$\Omega_1 = 0, \quad \Omega_2 = \frac{2\pi}{N}, \quad \Omega_3 = 2\left(\frac{2\pi}{N}\right), \quad \ldots, \quad \Omega_N = (N-1)\left(\frac{2\pi}{N}\right)$$

Therefore, from eq. (5.62), we see that the Fourier transform of $x[n]$ in eq. (5.63) is given by

$$X(\Omega) = a_0 \sum_{l=-\infty}^{+\infty} 2\pi\delta(\Omega - 2\pi l) + a_1 \sum_{l=-\infty}^{+\infty} 2\pi\delta\left(\Omega - \frac{2\pi}{N} - 2\pi l\right)$$

$$+ \ldots + a_{N-1} \sum_{l=-\infty}^{+\infty} 2\pi\delta\left(\Omega - (N-1)\frac{2\pi}{N} - 2\pi l\right) \tag{5.65}$$

This is illustrated in Figure 5.17. In Figure 5.17(a) we have depicted the first summation on the right-hand side of eq. (5.65), where we have used the fact that the Fourier series coefficients are periodic to write $2\pi a_0 = 2\pi a_N = 2\pi a_{-N}$. In Figure 5.17(b) we have illustrated the second term in eq. (5.63), and Figure 5.17(c) depicts the final term. Finally, Figure 5.17(d) depicts the entire expression for $X(\Omega)$. Note that because of the periodicity of the a_k, $X(\Omega)$ can be interpreted as a train of impulses occurring at multiples of the fundamental frequency $2\pi/N$, with the area of the

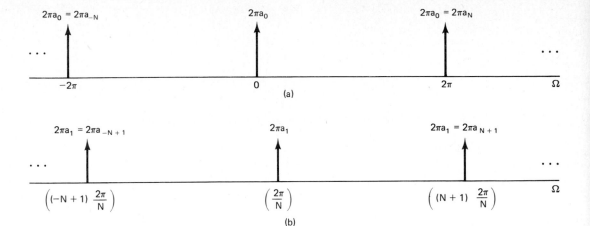

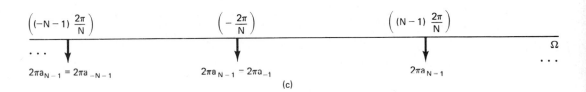

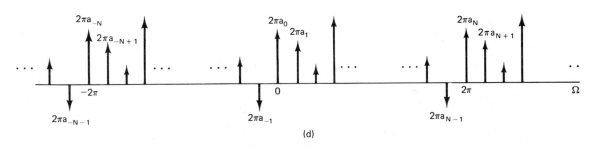

Figure 5.17 Fourier transform of a discrete-time periodic signal.

impulse located at $\Omega = 2\pi k/N$ being $2\pi a_k$. That is, an alternative and more convenient form for the Fourier transform of a periodic signal is

$$X(\Omega) = \sum_{k=-\infty}^{+\infty} 2\pi a_k \delta\left(\Omega - \frac{2\pi k}{N}\right) \tag{5.66}$$

Example 5.10

Consider the signal

$$x[n] = \cos \Omega_0 n = \tfrac{1}{2}e^{j\Omega_0 n} + \tfrac{1}{2}e^{-j\Omega_0 n} \tag{5.67}$$

From eq. (5.62) we can immediately write that

$$X(\Omega) = \sum_{l=-\infty}^{+\infty} \pi\{\delta(\Omega - \Omega_0 - 2\pi l) + \delta(\Omega + \Omega_0 - 2\pi l)\} \tag{5.68}$$

which is illustrated in Figure 5.18. This example will be very important in our discussion of discrete-time modulation in Chapter 7.

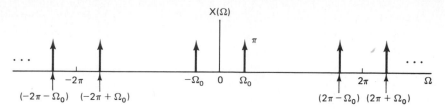

Figure 5.18 Discrete-time Fourier transform of $x[n] = \cos \Omega_0 n$.

Example 5.11

Consider again the periodic impulse train given in eq. (5.54). Using eqs. (5.55) and (5.66) we can calculate the Fourier transform of this signal:

$$X(\Omega) = \frac{2\pi}{N} \sum_{k=-\infty}^{+\infty} \delta\left(\Omega - \frac{2\pi k}{N}\right) \tag{5.69}$$

which is illustrated in Figure 5.19. Comparing Figures 5.15 and 5.19, we see another example of the inverse relationship between the time and frequency domains. As the spacing between the impulses in time (i.e., the period) gets longer, the spacing between impulses in frequency (the fundamental frequency) gets smaller.

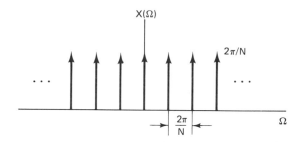

Figure 5.19 Fourier transform of a discrete-time periodic impulse train.

5.4.3 The DFT

In the introduction to this chapter we indicated that one of the reasons for the tremendous growth in the use of discrete-time methods for the analysis and synthesis of signals and systems was the development of exceedingly efficient tools for performing Fourier analysis for discrete-time sequences. At the heart of these methods is a technique that is very closely allied with the ideas that we have described and used in this and the preceding section and that is ideally suited for use on a digital computer or for implementation in digital hardware. This technique is the *discrete Fourier transform (DFT)*† for finite-duration signals. Although we will not spend a great deal of time discussing or using the DFT in this book, it is sufficiently important and is derived easily enough from what we have already done to deserve some mention.

Let $x[n]$ be a signal of finite duration; that is, there is an integer N_1 so that

$$x[n] = 0, \quad \text{outside the interval } 0 \leq n \leq N_1 - 1$$

†The DFT should not be confused with the *discrete-time Fourier transform*, with which we have been and will be almost exclusively concerned in this book. In order to remove possible confusion we will refer to the DFT only by its initials when we discuss it in this section, at the end of Section 5.6, and in several of the problems at the end of the chapter.

Much as we did in the preceding section we can construct a periodic signal $\tilde{x}[n]$ that is equal to $x[n]$ over one period. Specifically, let $N \geq N_1$ be a given integer, and let $\tilde{x}[n]$ be periodic with period N and such that

$$\tilde{x}[n] = x[n], \qquad 0 \leq n \leq N - 1 \tag{5.70}$$

The Fourier series coefficients for $\tilde{x}[n]$ are given by

$$a_k = \frac{1}{N} \sum_{n=\langle N \rangle} \tilde{x}[n] e^{-jk(2\pi/N)n}$$

Choosing the interval of summation to be that over which $\tilde{x}[n] = x[n]$, we obtain

$$a_k = \frac{1}{N} \sum_{n=0}^{N-1} x[n] e^{-jk(2\pi/N)n} \tag{5.71}$$

The set of coefficients defined by eq. (5.70) comprise the DFT of $x[n]$. Specifically, the DFT of $x[n]$ is usually denoted by $\tilde{X}(k)$, and is defined as

$$\tilde{X}(k) = a_k = \frac{1}{N} \sum_{n=0}^{N-1} x[n] e^{-jk(2\pi/N)n}, \qquad k = 0, 1, \ldots, N - 1 \tag{5.72a}$$

The importance of the DFT stems from several facts. First note that the original finite duration signal can be recovered from its DFT. Specifically, the synthesis equation (5.18) for the Fourier series representation of $\tilde{x}[n]$ allows us to compute $\tilde{x}[n]$ from the a_k. Then, using eqs, (5.70) and (5.72a), we have

$$x[n] = \sum_{k=0}^{N-1} \tilde{X}(k) e^{jk(2\pi/N)n}, \qquad n = 0, 1, \ldots, N - 1 \tag{5.72b}$$

Thus, the finite-duration signal can either be thought of as being specified by the finite set of nonzero values it assumes or by the finite set of values of $\tilde{X}(k)$ in its DFT. A second important feature of the DFT is that there is an extremely fast algorithm, called the *fast Fourier transform (FFT)*, for its calculation (see Problem 5.11 for an introduction to this extremely important technique). Also, because of its close relationship to the discrete-time Fourier series and transform, the DFT inherits some of their important properties.[†] As we will see, in Section 5.6, one of these properties, together with the FFT, provides an extremely efficient algorithm for computing the convolution of two finite-duration sequences. Finally, note that the choice of N in eq. (5.72) is not fixed, provided that N is chosen to be larger than the duration of $x[n]$. For this reason $\tilde{X}(k)$ in eq. (5.72a) is often referred to as an N-point DFT to make explicit the length of the summation in eq. (5.72a). As discussed in Problem 5.11, there are computational efficiencies to be gained from a judicious choice of N, such as choosing it to be a power of 2.

5.5 PROPERTIES OF THE DISCRETE-TIME FOURIER TRANSFORM

As with the continuous-time Fourier transform, there are a variety of properties of the discrete-time Fourier transform that provide further insight into the transform and in addition are often useful in reducing the complexity in the evaluation of transforms and inverse transforms. In this and the following sections we examine many

†See the texts on digital signal processing listed in the Bibliography at the end of the book for thorough discussions of the DFT and its properties and of the FFT.

The impulse train on the right-hand side of eq. (5.78) reflects the dc or average value that can result from summation. An important example of eq. (5.78) is the case when $x[n] = \delta[n]$. From Example 5.8 we have that $X(\Omega) - 1$, and eq. (5.78) yields

$$u[n] \quad \overset{\mathcal{F}}{\longleftrightarrow} \quad \frac{1}{1 - e^{-j\Omega}} + \pi \sum_{k=-\infty}^{+\infty} \delta(\Omega - 2\pi k) \tag{5.79}$$

The derivations of eqs. (5.78) and (5.79) are similar to our derivation of the integration property in Section 4.6, and both of these equations are derived in Problem 5.22.

5.5.6 Time and Frequency Scaling

Because of the discrete nature of the time index for discrete-time signals, the time and frequency scaling result in discrete time takes on a somewhat different form from its continuous-time counterpart. Let $x[n]$ be a signal with spectrum $X(\Omega)$. First let us consider the transform $Y(\Omega)$ of $y[n] = x[-n]$. From eq. (5.44),

$$Y(\Omega) = \sum_{n=-\infty}^{+\infty} y[n]e^{-j\Omega n} = \sum_{n=-\infty}^{+\infty} x[-n]e^{-j\Omega n} \tag{5.80}$$

Substituting $m = -n$ into eq. (5.80), we obtain

$$Y(\Omega) = \sum_{m=-\infty}^{+\infty} x[m]e^{-j(-\Omega)m} = X(-\Omega) \tag{5.81}$$

That is,

$$\boxed{x[-n] \quad \overset{\mathcal{F}}{\longleftrightarrow} \quad X(-\Omega)} \tag{5.82}$$

Although eq. (5.82) is analogous to the continuous-time case, differences arise when we try to scale time and frequency instead of simply reversing the time axis. Specifically in Section 4.6.5 we derived the continuous-time property

$$x(at) \quad \overset{\mathcal{F}}{\longleftrightarrow} \quad \frac{1}{|a|} X\left(\frac{\omega}{a}\right) \tag{5.83}$$

However, if we try to define the signal $x[an]$, we run into difficulties if a is not an integer. Therefore, we cannot slow down the signal by choosing $a < 1$. On the other hand, if we let a be an integer, for example if we consider $x[2n]$, we do not merely speed up the original signal. That is, since n can only take on integer values, the signal $x[2n]$ consists only of the even samples of $x[n]$.

The result that *does* closely parallel eq. (5.83) is the following. Let k be a positive integer, and define the signal

$$x_{(k)}[n] = \begin{cases} x[n/k], & \text{if } n \text{ is a multiple of } k \\ 0, & \text{if } n \text{ is not a multiple of } k \end{cases} \tag{5.84}$$

As illustrated in Figure 5.20 for $k = 3$, $x_{(k)}[n]$ is obtained from $x[n]$ by placing $(k - 1)$ zeros between successive values of the original signal. Intuitively, we can think of $x_{(k)}[n]$ as a slowed-down version of $x[n]$. If we calculate the Fourier transform of $x_{(k)}[n]$, we find, with the help of eq. (5.84), that

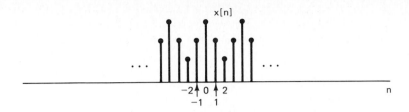

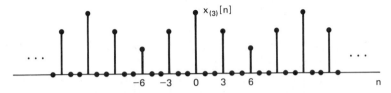

Figure 5.20 The signal $x_{(3)}[n]$ obtained from $x[n]$ by inserting two zeros between successive values of the original signal.

$$X_{(k)}(\Omega) = \sum_{n=-\infty}^{+\infty} x_{(k)}[n]e^{-j\Omega n} = \sum_{r=-\infty}^{+\infty} x_{(k)}[rk]e^{-j\Omega rk}$$

$$= \sum_{r=-\infty}^{+\infty} x[r]e^{-j(k\Omega)r} = X(k\Omega)$$

That is,

$$\boxed{x_{(k)}[n] \quad \overset{\mathcal{F}}{\longleftrightarrow} \quad X(k\Omega)} \tag{5.85}$$

Note that eq. (5.85) points out once again the inverse relationship between time and frequency. As the signal is spread out and slowed down in time by taking $k > 1$, its Fourier transform is compressed. For example, since $X(\Omega)$ is periodic with period 2π, $X(k\Omega)$ is periodic with period $2\pi/|k|$. This property is illustrated in Figure 5.21 for the example of a rectangular pulse.

5.5.7 Differentiation in Frequency

Again let

$$x[n] \quad \overset{\mathcal{F}}{\longleftrightarrow} \quad X(\Omega)$$

If we use the definition of $X(\Omega)$ in the analysis equation (5.44) and differentiate both sides, we obtain

$$\frac{dX(\Omega)}{d\Omega} = -\sum_{n=-\infty}^{+\infty} jnx[n]e^{-j\Omega n}$$

The right-hand side of this equation is nothing more than the Fourier transform of $-jnx[n]$. Therefore, multiplying both sides by j, we see that

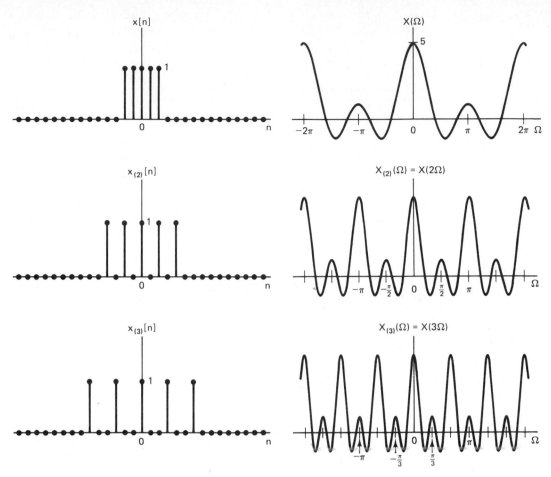

Figure 5.21 Inverse relationship between the time and frequency domains; as k increases $x_{(k)}[n]$ spreads out, while its transform is compressed.

$$nx[n] \quad \overset{\mathcal{F}}{\longleftrightarrow} \quad j\frac{dX(\Omega)}{d\Omega} \tag{5.86}$$

5.5.8 Parseval's Relation

If $x[n]$ and $X(\Omega)$ are a Fourier transform pair, then

$$\sum_{n=-\infty}^{+\infty} |x[n]|^2 = \frac{1}{2\pi} \int_{2\pi} |X(\Omega)|^2 \, d\Omega \tag{5.87}$$

We note that this is similar to eq. (4.110), and the derivation proceeds in a similar manner. In analogy with the continuous-time case, the left-hand side of eq. (5.87) is referred to as the *energy* in $x[n]$ and $|X(\Omega)|^2$ as the *energy-density spectrum*. Since the energy in a periodic sequence is infinite, eq. (5.87) is not useful in that case. For periodic signals a form of Parseval's relation can be derived, similar to eq. (4.111), that relates the energy in one period of the sequence to the energy in one period of the Fourier series coefficients. Specifically,

$$\frac{1}{N} \sum_{n=\langle N \rangle} |x[n]|^2 = \sum_{k=\langle N \rangle} |a_k|^2 \tag{5.88}$$

In the next few sections we consider several additional properties. The first two of these are the convolution and modulation properties, similar to those discussed in Sections 4.7 and 4.8. The third is the property of duality which is examined in Section 5.9, where we consider not only duality in the discrete-time domain but also the duality that exists *between* the continuous-time and discrete-time domains.

5.6 THE CONVOLUTION PROPERTY

In Section 4.7 we discussed the importance of the Fourier transform with regard to its effect on the operation of convolution and its use in dealing with LTI systems. An identical relation applies in discrete time, and for this reason we will find the discrete-time Fourier transform to be of great value in our examination of discrete-time LTI systems. Specifically, if $x[n]$, $h[n]$, and $y[n]$ are the input, impulse response, and output, respectively, of an LTI system, so that

$$y[n] = x[n] * h[n]$$

then

$$Y(\Omega) = X(\Omega)H(\Omega) \tag{5.89}$$

where $X(\Omega)$, $H(\Omega)$, and $Y(\Omega)$ are the Fourier transforms of $x[n]$, $h[n]$, and $y[n]$. The derivation parallels exactly that carried out in Section 4.7, eqs. (4.112)–(4.116). As in continuous time, the Fourier synthesis equation (5.43) for $x[n]$ can be interpreted as a decomposition of $x[n]$ into a linear combination of complex exponentials, with infinitesimal amplitudes proportional to $X(\Omega)$. Each of these exponentials is an eigenfunction of the system, and eq. (5.89) is in essence a statement of the fact that the change in complex amplitude experienced by each of these complex exponentials in passing through the system is $H(\Omega)$. As before, $H(\Omega)$ is referred to as the *frequency response* of the system. To illustrate the use of the convolution property, let us consider several examples.

Example 5.12

Consider an LTI system with impulse response

$$h[n] = \delta[n - n_0]$$

The frequency response $H(\Omega)$ is

$$H(\Omega) = \sum_{n=-\infty}^{+\infty} \delta[n - n_0] e^{-j\Omega n} = e^{-j\Omega n_0}$$

Thus, for any input $x[n]$ with Fourier transform $X(\Omega)$, the Fourier transform of the output is

$$Y(\Omega) = e^{-j\Omega n_0} X(\Omega) \tag{5.90}$$

We note that for this example, $y[n] = x[n - n_0]$ and eq. (5.90) is consistent with the time-shifting property, eq. (5.75).

Example 5.13

Let

$$h[n] = \alpha^n u[n]$$
$$x[n] = \beta^n u[n]$$

Evaluating the Fourier transforms of $h[n]$ and $x[n]$, we have

$$H(\Omega) = \frac{1}{1 - \alpha e^{-j\Omega}} \tag{5.91}$$

$$X(\Omega) = \frac{1}{1 - \beta e^{-j\Omega}} \tag{5.92}$$

so that

$$Y(\Omega) = H(\Omega)X(\Omega) = \frac{1}{(1 - \alpha e^{-j\Omega})(1 - \beta e^{-j\Omega})} \tag{5.93}$$

As with Example 4.20, determining the inverse transform of $Y(\Omega)$ is most easily done by expanding $Y(\Omega)$ in a partial fraction expansion. Specifically, $Y(\Omega)$ is a ratio of polynomials in powers of $e^{-j\Omega}$, and we would like to express this as a sum of simpler terms of this type so that we can recognize the inverse transform of each term by inspection (together, perhaps, with the use of the frequency differentiation property of Section 5.5.7). We discuss this at more length in Section 5.11 and in the Appendix and illustrate it now for this example.

If $\alpha \neq \beta$, the partial fraction expansion of $Y(\Omega)$ is of the form

$$Y(\Omega) = \frac{A}{1 - \alpha e^{-j\Omega}} + \frac{B}{1 - \beta e^{-j\Omega}} \tag{5.94}$$

Equating the right-hand sides of eqs. (5.93) and (5.94), we find that the correct values of the constants A and B are

$$A = \frac{\alpha}{\alpha - \beta}, \qquad B = -\frac{\beta}{\alpha - \beta}$$

Therefore, from eqs. (5.91) and (5.92), we can obtain the inverse transform of eq. (5.94) by inspection:

$$y[n] = \frac{\alpha}{\alpha - \beta} \alpha^n u[n] - \frac{\beta}{\alpha - \beta} \beta^n u[n]$$
$$= \frac{1}{\alpha - \beta} [\alpha^{n+1} u[n] - \beta^{n+1} u[n]] \tag{5.95}$$

For $\alpha = \beta$ the partial fraction expansion in eq. (5.94) is not valid. However, in this case

$$Y(\Omega) = \left(\frac{1}{1 - \alpha e^{-j\Omega}} \right)^2$$

which can be expressed as

$$Y(\Omega) = \frac{j}{\alpha} e^{j\Omega} \frac{d}{d\Omega} \left(\frac{1}{1 - \alpha e^{-j\Omega}} \right) \tag{5.96}$$

Exactly as we did in Example 4.20, we can use the frequency differentiation property, eq. (5.86), together with the Fourier transform pair

$$\alpha^n u[n] \quad \overset{\mathcal{F}}{\longleftrightarrow} \quad \frac{1}{1 - \alpha e^{-j\Omega}}$$

to deduce that

$$n\alpha^n u[n] \quad \overset{\mathcal{F}}{\longleftrightarrow} \quad j\frac{d}{d\Omega}\left(\frac{1}{1 - \alpha e^{-j\Omega}}\right)$$

To account for the factor $e^{j\Omega}$, we use the shifting property, eq. (5.75), to obtain

$$(n + 1)\alpha^{n+1} u[n + 1] \quad \overset{\mathcal{F}}{\longleftrightarrow} \quad je^{j\Omega}\frac{d}{d\Omega}\left(\frac{1}{1 - \alpha e^{-j\Omega}}\right)$$

and finally, accounting for the factor $1/\alpha$,

$$y[n] = (n + 1)\alpha^n u[n + 1] \quad \overset{\mathcal{F}}{\longleftrightarrow} \quad \frac{j}{\alpha}e^{j\Omega}\frac{d}{d\Omega}\left(\frac{1}{1 - \alpha e^{-j\Omega}}\right) = \left(\frac{1}{1 - \alpha e^{-j\Omega}}\right)^2 \qquad (5.97)$$

It is worth noting that although the left-hand side is multiplied by a step that begins at $n = -1$, the sequence $(n + 1)\alpha^n u[n + 1]$ is still zero prior to $n = 0$, since the factor $(n + 1)$ is zero at $n = -1$. Thus, we can alternatively express $y[n]$ as

$$y[n] = (n + 1)\alpha^n u[n] \qquad (5.98)$$

The frequency response $H(\Omega)$ plays the identical role for discrete-time LTI systems as that played by its continuous-time counterpart for continuous-time LTI systems. For example, the frequency response of the cascade of two discrete-time LTI systems is simply the product of the frequency responses of the two systems. It is also important to note that just as in continuous time, not every discrete-time LTI system has a frequency response. For example, the LTI system with impulse response $h[n] = 2^n u[n]$ does not have a finite response to sinusoidal inputs, which is reflected in the fact that the Fourier transform for $h[n]$ diverges. However, if an LTI system is stable, then from Section 3.4.4 we know that its impulse response is absolutely summable, that is,

$$\sum_{n=-\infty}^{+\infty} |h[n]| < \infty \qquad (5.99)$$

Referring to Section 5.3 and in particular to eq. (5.45) and the associated discussion, we see that eq. (5.99) guarantees the convergence of the Fourier transform of $h[n]$. Therefore, a stable LTI system has a well-defined frequency response $H(\Omega)$. For an illustration and further interpretation of these comments, see Problem 5.27.

5.6.1 Periodic Convolution

As in Section 4.7, the convolution property as considered thus far cannot be applied directly to two sequences that are periodic, since in that case the convolution summation will not converge. However, we can consider the *periodic convolution* of two sequences $\tilde{x}_1[n]$ and $\tilde{x}_2[n]$ which are periodic with common period N. The periodic convolution $\tilde{y}[n]$ is denoted by $\tilde{x}_1[n] \circledast \tilde{x}_2[n]$ and is defined as

$$\tilde{y}[n] = \tilde{x}_1[n] \circledast \tilde{x}_2[n] = \sum_{m=\langle N \rangle} \tilde{x}_1[m]\tilde{x}_2[n - m] \qquad (5.100)$$

which is the discrete-time counterpart of eq. (4.132). Equation (5.100) states that $\tilde{y}[n]$ is obtained by combining $\tilde{x}_1[n]$ and $\tilde{x}_2[n]$ in a manner that is reminiscent of the usual convolution of sequences, which is often referred to as *aperiodic convolution* to distinguish it from periodic convolution. In Figure 5.22 we have illustrated the pro-

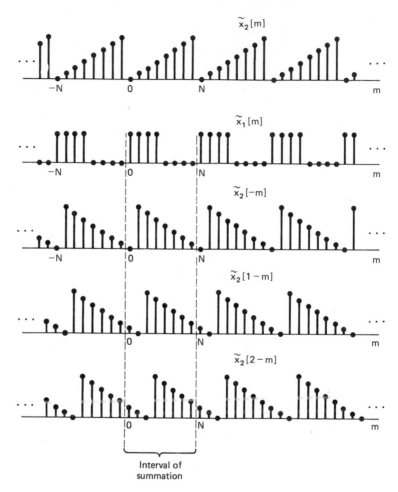

Figure 5.22 Procedure in forming the periodic convolution of two periodic sequences.

cedure used in calculating the periodic convolution. The signals $\tilde{x}_1[m]$ and $\tilde{x}_2[n - m]$ are multiplied together just as in aperiodic convolution. Note that both of these signals are periodic in m with period N and consequently so is their product. Also, the summation is carried out only over one period. To calculate $\tilde{y}[n]$ for successive values of n, $\tilde{x}_2[n - m]$ is shifted, and as one period slides out of the interval of summation, the next period slides in. Therefore, if n is increased by N, $\tilde{x}_2[n - m]$ will have been shifted one full period, and from this we can deduce that $\tilde{y}[n + N] = \tilde{y}[n]$ (i.e., that $\tilde{y}[n]$ is periodic with period N). Furthermore, it is a straightforward calculation to verify that the same result is obtained for any choice of the interval of summation in eq. (5.100).

For periodic convolution, the counterpart of the convolution property [eq. (5.89)] can be expressed directly in terms of the Fourier series coefficients. With $\{a_k\}$, $\{b_k\}$, and $\{c_k\}$ denoting the Fourier series coefficients of $\tilde{x}_1[n]$, $\tilde{x}_2[n]$, and $\tilde{y}[n]$, respectively, then

$$c_k = Na_k b_k \tag{5.101}$$

which is directly analogous to eq. (4.133).

The most important use of eq. (5.101) is in its application, together with the DFT, in the efficient calculation of the aperiodic convolution of two finite-duration sequences. Let $x_1[n]$ and $x_2[n]$ be two such sequences, and suppose that

$$
\begin{aligned}
x_1[n] &= 0, &&\text{outside the interval } 0 \leq n \leq N_1 - 1 \\
x_2[n] &= 0, &&\text{outside the interval } 0 \leq n \leq N_2 - 1
\end{aligned}
\tag{5.102}
$$

Let $y[n]$ be the aperiodic convolution of $x_1[n]$ and $x_2[n]$. Equation (5.102) then implies that

$$y[n] = x_1[n] * x_2[n] = 0, \qquad \text{outside the interval } 0 \leq n \leq N_1 + N_2 - 2 \tag{5.103}$$

(see Problem 3.11). Suppose we choose any integer $N \geq N_1 + N_2 - 1$ and define signals $\tilde{x}_1[n]$ and $\tilde{x}_2[n]$ that are periodic with period N and such that

$$
\begin{aligned}
\tilde{x}_1[n] &= x_1[n], &&0 \leq n \leq N - 1 \\
\tilde{x}_2[n] &= x_2[n], &&0 \leq n \leq N - 1
\end{aligned}
\tag{5.104}
$$

Let $\tilde{y}[n]$ be the periodic convolution of $\tilde{x}_1[n]$ and $\tilde{x}_2[n]$,

$$\tilde{y}[n] = \sum_{m=\langle N \rangle} \tilde{x}_1[m]\tilde{x}_2[n-m] \tag{5.105}$$

Then, as shown in Problem 5.12,

$$y[n] = \tilde{y}[n], \qquad 0 \leq n \leq N - 1 \tag{5.106}$$

The intuition behind eq. (5.106) can be gained by examining Figure 5.23. Essentially, we have "padded" $\tilde{x}_1[n]$ and $\tilde{x}_2[n]$ with enough zeros [i.e., we have chosen N long enough in eq. (5.104)] so that the periodic convolution $\tilde{y}[n]$ equals the aperiodic convolution $y[n]$ over one period. If we had chosen N too small, this would not be the case. Note also that from eq. (5.103) and the fact that $N \geq N_1 + N_2 - 1$, $y[n]$ is zero outside the interval $0 \leq n \leq N - 1$. Consequently, $y[n]$, the aperiodic convolution of $x_1[n]$ and $x_2[n]$ can be completely determined from the periodic convolution of $\tilde{x}_1[n]$ and $\tilde{x}_2[n]$. Also, from eq. (5.101), we note that we can calculate the Fourier series coefficients for $\tilde{y}[n]$ as the product of the Fourier series coefficients of $\tilde{x}_1[n]$ and $\tilde{x}_2[n]$. Since $\tilde{y}[n]$, $\tilde{x}_1[n]$, and $\tilde{x}_2[n]$ are identical to $y[n]$, $x_1[n]$, and $x_2[n]$, respectively, for $0 \leq n \leq N - 1$, we see from eq. (5.72a) that the Fourier series coefficients for the three periodic signals are equal to $\tilde{Y}(k)$, $\tilde{X}_1(k)$, and $\tilde{X}_2(k)$ the DFTs of $y[n]$, $x_1[n]$, and $x_2[n]$. Therefore, putting this all together we obtain an algorithm for the calculation of the aperiodic convolution of $x_1[n]$ and $x_2[n]$:

1. Calculate the DFTs $\tilde{X}_1(k)$ and $\tilde{X}_2(k)$ of $x_1[n]$ and $x_2[n]$ from eq. (5.72a)
2. Multiply these DFTs together to obtain the DFT of $y[n]$:

$$\tilde{Y}(k) = \tilde{X}_1(k)\tilde{X}_2(k) \tag{5.107}$$

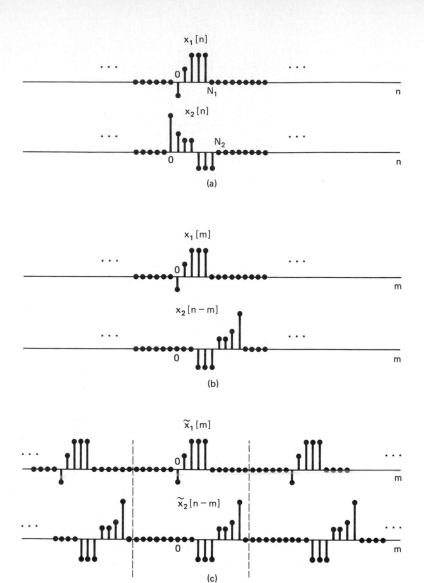

Figure 5.23 Calculation of the aperiodic convolution of two finite-duration signals as one period of the periodic convolution of two periodic signals, each equal to one of the original signals over one period: (a) original signals; (b) aperiodic convolution of $x_1[n]$ and $x_2[n]$; (c) periodic convolution where N is chosen large enough so that enough zeros are placed between the replicas of $x_1[n]$ and $x_2[n]$ so that the value resulting from evaluating the periodic convolution in (c) is identical to that for the periodic convolution in (b) for $0 \le n \le N - 1$.

3. Calculate the inverse DFT of $\tilde{Y}(k)$ from eq. (5.72b). The result is the desired convolution $y[n]$.

The only constraint in using this procedure is that of using an N-point DFT where $N \ge N_1 + N_2 - 1$. Since steps 1 and 3 can be performed efficiently using the fast

Fourier transform algorithm and since eq. (5.101) is computationally quite simple, the overall algorithm 1–3 represents an extremely efficient procedure for calculating the convolution of finite-duration signals.

5.7 THE MODULATION PROPERTY

In Section 4.8 we introduced the modulation property for continuous-time signals and indicated some of its applications through several examples. An analogous property exists for discrete-time signals and plays a similar role in applications. In this section we derive this result directly and give an example to illustrate it and to suggest possible applications.

To develop the modulation property, consider $y[n]$ equal to the product of $x_1[n]$ and $x_2[n]$, with $Y(\Omega), X_1(\Omega),$ and $X_2(\Omega)$ denoting the corresponding Fourier transforms. Then

$$Y(\Omega) = \sum_{n=-\infty}^{+\infty} y[n]e^{-j\Omega n} = \sum_{n=-\infty}^{+\infty} x_1[n]x_2[n]e^{-j\Omega n}$$

or, since

$$x_1[n] = \frac{1}{2\pi} \int_{2\pi} X_1(\theta)e^{j\theta n}\, d\theta \qquad (5.108)$$

then

$$Y(\Omega) = \sum_{n=-\infty}^{+\infty} x_2[n]\left\{\frac{1}{2\pi} \int_{2\pi} X_1(\theta)e^{j\theta n}\, d\theta\right\}e^{-j\Omega n} \qquad (5.109)$$

Interchanging the order of summation and integration, we obtain

$$Y(\Omega) = \frac{1}{2\pi} \int_{2\pi} X_1(\theta)\left[\sum_{n=-\infty}^{+\infty} x_2[n]e^{-j(\Omega-\theta)n}\right] d\theta \qquad (5.110)$$

The bracketed summation is $X_2(\Omega - \theta)$, and consequently eq. (5.110) becomes

$$\boxed{Y(\Omega) = \frac{1}{2\pi} \int_{2\pi} X_1(\theta)X_2(\Omega - \theta)\, d\theta} \qquad (5.111)$$

Equation (5.111) is in a form identical to eq. (4.132) and corresponds to a *periodic* convolution of $X_1(\Omega)$ and $X_2(\Omega)$.

The modulation property will be exploited in some detail when we discuss modulation and sampling in Chapters 7 and 8. Let us illustrate the use of eq. (5.111) with an example that we will see again in Chapter 7.

Example 5.14

Let $x_1[n]$ be the periodic sequence

$$x_1[n] = e^{j\pi n} = (-1)^n \qquad (5.112)$$

which is periodic with period 2. From eq. (5.60) we have that the Fourier transform of $x_1[n]$ is

$$X_1(\Omega) = 2\pi \sum_{r=-\infty}^{+\infty} \delta(\Omega - (2r + 1)\pi) \qquad (5.113)$$

as sketched in Figure 5.24(a). Taking $X_2(\Omega)$ to have the form shown in Figure 5.24(b),

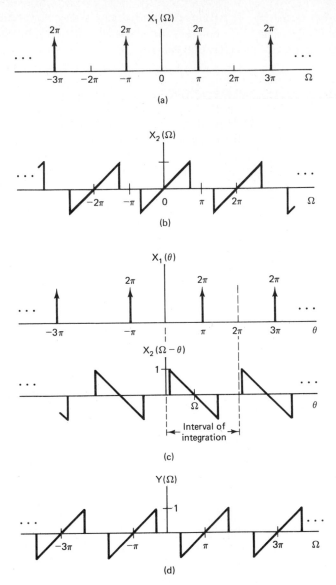

Figure 5.24 Discrete-time modulation property (a) Fourier transform of $x_1[n] = (-1)^n$; (b) Fourier transform of $x_2[n]$; (c) procedure for calculating the periodic convolution of eq. (5.111); (d) Fourier transform of $y[n] = x_1[n]\, x_2[n] = (-1)^n x_2[n]$.

we can carry out the convolution in eq. (5.111) graphically. In Figure 5.24(c) we have depicted $X_1(\theta)$ and $X_2(\Omega - \theta)$. In evaluating eq. (5.111), the integration can be carried out over any 2π interval in θ, which we choose here as $0 < \theta < 2\pi$. Thus, over the interval of integration,

$$X_1(\theta) X_2(\Omega - \theta) = 2\pi X_2(\Omega - \theta)\, \delta(\theta - \pi) = 2\pi X_2(\Omega - \pi)\, \delta(\theta - \pi)$$

so that

$$Y(\Omega) = \int_0^{2\pi} X_2(\Omega - \pi)\, \delta(\theta - \pi)\, d\theta = X_2(\Omega - \pi) \qquad (5.114)$$

For $X_2(\Omega)$ in Figure 5.24(b), $Y(\Omega)$ is then as sketched in Figure 5.24(d). Note that we could have obtained eq. (5.114) directly from the frequency-shifting property, eq. (5.76), with $\Omega_0 = \pi$.

Modulation of an arbitrary sequence $x_2[n]$ by $(-1)^n$ has the effect of changing the algebraic sign of every odd-numbered sequence value. From Figure 5.24(d) and more generally from eq. (5.114), we note that the effect in the frequency domain is to shift the periodic spectrum $X_2(\Omega)$ by one-half period (i.e., by π). Because of the periodicity in the spectrum, this has the effect of interchanging the high- and low-frequency regions of the spectrum. For this reason, modulation of a sequence by $(-1)^n$ has a number of useful applications, as explored more fully in Chapter 7.

5.8 TABLES OF FOURIER PROPERTIES AND OF BASIC FOURIER TRANSFORM AND FOURIER SERIES PAIRS

In Table 5.1 we summarize a number of important properties of the discrete-time Fourier transform. As we noted in Section 5.5, a number of these properties have counterparts for Fourier series, and these are summarized in Table 5.2. The derivation

TABLE 5.1 PROPERTIES OF THE DISCRETE-TIME FOURIER TRANSFORM

Aperiodic signal	Fourier transform
$x[n]$	$X(\Omega)$ periodic with
$y[n]$	$Y(\Omega)$ period 2π
$ax[n] + by[n]$	$aX(\Omega) + bY(\Omega)$
$x[n - n_0]$	$e^{-j\Omega n_0} X(\Omega)$
$e^{j\Omega_0 n} x[n]$	$X(\Omega - \Omega_0)$
$x^*[n]$	$X^*(-\Omega)$
$x[-n]$	$X(-\Omega)$
$x_{(k)}[n] = \begin{cases} x[n/k], & \text{if } n \text{ is a multiple of } k \\ 0, & \text{if } n \text{ is not a multiple of } k \end{cases}$	$X(k\Omega)$
$x[n] * y[n]$	$X(\Omega) Y(\Omega)$
$x[n]y[n]$	$\dfrac{1}{2\pi} \displaystyle\int_{2\pi} X(\theta) Y(\Omega - \theta)\, d\theta$
$x[n] - x[n - 1]$	$(1 - e^{-j\Omega}) X(\Omega)$
$\displaystyle\sum_{k=-\infty}^{n} x[k]$	$\dfrac{1}{1 - e^{-j\Omega}} X(\Omega) + \pi X(0) \displaystyle\sum_{k=-\infty}^{+\infty} \delta(\Omega - 2\pi k)$
$nx[n]$	$j\dfrac{dX(\Omega)}{d\Omega}$
$x[n]$ real	$\begin{cases} X(\Omega) = X^*(-\Omega) \\ \mathcal{R}e\{X(\Omega)\} = \mathcal{R}e\{X(-\Omega)\} \\ \mathcal{I}m\{X(\Omega)\} = -\mathcal{I}m\{X(-\Omega)\} \\ |X(\Omega)| = |X(-\Omega)| \\ \angle X(\Omega) = -\angle X(-\Omega) \end{cases}$
$x_e[n] = \mathcal{E}v\{x[n]\}$ $\quad [x[n] \text{ real}]$	$\mathcal{R}e\{X(\Omega)\}$
$x_o[n] = \mathcal{O}d\{x[n]\}$ $\quad [x[n] \text{ real}]$	$j\,\mathcal{I}m\{X(\Omega)\}$

Parseval's Relation for Aperiodic Signals

$$\sum_{n=-\infty}^{+\infty} |x[n]|^2 = \frac{1}{2\pi} \int_{2\pi} |X(\Omega)|^2\, d\Omega$$

of several of these are considered in Problems 5.4 and 5.6. Finally, in Table 5.3 we have assembled a list of some of the basic and most important discrete-time Fourier transform pairs. By comparing these tables with Tables 4.1 to 4.3, we can get a concise picture of some of the similarities and differences between continuous- and discrete-time Fourier transforms.

TABLE 5.2 PROPERTIES OF DISCRETE-TIME FOURIER SERIES

Periodic signal	Fourier series coefficients
$\left.\begin{array}{l} x[n] \\ y[n] \end{array}\right\}$ periodic with period N	$\left.\begin{array}{l} a_k \\ b_k \end{array}\right\}$ periodic with period N
$Ax[n] + By[n]$	$Aa_k + Bb_k$
$x[n - n_0]$	$a_k e^{-jk(2\pi/N)n_0}$
$e^{jM(2\pi/N)n}x[n]$	a_{k-M}
$x^*[n]$	a^*_{-k}
$x[-n]$	a_{-k}
$x_{(m)}[n] = \begin{cases} x[n/m] & \text{if } n \text{ is a multiple of } m \\ 0 & \text{if } n \text{ is not a multiple of } m \end{cases}$ (periodic with period mN)	$\dfrac{1}{m}a_k$ (viewed as periodic with period mN)
$\displaystyle\sum_{r=\langle N\rangle} x[r]y[n-r]$	$Na_k b_k$
$x[n]y[n]$	$\displaystyle\sum_{l=\langle N\rangle} a_l b_{k-l}$
$x[n] - x[n-1]$	$(1 - e^{-jk(2\pi/N)})a_k$
$\displaystyle\sum_{k=-\infty}^{n} x[k]$ (finite-valued and periodic only if $a_0 = 0$)	$\left(\dfrac{1}{1 - e^{-jk(2\pi/N)}}\right)a_k$
$x[n]$ real	$\begin{cases} a_k = a^*_{-k} \\ \mathcal{R}e\{a_k\} = \mathcal{R}e\{a_{-k}\} \\ \mathcal{I}m\{a_k\} = -\mathcal{I}m\{a_{-k}\} \\ |a_k| = |a_{-k}| \\ \sphericalangle a_k = -\sphericalangle a_{-k} \end{cases}$
$x_e[n] = \mathcal{E}v\{x[n]\}$ $\quad [x[n] \text{ real}]$	$\mathcal{R}e\{a_k\}$
$x_o[n] = \mathcal{O}d\{x[n]\}$ $\quad [x[n] \text{ real}]$	$j\,\mathcal{I}m\{a_k\}$

Parseval's Relation for Periodic Signals

$$\frac{1}{N}\sum_{n=\langle N\rangle} |x[n]|^2 = \sum_{k=\langle N\rangle} |a_k|^2$$

5.9 DUALITY

5.9.1 Discrete-Time Fourier Series

In considering the continuous-time Fourier transform, we observed a symmetry or duality between the analysis equation (4.61) and the synthesis equation (4.60). For the discrete-time Fourier transform a corresponding duality *does not* exist between the analysis equation (5.44) and the synthesis equation (5.43). However, there *is* a duality in the discrete-time Fourier *series* equations (5.18) and (5.19). More specifically, consider two periodic sequences with period N, related through the summation

$$f[m] = \frac{1}{N} \sum_{r=\langle N \rangle} g[r] e^{-jr(2\pi/N)m} \qquad (5.115)$$

If we let $m = k$ and $r = n$, eq. (5.115) becomes

$$f[k] = \frac{1}{N} \sum_{n=\langle N \rangle} g[n] e^{-jk(2\pi/N)n}$$

Comparing this to eq. (5.19), we see that the sequence $f[k]$ corresponds to the Fourier series coefficients of the signal $g[n]$. That is, if we adopt the notation

$$x[n] \overset{\mathcal{F}}{\longleftrightarrow} a_k$$

for a periodic discrete-time signal and its set of Fourier coefficients, we have that the two periodic sequences related through eq. (5.115) satisfy

$$g[n] \overset{\mathcal{F}}{\longleftrightarrow} f[k] \qquad (5.116)$$

Alternatively, if we let $m = n$ and $r = -k$, eq. (5.115) becomes

$$f[n] = \sum_{k=\langle N \rangle} \frac{1}{N} g[-k] e^{jk(2\pi/N)n}$$

Comparing with eq. (5.18), we find that $(1/N)g[-k]$ corresponds to the sequence of Fourier series coefficients of $f[n]$. That is,

$$f[n] \overset{\mathcal{F}}{\longleftrightarrow} \frac{1}{N} g[-k] \qquad (5.117)$$

Said another way, since the Fourier series coefficients a_k of a periodic signal $x[n]$ are themselves a periodic sequence, we can expand the a_k in a Fourier series. The duality property described above implies that the Fourier series coefficients for the periodic sequence a_k are the values $(1/N)x[-n]$ (i.e., are proportional to the original signal values reversed in time).

As in continuous time, this duality implies that every property of the discrete-time Fourier series has a dual. For example, referring to Table 5.2, the pair of properties

$$x[n - n_0] \overset{\mathcal{F}}{\longleftrightarrow} a_k e^{-jk(2\pi/N)n_0} \qquad (5.118)$$

$$e^{jM(2\pi/N)n} x[n] \overset{\mathcal{F}}{\longleftrightarrow} a_{k-M} \qquad (5.119)$$

are dual. Similarly, from this table we can extract another pair of dual properties,

$$\sum_{r=\langle N \rangle} x[r] y[n - r] \overset{\mathcal{F}}{\longleftrightarrow} N a_k b_k \qquad (5.120)$$

$$x[n] y[n] \overset{\mathcal{F}}{\longleftrightarrow} \sum_{l=\langle N \rangle} a_l b_{k-l} \qquad (5.121)$$

In addition to its consequences for the properties of discrete-time Fourier series, duality can often be useful in reducing the complexity of the calculations involved in determining Fourier series representations. Specifically, one evaluation of eq. (5.115) yields the Fourier series representation for *two* periodic sequences. This is illustrated in the next example.

TABLE 5.3 BASIC DISCRETE-TIME FOURIER TRANSFORM PAIRS

Signal	Fourier transform	Fourier series coefficient (if periodic)
$\displaystyle\sum_{k=\langle N\rangle} a_k e^{jk(2\pi/N)n}$	$\displaystyle 2\pi \sum_{k=-\infty}^{+\infty} a_k \delta\left(\Omega - \frac{2\pi k}{N}\right)$	a_k
$e^{j\Omega_0 N}$	$\displaystyle 2\pi \sum_{l=-\infty}^{+\infty} \delta(\Omega - \Omega_0 - 2\pi l)$	(a) $\Omega_0 = \dfrac{2\pi m}{N}$ $a_k = \begin{cases} 1, & k = m, m \pm N, m \pm 2N, \dots \\ 0, & \text{otherwise} \end{cases}$ (b) $\dfrac{\Omega_0}{2\pi}$ irrational $\Rightarrow$ The signal is aperiodic
$\cos \Omega_0 N$	$\displaystyle \pi \sum_{l=-\infty}^{+\infty} \{\delta(\Omega - \Omega_0 - 2\pi l) + \delta(\Omega + \Omega_0 - 2\pi l)\}$	(a) $\Omega_0 = \dfrac{2\pi m}{N}$ $a_k = \begin{cases} \frac{1}{2}, & k = \pm m, \pm m \pm N, \pm m \pm 2N, \dots \\ 0, & \text{otherwise} \end{cases}$ (b) $\dfrac{\Omega_0}{2\pi}$ irrational $\Rightarrow$ The signal is aperiodic
$\sin \Omega_0 N$	$\displaystyle \frac{\pi}{j} \sum_{l=-\infty}^{+\infty} \{\delta(\Omega - \Omega_0 - 2\pi l) - \delta(\Omega + \Omega_0 - 2\pi l)\}$	(a) $\Omega_0 = \dfrac{2\pi r}{N}$ $a_k = \begin{cases} \dfrac{1}{2j}, & k = r, r \pm N, r \pm 2N, \dots \\ -\dfrac{1}{2j}, & k = -r, -r \pm N, -r \pm 2N, \dots \\ 0, & \text{otherwise} \end{cases}$ (b) $\dfrac{\Omega_0}{2\pi}$ irrational $\Rightarrow$ The signal is aperiodic
$x[n] = 1$	$\displaystyle 2\pi \sum_{l=-\infty}^{+\infty} \delta(\Omega - 2\pi l)$	$a_k = \begin{cases} 1, & k = 0, \pm N, \pm 2N, \dots \\ 0, & \text{otherwise} \end{cases}$

TABLE 5.3 (Cont.)

Signal	Fourier transform	Fourier series coefficients (if periodic)
Periodic square wave $x[n] = \begin{cases} 1, & \|n\| \leq N_1 \\ 0, & N_1 < \|n\| \leq N/2 \end{cases}$ and $x[n+N] = x[n]$	$2\pi \sum\limits_{k=-\infty}^{+\infty} a_k \delta\left(\Omega - \dfrac{2\pi k}{N}\right)$	$a_k = \dfrac{\sin\left[(2\pi k/N)(N_1 + \frac{1}{2})\right]}{N \sin\left[2\pi k/2N\right]}, \quad k \neq 0, \pm N, \pm 2N, \dots$ $a_k = \dfrac{2N_1 + 1}{N}, \quad k = 0, \pm N, \pm 2N, \dots$
$\sum\limits_{k=-\infty}^{+\infty} \delta[n - kN]$	$\dfrac{2\pi}{N} \sum\limits_{k=-\infty}^{+\infty} \delta\left(\Omega - \dfrac{2\pi k}{N}\right)$	$a_k = \dfrac{1}{N}$ for all k
$a^n u[n], \quad \|a\| < 1$	$\dfrac{1}{1 - ae^{-j\Omega}}$	—
$x[n] = \begin{cases} 1, & \|n\| \leq N_1 \\ 0, & \|n\| > N_1 \end{cases}$	$\dfrac{\sin\left[\Omega(N_1 + \frac{1}{2})\right]}{\sin(\Omega/2)}$	—
$\dfrac{\sin Wn}{\pi n} = \dfrac{W}{\pi} \operatorname{sinc}\left(\dfrac{Wn}{\pi}\right)$ $0 < W < \pi$	$X(\Omega) = \begin{cases} 1, & 0 \leq \|\Omega\| \leq W \\ 0, & W < \|\Omega\| \leq \pi \end{cases}$ $X(\Omega)$ periodic with period 2π	—
$\delta[n]$	1	—
$u[n]$	$\dfrac{1}{1 - e^{-j\Omega}} + \sum\limits_{k=-\infty}^{+\infty} \pi\delta(\Omega - 2\pi k)$	—
$\delta[n - n_0]$	$e^{-j\Omega n_0}$	—
$(n + 1)a^n u[n], \quad \|a\| < 1$	$\dfrac{1}{(1 - ae^{-j\Omega})^2}$	—
$\dfrac{(n + r - 1)!}{n!\,(r - 1)!} a^n u[n], \quad \|a\| < 1$	$\dfrac{1}{(1 - ae^{-j\Omega})^r}$	—

Example 5.15

Consider the periodic sequence

$$x[n] = \frac{1}{N} \frac{\sin\left[(2\pi n/N)(N_1 + \frac{1}{2})\right]}{\sin\left(2\pi n/2N\right)} \tag{5.122}$$

From Example 5.3, we recognize this as the sequence of Fourier series coefficients of the discrete-time square wave of Figure 5.5. Thus, the Fourier series coefficients of $x[n]$ in eq. (5.122) will be $1/N$ times the sequence in Figure 5.5 reversed in order (which for this example has no effect since $x[n]$ is even). Consequently, the Fourier series coefficients for eq. (5.122) are as indicated in Figure 5.25.

Figure 5.25 Fourier series coefficients for the periodic sequence given in eq. (5.122).

5.9.2 Discrete-Time Fourier Transform and Continuous-Time Fourier Series

In addition to the duality for the discrete Fourier series, there is a duality between the *discrete-time* Fourier transform and the *continuous-time* Fourier series. Specifically, let us compare the continuous-time Fourier *series* equations (4.34) and (4.35) with the discrete-time Fourier *transform* equations (5.43) and (5.44). We repeat these equations here for convenience:

[eq. (5.43)]
$$x[n] = \frac{1}{2\pi} \int_{2\pi} X(\Omega) e^{j\Omega n}\, d\Omega \tag{5.123}$$

[eq. (5.44)]
$$X(\Omega) = \sum_{n=-\infty}^{+\infty} x[n] e^{-j\Omega n} \tag{5.124}$$

[eq. (4.34)]
$$x(t) = \sum_{k=-\infty}^{+\infty} a_k e^{jk\omega_0 t} \tag{5.125}$$

[eq. (4.35)]
$$a_k = \frac{1}{T_0} \int_{T_0} x(t) e^{-jk\omega_0 t}\, dt \tag{5.126}$$

Now let $f(u)$ represent a periodic function of a continuous variable with period 2π, and let $g[m]$ be a discrete sequence related to $f(u)$ by

$$f(u) = \sum_{m=-\infty}^{+\infty} g[m] e^{-jum} \tag{5.127}$$

Then with $u = \Omega$ and $m = n$, from eq. (5.124) we see that $f(\Omega)$ is the discrete-time Fourier transform of $g[n]$. That is,

$$g[n] \overset{\mathcal{F}}{\longleftrightarrow} f(\Omega) \tag{5.128}$$

Also, from this fact and from eq. (5.123), we see that we can recover $g[m]$ from $f(u)$ according to the relation

$$g[m] = \frac{1}{2\pi} \int_{2\pi} f(u) e^{jum}\, du \tag{5.129}$$

Alternatively, if we let $u = t$ and $m = -k$ in eq. (5.127), then $f(t)$ is periodic with period $T_0 = 2\pi$ and frequency $\omega_0 = 2\pi/T_0 = 1$, and, from eqs. (5.125) and (5.127) we see that $g[-k]$ is the sequence of Fourier series coefficients of $f(t)$. That is,

$$f(t) \overset{\mathcal{F}}{\longleftrightarrow} g[-k] \qquad (5.130)$$

where we have again adopted the notation

$$x(t) \overset{\mathcal{F}}{\longleftrightarrow} a_k$$

for a continuous-time periodic sequence and its Fourier series coefficients.

What eqs. (5.128) and (5.130) say is the following. Suppose that $x[n]$ is a discrete-time signal with Fourier transform $X(\Omega)$. Then, since this transform is a periodic function of a continuous variable we can expand it in a Fourier series using eq. (5.125) with $\omega_0 = 1$ and Ω, rather than t, as the continuous variables. From the duality relationship we can conclude that the Fourier series coefficients of $X(\Omega)$ will be the original sequence $x[n]$ reversed in order.

As with the other forms of duality, we can use the relationship just developed to translate properties of the continuous-time Fourier series into dual properties of the discrete-time Fourier transform. For example, in Section 4.7 we considered the periodic convolution of two periodic signals. In the case in which these signals have period $T_0 = 2\pi$, eqs. (4.132) and (4.133) become

$$\int_{2\pi} x_1(\tau)x_2(t - \tau)\, d\tau \overset{\mathcal{F}}{\longleftrightarrow} 2\pi a_k b_k$$

where a_k and b_k are the Fourier series coefficients of $x_1(t)$ and $x_2(t)$, respectively. The dual of this property is the modulation property for discrete-time signals

$$x[n]y[n] \overset{\mathcal{F}}{\longleftrightarrow} \frac{1}{2\pi} \int_{2\pi} X_1(\theta)X_2(\Omega - \theta)\, d\theta$$

In the next example we illustrate how the duality relationship can be useful in calculating the Fourier representation of signals.

Example 5.16

Consider the continuous-time signal $x(t)$, periodic with period 2π, and with Fourier series coefficients given by

$$a_k = \begin{cases} 1, & |k| \le N_1 \\ 0, & \text{otherwise} \end{cases}$$

From Example 5.7 we recognize this as the discrete-time rectangular pulse sequence with transform given by eq. (5.49). Therefore, from the duality property we have that

$$x(t) = \frac{\sin(N_1 + \frac{1}{2})t}{\sin(t/2)}$$

In a similar fashion, consider the discrete-time Fourier transform $X(\Omega)$ specified over the period $-\pi \le \Omega \le \pi$ as

$$X(\Omega) = \begin{cases} 1, & |\Omega| \le W \\ 0, & W < |\Omega| \le \pi \end{cases} \qquad (5.131)$$

From Example 4.5 we recognize this as a periodic square wave, with Fourier series coefficients given by eq. (4.40) with $T_1 = W$ and $\omega_0 = 1$. Using duality, we then have that the inverse transform of eq. (5.131) is

$$x[n] = \frac{\sin Wn}{\pi n} = \frac{W}{\pi} \operatorname{sinc}\left(\frac{Wn}{\pi}\right) \qquad (5.132)$$

In Table 5.4 we present a compact summary of the Fourier series and Fourier

TABLE 5.4 SUMMARY OF FOURIER SERIES AND TRANSFORM EXPRESSIONS

	Continuous-time		Discrete-time	
	Time domain	Frequency domain	Time domain	Frequency domain
Fourier Series	$x(t) = \sum\limits_{k=-\infty}^{+\infty} a_k e^{jk\omega_0 t}$ continuous time periodic in time	$a_k = \dfrac{1}{T_0}\displaystyle\int_{T_0} x(t) e^{-jk\omega_0 t}$ discrete frequency aperiodic in frequency	$x[n] = \sum\limits_{k=\langle N\rangle} a_k e^{jk(2\pi/N)n}$ discrete time periodic in time	$a_k = \dfrac{1}{N}\sum\limits_{n=\langle N\rangle} x[n] e^{-jk(2\pi/N)n}$ discrete frequency periodic in frequency
Fourier Transform	$x(t) = \dfrac{1}{2\pi}\displaystyle\int_{-\infty}^{+\infty} X(\omega) e^{j\omega t}\, d\omega$ continuous time aperiodic in time	$X(\omega) = \displaystyle\int_{-\infty}^{+\infty} x(t) e^{-j\omega t}\, dt$ continuous frequency aperiodic in frequency	$x[n] = \dfrac{1}{2\pi}\displaystyle\int_{2\pi} X(\Omega) e^{j\Omega n}$ discrete time aperiodic in time	$X(\Omega) = \sum\limits_{n=-\infty}^{+\infty} x[n] e^{-j\Omega n}$ continuous frequency periodic in frequency

duality

duality

duality

transform expressions for both continuous- and discrete-time signals, and we also indicate the duality relationships that apply in each case.

5.10 THE POLAR REPRESENTATION OF DISCRETE-TIME FOURIER TRANSFORMS

Let $x[n]$ be a discrete-time signal with transform $X(\Omega)$. Just as we did in Section 4.10, we can gain some further insight into properties of the discrete-time Fourier transform by examining its polar representation,

$$X(\Omega) = |X(\Omega)|e^{j \angle X(\Omega)} \tag{5.133}$$

Unlike the continuous-time case, both $|X(\Omega)|$ and $e^{j \angle X(\Omega)}$ are periodic with period 2π. However, the interpretations of these quantities are quite similar to those for their continuous-time counterparts. Specifically, $|X(\Omega)|$ contains the information about the relative magnitudes of the complex exponentials that make up $x[n]$, whereas the phase function $\angle X(\Omega)$ provides a description of the relative phases of the different complex exponentials in the Fourier representation of $x[n]$. For a given magnitude function $|X(\Omega)|$ we may obtain very different looking signals for different choices of the phase function $\angle X(\Omega)$. Therefore, a change in the phase function of $X(\Omega)$ leads to a distortion of the signal $x(t)$. As in continuous time, the case of *linear phase* is of particular importance. Specifically, suppose that we modify $\angle X(\Omega)$ by adding to it $m\Omega$, where m is an integer. Then the resulting transform is

$$X(\Omega)e^{jm\Omega}$$

and from the time-shifting property, eq. (5.75), the resulting signal is $x[n + m]$, which is simply a time-shifted version of the original signal.

The polar representation of Fourier transforms is often quite convenient for the examination of LTI systems. In particular, from the convolution property, the transforms of the input and output of an LTI system with frequency response $H(\Omega)$ are related by

$$Y(\Omega) = H(\Omega)X(\Omega)$$

which can also be written as

$$|Y(\Omega)| = |H(\Omega)||X(\Omega)| \tag{5.134}$$

$$\angle Y(\Omega) = \angle H(\Omega) + \angle X(\Omega) \tag{5.135}$$

Because of the multiplicative form of eq. (5.134) the magnitude of the frequency response of an LTI system is sometimes referred to as the *gain* of the system.

In the remaining sections of this chapter, as well as in our discussion of discrete-time filtering in Chapter 6, we will often find it useful to represent Fourier transforms graphically. Exactly as in continuous time, eqs. (5.134) and (5.135) suggest that for LTI analysis it is often convenient to graph $\angle X(\Omega)$ and the logarithm of $|X(\Omega)|$ as functions of Ω. In this way, the graph of $\angle Y(\Omega)$ and the logarithm of $|Y(\Omega)|$ can be obtained as the sums of the corresponding graphs for $X(\Omega)$ and $H(\Omega)$. Typical graphical representations of this type are illustrated in Figure 5.26. Here we have plotted $\angle H(\Omega)$ in radians and $|H(\Omega)|$ in decibels [i.e., $20 \log_{10} |H(\Omega)|$] as functions of Ω. Note

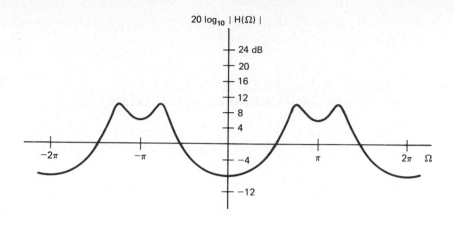

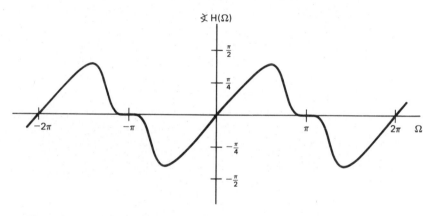

Figure 5.26 Typical graphical representations of the magnitude and phase of a discrete-time Fourier transform $H(\Omega)$.

that for $h[n]$ real, we actually need plot $H(\Omega)$ only for $0 \leq \Omega \leq \pi$, because in this case the symmetry property of the Fourier transform implies that we can then calculate $H(\Omega)$ for $-\pi \leq \Omega \leq 0$ using the relations $|H(\Omega)| = |H(-\Omega)|$ and $\sphericalangle H(-\Omega) = -\sphericalangle H(\Omega)$. Furthermore, we need consider no values of $|\Omega|$ greater than π because of the periodicity of $H(\Omega)$.

Recall that in continuous time we plotted $\sphericalangle H(\omega)$ and $20 \log_{10}|H(\omega)|$ versus ω, using a logarithmic frequency scale. This allowed a wider range of frequencies to be considered and also led to some useful simplifications for plotting frequency responses for continuous-time LTI systems specified by differential equations. In discrete time, however, the range of frequencies to be considered is limited, and the advantage found for differential equations (i.e., linear asymptotes) is not present for difference equations (see Section 5.11). Therefore, we have used a linear scale for Ω in Figure 5.26. Finally, just as in continuous time, in some cases it is more convenient to plot $|H(\Omega)|$ rather than its logarithm, especially for transforms for which $H(\Omega) = 0$ for some range of values of Ω. Thus, in the remainder of this book we use both logarithmic and linear plots for the magnitude of Fourier transforms and in any particular instance will use whichever is more appropriate.

5.11 THE FREQUENCY RESPONSE OF SYSTEMS CHARACTERIZED BY LINEAR CONSTANT-COEFFICIENT DIFFERENCE EQUATIONS

5.11.1 Calculation of Frequency and Impulse Responses for LTI Systems Characterized by Difference Equations

As defined in Chapter 3, a general linear, constant-coefficient difference equation for an LTI system with input $x[n]$ and output $y[n]$ is of the form

$$\sum_{k=0}^{N} a_k y[n - k] = \sum_{k=0}^{M} b_k x[n - k] \tag{5.136}$$

In this section we use the properties of the discrete-time Fourier transform to obtain an expression for the frequency response of the LTI system described by eq. (5.136). Having this expression, we can then use the technique of partial fraction expansion to determine the impulse response. This procedure closely parallels that used in Section 4.11 for continuous-time systems, and it provides an extremely useful method for analyzing discrete-time LTI systems described by difference equations.

As a first step in determining the frequency response for the LTI system specified by eq. (5.136), assume that the Fourier transforms of $x[n]$, $y[n]$, and the system impulse response $h[n]$ all exist, and denote their transforms by $X(\Omega)$, $Y(\Omega)$, and $H(\Omega)$, respectively. The convolution property (5.89) of the discrete-time Fourier transform then implies that

$$H(\Omega) = \frac{Y(\Omega)}{X(\Omega)}$$

Applying the Fourier transform to both sides of eq. (5.136) and using the linearity of the Fourier transform and the time-shifting property (5.75), we obtain the expression

$$\sum_{k=0}^{N} a_k e^{-jk\Omega} Y(\Omega) = \sum_{k=0}^{M} b_k e^{-jk\Omega} X(\Omega)$$

or, equivalently,

$$H(\Omega) = \frac{Y(\Omega)}{X(\Omega)} = \frac{\sum_{k=0}^{M} b_k e^{-jk\Omega}}{\sum_{k=0}^{N} a_k e^{-jk\Omega}} \tag{5.137}$$

Comparing eq. (5.137) with eq. (4.145), we see that as in the continuous-time case, $H(\Omega)$ is a ratio of polynomials, but in this case they are polynomials in the variable $e^{-j\Omega}$. As with eq. (4.145), the coefficients of the *numerator* polynomial are the same coefficients as appear on the *right* side of eq. (5.136), and the coefficients of the *denominator* polynomial are the same as appear on the *left* side of eq. (5.136). Therefore, the frequency response of the LTI system specified by eq. (5.136) can be written down by inspection.

Example 5.17

Consider the LTI system initially at rest that is characterized by

$$y[n] - ay[n-1] = x[n] \tag{5.138}$$

with $|a| < 1$. From eq. (5.137) the frequency response of this system is

$$H(\Omega) = \frac{1}{1 - ae^{-j\Omega}} \tag{5.139}$$

Comparing this with Example 5.5, we recognize this as the Fourier transform of the sequence $a^n u[n]$. Thus, the impulse response of the system is

$$h[n] = a^n u[n] \tag{5.140}$$

Example 5.18

Consider an LTI system initially at rest that is characterized by the difference equation

$$y[n] - \tfrac{3}{4} y[n-1] + \tfrac{1}{8} y[n-2] = 2x[n] \tag{5.141}$$

From eq. (5.137), the frequency response is

$$H(\Omega) = \frac{2}{1 - \tfrac{3}{4}e^{-j\Omega} + \tfrac{1}{8}e^{-j2\Omega}} \tag{5.142}$$

To determine the corresponding impulse response, we require the inverse transform of $H(\Omega)$. As in continuous time, an efficient procedure for doing this involves the technique of partial fraction expansion, which is discussed in detail in the Appendix (see Example A.3, in which the details of the calculations for this example are worked out; also see Example 5.13 earlier in this chapter). As a first step, we factor the denominator of eq. (5.142):

$$H(\Omega) = \frac{2}{(1 - \tfrac{1}{2}e^{-j\Omega})(1 - \tfrac{1}{4}e^{-j\Omega})} \tag{5.143}$$

Expanding eq. (5.143) in a partial fraction expansion yields

$$H(\Omega) = \frac{4}{1 - \tfrac{1}{2}e^{-j\Omega}} - \frac{2}{1 - \tfrac{1}{4}e^{-j\Omega}} \tag{5.144}$$

The inverse transform of each term can be recognized by inspection with the result that

$$h[n] = 4(\tfrac{1}{2})^n u[n] - 2(\tfrac{1}{4})^n u[n] \tag{5.145}$$

The procedure followed in Example 5.18 is identical in form to that used in continuous time. In particular, after expanding $H(\Omega)$ in a partial fraction expansion, the inverse transform of each term can be recognized by inspection. Thus, we can use this technique to invert the frequency response of any LTI system described by a linear constant-coefficient difference equation. Also, as illustrated in the next example, if the Fourier transform, $X(\Omega)$, of the input to such a system is a ratio of polynomials in $e^{-j\Omega}$, then so is $Y(\Omega)$. In this case we can use the same technique to find the response $y[n]$ to the input $x[n]$.

Example 5.19

Consider the LTI system of Example 5.18, and let the input to this system be

$$x[n] = (\tfrac{1}{4})^n u[n]$$

Then, using eq. (5.137) and Example 5.5 or 5.17,

$$Y(\Omega) = H(\Omega)X(\Omega) = \left[\frac{2}{(1 - \frac{1}{2}e^{-j\Omega})(1 - \frac{1}{4}e^{-j\Omega})}\right]\left[\frac{1}{1 - \frac{1}{4}e^{-j\Omega}}\right]$$

$$= \frac{2}{(1 - \frac{1}{2}e^{-j\Omega})(1 - \frac{1}{4}e^{-j\Omega})^2} \tag{5.146}$$

As described in the Appendix, the form of the partial fraction expansion in this case is

$$Y(\Omega) = \frac{B_{11}}{1 - \frac{1}{4}e^{-j\Omega}} + \frac{B_{12}}{(1 - \frac{1}{4}e^{-j\Omega})^2} + \frac{B_{21}}{1 - \frac{1}{2}e^{-j\Omega}} \tag{5.147}$$

where the constants B_{11}, B_{12}, and B_{21} can be determined using the techniques described in the Appendix. This particular example is worked out in detail in Example A.4, and the values obtained are

$$B_{11} = -4, \qquad B_{12} = -2, \qquad B_{21} = 8$$

so that

$$Y(\Omega) = -\frac{4}{1 - \frac{1}{4}e^{-j\Omega}} - \frac{2}{(1 - \frac{1}{4}e^{-j\Omega})^2} + \frac{8}{1 - \frac{1}{2}e^{-j\Omega}} \tag{5.148}$$

The first and third terms are of the same type as those encountered in Example 5.18, while the second term is of the same form as one seen in Example 5.13. Either from this example or from Table 5.3 we can invert this and higher-order terms of this type. The inverse transform of eq. (5.148) is then found to be

$$y[n] = \{-4(\tfrac{1}{4})^n - 2(n + 1)(\tfrac{1}{4})^n + 8(\tfrac{1}{2})^n\}u[n] \tag{5.149}$$

5.11.2 Cascade and Parallel-Form Structures

In Section 4.11 we saw that the use of Fourier transforms allowed us to develop two very important structures, the cascade and parallel forms, for the implementation of LTI systems characterized by differential equations. Following a similar procedure, we can now develop cascade and parallel-form structures for discrete-time LTI systems described by difference equations. For convenience in this discussion we will assume that $M = N$ in eq. (5.137). This can always be assumed to be true, with some of the b_k or a_k equal to zero.

The cascade structure is obtained by factoring the numerator and denominator of eq. (5.137) into products of first-order terms to obtain $H(\Omega)$ in the form

$$H(\Omega) = \frac{b_0 \displaystyle\prod_{k=1}^{N} (1 + \mu_k e^{-j\Omega})}{a_0 \displaystyle\prod_{k=1}^{N} (1 + \eta_k e^{-j\Omega})} \tag{5.150}$$

Some of the μ_k and η_k may be complex, but they then appear in complex-conjugate pairs. We can multiply each such pair of terms together to obtain a second-order term with real coefficients. Then, assuming that there are P such pairs in the numerator and Q in the denominator, we obtain

$$H(\Omega) = \frac{b_0 \displaystyle\prod_{k=1}^{P} (1 + \beta_{1k}e^{-j\Omega} + \beta_{2k}e^{-j2\Omega}) \displaystyle\prod_{k=1}^{N-2P} (1 + \mu_k e^{-j\Omega})}{a_0 \displaystyle\prod_{k=1}^{Q} (1 + \alpha_{1k}e^{-j\Omega} + \alpha_{2k}e^{-j2\Omega}) \displaystyle\prod_{k=1}^{N-2Q} (1 + \eta_k e^{-j\Omega})} \tag{5.151}$$

where we have renumbered the remaining real μ_k and η_k so that they run from 1 to

$N - 2P$ and $N - 2Q$, respectively. Therefore, the frequency response of any LTI system described by a linear constant-coefficient difference equation can be written as the product of first- and second-order terms. Consequently, this LTI system can be realized as the cascade of first- and second-order LTI systems each described by one of the terms in eq. (5.151). Thus, just as in continuous time, first- and second-order systems play an extremely important role in the analysis and synthesis of discrete-time systems, and we examine them further in the next section.

To illustrate the general form of the cascade structure, consider the case in which $H(\Omega)$ is represented as the product of second-order terms alone:

$$H(\Omega) = \frac{b_0}{a_0} \prod_{k=1}^{N/2} \frac{1 + \beta_{1k}e^{-j\Omega} + \beta_{2k}e^{-j2\Omega}}{1 + \alpha_{1k}e^{-j\Omega} + \alpha_{2k}e^{-j2\Omega}} \tag{5.152}$$

This corresponds to multiplying together pairs of the remaining first-order terms in eq. (5.151) and to assuming that N is even. If N is not even, we can add terms with zero coefficients to both the numerator and denominator. Each of the $N/2$ second-order terms in eq. (5.152) represents the frequency response of a system described by the difference equation

$$y[n] + \alpha_{1k}y[n-1] + \alpha_{2k}y[n-2] = x[n] + \beta_{1k}x[n-1] + \beta_{2k}x[n-2] \tag{5.153}$$

In Figure 5.27 we have illustrated the realization of $H(\Omega)$ in eq. (5.152) for $N = 6$. In this case $H(\Omega)$ is the product of three second-order terms, and consequently the system is the cascade of three subsystems as in eq. (5.153). In the figure we have used the direct form II realization (Figure 3.32) for each of these. As in continuous time, the cascade structure is not unique, since, for example, we have arbitrarily paired second-order numerator and denominator polynomials and have arbitrarily chosen the ordering for the second-order components in Figure 5.27. Other examples of cascade structures are considered in several of the problems at the end of the chapter.

The parallel-form structure in discrete time is obtained in the same manner as in continuous time, by performing a partial fraction expansion. Consider $H(\Omega)$ given in eq. (5.137), where again for convenience we will assume that $M = N$. Also, for simplicity, we will assume that all of the η_k are distinct. In this case a partial fraction expansion yields

$$H(\Omega) = \frac{b_N}{a_N} + \sum_{k=1}^{N} \frac{A_k}{1 + \eta_k e^{-j\Omega}} \tag{5.154}$$

As before, to obtain an implementation involving only real coefficients, we add together the pairs involving complex conjugate η_k's to obtain

$$H(\Omega) = \frac{b_N}{a_N} + \sum_{k=1}^{Q} \frac{\gamma_{0k} + \gamma_{1k}e^{-j\Omega}}{1 + \alpha_{1k}e^{-j\Omega} + \alpha_{2k}e^{-j2\Omega}} + \sum_{k=1}^{N-2Q} \frac{A_k}{1 + \eta_k e^{-j\Omega}} \tag{5.155}$$

Then using eq. (5.155), we can realize the LTI system with this frequency response as the parallel interconnection of LTI systems with frequency responses corresponding to each term in eq. (5.155). To illustrate the parallel-form structure, consider the case when N is even and $H(\Omega)$ is represented as a sum of second-order terms alone:

$$H(\Omega) = \frac{b_N}{a_N} + \sum_{k=1}^{N/2} \frac{\gamma_{0k} + \gamma_{1k}e^{-j\Omega}}{1 + \alpha_{1k}e^{-j\Omega} + \alpha_{2k}e^{-j2\Omega}} \tag{5.156}$$

To do this, we have added together pairs of the remaining first-order terms in eq.

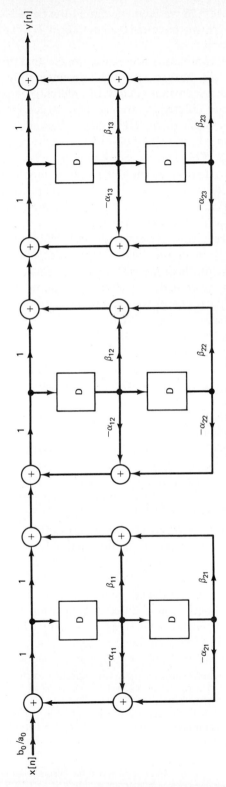

Figure 5.27 Cascade structure with a direct form II realization for each second-order subsystem.

(5.155). In Figure 5.28 we have illustrated the parallel-form realization of eq. (5.156) for $N = 6$, where we have used direct form II realizations for each of the three second-order terms.

As in the continuous-time case we see that there are a variety of different structures that can be used for the implementation of discrete-time LTI systems described by linear constant-coefficient difference equations. It is important to point out that there are differences among these structures that are of significance in the actual implementation of an LTI system. Cascade and parallel realizations consist of the interconnections of second-order systems which differ only in the values of their coefficients. This feature can have an impact on the cost of implementation, as it allows one to use essentially identical modules. Also, these structures offer other possibilities in terms of how they are implemented digitally. For example, in implementing a cascade structure we can imagine using a single, second-order digital system whose coefficients can be changed. Each successive value of the input is first processed by the digital system with its coefficients set equal to those of the first second-order system in the cascade structure; the result is then processed by the same digital system, this time with its coefficients set to those of the second second-order system in the cascade, etc. Here we require only a single second-order digital system, but the total processing time required to produce the next output is equal to the number of second-order systems in the cascade times the processing time of a single second-order system. On the other hand, one could implement a parallel structure using a number of second-order digital systems, one for each such system in the realization. These systems would then process $x[n]$ simultaneously, and consequently the total processing time to produce the next output is the processing time of a single second-order system.

In addition to issues such as those just described, the choice of a structure for implementation also involves the consideration of the fact that the actual implementations of different structures do not perform identically. Specifically, if we implement a discrete-time system digitally, we are confronted with the limitations of the finite register length of digital systems. Thus we cannot implement the desired coefficient values exactly. Also, numerical operations such as multiplication and addition are subject to roundoff, which is a nonlinear operation. It is a general fact that finite register length effects are different for different structures, and while we will not address this topic here, the methods we have developed in this chapter provide an essential ingredient for its investigation.†

In this section we have seen that the tools of discrete-time Fourier analysis are of great use in facilitating the examination of discrete-time LTI systems characterized by linear constant-coefficient difference equations. Just as in the continuous-time case, however, to use these techniques in examining any particular LTI system, the LTI system must have a frequency response. For example, the LTI system described by eq. (5.138) is unstable if $|a| > 1$ and consequently does not have a frequency response. Thus, whenever we apply the techniques described in this section, we will be

---

†See the texts on digital signal processing listed in the Bibliography at the end of the book for detailed treatments of the effects of finite register length on the digital implementation of discrete-time LTI systems.

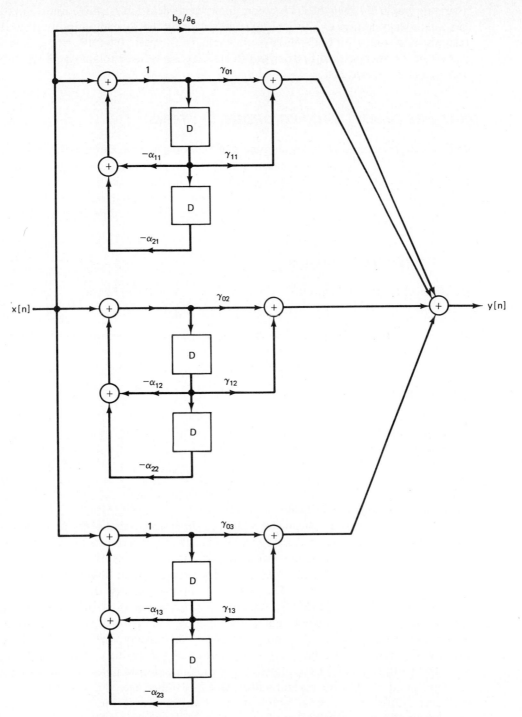

Figure 5.28 Parallel-form realization with a direct form II realization for each second-order subsystem.

assuming implicitly that the LTI system under consideration is stable and therefore has a frequency response. This can, of course, be checked by computing the impulse response and seeing if it is absolutely summable. In Chapter 10 we describe a generalization of the discrete-time Fourier transform that can be used to analyze both stable *and* unstable LTI systems.

5.12 FIRST-ORDER AND SECOND-ORDER SYSTEMS

In this section we parallel the development of Section 4.12 in examining the properties of first- and second-order systems. As in continuous time, these systems can be thought of as the building blocks from which we can construct systems with higher-order frequency responses. Therefore, by gaining some understanding of these basic systems, we will develop significant insight into the behavior of general LTI systems described by difference equations.

5.12.1 First-Order Systems

Consider the first-order causal LTI system described by the difference equation

$$y[n] - ay[n-1] = x[n] \tag{5.157}$$

with $|a| < 1$. From Example 5.17 we have that the frequency response of this system is

$$H(\Omega) = \frac{1}{1 - ae^{-j\Omega}} \tag{5.158}$$

and its impulse response is given by

$$h[n] = a^n u[n] \tag{5.159}$$

which is sketched in Figure 5.29 for several values of a. Also, the step response of this system is

$$s[n] = h[n] * u[n] = \frac{1 - a^{n+1}}{1 - a} u[n] \tag{5.160}$$

which is sketched in Figure 5.30.

The magnitude of the parameter a plays a role similar to that of the time constant τ of a continuous-time first-order system. Specifically, $|a|$ determines the rate at which the first-order system responds. For example, from eqs. (5.159) and (5.160) and Figures 5.29 and 5.30 we see that $h[n]$ and $s[n]$ converge to their final value at the rate at which $|a|^n$ converges to zero. Therefore, the impulse response decays sharply and the step response settles quickly for $|a|$ small. For $|a|$ nearer to 1, these responses are slower. Note that unlike its continuous-time analog, the first-order system described by eq. (5.157) *can* display oscillatory behavior. This occurs when $a < 0$ in which case the step response exhibits both overshoot of its final value and ringing.

In Figure 5.31(a) we have plotted the log magnitude and phase of the frequency response in eq. (5.158) for several values of $a > 0$. The case of $a < 0$ is illustrated in Figure 5.31(b). From these figures we see that for $a > 0$, the system attenuates high frequencies [i.e., $|H(\Omega)|$ is smaller for Ω near $\pm\pi$ than it is for Ω near 0], while when $a < 0$ the system amplifies high frequencies and attenuates low frequencies.

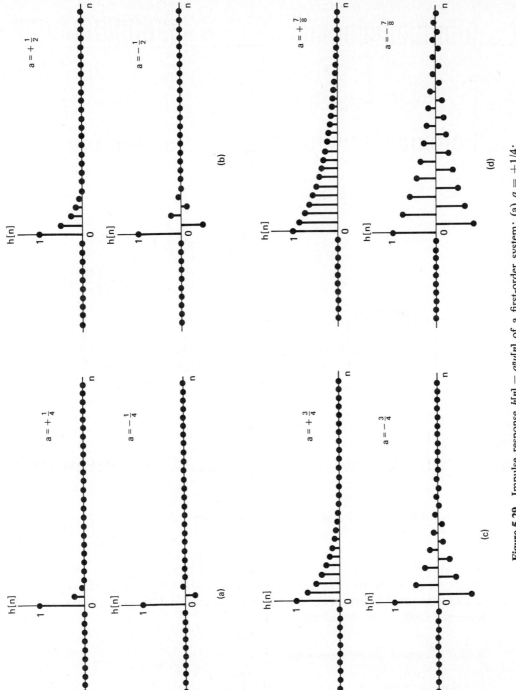

Figure 5.29 Impulse response $h[n] = a^n u[n]$ of a first-order system: (a) $a = \pm 1/4$; (b) $a = \pm 1/2$; (c) $a = \pm 3/4$; (d) $a = \pm 7/8$.

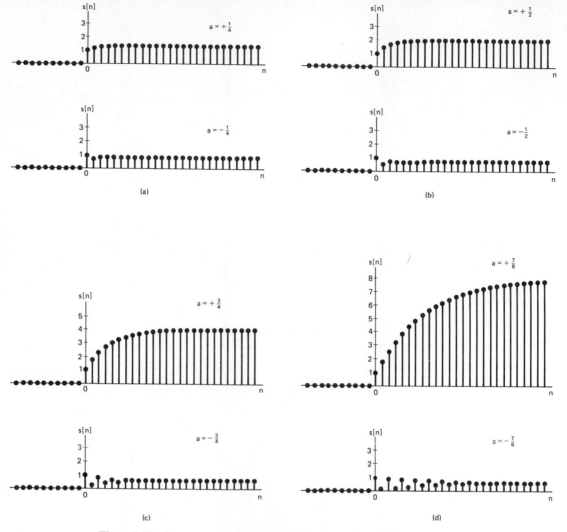

Figure 5.30 Step response $s[n]$ of a first-order system: (a) $a = \pm 1/4$; (b) $a = \pm 1/2$; (c) $a = \pm 3/4$; (d) $a = \pm 7/8$.

Note also that for $|a|$ small, the maximum and minimum values, $1/(1 + a)$ and $1/(1 - a)$, of $|H(\Omega)|$ are close together in value, and the graph of $|H(\Omega)|$ is relatively flat. On the other hand, for $|a|$ near 1, these quantities differ significantly, and consequently $|H(\Omega)|$ is more sharply peaked.

5.12.2 Second-Order Systems

Consider next the second-order causal LTI system described by

$$y[n] - 2r \cos \theta \, y[n - 1] + r^2 y[n - 2] = x[n] \qquad (5.161)$$

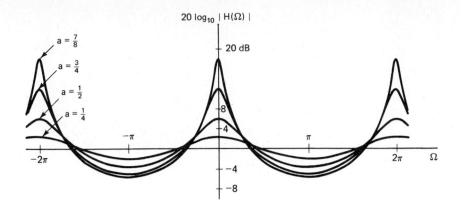

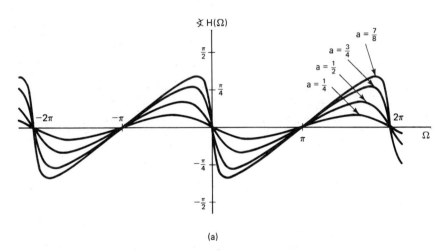

(a)

Figure 5.31 Magnitude and phase of the frequency response of eq. (5.158) for a first-order system: (a) plots for several values of $a > 0$; (b) plots for several values of $a < 0$.

with $0 < r < 1$ and $0 \leq \theta \leq \pi$. The frequency response for this system is

$$H(\Omega) = \frac{1}{1 - 2r \cos \theta \, e^{-j\Omega} + r^2 e^{-j2\Omega}} \tag{5.162}$$

The denominator of $H(\Omega)$ can be factored to obtain

$$H(\Omega) = \frac{1}{[1 - (re^{j\theta})e^{-j\Omega}][1 - (re^{-j\theta})e^{-j\Omega}]} \tag{5.163}$$

For $\theta \neq 0$ or π, the two factors in the denominator of $H(\Omega)$ are different, and a partial fraction expansion yields

$$H(\Omega) = \frac{A}{1 - (re^{j\theta})e^{-j\Omega}} + \frac{B}{1 - (re^{-j\theta})e^{-j\Omega}} \tag{5.164}$$

where

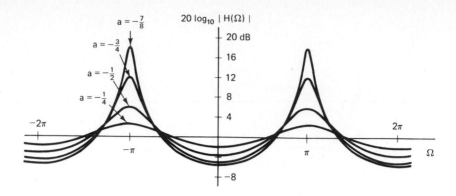

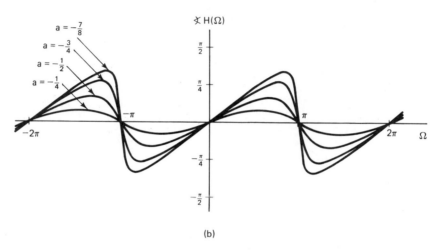

(b)

Figure 5.31 (cont.)

$$A = \frac{e^{j\theta}}{2j\sin\theta}, \qquad B = -\frac{e^{-j\theta}}{2j\sin\theta} \qquad (5.165)$$

In this case the impulse response of the system is

$$h[n] = [A(re^{j\theta})^n + B(re^{-j\theta})^n]u[n]$$
$$= r^n \frac{\sin[(n+1)\theta]}{\sin\theta}u[n] \qquad (5.166)$$

For $\theta = 0$ or π, the two factors in the denominator of eq. (5.163) are the same. In the case when $\theta = 0$,

$$H(\Omega) = \frac{1}{(1 - re^{-j\Omega})^2} \qquad (5.167)$$

and

$$h[n] = (n+1)r^n u[n] \qquad (5.168)$$

If $\theta = \pi$, then

$$H(\Omega) = \frac{1}{(1 + re^{-j\Omega})^2} \tag{5.169}$$

and

$$h[n] = (n + 1)(-r)^n u[n] \tag{5.170}$$

The impulse responses for second-order systems are plotted in Figure 5.32 for a range of values of r and θ. From this figure and from eq. (5.166) we see that the rate of decay of $h[n]$ is controlled by r—i.e. the closer r is to 1, the slower the decay in $h[n]$. Similarly, the value of θ determines the frequency of oscillation. For example, with $\theta = 0$ there is no oscillation in $h[n]$, while for $\theta = \pi$ the oscillations are rapid. The effect of different values of r and θ can also be seen by examining the step response of eq. (5.161). For $\theta \neq 0$ or π,

$$s[n] = h[n] * u[n] = \left[A\left(\frac{1 - (re^{j\theta})^{n+1}}{1 - re^{j\theta}}\right) + B\left(\frac{1 - (re^{-j\theta})^{n+1}}{1 - re^{-j\theta}}\right) \right] u[n] \tag{5.171}$$

Also, using the result of Problem 3.19, we find that for $\theta = 0$,

$$s[n] = \left[\frac{1}{(r-1)^2} - \frac{r}{(r-1)^2}r^n + \frac{r}{r-1}(n+1)r^n\right]u[n] \tag{5.172}$$

while for $\theta = \pi$,

$$s[n] = \left[\frac{1}{(r+1)^2} + \frac{r}{(r+1)^2}(-r)^n + \frac{r}{r+1}(n+1)(-r)^n\right]u[n] \tag{5.173}$$

The step response is plotted in Figure 5.33, again for a range of values of r and θ.

The second-order system given by eq. (5.161) is the counterpart of the *under-damped* second-order system in continuous time, while the special case of $\theta = 0$ is the critically damped case. That is, for any value of θ other than zero, the impulse response has a damped oscillatory behavior, and the step response exhibits ringing and overshoot. The frequency response of this system is depicted in Figure 5.34 for a number of values of r and θ. Here we see that a band of frequencies is amplified by this system. Note that θ essentially controls the location of the band that is amplified, and r then determines how sharply peaked the frequency response is within this band.

As we have just seen, the second-order system described by eq. (5.163) has factors with complex coefficients (unless $\theta = 0$ or π). It is also possible to consider second-order systems having factors with real coefficients. Specifically, consider $H(\Omega)$ of the form

$$H(\Omega) = \frac{1}{(1 - d_1 e^{-j\Omega})(1 - d_2 e^{-j\Omega})} \tag{5.174}$$

where d_1 and d_2 are both real numbers with $|d_1|, |d_2| < 1$. This is the frequency response for the difference equation

$$y[n] - (d_1 + d_2)y[n-1] + d_1 d_2 y[n-2] = x[n] \tag{5.175}$$

In this case

$$H(\Omega) = \frac{A}{1 - d_1 e^{-j\Omega}} + \frac{B}{1 - d_2 e^{-j\Omega}} \tag{5.176}$$

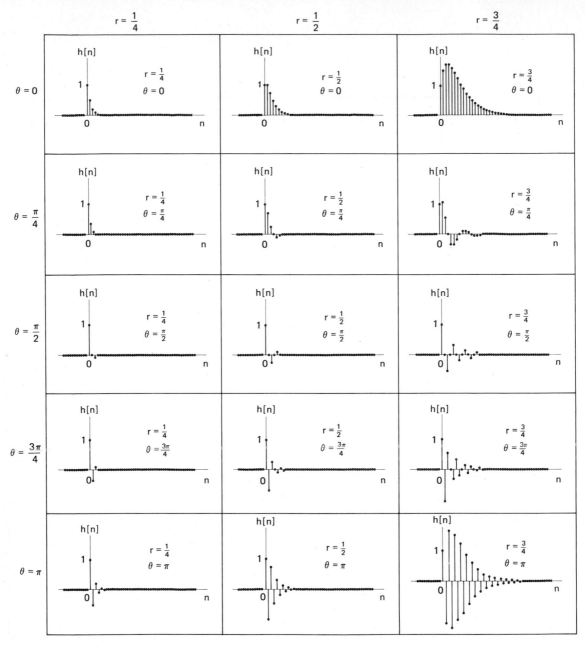

Figure 5.32 Impulse response of the second-order system of eq. (5.161) for a range of values of r and θ.

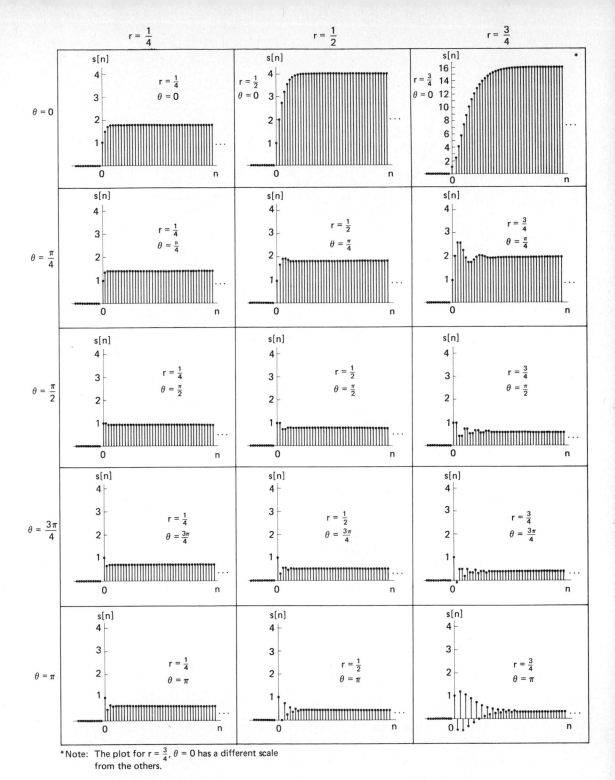

*Note: The plot for r = $\frac{3}{4}$, θ = 0 has a different scale from the others.

Figure 5.33 Step response of the second-order system of eq. (5.161) for a range of values of r and θ.

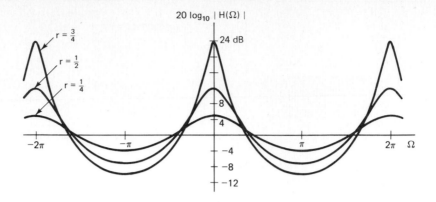

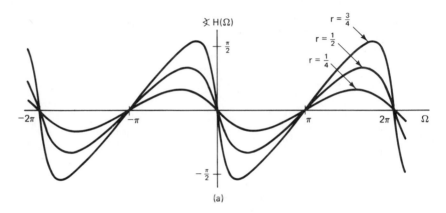

(a)

Figure 5.34 Magnitude and phase of the frequency response of the second-order system of eq. (5.161): (a) $\theta = 0$; (b) $\theta = \pi/4$; (c) $\theta = \pi/2$; (d) $\theta = 3\pi/4$; (e) $\theta = \pi$. Each plot contains curves corresponding to $r = 1/4, 1/2$, and $3/4$.

where

$$A = \frac{d_1}{d_1 - d_2}, \qquad B = \frac{d_2}{d_2 - d_1} \tag{5.177}$$

Thus,

$$h[n] = [Ad_1^n + Bd_2^n]u[n] \tag{5.178}$$

which is the sum of two decaying real exponentials. Also,

$$s[n] = \left[A\left(\frac{1 - d_1^{n+1}}{1 - d_1}\right) + B\left(\frac{1 - d_2^{n+1}}{1 - d_2}\right) \right]u[n] \tag{5.179}$$

The system with frequency response given by eq. (5.174) corresponds to the cascade of two first-order systems. Therefore, we can deduce most of its properties from our understanding of the first-order case. For example, the log-magnitude and

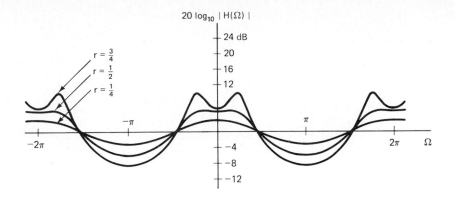

$$\theta = \frac{\pi}{4}$$

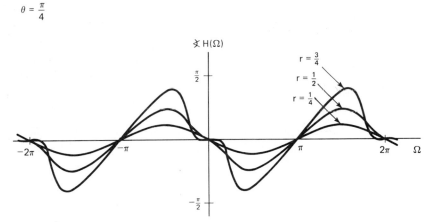

(b)

Figure 5.34 (cont.)

phase plots for eq. (5.174) can be obtained by adding together the plots for each of the two first-order terms. Also, as we saw for first-order systems, the response of the system is fast if $|d_1|$ and $|d_2|$ are small, but the system has a long settling time if either of these magnitudes is near 1. Furthermore, if d_1 and d_2 are negative, the response is oscillatory. The case when both d_1 and d_2 are positive is the counterpart of the overdamped case in continuous time, as the impulse and step responses settle without oscillation.

Finally, we note that, consistent with our observation at the close of Section 5.11, we have only examined those first- and second-order systems that are stable and consequently have frequency responses. In particular, the system described by eq. (5.157) is unstable for $|a| \geq 1$, the system of eq. (5.161) is unstable for $r \geq 1$, and the system of eq. (5.175) is unstable if either $|d_1|$ or $|d_2|$ exceeds 1. These statements can be verified directly by examining the impulse response in each case.

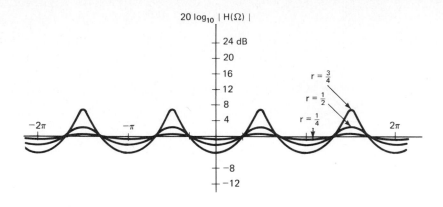

$$\theta = \frac{\pi}{2}$$

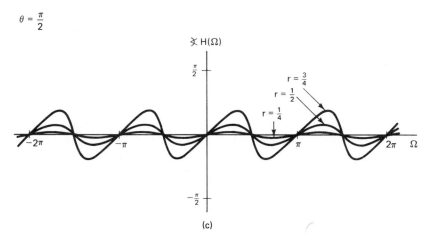

(c)

Figure 5.34 (cont.)

5.13 SUMMARY

In this chapter we have paralleled Chapter 4 as we developed the techniques of Fourier analysis for discrete-time signals and systems. As in Chapter 4, the primary motivation for this examination is the fact that complex exponential signals are eigenfunctions of discrete-time LTI systems. For this reason Fourier series and transform representations play a central role in the study of discrete-time signals and systems. We have now developed the basic tools, both in continuous time and in discrete time, and in Chapters 6 to 8 we apply these results to three topics of great practical importance: filtering, modulation, and sampling.

Throughout this chapter we have seen a great many similarities between continuous- and discrete-time Fourier analysis, and we have also seen some important differences. For example, the relationship between Fourier series and Fourier transforms in discrete time is exactly analogous to the continuous-time case, and many

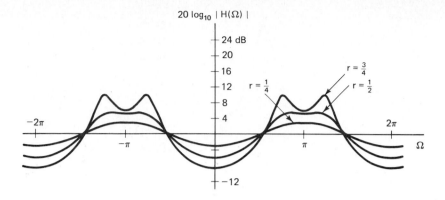

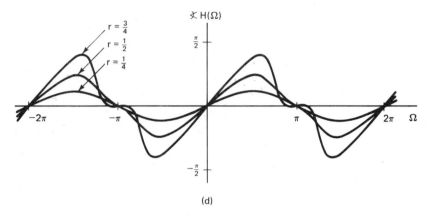

(d)

Figure 5.34 (cont.)

of the properties of continuous-time transforms have exact discrete-time counterparts. On the other hand, the discrete-time Fourier series for a periodic signal is a finite series, and the discrete-time Fourier transform of an aperiodic signal is always periodic with period 2π. Also, in addition to similarities and differences such as these, we have described the striking duality among the Fourier representations of continuous- and discrete-time signals.

Without a doubt the most important similarities between continuous- and discrete-time Fourier analysis are in their uses in examining signals and LTI systems. Specifically, the convolution property provides us with the basis for the frequency-domain analysis of LTI systems. We have already seen some of the utility of this approach in our examination of systems described by linear constant-coefficient differential or difference equations, and we will gain a further appreciation of its utility in Chapter 6 when we focus on the topic of frequency-selective filtering. In addition, the modulation properties in continuous and discrete time are the keys to our development of modulation systems in Chapter 7, and in Chapter 8 we use Fourier

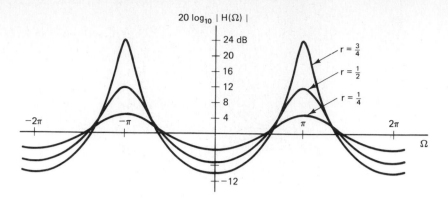

$\theta = \pi$

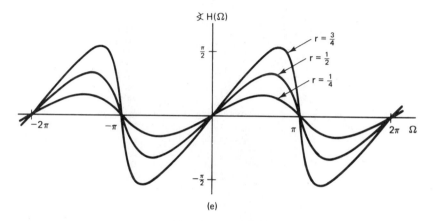

(e)

Figure 5.34 (cont.)

analysis in both continuous and discrete time to examine sampling systems and the discrete-time processing of continuous-time signals.

PROBLEMS

5.1. Determine the Fourier series coefficients for each of the periodic discrete-time signals given below. Plot the magnitude and phase of each set of coefficients a_k.

(a) $x[n] = \sin [\pi(n-1)/4]$

(b) $x[n] = \cos (2\pi n/3) + \sin (2\pi n/7)$

(c) $x[n] = \cos \left(\dfrac{11\pi n}{4} - \dfrac{\pi}{3}\right)$

(d) $x[n]$ is periodic with period 6 and

$$x[n] = (\tfrac{1}{2})^n \quad \text{for } -2 \le n \le 3$$

(e) $x[n] = \sin (2\pi n/3) \cos (\pi n/2)$

(f) $x[n]$ is periodic with period 4, and

$$x[n] = 1 - \sin \left(\dfrac{\pi n}{4}\right) \quad \text{for } 0 \le n \le 3$$

(g) $x[n]$ is periodic with period 12, and

$$x[n] = 1 - \sin\left(\frac{\pi n}{4}\right) \qquad \text{for } 0 \le n \le 11$$

(h) $x[n]$ as depicted in Figure P5.1(a)

(i) $x[n]$ as depicted in Figure P5.1(b)

(j) $x[n]$ as depicted in Figure P5.1(c)

(k) $x[n]$ as depicted in Figure P5.1(d)

(l) $x[n]$ as depicted in Figure P5.1(e)

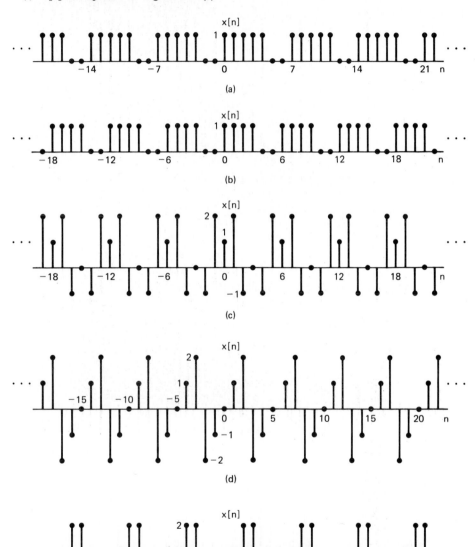

(a)

(b)

(c)

(d)

(e)

Figure P5.1

5.2. Consider the signal $x[n]$ depicted in Figure P5.2. This signal is periodic with period $N = 4$. The signal $x[n]$ can be expressed in terms of a discrete-time Fourier series:

$$x[n] = \sum_{k=0}^{3} a_k e^{jk(2\pi/4)n} \qquad \text{(P5.2-1)}$$

As mentioned in the text one way to determine the Fourier series coefficients is to treat eq. (P5.2-1) as a set of four linear equations [eq. (P5.2-1) for $n = 0, 1, 2, 3$] in the four unknowns (a_0, a_1, a_2, and a_3).

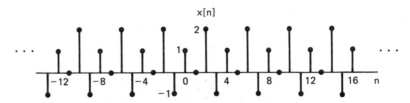

Figure P5.2

(a) Explicitly write out these four equations and solve them directly using any standard technique for solving four equations in four unknowns. (Be sure first to reduce the complex exponentials above to the simplest form.)

(b) Check your answer by calculating the a_k directly using the discrete-time Fourier series analysis equation

$$a_k = \tfrac{1}{4} \sum_{n=0}^{3} x[n] e^{-jk(2\pi/4)n}$$

5.3. In each of the following parts we specify the Fourier series coefficients of a signal that is periodic with period 8. Determine the signal $x[n]$ in each case.

(a) $a_k = \cos(k\pi/4) + \sin(3k\pi/4)$

(b) $a_k = \begin{cases} \sin(k\pi/3), & 0 \leq k \leq 6 \\ 0, & k = 7 \end{cases}$

(c) a_k as in Figure P5.3(a)

(d) a_k as in Figure P5.3(b)

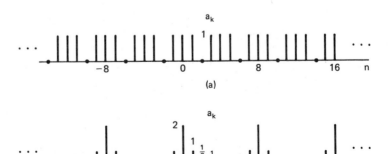

Figure P5.3

5.4. Let $x[n]$ be a periodic sequence with period N and Fourier series representation

$$x[n] = \sum_{k=\langle N \rangle} a_k e^{jk(2\pi/N)n} \qquad \text{(P5.4-1)}$$

(a) The Fourier series coefficients for each of the following signals can be expressed in terms of the a_k in eq. (P5.4-1). Derive these expressions.

(i) $x[n - n_0]$

(ii) $x[n] - x[n - 1]$

(iii) $x[n] - x\left[n - \dfrac{N}{2}\right]$ (assume that N is even)

(iv) $x[n] + x\left[n + \dfrac{N}{2}\right]$ (assume that N is even; note that this signal is periodic with period $N/2$)

(v) $x^*[-n]$

(vi) $x_{(m)}[n] = \begin{cases} x[n/m], & \text{if } n \text{ is a multiple of } m \\ 0, & \text{if } n \text{ is not a multiple of } m \end{cases}$

   $(x_{(m)}[n]$ is periodic with period $mN)$

(vii) $(-1)^n x[n]$ (assume that N is even)

(viii) $(-1)^n x[n]$ (assume that N is odd; note that this signal is periodic with period $2N$)

(ix) $y[n] = \begin{cases} x[n], & n \text{ even} \\ 0, & n \text{ odd} \end{cases}$

   (you will have to examine the cases N even and N odd separately)

(b) Suppose that N is even and that $x[n]$ in eq. (P5.4-1) satisfies

$$x[n] = -x\left[n + \frac{N}{2}\right] \qquad \text{for all } n$$

Show that $a_k = 0$ for all even integers k.

(c) Suppose that N is divisible by 4. Show that if

$$x[n] = -x\left[n + \frac{N}{4}\right] \qquad \text{for all } n$$

then $a_k = 0$ for every value of k that is a multiple of 4.

(d) More generally, suppose that N is divisible by the integer M. Show that if

$$\sum_{r=0}^{(N/M)-1} x\left[n + r\frac{N}{M}\right] = 0 \qquad \text{for all } n$$

then $a_k = 0$ for every value of k that is a multiple of M.

5.5. In this problem we consider the discrete-time counterpart of the concepts introduced in Problems 4.7 and 4.8. In analogy with the continuous-time case, two discrete-time signals $\phi_k[n]$ and $\phi_m[n]$ are said to be *orthogonal* over the interval (N_1, N_2) if

$$\sum_{n=N_1}^{N_2} \phi_k[n]\phi_m^*[n] = \begin{cases} A_k, & k = m \\ 0, & k \neq m \end{cases} \qquad \text{(P5.5-1)}$$

If the value of the constants A_k and A_m are both 1, then the signals are said to be *orthonormal*.

(a) Consider the signals

$$\phi_k[n] = \delta[n - k], \qquad k = 0, \pm 1, \pm 2, \ldots, \pm N$$

Show that these signals are orthonormal over the interval $(-N, N)$.

(b) Show that the signals

$$\phi_k[n] = e^{jk(2\pi/N)n}, \qquad k = 0, 1, \ldots, N-1$$

are orthogonal over any interval of length N.

(c) Show that if

$$x[n] = \sum_{i=1}^{M} a_i\phi_i[n]$$

where the $\phi_i[n]$ are orthogonal over the interval (N_1, N_2), then

$$\sum_{n=N_1}^{N_2} |x[n]|^2 = \sum_{i=1}^{M} |a_i|^2 A_i$$

(d) Let $\phi_i[n]$, $i = 0, 1, \ldots, M$, be a set of orthogonal functions over the interval (N_1, N_2), and let $x[n]$ be a given signal. Suppose that we wish to approximate $x[n]$ as a linear combination of the $\phi_i[n]$, that is,

$$\hat{x}[n] = \sum_{i=0}^{M} a_i\phi_i[n]$$

where the a_i are constant coefficients. Let

$$e[n] = x[n] - \hat{x}[n]$$

and show that if we wish to minimize

$$E = \sum_{m=N_1}^{N_2} |e[n]|^2$$

then the a_i are given by

$$a_i = \frac{1}{A_i} \sum_{n=N_1}^{N_2} x[n]\phi_i^*[n] \qquad \text{(P5.5-2)}$$

Hint: As in Problem 4.8, express E in terms of a_i, $\phi_i[n]$, A_i, and $x[n]$, write $a_i = b_i + jc_i$, and show that the equations

$$\frac{\partial E}{\partial b_i} = 0 \quad \text{and} \quad \frac{\partial E}{\partial c_i} = 0$$

are satisfied by a_i given by eq. (P5.5-2). [Note that applying this result when the $\phi_i[n]$ are as in part (b) yields eq. (5.19) for a_k.]

(e) Apply the result of part (d) when the $\phi_i[n]$ are as in part (a) to determine the coefficients a_i in terms of $x[n]$.

5.6. (a) Let $x[n]$ and $y[n]$ be periodic signals with

$$x[n] = \sum_{k=\langle N \rangle} a_k e^{jk(2\pi/N)n} \qquad \text{(P5.6-1)}$$

and

$$y[n] = \sum_{k=\langle N \rangle} b_k e^{jk(2\pi/N)n}$$

Derive the discrete-time modulation property. That is, show that

$$x[n]y[n] = \sum_{k=\langle N \rangle} c_k e^{jk(2\pi/N)n}$$

where

$$c_k = \sum_{l=\langle N \rangle} a_l b_{k-l} = \sum_{l=\langle N \rangle} a_{k-l} b_l$$

(b) Use the result of part (a) to find the Fourier series representation of the following signals, where $x[n]$ is given by eq. (P5.6-1).

(i) $x[n] \cos\left(\dfrac{6\pi n}{N}\right)$

(ii) $x[n] \displaystyle\sum_{r=-\infty}^{+\infty} \delta[n - rN]$

(iii) $x[n] \left(\displaystyle\sum_{r=-\infty}^{+\infty} \delta\left[n - \dfrac{rN}{3}\right]\right)$ (assume that N is divisible by 3)

(c) Find the Fourier series representation for the signal $x[n]y[n]$, where

$$x[n] = \cos(\pi n/3)$$
$$y[n] = \text{is periodic with period 12}$$
$$y[n] = \begin{cases} 1, & |n| \le 3 \\ 0, & 4 \le |n| \le 6 \end{cases}$$

(d) Use the result of part (a) to show that

$$\sum_{n=\langle N \rangle} x[n]y[n] = N \sum_{l=\langle N \rangle} a_l b_{-l}$$

and from this expression derive Parseval's relation for discrete-time periodic signals.

5.7. Let $x[n]$ be a real periodic signal, with period N and complex Fourier series coefficients a_k. Let the Cartesian form for a_k be denoted by

$$a_k = b_k + jc_k$$

where b_k and c_k are both real.

(a) Show that $a_{-k} = a_k^*$. What is the relation between b_k and b_{-k}? What is the relation between c_k and c_{-k}?

(b) Suppose that N is even. Show that $a_{N/2}$ is real.

(c) Show that $x[n]$ can also be expressed as a trigonometric Fourier series of the form

$$x[n] = a_0 + 2 \sum_{k=1}^{(N-1)/2} b_k \cos\left(\frac{2\pi kn}{N}\right) - c_k \sin\left(\frac{2\pi kn}{N}\right)$$

if N is odd, or

$$x[n] = (a_0 + a_{N/2}(-1)^n) + 2 \sum_{k=1}^{(N-2)/2} b_k \cos\left(\frac{2\pi kn}{N}\right) - c_k \sin\left(\frac{2\pi kn}{N}\right)$$

if N is even.

(d) Show that if the polar form of a_k is $A_k e^{j\theta_k}$, then the Fourier series representation for $x[n]$ can also be written as

$$x[n] = a_0 + 2 \sum_{k=1}^{(N-1)/2} A_k \cos\left(\frac{2\pi kn}{N} + \theta_k\right)$$

if N is odd, or

$$x[n] = (a_0 + a_{N/2}(-1)^n) + 2 \sum_{k=1}^{(N/2)-1} A_k \cos\left(\frac{2\pi kn}{N} + \theta_k\right)$$

if N is even.

(e) Suppose that $x[n]$ and $z[n]$, as depicted in Figure P5.7 have the sine–cosine series representations

$$x[n] = a_0 + 2 \sum_{k=1}^{3} \left\{ b_k \cos\left(\frac{2\pi kn}{7}\right) - c_k \sin\left(\frac{2\pi kn}{7}\right)\right\}$$

$$z[n] = d_0 + 2 \sum_{k=1}^{3} \left\{ d_k \cos\left(\frac{2\pi kn}{7}\right) - f_k \sin\left(\frac{2\pi kn}{7}\right)\right\}$$

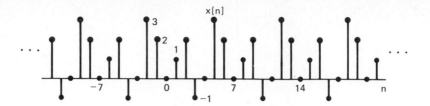

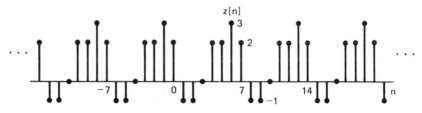

Figure P5.7

Sketch the signal.

$$y[n] = a_0 - d_0 + 2 \sum_{k=1}^{3} \left\{ d_k \cos\left(\frac{2\pi k n}{7}\right) + (f_k - c_k) \sin\left(\frac{2\pi k n}{7}\right) \right\}$$

5.8. (a) The triangular sequence $x_a[n]$ shown in Figure P5.8(a) is periodic with period $N = 6$ and thus has a discrete Fourier series representation given by

$$x_a[n] = \sum_{k=0}^{5} a_k e^{jk(2\pi/6)n}$$

Find and sketch the Fourier series coefficients a_k.

(b) The sequence $x_b[n]$ shown in Figure P5.8(b) is periodic with period 8 and is formed from $x_a[n]$ by inserting zeros between the triangular pulses. If the discrete Fourier series representation of $x_b[n]$ is given by

$$x_b[n] = \sum_{k=0}^{7} b_k e^{jk(2\pi/8)n}$$

determine and sketch the Fourier series coefficients b_k.

(c) The aperiodic triangular sequence $x_c[n]$ shown in Figure P5.8(c) corresponds to one period of the periodic sequence $x_a[n]$ (or $x_b[n]$). Determine $X_c(\Omega)$, the discrete Fourier transform of $x_c[n]$ and sketch your result.

(d) Show that the Fourier series coefficients a_k and b_k represent equally spaced samples of $X_c(\Omega)$:

$$a_k = c_1 X_c(k\Omega_1) \tag{P5.8-1}$$

$$b_k = c_2 X_c(k\Omega_2) \tag{P5.8-2}$$

Determine the values of the constants c_1, c_2, Ω_1, and Ω_2.

(e) Consider the aperiodic signal $x_d[n]$ in Figure P5.8(d). This signal is equal to $x_a[n]$ for $0 \leq n \leq 5$. Compute $X_d(\Omega)$ and show that

$$a_k = c_1 X_d(k\Omega_1)$$

where c_1 and Ω_1 have exactly the same values as in part (d).

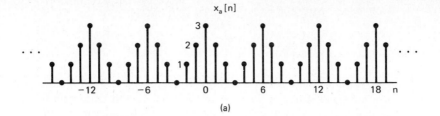

(a)

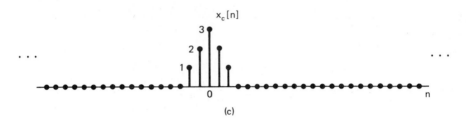

(b)

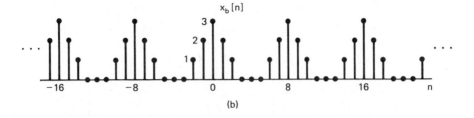

(c)

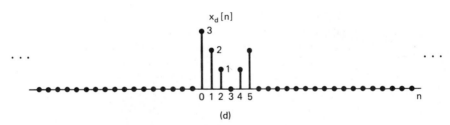

(d)

Figure P5.8

5.9. (a) Consider an LTI system with impulse response

$$h[n] = (\tfrac{1}{2})^{|n|}$$

Find the Fourier series representation of the output $y[n]$ for each of the following inputs.

(i) $x[n] = \sin(3\pi n/4)$

(ii) $x[n] = \displaystyle\sum_{k=-\infty}^{+\infty} \delta[n - 4k]$

(iii) $x[n]$ is periodic with period 6, and

$$x[n] = \begin{cases} 1, & n = 0, \pm 1 \\ 0, & n = \pm 2, \pm 3 \end{cases}$$

(iv) $x[n] = j^n + (-1)^n$

(b) Repeat part (a) for

$$h[n] = \begin{cases} 1, & 0 \le n \le 2 \\ -1, & -2 \le n \le -1 \\ 0, & \text{otherwise} \end{cases}$$

5.10. As mentioned in the text, it is often of interest to perform Fourier analysis of signals on a digital computer. Because of the discrete and finite nature of computers, any signal that is to be processed must necessarily be discrete and must also be of finite duration. Let $x[n]$ be such a signal, with

$$x[n] = 0 \qquad \text{outside the interval } 0 \le n \le N_1 - 1$$

The Fourier transform $X(\Omega)$ of $x[n]$ is a function of the continuous frequency variable Ω. However, because of physical limitations it is only possible to compute $X(\Omega)$ for a finite set of values of Ω.

Suppose that we evaluate $X(\Omega)$ at points that are equally spaced. Specifically, suppose that we choose an integer N and calculate $X(\Omega_k)$ for

$$\Omega_k = \frac{2\pi k}{N}, \qquad k = 0, 1, \ldots, N - 1$$

(a) Assume that $N \ge N_1$. Show that

$$\tilde{X}(k) = \frac{1}{N} X\left(\frac{2\pi k}{N}\right)$$

where $\tilde{X}(k)$ is the DFT of $x[n]$. From this, show that $x[n]$ can be completely determined from the samples of its Fourier transform

$$X\left(\frac{2\pi k}{N}\right), \qquad k = 0, 1, \ldots, N - 1$$

(b) In part (a) we have seen that $x[n]$ can be determined from N equally spaced samples of $X(\Omega)$ as long as $N \ge N_1$, the duration of $x[n]$. To illustrate that this condition is necessary, consider the two signals $x_1[n]$ and $x_2[n]$ depicted in Figure P5.10. Show that if we choose $N = 4$, we have that

$$X_1\left(\frac{2\pi k}{4}\right) = X_2\left(\frac{2\pi k}{4}\right)$$

for all values of k.

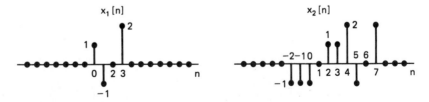

Figure P5.10

5.11. As indicated in the text and in Problem 5.10, there are many problems of practical importance in which one wishes to calculate the DFT of discrete-time signals. Often, these signals are of quite long duration, and in such cases it is very important to use computationally efficient procedures. In the text we mentioned that one of the reasons for the significant increase in the use of computerized techniques for the analysis of signals was the development of a very efficient technique, known as the fast Fourier transform (FFT) algorithm for the calculation of the DFT of finite-duration sequences. In this problem we develop the principle on which the FFT is based.

Let $x[n]$ be a signal that is 0 outside the interval $0 \leq n \leq N_1 - 1$. Recall from Section 5.4.3 that for $N \geq N_1$ the N-point DFT of $x[n]$ is given by

$$\tilde{X}(k) = \frac{1}{N} \sum_{n=0}^{N-1} x[n]e^{-jk(2\pi/N)n}, \qquad k = 0, 1, \ldots, N - 1 \qquad \text{(P5.11-1)}$$

It is convenient to rewrite eq. (P5.11-1) as

$$\tilde{X}(k) = \frac{1}{N} \sum_{k=0}^{N-1} x[n]W_N^{nk} \qquad \text{(P5.11-2)}$$

where

$$W_N = e^{-j2\pi/N}$$

(a) One method for calculating $\tilde{X}(k)$ is by direct evaluation of eq. (P5.11-2). A useful measure of the complexity of such a computation is the total number of complex multiplications required. Show that the number of complex multiplications required to evaluate eq. (P5.11-2) directly, for $k = 0, 1, \ldots, N - 1$, is N^2. Assume that $x[n]$ is complex and that the required values of W_N^{nk} have been precomputed and stored in a table. For simplicity, *do not* exploit the fact that, for certain values of n and k, W_N^{nk} is equal to ± 1 or $\pm j$ and hence does not, strictly speaking, require a full complex multiplication.

(b) Suppose that N is even. Let $f[n] = x[2n]$ represent the even-indexed samples of $x[n]$, and let $g[n] = x[2n + 1]$ represent the odd-indexed samples.

(i) Show that $f[n]$ and $g[n]$ are zero outside the interval $0 \leq n \leq (N/2) - 1$.

(ii) Show that the N-point DFT $\tilde{X}(k)$ of $x[n]$ can be expressed as

$$\tilde{X}(k) = \frac{1}{N} \sum_{n=0}^{(N/2)-1} f[n]W_{N/2}^{nk} + \frac{1}{N} W_N^k \sum_{n=0}^{(N/2)-1} g[n]W_{N/2}^{nk}$$

$$= \tfrac{1}{2}\tilde{F}(k) + \tfrac{1}{2}W_N^k \tilde{G}(k), \qquad k = 0, 1, \ldots, N - 1 \qquad \text{(P5.11-3)}$$

where

$$\tilde{F}(k) = \frac{2}{N} \sum_{n=0}^{(N/2)-1} f[n]W_{N/2}^{nk}$$

$$\tilde{G}(k) = \frac{2}{N} \sum_{n=0}^{(N/2)-1} g[n]W_{N/2}^{nk}$$

(iii) Show that

$$\tilde{F}\left(k + \frac{N}{2}\right) = \tilde{F}(k)$$

$$\tilde{G}\left(k + \frac{N}{2}\right) = \tilde{G}(k) \qquad \text{for all } k$$

Note that $\tilde{F}(k)$, $k = 0, 1, \ldots, (N/2) - 1$ and $\tilde{G}(k)$, $k = 0, 1, \ldots, (N/2) - 1$ are the $(N/2)$-point DFTs of $f[n]$ and $g[n]$, respectively. Thus, eq. (P5.11-3) indicates that the length N DFT of $x[n]$ can be calculated in terms of two DFTs of length $N/2$.

(iv) Determine the number of complex multiplications required to compute $\tilde{X}(k)$, $k = 0, 1, \ldots, N - 1$, from eq. (P5.11-3) by first computing $\tilde{F}(k)$ and $\tilde{G}(k)$. [Make the same assumptions about multiplications as in part (b), and do not count the multiplications by $\frac{1}{2}$ in eq. (P5.11-3).]

(c) If $N/2$ is also even then $f[n]$ and $g[n]$ can each be decomposed into sequences of even and odd indexed samples and therefore their DFTs can be computed using the same process as in eq. (P5.11-3). Furthermore, if N is an integer power of 2, we can continue to iterate this process, thus achieving a significant savings in computation time. Using this procedure, approximately how many complex multiplications are required for $N = 32$, 256, 1024, and 4096? Compare this to the direct method of calculation in part (a).

(d) Although the case in which N is a power of 2 is certainly the one used most commonly in practice, it is possible to obtain FFT algorithms for other cases. For example, suppose that N is divisible by 3, so that $N = 3M$. Show that the $\tilde{X}(k)$ can be calculated from the DFTs of length M of the following three signals.

$$f_1[n] = x[3n]$$
$$f_2[n] = x[3n + 1]$$
$$f_3[n] = x[3n + 2]$$

More generally, suppose that

$$N = p_1 p_2 \ldots p_R$$

where each p_i is a prime number (some of the p_i may be the same). Describe an efficient algorithm for calculating the DFT in this case.

(e) Consider the calculation of the inverse DFT of a signal $x[n]$, that is,

$$x[n] = \begin{cases} \sum_{k=\langle N \rangle} \tilde{X}(k) e^{jk(2\pi/N)n}, & 0 \leq n \leq N - 1 \\ 0, & \text{otherwise} \end{cases}$$

Suppose that N is a power of 2. Explain how the FFT algorithm can be applied to this inverse transform calculation.

5.12. In this problem we illustrate and expand upon the discussion in Section 5.6 on periodic convolution.

(a) Let $\tilde{x}_1[n]$ and $\tilde{x}_2[n]$ be periodic signals with common period N, and let $\tilde{y}[n]$ be their periodic convolution,

$$\tilde{y}[n] = \tilde{x}_1[n] \circledast \tilde{x}_2[n]$$

Verify that if a_k, b_k, and c_k are the Fourier coefficients of $\tilde{x}_1[n]$, $\tilde{x}_2[n]$, and $\tilde{y}[n]$, respectively, then

$$c_k = N a_k b_k$$

(b) Let $\tilde{x}_1[n]$ and $\tilde{x}_2[n]$ be two signals that are periodic with period 8 and are specified by

$$\tilde{x}_1[n] = \sin \left(\frac{3\pi n}{4} \right)$$

$$\tilde{x}_2[n] = \begin{cases} 1, & 0 \leq n \leq 3 \\ 0, & 4 \leq n \leq 7 \end{cases}$$

Find the Fourier series representation for the periodic convolution of these signals.

(c) Repeat part (b) for the following two periodic signals that also have period 8:

$$\tilde{x}_1[n] = \begin{cases} \sin \left(\frac{3\pi n}{4} \right), & 0 \leq n \leq 3 \\ 0, & 4 \leq n \leq 7 \end{cases}$$

$$\tilde{x}_2[n] = (\tfrac{1}{2})^n, \qquad 0 \leq n \leq 7$$

(d) Let $x_1[n]$ and $x_2[n]$ be finite-duration sequences with

$$x_1[n] = 0 \quad \text{for } n \text{ outside the interval } 0 \leq n \leq N_1 - 1$$
$$x_2[n] = 0 \quad \text{for } n \text{ outside the interval } 0 \leq n \leq N_2 - 1$$

$$\text{(P5.12-1)}$$

Let N be any integer at least as large as $N_1 + N_2 - 1$, and let $\tilde{x}_1[n]$ and $\tilde{x}_2[n]$ be periodic with period N, with

$$\tilde{x}_1[n] = x_1[n], \qquad 0 \le n \le N - 1$$
$$\tilde{x}_2[n] = x_2[n], \qquad 0 \le n \le N - 1$$

Show that

$$x_1[n] * x_2[n] = \begin{cases} \tilde{x}_1[n] \circledast \tilde{x}_2[n], & 0 \le n \le N - 1 \\ 0, & \text{otherwise} \end{cases}$$

(e) Let $x_1[n]$ and $x_2[n]$ be finite-duration signals as in part (d), satisfying eq. (P5.12-1), with $N_1 = N_2$. How many multiplications are required to calculate $x_1[n] * x_2[n]$ directly from the convolution sum? Let $N \ge 2N_1 - 1$ and assume that N is a power of 2. Suppose that we calculate $x_1[n] * x_2[n]$, as described in the text, by first using the FFT algorithm (see Problem 5.11) to compute the DFTs of $x_1[n]$ and $x_2[n]$, then multiplying these DFTs together, and then using the FFT to calculate the inverse transform. How many multiplies are required for this procedure? Compare the computational burden for the two methods for calculating $x_1[n] * x_2[n]$ when $N_1 = 100$ and $N = 256$. You may find the following identity useful:

$$\sum_{k=1}^{N_1} k = \frac{N_1(N_1 + 1)}{2}$$

5.13. Consider the following pairs of signals $x[n]$ and $y[n]$. For each pair determine if there is a discrete-time LTI system for which $y[n]$ is the output when the corresponding $x[n]$ is the input. If such a system exists, explain if the system is unique (i.e., whether there is more than one LTI system with the given input–output pair). Also determine the frequency response of an LTI system with the desired behavior. If no such LTI system exists for a given $x[n]$, $y[n]$ pair, explain why.

(a) $x[n] = (\frac{1}{2})^n$, $y[n] = (\frac{1}{4})^n$
(b) $x[n] = (\frac{1}{2})^n u[n]$, $y[n] = (\frac{1}{4})^n u[n]$
(c) $x[n] = (\frac{1}{2})^n u[n]$, $y[n] = 4^n u[-n]$
(d) $x[n] = e^{jn/8}$, $y[n] = 2e^{jn/8}$
(e) $x[n] = e^{jn/8} u[n]$, $y[n] = 2e^{jn/8} u[n]$
(f) $x[n] = j^n$, $y[n] = 2j^n(1 - j)$
(g) $x[n] = \cos(\pi n/3)$, $y[n] = \cos(\pi n/3) + \sqrt{3} \sin(\pi n/3)$
(h) $x[n]$ and $y_1[n]$ as in Figure P5.13
(i) $x[n]$ and $y_2[n]$ as in Figure P5.13

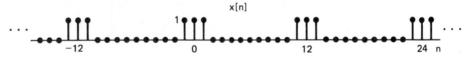

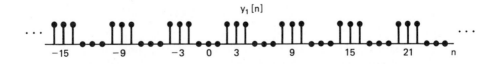

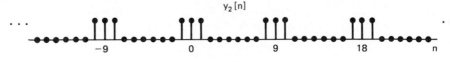

Figure P5.13

5.14. Compute the Fourier transform of each of the following signals.

(a) $x[n]$ as in Figure P5.14(a)

(b) $2^n u[-n]$

(c) $(\frac{1}{4})^n u[n+2]$

(d) $[a^n \sin \Omega_0 n] u[n], |a| < 1$

(e) $|a|^n \sin \Omega_0 n, |a| < 1$

(f) $(\frac{1}{2})^n \{u[n+3] - u[n-2]\}$

(g) $n\{u[n+N] - u[n-N-1]\}$

(h) $\cos (18\pi n/7) + \sin (2n)$

(i) $\sum\limits_{k=-\infty}^{+\infty} (\frac{1}{4})^n \, \delta[n - 3k]$

(j) $x[n]$ as in Figure P5.14(b)

(k) $\delta[4 - 2n]$

(l) $x[n] = \begin{cases} \cos (\pi n/3), & -4 \le n \le 4 \\ 0, & \text{otherwise} \end{cases}$

(m) $n(\frac{1}{2})^{|n|}$

(n) $\left[\dfrac{\sin (\pi n/2)}{\pi n}\right]\left[\dfrac{\sin (\pi n/4)}{\pi n}\right]$

(o) $x[n]$ as in Figure P5.14(c)

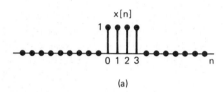

(a)

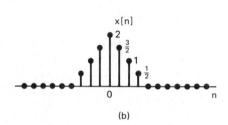

(b)

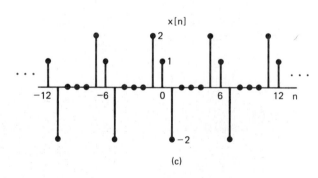

(c)

Figure P5.14

5.15. The following are Fourier transforms of discrete-time signals. Determine the signal corresponding to each transform.

(a) $X(\Omega) = \begin{cases} 0, & 0 \le |\Omega| \le W \\ 1, & W < |\Omega| \le \pi \end{cases}$

(b) $X(\Omega) = 1 - 2e^{-j3\Omega} + 4e^{j2\Omega} + 3e^{-j6\Omega}$

(c) $X(\Omega) = \sum\limits_{k=-\infty}^{+\infty} (-1)^k \, \delta\left(\Omega - \dfrac{\pi k}{2}\right)$

(d) $X(\Omega) = \cos^2 \Omega$

(e) $X(\Omega) = \cos(\Omega/2) + j\sin\Omega$ for $-\pi \le \Omega \le \pi$

(f) $X(\Omega)$ as in Figure P5.15(a)

(g) $X(\Omega)$ as in Figure P5.15(b)

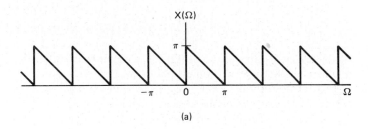

(a)

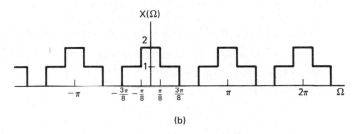

(b)

Figure P5.15

(h) $|X(\Omega)| = \begin{cases} 0, & 0 \le |\Omega| \le \pi/3 \\ 1, & \pi/3 < |\Omega| \le 2\pi/3 \\ 0, & 2\pi/3 < |\Omega| \le \pi \end{cases}$

$\sphericalangle X(\Omega) = 2\Omega$

(i) $X(\Omega) = \dfrac{e^{-j\Omega}}{1 + \frac{1}{6}e^{-j\Omega} - \frac{1}{6}e^{-j2\Omega}}$

5.16. In this chapter we showed that for $|\alpha| < 1$,

$$\alpha^n u[n] \quad \overset{\mathcal{F}}{\longleftrightarrow} \quad \frac{1}{1 - \alpha e^{-j\Omega}}$$

and

$$(n+1)\alpha^n u[n] \quad \overset{\mathcal{F}}{\longleftrightarrow} \quad \frac{1}{(1 - \alpha e^{-j\Omega})^2}$$

Use properties of the Fourier transform to show by induction that the inverse Fourier transform of

$$X(\Omega) = \frac{1}{(1 - \alpha e^{-j\Omega})^r}$$

is

$$x[n] = \frac{(n + r - 1)!}{n!\,(r-1)!}\alpha^n u[n]$$

5.17. (a) Let $X(\Omega)$ be the Fourier transform of $x[n]$. Derive expressions in terms of $X(\Omega)$ for the Fourier transforms of the following signals. (Do not assume that $x[n]$ is real.)

(i) $\mathcal{R}e\{x[n]\}$

(ii) $x^*[-n]$

(iii) $\mathcal{E}v\{x[n]\}$

(b) Let $X(\Omega)$ be the Fourier transform of a real signal $x[n]$. Show that $x[n]$ can be written as

$$x[n] = \int_0^\pi \{B(\Omega)\cos\Omega + C(\Omega)\sin\Omega\}\, d\Omega$$

by finding expressions for $B(\Omega)$ and $C(\Omega)$ in terms of $X(\Omega)$.

(c) Suppose that $x[n] = 0$ unless n is a multiple of a given integer M. Express the Fourier transform of $y[n] = x[nM]$ in terms of $X(\Omega) = \mathcal{F}\{x[n]\}$.

(d) Let $h_1[n]$ and $h_2[n]$ be the impulse responses of causal LTI systems, and let $H_1(\Omega)$ and $H_2(\Omega)$ be the corresponding frequency responses. Under these conditions, is the following equation true in general or not? Justify your answer.

$$\left[\frac{1}{2\pi}\int_{-\pi}^\pi H_1(\Omega)\, d\Omega\right]\left[\frac{1}{2\pi}\int_{-\pi}^\pi H_2(\Omega)\, d\Omega\right] = \frac{1}{2\pi}\int_{-\pi}^\pi H_1(\Omega)H_2(\Omega)\, d\Omega$$

(e) Is the following statement true or false in general? Again justify your answer.

If $X(\Omega) = \mathcal{F}\{x[n]\}$ is purely imaginary, the odd part of $x[n]$ is real.

(f) Derive the convolution property

$$x[n] * h[n] \quad \overset{\mathcal{F}}{\longleftrightarrow} \quad X(\Omega)H(\Omega)$$

(g) Let $x[n]$ and $h[n]$ be two signals, and let $y[n] = x[n] * h[n]$. Write two expressions for $y[0]$: one (using the convolution sum directly) in terms of $x[n]$ and $h[n]$, and one (using the convolution property of Fourier transforms) in terms of $X(\Omega)$ and $H(\Omega)$. Then, by judicious choice of $h[n]$, use these two expressions to show Parseval's relation, that is,

$$\sum_{n=-\infty}^{+\infty} |x[n]|^2 = \frac{1}{2\pi}\int_{-\pi}^\pi |X(\Omega)|^2\, d\Omega$$

In a similar fashion derive the following generalization of Parseval's relation,

$$\sum_{n=-\infty}^{+\infty} x[n]z^*[n] = \frac{1}{2\pi}\int_{-\pi}^\pi X(\Omega)Z^*(\Omega)\, d\Omega$$

(h) Let $x[n]$ be a discrete-time signal whose Fourier transform has a phase function given by

$$\sphericalangle X(\Omega) = \alpha\Omega, \qquad |\Omega| < \pi$$

Consider shifted versions of this signal, that is, signals of the form $x[n - n_0]$ for n_0 an integer. Are any of these signals even when $\alpha = 2$? What if $\alpha = \frac{1}{2}$?

(i) Let $x[n]$ be an aperiodic signal with Fourier transform $X(\Omega)$, and let $y[n]$ be given by

$$y[n] = \sum_{r=-\infty}^{+\infty} x[n + rN]$$

The signal $y[n]$ is periodic with period N. Show that its Fourier series coefficients are given by

$$a_k = \frac{1}{N} X\left(\frac{2\pi k}{N}\right)$$

5.18. Let $X(\Omega)$ denote the Fourier transform of the signal $x[n]$ depicted in Figure P5.18. Perform the following calculations without explicitly evaluating $X(\Omega)$.

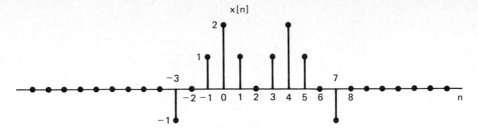

Figure P5.18

(a) Evaluate $X(0)$.

(b) Find $\sphericalangle X(\Omega)$.

(c) Evaluate

$$\int_{-\pi}^{\pi} X(\Omega)\,d\Omega$$

(d) Find $X(\pi)$.

(e) Determine and sketch the signal whose Fourier transform is $\mathcal{Re}\{X(\Omega)\}$.

(f) Evaluate:

(i) $\displaystyle\int_{-\pi}^{\pi} |X(\Omega)|^2\,d\Omega$

(ii) $\displaystyle\int_{-\pi}^{\pi} \left|\frac{dX(\Omega)}{d\Omega}\right|^2 d\Omega$

5.19. Determine which of the signals listed below, if any, have Fourier transforms that satisfy each of the following conditions.

    (i) $\mathcal{Re}\{X(\Omega)\} = 0$

    (ii) $\mathcal{Im}\{X(\Omega)\} = 0$

    (iii) There exists a real α such that

$$e^{j\alpha\Omega}X(\Omega) \qquad \text{is real}$$

    (iv) $\displaystyle\int_{-\pi}^{\pi} X(\Omega)\,d\Omega = 0$

    (v) $X(\Omega)$ periodic

    (vi) $X(0) = 0$

(a) $x[n]$ as in Figure P5.19(a)

(b) $x[n]$ as in Figure P5.19(b)

(c) $x[n] = (\frac{1}{2})^n u[n]$

(d) $x[n] = (\frac{1}{2})^{|n|}$

(e) $x[n] = \delta[n-1] + \delta[n+2]$

(f) $x[n] = \delta[n-1] + \delta[n+3]$

(g) $x[n]$ as in Figure P5.19(c)

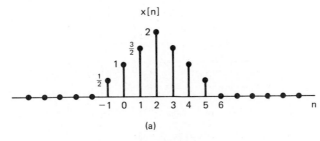

(a)

Figure P5.19

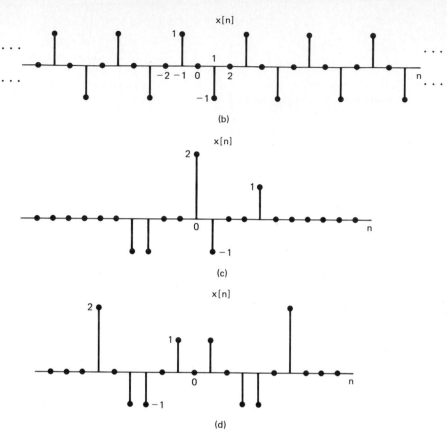

(b)

(c)

(d)

Figure P5.19 (cont.)

(h) $x[n]$ as in Figure P5.19(d)

(i) $x[n] = \delta[n - 1] - \delta[n + 1]$

5.20. Consider the signal depicted in Figure P5.20. Let the Fourier transform of this signal be written in rectangular form as

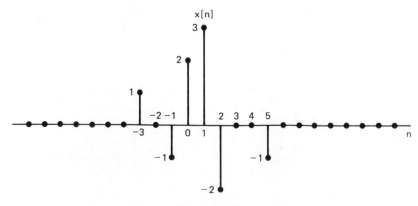

Figure P5.20

$$X(\Omega) = A(\Omega) + jB(\Omega)$$

Sketch the time function corresponding to the transform

$$Y(\Omega) = [B(\Omega) + A(\Omega)e^{j\Omega}]$$

5.21. Let $x_1[n]$ be the discrete-time signal whose Fourier transform, $X_1(\Omega)$, is depicted in Figure P5.21(a).

(a) Consider the signal $x_2[n]$ with Fourier transform $X_2(\Omega)$ as illustrated in Figure P5.21(b). Express $x_2[n]$ in terms of $x_1[n]$. [*Hint:* First express $X_2(\Omega)$ in terms of $X_1(\Omega)$, and then use properties of the Fourier transform.]

(b) Repeat part (a) for $x_3[n]$ with Fourier transform $X_3(\Omega)$ as shown in Figure P5.21(c).

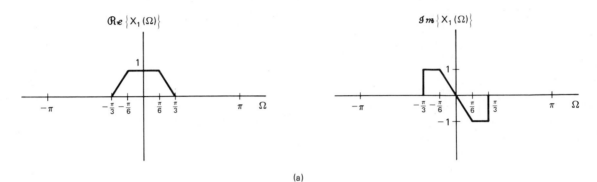

(a)

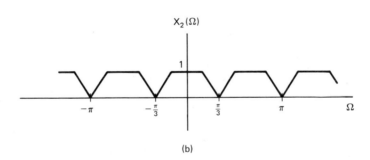

(b)

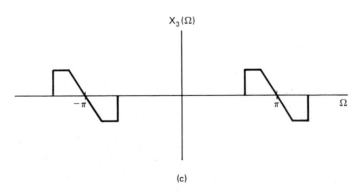

(c)

Figure P5.21

(c) Let

$$\alpha = \frac{\sum\limits_{n=-\infty}^{+\infty} n x_1[n]}{\sum\limits_{n=-\infty}^{+\infty} x_1[n]}$$

This quantity, which is the center of gravity of the signal $x_1[n]$, is usually referred to as the *delay time* for the signal $x_1[n]$. Find α (you can do this without first determining $x_1[n]$ explicitly).

(d) Consider the signal $x_4[n] = x_1[n] * h[n]$, where

$$h[n] = \frac{\sin(\pi n/6)}{\pi n}$$

Sketch $X_4(\Omega)$.

5.22. In this problem we derive the Fourier transform of the unit step $u[n]$.

(a) Use the fact that

$$\delta[n] = u[n] - u[n-1]$$

to deduce that

$$\mathcal{F}\{u[n]\} = \frac{1}{1 - e^{-j\Omega}} + g(\Omega)$$

where $g(\Omega)$ is zero unless $\Omega = 0, \pm 2\pi \pm 4\pi, \ldots$.

(b) Let $y[n] = \mathcal{O}d\{u[n]\}$ and show that $y[n]$ has the form

$$y[n] = u[n] + a + b\delta[n]$$

where a and b are constants that you should determine.

(c) Use the fact that $\mathcal{F}\{y[n]\}$ must be purely imaginary to deduce that

$$g(\Omega) = \pi \sum_{k=-\infty}^{+\infty} \delta(\Omega - 2\pi k)$$

(d) Use the result derived in parts (a)–(c) together with the convolution property to show that

$$\mathcal{F}\left\{ \sum_{m=-\infty}^{n} x[m] \right\} = \frac{X(\Omega)}{1 - e^{j\Omega}} + \pi X(0) \sum_{k=-\infty}^{+\infty} \delta(\Omega - 2\pi k)$$

5.23. (a) Let $x_1[n]$ be the signal

$$x_1[n] = \cos\left(\frac{\pi n}{3}\right) + \sin\left(\frac{\pi n}{2}\right)$$

and let $X_1(\Omega)$ denote its Fourier transform. Sketch this signal, together with the signals with the following Fourier transforms.

(i) $X_2(\Omega) = X_1(\Omega)e^{j\Omega}, \quad |\Omega| < \pi$

(ii) $X_3(\Omega) = X_1(\Omega)e^{-j3\Omega/2}, \quad |\Omega| < \pi$

(b) Let $z(t)$ be the continuous-time signal

$$z(t) = \cos\left(\frac{\pi t}{3T}\right) + \sin\left(\frac{\pi t}{2T}\right)$$

Note that $x_1[n]$ can be regarded as a sequence of evenly spaced samples of $z(t)$. Specifically,

$$x_1[n] = z(nT)$$

Show that $x_2[n]$ and $x_3[n]$ can also be regarded as evenly spaced samples of $z(t)$.

5.24. Let $x[m, n]$ be a signal depending upon the two independent, discrete variables m and n. In analogy with one dimension and with the continuous-time case treated in Problem

4.26, we can define the two-dimensional Fourier transform of $x[m, n]$ as

$$X(\Omega_1, \Omega_2) = \sum_{n=-\infty}^{+\infty} \sum_{m=-\infty}^{+\infty} x[m, n]e^{-j(\Omega_1 m + \Omega_2 n)} \tag{P5.24-1}$$

(a) Show that eq. (P5.24-1) can be calculated as two successive one-dimensional Fourier transforms, first in m, with n regarded as fixed, and then in n. Use this result to determine an expression for $x[m, n]$ in terms of $X(\Omega_1, \Omega_2)$.

(b) Suppose that

$$x[m, n] = a[m]b[n]$$

where $a[m]$ and $b[n]$ are each functions of only one independent variable. Let $A(\Omega)$ and $B(\Omega)$ denote the Fourier transforms of $a[m]$ and $b[n]$, respectively. Express $X(\Omega_1, \Omega_2)$ in terms of $A(\Omega)$ and $B(\Omega)$.

(c) Determine the two-dimensional Fourier transforms of the following signals.

   (i) $x[m, n] = \delta[m - 1]\,\delta[n + 4]$

   (ii) $x[m, n] = (\frac{1}{2})^{n-m}u[n - 2]u[-m]$

   (iii) $x[m, n] = (\frac{1}{2})^n \cos(2\pi m/3)u[n]$

   (iv) $x[m, n] = \begin{cases} 1, & \text{if } -2 < m < 2 \text{ and } -4 < n < 4 \\ 0, & \text{otherwise} \end{cases}$

   (v) $x[m, n] = \begin{cases} 1, & \text{if } -2 + n < m < 2 + n \text{ and } -4 < n < 4 \\ 0, & \text{otherwise} \end{cases}$

   (vi) $x[m, n] = \sin\left(\dfrac{\pi n}{3} + \dfrac{2\pi m}{5}\right)$

(d) Determine the signal $x[m, n]$ whose Fourier transform is given by

$$X(\Omega_1, \Omega_2) = \begin{cases} 1, & 0 < |\Omega_1| < \pi/4 \text{ and } 0 < |\Omega_2| < \pi/2 \\ 0, & \pi/4 < |\Omega_1| < \pi \text{ or } \pi/2 < |\Omega_2| < \pi \end{cases}$$

(e) Let $x[m, n]$ and $h[m, n]$ be two signals whose two-dimensional Fourier transforms are denoted by $X(\Omega_1, \Omega_2)$ and $H(\Omega_1, \Omega_2)$, respectively. Determine the transforms of the following signals in terms of $X(\Omega_1, \Omega_2)$ and $H(\Omega_1, \Omega_2)$.

   (i) $x[m, n]e^{jW_1 m}e^{jW_2 n}$

   (ii) $y[m, n] = \begin{cases} x[k, r], & \text{if } m = 2k \text{ and } n = 3r \\ 0, & \text{if } m \text{ is not a multiple of 2 or} \\ & \quad n \text{ is not a multiple of 3} \end{cases}$

   (iii) $y[m, n] = x[m, n]h[m, n]$

5.25. (a) Consider a discrete-time LTI system with impulse response

$$h[n] = (\tfrac{1}{2})^n u[n]$$

Use Fourier transforms to determine the response to each of the following input signals.

   (i) $x[n] = (\tfrac{3}{4})^n u[n]$

   (ii) $x[n] = (n + 1)(\tfrac{1}{4})^n u[n]$

   (iii) $x[n] = (-1)^n$

(b) Suppose that

$$h[n] = \left[\left(\frac{1}{2}\right)^n \cos\left(\frac{\pi n}{2}\right)\right]u[n]$$

Use Fourier transform techniques to determine the response to each of the following inputs.

   (i) $x[n] = (\tfrac{1}{2})^n u[n]$

   (ii) $x[n] = \cos(\pi n/2)$

(c) Let $x[n]$ and $h[n]$ be the signals with the following Fourier transforms:

$$X(\Omega) = 3e^{j\Omega} + 1 - e^{-j\Omega} + 2e^{-j3\Omega}$$

$$H(\Omega) = -e^{j\Omega} + 2e^{-j2\Omega} + e^{j4\Omega}$$

Determine $y[n] = x[n] * h[n]$.

5.26. Let $x[n]$ and $y[n]$ be two real signals with Fourier transforms $X(\Omega)$ and $Y(\Omega)$. The correlation function $\phi_{xy}[n]$ is given by

$$\phi_{xy}[n] = \sum_{k=-\infty}^{+\infty} x[k+n]y[k]$$

In a similar fashion we can define $\phi_{yx}[n]$, $\phi_{xx}[n]$, and $\phi_{yy}[n]$. Let $\Phi_{xy}(\Omega)$, $\Phi_{yx}(\Omega)$, $\Phi_{xx}(\Omega)$, and $\Phi_{yy}(\Omega)$ denote the Fourier transforms of $\phi_{xy}[n]$, $\phi_{yx}[n]$, $\phi_{xx}[n]$, and $\phi_{yy}[n]$, respectively.

(a) Find an expression for $\Phi_{xy}(\Omega)$ in terms of $X(\Omega)$ and $Y(\Omega)$.

(b) Show that $\Phi_{xx}(\Omega)$ is real and nonnegative for every Ω.

(c) Suppose that $x[n]$ is the input to an LTI system with real impulse response $h[n]$ and with corresponding frequency response $H(\Omega)$, and suppose that $y[n]$ is the output. Find expressions for $\Phi_{xy}(\Omega)$ and $\Phi_{yy}(\Omega)$ in terms of $\Phi_{xx}(\Omega)$ and $H(\Omega)$.

(d) Let $x[n] = (\frac{1}{2})^n u[n]$, and let the LTI system impulse response be $h[n] = (\frac{1}{4})^n u[n]$. Compute $\Phi_{xx}(\Omega)$, $\Phi_{xy}(\Omega)$, and $\Phi_{yy}(\Omega)$ using the results of the previous parts of this problem.

(e) Repeat part (d) for $x[n] = 2^n u[-n]$ and $h[n] = 4^n u[-n]$.

(f) Repeat part (d) for $x[n]$ and $h[n]$ shown in Figure P5.26.

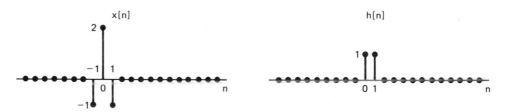

Figure P5.26

5.27. In our discussion of frequency responses for continuous-time and discrete-time LTI systems we pointed out that not all LTI systems have finite-valued responses to complex exponential inputs. In this problem we provide a bit more insight into this issue.

(a) Consider the LTI system with unit sample response

$$h[n] = (\tfrac{1}{4})^n u[n]$$

Compute the response $y_1[n]$ of this system to the input

$$x_1[n] = (-1)^n$$

(b) Use the convolution sum to determine the output $y_2[n]$ when the input is

$$x_2[n] = (-1)^n u[n]$$

Show that

$$y_2[n] = y_1[n] + y_T[n]$$

where

$$\lim_{n \to \infty} y_T[n] = 0$$

Often $y_T[n]$ is referred to as the *transient response* of the LTI system, and $y_1[n]$ as the *steady-state* response. That is, as $n \longrightarrow \infty$, the system response to $x_2[n]$ approaches the response that would have resulted if the sinusoidal signal $(-1)^n$ had been applied for all time. The transient response thus is due to the fact that the signal $x_2[n]$ is "started" at $n = 0$, and the effect of this initial start-up dies out asymptotically as $n \longrightarrow \infty$. This result extends quite generally to the response of any *stable* LTI system. Specifically, if $x_3[n]$ is a periodic input, then the response of a stable LTI system to $x_4[n] = x_3[n]u[n]$ converges to a steady state which is the response to $x_3[n]$. On the other hand, as we mentioned in the text, an unstable LTI system may have an infinite response to a sinusoidal input. To gain some understanding of this, let us consider applying the input of part (b) to an unstable LTI system.

(c) Consider the LTI system with impulse response

$$h[n] = 2^n u[n]$$

Compute the response of this system to $x_2[n]$. Does this response approach a steady state as $n \longrightarrow \infty$?

5.28. In Chapter 4 we indicated that the continuous-time LTI system with impulse response

$$h(t) = \frac{W}{\pi} \operatorname{sinc}\left(\frac{Wt}{\pi}\right) = \frac{\sin Wt}{\pi t}$$

plays a very important role in LTI system analysis. The same is true of the discrete-time LTI system with frequency response

$$h[n] = \frac{W}{\pi} \operatorname{sinc}\left(\frac{Wn}{\pi}\right) = \frac{\sin Wn}{\pi n}$$

In this problem we illustrate some of the properties of this system and others that are closely related to it.

(a) Consider the signal

$$x[n] = \sin\left(\frac{\pi n}{8}\right) - 2 \cos\left(\frac{\pi n}{4}\right)$$

Suppose that this signal is the input to each of the LTI systems with impulse responses given below. Determine the output in each case.

(i) $h[n] = \dfrac{\sin (\pi n/6)}{\pi n}$

(ii) $h[n] = \dfrac{\sin (\pi n/6)}{\pi n} + \dfrac{\sin (\pi n/2)}{\pi n}$

(iii) $\dfrac{\sin (\pi n/6) \sin (\pi n/3)}{\pi^2 n^2}$

(iv) $\dfrac{\sin (\pi n/6) \sin (\pi n/3)}{\pi n}$

(b) Consider an LTI system with unit sample response

$$h[n] = \frac{\sin (\pi n/3)}{\pi n}$$

Determine the output for each of the inputs listed below.

(i) $x[n] =$ the square wave depicted in Figure P5.28

(ii) $x[n] = \displaystyle\sum_{k=-\infty}^{+\infty} \delta[n - 8k]$

(iii) $x[n] = (-1)^n$ times the square wave in Figure P5.28

(iv) $x[n] = \delta[n + 1] + \delta[n - 1]$

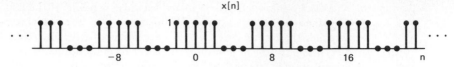

$$x[n]$$

Figure P5.28

5.29. (a) Let $h[n]$ and $g[n]$ be the impulse responses of two stable discrete-time LTI systems that are inverses of each other. What is the relationship between the frequency responses of these two systems?

(b) Consider causal LTI systems described by the following difference equations. In each case determine the impulse response of the inverse system and the difference equation that characterizes the inverse.

(i) $y[n] = x[n] - \frac{1}{4}x[n-1]$

(ii) $y[n] + \frac{1}{2}y[n-1] = x[n]$

(iii) $y[n] + \frac{1}{2}y[n-1] = x[n] - \frac{1}{4}x[n-1]$

(iv) $y[n] + \frac{5}{4}y[n-1] - \frac{1}{8}y[n-2] = x[n] - \frac{1}{4}x[n-1] - \frac{1}{8}x[n-2]$

(v) $y[n] + \frac{5}{4}y[n-1] - \frac{1}{8}y[n-2] = x[n] - \frac{1}{2}x[n-1]$

(vi) $y[n] + \frac{5}{4}y[n-1] - \frac{1}{8}y[n-2] = x[n]$

(c) Consider stable LTI systems which are inverses of each other. Suppose that these systems are causal and that one of them has a cascade structure implementation as in Figure P5.29-1 (the ρ_i here are given coefficients). Find a cascade structure for the inverse of this system (including a specification of the coefficient values in the structure).

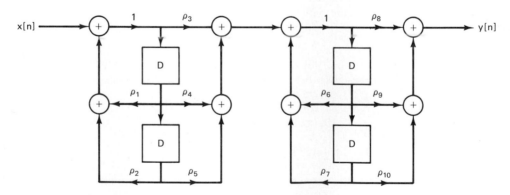

Figure P5.29-1

(d) Consider the causal, discrete-time LTI system described by the difference equation

$$y[n] + y[n-1] + \frac{1}{4}y[n-2] = x[n-1] - \frac{1}{2}x[n-2] \qquad \text{(P5.29-1)}$$

What is the inverse of this system? Show that the inverse is not causal. Find another causal LTI system that is an "inverse with delay" of the system described by eq. (P5.29-1). Specifically, find a causal LTI system so that the output $w[n]$ in Figure P5.29-2 equals $x[n-1]$.

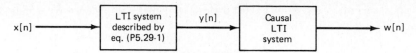

Figure P5.29-2

5.30. (a) Let $h[n]$ be the impulse response of a real, causal, discrete-time LTI system. Show that this system is completely specified by the real part of its frequency response. (*Hint:* Show how $h[n]$ can be recovered from $\mathcal{E}v\{h[n]\}$. What is the Fourier transform of $\mathcal{E}v\{h[n]\}$?) This is the discrete-time counterpart of the *real-part sufficiency* property of causal LTI systems considered in Problem 4.32 for continuous-time systems.

(b) Let $h[n]$ be real and causal. If

$$\mathcal{R}e\{H(\Omega)\} = 1 + \alpha \cos 2\Omega \qquad (\alpha \text{ real})$$

determine $h[n]$ and $H(\Omega)$.

(c) Show that $h[n]$ can be completely recovered from knowledge of $\mathcal{I}m\{H(\Omega)\}$ and $h[0]$.

(d) Find two real, causal LTI systems whose frequency responses have imaginary parts equal to $\sin \Omega$.

5.31. (a) Let $x[n]$ be a discrete-time sequence with Fourier transform $X(\Omega)$, which is illustrated in Figure P5.31. Sketch the Fourier transform of

$$z[n] = x[n]p[n]$$

for each of the following signals $p[n]$.

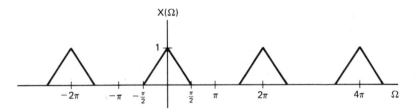

Figure P5.31

(i) $p[n] = \cos \pi n$

(ii) $p[n] = \cos (\pi n/2)$

(iii) $p[n] = \sin (\pi n/2)$

(iv) $p[n] = \displaystyle\sum_{k=-\infty}^{+\infty} \delta[n - 2k]$

(v) $p[n] = \displaystyle\sum_{k=-\infty}^{+\infty} \delta[n - 4k]$

(b) Suppose that the signal $z[n]$ of part (a) is applied as the input to an LTI system with unit impulse response

$$h[n] = \frac{\sin (\pi n/2)}{\pi n}$$

Determine the output $y[n]$ for each of the choices of $p[n]$ in part (a).

5.32. In this problem we introduce the concept of *windowing*, which is of great importance both in the design of LTI systems and in the spectral analysis of signals. Windowing is the operation of taking a signal $x[n]$ and multiplying it by a finite-duration *window signal* $w[n]$. That is,

$$z[n] = x[n]w[n]$$

Note that $z[n]$ is also of finite duration.

The importance of windowing in spectral analysis stems from the fact that in numerous applications one wishes to compute the Fourier transform of a measured signal. Since in actual practice we can only measure a signal $x[n]$ over a finite time interval (the *time window*), the actual signal available for spectral analysis is

$$z[n] = \begin{cases} x[n], & -M \le n \le M \\ 0, & \text{otherwise} \end{cases}$$

where $-M \le n \le M$ is the time window. Thus,

$$z[n] = x[n]w[n]$$

where $w[n]$ is the *rectangular window*

$$w[n] = \begin{cases} 1, & -M \le n \le M \\ 0, & \text{otherwise} \end{cases} \tag{P5.32-1}$$

Windowing also plays a role in LTI system design. Specifically, for a variety of reasons (such as the potential utility of the FFT algorithm—see Problem 5.11) it is often advantageous to design a system that has an impulse response of finite duration to achieve some desired signal processing objective. That is, we often begin with a desired frequency response $H(\Omega)$, whose inverse transform $h[n]$ is an impulse response of infinite (or at least excessively long) duration. What is required, then, is the construction of an impulse response $g[n]$ of finite duration whose transform $G(\Omega)$ adequately approximates $H(\Omega)$. One general approach to choosing $g[n]$ is to find a window function $w[n]$ such that the transform of $h[n]w[n]$ meets the desired specifications on $G(\Omega)$.

Clearly, the windowing of a signal has an effect on the resulting spectrum. In this problem we illustrate this effect.

(a) To gain some understanding of the effect of windowing, consider windowing the signal

$$x[n] = \sum_{k=-\infty}^{+\infty} \delta[n - k]$$

using the rectangular window signal given in eq. (P5.32-1).
  (i)   What is $X(\Omega)$?
  (ii) Sketch the transform of $z[n] = x[n]w[n]$ when $M = 1$.
  (iii) Do the same for $M = 10$.

(b) Consider next a signal $x[n]$ whose Fourier transform is specified by

$$X(\Omega) = \begin{cases} 1, & |\Omega| < \pi/4 \\ 0, & \pi/4 < |\Omega| \le \pi \end{cases}$$

Let $z[n] = x[n]w[n]$, where $w[n]$ is the rectangular window of eq. (P5.32-1). Roughly sketch $Z(\Omega)$ for $M = 4, 8, 16$.

One of the problems with the use of a rectangular window is that it introduces ripples in the transform $Z(\Omega)$ (this is in fact directly related to the Gibbs phenomenon). For this reason, a variety of other window signals have been developed. These signals are tapered, that is, they go from 0 to 1 more gradually than the abrupt transition of the rectangular window. What this does is to reduce the ripples in $Z(\Omega)$ at the expense of adding a bit of distortion in terms of further smoothing of $X(\Omega)$.

(c) To illustrate the points made above, consider the signal $x[n]$ described in part (b), and let $z[n] = x[n]w[n]$, where $w[n]$ is the *triangular* or *Bartlett window*:

$$w[n] = \begin{cases} 1 - \dfrac{|n|}{M+1}, & -M \le n \le M \\ 0, & \text{otherwise} \end{cases}$$

Roughly sketch the Fourier transform of $z[n] = x[n]w[n]$ for $M = 4, 8, 16$. [*Hint:* Note that the triangular signal can be obtained as the convolution of a rectangular signal with itself. This fact leads to a convenient expression for $W(\Omega)$.]

(d) Let $z[n] = x[n]w[n]$ where $w[n]$ is a raised cosine signal known as the *Hanning window*:

$$w[n] = \begin{cases} \frac{1}{2}[1 + \cos(\pi n/M)], & -M \le n \le M \\ 0, & \text{otherwise} \end{cases}$$

Roughly sketch $Z(\Omega)$ for $M = 4, 8, 16$.

5.33. (a) Consider the interconnected system depicted in Figure P5.33-1(a), where

$$h_1[n] = \delta[n] - \frac{\sin(\pi n/2)}{\pi n}$$

and where $H_2(\Omega)$ and $H_3(\Omega)$ are as shown in Figure P5.33-2. Find the output if the input has the Fourier transform depicted in Figure P5.33-3.

(b) Repeat part (a) for the interconnected system of Figure P5.33-1(b), using the same

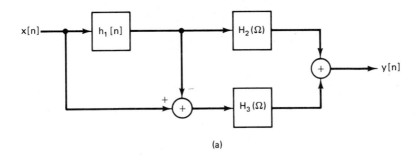

(a)

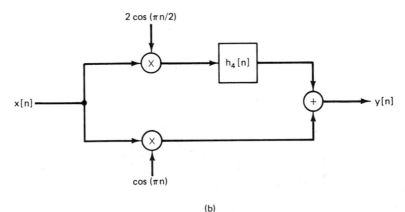

(b)

Figure P5.33-1

input as in part (a). Here

$$h_4[n] = \frac{\sin(\pi n/2)}{\pi n}$$

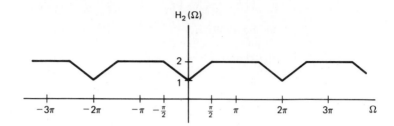

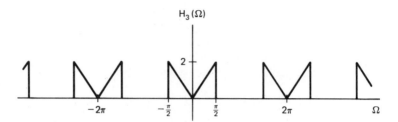

Figure P5.33-2

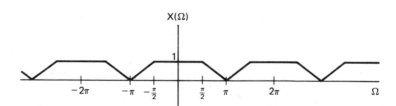

Figure P5.33-3

5.34. Consider a discrete-time signal $x[n]$ with Fourier transform as illustrated in Figure P5.34. Provide dimensioned sketches of the following continuous-time signals.

(a) $x_1(t) = \displaystyle\sum_{n=-\infty}^{+\infty} x[n]e^{j(2\pi/10)nt}$

(b) $x_2(t) = \displaystyle\sum_{n=-\infty}^{+\infty} x[-n]e^{j(2\pi/10)nt}$

(c) $x_3(t) = \displaystyle\sum_{n=-\infty}^{+\infty} \mathcal{O}d\{x[n]\}e^{j(2\pi/8)nt}$

(d) $x_4(t) = \displaystyle\sum_{n=-\infty}^{+\infty} \mathcal{R}e\{x[n]\}e^{j(2\pi/6)nt}$

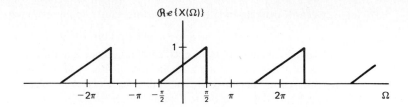

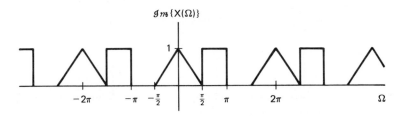

Figure P5.34

5.35. (a) A particular discrete-time system has input $x[n]$ and output $y[n]$. The Fourier transforms of these signals are related by the following equation

$$Y(\Omega) = 2X(\Omega) + e^{-j\Omega}X(\Omega) - \frac{dX(\Omega)}{d\Omega}$$

(i) Is the system linear? Clearly justify your answer.
(ii) Is the system time invariant? Clearly justify your answer.
(iii) What is $y[n]$ if $x[n] = \delta[n]$?

(b) Consider a discrete-time system for which the transform $Y(\Omega)$ of the output is related to the transform of the input through the relation

$$Y(\Omega) = \int_{\Omega - \pi/4}^{\Omega + \pi/4} X(\Omega) \, d\Omega$$

Find an expression for $y[n]$ in terms of $x[n]$.

5.36. (a) We want to design a discrete-time LTI system that has the property that if the input is

$$x[n] = (\tfrac{1}{2})^n u[n] - \tfrac{1}{4}(\tfrac{1}{2})^{n-1} u[n-1]$$

then the output is

$$y[n] = (\tfrac{1}{3})^n u[n]$$

(i) Find the impulse *and* frequency responses of a discrete time, LTI system that has this property.
(ii) Find a difference equation relating $x[n]$ and $y[n]$ that characterizes this system.
(iii) Determine a realization of the system in terms of adders, coefficient multipliers, and unit delays. Use as few unit delays as possible.

(b) Suppose that a system has the response $(\tfrac{1}{4})^n u[n]$ to the input $(n + 2)(\tfrac{1}{2})^n u[n]$. If the output of this system is $\delta[n] - (-\tfrac{1}{2})^n u[n]$, what is the input?

5.37. (a) Consider a discrete-time LTI system with unit sample response

$$h[n] = (\tfrac{1}{2})^n u[n] + \tfrac{1}{2}(\tfrac{1}{4})^n u[n]$$

Determine a linear constant-coefficient difference equation relating the input and output of the system. Find a realization of this system as a cascade of first-order systems, using adders, coefficient multipliers, and delay elements.

(b) Figure P5.37 depicts a block diagram implementation of a causal LTI system.
  (i)   Find a difference equation relating $x[n]$ and $y[n]$ for this system.
  (ii) What is the frequency response of the system?
  (iii) Determine the system's unit sample response.

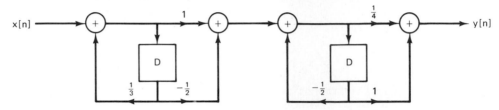

Figure P5.37

5.38. We are given a discrete-time, linear, time-invariant, causal system with input denoted by $x[n]$ and output denoted by $y[n]$. This system is specified by a *pair* of difference equations, involving an intermediate signal $w[n]$.

$$y[n] + \tfrac{1}{4}y[n-1] + \ w[n] + \tfrac{1}{2}w[n-1] = \tfrac{2}{3}x[n]$$
$$y[n] - \tfrac{5}{4}y[n-1] + 2w[n] - 2w[n-1] = -\tfrac{5}{3}x[n]$$

(a) Find the frequency response and the unit sample response of this system.
(b) Find a single difference equation relating $x[n]$ and $y[n]$ for this system.
(c) Determine a realization of the system in terms of adders, coefficient multipliers, and unit delays. Use as few unit delays as possible.

5.39. Consider a causal LTI system described by the difference equation

$$y[n] + \tfrac{1}{2}y[n-1] = x[n]$$

(a) Determine the frequency response $H(\Omega)$ of this system.
(b) What is the response of this system to the following inputs?
  (i) $x[n] = (\tfrac{1}{2})^n u[n]$
  (ii) $x[n] = (-\tfrac{1}{2})^n u[n]$
  (iii) $x[n] = \delta[n] + \tfrac{1}{2}\delta[n-1]$
  (iv) $x[n] = \delta[n] - \tfrac{1}{2}\delta[n-1]$
(c) Find the response to the inputs with the following Fourier transforms.
  (i) $X(\Omega) = \dfrac{1 - \tfrac{1}{4}e^{-j\Omega}}{1 + \tfrac{1}{2}e^{-j\Omega}}$
  (ii) $X(\Omega) = \dfrac{1 + \tfrac{1}{2}e^{-j\Omega}}{1 - \tfrac{1}{4}e^{-j\Omega}}$
  (iii) $X(\Omega) = \dfrac{1}{(1 - \tfrac{1}{4}e^{-j\Omega})(1 + \tfrac{1}{2}e^{-j\Omega})}$
  (iv) $X(\Omega) = 1 + 2e^{-3j\Omega}$

5.40. Consider an LTI system initially at rest and described by the difference equation

$$y[n] - \tfrac{1}{9}y[n-2] = x[n]$$

(a) Find the impulse response of this system.
(b) What is the response of the system for each of the following inputs?
   (i) $x[n] = (\frac{1}{3})^n u[n]$
   (ii) $x[n] = [(\frac{1}{3})^n + (-\frac{1}{3})^n]u[n]$
   (iii) $x[n] = (n+1)(\frac{1}{3})^n u[n]$
(c) Determine realizations of the following types for this system.
   (i) Direct form II
   (ii) Cascade of first-order systems
   (iii) Parallel interconnection of first-order systems

5.41. Sketch the log magnitude and phase of each of the following frequency responses.

(a) $1 + \frac{1}{2}e^{-j\Omega}$

(b) $1 + 2e^{-j\Omega}$

(c) $1 - 2e^{-j\Omega}$

(d) $1 + 2e^{-j2\Omega}$

(e) $\dfrac{1}{(1 + \frac{1}{2}e^{-j\Omega})^3}$

(f) $\dfrac{1 + \frac{1}{2}e^{-j\Omega}}{1 - \frac{1}{2}e^{-j\Omega}}$

(g) $\dfrac{1 + 2e^{-j\Omega}}{1 + \frac{1}{2}e^{-j\Omega}}$

(h) $\dfrac{1 - 2e^{-j\Omega}}{1 + \frac{1}{2}e^{-j\Omega}}$

(i) $\dfrac{1}{(1 - \frac{1}{4}e^{-j\Omega})(1 - \frac{3}{4}e^{-j\Omega})}$

(j) $\dfrac{1}{(1 - \frac{1}{4}e^{-j\Omega})(1 + \frac{3}{4}e^{-j\Omega})}$

(k) $\dfrac{1 + 2e^{-2j\Omega}}{(1 - \frac{1}{2}e^{-j\Omega})^2}$

5.42. A causal LTI system is described by the difference equation

$$y[n] - ay[n-1] = bx[n] + x[n-1]$$

where a is real and less than 1 in magnitude.

(a) Find a value of b so that the frequency response of the system satisfies

$$|H(\Omega)| = 1 \qquad \text{for all } \Omega$$

Such a system is called an *all-pass system*, as the system does not attenuate the input $e^{j\Omega n}$ for *any* value of Ω. Use this value of b in the rest of the problems.
(b) Roughly sketch $\sphericalangle H(\Omega)$, $0 \le \Omega \le \pi$, when $a = \frac{1}{2}$.
(c) Roughly sketch $\sphericalangle H(\Omega)$, $0 \le \Omega \le \pi$, when $a = -\frac{1}{2}$.
(d) Find and plot the output of this system with $a = -\frac{1}{2}$ when the input is

$$x[n] = (\frac{1}{2})^n u[n]$$

From this example we see that a nonlinear change in phase can have a significantly different effect on a signal than the time shift that results from linear phase.

5.43. A particular causal LTI system is described by the difference equation

$$y[n] - \frac{\sqrt{2}}{2}y[n-1] + \frac{1}{4}y[n-2] = x[n] - x[n-1]$$

(a) Find the impulse response of this system.
(b) Sketch the log magnitude and phase of the system frequency response.

5.44. A causal LTI system is characterized by the difference equation

$$y[n] - \frac{1}{4}y[n-1] - \frac{1}{8}y[n-2] = x[n] - 2x[n-2] + x[n-4]$$

(a) Construct the direct form II realization for this system.
(b) Construct a cascade realization (using unit delays, adders, and coefficient multipliers) consisting of the series interconnection of two second-order systems. Note that one of these second-order systems is an FIR system.
(c) Construct a cascade realization consisting of four first-order systems, two of which are FIR systems.

(d) Using the technique described in the Appendix on partial fraction expansion write the frequency response $H(\Omega)$ of the system in the form

$$H(\Omega) = c_0 + c_1 e^{-j\Omega} + c_2 e^{-j2\Omega} + \frac{\alpha_0 + \alpha_1 e^{-j\Omega}}{1 - \frac{1}{4}e^{-j\Omega} - \frac{1}{8}e^{-j2\Omega}}$$

where c_0, c_1, c_2, α_0, and α_1 are coefficients that you should determine. Use this expression in order to help you:

(i) Determine the unit sample response of the system.

(ii) Construct a realization of the system consisting of the parallel interconnection of a recursive second-order system and an FIR system.

5.45. (a) Consider two LTI systems with the following frequency responses:

$$H_1(\Omega) = \frac{1 + \frac{1}{2}e^{-j\Omega}}{1 + \frac{1}{4}e^{-j\Omega}}$$

$$H_2(\Omega) = \frac{\frac{1}{2} + e^{-j\Omega}}{1 + \frac{1}{4}e^{-j\Omega}}$$

Show that both of these frequency responses have the same magnitude function [i.e., $|H_1(\Omega)| = |H_2(\Omega)|$] but that the phase of $H_2(\Omega)$ is larger in magnitude then the phase of $H_1(\Omega)$.

(b) Determine and sketch the impulse and step responses of these two systems.

(c) Show that

$$H_2(\Omega) = G(\Omega)H_1(\Omega)$$

where $G(\Omega)$ is an *all-pass system* [i.e., $|G(\Omega)| = 1$ for all Ω] (see Problem 5.42).

The system with frequency response $H_1(\Omega)$ is usually referred to as a minimum phase system, and its unit sample response is called a minimum phase signal, while the system with frequency response $H_2(\Omega)$ is called a nonminimum phase system. The reason for this terminology can be seen in the result of part (a). These systems and their higher-order generalizations are the discrete-time counterparts of those introduced in Problem 4.45.

5.46. Consider a system consisting of the cascade of two LTI systems with frequency responses

$$H_1(\Omega) = \frac{2 - e^{-j\Omega}}{1 + \frac{1}{2}e^{-j\Omega}}$$

$$H_2(\Omega) = \frac{1}{1 - \frac{1}{2}e^{-j\Omega} + \frac{1}{4}e^{-j2\Omega}}$$

(a) Find the difference equation describing the overall system.

(b) Find a realization of the overall system consisting of the cascade of a first-order and a second-order system, where the first-order system has as its frequency response

$$\frac{1}{1 + \frac{1}{2}e^{-j\Omega}}$$

(c) Determine the impulse response of the overall system.

(d) Find a realization of the overall system, using unit delays, adders, and coefficient multipliers, consisting of the parallel interconnection of a first-order and a second-order system.

5.47. In our description of the cascade structure in Section 5.11, we implicitly assumed that $b_0 \neq 0$ [see eq. (5.150)]. This was done to simplify the discussion in the text, and as we will see in this problem, cascade structures can be readily constructed for systems for which $b_0 = 0$.

(a) Construct the direct form II realization for the causal LTI system with frequency response

$$H(\Omega) = \frac{e^{-j\Omega}}{1 - \frac{1}{6}e^{-j\Omega} - \frac{1}{6}e^{-j2\Omega}}$$

(b) Consider the causal LTI system with frequency response

$$H(\Omega) = \frac{e^{-j\Omega} + 2e^{-j3\Omega}}{(1 - \frac{1}{6}e^{-j\Omega} - \frac{1}{6}e^{-j2\Omega})(1 - e^{-j\Omega} + \frac{1}{4}e^{-j2\Omega})}$$

Construct a realization of this system consisting of the cascade of two second-order systems.

(c) Let

$$H(\Omega) = \frac{e^{-j4\Omega}}{(1 - \frac{1}{6}e^{-j\Omega} - \frac{1}{6}e^{-j2\Omega})}$$

Construct a realization consisting of the cascade of two second-order systems (one of which will be an FIR system).

(d) Consider the frequency response of part (c). Use the method described in the Appendix on partial fraction expansion to construct a realization consisting of the parallel interconnection of a recursive second-order system and a nonrecursive second-order system.

In a variety of important applications it is of interest to change the relative amplitudes of the frequency components in a signal or perhaps eliminate some frequency components entirely, a process referred to as *filtering*. For linear time-invariant systems, the spectrum of the output is that of the input multiplied by the frequency response of the system. Consequently, filtering can be conveniently accomplished through the use of such systems with an appropriately chosen frequency response. This represents one of the very important applications of linear time-invariant systems.

One example in which linear time-invariant filtering is encountered is in audio systems. In such systems, a filter is typically included to permit the listener to modify the relative amounts of low-frequency energy (bass) and high-frequency energy (treble). The filter corresponds to a linear time-invariant system whose frequency response is changed by manipulating the tone controls. Also, in high-fidelity audio systems, a filter is often included in the preamplifier to compensate for the frequency-response characteristics of the speakers. An example, shown in Figure 6.1, is the frequency response of the equalizer circuits used for one particular series of audio speakers. The equalizing circuits are designed to compensate for the frequency response of the speakers and the listening room and consist of a cascade of two stages: a control stage and an equalizer stage. The equalizer has the fixed frequency response indicated in Figure 6.1(c). The control stage consists of the cascade of two filters, as indicated in Figure 6.1(a) and (b). The first filter is a low-frequency filter controlled by a two-position switch to provide one of the two frequency responses indicated. The second filter in the control stage has two continuously adjustable slider switches to vary the frequency response within the limits indicated in Figure 6.1(b).

Filtering

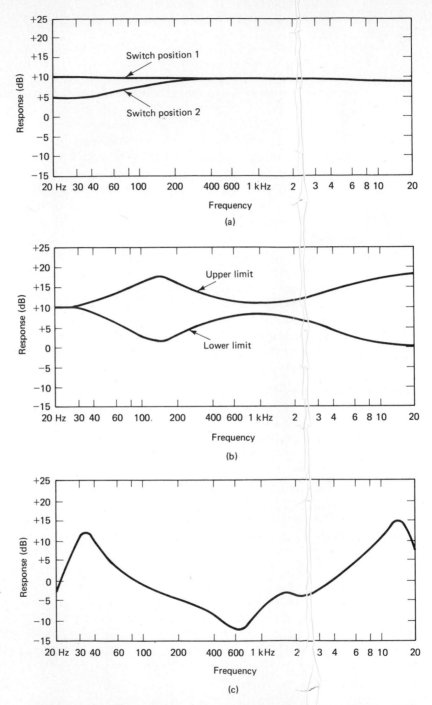

Figure 6.1 Frequency response of the equalizer circuits for one particular series of audio speakers: (a) low-frequency filter controlled by a two-position switch; (b) upper and lower frequency limits on a continuously adjustable shaping filter; (c) fixed frequency response of the equalizer stage.

Another class of LTI filters often encountered is that for which the filter output approximates the derivative of the filter input. It follows from the differentiation property of Fourier transforms, discussed in Chapters 4 and 5, that the frequency response of a differentiating filter is of the form $H(\omega) = j\omega$ (i.e., amplitude that varies linearly with frequency). The frequency-response characteristics of a differentiating filter are shown in Figure 6.2. Differentiating filters are useful in enhancing rapid transitions in

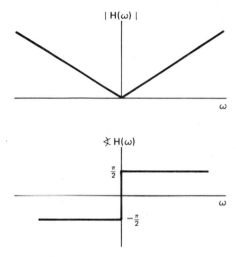

Figure 6.2 Frequency-response characteristics of a filter for which the output is the derivative of the input.

a signal, and one application in which they often arise is in enhancement of edges in picture processing. Figure 6.3 illustrates the effect of a differentiating filter on an image. Figure 6.3(a) shows two original images and in Figure 6.3(b), the result of processing with a filter having a linear frequency response. Since the derivative at edges in a picture is greater than in regions where the brightness varies slowly spatially, the effect of the differentiating filter is to enhance the edges.

Frequency-selective filters which pass signals undistorted in one or a set of frequency bands and attenuate or totally eliminate signals in the remaining frequency bands are another important class of LTI filters. The use of frequency-selective filters arises in a variety of situations. For example, if surface noise in an audio recording is in a higher frequency band than the music or voice on the recording, it can be removed by frequency-selective filtering. Another important application of frequency-selective filters is in communications systems. As we discuss in detail in Chapter 7, the basis for amplitude modulation (AM) systems is the transmission of information from many different sources simultaneously by putting the information from each channel into a separate frequency band and extracting the individual channels or bands at the receiver using frequency-selective filters. Frequency-selective filters for separating the individual channels and frequency-shaping filters for adjusting the tone quality form a major part of any home radio and television receiver.

Analysis of economic data sequences such as the stock market average commonly utilizes discrete-time filters. Often the long-term variations (which correspond to low frequencies) have a different significance than the short-term variations (which correspond to high frequencies), and it is useful to analyze these components sepa-

(a)

(b)

Figure 6.3 Effect of a differentiating filter on an image: (a) two original images; (b) the result of processing the original images with a differentiating filter.

rately. The separation of these components is typically accomplished using discrete-time frequency-selective filters. Filtering of economic data sequences is also used to *smooth* the data to remove random fluctuations (which are generally high frequency) superimposed on the meaningful data.

These are only a few of the many applications of filtering using linear time-invariant systems, and in this and the next several chapters we see other examples of both continuous-time and discrete-time filters. In detail, the topic of filtering encompasses many issues, such as those involving design and implementation. The selection of an appropriate filter for a specific application often becomes very much of an art. Our principal goal in this chapter is to *introduce* the basic concept of filtering, not to attempt to cover the topic in detail. As we will see, the basic concepts stem directly from the notions and properties of the Fourier transform as developed in Chapters 4 and 5. Thus, in addition to introducing an important area of application, the discussion will serve as a focus for further understanding of the properties and importance of the Fourier transform.

6.1 IDEAL FREQUENCY-SELECTIVE FILTERS

6.1.1 Frequency-Domain Characteristics of Ideal Frequency-Selective Filters

The mathematical basis for filtering using LTI systems is the fact that for such systems the Fourier transform of the output is that of the input multiplied by the frequency response of the system. An *ideal frequency-selective* filter is one that exactly passes complex exponentials at one set of frequencies and completely rejects the rest. The frequency response $H(\omega)$ of a continuous-time filter that passes complex exponentials $e^{j\omega t}$ for values of ω in the range $-\omega_c \leq \omega \leq \omega_c$ and rejects all others is

$$H(\omega) = \begin{cases} 1, & |\omega| \leq \omega_c \\ 0, & |\omega| > \omega_c \end{cases} \tag{6.1}$$

and is shown in Figure 6.4. A filter with this frequency response is referred to as an

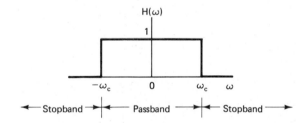

Figure 6.4 Frequency response of an ideal lowpass filter.

ideal lowpass filter since the band of frequencies that it passes is centered around $\omega = 0$. The band of frequencies passed by the filter is referred to as the *passband*, and the band of frequencies rejected is called the *stopband* of the filter. The frequency ω_c is referred to as the *cutoff frequency*. Two other ideal continuous-time frequency-selective filters are shown in Figure 6.5. Figure 6.5(a) corresponds to a *highpass filter* and Figure 6.5(b) to a *bandpass* filter in reference to the frequency range that the class of filters selects.

Figure 6.6 shows the frequency response of several discrete-time ideal filters, with Figure 6.6(a) corresponding to a lowpass filter, 6.6(b) to a highpass filter, and 6.6(c) to a bandpass filter. The frequency-response characteristics of the continuous-time and discrete-time ideal filters differ by virtue of the fact that for discrete-time filters, the frequency response $H(\Omega)$ must be periodic with period 2π. As discussed in Chapter 2, frequencies around even multiples of π ($\Omega = 0$, $\Omega = \pm 2\pi$, etc.) are inter-

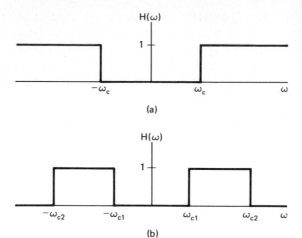

Figure 6.5 (a) Frequency response of an ideal highpass filter; (b) frequency response of an ideal bandpass filter.

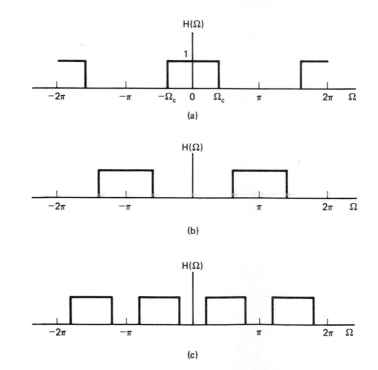

Figure 6.6 Discrete-time ideal frequency-selective filters: (a) lowpass; (b) highpass; (c) bandpass.

preted as low frequencies, whereas those around odd multiples of π ($\Omega = \pm\pi, \pm3\pi$, etc.) are interpreted as high frequencies.

Each of the ideal filters in Figures 6.4, 6.5, and 6.6 has a frequency response that is real and nonnegative, in other words, has a zero phase characteristic. This phase characteristic stems from our requirement that any signal whose frequency components are totally within the passband of the filter be passed undistorted. We have seen in a number of examples in Chapters 4 and 5 how modification of the phase can lead to severe distortion in the shape of a signal even when the magnitude of the spectrum is undisturbed. A special case in which a nonzero phase characteristic results in a "distortion" that is often acceptable is that of *linear phase*, as indicated in Figure 6.7 for an ideal continuous-time lowpass filter. As a consequence of property 4.6.3, eq. (4.85), a linear phase characteristic introduces a *time shift*. A similar statement applies to other ideal filters, both continuous-time and discrete-time.

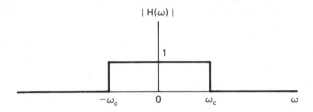

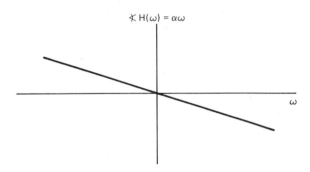

Figure 6.7 Ideal lowpass filter with phase proportional to ω.

6.1.2 Time-Domain Characteristics of Ideal Frequency-Selective Filters

So far, the discussion has focused on the frequency-domain characteristics of ideal frequency-selective filters. Often in designing and utilizing filters, it is also important to take into account the time-domain characteristics, such as the impulse response and step response. In the following discussion we will focus on the ideal lowpass filter for both continuous and discrete time.

The impulse response of the ideal lowpass filter corresponds to the inverse Fourier transform of the frequency response of Figure 6.4 and as developed in Example 4.11 is given by

$$h_{lp}(t) = \frac{\omega_c}{\pi} \, \text{sinc} \left(\frac{\omega_c t}{\pi} \right) \tag{6.2}$$

This impulse response is sketched in Figure 6.8. Here the width of the filter passband

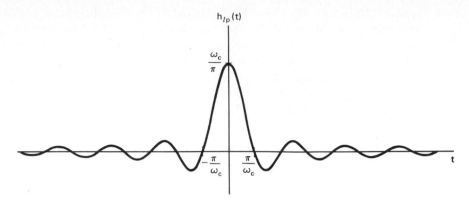

Figure 6.8 Impulse response of ideal lowpass filter.

is proportional to ω_c and the width of the main lobe of the impulse response is proportional to $1/\omega_c$. As the filter bandwidth increases, the impulse response becomes narrower, and vice versa. This is, of course, consistent with the scaling property for Fourier transforms discussed in Section 4.6.5. For the ideal lowpass filter with linear phase corresponding to Figure 6.7, the impulse response is simply delayed by α, as indicated in Figure 6.9.

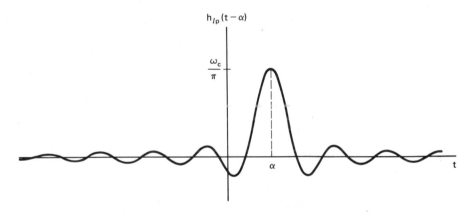

Figure 6.9 Impulse response of the filter of Figure 6.7.

For the discrete-time ideal lowpass filter, the impulse response can be obtained from Example 5.16 as

$$h_{lp}[n] = \frac{\Omega_c}{\pi} \operatorname{sinc}\left(\frac{\Omega_c n}{\pi}\right) \tag{6.3}$$

and is sketched in Figure 6.10 for $\Omega_c = \pi/4$.

The step responses of the ideal lowpass filter for continuous time and discrete time are illustrated in Figure 6.11. In both cases we see that in the time domain, the response of an ideal lowpass filter exhibits overshoot and ringing, much like the under-damped second-order systems discussed in Section 4.12 for continuous time and in Section 5.12 for discrete time. In some contexts this time-domain behavior may be

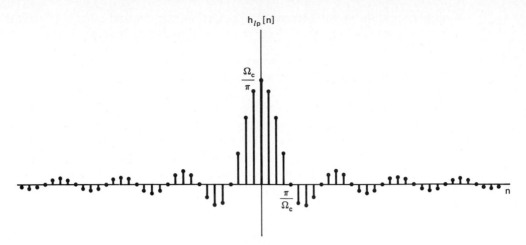

Figure 6.10 Impulse response of discrete-time ideal lowpass filter with $\Omega_c = \frac{\pi}{4}$.

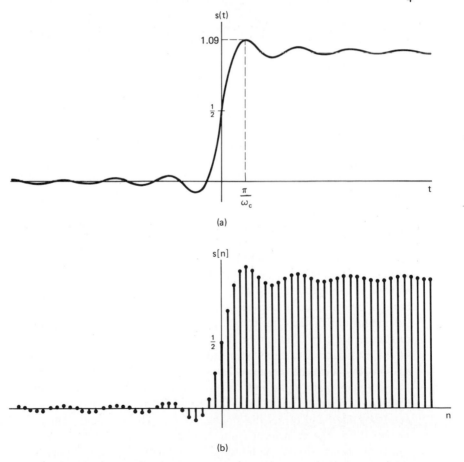

(a)

(b)

Figure 6.11 (a) Step response of continuous-time ideal lowpass filter; (b) step response of discrete-time ideal lowpass filter.

undesirable. As we consider in the next section, for this and other reasons it is often of interest to design filters with a more gradual transition from passband to stopband.

6.2 NONIDEAL FREQUENCY-SELECTIVE FILTERS

The class of filters discussed in Section 6.1 are referred to as ideal filters because they *exactly* pass one set of frequencies and *completely* reject others. However, this is not necessarily desirable. For example, in many filtering contexts, the signals to be separated do not lie in totally disjoint frequency bands. A typical situation might be that depicted in Figure 6.12, where the spectra of two signals overlap slightly. A filter with

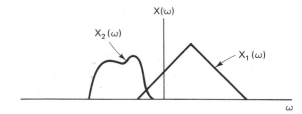

Figure 6.12 Two spectra that are slightly overlapping.

a gradual transition from passband to stopband is generally preferable when filtering the sum of signals with overlapping spectra.

As we indicated in Section 6.1.2, another consideration is suggested by examination of the step response, shown in Figure 6.11, of an ideal lowpass filter. For both continuous time and discrete time, the step response sufficiently far away from the discontinuity is approximately equal to the value of the step. In the vicinity of the discontinuity, however, it overshoots this value and exhibits ringing. In some situations this time-domain behavior may be undesirable. Since in many cases, the characteristics of the "ideal" frequency-selective filter are undesirable it is often preferable to allow some flexibility in the behavior of the filter in the passband and in the stopband as well as to permit a more gradual transition between the passband and stopband as opposed to the abrupt transition characteristic of the "ideal" filters. For example, in the case of continuous-time lowpass filters, the specifications may allow some deviation from unity gain in the passband and from zero gain in the stopband as well as the specifications of a passband edge and stopband edge with a transition band between them. Thus, specifications for a lowpass filter are often stated to require the magnitude of the filter frequency response to lie in the nonshaded area indicated in Figure 6.13. In this figure a deviation from unity of plus and minus δ_1 is allowed in the passband and a deviation of δ_2 from zero is allowed in the stopband. The amount by which the frequency response differs from unity in the passband is referred to as the *passband ripple* and the amount by which it deviates from zero in the stopband is referred to as the *stopband ripple*. The frequency ω_p is referred to as the *passband edge* and ω_s as the *stopband edge*. The frequency range from ω_p to ω_s is provided for the transition from passband to stopband and is referred to as the *transition band*. Similar definitions apply to discrete-time lowpass filters as well as other continuous- and discrete-time frequency-selective filters.

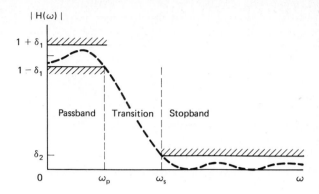

Figure 6.13 Tolerances for a lowpass filter. Allowable passband ripple is $\pm\delta_1$, and allowable stopband ripple is δ_2. The dashed curve illustrates one possible frequency-response curve that stays within the tolerance limits.

Even in cases when the ideal frequency-selective characteristics are desirable, they may not be attainable. For example, from eqs. (6.2) and (6.3) it is evident that the ideal lowpass filter is noncausal and consequently must be approximated for real-time filtering by a causal system. It can similarly be shown that the ideal highpass and bandpass filters are noncausal. When filtering is to be carried out in real time, causality is a necessary constraint, and thus a causal approximation to the ideal characteristics would be required. A further consideration that motivates providing some flexibility in the filter characteristics is ease of implementation. In general, the more precisely we try to approximate or implement an ideal frequency-selective filter, the more complicated or costly the implementation becomes whether in terms of components such as resistors, capacitors, and operational amplifiers, in continuous time, or in terms of memory registers, multipliers, and adders, in discrete time. In many filtering contexts a precise filter characteristic may not be essential and a simple filter will suffice.

Some of the points made above can be summarized in the interpretation of an automobile suspension system as a lowpass filter. In Figure 6.14 is shown a diagrammatic representation of a simple suspension system comprised of a spring and a dashpot (shock absorber). The road surface can be thought of as a superposition of rapid small-amplitude changes in elevation (high frequencies) representing the roughness of the road surface and gradual changes in elevation (low frequencies) due to the general topography. The automobile suspension system is generally intended to filter out rapid variations in the ride due to the road surface (i.e., to act as a lowpass filter).

The basic intent of the lowpass filter represented by the suspension system is to provide a smooth ride and there is no sharp natural division between the frequencies to be passed and those to be rejected. Thus, it is reasonable to accept a lowpass filter that has a gradual transition from passband to stopband. Furthermore, the time-domain characteristics of the system are important. If the impulse response or step response of the suspension system exhibits ringing, then a large bump in the road (modeled as an impulse input) or a curb (modeled as a step input) will result in an uncomfortable oscillatory response. In fact, a common test for a suspension system

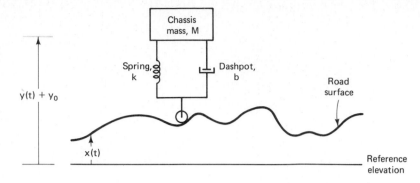

Figure 6.14   Diagrammatic representation of an automotive suspension system. y_0 represents the distance between the chassis and the road surface when the automobile is at rest, $y(t) + y_0$ the position of the chassis above the reference elevation, and $x(t)$ the elevation of the road above the reference elevation.

is to introduce an excitation by depressing and then releasing the chassis. If the response exhibits ringing, it is an indication that the shock absorbers need to be replaced.

Cost and ease of implementation also play an important role in the design of automobile suspension systems. Many studies have been carried out to determine the most desirable frequency-response characteristics for suspension systems from the point of view of passenger comfort. In situations where the cost may be warranted, such as for passenger railway cars, intricate and costly suspension systems are used. For the automotive industry, cost is an important factor and simple, less costly suspension systems are generally used. A typical automotive suspension system consists simply of the chassis connected to the wheels through a spring and a dashpot.

6.3 EXAMPLES OF CONTINUOUS-TIME FREQUENCY-SELECTIVE FILTERS DESCRIBED BY DIFFERENTIAL EQUATIONS

The automobile suspension system is an example of the approximation of a frequency-selective filter using components that are describable by linear constant-coefficient differential equations. More generally, in a variety of electrical, mechanical, and numerical contexts, systems describable by linear constant-coefficient differential or difference equations provide a particularly convenient way of implementing nonideal filters. In this and the next section we consider several examples, including the automobile suspension system, that illustrate the implementation of continuous-time and discrete-time frequency-selective filters through the use of differential and difference equations.

6.3.1 RC Lowpass and Highpass Filters

As an example of a simple continuous-time lowpass filter, consider the first-order *RC* circuit in Figure 6.15. The capacitor voltage is considered to be the system output and the source voltage is the system input. The output voltage is related to the input voltage through the linear constant-coefficient differential equation

$$RC\frac{dv_c(t)}{dt} + v_c(t) = v_s(t) \tag{6.4}$$

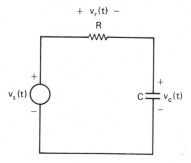

Figure 6.15 First-order *RC* filter.

Thus, this corresponds to a first-order system of the form considered in Section 4.12 with time constant τ equal to *RC*. The impulse and step response were illustrated in Figure 4.39 and are shown again in Figure 6.16. The Bode plot for the frequency

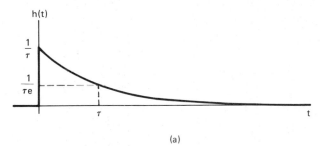

(a)

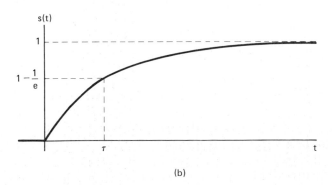

(b)

Figure 6.16 (a) Impulse response of *RC* lowpass filter with $\tau = RC$; (b) step response of *RC* lowpass filter with $\tau = RC$.

response was shown in Figure 4.40 and is reproduced in Figure 6.17. From Figure 6.17 we can see that this filter acts as an approximation to an ideal lowpass filter. As an alternative to choosing the capacitor voltage as the output, we can instead choose the voltage across the resistor. In this case, the differential equation relating input and output is given by

$$RC\frac{dv_R(t)}{dt} + v_R(t) = RC\frac{dv_s(t)}{dt} \tag{6.5}$$

and the frequency response $G(\omega)$ is given by

$$G(\omega) = \frac{V_R(\omega)}{V_S(\omega)} = \frac{j\omega RC}{1 + j\omega RC} \tag{6.6}$$

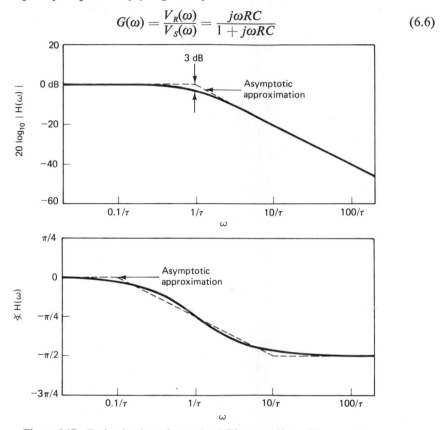

Figure 6.17 Bode plot for a first-order RC lowpass filter with $\tau = RC$.

The step response is shown in Figure 6.18 and the Bode plot for the frequency response is shown in Figure 6.19. From Figure 6.19 we note that with the resistor voltage taken as the output, the RC circuit behaves as an approximation to a highpass filter.

We see from this example that a simple RC circuit can serve as a highpass or a lowpass filter, depending upon the choice of the physical output variable. These filters do not have a sharp transition from passband to stopband. If desired, more complex filters with a sharper transition can be implemented by using more energy storage elements (capacitances and inductances in electrical filters and springs and masses in mechanical filters), leading to higher-order differential equations. The

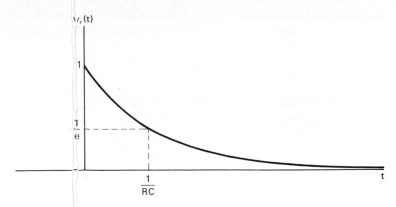

Figure 6.18   Step response of RC circuit with output $v_r(t)$.

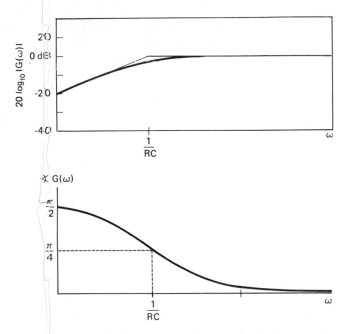

Figure 6.19   Bode plot for frequency response of RC circuit with output $v_r(t)$.

automobile suspension system, which we introduced earlier and now explore further, represents one such example.

6.3.2 The Automobile Suspension System as a Lowpass Filter

As introduced earlier, an automobile suspension system can be interpreted as a lowpass filter. In the diagrammatic representation in Figure 6.14, y_0 represents the distance between the chassis and the road surface when the automobile is at rest, $y(t) + y_0$

the position of the chassis above a reference elevation, and $x(t)$ the elevation of the road above a reference elevation. The differential equation governing the motion of the chassis is then

$$M \frac{d^2 y(t)}{dt} + b \frac{dy(t)}{dt} + ky(t) = kx(t) + b \frac{dx(t)}{dt} \tag{6.7}$$

where M is the mass of the chassis and k and b are the spring and shock absorber constants. The frequency response of the system is

$$H(\omega) = \frac{k + bj\omega}{(j\omega)^2 M + b(j\omega) + k}$$

or

$$H(\omega) = \frac{\omega_n^2 + 2\zeta\omega_n(j\omega)}{(j\omega)^2 + 2\zeta\omega_n(j\omega) + \omega_n^2} \tag{6.8}$$

where

$$\omega_n = \sqrt{\frac{k}{M}} \quad \text{and} \quad 2\zeta\omega_n = \frac{b}{M}$$

We note that the denominator is identical to the second-order system discussed in Section 4.12, with ω_n representing the natural frequency and ζ the damping ratio. The Bode plot representing the magnitude of the frequency response is sketched in Figure 6.20 for several different values of damping ratio. Figure 6.21 illustrates the step response for several different values of damping ratio.

The filter cutoff frequency is controlled primarily through ω_n, or equivalently for a fixed chassis mass, by an appropriate choice of spring constant k. For a given ω_n, the damping ratio is then adjusted through the damping factor b associated with

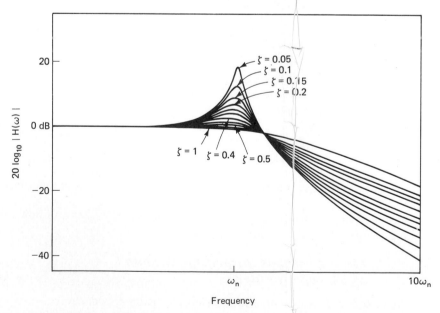

Figure 6.20 Log magnitude of frequency response of automobile suspension system.

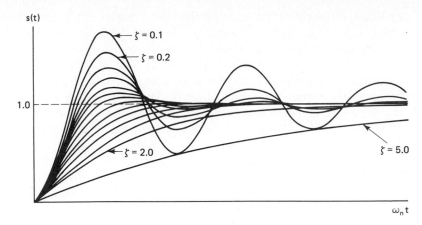

Figure 6.21 Step response of automobile suspension system for various values of damping ratio ($\zeta = 0.1, 0.2, 0.3, 0.4, 0.5, 0.6, 0.7, 0.8, 0.9, 1.0, 1.2, 1.5, 2.0, 5.0$).

the shock absorbers. As the natural frequency ω_n is decreased, the suspension will tend to filter out slower road variations, thus providing a smoother ride. On the other hand, we see from Figure 6.21 that the rise time of the system increases, and thus the system will feel more sluggish. On the one hand, it would be desirable to keep ω_n small to improve the lowpass filtering and on the other to have ω_n large for rapid time response. These, of course, are conflicting requirements and illustrate again the need for a trade-off between time-domain and frequency-domain characteristics. Typically, a suspension system with a low value of ω_n, so that the rise time is long, is referred to as "soft" and one with a high value of ω_n, so that the rise time is short, is referred to as "hard." From Figures 6.20 and 6.21 we observe also that as the damping ratio decreases, the frequency response of the system cuts off more sharply, but the overshoot and ringing in the step response tend to increase, another trade-off between the time and frequency domains. Generally, the shock absorber damping is chosen to have a rapid rise time and yet avoid overshoot and ringing. This choice corresponds to the critically damped case considered in Section 4.12. As mentioned previously, a common test for shock absorbers is to apply a step excitation to the suspension system by depressing and releasing the chassis and observing whether or not ringing occurs. If it does, this is an indication that the dashpot constant has decreased, representing a deterioration of the shock absorber, and the system has become underdamped.

6.4 EXAMPLES OF DISCRETE-TIME FREQUENCY-SELECTIVE FILTERS DESCRIBED BY DIFFERENCE EQUATIONS

As we saw in Chapter 5, discrete-time systems characterized by linear constant-coefficient difference equations, are conveniently implemented using coefficient multipliers, storage or delay registers, and adders. Thus, filters characterized by difference equations take on particular importance. In Chapter 3 we introduced the two basic classes of difference equations: nonrecursive difference equations, for which the impulse response is of finite length, and recursive difference equations, for which the impulse response is of infinite length. There are specific advantages and disadvantages to recursive and nonrecursive filters. For example, it is often desirable for the phase characteristics of a filter to be zero or linear, so that the phase affects the shape of the output signal by at most a time delay. As developed in Problem 6.7, if a filter is to be *causal* and have exactly linear phase, its impulse response must be of finite length, and consequently the difference equation must be nonrecursive. On the other hand, it is generally true that the same filter specifications require a higher-order equation and consequently more coefficients and delays when implemented using a nonrecursive difference equation compared with using a recursive difference equation. In the following discussion we consider separately the two classes of discrete-time filters.

6.4.1 Nonrecursive Discrete-Time Filters

As we have emphasized in several discussions, lowpass filtering can be thought of as a smoothing operation. For discrete-time sequences a common smoothing operation is one referred to as a *moving average*, where the smoothed value $y[n]$ for any n, say n_0, is an average of values of $x[n]$ in the vicinity of n_0. The basic idea is that by averaging values locally, rapid variations from point to point will be averaged out and slow variations will be retained, corresponding to smoothing or lowpass filtering the original sequence. As an example, a three-point moving average of an input sequence $x[n]$ is of the form

$$y[n] = \tfrac{1}{3}(x[n-1] + x[n] + x[n+1]) \tag{6.9}$$

so that each output $y[n]$ is the average of three consecutive input values. Equation (6.9) is a nonrecursive linear constant-coefficient difference equation, and using the procedure developed in Section 5.11 the filter transfer function is

$$H(\Omega) = \tfrac{1}{3}\{1 + 2\cos\Omega\} \tag{6.10}$$

The magnitude and phase of $H(\Omega)$ are sketched in Figure 6.22. We observe that it has the general characteristics of a lowpass filter although, as with the simple *RC* circuit, it does not have a sharp transition from passband to stopband.

A similar approach can also be used to approximate a highpass filter as well as a lowpass filter. To illustrate this, again with a simple example, consider the difference equation

$$y[n] = \frac{x[n] - x[n-1]}{2} \tag{6.11}$$

For input signals that are approximately constant, the value of $y[n]$ is close to zero. For input signals that vary greatly from sample to sample, the values of $y(n)$ can be expected to have large amplitude. We would thus expect this equation to approximate a highpass filter, since high-frequency components are reflected in large variations between adjacent sequence values. The frequency response associated with eq. (6.11) is

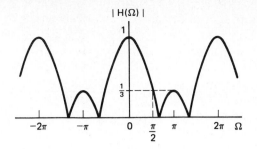

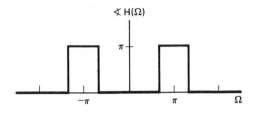

Figure 6.22 Frequency response of three-point moving-average lowpass filter.

$$H(\Omega) = \frac{1}{2}[1 - e^{-j\Omega}] = je^{-j\Omega/2} \sin\left(\frac{\Omega}{2}\right) \tag{6.12}$$

and its magnitude is shown in Figure 6.23.

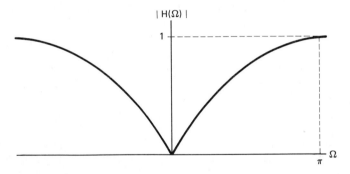

Figure 6.23 Frequency response of two-point moving-average highpass filter.

The three-point moving-average filter in eq. (6.9) has no parameters that can be changed to adjust the effective cutoff frequency. As a generalization of this moving-average filter, we can consider averaging over $N + M - 1$ neighboring points, that is, to use a difference equation of the form

$$y[n] = \frac{1}{N + M + 1} \sum_{k=-N}^{M} x[n - k] \tag{6.13}$$

The corresponding impulse response is a rectangular pulse. If $N = 0$ or is negative the moving average filter of eq. (6.13) is causal. The filter frequency response is

$$H(\Omega) = \frac{1}{N+M+1} \sum_{k=-N}^{M} e^{-j\Omega k} \tag{6.14}$$

or

$$H(\Omega) = \frac{1}{N+M+1} e^{j\Omega[(N-M)/2]} \frac{\sin\left[\Omega\left(\frac{M+N+1}{2}\right)\right]}{\sin(\Omega/2)} \tag{6.15}$$

The log magnitude of $H(\Omega)$ is shown in Figure 6.24 for $M + N + 1 = 33$ and $M + N + 1 = 65$.

It is common to apply a moving-average filter to many economic indicators to attenuate the short-term fluctuations in relation to longer-term trends. In Figure 6.25 we illustrate the use of a moving-average filter of the form of eq. (6.13), on the weekly Dow Jones stock market index for a 7-year period. In Figure 6.25(b) is a 51-day mov-

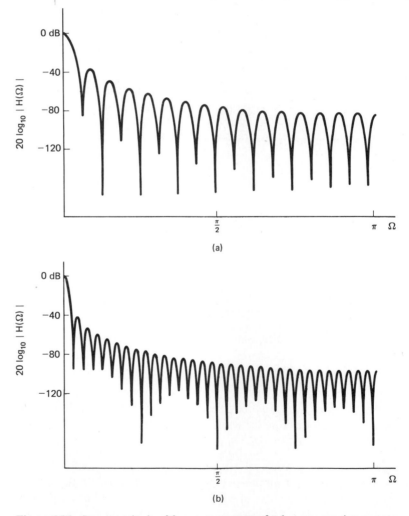

Figure 6.24 Log magnitude of frequency response for lowpass moving-average filter of eq. (6.13): (a) $M = N = 16$; (b) $M = N = 32$.

Filtering   Chap. 6

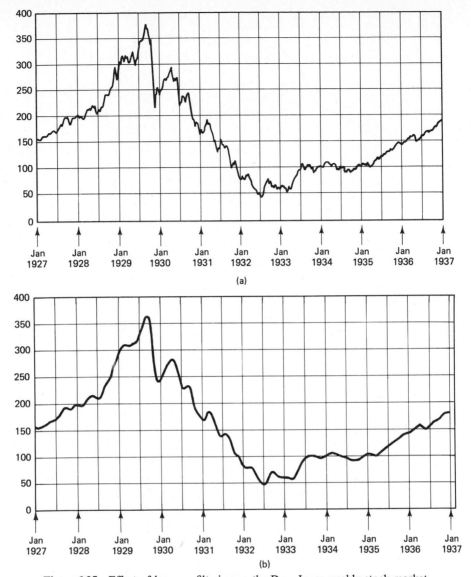

Figure 6.25 Effect of lowpass filtering on the Dow Jones weekly stock market index over a 10-year period: (a) weekly index; (b) 51-day moving average applied to (a); (c) 201-day moving average applied to (a). The weekly stock market index and the two moving averages are discrete-time sequences. For clarity in the graphical display the three sequences are shown here with the individual sequence values connected by straight lines to form a continuous curve.

ing average (i.e., $N = M = 25$) and in Figure 6.25(c) is a 201-day moving average (i.e., $N = M = 100$) applied to the weekly Dow Jones index in Figure 6.25(a). Both are considered to be useful averages, with the 51-day average tracking cyclical trends that occur during the course of a year and the 201-day average primarily emphasizing trends over a longer time frame.

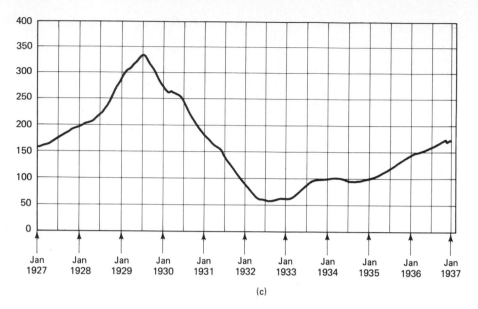

(c)

Figure 6.25 (cont.)

A further generalization of the moving-average filter can be made by forming a *weighted* average of $(N + M + 1)$ neighboring points, that is, by using a difference equation of the form

$$y[n] = \sum_{k=-N}^{M} b_k x[n - k] \qquad (6.16)$$

where the coefficients b_k can be selected to achieve the prescribed filter characteristics. Equation (6.16) is then in the form of a general nonrecursive difference equation.

There are a variety of techniques available for choosing the coefficients in eq. (6.16) to meet certain specifications on the filter. These procedures are discussed in detail in a number of texts,† and although we do not discuss these procedures here, it is worth emphasizing that they rely heavily on the basic concepts and tools developed thus far in this book. To illustrate how adjustment of these coefficients can influence the filter response, let us consider a filter of the form of eq. (6.16), with $N = M = 16$ and the filter coefficients chosen as

$$b_k = \begin{cases} \dfrac{2}{33} \, \text{sinc} \left(\dfrac{2k}{33} \right), & |k| \leq 32 \\ 0, & |k| > 32 \end{cases} \qquad (6.17)$$

The impulse response of this filter is

$$h[n] = \begin{cases} \dfrac{2}{33} \, \text{sinc} \left(\dfrac{2n}{33} \right), & |n| \leq 32 \\ 0, & |n| > 32 \end{cases} \qquad (6.18)$$

†See, for example, R. W. Hamming, *Digital Filters* (Englewood Cliffs, N.J.: Prentice-Hall, Inc., 1977); A. V. Oppenheim and R. W. Schafer, *Digital Signal Processing* (Englewood Cliffs, N.J.: Prentice-Hall, Inc., 1975); and L. R. Rabiner and B. Gold, *Theory and Application of Digital Signal Processing* (Englewood Cliffs, N.J.: Prentice-Hall, Inc., 1975).

Comparing with eq. (6.3), we see that this corresponds to truncating for $|n| > 32$ the impulse response for the ideal lowpass filter with cutoff frequency $\Omega_c = 2\pi/33$.

In general, the coefficients b_k can be adjusted so that the cutoff is at a desired frequency. For the example of Figure 6.26, the cutoff frequency was chosen to match approximately the cutoff frequency of Figure 6.24 for $N = M = 16$. Figure 6.26(a) shows the impulse response and Figure 6.26(b) shows the log magnitude of the frequency response. Comparing this frequency response to Figure 6.24, we observe that the passband of the filter has approximately the same width but that the transition to the stopband is sharper. In Figure 6.27(a) and (b) the magnitudes (on a linear vertical scale) of the two filters are shown for comparison. It should be clear from the comparison of the two examples that by intelligent choice of the weighting coefficients, the transition band can be sharpened. An example of a higher-order lowpass filter ($N = M = 125$), with the coefficients determined through a numerical algorithm referred to as the Parks-McClellan algorithm,[†] to obtain the sharpest cutoff possible, is shown in Figure 6.28.

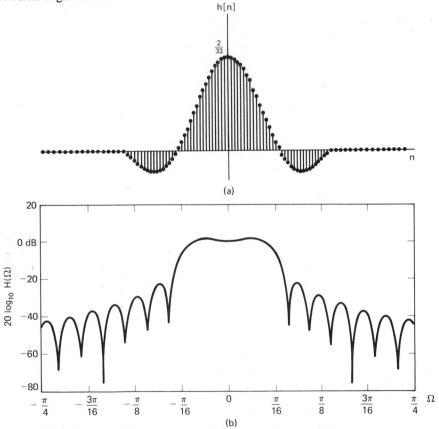

(a)

(b)

Figure 6.26 (a) Impulse response for the nonrecursive filter of eq. (6.18); (b) log magnitude of frequency response.

†A. V. Oppenheim and R. W. Schafer, *Digital Signal Processing* (Englewood Cliffs, N.J.: Prentice-Hall, Inc., 1975), chap. 5.

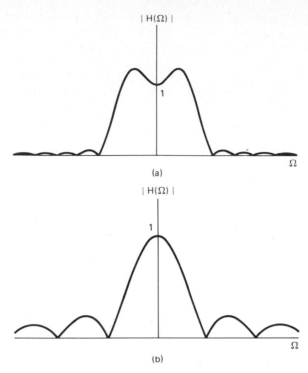

Figure 6.27 Comparison on a linear amplitude scale of the frequency response of (a) Figure 6.26 and (b) Figure 6.24(a).

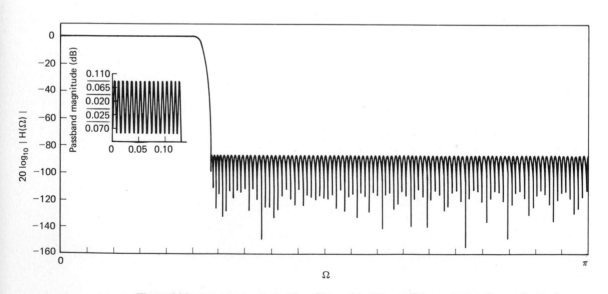

Figure 6.28 Lowpass nonrecursive filter with 251 coefficients designed to obtain the sharpest possible cutoff.

6.4.2 Recursive Discrete-Time Filters

In Section 6.4.1 we considered moving-average or nonrecursive filters. Another important class of discrete-time filters are those described by the class of recursive difference equations. The use of the discrete-time Fourier transform to obtain the frequency response from difference equations of this type was considered in Section 5.11. As a simple illustration of how such a difference equation can act as a lowpass or highpass filter, consider the discrete-time system described by the difference equation

$$y[n] - ay[n-1] = x[n] \tag{6.19}$$

The corresponding frequency response is

$$H(\Omega) = \frac{1}{1 - ae^{-j\Omega}} \tag{6.20}$$

The magnitude and phase of $H(\Omega)$ is shown in Figure 6.29(a) for $a = 0.6$ and in Figure 6.29(b) for $a = -0.6$. We observe that for the coefficient a positive, the difference equation (6.19) behaves as a lowpass filter, whereas for a negative, it behaves as a highpass filter. Just as with differential equations, higher-order recursive difference equations can be used to provide sharper filter characteristics and to provide more flexibility in balancing time-domain and frequency-domain constraints. There are a number of specific classes of continuous- and discrete-time filters for which standard procedures have been developed to determine the coefficients of the as-

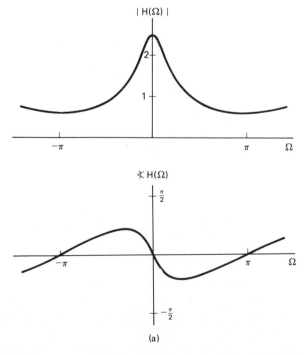

(a)

Figure 6.29 Frequency response of first-order recursive discrete-time filter: (a) coefficient equal to 0.6; (b) coefficient equal to −0.6.

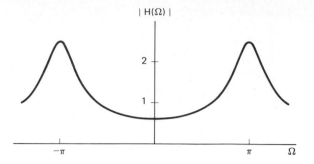

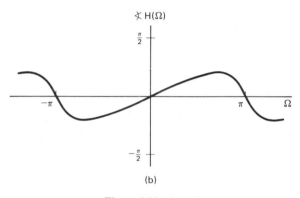

(b)

Figure 6.29 (cont.)

sociated differential or difference equation. In the next section we introduce one such class, referred to as Butterworth filters.

6.5 THE CLASS OF BUTTERWORTH FREQUENCY-SELECTIVE FILTERS

As we saw in Section 6.4, useful filters can be implemented in terms of differential equations in the continuous-time case and difference equations in the discrete-time case. As the order of the equation is allowed to increase, better approximations to ideal filters can result. There are several classes of procedures commonly applied to filter design and there is a rich and detailed literature on the subject.† In this section we introduce one of these classes, referred to as the Butterworth filter designs, to illustrate some of the issues and trade-offs involved. In this discussion we focus only on the magnitude of the frequency response of these filters.

To introduce the class of continuous-time Butterworth filters, let us return to the *RC* lowpass filter considered in Section 6.3.1. As we discussed, it is of the form of a first-order system as was considered in Section 4.12 with frequency response given by

†See, for example, references on page 425.

$$H(\omega) = \frac{1}{j\omega\tau + 1}, \qquad \tau = RC \tag{6.21}$$

A Bode plot of $|H(\omega)|$ is shown in Figures 4.40 and 6.17. The low-frequency asymptote is a constant and the high-frequency asymptote has a slope of -1. The slope of the high-frequency asymptote is controlled by the power-law dependence of ω in the denominator, and if ω appeared in the denominator raised to a higher integer power, the high-frequency asymptote would fall off more sharply and consequently the stopband attenuation would be greater. This, in fact, is how the frequency response is controlled in the class of filters referred to as *Butterworth filters*. Specifically, the class of Butterworth filters is that for which the magnitude squared of the frequency response $B(\omega)$ is of the form

$$|B(\omega)|^2 = \frac{1}{1 + (\omega/\omega_c)^{2N}} \tag{6.22}$$

The parameter N is referred to as the filter order. In Chapter 9 we will see how to obtain the differential equation that implements the transfer function of eq. (6.22) and will see, in particular, that the parameter N corresponds to the order of the differential equation. The parameter ω_c is the frequency at which $|B(\omega)|$ is at $1/\sqrt{2}$ times its value at $\omega = 0$; equivalently, on a Bode plot, the gain is 3 dB down from its maximum value at $\omega = 0$. In Figure 6.30 is shown the Bode plot for the gain of a Butterworth filter for $N = 1, 2, 4,$ and 8, and in Figure 6.31 is shown $|B(\omega)|$ on a

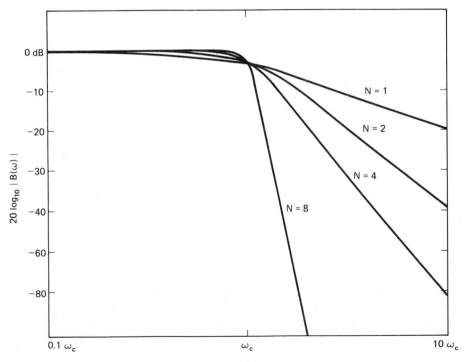

Figure 6.30 Gain on a logarithmic amplitude scale for Butterworth filters with $N = 1, 2, 4, 8$.

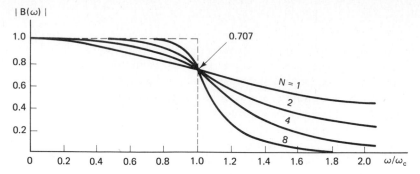

Figure 6.31 Gain on a linear amplitude scale for Butterworth filters with $N = 1, 2, 4, 8$.

linear scale. As is also evident from eq. (6.22), the gain at $\omega = \omega_c$ is $1/\sqrt{2}$ independent of the filter order; and the higher the order, the sharper the transition from passband to stopband. The simple RC circuit of Section 6.3.1 represents an implementation of a first-order Butterworth filter with $RC = 1/\omega_c$. In Figure 6.32 are

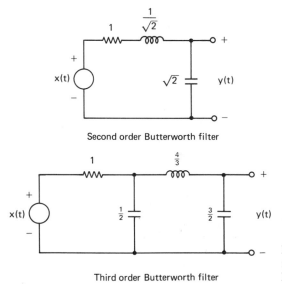

Second order Butterworth filter

Third order Butterworth filter

Figure 6.32 Typical circuits for implementing a second- and third-order Butterworth filter.

shown typical circuits for implementing second- and third-order Butterworth filters. Shown in Figure 6.33 are the impulse response and step response of second- and third-order Butterworth filters.

Butterworth filters are a very commonly used class of filters and the parameters and circuits for implementing them to meet specific passband, stopband, and transition-band characteristics are extensively tabulated and readily available. From Figures 6.30 and 6.31 we see that a characteristic of the class of Butterworth filters is the *monotonic* shape of the magnitude curve in the passband and stopband. Butterworth filters are also referred to as having a *maximally flat* frequency response. This characteristic is defined and explored in Problem 6.26. There are other classes of

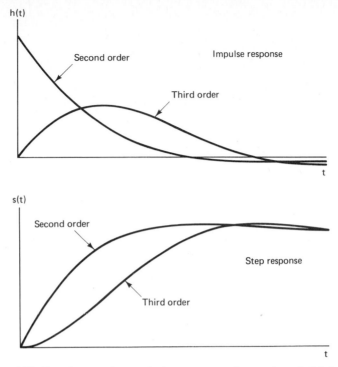

Figure 6.33 Impulse response and step response of second- and third-order Butterworth filters.

commonly used filters which have what is referred to as an *equiripple* characteristic in the passband, the stopband, or both. As illustrated in Figure 6.34, an equiripple characteristic in a frequency band corresponds to a characteristic that oscillates between the maximum and minimum values in the band. Figure 6.34(a) corresponds to an equiripple characteristic in the passband and a monotonic behavior in the stopband, Figure 6.34(b) to an equiripple characteristic in the stopband and a monotonic characteristic in the passband, and Figure 6.34(c) to an equiripple characteristic in both bands. The class of filters that provides the characteristics shown in Figure 6.34(a) and (b) is referred to as the class of *Chebychev* filters, and that which provides the characteristic in Figure 6.34(c) is the class of *elliptic* filters. As with Butterworth filters, parameters and circuits for implementing Chebychev and elliptic filters are available in tabulated form.† Although the discussion above was phrased in terms of

†Tables of parameters and circuits for Butterworth, Chebychev, and elliptic filters can be found, for example, in Erich Christian and Egon Eisenmann, *Filter Design Tables and Graphs* (New York: John Wiley and Sons, Inc., 1966); Anatol I. Zverev, *Handbook of Filter Synthesis* (New York: John Wiley and Sons, Inc., 1967); and D. E. Johnson, J. R. Johnson, and H. P. Moore, *A Handbook of Active Filters* (Englewood Cliffs, N.J.: Prentice-Hall, Inc., 1980). A more complete introduction to these classes of filters can be found in Richard W. Daniels, *Approximation Methods for Electronic Filter Design* (New York: McGraw-Hill Book Company, 1974); David E. Johnson, *Introduction to Filter Theory* (Englewood Cliffs, N.J.: Prentice-Hall, Inc., 1976); and L. P. Huelsman and P. E. Allen, *Introduction to the Theory and Design of Active Filters* (New York: McGraw-Hill Book Company, 1980).

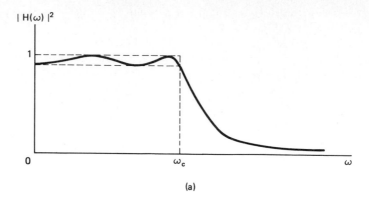

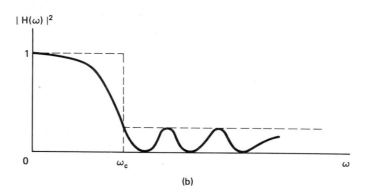

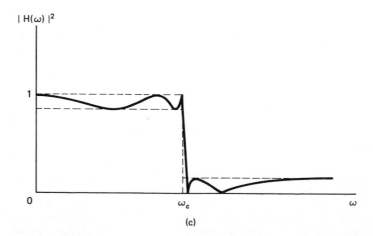

Figure 6.34 Chebychev and elliptic filters: (a) Chebychev filter with equiripple passband and monotonic stopband characteristics; (b) Chebychev filter with monotonic passband and equiripple stopband; (c) elliptic filter having equiripple behavior in both the passband and the stopband.

continuous-time lowpass filters, there are also corresponding designs for highpass and bandpass filters.

Each of the foregoing classes of continuous-time filters also has a direct counterpart in discrete-time filters. The class of discrete-time Butterworth filters has a frequency response $B(\Omega)$ for which the magnitude squared is of the form

$$| B(\Omega)|^2 = \frac{1}{1 + \left(\dfrac{\tan(\Omega/2)}{\tan(\Omega_c/2)}\right)^{2N}} \tag{6.23}$$

where, again, N is referred to as the filter order. Figure 6.35 shows $|B(\Omega)|$ on a linear scale for several values of N and we observe that the discrete-time Butterworth filter has characteristics similar to the continuous-time one, namely a monotonic frequency response in the passband and the stopband. At $\Omega = \Omega_c$, $|B(\Omega)|$ is $1/\sqrt{2}$ times the gain at $\Omega = 0$ independent of the order of the Butterworth filter (i.e., independent of N). In Problem 6.28 the difference equation for a first-order Butterworth filter is considered. In Chapter 10 we discuss how the *differential* equation that implements an Nth-order continuous-time Butterworth filter can be transformed to a *difference* equation that implements an Nth-order discrete-time Butterworth filter.

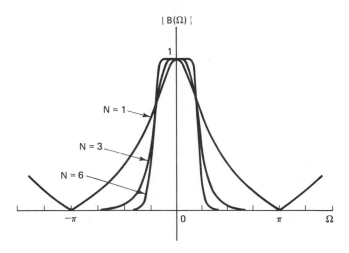

Figure 6.35 Gain on a linear amplitude scale for discrete-time Butterworth filters for several values of N.

6.6 SUMMARY

In this chapter we have discussed the application of linear time-invariant continuous- and discrete-time systems to filtering. Although this represents only a preliminary introduction to a topic of considerable importance, the basic concepts stem directly from our discussions in Chapters 4 and 5. For frequency-selective filters we first introduced the notion of ideal filters, that is, filters that exactly pass some bands of frequency and totally reject others. Then, motivated in part by considerations of time-domain and frequency-domain trade-offs and in part by considerations such as causality and ease of implementation, we introduced the concept of frequency-selective filters with transition bands and with tolerance limits in the passbands and stopbands. A variety of these issues were explored and developed in the context of an automobile suspension system and simple RC filters in continuous time, and moving-average and recursive filters in discrete time. The chapter concluded with the introduction of the class of Butterworth lowpass filters.

PROBLEMS

6.1. A causal LTI filter has the frequency response $H(\omega)$ shown in Figure P6.1. For each of the input signals given below, determine the filtered output signal $y(t)$.

(a) $x(t) = e^{jt}$　　　　　　　　　　　　(b) $x(t) = (\sin \omega_0 t)u(t)$

(c) $X(\omega) = \dfrac{1}{(j\omega)(6 + j\omega)}$　　　　(d) $X(\omega) = \dfrac{1}{2 + j\omega}$

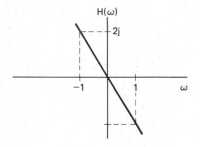

Figure P6.1

6.2. Shown in Figure P6.2-1 is the frequency response $H(\omega)$ of a continuous-time filter referred to as a lowpass differentiator. For each of the input signals $x(t)$ below, determine the filtered output signal $y(t)$.

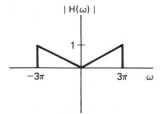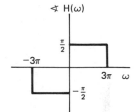

Figure P6.2-1

(a) $x(t) = \cos (2\pi t + \theta)$
(b) $x(t) = \cos (4\pi t + \theta)$
(c) $x(t)$ is a half-wave-rectified sine wave of period 1, as sketched in Figure P6.2-2.

$$x(t) = \begin{cases} \sin 2\pi t, & m \le t \le (m + \tfrac{1}{2}) \\ 0, & (m + \tfrac{1}{2}) \le t \le m \end{cases} \quad \text{for any integer } m$$

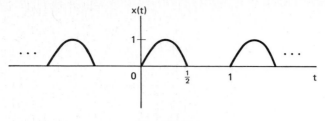

Figure P6.2-2

6.3. Shown in Figure P6.3 is the frequency response $H(\Omega)$ of a discrete-time differentiator. Determine the output signal $y[n]$ as a function of Ω_0 if the input $x[n]$ is

$$x[n] = \cos [\Omega_0 n + \theta]$$

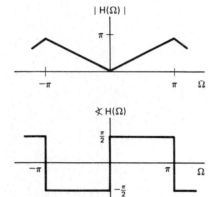

Figure P6.3

6.4. Shown in Figure P6.4 is $|H(\omega)|$ for a lowpass filter. Determine and sketch the impulse response of the filter for each of the following phase characteristics:
(a) $\sphericalangle H(\omega) = 0$
(b) $\sphericalangle H(\omega) = \omega T$, where T is a constant
(c) $\sphericalangle H(\omega) = \begin{cases} \dfrac{\pi}{2}, & \omega > 0 \\ -\dfrac{\pi}{2}, & \omega < 0 \end{cases}$

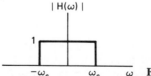

Figure P6.4

6.5. Consider an ideal discrete-time lowpass filter with impulse response $h[n]$ and for which the frequency response $H(\Omega)$ is that shown in Figure P6.5. Let us consider obtaining a new filter with impulse response $h_1[n]$ and frequency response $H_1(\Omega)$ as follows:

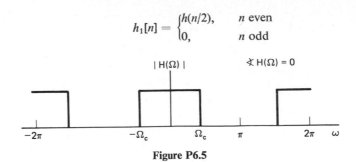

$$h_1[n] = \begin{cases} h(n/2), & n \text{ even} \\ 0, & n \text{ odd} \end{cases}$$

Figure P6.5

This corresponds to inserting a sequence value of zero between each sequence value of $h[n]$. Determine and sketch $H_1(\Omega)$ and state the class of ideal filters to which it belongs (e.g., lowpass, highpass, bandpass, multiband, etc.).

6.6. In the discussion in Section 4.6.7 of Parseval's theorem for continuous-time signals, we showed that

$$\int_{-\infty}^{+\infty} |x(t)|^2 \, dt = \frac{1}{2\pi} \int_{-\infty}^{+\infty} |X(\omega)|^2 \, d\omega$$

This says that the total energy in a signal can be obtained by integrating $|X(\omega)|^2$ over all frequencies. Now consider a real-valued signal $x(t)$ processed by the ideal bandpass filter $H(\omega)$ shown in Figure P6.6.

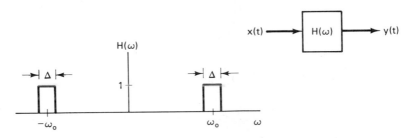

Figure P6.6

(a) Express the energy in the output signal $y(t)$ as an integration over frequency of $|X(\omega)|^2$. For Δ sufficiently small so that $|X(\omega)|$ is approximately constant over a frequency interval of width Δ, show that the energy in the output $y(t)$ of the bandpass filter is approximately proportional to $\Delta|X(\omega_0)|^2$.

On the basis of the foregoing result, $\Delta|X(\omega_0)|^2$ is proportional to the energy in the signal in a bandwidth Δ around the frequency ω_c. For this reason $|X(\omega)|^2$ is often referred to as the *energy-density spectrum* of the signal $x(t)$.

(b) Derive the corresponding result for discrete-time signals. Specifically, show that if a real discrete-time signal $x[n]$ is filtered with an ideal bandpass filter with center frequency Ω_0 and bandwidth Δ, then if Δ is sufficiently small so that $|X(\Omega)|$ is approximately constant over a frequency interval of width Δ, the energy in the output of the bandpass filter is approximately proportional to $\Delta|X(\Omega_0)|^2$.

6.7. It was stated in Section 6.4 that for a discrete-time filter to be *causal* and have exactly linear phase, its impulse response must be of finite length and consequently the difference equation must be nonrecursive. To focus on the insight behind this statement,

430   Filtering   Chap. 6

we consider a particular case, that of a linear phase characteristic for which the slope of the phase is an integer. Thus, the frequency response is assumed to be of the form

$$H(\Omega) = H_r(\Omega)e^{-j M \Omega}, \qquad -\pi < \Omega < \pi \qquad \text{(P6.7-1)}$$

where $H_r(\Omega)$ is real and even.

Let $h[n]$ denote the impulse response of the filter with frequency response $H(\Omega)$ and $h_r[n]$ the impulse response of the filter with frequency response $H_r(\Omega)$.

(a) By using the appropriate properties in Table 5.1, show that:

    1. $h_r[n] = h_r[-n]$ (i.e., $h_r[n]$ is symmetric about $n = 0$).
    2. $h[n] = h_r[n - M]$.

(b) Using your result in part (a), show that with $H(\Omega)$ of the form shown in eq. (P6.7-1), $h[n]$ is symmetric about $n = M$, that is,

$$h[M + n] = h[M - n] \qquad \text{(P6.7-2)}$$

(c) According to the result in part (b), the linear phase characteristic in eq. (P6.7-1) imposes a symmetry in the impulse response. Show that if $h[n]$ is causal and has the symmetry in eq. (P6.7-2), then

$$h[n] = 0, \qquad n < 0 \text{ and } n > 2M$$

(i.e., it must be of finite length).

6.8. Let $h[n]$ be the unit sample response of an FIR filter so that $h[n] = 0$ for $n < 0$, $n \geq N$. Assume that $h[n]$ is real. We can guarantee that the filter will have linear phase by imposing certain symmetry conditions on its unit sample response $h[n]$.

    The frequency response of this filter can be represented in the form

$$H(\Omega) = \hat{H}(\Omega)e^{j\theta(\Omega)}$$

where

$$\hat{H}(\Omega) \text{ is real}$$

(a) Find $\theta(\Omega)$ for $0 \leq \Omega \leq \pi$ when $h(n)$ satisfies the condition

$$h[n] = h[N - 1 - n]$$

(b) Find $\theta(\Omega)$ for $0 \leq \Omega \leq \pi$ when

$$h[n] = -h[N - 1 - n]$$

(Be careful; it may be necessary to treat cases N even and N odd separately.)

6.9. In the design of either analog or digital filters, we often approximate a specified magnitude characteristic without particular regard to the phase. For example, standard design techniques for lowpass and bandpass filters are typically derived from consideration of the magnitude characteristics only.

    In many filtering problems, one would ideally like the phase characteristics to be zero or linear. For causal filters it is impossible to have zero phase. However, for many digital filtering applications, it is not necessary that the unit sample response of the filter be zero for $n < 0$ if the processing is not to be carried out in real time.

    One technique commonly used in digital filtering when the data to be filtered are of finite duration and stored, for example, on a disc or magnetic tape, is to process the data forward and then backward through the same filter.

    Let $h[n]$ be the unit sample response of a causal filter with an arbitrary phase characteristic. Assume that $h[n]$ is real and denote its Fourier transform by $H(\Omega)$. Let $x[n]$ be the data that we want to filter. The filtering operation is performed as follows:

(a) *Method A:* Process $x[n]$ to get $s[n]$ as indicated in Fig. P6.9-1.

1. Determine the overall unit sample response $h_1[n]$ that relates $x[n]$ and $s[n]$, and show that it has a zero phase characteristic.
2. Determine $|H_1(\Omega)|$ and express it in terms of $|H(\Omega)|$ and $\measuredangle H(\Omega)$.

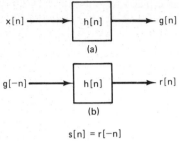

(a)

(b)

$s[n] = r[-n]$

(c)

Figure P6.9-1

(b) *Method B:* Process $x[n]$ through the filter $h[n]$ to get $g[n]$ (Fig. P6.9-2). Also process $x[n]$ backward through $h[n]$ to get $r[n]$. The output $y[n]$ is then taken as the sum of $g[n]$ and $r[-n]$. This composite set of operations can be represented by a filter, with input $x[n]$, output $y[n]$, and unit sample response $h_2[n]$.

1. Show that the composite filter $h_2[n]$ has a zero phase characteristic.
2. Determine $|H_2(\Omega)|$ and express it in terms of $|H(\Omega)|$ and $\measuredangle H(\Omega)$.

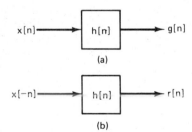

(a)

(b)

Figure P6.9-2

(c) Suppose that we are given a sequence of finite duration, on which we would like to perform a bandpass, zero-phase filtering operation. Furthermore, assume that we are given the bandpass filter $h[n]$, with frequency response as specified in Figure P6.9-3, which has the magnitude characteristic that we desire but linear phase.

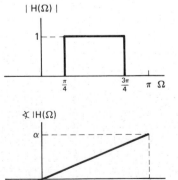

Figure P6.9-3

To achieve zero phase, we could use either method A or B. Determine and sketch $|H_1(\Omega)|$ and $|H_2(\Omega)|$. From these results, which method would you use to achieve the desired bandpass filtering operation? Explain why. More generally, if $h[n]$ has the desired magnitude but a nonlinear phase characteristic, which method is preferable to achieve a zero phase characteristic?

6.10. Figure P6.10 shows a system commonly used to obtain a highpass filter from a lowpass filter, and vice versa.

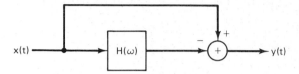

Figure P6.10

(a) If $H(\omega)$ is an ideal lowpass filter with cutoff frequency ω_{lp}, show that the overall system corresponds to an ideal highpass filter. Determine its cutoff frequency, and sketch its impulse response.

(b) If $H(\omega)$ is an ideal highpass filter with cutoff frequency ω_{hp}, show that the overall system corresponds to an ideal lowpass filter and determine its cutoff frequency.

(c) Now suppose that $H(\omega)$ corresponds to the RC lowpass filter discussed in Section 6.3.1. Determine and sketch the frequency response of the resulting highpass filter of Figure P6.10.

(d) Determine how the highpass filter obtained in part (b) compares with the highpass RC filter discussed in Section 6.3.1.

(e) If the interconnection of Figure P6.10 is applied to an ideal discrete-time lowpass filter, will the resulting system be an ideal discrete-time highpass filter?

6.11. In Problem 6.10 we considered a system (Figure P6.10) commonly used to obtain a highpass filter from a lowpass filter, and vice versa. In this problem we explore the system further and in particular consider a potential difficulty if the phase of $H(\omega)$ is not properly chosen.

(a) Referring to Figure P6.10, let us first assume that $H(\omega)$ is real and is as shown in Figure P6.11. In particular, then,

$$1 - \delta_1 < H(\omega) < 1 + \delta_1, \qquad 0 \le \omega \le \omega_1$$
$$-\delta_2 < H(\omega) < +\delta_2, \qquad \omega_2 < \omega$$

Determine and sketch the resulting frequency response for the overall system of Figure P6.10. Does the resulting system correspond to an approximation to a highpass filter?

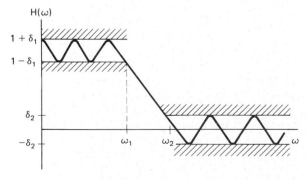

Figure P6.11

Now let $H(\omega)$ in Figure P6.10 be of the form

$$H(\omega) = H_1(\omega)e^{j\theta(\omega)} \tag{P6.11}$$

where $H_1(\omega)$ is identical to Figure P6.11 and $\theta(\omega)$ is an unspecified phase characteristic.

(b) With $H(\omega)$ in the more general form of eq. (P6.11), does it still correspond to an approximation to a lowpass filter?

(c) Without making any assumptions about $\theta(\omega)$, determine and sketch the tolerance limits on the magnitude of the frequency response of the overall system of Figure P6.10.

(d) If $H(\omega)$ in Figure P6.10 is an approximation to a lowpass filter with unspecified phase characteristics, will the overall system in Figure P6.10 necessarily correspond to an approximation to a highpass filter?

(e) Do the same conclusions reached in this problem apply in the use of the interconnection of Figure P6.10 for discrete-time filters?

6.12. Let $h_d[n]$ denote the unit sample response of an ideal desired system with frequency response $H_d(\Omega)$ and let $h[n]$ denote the unit sample response for an FIR system of length N and with frequency response $H(\Omega)$. In this problem we show that a rectangular window of length N samples applied to $h_d[n]$ will produce a unit sample response $h[n]$ such that the mean-square error

$$\epsilon^2 = \frac{1}{2\pi} \int_{-\pi}^{\pi} |H_d(\Omega) - H(\Omega)|^2 \, d\Omega$$

is minimized.

(a) The error function $E(\Omega) = H_d(\Omega) - H(\Omega)$ can be expressed as the power series

$$E(\Omega) = \sum_{n=-\infty}^{\infty} e[n]e^{-j\Omega n}$$

Find the coefficients $e[n]$ in terms of $h_d[n]$ and $h[n]$.

(b) Using Parseval's relation as stated in Section 5.5.8, express the mean-square error ϵ^2 in terms of the coefficients $e[n]$.

(c) Show that for a unit sample response $h[n]$ of length N samples, ϵ^2 is minimized when

$$h[n] = \begin{cases} h_d[n], & 0 \le n \le N - 1 \\ 0, & \text{otherwise} \end{cases}$$

That is, simple truncation gives the best mean-square approximation to a desired frequency response for a fixed value of N.

6.13. As indicated in Section 6.2, the concepts of frequency-selective filtering are often used to separate two signals that have been added. If the spectra of the two signals do not overlap, ideal frequency-selective filters are desirable. However, if the spectra overlap, it is often preferable to design the filter to have a gradual transition from passband to stopband. In this problem we explore one approach for determining the frequency response of a filter to be used for approximately separating signals with overlapping spectra. Let $x(t)$ denote a composite continuous-time signal, consisting of the sum of two signals $s(t) + w(t)$. As indicated in Figure P6.13-1, we would like to design an LTI filter to recover $s(t)$ from $x(t)$. The filter frequency response $H(\omega)$ is to be chosen so that in some sense $y(t)$ is a "good" approximation to $s(t)$.

Let us define a measure of the error between $y(t)$ and $s(t)$ at each frequency ω as

$$\epsilon(\omega) \triangleq |S(\omega) - Y(\omega)|^2$$

where $S(\omega)$ and $Y(\omega)$ are the Fourier transforms of $s(t)$ and $y(t)$, respectively.

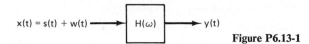

Figure P6.13-1

(a) Express $\epsilon(\omega)$ in terms of $S(\omega)$, $H(\omega)$, and $W(\omega)$, where $W(\omega)$ is the Fourier transform of $w(t)$.

(b) Let us restrict $H(\omega)$ to be real, so that $H(\omega) = H^*(\omega)$. By setting the derivative of $\epsilon(\omega)$ with respect to $H(\omega)$ to zero, determine $H(\omega)$ to minimize the error $\epsilon(\omega)$.

(c) Show that if the spectra of $S(\omega)$ and $W(\omega)$ are nonoverlapping, the result in part (b) reduces to an ideal frequency-selective filter.

(d) From your result in part (b), determine and sketch $H(\omega)$ if $S(\omega)$ and $W(\omega)$ are as shown in Figure P6.13-2.

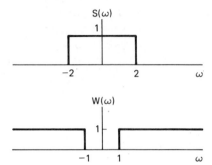

Figure P6.13-2

6.14. In Problem 6.13 we considered one specific criterion for the determination of the frequency response of a continuous-time filter for recovering a signal from the sum of two signals when their spectra overlap in frequency. Develop for the discrete-time case the result corresponding to that obtained in part (b) of Problem 6.13.

6.15. As defined in Section 6.2, an ideal bandpass filter is one that passes only a range of frequencies without any change in amplitude or phase. As shown in Figure P6.15-1, let the passband be

$$\omega_0 - \frac{W}{2} \le |\omega| \le \omega_0 + \frac{W}{2}$$

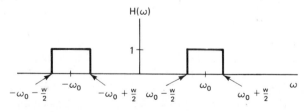

Figure P6.15-1

(a) What is the impulse response $h(t)$ of this filter?

(b) We can approximate an ideal bandpass filter by cascading a first-order lowpass and a first-order highpass section as shown in Figure P6.15-2. Sketch the Bode diagrams for each of the two filters $H_1(\omega)$ and $H_2(\omega)$.

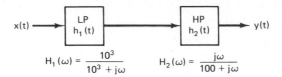

$$H_1(\omega) = \frac{10^3}{10^3 + j\omega} \qquad H_2(\omega) = \frac{j\omega}{100 + j\omega}$$

Figure P6.15-2

(c) Determine the Bode diagram for the overall bandpass filter in terms of your results from part (b).

6.16. In many situations we have available an analog or digital filter module, such as a basic hardware element or a computer subroutine. By using it repetitively or by combining identical modules it is possible to implement a new filter with improved passband or stopband characteristics. In this and the next problem we consider two procedures for doing this. Although the discussion is phrased in terms of discrete-time filters, much of the discussion applies directly to continuous-time filters as well.

Consider a lowpass filter with frequency response $H(\Omega)$ for which $|H(\Omega)|$ falls within the tolerance limits shown in Figure P6.16, that is,

$$1 - \delta_1 \leq |H(\Omega)| \leq 1 + \delta_1, \qquad 0 \leq \Omega \leq \Omega_1$$
$$0 \leq |H(\Omega)| \leq \delta_2, \qquad \Omega_2 \leq \Omega \leq \pi$$

A new filter with frequency response $G(\Omega)$ is formed by cascading two identical filters, both with frequency response $H(\Omega)$.

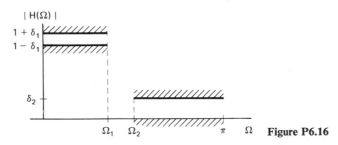

Figure P6.16

(a) Determine the tolerance limits on $|G(\Omega)|$.
(b) Assuming that $H(\Omega)$ is a good approximation to a lowpass filter so that $\delta_1 \ll 1$ and $\delta_2 \ll 1$, determine whether the passband ripple for $G(\Omega)$ is larger or smaller than the passband ripple for $H(\Omega)$. Also, determine whether the stopband ripple for $G(\Omega)$ is larger or smaller than the stopband ripple for $H(\Omega)$.
(c) If N identical filters with frequency response $H(\Omega)$ are cascaded to obtain a new frequency response $G(\Omega)$ and again assuming that $\delta_1 \ll 1$ and $\delta_2 \ll 1$, determine the approximate tolerance limits on $|G(\Omega)|$.

6.17. In Problem 6.16 we considered one method for using a basic filter module repetitively to implement a new filter with improved characteristics. In this problem we consider an alternative approach, proposed by J. W. Tukey in the book *Exploratory Data Analysis* (Reading, Mass.: Addison-Wesley Publishing Co., Inc., 1976). The procedure is shown in block-diagram form in Figure P6.17-1.

(a) Suppose that $H(\Omega)$ is real and has a passband ripple of $\pm\delta_1$ and a stopband ripple of $\pm\delta_2$ (i.e., it falls within the tolerance limits indicated in Figure P6.17-2). The frequency response $G(\Omega)$ of the overall system in Figure 6.17-1 falls within the tolerance limits indicated on Figure P6.17-3. Determine $A, B, C,$ and D in terms of δ_1 and δ_2.

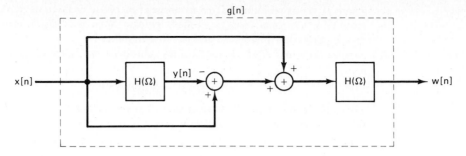

Figure P6.17-1

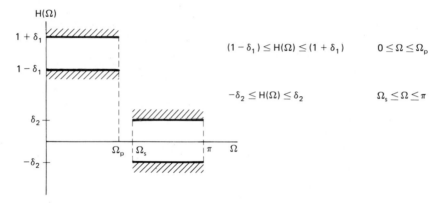

$$(1 - \delta_1) \le H(\Omega) \le (1 + \delta_1) \qquad 0 \le \Omega \le \Omega_p$$

$$-\delta_2 \le H(\Omega) \le \delta_2 \qquad \Omega_s \le \Omega \le \pi$$

Figure P6.17-2

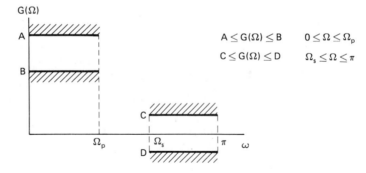

$$A \le G(\Omega) \le B \qquad 0 \le \Omega \le \Omega_p$$
$$C \le G(\Omega) \le D \qquad \Omega_s \le \Omega \le \pi$$

Figure P6.17-3

(b) If $\delta_1 \ll 1$ and $\delta_2 \ll 1$, what are the approximate passband and stopband ripple associated with $G(\Omega)$? Indicate in particular whether the passband ripple for $G(\Omega)$ is larger or smaller than the passband ripple for $H(\Omega)$. Also indicate whether the stopband ripple for $G(\Omega)$ is larger or smaller than the stopband ripple for $H(\Omega)$.

(c) In parts (a) and (b) we assumed that $H(\Omega)$ is real. Now consider $H(\Omega)$ to have the more general form

$$H(\Omega) = H_1(\Omega)e^{j\theta(\Omega)}$$

where $H_1(\Omega)$ is real and $\theta(\Omega)$ is an unspecified phase characteristic. If $|H(\Omega)|$ is a reasonable approximation to an ideal lowpass filter, will $|G(\Omega)|$ necessarily be a reasonable approximation to an ideal lowpass filter?

(d) Now assume that $H(\Omega)$ is an FIR linear-phase lowpass filter so that $H(\Omega)$ has the form

$$H(\Omega) = H_1(\Omega)e^{jM\Omega}$$

where $H_1(\Omega)$ is real and M is an integer. Show how to modify the system in Figure P6.17-1 so that the overall system will approximate a lowpass filter.

6.18. In Figure P6.18-1 we show the magnitude of the frequency response for an ideal continuous-time differentiator. A nonideal differentiator would have a frequency response that is some approximation to the frequency response of Figure P6.18-1.

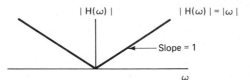

Figure P6.18-1

(a) Consider a nonideal differentiator with frequency response $G(\omega)$ for which $|G(\omega)|$ is constrained to be within $\pm 10\%$ of the magnitude of the frequency response of the ideal differentiator at all frequencies, that is,

$$-0.1\,|H(\omega)| \leq [|G(\omega)| - |H(\omega)|] \leq 0.1\,|H(\omega)|$$

Indicate with a sketch the region in a plot of $|G(\omega)|$ vs. ω, where $|G(\omega)|$ must be confined to meet this specification.

(b) The system in Figure P6.18-2, incorporating the use of an ideal delay of T seconds, is sometimes used to approximate a continuous-time differentiator. For $T = 10^{-2}$ seconds, determine the frequency range over which the magnitude of the frequency response of the system in Figure P6.18-2 is within $\pm 10\%$ of that for an ideal differentiator.

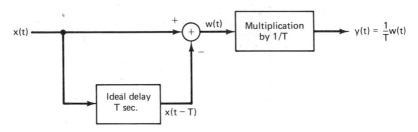

Figure P6.18-2

6.19. In Section 6.4.1, eq. (6.9), we considered a three-point moving-average filter of the form

$$y[n] = \tfrac{1}{3}\{x[n-1] + x[n] + x[n+1]\}$$

with the frequency response shown in Figure 6.22. A more general three-point symmetric moving average, referred to as a weighted moving average, is of the form

$$y[n] = b\{ax[n-1] + x[n] + ax[n+1]\} \quad\quad\quad \text{(P6.19)}$$

(a) Determine as a function of a and b the frequency response $H(\Omega)$ of the three-point moving average in eq. (P6.19).

(b) Determine the scaling factor b so that $H(\Omega)$ has unity gain at zero frequency.

(c) In many time-series analysis problems, a common choice for the coefficient a in the weighted moving average in eq. (P6.19) is $a = \frac{1}{2}$. Determine and sketch the frequency response of the resulting filter.

6.20. Consider a four-point moving-average discrete-time filter for which the difference equation is

$$y[n] = b_0 x[n] + b_1 x[n-1] + b_2 x[n-2] + b_3 x[n-3]$$

Determine and sketch the magnitude of the frequency response for each of the following cases:

(a) $b_0 = b_3 = 0, \quad b_1 = b_2$
(b) $b_1 = b_2 = 0, \quad b_0 = b_3$
(c) $b_0 = b_1 = b_2 = b_3$
(d) $b_0 = -b_1 = b_2 = -b_3$

6.21. In Figure P6.21-1 we show a discrete-time system consisting of a parallel combination of N LTI filters with impulse response $h_k[n]$, $k = 0, 1, \ldots, N-1$. For any k, $h_k[n]$ is related to $h_0[n]$ by the expression

$$h_k[n] = e^{j(2\pi nk/N)} h_0[n]$$

(a) If $h_0[n]$ is an ideal discrete-time lowpass filter with frequency response $H_0(\Omega)$ as shown in Figure P6.21-2, sketch the Fourier transforms of $h_1[n]$ and $h_{N-1}[n]$ for Ω in the range $-\pi < \Omega \leq +\pi$.

(b) Determine the value of the cutoff frequency Ω_c in Figure P6.21-2 in terms of N $(0 < \Omega_c \leq \pi)$ such that the system of Figure P6.21-1 is an identity system; that is, $y[n] = x[n]$ for all n and any input $x[n]$.

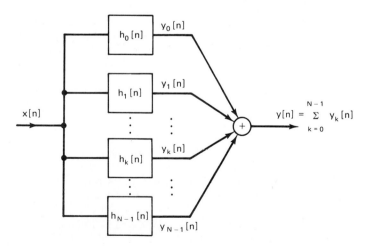

Figure P6.21-1

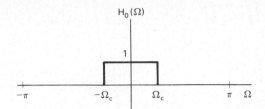

Figure P6.21-2

(c) Suppose that $h[n]$ is no longer restricted to be an ideal lowpass filter. If $h[n]$ denotes the impulse response of the entire system in Figure P6.21-1 with input $x[n]$ and output $y[n]$, then $h[n]$ can be expressed in the form

$$h[n] = r[n]h_0[n]$$

Determine and sketch $r[n]$.

(d) From your result of part (c), determine a necessary and sufficient condition on $h_0[n]$ to ensure that the overall system will be an identity system (i.e., such that for any input $x[n]$, the output $y[n]$ will be identical to $x[n]$). Your answer should not contain any sums.

6.22. In many filtering applications it is often undesirable for the step response of a filter to overshoot its final value. In picture processing, for example, the overshoot in the step response of a linear filter may produce flare, or intensity increase, at sharp boundaries. It is possible, however, to eliminate overshoot by requiring that the impulse response of the filter be positive for all time.

(a) Show that if $h(t)$, the impulse response of a continuous-time LTI filter, is always greater than or equal to zero [$h(t) \geq 0$], the step response of the filter is a monotonically nondecreasing function and therefore will not have overshoot.

(b) Show that if $h[n]$, the impulse response of a discrete-time LTI filter, is always greater than or equal to zero, the step response of the filter is a monotonically nondecreasing function and therefore will not have overshoot.

6.23. Using a specific filter design procedure, a nonideal continuous-time lowpass filter with frequency response $H_0(\omega)$, impulse response $h_0(t)$, and step response $s_0(t)$ has been designed. The cutoff frequency of the filter is at $\omega = 2\pi \times 10^2$ rad/sec and the step response rise time τ_r, defined as the time required for the step response to go from 10% of its final value to 90% of its final value, is $\tau_r = 10^{-2}$ second. From this design we can obtain a new filter with an arbitrary cutoff frequency ω_c by the use of frequency scaling. The frequency response of the resulting filter $H_{lp}(\omega)$ is then of the form

$$H_{lp}(\omega) = H_0(a\omega)$$

where a is an appropriate scale factor.

(a) Determine the scale factor a so that $H_{lp}(\omega)$ has a cutoff frequency of ω_c.

(b) Determine in terms of ω_c and $h_0(t)$ the impulse response $h_{lp}(t)$ of the new filter.

(c) Determine in terms of ω_c and $s_0(t)$ the step response $s_{lp}(t)$ of the new filter.

(d) Determine and sketch the rise time of the new filter as a function of its cutoff frequency ω_c.

This is one illustration of the trade-off between time-domain and frequency-domain characteristics. In particular as the cutoff frequency decreases, the rise time tends to increase.

(e) Is your result in part (d) dependent on the specific definition chosen for the rise time? (A different definition of rise time was considered in Problem 4.35.)

6.24. In the design of digital filters we often would like to choose a filter with a specified magnitude characteristic that has the shortest time duration. That is, the impulse response, which is the Fourier transform of the complex frequency spectrum, should be as narrow as possible. Assuming that $h[n]$ is real, we wish to show that if the phase $\theta(\Omega)$ associated with the frequency response $H(\Omega)$ is zero, the duration of the impulse response is minimum. Let the frequency response be expressed as

$$H(\Omega) = |H(\Omega)| e^{j\theta(\Omega)}$$

and let us consider the quantity D to be a measure of the duration of the associated impulse response $h[n]$, where

$$D = \sum_{n=-\infty}^{+\infty} n^2 h^2[n] = \sum_{n=-\infty}^{+\infty} (nh[n])^2$$

(a) Using the derivative property of the Fourier transform and Parseval's relation, express D in terms of $H(\Omega)$.

(b) By expressing $H(\Omega)$ in terms of its magnitude $|H(\Omega)|$ and phase $\theta(\Omega)$, use your result from part (a) to show that D is minimized when the phase $\theta(\Omega) = 0$.

6.25. As discussed in Section 6.5, the magnitude squared of the frequency response of the class of continuous-time Butterworth filters is

$$|B(\omega)|^2 = \frac{1}{1 + (\omega/\omega_c)^{2N}}$$

Let us define the passband edge frequency ω_p as the frequency below which $|B(\omega)|^2$ is greater than one-half of its value at $\omega = 0$; that is,

$$|B(\omega)|^2 \geq \tfrac{1}{2}|B(0)|^2, \qquad |\omega| < \omega_p$$

Let us define the stopband edge frequency ω_s as the frequency above which $|B(\omega)|^2$ is less than 10^{-2} times its value at $\omega = 0$, that is,

$$|B(\omega)|^2 \leq 10^{-2}|B(0)|^2, \qquad |\omega| > \omega_s$$

The transition band is then the frequency range between ω_p and ω_s. The ratio ω_s/ω_p is referred to as the transition ratio.

For fixed ω_p, and making reasonable approximations, determine and sketch the transition ratio as a function of N for the class of Butterworth filters.

6.26. The class of continuous-time Butterworth lowpass filters has the squared-magnitude function given in eq. (6.22):

$$|B(\omega)|^2 = \frac{1}{1 + (\omega/\omega_c)^{2N}}$$

In Figure 6.30 the Bode plot for $|B(\omega)|^2$ for $N = 1, 2, 4$, and 8 is shown. The class of discrete-time Butterworth lowpass filters has the squared-magnitude function given in eq. (6.23):

$$|B(\Omega)|^2 = \frac{1}{1 + \left(\dfrac{\tan{(\Omega/2)}}{\tan{(\Omega_c/2)}}\right)^{2N}}$$

In the passband an ideal lowpass filter is completely flat. As a consequence, all derivatives of the gain of an ideal lowpass filter around $\omega = 0$ are zero.

(a) Show that for the Nth-order continuous-time Butterworth filter frequency response in eq. (6.22), the first $(2N - 1)$ derivatives of $|B(\omega)|^2$ are zero at $\omega = 0$. This property of an Nth-order transfer function is referred to as *maximally flat*.

(b) Determine whether or not the Nth-order discrete-time Butterworth filter is maximally flat.

6.27. The magnitude squared of the transfer functions of a continuous-time Butterworth filter and a discrete-time Butterworth filter are given in eqs. (6.22) and (6.23).

(a) Determine the parameters ω_c and N for the *lowest*-order continuous-time Butterworth filter for which the passband frequency response is constant to within 0.75 dB for frequencies below $\omega = 2\pi \times 100$ and for which the stopband attenuation is at least 20 dB for frequencies greater than $\omega = 2\pi \times 110$. *Note that N must be an integer.*

(b) Determine the parameters Ω_c and N for the *lowest*-order discrete-time Butterworth filter for which the passband frequency response is constant to within 0.75 dB for frequencies below $\Omega = 0.26\pi$ and for which the stopband attenuation is at least 20 dB in the frequency range $0.4\pi \leq \Omega \leq \pi$.

6.28. For discrete-time filters the lowpass Butterworth filter is characterized by a squared-magnitude characteristic

$$|B(\Omega)|^2 = \frac{1}{1 + \left[\dfrac{\tan{(\Omega/2)}}{\tan{(\Omega_c/2)}}\right]^{2N}}$$

where Ω_c is the cutoff frequency (which we shall take to be $\pi/2$) and N is the order of the filter (which we shall consider as $N = 1$). Thus, we have

$$|B(\Omega)|^2 = \frac{1}{1 + \tan^2{(\Omega/2)}}$$

(a) Using the trigonometric identities, show that $|B(\Omega)|^2 = \cos^2{(\Omega/2)}$.

(b) Let $B(\Omega) = a \cos{(\Omega/2)}$. For what complex values of a is $|B(\Omega)|^2$ the same as in part (a)?

(c) Show that $B(\Omega)$ from part (b) is the transfer function corresponding to a difference equation of the form

$$y[n] = \alpha x[n] + \beta x[n - \gamma]$$

In particular, determine α, β, and γ.

6.29. In Figure 6.32 are shown typical circuits for implementing a second-order and a third-order Butterworth filter. Determine the frequency response of each of these circuits and demonstrate that they in fact correspond to the appropriate-order Butterworth filters.

6.30. In Figure P6.30 are shown digital networks for implementing a second-order and a third-order Butterworth filter.

(a) Determine the difference equation associated with each of these two networks.

(b) Determine the frequency response of each and demonstrate that they correspond to a second-order and a third-order Butterworth filter.

6.31. Consider a discrete-time signal $x[n]$ that is composed of the sum of two signals $s[n]$ and $g[n]$, that is, $x[n] = s[n] + g[n]$, where

$$s[n] = A \cos\left(\frac{2\pi n}{6} + \frac{\pi}{8}\right)$$

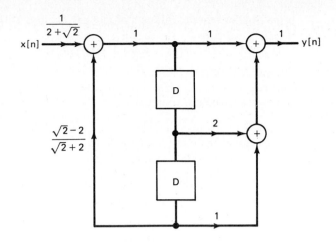

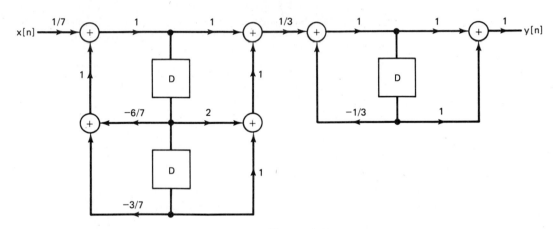

Figure P6.30

and

$$g[n] = B(-1)^n$$

Let us assume that the sinusoidal tone $s[n]$ is the desired component (signal) and $g[n]$ is the undesired component (noise). Since $s[n]$ and $g[n]$ are periodic signals, we can define the signal-to-noise ratio R_x for $x[n]$ as the average power P_s in $s[n]$ divided by the average power P_g in $g[n]$. The average power for a periodic discrete-time signal $f[n]$ with period N is defined as

$$P_f = \frac{1}{N} \sum_{n=0}^{N-1} f^2[n]$$

If the Fourier series coefficients of $f[n]$ are F_k, then Parseval's relation gives

$$P_f = \frac{1}{N}\sum_{n=0}^{N-1} f^2[n] = \sum_{k=0}^{N-1} |F_k|^2$$

(a) Determine $R_x = P_s/P_g$.

(b) To increase the signal-to-noise ratio, we shall put $x[n]$ through an LTI filter characterized by the difference equation

$$y[n] - ay[n-1] = x[n]$$

Let $y_s[n]$ denote the output due to $s[n]$, that is,

$$y_s[n] - ay_s[n-1] = s[n]$$

Let $y_g[n]$ denote the output due to $g[n]$, that is,

$$y_g[n] - ay_g[n-1] = g[n]$$

Thus, when $x[n] = s[n] + g[n]$, we have $y[n] = y_s[n] + y_g[n]$. Determine $R_y = P_{y_s}/P_{y_g}$ in terms of R_x and the filter parameter a.

(c) Show that the output signal-to-noise ratio R_y is greater than the input signal-to-noise ratio R_x for any value of a in the range $0 < a < 1$.

6.32. In this problem we explore some of the filtering issues involved in the commercial version of a typical system that is used in most modern cassette tape decks to reduce noise. The primary source of noise is the high-frequency hiss in the tape playback process, which, in some part, is due to the friction between the tape and the playback head. Let us assume that the noise hiss which is added to the signal upon playback has the spectrum of Figure P6.32-1 when measured in decibels, with 0 dB equal to the signal level at 100 Hz. The signal spectrum $S(\omega)$ has the shape shown in Figure P6.32-2.

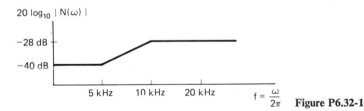

Figure P6.32-1

Figure P6.32-2

The system that we analyze has a filter $H_1(\omega)$ that conditions the signal $s(t)$ before it is recorded. Upon playback, the hiss $n(t)$ is added to the signal. The system is represented schematically in Figure P6.32-3.

We would like the overall system to have a signal-to-noise ratio of 40 dB over the frequency range 50 Hz $< \omega/2\pi < 20$ kHz.

(a) Determine the transfer characteristic of the filter $H_1(\omega)$. Sketch a Bode plot of $H_1(\omega)$.

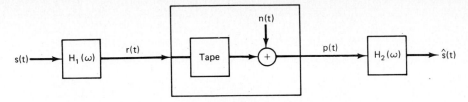

Figure P6.32-3

(b) If we were to listen to the signal $p(t)$, assuming that the playback process does nothing more than add hiss to the signal, how do you think it would sound?

(c) What should the Bode plot and transfer characteristic of the filter $H_2(\omega)$ be such that the signal $\hat{s}(t)$ sounds similar to $s(t)$?

(d) Using the techniques discussed in Chapters 3 and 4, determine the differential equation to realize $H_2(\omega)$.

In a wide variety of engineering systems a concept referred to as modulation plays a central role. In general, a modulation system is one in which one signal is used to control some parameter of another signal. In Sections 4.8 and 5.7, in the context of properties of the Fourier transform, we introduced one example of modulation, commonly referred to as *amplitude modulation*, whereby one signal is used to vary the amplitude of a second signal. In this chapter we discuss the principle of amplitude modulation in considerably more detail and as we will see, the modulation property developed in Sections 4.8 and 5.7 will play a central role. One of the particularly important classes of applications of amplitude modulation is in communications systems. Any specific communications channel typically has associated with it a particular frequency range over which it is best suited for signal transmission. For example, the atmosphere will rapidly attenuate signals in the audible frequency range (10 Hz to 20 kHz), whereas it will propagate signals at a higher-frequency range over longer distances. Thus, in transmitting audio signals such as speech or music over a communications channel that relies on propagation through the atmosphere, a modulation system is used to embed the signal to be transmitted in a higher-frequency carrier signal. One common modulation system for this purpose is *sinusoidal amplitude modulation*, in which the information-bearing signal, speech or music for example, is used to vary the amplitude of a sinusoidal carrier signal whose frequency is in the appropriate range. As we will see, through the use of modulation systems it is also possible to transmit simultaneously more than one signal with overlapping spectra over the same channel, through a concept referred to as *multiplexing*.

Modulation

Another application of the principles of amplitude modulation is a process whereby a train of rectangular pulses with equal spacing and amplitude is multiplied by the information-bearing signal. This process is referred to as *pulse amplitude modulation*. In addition to being an important modulation technique in communications systems, it is closely related to the concept of sampling, whereby under certain conditions a signal can be represented by equally spaced time samples. Sampling, which is the focus of Chapter 8, provides an important bridge between continuous-time and discrete-time signals and systems.

Our primary goal in this chapter is to introduce the basic concepts of amplitude modulation, using the properties and insights developed in Chapters 4, 5, and 6. As we have seen throughout this book, there are strong conceptual and analytical similarities between the properties of continuous-time and discrete-time signals and systems, and throughout this chapter we continue to emphasize these relationships. Major applications of amplitude-modulation systems are in the realm of continuous-time communications systems and in the conversion of continuous-time to discrete-time signals. As with any concept that is closely tied to a wide variety of important applications, there are a large number of detailed issues to be considered and there are many excellent books on the subject.† As was the case in Chapter 6 with regard to filtering, it is our intent to introduce the concept of amplitude modulation. While we will not be developing the topic in depth, it should be stressed that the fundamentals of signal and system analysis developed thus far in this text provide all the basic tools for a detailed understanding of modulation systems.

Although the principal focus of the chapter is amplitude modulation, there are other important types of modulation, such as sinusoidal frequency or phase modulation, in which the information-bearing signal is used to vary either the frequency of a sinusoidal carrier in the neighborhood of a center frequency, or to vary its phase. Frequency modulation is not as straightforward to analyze as is amplitude modulation. However there are a number of special cases that are relatively easy to develop and that serve to illustrate many of the basic principles of this form of modulation and its relation to amplitude modulation. These cases are considered at the end of the chapter.

We begin with the analysis and applications of sinusoidal amplitude modulation for continuous-time signals, followed by a discussion of pulse amplitude modulation, both of which play a major role in modern communications systems. Initially the discussion is carried out in the context of continuous-time signals. However, all of the basic results have a direct counterpart for discrete-time signals, and in Section 7.5 these relations and some of their applications are discussed.

---

†See, for example, Harold S. Black, *Modulation Theory* (New York: D. Van Nostrand Company, 1953); Mischa Schwartz, *Information Transmission, Modulation and Noise*, 3rd ed. (New York: McGraw-Hill Book Company, 1980); J. S. Spilker, *Digital Communications by Satellite* (Englewood Cliffs, N.J.: Prentice-Hall, Inc., 1977); and Simon Haykin, *Communication Systems* (New York: John Wiley and Sons, 1978).

7.1 CONTINUOUS-TIME SINUSOIDAL AMPLITUDE MODULATION

Many systems rely on the concept of sinusoidal amplitude modulation, in which a complex exponential or sinusoidal signal $c(t)$ has its amplitude multiplied (modulated) by the information-bearing signal $x(t)$. The signal $x(t)$ is typically referred to as the *modulating signal* and the signal $c(t)$ as the *carrier signal*. A system for sinusoidal amplitude modulation is depicted in Figure 7.1.

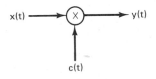

x(t) ⟶ × ⟶ y(t)

c(t)

Figure 7.1 Amplitude-modulation system.

There are two common forms of sinusoidal amplitude modulation, one in which the carrier signal $c(t)$ is a complex exponential of the form

$$c(t) = e^{j(\omega_c t + \theta_c)} \tag{7.1}$$

and the second in which the carrier signal is sinusoidal of the form

$$c(t) = \cos(\omega_c t + \theta_c) \tag{7.2}$$

In both cases, the frequency ω_c is referred to as the *carrier frequency*. Let us consider first the case of a complex exponential carrier and for convenience let us choose $\theta_c = 0$ so that the modulated signal $y(t)$ is

$$y(t) = x(t)e^{j\omega_c t} \tag{7.3}$$

From the modulation property (Section 4.8), and with $X(\omega)$, $Y(\omega)$, and $C(\omega)$ denoting the Fourier transforms of $x(t)$, $y(t)$, and $c(t)$, respectively,

$$Y(\omega) = \frac{1}{2\pi} X(\omega) * C(\omega) \tag{7.4}$$

For $c(t)$ a complex exponential as given in eq. (7.1),

$$C(\omega) = 2\pi\delta(\omega - \omega_c) \tag{7.5}$$

and hence, from Section 4.5.2,

$$Y(\omega) = X(\omega - \omega_c) \tag{7.6}$$

Thus, the spectrum of the modulated output $y(t)$ is simply that of the input, shifted in frequency by an amount equal to the carrier frequency ω_c. For example, with $X(\omega)$ bandlimited with highest frequency ω_M (and bandwidth $2\omega_M$) as depicted in Figure 7.2(a), the output spectrum $Y(\omega)$ is that shown in Figure 7.2(c).

From eq. (7.3) it is clear that $x(t)$ can be recovered from the modulated signal $y(t)$ by multiplying by the complex exponential $e^{-j\omega_c t}$, that is,

$$x(t) = y(t)e^{-j\omega_c t} \tag{7.7}$$

In the frequency domain this has the effect of shifting the spectrum of the modulated

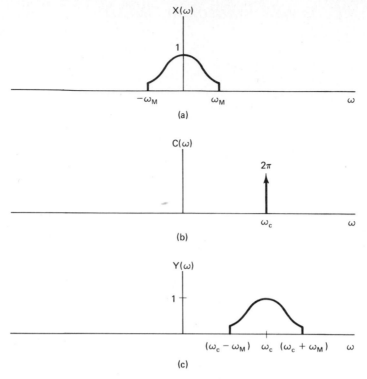

(a)

(b)

(c)

Figure 7.2 Effect in the frequency domain of amplitude modulation with a complex exponential carrier: (a) spectrum of modulating signal $x(t)$; (b) spectrum of carrier $c(t) = e^{j\omega_c t}$; (c) spectrum of amplitude-modulated signal.

signal back to its original position on the frequency axis. The process of recovering the original signal from the modulated signal is referred to as *demodulation*.

Since $e^{j\omega_c t}$ is a complex signal, eq. (7.3) can be rewritten as

$$y(t) = x(t) \cos \omega_c t + j x(t) \sin \omega_c t \qquad (7.8)$$

A representation of the system of Figure 7.1 with $x(t)$ real and $c(t)$ a complex exponential signal in the form of eq. (7.1) is shown in Figure 7.3. This then requires two separate multipliers and two sinusoidal carrier signals which have a phase

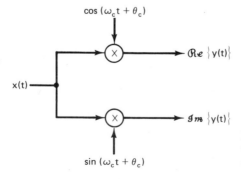

Figure 7.3 Implementation of amplitude modulation with a complex exponential carrier: $c(t) = e^{j(\omega_c t + \theta_c)}$.

difference of $\pi/2$. In Section 7.3 we study some cases in which there are particular advantages to using a complex exponential carrier signal and retaining or transmitting both the real and imaginary parts of $y(t)$, as generated by the system of Figure 7.3. However, in many situations it is often simpler and equally effective to use a sinusoidal carrier of the form of eq. (7.2). In effect, this corresponds to retaining only the real or imaginary part of the output of Figure 7.3. The system using a sinusoidal carrier is depicted in Figure 7.4.

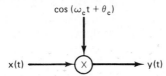

Figure 7.4 Amplitude modulation with a sinusoidal carrier.

The effect of sinusoidal amplitude modulation using a sinusoidal carrier in the form of eq. (7.2) can be analyzed in a manner identical to that used in the preceding case. Again, for convenience we choose $\theta_c = 0$. In this case, the spectrum of the carrier signal is

$$C(\omega) = \pi[\delta(\omega - \omega_c) + \delta(\omega + \omega_c)] \tag{7.9}$$

and thus from eq. (7.4),

$$Y(\omega) = \tfrac{1}{2}[X(\omega - \omega_c) + X(\omega + \omega_c)] \tag{7.10}$$

With $X(\omega)$ depicted in Figure 7.5(a), the spectrum of $y(t)$ is that shown in Figure

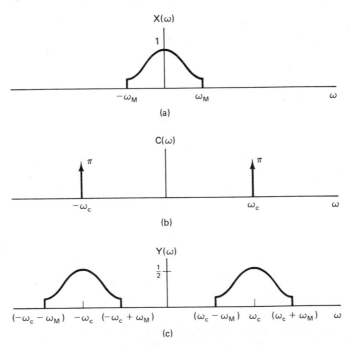

Figure 7.5 Effect in the frequency domain of amplitude modulation with a sinusoidal carrier: (a) spectrum of modulating signal $x(t)$; (b) spectrum of carrier $c(t) = \cos \omega_c t$; (c) spectrum of amplitude-modulated signal.

7.5(c). Note that there is now a replication of the spectrum of the original signal, centered around both $+\omega_c$ and $-\omega_c$, compared with the previous case of a complex exponential carrier, for which a replication of the spectrum of the original signal is centered only around ω_c. Also, with a complex exponential carrier, the basic shape of the spectrum of $x(t)$ is preserved in $Y(\omega)$ for any choice of carrier frequency ω_c. Thus with a complex exponential carrier $x(t)$ can always be recovered from $y(t)$ for any choice of ω_c by shifting the spectrum back to its original location. With a sinusoidal carrier, on the other hand, as we see from Figure 7.5, if $\omega_c < \omega_M$, then there will be an overlap between the two replications of $X(\omega)$. For example, Figure 7.6 depicts $Y(\omega)$ for $\omega_c = \omega_M/2$. Clearly, the spectrum of $x(t)$ is no longer replicated in $Y(\omega)$, and thus it may no longer be possible to recover $x(t)$ from $y(t)$.

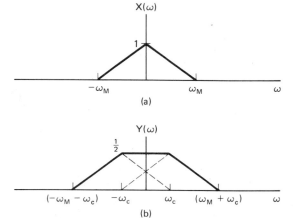

Figure 7.6 Sinusoidal amplitude modulation with carrier $\cos \omega_c t$ for which $\omega_c = \omega_M/2$: (a) spectrum of modulating signal; (b) spectrum of modulated signal.

Assuming that $\omega_c > \omega_M$, demodulation of a signal that was modulated with a sinusoidal carrier is relatively straightforward. Specifically, consider

$$y(t) = x(t) \cos \omega_c t \tag{7.11}$$

As was suggested in Example 4.23, the original signal can be recovered by modulating $y(t)$ with the same sinusoidal carrier and applying a lowpass filter to the result. To see this, consider

$$w(t) = y(t) \cos \omega_c t \tag{7.12}$$

Figure 7.7 shows the spectra of $y(t)$ and $w(t)$, and we observe that $x(t)$ can be recovered from $w(t)$ by applying an ideal lowpass filter with a gain of 2 and a cutoff frequency which is greater than ω_M and less than $(2\omega_c - \omega_M)$. The frequency response of the lowpass filter is indicated by the dashed line in Figure 7.7(c).

The basis for using eq. (7.12) and a lowpass filter to demodulate $y(t)$ can also be seen algebraically. From eqs. (7.11) and (7.12) it follows that

$$w(t) = x(t) \cos^2 \omega_c t$$

or, using the trigonometric identity

$$\cos^2 \omega_c t = \tfrac{1}{2} + \tfrac{1}{2} \cos 2\omega_c t$$

we can rewrite $w(t)$ as

$$w(t) = \tfrac{1}{2}x(t) + \tfrac{1}{2}x(t) \cos 2\omega_c t \tag{7.13}$$

Modulation   Chap. 7

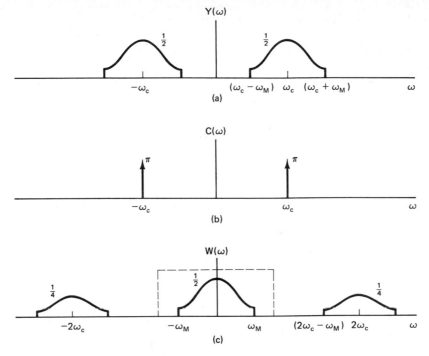

Figure 7.7 Demodulation of an amplitude-modulated signal with a sinusoidal carrier: (a) spectrum of modulated signal; (b) spectrum of carrier signal; (c) spectrum of modulated signal multiplied by the carrier. The dashed line indicates the frequency response of a lowpass filter used to extract the demodulated signal.

Thus, $w(t)$ consists of the sum of two terms, one being one-half the original signal and the second one-half the original signal modulated with a sinusoidal carrier at twice the original carrier frequency ω_c. Both of these terms are apparent in the spectrum shown in Figure 7.7(c). Applying the lowpass filter to $w(t)$ corresponds to retaining the first term on the right-hand side of eq. (7.13) and eliminating the second term.

The overall system for amplitude modulation and demodulation using a complex exponential carrier is depicted in Figure 7.8 and the overall system for modulation and demodulation using a sinusoidal carrier is depicted in Figure 7.9. In these figures we have indicated the more general case in which for both the complex exponential and sinusoidal carriers a carrier phase θ_c is included. The modification of the analysis above to include θ_c is straightforward and is considered in Problem 7.1.

In the systems of Figures 7.8 and 7.9, the demodulating signal is assumed to be synchronized in phase with the modulating signal. Let us consider the demodulated result for both systems when the modulator and demodulator are not synchronized. For the case of the complex exponential carrier, with θ_c denoting the phase of the modulating carrier and ϕ_c the phase of the demodulating carrier,

$$y(t) = e^{j(\omega_c t + \theta_c)} x(t) \tag{7.14a}$$

$$w(t) = e^{-j(\omega_c t + \phi_c)} y(t) \tag{7.14b}$$

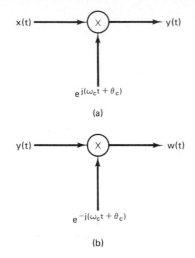

(a)

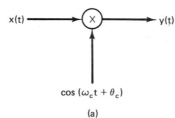

(b)

Figure 7.8 System for amplitude modulation and demodulation using a complex exponential carrier: (a) modulation; (b) demodulation.

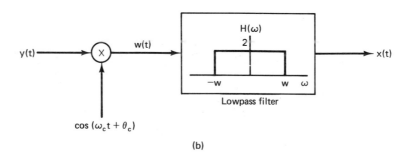

Figure 7.9 Amplitude modulation and demodulation with a sinusoidal carrier: (a) modulation system; (b) demodulation system. The lowpass filter cutoff frequency w is greater than ω_M and less than $(2\omega_C - \omega_M)$.

or

$$w(t) = e^{j(\theta_c - \phi_c)}x(t) \tag{7.15}$$

Thus, if $\theta_c \neq \phi_c$, $w(t)$ will have a complex amplitude factor. For the particular case in which $x(t)$ is positive, $x(t) = |w(t)|$ and thus $x(t)$ can be recovered by taking the absolute value of the demodulated signal.

For the sinusoidal carrier, again let θ_c and ϕ_c denote the phase of the modulating and demodulating carrier, respectively, as indicated in Figure 7.10. The input $w(t)$ to the lowpass filter is now

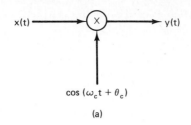

(a)

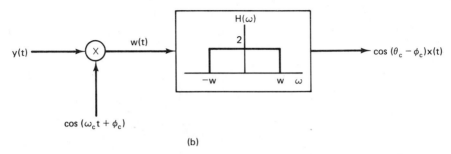

(b)

Figure 7.10 Sinusoidal amplitude modulation and demodulation system for which the carrier signals in the modulator and demodulator are not synchronized: (a) modulator; (b) demodulator.

$$w(t) = x(t) \cos(\omega_c t + \theta_c) \cos(\omega_c t + \phi_c) \tag{7.16}$$

or, using the trigonometric identity

$$\cos(\omega_c t + \theta_c) \cos(\omega_c t + \phi_c) = \tfrac{1}{2}\cos(\theta_c - \phi_c) + \tfrac{1}{2}\cos(2\omega_c t + \theta_c + \phi_b) \tag{7.17}$$

we have

$$w(t) = \tfrac{1}{2}\cos(\theta_c - \phi_c)x(t) + \tfrac{1}{2}x(t)\cos(2\omega_c t + \theta_c + \phi_c) \tag{7.18}$$

and the output of the lowpass filter is then $x(t)$ multiplied by the amplitude factor $\cos(\theta_c - \phi_c)$. If the oscillators in the modulator and demodulator are in phase, $\theta_c = \phi_c$ and the output of the lowpass filter is $x(t)$. On the other hand, if these oscillators have a phase difference of $\pi/2$, the output will be zero. In general, for maximum output signal the oscillators should be in phase. Of even more importance, the phase relation between these two oscillators must be maintained over time so that the amplitude factor $\cos(\theta_c - \phi_c)$ does not vary. This requires careful synchronization between the modulator and demodulator, which is often difficult, particularly when they are geographically separated as is typical in a communications system. The corresponding effects of and the need for synchronization not only between the phase of the modulator and demodulator but between the frequencies of the carrier signals present in both is explored in detail in Problem 7.3.

7.1.1 Asynchronous Demodulation

Demodulation using a carrier signal synchronized in phase with the modulator carrier is referred to as *synchronous demodulation*. In many systems that employ sinusoidal amplitude modulation an alternative demodulation procedure, referred to as *asynchronous demodulation*, is commonly used which avoids the need for synchronization between the modulator and demodulator. Assume that $x(t)$ is always positive and that the carrier frequency ω_c is much higher than ω_M, the highest frequency in the modulating signal. The modulated signal $y(t)$ will then have the general form illustrated in Figure 7.11. In particular, the *envelope* of $y(t)$, that is, a smooth curve con-

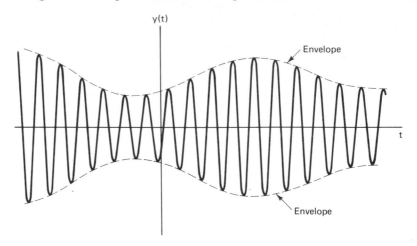

Figure 7.11 Amplitude-modulated signal for which the modulating signal is positive. The dashed curve represents the envelope of the modulated signal.

necting the peaks in $y(t)$, would appear to be a reasonable approximation to $x(t)$. Thus, $x(t)$ could be approximately recovered through the use of a system that tracks these peaks to extract the envelope. Such a system is referred to as an *envelope detector*. One example of a simple circuit that acts as an envelope detector is shown in Figure 7.12(a). This circuit is generally followed by a lowpass filter to reduce the variations at the carrier frequency which are evident in Figure 7.12(b) and which will generally be present in the output of an envelope detector of the type indicated in Figure 7.12(a).

    The two basic assumptions required for asynchronous demodulation are that $x(t)$ is positive and that $x(t)$ varies slowly compared to ω_c, so that the envelope is easily tracked. This second condition is satisfied, for example, in audio transmission over a radio-frequency (RF) channel where the highest frequency present in $x(t)$ is typically 15 to 20 kHz and $\omega_c/2\pi$ is in the range 500 to 2000 kHz. The first condition, that $x(t)$ be positive, can be satisfied by simply adding an appropriate constant value to $x(t)$ or equivalently by a simple change in the modulator, as shown in Figure 7.13. The output of the envelope detector then approximates $x(t) + A$, from which $x(t)$ is easily obtained.

    To use the envelope detector for demodulation, we require that A be sufficiently large so that $x(t) + A$ is positive. Let K denote the maximum amplitude of $x(t)$, that is, $|x(t)| \leq K$. For $x(t) + A$ to be positive we require that $A > K$. The ratio K/A is commonly referred to as the *modulation index m*. Expressed in percent, this is referred to as the *percent modulation*. An illustration of the output of the modulator of Figure 7.13 for $x(t)$ sinusoidal and two values of m, specifically $m = 0.5$ (50% modulation) and $m = 1$ (100% modulation), is shown in Figure 7.14.

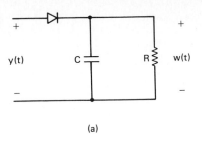

(a)

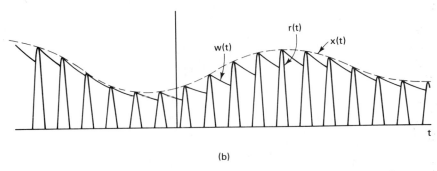

(b)

Figure 7.12 Demodulation by envelope detection: (a) circuit for envelope detection using half-wave rectification; (b) waveforms associated with the envelope detector in (a): $r(t)$ is the half-wave rectified signal, $x(t)$ is the true envelope, and $w(t)$ is the envelope obtained from the circuit in (a). The relationship between $x(t)$ and $w(t)$ has been exaggerated in (b) for purposes of illustration. In a practical asynchronous demodulation system $w(t)$ would typically be a much closer approximation to $x(t)$ than depicted here.

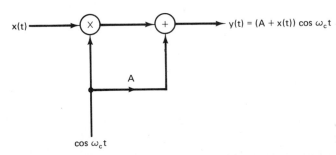

Figure 7.13 Modulator for an asynchronous modulation/demodulation system.

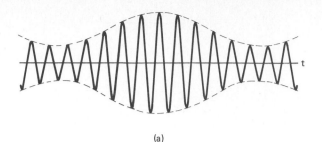

(a)

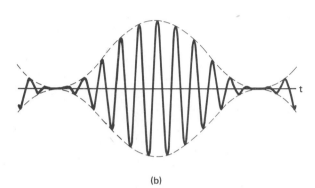

(b)

Figure 7.14 Output of the amplitude-modulation system of Figure 7.13: (a) modulation index $m = 0.5$; (b) modulation index $m = 1.0$.

In Figure 7.15 we show a comparison of the spectra associated with the modulated signal when synchronous demodulation and when asynchronous demodulation are to be used. We note in particular that the output of the modulator for the asynchronous system in Figure 7.13 has the additional component $A \cos \omega_c t$, which is not present or necessary in the synchronous system. This is represented in the spectrum of Figure 7.15(c) by the presence of impulses at $+\omega_c$ and $-\omega_c$. For a fixed maximum modulating signal amplitude K, as A is decreased the modulation index m increases and the relative amount of carrier present in the modulated output decreases. Since the carrier component in the output contains no information, its presence represents an inefficiency, for example in the amount of power required to transmit the modulated signal, and thus in one sense it is desirable to make m as large as possible. On the other hand, the ability of a simple envelope detector such as that in Figure 7.12 to follow the envelope and thus extract $x(t)$ improves as the modulation index decreases. Thus, there is a trade-off between the efficiency of the system in terms of the power in the output of the modulator, and the "quality" of the demodulated signal.

There are a number of advantages and disadvantages to the asynchronous modulation-demodulation system of Figures 7.12 and 7.13 compared with the synchronous system of Figure 7.9. The synchronous system requires a more sophisticated demodulator because the oscillator in the demodulator must be synchronized

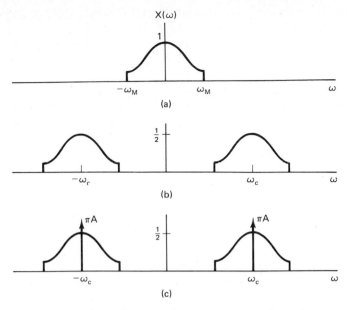

Figure 7.15 Comparison of spectra for synchronous and asynchronous sinusoidal amplitude-modulation systems: (a) spectrum of modulating signal; (b) spectrum of $x(t) \cos \omega_c t$ representing modulated signal in a synchronous system; (c) spectrum of $[x(t) + A] \cos \omega_c t$ representing modulated signal in an asynchronous system.

with the oscillator in the modulator both in phase and in frequency. On the other hand, the asynchronous modulator in general requires more output power than the synchronous modulator since, for the envelope detector to operate properly, the envelope must be positive, or equivalently there must be a carrier component present in the transmitted signal. This is often preferable in cases such as that associated with public radio broadcasting, in which it is desirable to mass-produce large numbers of receivers (demodulators) at moderate cost. The additional cost in transmitted power is then offset by the savings in receiver expense. On the other hand, in situations in which transmitter power requirements are at a premium, as in satellite communications, the cost of implementing a more sophisticated synchronous receiver is warranted.

7.2 SOME APPLICATIONS OF SINUSOIDAL AMPLITUDE MODULATION

7.2.1 Frequency-Selective Filtering with Variable Center Frequency

One important application of modulation is in the implementation of bandpass filters for which the center frequency can be varied. In building a frequency-selective bandpass filter with elements such as resistors, inductors, and capacitors, the center frequency will depend on a number of the element values, all of which must be varied simultaneously in the correct way. This is generally difficult and cumbersome in comparison with building a filter whose characteristics are fixed. An alternative to varying the filter characteristics is to use a fixed frequency-selective filter and shift the signal spectrum appropriately, using the principles of sinusoidal amplitude modulation discussed in Section 7.1.

For example, consider the system shown in Figure 7.16. The spectra of the signals $x(t)$, $y(t)$, $w(t)$, and $f(t)$ are illustrated in Figure 7.17. We observe that the

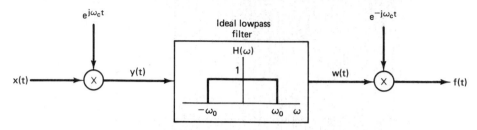

Figure 7.16 Implementation of a bandpass filter using amplitude modulation with a complex exponential carrier.

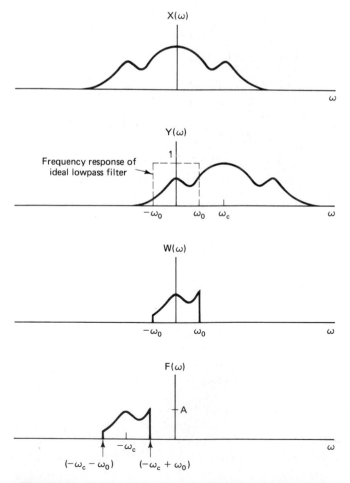

Figure 7.17 Spectra of the signals in the system of Figure 7.16.

overall system of Figure 7.16 is equivalent to an ideal bandpass filter with center frequency $-\omega_c$ and bandwidth $2\omega_0$, as illustrated in Figure 7.18. As the frequency ω_c of the complex exponential oscillator is varied, the center frequency of the bandpass filter varies.

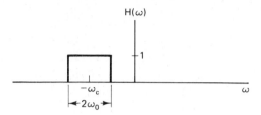

Figure 7.18 Bandpass filter equivalent of Figure 7.16.

In the system of Figure 7.16, with $x(t)$ real, the signals $y(t)$, $w(t)$, and $f(t)$ are all complex. If we retain only the real part of $f(t)$, the resulting spectrum is that shown in Figure 7.19 and the equivalent bandpass filter passes bands of frequencies

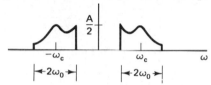

Figure 7.19 Spectrum of $\mathcal{R}e\{f(t)\}$ associated with Figure 7.16.

centered around ω_c and $-\omega_c$, as indicated in Figure 7.20. Under certain conditions it is also possible to use sinusoidal rather than complex exponential modulation to implement the system of Figure 7.20. This is explored further in Problem 7.12.

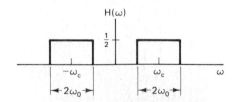

Figure 7.20 Equivalent bandpass filter for $\mathcal{R}e\{f(t)\}$ in Figure 7.19.

7.2.2 Sinusoidal Amplitude Modulation for Communications: Frequency-Division Multiplexing

One of the most widespread areas of application of sinusoidal amplitude modulation is in communications and information transmission systems. The basic need for modulation arises for two reasons. Different transmission media that are used for transmission of signals are generally suited to a particular frequency range that might not match the frequency range of the signals to be transmitted. In telephone transmission systems, for example, long-distance transmission is often accomplished over microwave or satellite links. The individual voice signals are in the frequency range 200 Hz to 4 kHz, whereas a microwave link requires signals in the range 300 megahertz (MHz) to 300 gigahertz (GHz), and communication satellite links operate in the frequency range 1.5 to 20 GHz. Thus, for transmission over these channels, the voice signal must be modulated onto a carrier at these higher frequencies.

An additional role that modulation plays in transmission of signals stems from the fact that many systems used for signal transmission provide more bandwidth than is required for any one signal. For example, a typical microwave link has a total bandwidth of several gigahertz, which is considerably greater than the bandwidth required for one voice channel. If the individual voice signals, which are overlapping in frequency, have their frequency content shifted through sinusoidal amplitude modulation so that the spectra of the modulated signals no longer overlap, they can be transmitted *simultaneously* over a single wideband channel. The resulting concept is referred to as *frequency-division multiplexing* (FDM). Frequency-division multiplexing using a sinusoidal carrier is illustrated in Figure 7.21. The individual

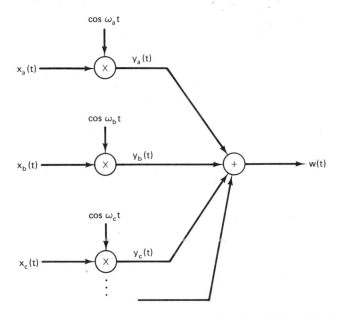

Figure 7.21 Frequency-division multiplexing using sinusoidal amplitude modulation.

signals to be transmitted are assumed to be bandlimited and are modulated with different carrier frequencies. The modulated signals are then summed and transmitted simultaneously over the same communications channel. The spectra of the individual subchannels and the composite multiplexed signal are illustrated in Figure 7.22. Through this multiplexing process, the individual input signals are allocated distinct segments of the frequency band. To recover the individual channels in the demultiplexing process, there are two basic steps: bandpass filtering to extract the modulated signal corresponding to a specific channel, followed by demodulation to recover the original signal. This is illustrated in Figure 7.23 to recover channel *a*, where synchronous demodulation is assumed.

Telephone communications is one important application of frequency-division multiplexing. Another is in the transmission of signals through the atmosphere in the

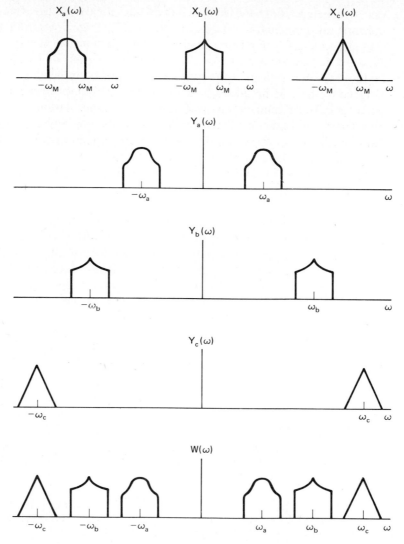

Figure 7.22 Spectra associated with the frequency-division multiplexing system of Figure 7.21.

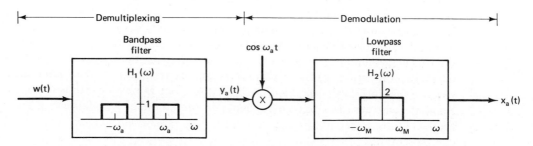

Figure 7.23 Demultiplexing and demodulation for a frequency-division multiplexed signal.

radio-frequency (RF) band. In the United States the use of these frequencies for signal transmission covering the range 10 kHz to 275 GHz is controlled by the Federal Communications Commission, and different portions of the frequency range are allocated for different purposes. The current allocation of frequencies is shown in Figure 7.24. As indicated, the frequency range in the neighborhood of 1 MHz is assigned to the AM broadcast band, where AM refers specifically to the use of sinusoidal amplitude modulation. Individual AM radio stations are assigned specific frequencies within the AM band and thus many stations can broadcast simultaneously through this use of frequency-division multiplexing. In principle, at the receiver, an individual radio station can be selected by demultiplexing and demodulating as illustrated in Figure 7.23. The tuning dial on the receiver would then control both the center frequency of the bandpass filter and the frequency of the demodulating oscillator. In fact, for public broadcasting, asynchronous modulation and demodulation are used to simplify the receiver and reduce its cost. Furthermore, the demultiplexing in Figure 7.23 requires a sharp cutoff bandpass filter with variable center frequency. As discussed in Section 7.2.1, variable frequency-selective filters are difficult to implement, and consequently a fixed filter is implemented instead and an intermediate stage of modulation and filtering [referred to in a radio receiver as the intermediate-frequency (IF) stage] is used. The use of modulation to slide the signal spectrum past a fixed bandpass filter replaces the use of a variable bandpass filter in a manner similar to the procedure discussed in Section 7.2.1. This basic procedure is incorporated in typical home AM radio receivers. Some of the more detailed issues involved are considered in Problem 7.17.

As illustrated in Figure 7.22, in the frequency-division multiplexing system of Figure 7.21, the spectrum of each individual signal is replicated at both positive and negative frequencies, and thus the modulated signal occupies twice the bandwidth of the original. This represents an inefficient use of bandwidth. In the next section we consider an alternative form of sinusoidal amplitude modulation, which leads to more efficient use of bandwidth at the cost of a more complicated modulation and demodulation system.

7.3 SINGLE-SIDEBAND AMPLITUDE MODULATION

For the sinusoidal amplitude modulation systems discussed in Section 7.2, the total bandwidth of the original signal $x(t)$ is $2\omega_M$, including both positive and negative frequencies, where ω_M is the highest frequency present in $x(t)$. With the use of a complex exponential carrier, this spectrum is translated to ω_c and the total width of the frequency band over which there is signal energy is still $2\omega_M$, although the modulated signal is now complex. With a sinusoidal carrier, on the other hand, the signal spectrum is shifted to $+\omega_c$ and $-\omega_c$, and thus twice the bandwidth is required. This suggests that there is a basic redundancy in the modulated signal with a sinusoidal carrier. Using a technique referred to as *single-sideband modulation*, this redundancy can be removed.

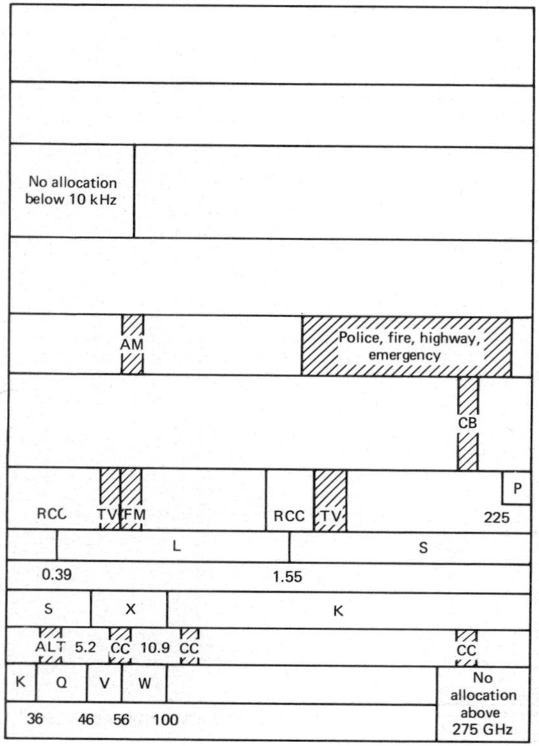

Frequency range	Designation	Wavelength range	Service	Propagation method	Channel features
30–300 Hz	ELF (extremely low frequency)	10^4–10^3 km	Macrowave, submarine communication	Megametric waves	Penetration of conducting earth and seawater
0.3–3 kHz	VF (voice frequency)	1000–100 km	Data terminals, telephony		
3–30 kHz	VLF (very low frequency)	100–10 km	Navigation, telephony telegraphy, frequency and timing standards	Surface ducting (ground wave)	Low attenuation, little fading, extremely stable phase and frequency, large antennas
30–300 kHz	LF (low frequency)	10–1 km	Industrial (power line) communication, aeronautical and maritime long-range navigation	Mostly surface ducting	Slight fading, high atmospheric noise
0.3–3 MHz	MF (medium frequency)	1 km–100 m	Mobile, AM broadcasting, amateur, public safety	Ducting and ionospheric reflection (sky wave)	Increased fading, but reliable
3–30 MHz	HF (high frequency)	100–10 m	Military communication, aeronautical fixed, amateur and citizens band, industrial	Ionospheric reflecting sky wave, 50–400 km layer altitudes	Intermittent and frequency-selective fading, multipath
30–300 MHz	VHF (very high frequency)	10–1 m	FM and TV broadcast, land transportation (taxis, buses, railroad)	Sky wave (ionospheric and tropospheric scatter)	Fading, scattering and multipath
0.3–3 GHz	UHF (ultra high frequency)	1 m–10 cm	UHF TV, space telemetry, radar, military	Transhorizon tropospheric scatter and line-of-sight relaying	
3–30 GHz	SHF (super high frequency)	10–1 cm	Satellite and space communication, common carrier (CC), microwave	Line-of-sight ionosphere penetration	Ionospheric penetration, extraterrestrial noise, high directivity
30–300 GHz	EHF (extremely high frequency)	1 cm–1 mm	Reasearch, government, radio astronomy	Line-of-sight	Water vapor and oxygen absorption

No allocation below 10 kHz

Police, fire, highway, emergency

AM
CB
P
225
RCC TV FM TV RCC S
0.39 L 1.55
S X K
ALT 5.2 CC 10.9 CC
K Q V W
36 46 56 100

CC

No allocation above 275 GHz

Figure 7.24 Allocation of frequencies in the RF spectrum. After W. David Gregg, *Analog and Digital Communication* (New York: John Wiley & Sons, Inc., 1977).

465

The spectrum of $x(t)$ is illustrated in Figure 7.25(a), where we have shaded the positive and negative frequency components differently. The spectrum in Figure 7.25(b) results from modulation with a sinusoidal carrier, where we identify an upper and lower sideband for the portion of the spectrum centered at $+\omega_c$ and that centered at $-\omega_c$. Comparing Figure 7.25(a) and (b) it should be evident that $X(\omega)$ can be recovered if only the upper sidebands at positive and negative frequencies are retained, or alternatively if only the lower sidebands at positive and negative frequencies are retained. The resulting spectrum if only the upper sidebands are retained is shown in Figure 7.25(c), and the resulting spectrum if only the lower sidebands are retained is shown in Figure 7.25(d). The conversion of $x(t)$ to the form corresponding to Figure 7.25(c) or (d) is referred to as *single-sideband modulation* (SSB), in contrast to the *double-sideband modulation* (DSB) of Figure 7.25(b), in which both sidebands are retained.

There are several methods by which the single-sideband signal can be obtained. One is to apply a sharp cutoff bandpass or highpass filter to the double-sideband

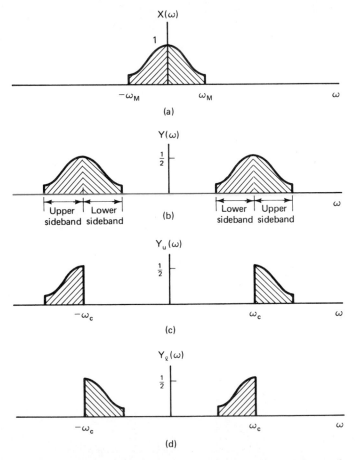

Figure 7.25 Double- and single-sideband modulation: (a) spectrum of modulating signal; (b) spectrum after modulation with a sinusoidal carrier; (c) spectrum with only the upper sidebands; (d) spectrum with only the lower sidebands.

signal of Figure 7.25(b) to remove the unwanted sideband as illustrated in Figure 7.26. The second is to use a procedure that utilizes phase shifting. The resulting system to retain the lower sidebands is shown in Figure 7.27. The system $H(\omega)$ is a "90° phase-shift network," of the form

$$H(\omega) = \begin{cases} -j, & \omega > 0 \\ j, & \omega < 0 \end{cases} \tag{7.19}$$

The spectra of $x(t)$, $y_1(t) = x(t) \cos \omega_c t$, $y_2(t) = x_p(t) \sin \omega_c t$, and $y(t)$ are illustrated in Figure 7.28. As considered in Problem 7.20, to retain the upper sidebands instead of the lower sidebands, the phase characteristic of $H(\omega)$ is reversed so that

$$H(\omega) = \begin{cases} j, & \omega > 0 \\ -j, & \omega < 0 \end{cases} \tag{7.20}$$

As explored in Problem 7.21, synchronous demodulation of single-sideband systems can be accomplished in a manner identical to synchronous demodulation of double-sideband systems. The price paid for the increased efficiency of single-sideband systems is added complexity in the modulator.

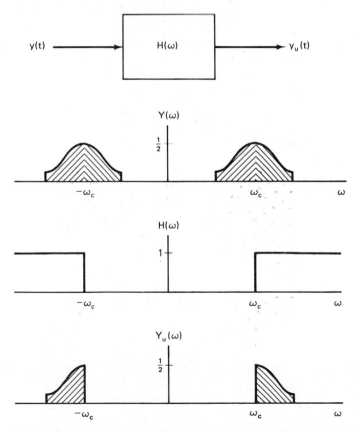

Figure 7.26 System for retaining the upper sidebands using ideal highpass filtering.

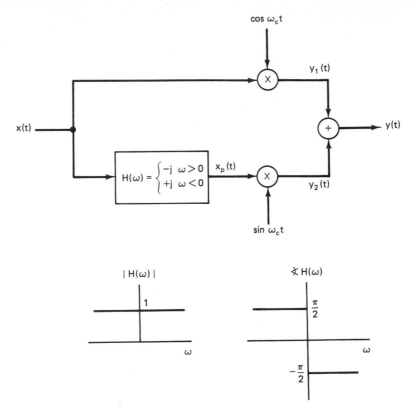

Figure 7.27 System for single-sideband amplitude modulation in which only the lower sidebands are retained.

In summary, in Sections 7.1 to 7.3 we have seen a number of variations of sinusoidal amplitude modulation. A carrier signal can be included in the modulated output, which requires more power for transmission but results in a simpler demodulator. Alternatively, only the upper or lower sidebands in the modulated output can be retained, which makes more efficient use of bandwidth and transmitter power but requires a more sophisticated modulator. Sinusoidal amplitude modulation with both sidebands and the presence of a carrier is typically abbreviated as AM-DSB/WC (amplitude modulation double sideband/with carrier) and when the carrier is suppressed or absent as AM-DSB/SC (amplitude modulation double sideband/suppressed carrier). The corresponding single-sideband systems are abbreviated as AM-SSB/WC and AM-SSB/SC.

The previous sections of this chapter were intended to provide an introduction to many of the basic concepts associated with sinusoidal amplitude modulation. There are many variations in details and implementation and the reader is referred to the Bibliography for an indication of the many excellent books that explore this topic further.

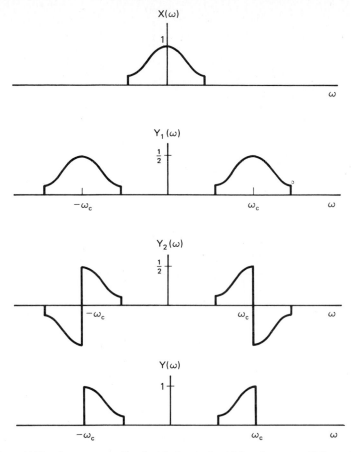

Figure 7.28 Spectra associated with the single-sideband system of Figure 7.27.

7.4 PULSE AMPLITUDE MODULATION AND TIME-DIVISION MULTIPLEXING

In previous sections, we considered amplitude modulation with a sinusoidal carrier. Another important class of amplitude-modulation techniques corresponds to the use of a carrier signal which is a pulse train, as illustrated in Figure 7.29. Amplitude modulation of this type is referred to as *pulse amplitude modulation* (PAM). One of the important applications of pulse amplitude modulation is for transmission of several signals over a single channel. As indicated in Figure 7.29, the modulated output signal $y(t)$ is nonzero only while the carrier signal $p(t)$ is "on" (i.e., is nonzero). During the off intervals other pulse-amplitude-modulated signals could be transmitted. Two equivalent representations of this process are shown in Figure 7.30. In this technique for transmitting several signals over a single channel, each signal is in

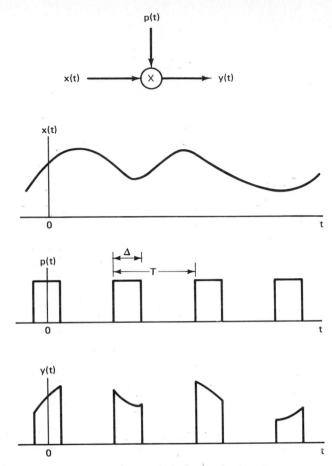

Figure 7.29 Pulse amplitude modulation.

effect assigned a set of time slots of duration Δ which repeat every T seconds and which do not overlap with the slots assigned to other signals. The smaller the ratio Δ/T, the larger the number of signals that can be transmitted over the channel. This procedure is referred to as *time-division multiplexing* (TDM). Whereas frequency-division multiplexing, as discussed in Section 7.2.2, assigns different *frequency* intervals to individual signals, time-division multiplexing assigns different *time* intervals to individual signals. Demultiplexing of the individual PAM signals from the composite signal in Figure 7.30 is accomplished by time gating to select the particular time slots associated with each individual signal.

In general, we would not expect that an arbitrary continuous-time signal could be recovered after pulse amplitude modulation. However, somewhat surprisingly, under certain conditions, the original signal $x(t)$ can be exactly recovered by lowpass filtering. To develop these conditions, we note that as with sinusoidal amplitude modulation, if

$$y(t) = x(t)p(t)$$

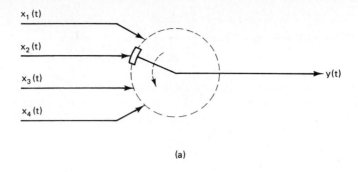

(a)

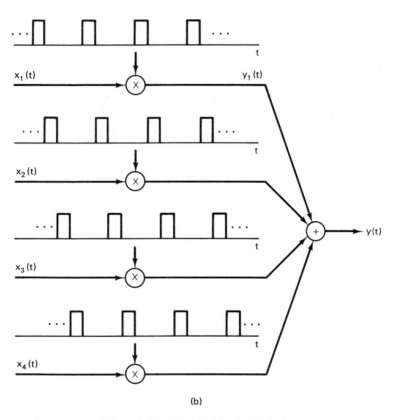

(b)

Figure 7.30   Time-division multiplexing.

then

$$Y(\omega) = \frac{1}{2\pi} X(\omega) * P(\omega) \qquad (7.21)$$

Since $p(t)$ is periodic with period T, $P(\omega)$ consists of impulses in frequency spaced by $2\pi/T$, that is,

$$P(\omega) = 2\pi \sum_{k=-\infty}^{-\infty} a_k \, \delta(\omega - k\omega_p) \qquad (7.22)$$

where $\omega_p = 2\pi/T$ and the coefficients a_k are the Fourier series coefficients of $p(t)$, which, from Example 4.13, are

$$a_k = \frac{\sin (k\omega_p \, \Delta/2)}{\pi k} \tag{7.23}$$

The spectrum of $p(t)$ is thus as indicated in Figure 7.31(b). With the spectrum of $x(t)$ as illustrated in Figure 7.31(a), the resulting spectrum of the PAM signal $y(t)$ is shown in Figure 7.31(c). In general, from eqs. (7.21) and (7.22), $Y(\omega)$ is a sum of scaled replications of $X(\omega)$ shifted in frequency, that is,

$$Y(\omega) = \sum_{k=-\infty}^{+\infty} a_k X(\omega - k\omega_p) \tag{7.24}$$

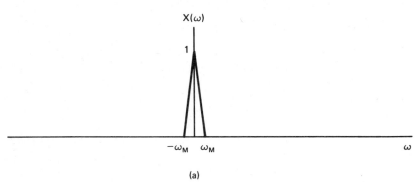

(a)

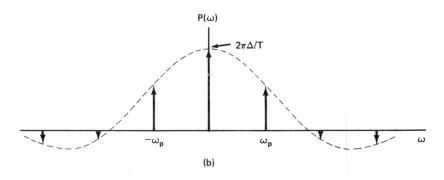

(b)

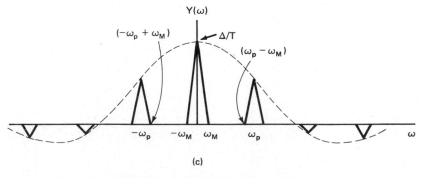

(c)

Figure 7.31   Spectra associated with pulse amplitude modulation.

From Figure 7.31(c) and eq. (7.24) it should be clear that $X(\omega)$ can be extracted from $Y(\omega)$, and equivalently $x(t)$ can be recovered from $y(t)$, provided that

$$(\omega_p - \omega_M) > \omega_M \tag{7.25a}$$

or

$$\omega_p > 2\omega_M \tag{7.25b}$$

where $X(\omega)$ is zero for $|\omega| > \omega_M$. Equations (7.25) place a constraint on the relation between the period T of the carrier and the bandwidth of the input signal. If this constraint is satisfied, then $x(t)$ can be recovered from $y(t)$ through the use of a lowpass filter with cutoff frequency greater than ω_M and less than $(\omega_p - \omega_M)$.

It is interesting to note that the constraint in eq. (7.25) that dictates our ability to recover $x(t)$ from $y(t)$ depends on the period T but does not depend on the duration Δ (i.e., the length of time that the carrier is "on" in each period). Thus, as long as eq. (7.25b) is satisfied and we are willing to supply the necessary amplification (T/Δ) in the lowpass filter used to recover $x(t)$, we can in principle select time slices of $x(t)$ that are arbitrarily narrow. In practice, as the width of the time slices decreases, the amplitude of $p(t)$ is generally increased so that the pulse-amplitude-modulated signal is not unduly sensitive to errors and noise encountered in transmission over a channel. In the limit as Δ approaches zero and the amplitude of $p(t)$ is increased in proportion to $1/\Delta$, $p(t)$ becomes a periodic impulse train and $x(t)$ is represented by instantaneous values or *samples* at periodic time instants. This specific case is the subject of Chapter 8.

7.5 DISCRETE-TIME AMPLITUDE MODULATION

The concepts of amplitude modulation developed in the previous sections for continuous-time signals apply as well to discrete-time signals. By far the most widespread application for amplitude modulation is in communications systems. In many communications systems the continuous-time signal is converted to a discrete-time signal through the concept of sampling, which was introduced briefly in Section 7.4 and which is developed in more detail in Chapter 8. In these systems the modulation may then be carried out in discrete time. Furthermore, in many of these discrete-time communications systems, there is the need to convert from frequency-division multiplexing to time-division multiplexing, or vice versa, entirely within the discrete-time domain. Such discrete-time systems, referred to as transmultiplexing or transmodulation systems, represent an extremely important application of discrete-time amplitude modulation, and are considered again in Chapter 8, after the relation between continuous-time and discrete-time signals and systems has been developed. A second important application for discrete-time amplitude modulation, as was the case for continuous-time amplitude modulation, is in the implementation of frequency-selective bandpass filters with variable center frequency.

A discrete-time amplitude-modulation system is shown in Figure 7.32, where $c[n]$ is the carrier and $x[n]$ is the modulating signal. The basis for our analysis of continuous-time amplitude modulation was the modulation property for Fourier transforms, specifically the fact that multiplication in the time domain leads to a

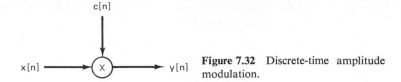

Figure 7.32 Discrete-time amplitude modulation.

convolution in the frequency domain. As discussed in Section 5.7, there is a corresponding property for discrete-time signals. Specifically, consider

$$y[n] = x[n]c[n]$$

With $X(\Omega)$, $Y(\Omega)$, and $C(\Omega)$ denoting the Fourier transforms of $x[n]$, $y[n]$, and $c[n]$, respectively, $Y(\Omega)$ is proportional to the periodic convolution of $X(\Omega)$ and $C(\Omega)$, that is,

$$Y(\Omega) = \frac{1}{2\pi} \int_{2\pi} X(\theta)C(\Omega - \theta) \, d\theta \qquad (7.26)$$

Since $X(\Omega)$ and $C(\Omega)$ are periodic with a period of 2π, the integration can be performed over any frequency interval of length 2π.

Let us first consider sinusoidal amplitude modulation with a complex exponential carrier, so that

$$c[n] = e^{j\Omega_c n} \qquad (7.27)$$

As we developed in Sec. 5.4.2, the Fourier transform of $c[n]$ is a periodic impulse train, that is,

$$C(\Omega) = \sum_{k=-\infty}^{+\infty} 2\pi \, \delta(\Omega - \Omega_c + k2\pi) \qquad (7.28)$$

Thus, $C(\Omega)$ is as sketched in Figure 7.33(b). With $X(\Omega)$ as illustrated in Figure 7.33(a), the spectrum of the modulated signal will be that shown in Figure 7.33(c). In particular, we note that $Y(\Omega) = X(\Omega - \Omega_c)$. This is the discrete-time counterpart to Figure 7.2 and here again, with $x[n]$ real the modulated signal will be complex. Demodulation is accomplished by multiplying by $e^{-j\Omega_c n}$ to translate the spectrum back to its original location on the frequency axis, so that

$$x[n] = y[n]e^{-j\Omega_c n} \qquad (7.29)$$

When $\Omega_c = \pi$, $c[n]$ is the sequence $(-1)^n$ and the high and low frequencies in $x[n]$ become interchanged in $y[n]$, as illustrated in Figure 7.34. For this specific choice of $c[n]$, $y[n]$ is obtained from $x[n]$ by changing the algebraic sign of the values in $x[n]$ for odd values of n, and demodulation corresponds to changing the algebraic sign of the values of $y[n]$ for odd values of n.

One example in which amplitude modulation with $\Omega_c = \pi$ is often useful is in utilizing a lowpass filter to achieve highpass filtering, and vice versa. A system to accomplish this is illustrated in Figure 7.35. By modulating with the carrier $(-1)^n$, the high frequencies are shifted to low frequencies, and after lowpass filtering, the signal is demodulated.

As an alternative to the use of a complex exponential carrier we can use a sinusoidal carrier, in which case with $x[n]$ real, the modulated signal $y[n]$ will also be real. With $c[n] = \cos \Omega_c n$, the spectrum of the carrier consists of periodically

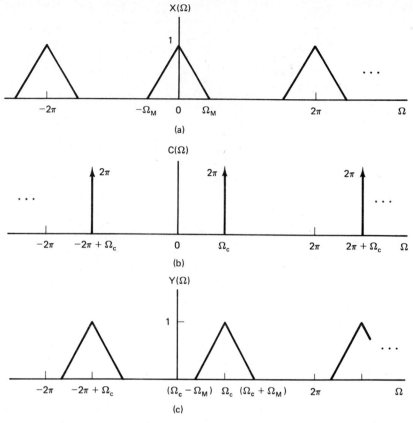

Figure 7.33 (a) Spectrum of $x[n]$; (b) spectrum of $c[n] = e^{j\Omega_c n}$; (c) spectrum of $y[n] = x[n]c[n]$.

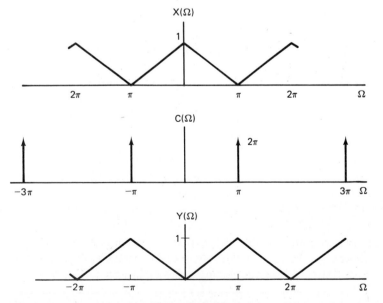

Figure 7.34 Spectra for amplitude modulation with carrier $e^{j\pi n} = (-1)^n$.

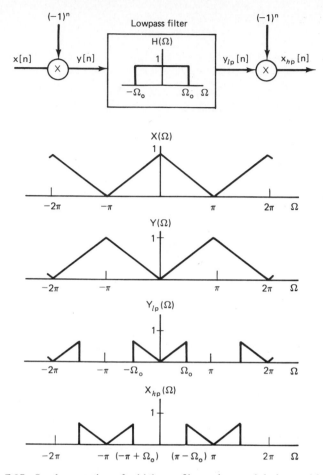

Figure 7.35 Implementation of a highpass filter using modulation and lowpass filtering.

repeated pairs of impulses at $\Omega = \pm\Omega_c + k2\pi$, as illustrated in Figure 7.36(b). With $X(\Omega)$ as shown in Figure 7.36(a), the resulting spectrum for the modulated signal is that shown in Figure 7.36(c), and in particular corresponds to replicating $X(\Omega)$ at the frequencies $\Omega = \pm\Omega_c + k2\pi$. In order that the individual replications of $X(\Omega)$ not overlap, we require that

$$\Omega_c > \Omega_M \tag{7.30}$$

and

$$(2\pi - \Omega_c - \Omega_M) > (\Omega_c + \Omega_M)$$

or, equivalently,

$$\Omega_c < (\pi - \Omega_M) \tag{7.31}$$

The first condition is identical to that in Section 7.1 for continuous-time sinusoidal amplitude modulation, while the second results from the inherent periodicity of discrete-time spectra. Combining eqs. (7.30) and (7.31), with a sinusoidal carrier, we restrict Ω_c so that

$$\Omega_M < \Omega_c < (\pi - \Omega_M) \qquad (7.32)$$

Demodulation can be accomplished in a manner similar to that used in continuous time. Specifically, as illustrated in Figure 7.37, multiplication of $y[n]$ with the same carrier used in the modulator results in several replications of the original signal spectrum, one of which is centered about $\Omega = 0$. By lowpass filtering to eliminate the unwanted replications of $X(\Omega)$, the demodulated signal is obtained.

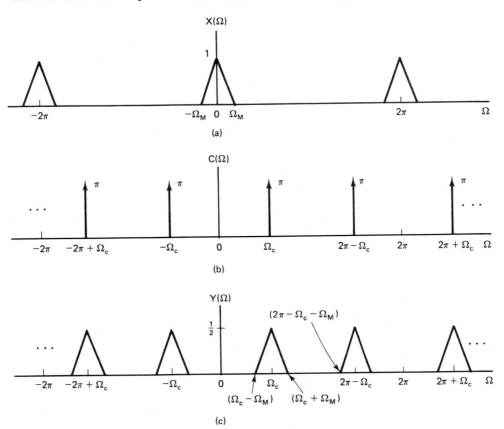

Figure 7.36 Spectra associated with discrete-time amplitude modulation using a sinusoidal carrier.

As should be evident from the foregoing discussion, an analysis of discrete-time amplitude modulation proceeds in a manner very similar to continuous-time amplitude modulation with only slight differences. For example, as explored in Problem 7.24, in the synchronous modulation and demodulation system, the effect of a phase difference or a frequency difference between the sinusoidal carrier in the modulator and demodulator is identical in both discrete and continuous time.

As in Section 7.4, we can also consider discrete-time pulse amplitude modulation and time-division multiplexing. A single channel of such a system is shown in Figure 7.38. Here we retain $(M + 1)$ consecutive values of $x[n]$ in each period of interval

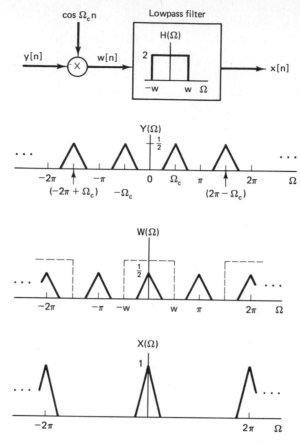

Figure 7.37 System for discrete-time synchronous demodulation and associated spectra.

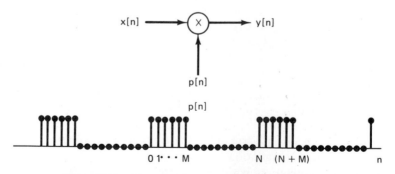

Figure 7.38 Discrete-time pulse amplitude modulation.

N. The analysis of this system is very similar to the continuous-time case and is explored in greater detail in Problem 7.33. In that exercise it is shown that if the bandwidth of $X(\Omega)$ is sufficiently small relative to $2\pi/N$, the fundamental frequency of $p[n]$, then $x[n]$ can be recovered from $y[n]$ by lowpass filtering. As in continuous time, this result depends only on N and the bandwidth of $X(\Omega)$ and *not* on the duration $M + 1$ of each pulse. For example, we could take the case $M = 0$ (i.e., we could consider transmitting a single sample of $x[n]$ each period), and if the bandwidth constraint is satisfied, we can recover *all* of $x[n]$ by lowpass filtering. This is the basic result behind discrete-time sampling, a topic which, together with continuous-time sampling, we explore in Chapter 8.

7.6 CONTINUOUS-TIME FREQUENCY MODULATION

In the preceding sections we discussed a number of specific amplitude-modulation systems in which the modulating signal was used to vary the amplitude of a sinusoidal or a pulse carrier. As we have seen, such systems are amenable to detailed analysis using the frequency-domain techniques that we have developed in preceding chapters. Another very important class of modulation techniques is referred to as *angle modulation*, in which the modulating signal is used to control the frequency or phase of a sinusoidal carrier. Modulation systems of this type have a number of advantages over sinusoidal amplitude modulation. For example, the modulated signal has a constant peak amplitude and thus the transmitter can always operate at peak power. In addition, amplitude variations introduced over a transmission channel due to additive disturbances or fading can, to a large extent, be eliminated at the receiver. For this reason, in public broadcasting and a variety of other contexts, frequency modulation (FM) reception is typically better than AM reception. On the other hand, as we will see, frequency modulation generally requires greater bandwidth than does sinusoidal amplitude modulation.

Angle modulation systems are highly nonlinear and consequently are not as straightforward to analyze as are the amplitude modulation systems discussed in preceding sections. However, the frequency-domain methods we have developed do allow us to gain some understanding of the nature and operation of angle modulation systems. To introduce the analysis of angle modulation, consider a sinusoidal carrier signal $c(t)$, expressed in the form

$$c(t) = A \cos(\omega_c t + \theta_c) \tag{7.33}$$

where ω_c is the frequency and θ_c the phase of the carrier. Angle modulation, in general, corresponds to using the modulating signal to vary the angle $\theta(t) = (\omega_c t + \theta_c)$. One form that this sometimes takes is to use the modulating signal $x(t)$ to vary the *phase* θ_c so that the modulated signal $y(t)$ takes the form

$$y(t) = A \cos[\omega_c t + \theta_c(t)] \tag{7.34}$$

where θ_c is now a function of time, specifically of the form

$$\theta_c(t) = \theta_0 + k_p x(t) \tag{7.35}$$

If $x(t)$ is constant, for example, the phase of $y(t)$ will be constant and proportional to the amplitude of $x(t)$. Angle modulation of the form of eq (7.34) is referred to as *phase modulation*. Another form of angle modulation corresponds to varying the *derivative* of the angle linearly with the modulating signal, that is,

$$y(t) - A \cos \theta(t) \tag{7.36}$$

where

$$\frac{d\theta(t)}{dt} = \omega_c + k_f x(t) \tag{7.37}$$

For $x(t)$ constant, $y(t)$ is sinusoidal with a frequency that is proportional to the amplitude of $x(t)$. For that reason, angle modulation of the form of eqs. (7.36) and (7.37) is commonly referred to as *frequency modulation.*

Although phase modulation and frequency modulation are different forms of angle modulation, the difference is not a significant one since they can be easily related. Specifically, from eqs. (7.34) and (7.35), for phase modulation

$$\frac{d\theta(t)}{dt} = \omega_c + k_p \frac{dx(t)}{dt} \tag{7.38}$$

and thus, comparing eqs. (7.37) and (7.38), phase modulating with $x(t)$ is identical to frequency modulating with the derivative of $x(t)$. Likewise frequency modulating with $x(t)$ is identical to phase modulating with the integral of $x(t)$. An illustration of phase modulation and frequency modulation is shown in Figure 7.39(a) and (b). In both cases, the modulating signal is $x(t) = tu(t)$ (i.e., a linear increase with time for $t > 0$). In Figure 7.39(c) an example of frequency modulation is shown with a step (the derivative of a ramp) as the modulating signal [i.e., $x(t) = u(t)$]. The correspondence between Figure 7.39(a) and (c) should be evident. Frequency modulation with a step corresponds to the frequency of the sinusoidal carrier changing instantaneously from one value to another when $x(t)$ changes value at $t = 0$, much as the frequency of a sinusoidal oscillator changes when the frequency setting is switched instantaneously. When the frequency modulation is a ramp, as in Figure 7.39(b), "frequency" is changing linearly with time. This notion of a time-varying frequency is often best expressed in terms of the concept of *instantaneous frequency*. For $y(t)$ of the form

$$y(t) = A \cos \theta(t) \tag{7.39}$$

the instantaneous frequency ω_i of the sinusoid is defined as

$$\omega_i(t) = \frac{d\theta(t)}{dt} \tag{7.40}$$

Thus, for $y(t)$ truly sinusoidal [i.e., $\theta(t) = (\omega_c t + \theta_0)$], the instantaneous frequency is ω_c, as we would expect. For phase modulation as expressed in eqs. (7.34) and (7.35), the instantaneous frequency is $[\omega_c + k_p(dx(t)/dt)]$ and for frequency modulation as expressed in eqs. (7.36) and (7.37), the instantaneous frequency is $[\omega_c + k_f x(t)]$.

Since frequency modulation and phase modulation are easily related, we will phrase the remaining discussion in terms of only one of these, specifically frequency modulation. To gain some insight into how the spectrum of the frequency-modulated signal is affected by the modulating signal $x(t)$, it is useful to consider two cases in which the modulating signal is sufficiently simple so that some of the essential properties of frequency modulation become evident.

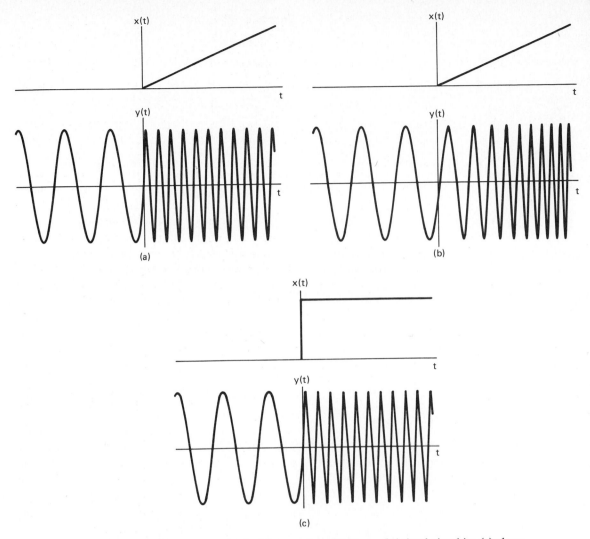

Figure 7.39 Phase modulation and frequency modulation and their relationship: (a) phase modulation with a ramp as the modulating signal; (b) frequency modulation with a ramp as the modulating signal; (c) frequency modulation with a step (the derivative of a ramp) as the modulating signal.

7.6.1 Narrowband Frequency Modulation

Let us consider the case of frequency modulation with $x(t)$ sinusoidal of the form

$$x(t) = A \cos \omega_m t \tag{7.41}$$

From eq. (7.40), the instantaneous frequency is

$$\omega_i(t) = \omega_c + k_f A \cos \omega_m t \tag{7.42}$$

and in particular varies sinusoidally between $\omega_c + k_f A$ and $\omega_c - k_f A$. With $\Delta\omega$ defined as

$$\Delta\omega = k_f A$$

we have

$$\omega_i(t) = \omega_c + \Delta\omega \cos \omega_m t$$

and

$$y(t) = \cos [\omega_c t + \int x(t) \, dt] \tag{7.43}$$
$$= \cos \left(\omega_c t + \frac{\Delta\omega}{\omega_m} \sin \omega_m t + \theta_0\right)$$

where θ_0 is a constant of integration. For convenience we will choose $\theta_0 = 0$, so that

$$y(t) = \cos [\omega_c t + \frac{\Delta\omega}{\omega_m} \sin \omega_m t] \tag{7.44}$$

The factor $\Delta\omega/\omega_m$, which we denote by m, is defined as the *modulation index* for frequency modulation. The properties of FM systems tend to be different depending on whether the modulation index m is small or large. The case in which m is small is referred to as *narrowband FM*. In general, we can rewrite eq. (7.44) as

$$y(t) = \cos (\omega_c t + m \sin \omega_m t) \tag{7.45a}$$

or

$$y(t) = \cos \omega_c t \cos (m \sin \omega_m t) - \sin \omega_c t \sin (m \sin \omega_m t) \tag{7.45b}$$

When m is sufficiently small ($\ll \pi/2$), we can make the approximations

$$\cos (m \sin \omega_m t) \simeq 1 \tag{7.46a}$$
$$\sin (m \sin \omega_m t) \simeq m \sin \omega_m t \tag{7.46b}$$

so that eq. (7.45) becomes

$$y(t) \simeq \cos \omega_c t - m(\sin \omega_m t)(\sin \omega_c t) \tag{7.47}$$

The spectrum of $y(t)$ based on this approximation is shown in Figure 7.40. We note

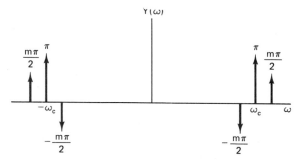

Figure 7.40 Approximate spectrum for narrowband FM.

that it has a similarity to AM-DSB/WC in that the carrier frequency is present in the spectrum and there are sidebands representing the spectrum of the modulating signal. However, in AM-DSB/WC the additional carrier injected is in phase with the modulated carrier, whereas as we see in eq. (7.47) for the case of narrowband FM, the carrier signal has a phase difference of $\pi/2$ in relation to the amplitude-modulated carrier. The waveforms corresponding to AM-DSB/WC and FM are also very different. Figure 7.41(a) illustrates the narrowband FM time waveform corresponding to eq. (7.47). For comparison, in Figure 7.41(b) is shown the AM-DSB/WC signal

$$y_2(t) = \cos \omega_c t + m(\cos \omega_m t)(\cos \omega_c t) \tag{7.48}$$

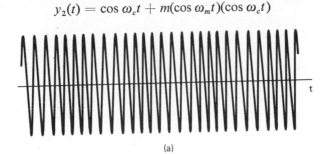

(a)

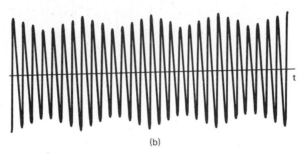

(b)

Figure 7.41 Comparison of narrowband FM and AM-DSB/WC: (a) narrowband FM; (b) AM-DSB/WC.

For the narrowband FM signal of eq. (7.47), the bandwidth of the sidebands is equal to the bandwidth of the modulating signal, and in particular, although the approximation in eq. (7.47) is based on assuming that $m \ll \pi/2$, the bandwidth of the sidebands is otherwise independent of the modulation index m (i.e., it depends only on the bandwidth of the modulating signal, not on its amplitude). A similar statement applies for narrowband FM with a more general modulating signal.

7.6.2 Wideband Frequency Modulation

When m is large, the approximation leading to eq. (7.47) no longer applies and the spectrum of $y(t)$ depends on both the amplitude and spectrum of the modulating signal $x(t)$. With $y(t)$ expressed in the form of eq. (7.45), we note that the terms $\cos [m \sin \omega_m t]$ and $\sin [m \sin \omega_m t]$ are periodic signals with fundamental frequency ω_m. Thus, the Fourier transform of each of these signals is an impulse train with impulses at integer multiples of ω_m and amplitudes proportional to the Fourier series coefficients. The Fourier series coefficients for these two periodic signals involve a class of functions referred to as Bessel functions of the first kind. The first term in eq. (7.45)b corresponds to a sinusoidal carrier of the form $\cos \omega_c t$ amplitude-modulated by the periodic signal $\cos{[m \sin{\omega_m t}]}$ and the second term a sinusoidal carrier $\sin{\omega_c t}$ amplitude-modulated by the periodic signal $\sin{[m \sin{\omega_m t}]}$. Multiplication by the carrier signals has the effect in the frequency domain of translating the spectrum of eq. (7.45)b to the carrier frequency ω_c. In Figure 7.42(a) and (b) we illustrate the spectra of the two individual terms in eq. (7.45), and in Figure 7.42(c) the combined spectrum representing the modulated signal $y(t)$. In particular, the spectrum of $y(t)$ consists of impulses at frequencies $\pm \omega_c + n\omega_m$, $n = 0, \pm 1, \pm 2, \ldots$ and is not, strictly speaking, bandlimited around $\pm \omega_c$. However, the behavior of the Fourier series coefficients of $\cos{[m \sin{\omega_m t}]}$ and $\sin{[m \sin{\omega_m t}]}$ are such that the amplitude of the nth harmonic for $|n| > m$ can be considered to be negligible, and thus the total bandwidth B of each sideband centered around $+\omega_c$ and $-\omega_c$ is effectively limited to $2m\omega_m$, that is,

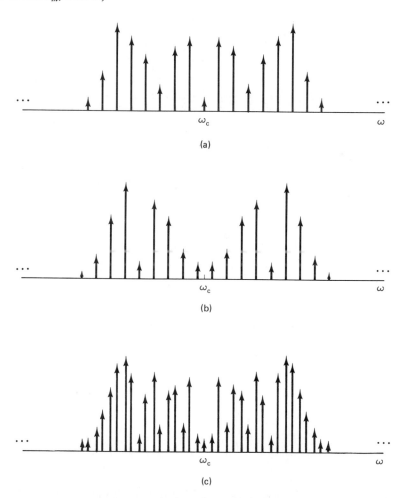

Figure 7.42 Magnitude of spectrum of wideband frequency modulation with $m = 12$: (a) magnitude of spectrum of $\cos{\omega_c t} \cos{[m \sin{\omega_m t}]}$; (b) spectrum of $\sin{\omega_c t} \sin{[m \sin{\omega_m t}]}$; (c) combined spectral magnitude of $\cos{[\omega_c t + m \sin{\omega_m t}]}$.

$$B \simeq 2m\omega_m \tag{7.49}$$

or, in terms of k_f and A, since $m = k_f A/\omega_m, = \Delta\omega/\omega_m$

$$B \simeq 2k_f A = 2\Delta\omega \tag{7.50}$$

In particular, comparing eqs. (7.42) and (7.50), we note that the effective bandwidth of each sideband is equal to the total excursion of the instantaneous frequency around the carrier frequency. Therefore, for wideband FM, since we assume that m is large, the bandwidth of the modulated signal is much larger than the bandwidth of the modulating signal and, in contrast to the narrowband case, the bandwidth of the transmitted signal in wideband FM is directly proportional to the modulating signal amplitude and the gain factor k_f.

7.6.3 Periodic-Square-Wave Modulating Signal

Another example that serves to lend insight into the properties of frequency modulation is that of a modulating signal which is a periodic square wave. Referring to eq. (7.42), let $k_f = 1$ so that $\Delta\omega = A$ and let $x(t)$ be given by Figure 7.43. The

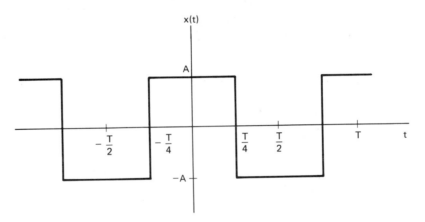

Figure 7.43 Symmetric periodic square wave.

modulated signal $y(t)$ is illustrated in Figure 7.44. The instantaneous frequency is $\omega_c + \Delta\omega$ when $x(t)$ is positive, and $\omega_c - \Delta\omega$ when $x(t)$ is negative. Thus, $y(t)$ can also be written as

$$y(t) = r(t) \cos{[(\omega_c + \Delta\omega)t]} + r\left(t - \frac{T}{2}\right) \cos{[(\omega_c - \Delta\omega)t]} \tag{7.51}$$

where $r(t)$ is the symmetric square wave shown in Figure 7.45. Thus, for this particular modulating signal, we are able to recast the problem of determining the spectrum of the FM signal $y(t)$ as the determination of the spectrum of the sum of the two AM signals in eq. (7.51). Specifically,

$$Y(\omega) = \tfrac{1}{2}[R(\omega + \omega_c + \Delta\omega) + R(\omega - \omega_c - \Delta\omega)]$$
$$+ \tfrac{1}{2}[R_T(\omega + \omega_c - \Delta\omega) + R_T(\omega - \omega_c + \Delta\omega)] \tag{7.52}$$

where $R(\omega)$ is the Fourier transform of the periodic square wave $r(t)$ in Figure 7.45

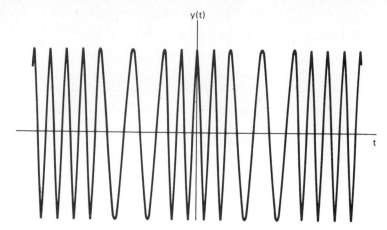

Figure 7.44 Frequency modulation with a periodic-square-wave modulating signal.

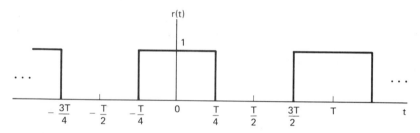

Figure 7.45 Symmetric square-wave term in eq. (7.51).

and $R_T(\omega)$ is the Fourier transform of $r(t - T/2)$. From Example 4.13

$$R(\omega) = \sum_{k=-\infty}^{\infty} \frac{2}{2k+1}(-1)^k \,\delta\!\left[\omega - \frac{2\pi(2k+1)}{T}\right] + \pi\delta(\omega) \qquad (7.53)$$

and

$$R_T(\omega) = R(\omega)e^{-j\omega T/2} \qquad (7.54)$$

The spectrum of $Y(\omega)$ is illustrated in Figure 7.46. As with wideband FM, the spectrum has the general appearance of two sidebands, centered around $\omega_c \pm \Delta\omega$, which

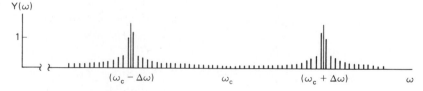

Figure 7.46 Spectrum for $\omega > 0$ corresponding to frequency modulation with a periodic square-wave modulating signal.

Modulation   Chap. 7

decay for $\omega < (\omega_c - \Delta\omega)$ and $\omega > (\omega_c + \Delta\omega)$. Systems for demodulation of FM signals typically are of two types, one corresponding to converting the FM signal to an AM signal through differentiation and the second corresponding to tracking directly the phase or frequency of the modulated signal.

The foregoing discussion provides only a brief introduction to the characteristics of frequency modulation, and again we have seen how the basic techniques developed in the earlier chapters can be exploited to analyze and develop insight into an important class of systems.

7.7 SUMMARY

In this chapter we have introduced the concept of amplitude modulation and discussed a number of its applications. The properties of amplitude modulation for both continuous time and discrete time are most easily interpreted in the frequency domain through the modulation property of Fourier transforms as discussed in Chapters 4 and 5. Amplitude modulation with a complex exponential or sinusoidal carrier is typically used to shift the spectrum of a signal in frequency and is applied, for example, in communications systems to place the signal spectrum in a frequency range suitable for transmission and to permit frequency-division multiplexing. Variations of sinusoidal amplitude modulation, such as insertion of a carrier signal for asynchronous systems and single- and double-sideband systems, were discussed. Pulse amplitude modulation, used, for example, in time-division multiplexing systems, was discussed. As the pulse width becomes small, so that the pulse train approaches an impulse train, pulse amplitude modulation corresponds to instantaneous time sampling.

In the present chapter we also briefly introduced the concepts of frequency and phase modulation. Although these forms of modulation are more difficult to analyze in detail, it is possible to gain significant insight through the frequency domain.

PROBLEMS

7.1. In Section 7.1 we analyzed the sinusoidal amplitude modulation and demodulation system of Figure 7.9, assuming that the phase θ_c of the carrier signal was zero.

(a) For the more general case of arbitrary phase θ_c in Figure 7.9, show that the signal $w(t)$ in the demodulation system can be expressed as

$$w(t) = \tfrac{1}{2}x(t) + \tfrac{1}{2}x(t)\cos(2\omega_c t + 2\theta_c)$$

(b) If $x(t)$ has a spectrum that is zero for $|\omega| \geq \omega_M$, determine the relationships required among W [the cutoff frequency of the ideal lowpass filter in Figure 7.9(b)], ω_c (the carrier frequency), and ω_M so that the output of the lowpass filter is proportional to $x(t)$. Does your answer depend on the carrier phase θ_c?

7.2. In Figure P7.2-1 a system is shown with input signal $x(t)$ and output signal $y(t)$. The input signal has the Fourier transform spectrum $X(\omega)$ shown in Figure P7.2-2. Determine and sketch $Y(\omega)$, the spectrum of $y(t)$.

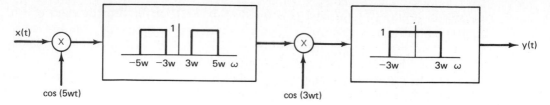

x(t)

cos (5wt)

cos (3wt)

y(t)

Figure P7.2-1

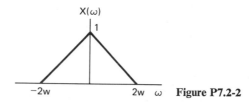

$X(\omega)$

1

$-2w$ $2w$ ω **Figure P7.2-2**

7.3. In Section 7.1 we discussed the effect of a loss in synchronization in phase between the carrier signals in the modulator and demodulator for sinusoidal amplitude modulation. Specifically, we showed that the output of the demodulation is attenuated by the cosine of the phase difference, and in particular, when the modulator and demodulator have a phase difference of $\pi/2$, the demodulator output is zero. As we demonstrate in this problem, it is also important to have *frequency* synchronization between the modulator and demodulator.

Consider the amplitude modulation and demodulation systems in Figure 7.9 with $\theta_c = 0$ and with a change in the *frequency* of the demodulator carrier so that

$$w(t) = y(t) \cos \omega_d t$$

where

$$y(t) = x(t) \cos \omega_c t$$

Let us denote the frequency difference between the modulator and demodulator as $\Delta\omega$ (i.e., $(\omega_d - \omega_c) = \Delta\omega$). Also, assume that $x(t)$ is bandlimited with $X(\omega) = 0$ for $|\omega| \geq \omega_M$ and assume that the cutoff frequency W of the lowpass filter in the demodulator satisfies the inequality

$$(\omega_M + \Delta\omega) < W < (2\omega_c + \Delta\omega - \omega_M)$$

(a) Show that the output of the demodulator lowpass filter is proportional to $x(t) \cos (\Delta\omega t)$.

(b) If the spectrum of $x(t)$ is that shown in Figure P7.3, sketch the spectrum of the output of the demodulator.

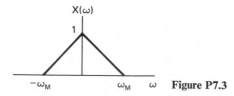

$X(\omega)$

1

$-\omega_M$ ω_M ω **Figure P7.3**

7.4. In discussing amplitude-modulation systems, the modulation and demodulation were carried out through the use of a multiplier. Since multipliers are often difficult to implement, many practical systems use a nonlinearity. In this problem we illustrate the basic concept.

In Figure P7.4 we show one such system for amplitude modulation. The system consists of squaring the *sum* of the modulating signal and the carrier, and then band-pass-filtering to obtain the amplitude-modulated signal.

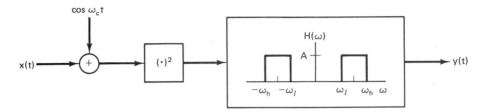

Figure P7.4

Assume that $x(t)$ is bandlimited so that $X(\omega) = 0$, $|\omega| > \omega_M$. Determine the bandpass filter parameters A, ω_ℓ, and ω_h so that $y(t)$ is an amplitude-modulated version of $x(t)$ [i.e., so that $y(t) = x(t) \cos \omega_c t$]. Specify the necessary constraints, if any, on ω_c and ω_M.

7.5. The following scheme has been proposed to perform amplitude modulation. The input signal $x(t)$ is added to the carrier signal $\cos \omega_c t$ and then put through a nonlinear device so that the output $z(t)$ is related to the input by

$$z(t) = e^{y(t)} - 1$$
$$y(t) = x(t) + \cos \omega_c t$$

This is illustrated in Figure P7.5-1. Such a nonlinear relation can be implemented

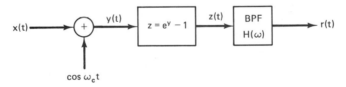

Figure P7.5-1

through the current/voltage characteristics of a diode where, with $i(t)$ and $v(t)$ the diode current and voltage, respectively,

$$i(t) = I_0 e^{av(t)} - 1 \qquad (a \text{ real})$$

To study the effects of this nonlinearity, we can examine the spectrum of $z(t)$ and how it relates to $X(\omega)$ and ω_c. To accomplish this, we use the power series for e^y, which is

$$e^y = 1 + y + \tfrac{1}{2}y^2 + \tfrac{1}{6}y^3 + \cdots$$

(a) If the spectrum of $x(t)$ is given by Figure P7.5-2, and if $\omega_c = 100\omega_1$, sketch and label $Z(\omega)$, the spectrum of $z(t)$, using the first four terms in the power series for e^y.

(b) The bandpass filter (BPF) has parameters as shown in Figure P7.5-3. Determine the range of α and the range of β such that $r(t)$ is an amplitude-modulated version of $x(t)$.

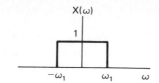

Figure P7.5-2

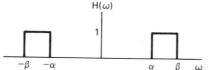

Figure P7.5-3

7.6. In Figure P7.6 is shown a system to be used for sinusoidal amplitude modulation, where $x(t)$ is bandlimited with maximum frequency ω_M so that $X(\omega) = 0 \; |\omega| > \omega_M$. As indicated, the signal $s(t)$ is a periodic impulse train with period T and with an offset from $T = 0$ of Δ. The system $H(\omega)$ is a bandpass filter.

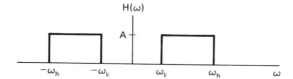

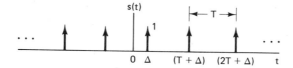

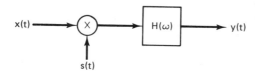

Figure P7.6

(a) With $\Delta = 0$, $\omega_M = \pi/2T$, $\omega_\ell = \pi/T$, and $\omega_h = 3\pi/T$, show that $y(t)$ is proportional to $x(t) \cos \omega_c t$, where $\omega_c = 2\pi/T$.

(b) If ω_M, ω_ℓ, and ω_h are the same as given in part (a) but Δ is not necessarily zero, show that $y(t)$ is proportional to $x(t) \cos (\omega_c t + \theta_c)$, and determine ω_c and θ_c as a function of T and Δ.

(c) Determine the maximum allowable value of ω_M relative to T so that $y(t)$ is proportional to $x(t) \cos (\omega_c t + \theta_c)$.

7.7. The modulation/demodulation scheme proposed below is similar to sinusoidal amplitude modulation except that the demodulation is done with a square wave with the same zero crossings as $\cos \omega_c t$. The system is shown in Figure P7.7-1, where the relation between $\cos \omega_c t$ and $p(t)$ is shown in Figure P7.7-2. Let the input signal $x(t)$ be a bandlimited signal with maximum frequency $\omega_M < \omega_c$, as shown in Figure P7.7-3.

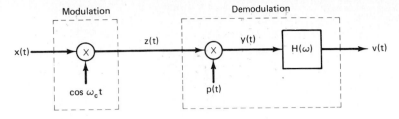

Figure P7.7-1

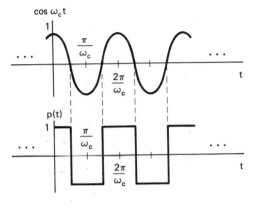

Figure P7.7-2

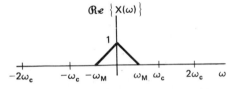

Figure P7.7-3

(a) Sketch and dimension the real and imaginary parts of $Z(\omega)$, $P(\omega)$, and $Y(\omega)$, the Fourier transforms of $z(t)$, $p(t)$, and $y(t)$, respectively.

(b) Sketch and dimension a filter $H(\omega)$ so that $v(t) = x(t)$.

7.8. In Figure P7.8-1 a communication system is shown that transmits a bandlimited signal

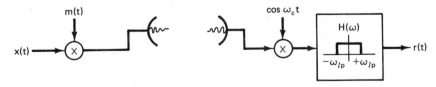

Figure P7.8-1

$x(t)$ as periodic bursts of high-frequency energy. Assume that $X(\omega) = 0$ for $|\omega| > \omega_M$. Two possible choices are to be considered for the modulating signal $m(t)$, which we denote as $m_1(t)$ and $m_2(t)$. $m_1(t)$ is a periodic train of sinusoidal pulses, each of duration D, as shown in Figure P7.8-2, that is,

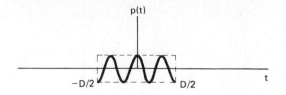

$p(t)$

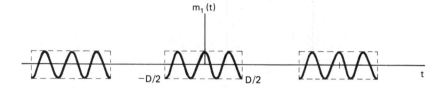

$m_1(t)$

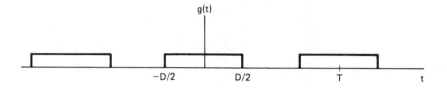

$g(t)$

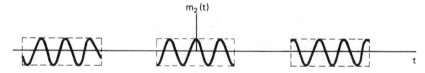

$m_2(t)$

Figure P7.8-2

$$m_1(t) = \sum_{k=-\infty}^{+\infty} p(t - kT)$$

where

$$p(t) = \begin{cases} \cos \omega_c t & \text{for } |t| < \dfrac{D}{2} \\ 0 & \text{for } |t| > \dfrac{D}{2} \end{cases}$$

$m_2(t)$ is $\cos \omega_c t$ periodically blanked or gated, that is, $m_2(t) = g(t) \cos \omega_c t$, where $g(t)$ is shown in Figure P7.8-2.

The following relationships between the parameters T, D, ω_c, and ω_M are to be assumed:

$$D < T$$

$$\omega_c \gg \frac{2\pi}{D}$$

$$\frac{2\pi}{T} > 2\omega_M$$

Also, assume that sinc (x) is negligible for $x \gg 1$.

Determine whether for some choice of ω_{1p} either $m_1(t)$ or $m_2(t)$ will result in a demodulated signal $r(t)$ that is proportional to the signal $x(t)$. For each case in which your answer is yes, determine an acceptable range for ω_{1p}.

7.9. In Section 7.1.1 we discussed the use of an envelope detector for asynchronous demodulation of an AM signal of the form $y(t) = [x(t) + A] \cos(\omega_c t + \theta_c)$. An alternative demodulation system, which also does not require phase synchronization but does require frequency synchronization, is shown in block-diagram form in Figure P7.9. The lowpass filters both have a cutoff frequency of ω_c. $y(t)$ is of the form $y(t) = [x(t) + A] \cos(\omega_c t + \theta_c)$ with θ_c constant but unknown. $x(t)$ is band-limited with $X(\omega) = 0$, $|\omega| > \omega_M$ and with $\omega_M < \omega_c$. As was required for the use of an envelope detector, $x(t) + A > 0$ for all t.

Show that the system in Figure P7.9 can be used to recover $x(t)$ from $y(t)$ without knowledge of the modulator phase θ_c.

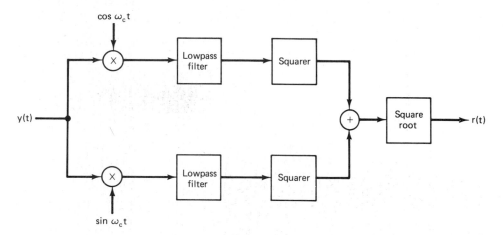

Figure P7.9

7.10. As discussed in Section 7.1.1, asynchronous modulation–demodulation requires the injection of the carrier signal so that the modulated signal is of the form

$$y(t) = [A + x(t)] \cos(\omega_c t + \theta_c) \qquad (\text{P } 7.10)$$

where $[A + x(t)] > 0$ for all t. The presence of the carrier means that more transmitter power is required, representing an inefficiency.

(a) Let $x(t)$ be given by $x(t) = \cos \omega_M t$ with $\omega_M < \omega_c$ and $[A + x(t)] > 0$. For a periodic signal $y(t)$ with period T, the time average power P_y is defined as $P_y = (1/T) \int_T y^2(t) \, dt$. Determine and sketch P_y for $y(t)$ in eq. (P7.10). Express your answer as a function of the modulation index m, defined as the maximum absolute value of $x(t)$, divided by A.

(b) The efficiency of transmission of an amplitude-modulated signal is defined to be the ratio of the power in the sidebands of the signal to the total power in the signal. With $x(t) = \cos \omega_M t$ and with $\omega_M < \omega_c$ and $[A + x(t)] > 0$, determine and sketch the efficiency of the modulated signal as a function of the modulation index m.

7.11. In Figure P7.11 a system is proposed whereby a bandpass filter can be synthesized out of two lowpass filters and an oscillator. Let the input signal be bandlimited with the spectrum $X(\omega)$ shown in Figure P7.11, where we have shaded the negative-frequency portion to distinguish it from the positive-frequency portion.

With $\omega_M < \omega_c$, $\omega_1 < \omega_c$, and $\omega_2 > (\omega_c - \omega_1)$, sketch the spectra of $r_1(t)$, $r_2(t)$, $r_3(t)$, and $y(t)$. Demonstrate, in particular, that the overall system is equivalent

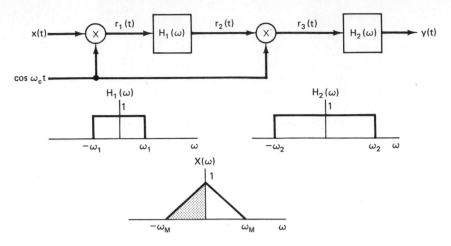

Figure P7.11

to a bandpass filter and determine the upper and lower cutoff frequencies of the bandpass filter in terms of ω_c, ω_1, and ω_2.

7.12. In Section 7.2.1 we discussed the use of amplitude modulation with a complex exponential carrier to implement a bandpass filter. The specific system was shown in Figure 7.16, and if only the real part of $f(t)$ is retained, the equivalent bandpass filter is that shown in Figure 7.20.

In Figure P7.12 we indicate an implementation of a bandpass filter using sinu-

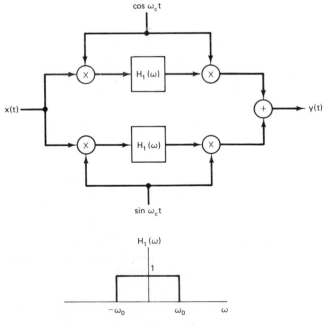

Figure P7.12

soidal modulation and lowpass filters. Show that the output $y(t)$ of this system is identical to that which would be obtained by retaining only $\mathcal{R}e\{f(t)\}$ in Figure 7.16.

7.13. The system shown in Figure P7.13-1 is proposed to implement a lowpass filter. The system $H(\omega)$ is an ideal "90° phase shifter" for which the frequency response is

$$H(\omega) = \begin{cases} e^{+j(\pi/2)} \, (=j), & \omega > 0 \\ e^{-j(\pi/2)} \, (=-j), & \omega < 0 \end{cases}$$

The system output $y(t)$ is the imaginary part of the complex signal $r_3(t)$.

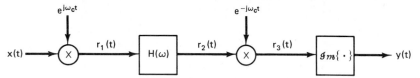

Figure P7.13-1

(a) With $X(\omega)$ real and as sketched in Figure P7.13-2, sketch the spectra of $r_1(t)$, $r_2(t)$, $r_3(t)$, and $y(t)$. Demonstrate in particular that the system implements an ideal lowpass filter and determine the cutoff frequency of the lowpass filter as a function of the modulating frequency ω_c.

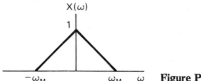

Figure P7.13-2

(b) Show that the system in Figure P7.13-3 is equivalent to the system in Figure P7.13-1, where the system $H(\omega)$ is identical in both cases, and $x(t)$ is real.

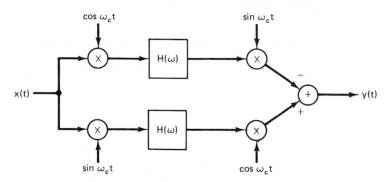

Figure P7.13-3

7.14. A commonly used system to maintain privacy in voice communications is a speech scrambler. As illustrated in Figure P7.14-1, the input to the system is a normal speech signal $x(t)$ and the output is the scrambled version $y(t)$. The signal $y(t)$ is transmitted and then unscrambled at the receiver.

    We assume that all inputs to the scrambler are real and bandlimited to frequency ω_M, that is, $X(\omega) = 0$ for $|\omega| > \omega_M$. Given any such input, our proposed scrambler permutes different bands of the input signal spectrum. In addition, the output signal

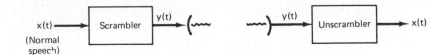

Figure P7.14-1

is real and bandlimited to the same frequency band, that is, $Y(\omega) = 0$ for $|\omega| > \omega_M$. The specific permuting algorithm for our scrambler is

$$Y(\omega) = X(\omega - \omega_M), \qquad \omega > 0$$
$$Y(\omega) = X(\omega + \omega_M), \qquad \omega < 0$$

(a) If $X(\omega)$ is given by the spectrum shown in Figure P7.14-2, sketch the spectrum of the scrambled signal $y(t)$.

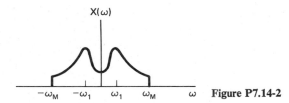

Figure P7.14-2

(b) Using amplifiers, multipliers, adders, oscillators, and whatever ideal filters you find necessary, draw the block dragram for such an ideal scrambler.

(c) Again, using amplifiers, multipliers, adders, oscillators, and ideal filters, draw a block diagram for the associated unscrambler.

7.15. In practice it is often very difficult to build an amplifier at very low frequencies. Consequently, low-frequency amplifiers typically exploit the principles of amplitude modulation to shift the signal into a higher-frequency band. Such an amplifier is referred to as a chopper amplifier and is illustrated in block-diagram form in Figure P7.15.

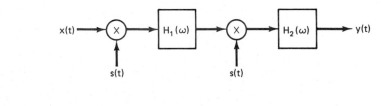

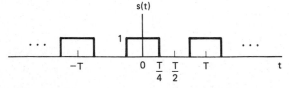

Figure P7.15

(a) Determine in terms of T the highest allowable frequency present in $x(t)$ if $y(t)$ is to be proportional to $x(t)$ (i.e., if the overall system is to be equivalent to an amplifier).

(b) With $x(t)$ bandlimited as specified in part (a), determine the gain of the overall system in Figure P7.15 in terms of A and T.

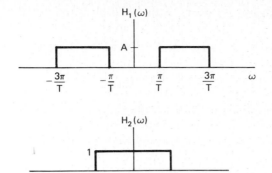

Figure P7.15 (cont.)

7.16. We wish to communicate one of two possible messages, message m_0 or message m_1. To do so we will send a burst of one of two frequencies over a time interval of length T. Note that T is independent of which message is being transmitted. For message m_0 we will send $\cos \omega_0 t$ and for message m_1 we will send $\cos \omega_1 t$. Thus, a burst $b(t)$ will look as shown in Figure P7.16-1. Such a communications system is called *frequency shift keying* (FSK). When the burst of frequency $b(t)$ is received, we wish to determine if it represents message m_0 or message m_1. To accomplish this we do as shown in Figure P7.16-2.

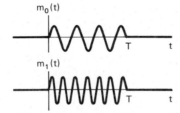

Figure P7.16-1

(a) Show that the maximum difference between the absolute values of the two lines in Figure P7.16-2 occurs when $\cos \omega_0 t$ and $\cos \omega_1 t$ have the relationship

$$\int_0^T \cos \omega_0 t \cos \omega_1 t \, dt = 0$$

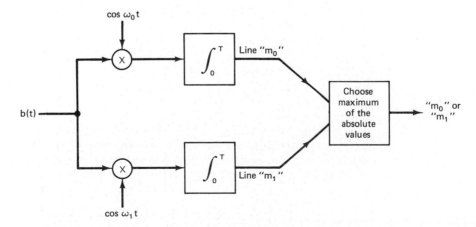

Figure P7.16-2

(b) Is it possible to choose ω_0 and ω_1 such that there is *no* interval of length T for which

$$\int_0^T \cos \omega_0 t \cos \omega_1 t \, dt = 0 ?$$

7.17. As described in the text, it is quite difficult to build a filter with good frequency selectivity and gain that can be tuned over its entire range of operation. Since the accurate demultiplexing and demodulation of radio and television signals require this type of tunable filter, a system called the superheterodyne receiver has been designed to be equivalent to this tunable filter. The basic system is shown in Figure P7.17-1.

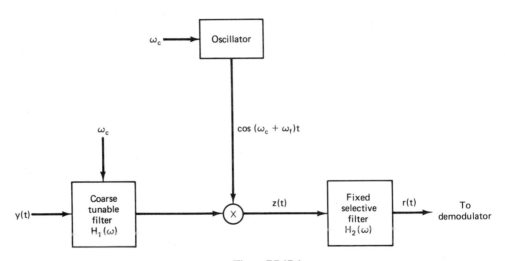

Figure P7.17-1

(a) The input signal $y(t)$ consists of the superposition of many amplitude modulated signals which have been multiplexed using frequency division multiplexing so that they each occupy different frequency channels. Let us consider one such channel which contains the amplitude modulated signal $y_1(t) = x_1(t) \cos \omega_c t$ with spectrum $Y_1(\omega)$ as depicted in Fig. P7.17-2. We want to demultiplex and demodulate to recover the modulating signal $x_1(t)$, using the system of Fig. P7.17-1. The coarse tunable filter has the spectrum $H_1(\omega)$ shown in Figure P7.17-2(b). Determine the spectrum $Z(\omega)$ of the input signal to the fixed selective filter $H_2(\omega)$. Sketch and label the spectrum $Z(\omega)$ for $\omega > 0$.

(b) The fixed frequency selective filter is a bandpass type centered around fixed frequency ω_f, as shown in Figure P7.17-3. We would like the output of filter $H_2(\omega)$ to be $r(t) = x_1(t) \cos \omega_f t$. In terms of ω_c and ω_M, what constraint must ω_T satisfy to guarantee that an undistorted spectrum of $x_1(t)$ is centered around $\omega = \omega_f$?

(c) What must G, α, and β be in Figure P7.17-3 so that $r(t) = x_1(t) \cos \omega_f t$?

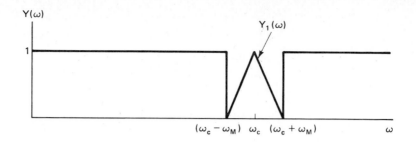

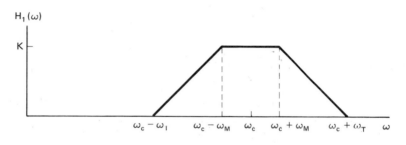

(a) Input signal

(b) Coarse tunable filter

Figure P7.17-2

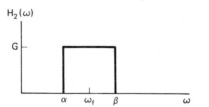

Figure P7.17-3

7.18. In Section 7.2.2 we discussed the use of sinusoidal amplitude modulation for frequency-division multiplexing whereby several signals are shifted into different frequency bands and then summed for simultaneous transmission. In this problem we explore another multiplexing concept referred to as *quadrature multiplexing*. In this multiplexing procedure two signals can be transmitted simultaneously in the same frequency band if the two carrier signals are 90° out of phase. The multiplexing system is shown in Figure P7.18(a) and the demultiplexing system in Figure P7.18(b). $x_1(t)$ and $x_2(t)$ are both assumed to be bandlimited with maximum frequency ω_M, so that $X_1(\omega) = X_2(\omega) = 0$ for $|\omega| > \omega_M$. The carrier frequency ω_c is assumed to be greater than ω_M. Show that $y_1(t) = x_1(t)$ and $y_2(t) = x_2(t)$.

7.19. In Figure 7.27 we indicated a system for implementing single-sideband modulation in which only the lower sidebands are retained and in Figure 7.28 illustrated the corresponding spectra, assuming that $X(\omega)$ is real and symmetric. By sketching the spectra of $y_1(t)$, $y_2(t)$, and $y(t)$ for $X(\omega)$, as indicated in Figure P7.19, show that for $X(\omega)$ imaginary and antisymmetric the system also retains only the lower sidebands.

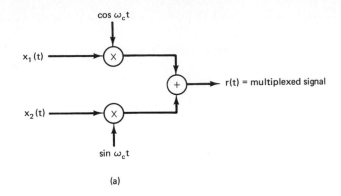

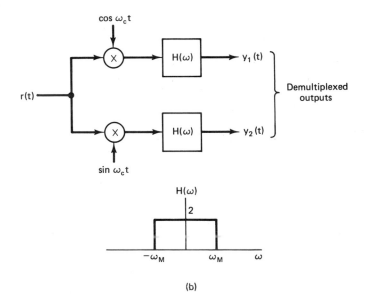

(a)

(b)

Figure P7.18

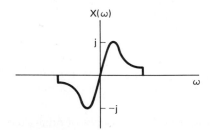

Figure P7.19

7.20. In Section 7.3 we discussed the implementation of single-sideband modulation using 90° phase-shift networks and in Figures 7.27 and 7.28 specifically illustrated the system and associated spectra to retain the lower sidebands.

In Figure P7.20-1 is shown the corresponding system for retaining the upper sidebands.

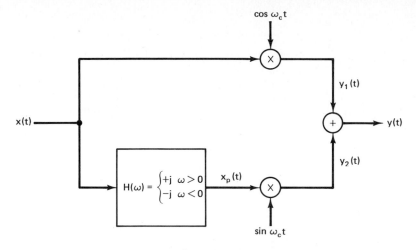

Figure P7.20-1

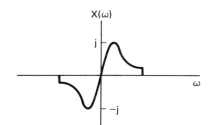

Figure P7.20-2

(a) With the same $X(\omega)$ illustrated in Figure 7.28, sketch $Y_1(\omega)$, $Y_2(\omega)$, and $Y(\omega)$ for the system in Figure P7.20-1, and demonstrate specifically that only the upper sidebands are retained.

(b) For $X(\omega)$ imaginary, as illustrated in Figure P7.20-2, sketch $Y_1(\omega)$, $Y_2(\omega)$, and $Y(\omega)$ for the system in Figure P7.20-1, and demonstrate that for this case also, only the upper sidebands are retained.

7.21. Single-sideband modulation discussed in Section 7.3 is commonly used in point-to-point voice communication. It offers many advantages, including effective use of available power, bandwidth conservation, and insensitivity to some forms of random fading in the channel. In double-sideband suppressed carrier (DSB/SC) systems the spectrum of the modulating signal appears in its entirety in two places in the transmitted spectrum. Single-sideband modulation eliminates this redundancy, thus conserving bandwidth and increasing the signal-to-noise ratio within the remaining portion of the spectrum that is transmitted.

In Figure P7.21-1 two methods of generating an amplitude-modulated single-sideband signal are shown. The system in Figure P7.21-1(a) can be used to generate a single-sideband signal for which the lower sideband is retained and the system in Figure P7.21-1(b) a single-sideband signal for which the upper sideband is retained.

(a) For $X(\omega)$ as shown in Figure P7.21-2, determine and sketch $S(\omega)$, the Fourier transform of the lower-sideband modulated signal and $R(\omega)$, the Fourier transform of the upper-sideband modulated signal. Assume that $\omega_c > \omega_3$.

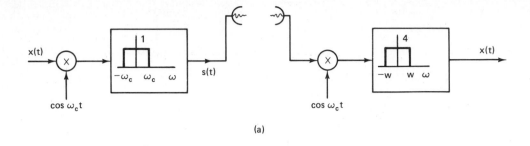

(a)

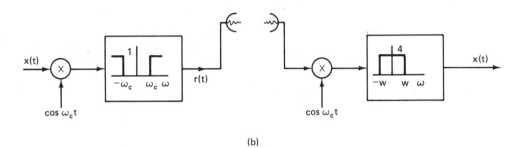

(b)

Figure P7.21-1

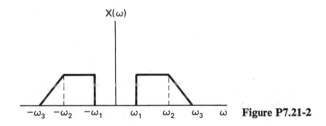

Figure P7.21-2

The upper-sideband modulation scheme is particularly useful with voice communications, as any real filter has a finite transition region for the cutoff (i.e., near ω_c). This region can be accommodated with negligible distortion, as the voice signal does not have any significant energy near $\omega = 0$ (i.e., for $|\omega| < \omega_1 = 2\pi \times 40$ Hz).

(b) Another (third) procedure for generating a single-sideband signal is termed the phase-shift method and is illustrated in Figure P7.21-3. Show that the single-sideband signal generated is proportional to that generated by the lower-sideband modulation scheme of Figure P7.21-1(a) [i.e., $p(t)$ is proportional to $s(t)$].

(c) All three AM-SSB signals can be demodulated using the scheme shown on the right-hand side of Figure P7.21-1. Show that whether the received signal is $s(t)$, $r(t)$, or $p(t)$, as long as the oscillator at the receiver is in phase with oscillators at the transmitter, and $w = \omega_c$, the output of the demodulator is $x(t)$.

The distortion that results when the oscillator is not in phase with the transmitter, called quadrature distortion, can be particularly troublesome in data communications.

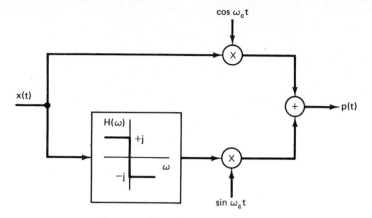

Figure P7.21-3

7.22. A pulse-amplitude-modulated (PAM) system for communications may be modeled as in Figure P7.22-1. The output of the system is $q(t)$, the PAM signal.

    (a) Let $x(t)$ be a bandlimited signal [i.e., $X(\omega) = 0\ |\omega| \geq \pi/T$], as shown in Figure P7.22-2. Determine and sketch $R(\omega)$ and $Q(\omega)$.

    (b) Find the maximum value of Δ such that $w(t) = x(t)$ with an appropriate filter $M(\omega)$.

    (c) Determine and sketch a compensating filter $M(\omega)$ such that $w(t) = x(t)$.

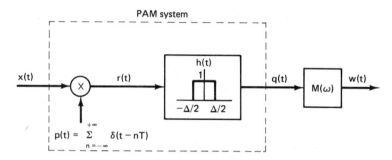

Figure P7.22-1

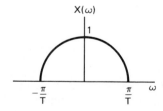

Figure P7.22-2

7.23. Consider a discrete-time signal $x[n]$ with Fourier transform shown in Figure P7.23-1. The signal is amplitude-modulated by a sinusoidal sequence, as indicated in Figure P7.23-2.

    (a) Determine and sketch $Y(\Omega)$, the Fourier transform of $y[n]$.

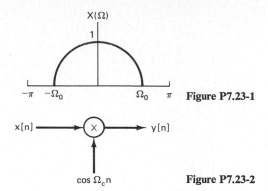

Figure P7.23-1

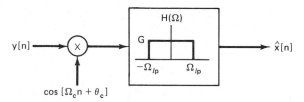

Figure P7.23-2

(b) A proposed demodulation system is shown in Figure P7.23-3. For what value of θ_c, Ω_{1p}, and G will $\hat{x}[n] = x[n]$? Are any restrictions on Ω_c and Ω_{1p} necessary to guarantee that $x[n]$ is recoverable from $y[n]$?

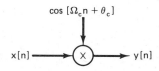

Figure P7.23-3

7.24. In Section 7.5 we considered synchronous discrete-time modulation and demodulation with a sinusoidal carrier. In this problem we want to consider the effect of a loss in synchronization in phase and/or frequency. The modulation and demodulation systems are shown in Figure P7.24-1, where both a phase and frequency difference between

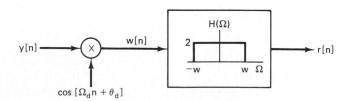

Figure P7.24-1

Modulation   Chap. 7

the modulator and demodulator carriers is indicated. Let the frequency difference $\Omega_d - \Omega_c$ be denoted as $\Delta\Omega$ and the phase difference $\theta_d - \theta_c$ as $\Delta\theta$.

(a) If the spectrum of $x[n]$ is that shown in Figure P7.24-2, sketch the spectrum of $w[n]$, assuming $\Delta\Omega = 0$.

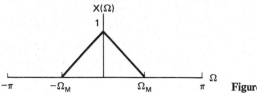

Figure P7.24-2

(b) If $\Delta\Omega = 0$, show that w can be chosen so that the output $r[n]$ is $r[n] = x[n] \cos \Delta\theta$. In particular what is $r[n]$ if $\Delta\theta = \pi/2$?

(c) For $\Delta\theta = 0$, and $w = \Omega_M + \Delta\Omega$, show that the output $r[n] = x[n] = x[n] \cos[\Delta\Omega n]$ (assume that $\Delta\Omega$ is small).

7.25 Let $x[n]$ be a discrete-time signal with spectrum $X(\Omega)$ and let $p(t)$ be a continuous-time pulse function with spectrum $P(\omega)$. We form the pulse-amplitude-modulated signal $y(t)$ as follows:

$$y(t) = \sum_{n=-\infty}^{+\infty} x[n]p(t - n)$$

(a) Determine the spectrum $Y(\omega)$ in terms of $X(\Omega)$ and $P(\omega)$.

(b) If

$$p(t) = \begin{cases} \cos 8\pi t, & 0 \le t \le 1 \\ 0, & \text{elsewhere} \end{cases}$$

determine $P(\omega)$ and $Y(\omega)$.

7.26. When designing filters with highpass or bandpass characteristics, it is often convenient to first design a lowpass filter with the desired passband and stopband specifications and then transform this prototype filter to the desired highpass or bandpass filter. Such transformations are called lowpass-to-highpass or lowpass-to-bandpass transformations. Designing filters in this manner is convenient because it only requires us to formulate our filter design algorithms for the class of filters with lowpass characteristics. As one example of such a procedure, consider a discrete-time lowpass filter with impulse response $h_{lp}[n]$ and frequency response $H_{lp}(\Omega)$ as sketched in Figure P7.26. The impulse response is modulated with the sequence $(-1)^n$ to obtain $h_{hp}[n] =$. $(-1)^n h_{lp}[n]$.

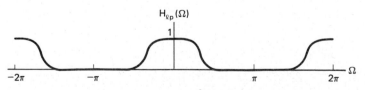

Figure P7.26

(a) Determine and sketch $H_{\mathrm{hp}}(\Omega)$ in terms of $H_{\mathrm{lp}}(\Omega)$. Show in particular that for $H_{\mathrm{lp}}(\Omega)$ as shown in Figure P7.26, $H_{\mathrm{hp}}(\Omega)$ corresponds to a highpass filter.

(b) Show that modulation of the impulse response of a discrete-time highpass filter by $(-1)^n$ will transform it to a lowpass filter.

7.27. In Problem 7.26 we showed that modulation of the impulse response of a lowpass filter by $(-1)^n$ will convert it to a highpass filter, and vice versa. In this problem we consider how to effect this modulation by appropriately modifying the coefficients in the difference equation which implements the filter.

Consider a lowpass filter with impulse response $h_{\mathrm{lp}}[n]$, frequency response $H_{\mathrm{lp}}(\Omega)$, and for which the difference equation relating the input $x[n]$ and output $y[n]$ is

$$\sum_{k=0}^{N} a_k y[n-k] = \sum_{k=0}^{M} b_k x[n-k] \tag{P7.27-1}$$

(a) Express the frequency response $H_{\mathrm{lp}}(\Omega)$ in terms of the difference-equation coefficients.

(b) The lowpass filter $h_{\mathrm{lp}}[n]$ is to be transformed to a highpass filter $h_{\mathrm{hp}}[n]$ by modulation with $(-1)^n$, that is,

$$h_{\mathrm{hp}}[n] = (-1)^n h_{\mathrm{lp}}[n]$$

Express the frequency response $H_{\mathrm{hp}}(\Omega)$ in terms of $H_{\mathrm{lp}}(\Omega)$.

(c) By combining your results in parts (a) and (b), show that if the difference equation (P7.27-1) implements $H_{\mathrm{lp}}(\Omega)$, then the difference equation

$$\sum_{k=0}^{N} (-1)^k a_k y[n-k] = \sum_{k=0}^{M} (-1)^k b_k x[n-k] \tag{P7.27-2}$$

implements $H_{\mathrm{hp}}(\Omega)$.

7.28. In Problem 7.13 we considered a continuous-time system that implements a lowpass filter using modulators and 90° phase shifters. In Figures P7.28-1 and P7.28-2 are shown the discrete-time counterparts to the systems in Figures P7.13-1 and P7.13-3.

(a) Show that the systems of Figures P7.28-1 and P7.28-2 are equivalent.

(b) Determine whether or not the foregoing systems are equivalent to an ideal lowpass filter. If so, determine the filter cutoff frequency as a function of Ω_c.

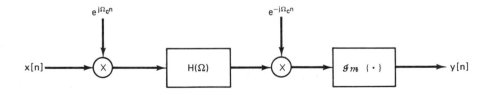

$$H(\Omega) = \begin{cases} e^{+j\frac{\pi}{2}} & 0 < \Omega < \pi \\ e^{-j\frac{\pi}{2}} & -\pi < \Omega < 0 \end{cases}$$

Figure P7.28-1

Modulation   Chap. 7

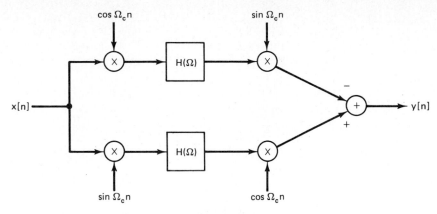

Figure P7.28-2

7.29. Consider the discrete-time system shown in Figure P7.29-1. The input sequence $x[n]$ is multiplied by $\phi_1[n]$ and the product is taken as the input to an LTI system. The final output $y[n]$ is then obtained as the product of the output of the LTI system multiplied by $\phi_2[n]$.

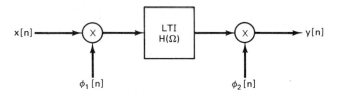

Figure P7.29-1

(a) In general, is the overall system linear? Is it shift-invariant?

(b) If $\phi_1[n] = z^{-n}$ and $\phi_2[n] = z^n$, where z is any complex number, show that the overall system is LTI.

(c) If $\phi_1[n] = (j)^{-n}$, $\phi_2[n] = j^n$, and $H(\Omega)$ is as shown in Figure P7.29-2, sketch the frequency response (magnitude and phase) of the overall system; that is, sketch $Y(\Omega)/X(\Omega)$ as a function of Ω.

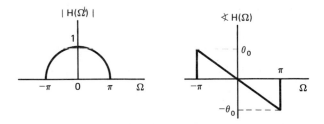

Figure P7.29-2

7.30. A discrete-time filter bank is to be implemented by using a basic lowpass filter and appropriate complex exponential amplitude modulation as indicated in Figure P7.30-1.

(a) With $H(\Omega)$ an ideal lowpass filter, as shown in Figure P7.30-2(a), the ith channel

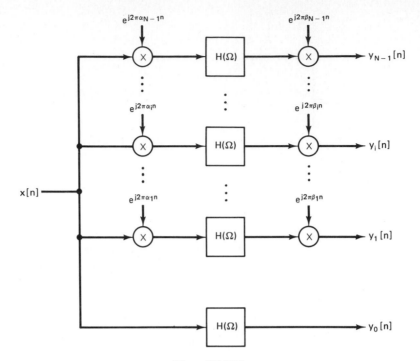

Figure P7.30-1

of the filter bank is to be equivalent to a bandpass filter with frequency response shown in Figure P7.30-2(b). Determine the values of α_i and β_i to accomplish this.

(b) Again with $H(\Omega)$ as in Figure P7.30-2, and with $\Omega_i = 2\pi i/N$, determine the value of Ω_0 in terms of N so that the filter bank covers the entire frequency band without any overlap.

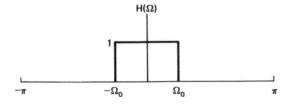

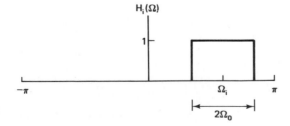

Figure P7.30-2

Modulation   Chap. 7

7.31. The continuous-time system illustrated in Figure P7.31-1 and the discrete-time system illustrated in Figure P7.31-2 are similar in the sense that both use the same configuration to make a lowpass filter appear in the overall system response like a bandpass filter.

(a) Show that in both cases, whatever the input is to the lowpass filter, the output $y(t)$ or $y[n]$ will be a bandpass signal.

(b) Determine and sketch the frequency response of the overall system of Figures P7.31-1 and P7.31-2.

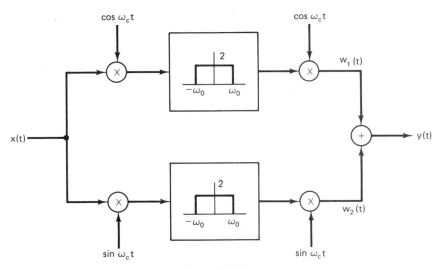

Figure P7.31-1

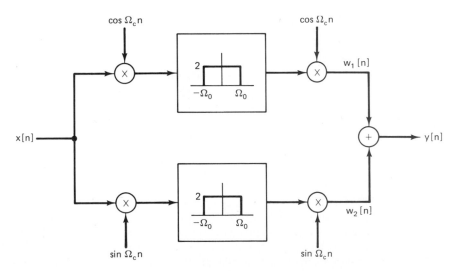

Figure P7.31-2

7.32. In Problem 7.18 we introduced the concept of *quadrature multiplexing*, whereby two signals are summed after modulating each with carrier signals of identical frequency but with a phase difference of 90°. The corresponding discrete-time multiplexer and demultiplexer are shown in Figure P7.32. The signals $x_1[n]$ and $x_2[n]$ are both assumed to be bandlimited with maximum frequency Ω_M so that

$$X_1(\Omega) = X_2(\Omega) = 0 \qquad \text{for } \Omega_M < \Omega < (2\pi - \Omega_M)$$

(a) Determine the range of values for Ω_c so that $x_1[n]$ and $x_2[n]$ can be recovered from $r[n]$.

(b) With Ω_c satisfying the conditions in part (a), determine $H(\Omega)$ so that $y_1[n] = x_1[n]$ and $y_2[n] = x_2[n]$.

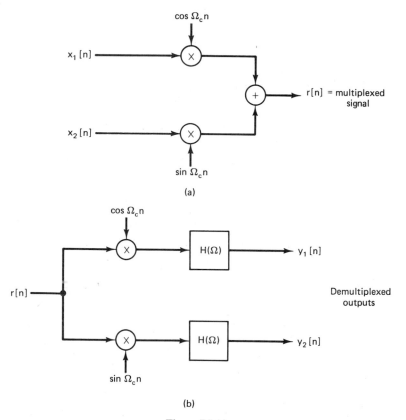

(a)

(b)

Figure P7.32

7.33. In Section 7.5 we introduced the notion of discrete-time pulse amplitude modulation (PAM) and indicated that its analysis paralleled very closely that of continuous-time pulse amplitude modulation as discussed in Section 7.4. In this problem, we consider the analysis of discrete-time PAM. The system to be considered is that shown in Figure 7.38.

(a) Determine and sketch the discrete-time Fourier transform of the periodic square-wave signal $p[n]$ in Figure 7.38.

(b) Assume that $x[n]$ has the spectrum shown in Figure P7.33. With $\Omega_M = \pi/2N$ and with $M = 1$ in Figure 7.38, sketch $Y(\Omega)$, the Fourier transform of $y[n]$.

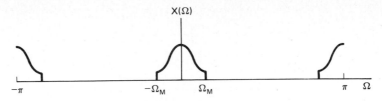

Figure P7.33

(c) Now assume that $X(\Omega)$ is known to be bandlimited with $X(\Omega) = 0$, $\Omega_M < \Omega$ $< (2\pi - \Omega_M)$, but is otherwise unspecified. For the system of Figure 7.38 determine, as a function of N, the maximum allowable value of Ω_M that will permit $x[n]$ to be recovered from $y[n]$. Indicate whether your result depends on M.

(d) With Ω_M and N satisfying the conditions determined in part (c), state or show in block-diagram form how to recover $x[n]$ from $y[n]$.

7.34. A bandlimited signal $x(t)$ is to be transmitted using narrowband FM techniques. By narrowband we mean that the modulation index m, as defined in Section 7.6, is much less than $\pi/2$. Before $x(t)$ is transmitted to the modulator, it is processed so that $X(\omega)|_{\omega=0} = 0$ and $|x(t)| < 1$. This normalized $x(t)$ is now used to angle-modulate a carrier to form the FM signal

$$y(t) = \cos\left(\omega_c t + m \int_{-\infty}^{t} x(\tau)\, d\tau\right)$$

(a) Determine the instantaneous frequency ω_i.
(b) Using eqs. (7.46), the narrowband assumption ($m \ll \pi/2$), and the normalization conditions cited above, show that

$$y(t) \simeq \cos \omega_c t - \left(m \int_{-\infty}^{t} x(\tau)\, d\tau\right) \sin \omega_c t$$

(c) What is the relationship among the bandwidth of $y(t)$, the bandwidth of $x(t)$, and the carrier frequency ω_c?

7.35. Consider the complex exponential time function

$$s(t) = e^{j\theta(t)} \tag{P7.35}$$

where $\theta(t) = \omega_0 t^2/2$.

Since the instantaneous frequency $\omega_i = d\theta/dt$ is a function of time, this signal may be regarded as an FM signal. In particular, since the signal sweeps linearly through the frequency spectrum with time, it is often called a frequency "chirp" or "chirp signal."

(a) Determine the instantaneous frequency.
(b) Determine and sketch the magnitude and phase of the Fourier transform of the chirp signal. To evaluate the Fourier transform integral you may find it helpful to "complete the square" in the exponent in the integrand and to use the relation

$$\int_{-\infty}^{+\infty} e^{jz^2}\, dz = \sqrt{\frac{\pi}{2}}\,(1 + j)$$

(c) Consider the system in Figure P7.35, where $s(t)$ is the chirp signal in eq. (P7.35). Show that $y(t) = X(\omega_0 t)$, where $X(\omega)$ is the Fourier transform of $x(t)$.

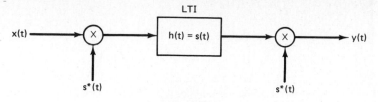

Figure P7.35

Note: The system in Figure P7.35 is referred to as the chirp transform algorithm and is often used in practice to obtain the Fourier transform of a signal.

Under certain conditions a continuous-time signal can be completely represented by and recoverable from knowledge of its instantaneous values or *samples* equally spaced in time. This somewhat surprising property follows from a basic result which is referred to as the *sampling theorem*. This theorem is extremely important and useful. It is exploited, for example, in moving pictures, which consist of a time sequence of individual frames, each of which represents an instantaneous view (i.e., a time sample) of a continuously changing scene. When these samples are viewed in time sequence at a sufficiently fast rate, we perceive an accurate representation of the original continuously moving scene. As another example, printed pictures typically consist of a very fine grid of points, each corresponding to a sample of the spatially continuous picture to be represented. If the samples are sufficiently close together the picture appears to be spatially continuous, although under a magnifying glass its representation in terms of samples becomes evident.

Much of the importance of the sampling theorem also lies in its role as a bridge between continuous-time signals and discrete-time signals. As we develop in some detail, the ability under certain conditions to completely represent a continuous-time signal by a sequence of instantaneous samples provides a mechanism for representing a continuous-time signal by a discrete-time signal. In many contexts, processing of discrete-time signals is more flexible and is often preferable to processing of continuous-time signals, in part because of the increasing availability of inexpensive, lightweight, programmable and easily reproducible digital and discrete-time systems. This technology also offers the possibility of exploiting the concept of sampling to convert

Sampling

a continuous-time signal to a discrete-time signal. After processing the discrete-time signal using a discrete-time system, we can then convert back to continuous time.

In the following discussion, we first introduce and develop the concept of sampling and the process of reconstructing a continuous-time signal from its samples. We then explore the processing of continuous-time signals that have been converted to discrete-time signals through sampling. Next we consider the dual concept to time-domain sampling, specifically sampling in the frequency domain. Finally, we develop the concept and some applications of sampling applied to discrete-time signals.

8.1 REPRESENTATION OF A CONTINUOUS-TIME SIGNAL BY ITS SAMPLES: THE SAMPLING THEOREM

In general, we could not expect that in the absence of any additional conditions or information, a signal could be uniquely specified by a sequence of equally spaced samples. For example, in Figure 8.1 we illustrate three different continuous-time signals, all of which have identical values at integer multiples of T, that is,

$$x_1(kT) = x_2(kT) = x_3(kT)$$

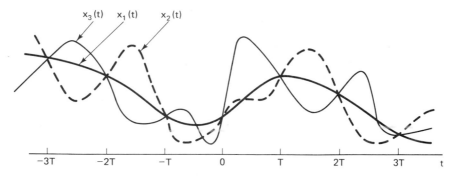

Figure 8.1 Three continuous-time signals with identical values at integer multiples of T.

In general, there are an infinite number of signals that can generate a given set of samples. As we will see, however, if a signal is bandlimited and if the samples are taken sufficiently close together, in relation to the highest frequency present in the signal, then the samples *uniquely* specify the signal and we can reconstruct it perfectly. The basic result was suggested in Section 7.4 in the context of pulse amplitude modulation. Specifically, if a bandlimited signal $x(t)$ is amplitude-modulated with a periodic pulse train, corresponding to extracting equally spaced time segments, it can be recovered exactly by lowpass filtering if the fundamental frequency of the modulating pulse train is greater than twice the highest frequency present in $x(t)$. Furthermore, the ability to recover $x(t)$ is independent of the time duration of the individual pulses. Thus, as suggested by Figures 8.2 and 8.3 as this duration becomes arbitrarily small, pulse amplitude modulation is, in effect, representing $x(t)$ by instantaneous samples equally spaced in time. In the pulse-amplitude-modulation system in Figure 8.2, we

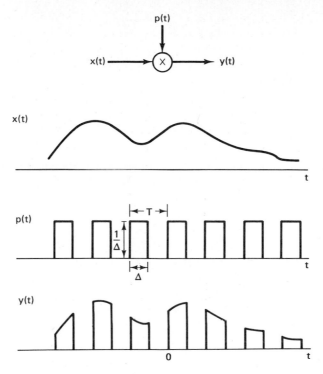

Figure 8.2 Pulse amplitude modulation. As $\Delta \longrightarrow 0$, $p(t)$ approaches an impulse train.

have scaled the amplitude of the pulse train to be inversely proportional to the pulse width Δ. In any practical pulse-amplitude-modulation system, it is particularly important as Δ becomes small to maintain a constant time-average power in the modulated signal. As illustrated in Figure 8.3, as Δ approaches zero the modulated signal then becomes an impulse train for which the individual impulses have values corresponding to instantaneous samples of $x(t)$ at time instants spaced T seconds apart.

8.1.1 Impulse-Train Sampling

In a manner identical to that used to analyze the more general case of pulse amplitude modulation, let us consider the specific case of impulse-train sampling depicted in Figure 8.3. The impulse train $p(t)$ is referred to as the *sampling function*, the period T as the *sampling period*, and the fundamental frequency of $p(t)$, $\omega_s = 2\pi/T$, as the *sampling frequency*. In the time domain we have

$$x_p(t) = x(t)p(t) \tag{8.1a}$$

where

$$p(t) = \sum_{n=-\infty}^{+\infty} \delta(t - nT) \tag{8.1b}$$

$x_p(t)$ is an impulse train with the amplitudes of the impulses equal to the samples of $x(t)$ at intervals spaced by T, that is,

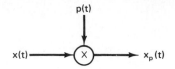

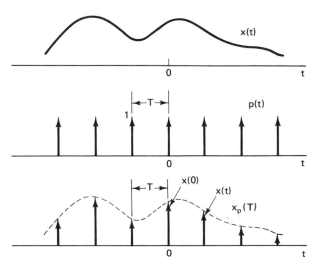

Figure 8.3 Pulse amplitude modulation with an impulse train.

$$x_p(t) = \sum_{n=-\infty}^{+\infty} x(nT)\,\delta(t - nT) \tag{8.2}$$

From the modulation property [Sec. 4.8],

$$X_p(\omega) = \frac{1}{2\pi}[X(\omega) * P(\omega)] \tag{8.3}$$

and from Example 4.15,

$$P(\omega) = \frac{2\pi}{T} \sum_{k=-\infty}^{+\infty} \delta(\omega - k\omega_s) \tag{8.4}$$

so that

$$X_p(\omega) = \frac{1}{T} \sum_{k=-\infty}^{+\infty} X(\omega - k\omega_s) \tag{8.5}$$

That is, $X_p(\omega)$ is a periodic function of frequency consisting of a sum of shifted replicas of $X(\omega)$, scaled by $1/T$ as illustrated in Figure 8.4. In Figure 8.4(c), $\omega_M < (\omega_s - \omega_M)$ or equivalently $\omega_s > 2\omega_M$, and thus there is no overlap between the shifted replicas of $X(\omega)$, whereas in Figure 8.4(d) with $\omega_s < 2\omega_M$, there is overlap. For the case illustrated in Figure 8.4(c), $X(\omega)$ is faithfully reproduced at integer multiples of the sampling frequency. Consequently, if $\omega_s > 2\omega_M$, $x(t)$ can be recovered exactly from $x_p(t)$ by means of a lowpass filter with gain T and a cutoff

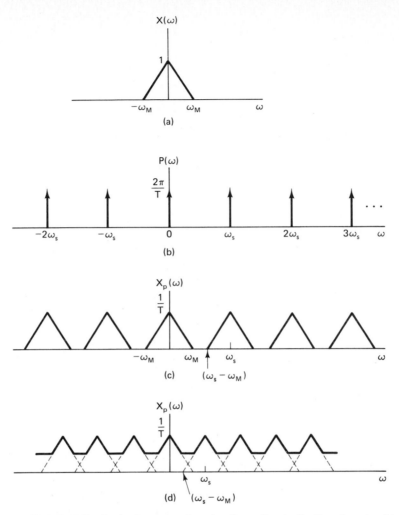

Figure 8.4 Effect in the frequency domain of sampling in the time domain: (a) spectrum of original signal; (b) spectrum of sampling function; (c) spectrum of sampled signal with $\omega_s > 2\omega_M$; (d) spectrum of sampled signal with $\omega_s < 2\omega_M$.

frequency greater than ω_M and less than $\omega_s - \omega_M$, as indicated in Figure 8.5. This basic result, referred to as the *sampling theorem*, can be stated as follows:[†]

†This important and elegant theorem was available for many years in a variety of forms in the mathematics literature. See, for example, J. M. Whittaker, "Interpolatory Function Theory," *Cambridge Tracts in Mathematics and Mathematical Physics*, no. 33 (Cambridge, 1935), chap. 4. It did not appear explicitly in the literature of communication theory until the publication in 1949 of the classic paper by Shannon entitled "Communication in the Presence of Noise" (*Proceedings of the IRE*, January, 1949, pp. 10–21). However, H. Nyquist in 1928 and D. Gabor in 1946 had pointed out, based on the use of Fourier Series, that 2TW numbers are sufficient to represent a function of time duration T and highest frequency W. [H. Nyquist, "Certain Topics in Telegraph Transmission Theory," *AIEE Transactions*, 1946, p. 617; D. Gabor, "Theory of Communication," *Journal of IEE* 93, no. 26 (1946): 429.]

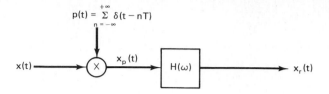

$$p(t) = \sum_{n=-\infty}^{+\infty} \delta(t - nT)$$

$x(t) \longrightarrow \bigotimes \xrightarrow{\ x_p(t)\ } \boxed{H(\omega)} \longrightarrow x_r(t)$

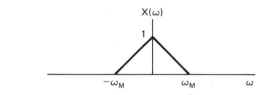

$X(\omega)$

1

$-\omega_M$ $\qquad$ ω_M $\qquad$ ω

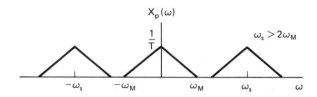

$X_p(\omega)$

$\dfrac{1}{T}$ $\qquad\qquad$ $\omega_s > 2\omega_M$

$-\omega_s$ $\quad$ $-\omega_M$ $\qquad$ ω_M $\quad$ ω_s $\qquad$ ω

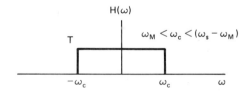

$H(\omega)$

$\omega_M < \omega_c < (\omega_s - \omega_M)$

T

$-\omega_c$ $\qquad$ ω_c $\qquad$ ω

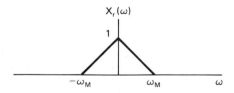

$X_r(\omega)$

1

$-\omega_M$ $\qquad$ ω_M $\qquad$ ω

Figure 8.5 Exact recovery of a continuous-time signal from its samples using an ideal lowpass filter.

Sampling Theorem:

Let $x(t)$ be a bandlimited signal with $X(\omega) = 0$ for $|\omega| > \omega_M$. Then $x(t)$ is uniquely determined by its samples $x(nT)$, $n = 0, \pm1, \pm2, \ldots$ if

$$\omega_s > 2\omega_M$$

where

$$\omega_s = \frac{2\pi}{T}$$

Given these samples, we can reconstruct $x(t)$ by generating a periodic impulse train in which successive impulses have amplitudes that are successive sample values. This impulse train is then processed through an ideal lowpass filter with gain T and cutoff frequency greater than ω_M and less than $(\omega_s - \omega_M)$. The resulting output signal will exactly equal $x(t)$.

The sampling frequency ω_s is also referred to as the *Nyquist frequency*. The frequency $2\omega_M$, which, under the sampling theorem, must be exceeded by the sampling frequency, is commonly referred to as the *Nyquist rate*.

8.1.2 Sampling with a Zero-Order Hold

The sampling theorem establishes the fact that a bandlimited signal is uniquely represented by its samples, and is motivated on the basis of impulse-train sampling. In practice, narrow large-amplitude pulses, which approximate impulses, are relatively difficult to generate and transmit, and it is often more convenient to generate the sampled signal in a form referred to as a *zero-order hold*. Such a system samples $x(t)$ at a given sampling instant and holds that value until the succeeding sampling instant, as illustrated in Figure 8.6. Reconstruction of $x(t)$ from the output of a zero-order hold

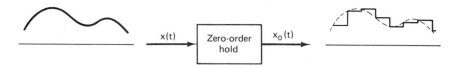

Figure 8.6 Sampling utilizing a zero-order hold.

can again be carried out by lowpass filtering. However, in this case, the required filter no longer has constant gain in the passband. To develop the required filter characteristic, we first note that the output $x_0(t)$ of the zero-order hold can in principle be generated by impulse-train sampling followed by an LTI system with a rectangular impulse response as depicted in Figure 8.7. To reconstruct $x(t)$ from $x_0(t)$, we consider processing $x_0(t)$ with an LTI system with impulse response $h_r(t)$ and frequency response $H_r(\omega)$. The cascade of this system with the system of Figure 8.7 is shown in Figure 8.8, where we wish to specify $H_r(\omega)$ so that $r(t) = x(t)$. Comparing the system in Figure 8.8 with that in Figure 8.5, we see that $r(t) = x(t)$ if the cascade combination of $h_0(t)$ and $h_r(t)$ is the ideal lowpass filter $H(\omega)$ used in Figure 8.5. Since, from

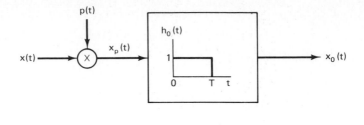

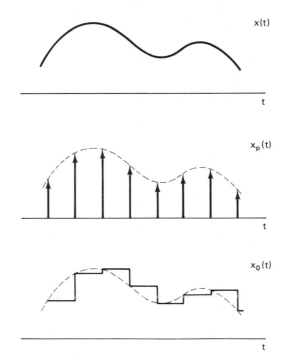

Figure 8.7 Zero-order hold as impulse train sampling followed by convolution with a rectangular pulse.

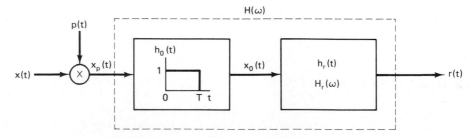

Figure 8.8 Cascade of the representation of a zero-order hold (Figure 8.7) with a reconstruction filter.

Example 4.10 and the time-shifting property 4.6.3

$$H_0(\omega) = e^{-j\omega T/2}\left[\frac{2\sin(\omega T/2)}{\omega}\right] \tag{8.6}$$

This requires that

$$H_r(\omega) = \frac{e^{j\omega T/2} H(\omega)}{\left[\dfrac{2\sin(\omega T/2)}{\omega}\right]} \tag{8.7}$$

For example with the cutoff frequency of $H(\omega)$ as $\omega_s/2$, the ideal magnitude and phase for the reconstruction filter following a zero-order hold is that shown in Figure 8.9.

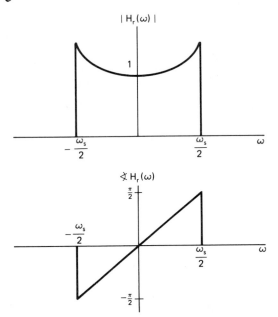

Figure 8.9 Magnitude and phase for reconstruction filter for zero-order hold.

In many situations the zero-order hold is considered to be an adequate approximation to the original signal without any additional lowpass filtering and in essence represents a possible, although admittedly very coarse, interpolation between the sample values. In the next section we explore in more detail the general concept of interpreting the reconstruction of a signal from its samples as a process of interpolation.

8.2 RECONSTRUCTION OF A SIGNAL FROM ITS SAMPLES USING INTERPOLATION

Interpolation is a commonly used procedure for reconstructing a function either approximately or exactly from samples. One simple interpolation procedure is the zero-order hold discussed in Section 8.1. Another simple and useful form of interpolation is *linear interpolation*, whereby adjacent sample points are connected by a straight line as illustrated in Figure 8.10. In more complicated interpolation for-

Figure 8.10 Linear interpolation between sample points. The dashed curve represents the original signal and the solid curve the linear interpolation.

mulas, sample points may be connected by higher-order polynomials or other mathematical functions.

As we have seen in Section 8.1, for a bandlimited signal, if the sampling instants are sufficiently close, then the signal can be reconstructed exactly, i.e., through the use of a lowpass filter exact interpolation can be carried out between the sample points. The interpretation of the reconstruction of $x(t)$ as a process of interpolation becomes evident when we consider the effect in the time domain of the lowpass filter in Figure 8.5. In particular, the output $x_r(t)$ is

$$x_r(t) = x_p(t) * h(t)$$

or with $x_p(t)$ given by eq. (8.2),

$$x_r(t) = \sum_{n=-\infty}^{+\infty} x(nT)h(t - nT) \tag{8.8}$$

Equation (8.8) represents an interpolation formula since it describes how to fit a continuous curve between the sample points. For the ideal lowpass filter $H(\omega)$ in Figure 8.5, $h(t)$ is given by

$$h(t) = T\frac{\omega_c}{\pi} \text{ sinc}\left(\frac{\omega_c t}{\pi}\right) \tag{8.9}$$

so that

$$x_r(t) = \sum_{n=-\infty}^{+\infty} x(nT)T\frac{\omega_c}{\pi} \text{ sinc}\left[\frac{\omega_c(t - nT)}{\pi}\right] \tag{8.10}$$

The reconstruction according to eq. (8.10) with $\omega_c = \omega_s/2$ is illustrated in Figure 8.11.

Interpolation using the sinc function as in eq. (8.10) is commonly referred to as *bandlimited interpolation*, since it implements exact reconstruction if $x(t)$ is bandlimited and the sampling frequency satisfies the conditions of the sampling theorem. Since a very good approximation to an ideal lowpass filter is relatively difficult to implement, in many cases it is preferable to use a less accurate but simpler filter (or equivalently interpolating function) $h(t)$. For example, as we previously indicated, the zero-order hold can be viewed as a form of interpolation between sample values in which the interpolating function $h(t)$ is the impulse response $h_0(t)$ depicted in Figure 8.7. In that sense, with $x_0(t)$ in Figure 8.7 corresponding to the approximation to $x(t)$, the system $h_0(t)$ represents an approximation to the ideal lowpass filter required for the exact interpolation. Figure 8.12 shows the magnitude of the transfer function of the zero-order-hold interpolating filter, superimposed on the desired transfer function of the exact interpolating filter. Both from Figure 8.12 and from Figure 8.7 we see

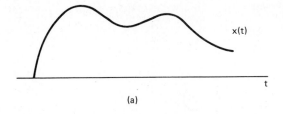

(a)

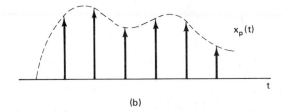

(b)

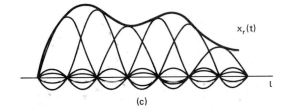

(c)

Figure 8.11 Ideal bandlimited interpolation using the sinc function.

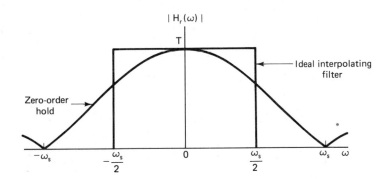

Figure 8.12 Transfer function for the zero-order hold and for the ideal interpolating filter.

that the zero-order hold is a very rough approximation, although in some cases it is sufficient. For example, if, in a given application, there is additional lowpass filtering that is naturally applied, this will tend to improve the overall interpolation. This is illustrated in the case of pictures in Figure 8.13. Figure 8.13(a) shows a picture with "impulse" sampling (i.e., sampling with spatially narrow pulses). Figure 8.13(b) is the result of applying a two-dimensional zero-order hold to Figure 8.13(a) with a resulting mosaic effect when viewed at close range. However, the human visual system inherently

(a)

(b)

Figure 8.13 (a) The original pictures of Figs. 2.2 and 4.2 with impulse sampling; (b) zero-order hold applied to the pictures in (a). The visual system naturally introduces lowpass filtering with a cutoff frequency that increases with distance. Thus, when viewed at a distance, the discontinuities in the mosaic in Figure 8.13(b) are not resolved; (c) result of applying a zero-order hold after impulse sampling with one-half the horizontal and vertical spacing used in (a) and (b).

524

(c)

Figure 8.13 (cont.)

imposes lowpass filtering, and consequently when viewed at a distance, the discontinuities in the mosaic are not resolved. In Figure 8.13(c) a zero-order hold is again used, but here the sample spacing in each direction is half that in Figure 8.13(a). With normal viewing, considerable lowpass filtering is naturally applied although, particularly with a magnifying glass, the mosaic effect is still somewhat evident.

Another approximate form of interpolation often used is linear interpolation, for which the reconstructed signal is continuous, although its derivative is not. Linear interpolation, sometimes referred to as a first-order hold, was illustrated in Figure 8.10 and can also be viewed as an interpolation in the form of Figure 8.5 and eq. (8.8) with $h(t)$ triangular, as illustrated in Figure 8.14. The associated transfer function $H(\omega)$ is also shown in Figure 8.14 and is given by

$$H(\omega) = \frac{1}{T}\left[\frac{\sin(\omega T/2)}{\omega/2}\right]^2 \tag{8.11}$$

The transfer function of the first-order hold in Figure 8.14 is shown superimposed on the transfer function for the ideal interpolating filter. Figure 8.15 corresponds to the same pictures as in Figure 8.13 but with a first-order hold applied to the sampled picture.

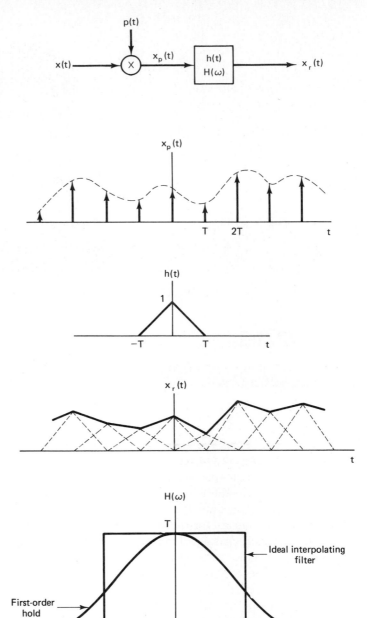

Figure 8.14 Linear interpolation (first-order hold) as impulse-train sampling followed by convolution with a triangular impulse response.

Figure 8.15 Figure 8.13 with a first-order hold applied to the sampled pictures.

8.3 THE EFFECT OF UNDERSAMPLING: ALIASING

In the discussion in previous sections, it was assumed that the sampling frequency was sufficiently high so that the conditions of the sampling frequency were met. As was illustrated in Figure 8.4, with $\omega_s > 2\omega_M$, the spectrum of the sampled signal consists of exact replications of the spectrum of $x(t)$, and this forms the basis for the sampling theorem. When $\omega_s < 2\omega_M$, $X(\omega)$, the spectrum of $x(t)$, is no longer replicated in $X_p(\omega)$ and thus is no longer recoverable by lowpass filtering. This effect, in which the individual terms in eq. (8.5) overlap, is referred to as *aliasing*, and in this section we explore its effect and consequences.

Clearly, if the system of Figure 8.5 is applied to a signal with $\omega_s < 2\omega_M$, the reconstructed signal $x_r(t)$ will no longer be equal to $x(t)$. However, as explored in Problem 8.4 the original signal and the signal $x_r(t)$ which is reconstructed using bandlimited interpolation will always be equal at the sampling instants; that is, for any choice of ω_s,

$$x_r(nT) = x(nT), \qquad n = 0, \pm 1, \pm 2, \dots \qquad (8.12)$$

Some insight into the relationship between $x(t)$ and $x_r(t)$ when $\omega_s < 2\omega_M$ is provided by considering in more detail the comparatively simple case of a sinusoidal signal. Thus, let $x(t)$ be given by

$$x(t) = \cos \omega_0 t \qquad (8.13)$$

with Fourier transform $X(\omega)$ as indicated in Figure 8.16(a). In this figure, we have graphically distinguished the impulse at ω_0 from that at $-\omega_0$ for convenience as the discussion proceeds. Let us consider $X_p(\omega)$, the spectrum of the sampled signal and focus in particular on the effect of a change in the frequency ω_0 with the sampling frequency ω_s fixed. In Figure 8.16(b) — (e) we illustrate $X_p(\omega)$ for several values of

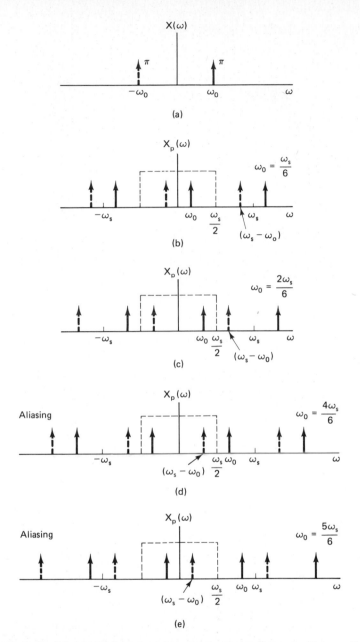

Figure 8.16 Effect in the frequency domain of oversampling and undersampling: (a) spectrum of original sinusoidal signal; (b), (c) spectrum of sampled signal with $\omega_s > 2\omega_0$; (d), (e) spectrum of sampled signal with $\omega_s < 2\omega_0$.

ω_0. Also indicated by a dashed line is the passband of the lowpass filter of Figure 8.5 with $\omega_c = \omega_s/2$. Note that no aliasing occurs in (b) or (c), since $\omega_0 < \omega_s/2$, whereas aliasing does occur in (d) and (e). For each of the four cases, the lowpass filtered output $x_r(t)$ is given by:

(a) $\omega_0 = \dfrac{\omega_s}{6};$ $\quad x_r(t) = \cos \omega_0 t = x(t)$

(b) $\omega_0 = \dfrac{2\omega_s}{6};$ $\quad x_r(t) = \cos \omega_0 t = x(t)$

(c) $\omega_0 = \dfrac{4\omega_s}{6};$ $\quad x_r(t) = \cos (\omega_s - \omega_0)t \neq x(t)$

(d) $\omega_0 = \dfrac{5\omega_s}{6};$ $\quad x_r(t) = \cos (\omega_s - \omega_0)t \neq x(t)$

When aliasing occurs, the original frequency ω_0 takes on the identity or "alias" of a lower frequency $(\omega_s - \omega_0)$. For $\omega_s/2 < \omega_0 < \omega_s$, as ω_0 increases relative to ω_s, the output frequency $(\omega_s - \omega_0)$ decreases. When $\omega_s = \omega_0$, for example, the reconstructed signal is a constant. This is consistent with the fact that when sampling once per cycle, the samples are all equal and would be identical to those obtained by sampling a constant signal $(\omega_0 = 0)$.

In Figure 8.17 we have depicted for each of these cases the signal $x(t)$, its samples, and the reconstructed signal $x_r(t)$. From these figures we can see how the lowpass filter interpolates between the samples, in particular, always fitting a sinusoid of frequency less than $\omega_s/2$ to the samples of $x(t)$.

The effect of undersampling, whereby higher frequencies are reflected into lower frequencies, is the principle on which the stroboscopic effect is based. Consider, for example, the situation depicted in Figure 8.18, in which we have a disc, rotating at a constant rate, with a single radial line marked on the disc. The flashing strobe acts as a sampling system since it illuminates the disc for extremely brief time intervals

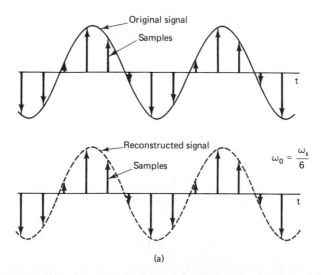

(a)

Figure 8.17 Effect of aliasing on a sinusoidal signal. For each of four values of ω_0, the original sinusoidal signal (solid curve), its samples, and the reconstructed signal (dashed curve) are illustrated: (a) $\omega_0 = \omega_s/6$; (b) $\omega_0 = 2\omega_s/6$; (c) $\omega_0 = 4\omega_s/6$; (d) $\omega_0 = 5\omega_s/6$. In (a) and (b) no aliasing occurs, whereas in (c) and (d) there is aliasing.

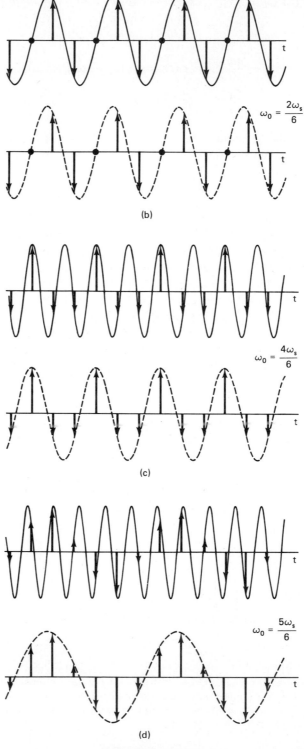

$$\omega_0 = \frac{2\omega_s}{6}$$

(b)

$$\omega_0 = \frac{4\omega_s}{6}$$

(c)

$$\omega_0 = \frac{5\omega_s}{6}$$

(d)

Figure 8.17 (cont.)

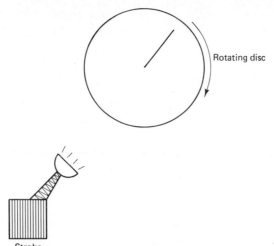

Rotating disc

Strobe

Figure 8.18 Strobe effect.

at a periodic rate. When the strobe frequency is much higher than the rotational speed of the disc, the speed of rotation of the disc is perceived correctly. When the strobe frequency becomes less than twice the rotational frequency of the disc, the disc rotation appears to be at a lower frequency than is actually the case (and, incidentally, in the wrong direction). At one flash per revolution, corresponding to $\omega_s = \omega_0$, the radial line appears stationary (i.e., the rotational frequency of the disc and its harmonics have been aliased to zero frequency). A similar effect is commonly observed in western movies, where the stagecoach wheels appear to be rotating more slowly than would be consistent with the forward motion, and sometimes in the wrong direction. In this case the sampling process corresponds to the fact that moving pictures are a sequence of individual frames, with the frame rate (usually between 18 and 24 frames per second) corresponding to the sampling frequency.

The preceding discussion is generally suggestive of the interpretation of the stroboscopic effect as an example of aliasing due to undersampling, and in many cases represents a useful application of aliasing. Another useful application of aliasing arises in a measurement instrument referred to as the sampling oscilloscope. It is intended for observing very high frequency waveforms and exploits the principles of sampling to alias these frequencies into ones that are more easily displayed. This application is explored in more detail in Problem 8.9.

8.4 DISCRETE-TIME PROCESSING OF CONTINUOUS-TIME SIGNALS

In many applications there is a significant advantage offered in processing a continuous-time signal by first converting it to a discrete-time signal and after processing, converting back to a continuous-time signal. The discrete-time signal processing can then be implemented with a general- or special-purpose computer, with microprocessors or with any of the variety of devices that are specifically oriented toward discrete-time signal processing.

In broad terms, this approach to continuous-time signal processing can be viewed as the cascade of three operations as indicated in Figure 8.19, where $x_c(t)$

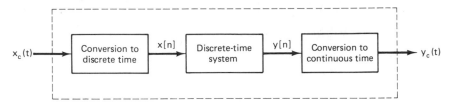

Figure 8.19 Discrete-time processing of continuous-time signals.

and $y_c(t)$ are continuous-time signals and $x[n]$ and $y[n]$ are the discrete-time signals representing $x_c(t)$ and $y_c(t)$. The overall system in Figure 8.19 is, of course, a continuous-time system in the sense that the system input and output are both continuous-time signals. The theoretical basis for conversion of a continuous-time signal to a discrete-time signal and the reconstruction of a continuous-time signal from its discrete-time representation lies in the sampling theorem, as discussed in Section 8.1. Through the process of periodic sampling with the sampling frequency consistent with the conditions of the sampling theorem, the continuous-time signal $x_c(t)$ is exactly represented by a sequence of instantaneous sample values $x_c(nT)$; that is, the discrete-time sequence $x[n]$ is related to $x_c(t)$ by

$$x[n] = x_c(nT) \tag{8.14}$$

The transformation of $x_c(t)$ to $x[n]$ corresponding to the first system in Figure 8.19 will be referred to as continuous-to-discrete-time conversion and will be abbreviated C/D. The reverse operation corresponding to the third system in Figure 8.19 will be abbreviated D/C, representing discrete-time to continuous-time conversion. In systems such as digital computers and digital systems for which the discrete-time signal is represented in digital form, the device commonly used to implement the C/D conversion is referred to as an analog-to-digital (A-to-D) converter and the device used to implement the D/C conversion is referred to as a digital-to-analog (D-to-A) converter.

To understand further the relationship between the continuous-time signal $x_c(t)$ and its discrete-time representation $x[n]$, it is helpful to represent the continuous-time to discrete-time transformation as a process of periodic sampling followed by a mapping of the impulse train to a sequence. These two steps are illustrated in Figure 8.20. In the first step, representing the sampling process, the impulse train $x_p(t)$ corresponds to a sequence of impulses with amplitudes corresponding to the samples of $x_c(t)$ and with a time spacing equal to the sampling period T. In the conversion from the impulse train to the discrete-time sequence, we obtain $x[n]$, corresponding to the same sequence of samples of $x_c(t)$, but with unity spacing in terms of the new independent variable n. Thus, in effect, the conversion from the impulse train sequence of samples to the discrete-time sequence of samples can be thought of as a time normalization. The time normalization in converting $x_p(t)$ to $x[n]$ is evident in Figure 8.20(b) and (c), in which $x_p(t)$ and $x[n]$ are illustrated for a sampling rate of $T = T_1$ and $T = 2T_1$.

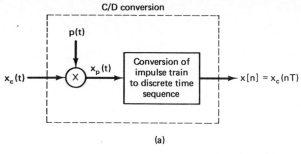

(a)

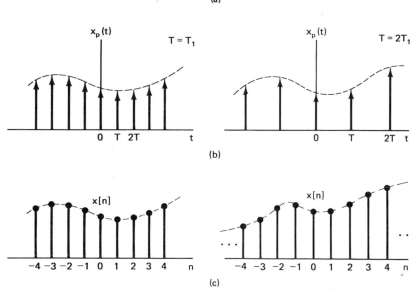

(b)

(c)

Figure 8.20 Sampling with a periodic impulse train followed by conversion to a discrete-time sequence: (a) overall system; (b) $x_p(t)$ for two sampling rates. The dashed envelope represents $x_c(t)$; (c) the output sequence for the two different sampling rates.

In the frequency domain, the relationship between $x_c(t)$ and $x_p(t)$ is that developed in eq. (8.5) and illustrated in Figure 8.4:

$$X_p(\omega) = \frac{1}{T} \sum_{k=-\infty}^{+\infty} X_c(\omega - k\omega_s) \qquad (8.15)$$

Alternatively, we may express $X_p(\omega)$ in terms of the sample values of $x_c(t)$ by applying the Fourier transform to eq. (8.2). Thus,

$$x_p(t) = \sum_{n=-\infty}^{+\infty} x_c(nT)\,\delta(t - nT) \qquad (8.16)$$

$$X_p(\omega) = \sum_{n=-\infty}^{+\infty} x_c(nT)e^{-j\omega nT} \qquad (8.17)$$

Now consider the discrete-time Fourier transform of $x[n]$ given by

$$X(\Omega) = \sum_{n=-\infty}^{+\infty} x[n]e^{-j\Omega n} \qquad (8.18)$$

or, using eq. (8.14),

$$X(\Omega) = \sum_{n=-\infty}^{+\infty} x_c(nT)e^{-j\Omega n} \qquad (8.19)$$

Comparing eqs. (8.17) and (8.19), we see that $X(\Omega)$ and $X_p(\omega)$ are related through

$$X(\Omega) = X_p\left(\frac{\Omega}{T}\right) \qquad (8.20)$$

The relationship among $X_c(\omega)$, $X_p(\omega)$, and $X(\Omega)$ is illustrated in Figure 8.21 for two

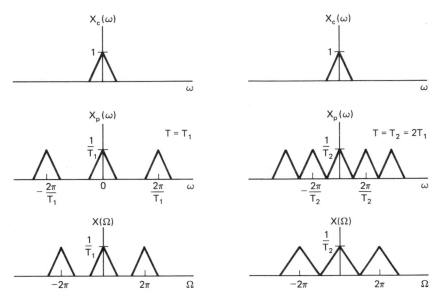

Figure 8.21 Relationship between $X_c(\omega)$, $X_p(\omega)$, and $X(\Omega)$ for two different sampling rates.

different sampling rates. From this figure, we note that $X(\Omega)$ is a frequency-scaled replication of $X_p(\omega)$, and in particular, is periodic in Ω with period 2π. This periodicity is, of course, characteristic of any discrete-time Fourier transform. The spectrum of $x[n]$ is related to that of $x_c(t)$ through periodic replication represented by eq. (8.15) followed by linear frequency scaling represented by eq. (8.20). The periodic replication is a consequence of the first step in the conversion process in Figure 8.20, namely the impulse train sampling. The linear frequency scaling in eq. (8.20) can be thought of as a consequence of the time normalization introduced by converting from the impulse train $x_p(t)$ to the discrete-time sequence $x[n]$. From the time-scaling property of the Fourier transform in Section 4.6.5, scaling of the time axis by $1/T$ will introduce a scaling of the frequency axis by T. Thus, the relationship $\Omega = \omega T$ is consistent with the notion that in converting from $x_p(t)$ to $x[n]$ the time axis is scaled by $1/T$.

In the overall system of Figure 8.19, after processing with a discrete-time system, the resulting sequence is converted back to a continuous-time signal. This process is the reverse of the steps in Figure 8.20. Specifically, from the sequence $y[n]$ a continuous-time impulse train $y_p(t)$ can be generated. Recovery of the continuous-time signal $y_c(t)$ from this impulse train is then accomplished by means of lowpass filtering, as illustrated in Figure 8.22.

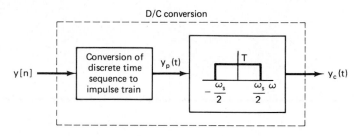

Figure 8.22   Conversion of a discrete-time sequence to a continuous-time signal.

Now let us consider the overall system of Figure 8.19, represented as shown in Figure 8.23. Clearly, if the discrete-time system is an identity system (i.e., $x[n] = y[n]$),

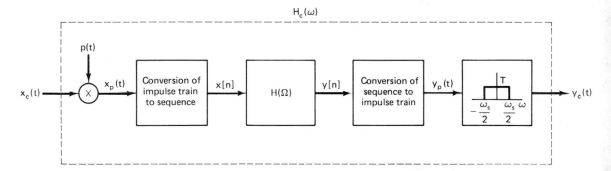

Figure 8.23   Overall system for filtering a continuous-time signal using a discrete-time filter.

then, assuming that the conditions of the sampling theorem are met, the overall system will be an identity system. The characteristics of the overall system with a more general frequency response $H(\Omega)$ is perhaps best understood by examining the representative example depicted in Figure 8.24. On the left-hand side of the figure are the representative spectra $X_c(\omega)$, $X_p(\omega)$, and $X(\Omega)$. The spectrum $Y(\Omega)$ corresponding to the output of the discrete-time filter is the product of $X(\Omega)$ and $H(\Omega)$, and this is depicted in Figure 8.24(d) by overlaying $H(\Omega)$ and $X(\Omega)$. The transformation to $Y_c(\omega)$ then corresponds to applying a frequency scaling and lowpass filtering, resulting in the spectra indicated in Figure 8.24(e) and (f). Since $Y(\Omega)$ is the product of the two overlaid spectra in 8.24(d), the scaling and filtering are applied to both. In comparing 8.24(a) and (f), it should be clear that

$$Y_c(\omega) = X_c(\omega)H(\omega T) \tag{8.21}$$

Consequently, the overall system of Figure 8.23 is, in fact, equivalent to a continuous-time LTI system with frequency response $H_c(\omega)$, which is related to the discrete-time frequency response $H(\Omega)$ through

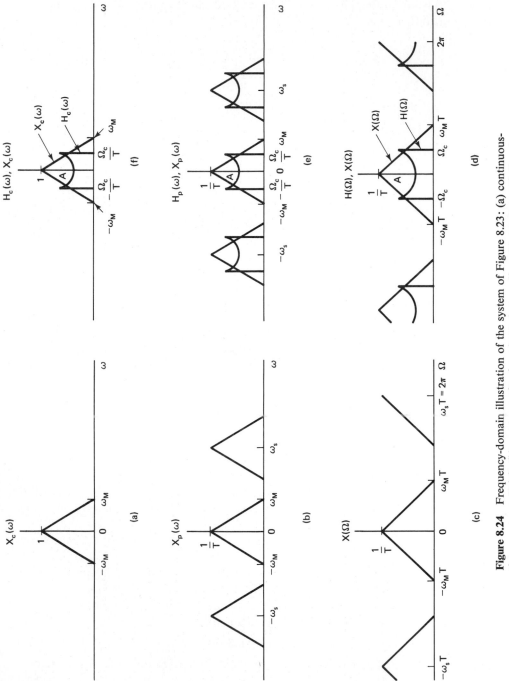

Figure 8.24 Frequency-domain illustration of the system of Figure 8.23: (a) continuous-time spectrum $X_c(\omega)$; (b) spectrum after impulse-train sampling; (c) spectrum of discrete-time sequence; (d) $H(\Omega)$ and $X(\Omega)$ that are multiplied to form $Y(\Omega)$; (e) spectra that are multiplied to form $Y_p(\omega)$; (f) spectra that are multiplied to form $Y_c(\omega)$.

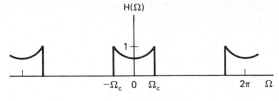

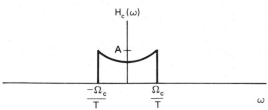

Figure 8.25 Discrete-time frequency response and the equivalent continuous-time frequency response for the system of Figure 8.23.

$$H_c(\omega) = \begin{cases} H(\omega T), & |\omega| < \dfrac{\omega_s}{2} \\ 0, & |\omega| > \dfrac{\omega_s}{2} \end{cases} \tag{8.22}$$

The equivalent continuous-time filter frequency response is one period of the discrete-time filter characteristics with a linear scale change applied to the frequency axis. This relationship between the discrete-time frequency response and the equivalent continuous-time frequency response is illustrated in Figure 8.25.

The equivalence of the overall system of Figure 8.23 to an LTI system is somewhat surprising in view of the fact that the impulse train modulator is clearly *not* a time-invariant component. In fact, the overall system of Figure 8.23 is not time-invariant for arbitrary inputs. For example, if $x_c(t)$ was a narrow rectangular pulse, of duration less that T, then a time shift of $x_c(t)$ could generate a sequence $x[n]$ that either had all zero sequence values or had one nonzero sequence value, depending on the alignment of the rectangular pulse relative to the sampling impulse train. However, as suggested by the spectra of Figure 8.24, for *bandlimited input signals* with a sampling rate sufficiently high to avoid aliasing, the system of Figure 8.23 *is* equivalent to a continuous-time time-invariant system. For such inputs, Figure 8.23 and eq. (8.22) provide the conceptual basis for continuous-time processing using discrete-time filters. This is now explored further in the context of some specific examples.

8.4.1 Digital Differentiator

Let us consider the discrete-time implementation of a continuous-time bandlimited differentiating filter. As discussed in Chapter 5, the frequency response of a continuous-time differentiating filter is

$$H_c(\omega) = j\omega \tag{8.23}$$

and that of a bandlimited differentiator with cutoff frequency ω_c is

$$H_c(\omega) = \begin{cases} j\omega, & |\omega| < \omega_c \\ 0, & |\omega| > \omega_c \end{cases} \tag{8.24}$$

as sketched in Figure 8.26. Using eq. (8.22) with a sampling frequency $\omega_s = 2\omega_c$, the corresponding discrete-time transfer function $H(\Omega)$ is

$$H(\Omega) = j\left(\frac{\Omega}{T}\right), \qquad |\Omega| < \pi \tag{8.25}$$

as sketched in Figure 8.27. With this discrete-time transfer function, $y_c(t)$ in figure 8.23 will be the derivative of $x_c(t)$ as long as there is no aliasing in sampling $x_c(t)$.

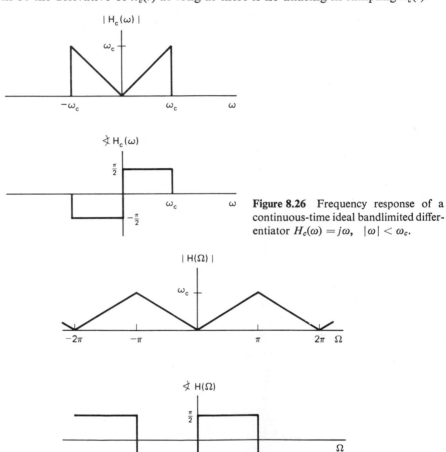

Figure 8.26 Frequency response of a continuous-time ideal bandlimited differentiator $H_c(\omega) = j\omega$, $|\omega| < \omega_c$.

Figure 8.27 Frequency response of discrete-time filter to implement a continuous-time bandlimited differentiator.

8.4.2 Half-Sample Delay

Let us consider the implementation of a time shift (delay) of a continuous-time signal through the use of a system in the form of Figure 8.19. Thus, we require that the input and output of the overall system be related by

$$y_c(t) = x_c(t - \Delta) \tag{8.26}$$

where Δ represents the delay time. From the time-shifting property derived in Section 4.6.3,

$$Y_c(\omega) = e^{-j\omega\Delta}X_c(\omega)$$

Since $x_c(t)$ must be bandlimited to be processed with the system of Figure 8.19, and since the equivalent continuous-time system to be implemented must be bandlimited, choose

$$H_c(\omega) = \begin{cases} e^{-j\omega\Delta}, & |\omega| < \omega_c \\ 0, & \text{otherwise} \end{cases} \tag{8.27}$$

where ω_c is the cutoff frequency of the continuous-time filter. The magnitude and phase of the frequency response is shown in Figure 8.28(a). With the sampling fre-

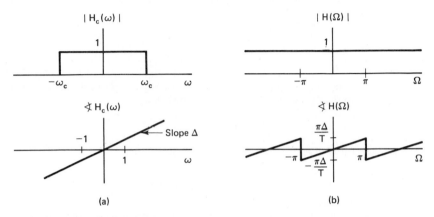

Figure 8.28 (a) Magnitude and phase of the frequency response for a continuous-time delay; (b) magnitude and phase of the frequency response for the corresponding discrete-time delay.

quency ω_s taken as $\omega_s = 2\omega_c$, the corresponding discrete-time frequency response $H(\Omega)$ is

$$H(\Omega) = e^{-j\Omega\Delta/T}, \qquad |\Omega| < \pi \tag{8.28}$$

and is shown in Figure 8.28(b).

For bandlimited inputs the output of the system of Figure 8.23 with $H(\Omega)$ as in eq. (8.28) is a delayed replica of the input. For Δ/T an integer the sequence $y[n]$ is a delayed replica of $x[n]$, that is,

$$y[n] = x\left[n - \frac{\Delta}{T}\right] \tag{8.29}$$

For Δ/T not an integer, eq. (8.29) has no meaning since sequences are defined only at integer values of the index. However, we can interpret the relationship between $x[n]$ and $y[n]$ in these cases in terms of bandlimited interpolation. The signals $x_c(t)$ and $x[n]$ are related through sampling and bandlimited interpolation, as are $y_c(t)$ and $y[n]$. With $H(\Omega)$ in eq. (8.28), $y[n]$ is equal to samples of the bandlimited interpolation of the sequence $x[n]$. This is illustrated in Figure 8.29 with $(\Delta/T) = \frac{1}{2}$, which is sometimes referred to as a one-half-sample delay.

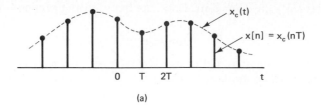

(a)

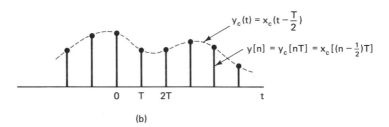

(b)

Figure 8.29 (a) Sequence of samples of a continuous-time signal $x_c(t)$; (b) sequence in (a) with a "half-sample" delay.

8.5 SAMPLING IN THE FREQUENCY DOMAIN

The sampling theorem was developed in Section 8.1 for time-domain sampling of bandlimited signals. As was discussed in Section 4.6.6, there is a duality between the time domain and frequency domain for continuous-time signals. In this section we develop the dual to the time-domain sampling theorem, whereby a time-limited signal can be reconstructed from *frequency-domain* samples.

To develop this result, consider the frequency-domain sampling operation in Figure 8.30, which is the frequency-domain dual of the time-domain impulse train sampling in Figure 8.3. Since in the frequency domain,

$$\tilde{X}(\omega) = X(\omega)P(\omega) \tag{8.30}$$

then in the time domain,

$$\tilde{x}(t) = x(t) * p(t) \tag{8.31}$$

where

$$p(t) = \frac{1}{\omega_0} \sum_{k=-\infty}^{+\infty} \delta\left(t - \frac{2\pi}{\omega_0}k\right) \tag{8.32}$$

Thus,

$$\tilde{x}(t) = \frac{1}{\omega_0} \sum_{k=-\infty}^{+\infty} x\left(t - \frac{2\pi k}{\omega_0}\right) \tag{8.33}$$

Equation (8.33) is the dual of eq. (8.5) and the relation between $x(t)$ and $\tilde{x}(t)$ in eq. (8.33) is similar to the relationship between $x(t)$ and $\tilde{x}(t)$ in Figure 4.12. If $x(t)$ is time-limited with

$$x(t) = 0, \qquad |t| > T_m \tag{8.34}$$

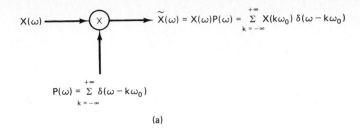

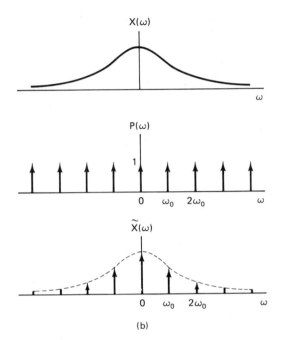

Figure 8.30 Frequency-domain impulse-train sampling: (a) overall operation; (b) associated spectra.

then, as illustrated in Figure 8.31, with

$$\frac{2\pi}{\omega_0} > 2T_m \tag{8.35}$$

$\tilde{x}(t)$ as given by eq. (8.33) consists of nonoverlapping periodic replicas of $x(t)$ spaced at integer multiples of $T_0 = 2\pi/\omega_0$. In this case, the original signal $x(t)$ [and thus, of course, its transform $X(\omega)$] can be reconstructed by "low-time windowing" of $\tilde{x}(t)$, that is,

$$x(t) = \tilde{x}(t)w(t) \tag{8.36}$$

where

$$w(t) = \begin{cases} \omega_0, & |t| \leq \dfrac{\pi}{\omega_0} \\ 0, & |t| > \dfrac{\pi}{\omega_0} \end{cases} \tag{8.37}$$

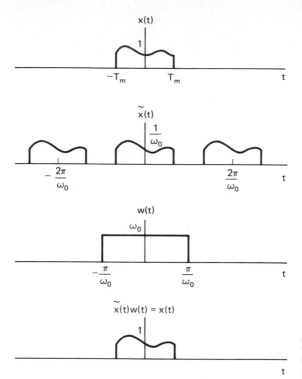

Figure 8.31 Time-domain waveforms associated with frequency-domain sampling.

as illustrated in Figure 8.31. If the inequality in eq. (8.35) is not satisfied, then the replications of $x(t)$ in eq. (8.33) and Figure 8.31 will overlap and $x(t)$ will no longer be recoverable from $\tilde{x}(t)$. This is the dual of frequency-domain aliasing, as discussed in Section 8.1, and is commonly referred to as *time-domain aliasing*.[†]

In analogy with the discussion in Section 8.2, low-time windowing of $\tilde{x}(t)$ to recover $x(t)$ can be interpreted as interpolation of the frequency-domain samples of $X(\omega)$. Specifically, from eq. (8.36),

$$X(\omega) = \frac{1}{2\pi}\tilde{X}(\omega) * W(\omega) \tag{8.38}$$

where

$$\tilde{X}(\omega) = \sum_{k=-\infty}^{+\infty} X(k\omega_0)\,\delta(\omega - k\omega_0) \tag{8.39}$$

and $W(\omega)$ is the Fourier transform of $w(t)$, that is,

$$W(\omega) = 2\pi \, \text{sinc}\left(\frac{\omega}{\omega_0}\right) \tag{8.40}$$

Combining eqs. (8.38), (8.39), and (8.40),

$$X(\omega) = \sum_{k=-\infty}^{+\infty} X(k\omega_0)\,\text{sinc}\left[\frac{\omega - k\omega_0}{\omega_0}\right] \tag{8.41}$$

[†] In eq. (8.34) we have assumed that the time interval over which $x(t)$ is nonzero is centered about $t = 0$. As developed in Problem 8.20, the results in this section are easily modified to apply if $x(t)$ is nonzero over *any* specified time interval of length $2T_m$.

which is the dual of eq. (8.10). In particular, the sinc function provides exact interpolation between equally spaced frequency samples for a time-limited signal just as it does between equally spaced time samples for a bandlimited signal.

8.6 SAMPLING OF DISCRETE-TIME SIGNALS

Thus far in this chapter we have considered the sampling of continuous-time signals and in addition to developing the analysis necessary to understand continuous-time sampling, we have introduced a number of its applications. As we will see in this section, a very similar set of properties and results with a number of important applications can be developed for sampling of discrete-time signals.

In analogy with continuous-time sampling as carried out using the system of Figure 8.3, sampling of a discrete-time signal can be represented as shown in Figure 8.32. Here the new sequence $x_p[n]$ resulting from the sampling process is equal to

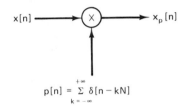

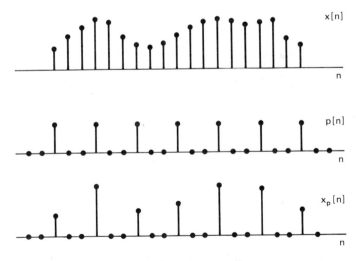

Figure 8.32 Discrete-time sampling.

the original sequence $x[n]$ at integer multiples of the sampling period N and is zero at the intermediate samples, that is,

$$x_p[n] = \begin{cases} x[n], & \text{if } n = \text{an integer multiple of } N \\ 0, & \text{otherwise} \end{cases} \qquad (8.42)$$

Paralleling the development in Section 8.1, the effect in the frequency domain of discrete-time sampling is seen by using the modulation property developed in Section 5.7. Thus, with

$$x_p[n] = x[n]p[n] = \sum_{k=-\infty}^{+\infty} x[kN]\delta[n - kN] \tag{8.43}$$

we have in the frequency domain that

$$X_p(\Omega) = \frac{1}{2\pi} \int_{2\pi} P(\theta)X(\Omega - \theta)\, d\theta \tag{8.44}$$

As developed in Example 5.11, the Fourier transform of the sampling sequence $p[n]$ is

$$P(\Omega) = \frac{2\pi}{N} \sum_{k=-\infty}^{+\infty} \delta(\Omega - k\Omega_s) \tag{8.45}$$

where Ω_s, the sampling frequency, is $2\pi/N$. Combining eqs. (8.44) and (8.45), we have

$$X_p(\Omega) = \frac{1}{N} \sum_{k=0}^{N-1} X(\Omega - k\Omega_s) \tag{8.46}$$

Equation (8.46) is the counterpart for discrete-time sampling to eq. (8.5) for continuous-time sampling and is illustrated in Figure 8.33. In Figure 8.33(c), with $(\Omega_s - \Omega_M) > \Omega_M$ or equivalently $\Omega_s > 2\Omega_M$, there is no aliasing [i.e., the nonzero portions of the replicas of $X(\Omega)$ do not overlap], whereas with $\Omega_s < 2\Omega_M$, as in Figure 8.33(d), frequency-domain aliasing results. In the absence of aliasing $X(\Omega)$ is faithfully reproduced around $\Omega = 0$ and integer multiples of 2π. Consequently, $x[n]$ can be recovered from $x_p[n]$ by means of a lowpass filter with gain N and a cutoff frequency greater than Ω_M and less than $\Omega_s - \Omega_M$, as illustrated in Figure 8.34, where we have specified the cutoff frequency of the lowpass filter as $\Omega_s/2$. If the overall system of Figure 8.34(a) is applied to a sequence for which $\Omega_s < 2\Omega_M$ so that aliasing results, $x_r[n]$ will no longer be equal to $x[n]$. However, as with continuous-time sampling, the two sequences *will* be equal at multiples of the sampling period; that is, corresponding to eq. (8.12) we have

$$x_r[kN] = x[kN], \qquad k = 0, \pm 1, \pm 2, \ldots \tag{8.47}$$

independent of whether aliasing occurs (see Problem 8.23).

The reconstruction of $x[n]$ through the use of a lowpass filter applied to $x_p[n]$ can be interpreted in the time domain as an interpolation formula similar to eq. (8.10). Specifically, with $h[n]$ denoting the impulse response of the lowpass filter, we have

$$h[n] = \frac{N\Omega_c}{2\pi} \operatorname{sinc}\left(\frac{\Omega_c n}{\pi}\right) \tag{8.48}$$

The reconstructed sequence $x_r[n]$ is then

$$x_r[n] = x_p[n] * h[n] \tag{8.49}$$

or, equivalently,

$$x_r[n] = \sum_{k=-\infty}^{+\infty} x[kN]\frac{N\Omega_c}{2\pi} \operatorname{sinc}\left[\frac{\Omega_c}{\pi}(n - kN)\right] \tag{8.50}$$

Equation (8.50) represents ideal bandlimited interpolation and requires the implementation of an ideal lowpass filter. In typical applications a suitable approximation for

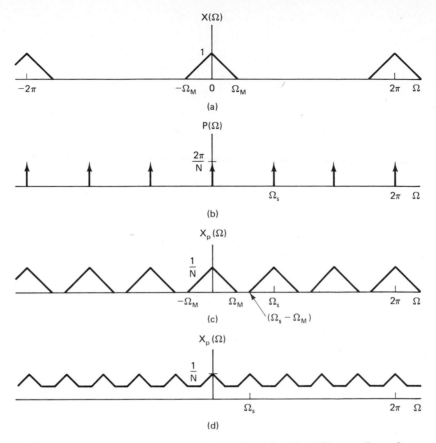

Figure 8.33 Effect in the frequency domain of impulse-train sampling of a discrete-time signal: (a) spectrum of original signal; (b) spectrum of sampling sequence; (c) spectrum of sampled signal with $\Omega_s > 2\Omega_M$; (d) spectrum of sampled signal with $\Omega_s < 2\Omega_M$. Note that aliasing occurs.

the lowpass filter in Fig. 8.34 is used, in which case the equivalent interpolation formula is of the form

$$x_r[n] = \sum_{k=-\infty}^{+\infty} x[kN]h_r[n - kN] \tag{8.51}$$

where $h_r[n]$ is the impulse response of the interpolating filter. Some specific examples including the discrete-time counterparts of the zero-order hold and first-order hold discussed in Section 8.2 for continuous-time interpolation are considered in Problem 8.22.

In Section 8.5 we discussed the continuous-time dual to time-domain sampling of bandlimited signals, specifically frequency-domain sampling of time-limited signals. A similar result also applies to sampling the discrete-time Fourier transform for a time-limited sequence. Again, consider multiplying the discrete-time Fourier transform of a sequence by an impulse train in the frequency domain as shown in Figure 8.35. This is identical to the procedure illustrated in Figure 8.30 with the exception that the spacing Ω_0 between frequency samples is restricted such that

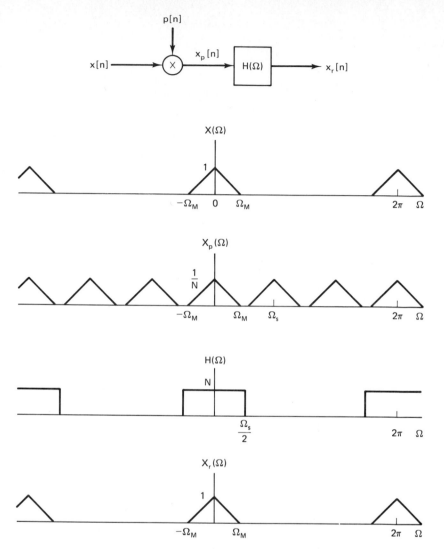

Figure 8.34 Exact recovery of a discrete-time signal from its samples using an ideal lowpass filter.

$2\pi/\Omega_0$ is an integer N to ensure that $P(\Omega)$ and $\tilde{X}(\Omega)$ are both periodic in Ω with period 2π. Then, in the time domain

$$\tilde{x}[n] = x[n] * p[n] \tag{8.52}$$

From Example 5.11,

$$p[n] = \frac{1}{\Omega_0} \sum_{k=-\infty}^{+\infty} \delta\left[n - k\frac{2\pi}{\Omega_0}\right] \tag{8.53}$$

or, with $\Omega_0 = \dfrac{2\pi}{N}$,

$$\tilde{x}[n] = \frac{N}{2\pi} \sum_{k=-\infty}^{+\infty} x[n - kN] \tag{8.54}$$

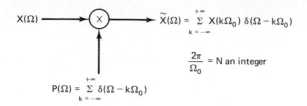

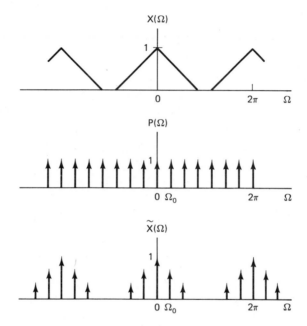

Figure 8.35 Impulse-train sampling of the discrete-time Fourier transform.

Equations (8.52) and (8.54) are the counterparts to eqs. (8.31) and (8.33). Assuming that $x[n]$ is time-limited with

$$x[n] = 0, \qquad n < 0, \, n > N - 1 \tag{8.55}$$

then, as illustrated in Figure 8.36, $\tilde{x}[n]$ consists of nonoverlapping periodic replications of $x[n]$ spaced at integer multiples of N. Thus, the original signal $x[n]$ can be reconstructed by "low-time windowing" of $\tilde{x}[n]$, that is,

$$x[n] = \tilde{x}[n]w[n] \tag{8.56}$$

where

$$w[n] = \begin{cases} \Omega_0, & 0 \leq n \leq N - 1 \\ 0, & \text{otherwise} \end{cases} \tag{8.57}$$

as illustrated in Figure 8.36. As in the continuous-time case, if eq. (8.55) is not satisfied, then the periodic replications of $x[n]$ in eq. (8.52) overlap and time-domain aliasing results.

It is useful to note that, to within the constant scale factor $N/2\pi$, the periodic sequence $\tilde{x}[n]$ in eq. (8.54) is identical to the periodic replication of a finite-duration

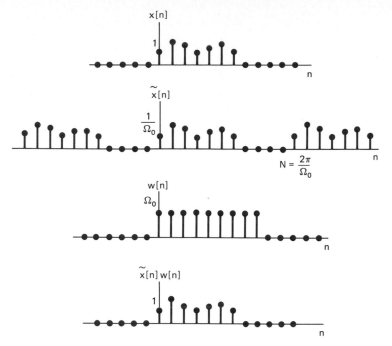

Figure 8.36 Time-domain sequences associated with frequency-domain sampling.

sequence as considered in Section 5.4.3. In fact, the DFT of a sequence $x[n]$ of length N is proportional to samples of the discrete-time Fourier transform equally spaced in frequency by $2\pi/N$.

8.7 DISCRETE-TIME DECIMATION AND INTERPOLATION

There are a variety of important applications of the principles of discrete-time sampling such as in signal multiplexing and filter design and implementation. In many of these applications it is inefficient to represent, transmit, or store the sampled sequence $x_p[n]$ directly in the form depicted in Fig. 8.32 since, in between the sampling instants, the sampled sequence $x_p[n]$ is *known* to be zero. Thus, often the sampled sequence is replaced by a new sequence $x_d[n]$, which is simply every Nth value of $x_p[n]$ or equivalently $x[n]$, that is,

$$x_d[n] = x[nN] = x_p[nN] \tag{8.58}$$

The operation of extracting every Nth sample is commonly referred to as *decimation*.† The relationship between $x[n]$, $x_p[n]$, and $x_d[n]$ is illustrated in Figure 8.37. To determine the effect in the frequency domain of decimation, we wish to determine $X_d(\Omega)$, the Fourier transform of $x_d[n]$. To this end, we note that

---

†Technically, decimation would correspond to extracting every *tenth* sample. However, it has become common terminology to refer to the operation as decimation even when N is not equal to 10.

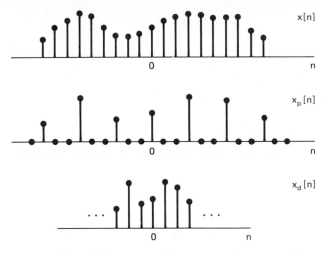

Figure 8.37 Relationship between $x_p[n]$ corresponding to sampling and $x_d[n]$ corresponding to decimation.

$$X_d(\Omega) = \sum_{n=-\infty}^{+\infty} x_d[n]e^{-j\Omega n} \tag{8.59}$$

or, using eq. (8.58),

$$X_d(\Omega) = \sum_{n=-\infty}^{+\infty} x_p[nN]e^{-j\Omega n} \tag{8.60}$$

Since $x_p[n]$ is zero except at integer multiples of N, we can equivalently write that

$$X_d(\Omega) = \sum_{n=-\infty}^{+\infty} x_p[n]e^{-j\Omega n/N} \tag{8.61}$$

The individual terms in the summation will only be nonzero for n equal to integer multiples of N. Furthermore, we recognize the right-hand side of eq. (8.61) as the Fourier transform of $x_p[n]$, specifically,

$$\sum_{n=-\infty}^{+\infty} x_p[n]e^{-j\Omega n/N} = X_p\left(\frac{\Omega}{N}\right) \tag{8.62}$$

Thus, from eqs. (8.61) and (8.62), we conclude that

$$X_d(\Omega) = X_p\left(\frac{\Omega}{N}\right) \tag{8.63}$$

This is illustrated in Figure 8.38 and we observe that from eq. (8.63) the spectra for the sampled sequence and the decimated sequence differ only in a frequency scaling or normalization. If the original spectrum $X(\Omega)$ is appropriately bandlimited so that there is no aliasing present in $X_p(\Omega)$, then as illustrated in Figure 8.38, the effect of decimation is to spread the spectrum of the original sequence over a larger portion of the frequency band.

When the original sequence $x[n]$ is obtained by sampling a continuous-time signal, the process of decimation can be viewed as reducing the sampling rate on the continuous-time signal by a factor of N. To avoid aliasing in the decimation process, $X(\Omega)$ cannot occupy the full frequency band. This requirement is indicative of the

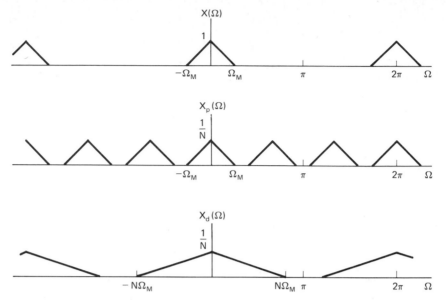

Figure 8.38 Frequency-domain illustration of the relationship between sampling and decimation.

fact that the original continuous-time signal was oversampled, and thus that the sampling rate can be reduced. With the interpretation of the sequence $x[n]$ as samples of a continuous-time signal, the process of decimation is often referred to as *downsampling*.

In some applications in which a sequence is obtained by sampling a continuous-time signal, the original sampling rate may be as low as possible without introducing aliasing, but after additional processing the bandwidth of the sequence may be reduced. An example of such a situation is shown in Figure 8.39. Since the output of the discrete-time filter is bandlimited, downsampling or decimation can be applied.

Just as in some applications it is useful to downsample, there are situations in which it is useful to convert a sequence to a *higher* equivalent sampling rate, a process referred to as *upsampling* or *interpolating*. Interpolation or upsampling is basically the reverse process to that of decimation or downsampling. As was illustrated in Figures 8.37 and 8.38, we analyzed decimation by first sampling and then retaining only the sequence values at the sampling instants. To upsample, we reverse the process. Thus, referring to Figure 8.37, consider the process of upsampling the sequence $x_d[n]$ to obtain $x[n]$. From $x_d[n]$ we form the sequence $x_p[n]$ by "inserting" $(N - 1)$ sequence points with zero amplitude between each of the sequence values in $x_d[n]$. The interpolated sequence $x[n]$ is then obtained from $x_p[n]$ by lowpass filtering. The overall procedure is summarized in Figure 8.40.

As one example of an application which requires upsampling, consider a set of sequences which we wish to frequency-division-multiplex. With M channels, it is required that the spectral energy for each input channel $x_i[n]$ be bandlimited, that is,

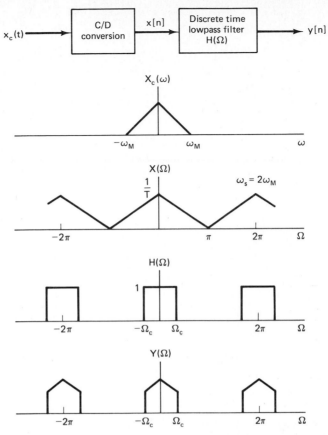

Figure 8.39 Continuous-time signal that was originally sampled at the Nyquist rate. After discrete-time filtering, the resulting sequence can be further down-sampled.

$$X_i(\Omega) = 0, \qquad \frac{\pi}{M} < |\Omega| < \pi \qquad (8.64)$$

If the sequences originally occupied the entire frequency band corresponding, for example, to having sampled a set of continuous-time signals at the Nyquist rate, then they would first have to be converted to a higher sampling rate (i.e., upsampled) before frequency-division multiplexing.

8.7.1 Discrete-Time Transmodulation

One context in which discrete-time modulation, decimation, and interpolation are widely used is that of digital communications systems. Typically, in such systems continuous-time signals are transmitted over communication channels in the form of discrete-time signals, utilizing the concepts of sampling as presented in this chapter. In such communication systems, the continuous-time signals are often in the form of

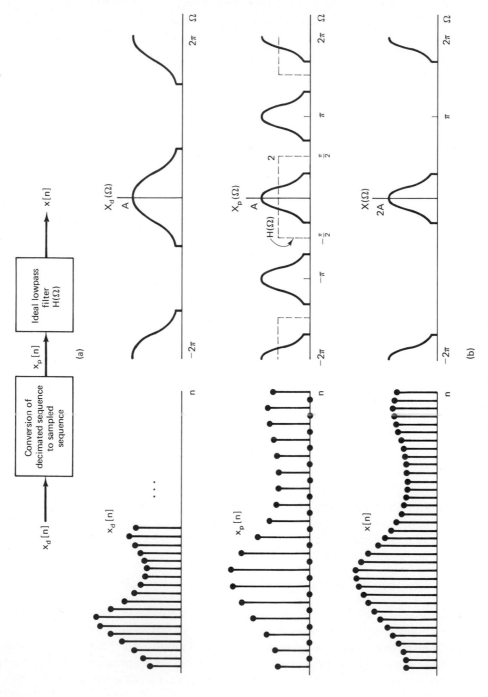

Figure 8.40 Upsampling: (a) overall system; (b) associated sequences and spectra for upsampling by a factor of 2.

time-division-multiplexed (TDM) or frequency-division-multiplexed (FDM) signals. These signals are then converted to discrete-time sequences with the sequence values represented digitally, for storage or long-distance transmission. In some systems, because of different constraints or requirements at the transmitting end and the receiving end, or because sets of signals that have been individually multiplexed by different methods are then multiplexed together there is often the requirement for converting from sequences representing TDM signals to sequences representing FDM signals, or vice versa. This conversion from one modulation or multiplexing scheme to another is referred to as *transmodulation* or *transmultiplexing*. In the context of digital communication systems, one obvious way of implementing transmultiplexing is to first convert back to continuous-time signals, demultiplex and demodulate, and then modulate and multiplex as required. However, if the new signal is then to be converted back to discrete time, it is clearly more efficient for the entire process to be carried out directly in the discrete-time domain. Figure 8.41 shows in block-diagram form the steps involved in converting a discrete-time TDM signal to a discrete-time FDM signal. Note that after demultiplexing the TDM signal, each channel must be upsampled in preparation for frequency-division multiplexing.

8.8 SUMMARY

In this chapter we have developed the concept of sampling, whereby a continuous-time or discrete-time signal is represented by a sequence of equally spaced samples. The conditions under which the signal is exactly recoverable from the samples is embodied in the sampling theorem. This theorem requires, for exact reconstruction, that the signal to be sampled be bandlimited and that the sampling frequency be greater than twice the highest frequency in the signal to be sampled. Under these conditions reconstruction of the original signal is carried out by means of ideal lowpass filtering. The time-domain interpretation of this ideal reconstruction procedure is interpolation with a sinc function, often referred to as ideal bandlimited interpolation. In practical implementations, the lowpass filter is approximated and the interpolation in the time domain is no longer exact. In some instances, simple interpolation procedures such as a zero-order hold or linear interpolation (first-order hold) suffice.

If a signal is undersampled (i.e., the sampling frequency is less than that required by the sampling theorem), then the signal reconstructed by ideal bandlimited interpolation will be related to the original signal through a form of distortion referred to as aliasing. In many instances it is important to choose the sampling rate to avoid aliasing. However, there are a variety of important examples, such as the stroboscope, where the presence of aliasing is important and is exploited.

Sampling has a number of important applications. One particularly significant set of applications relates to using the concept of sampling to process continuous-time signals with discrete-time systems, using minicomputers or microprocessors, or any of a variety of devices specifically oriented toward discrete-time signal processing.

The basic theory of sampling is similar for both continuous-time and discrete-time signals. In the discrete-time case there is the closely related concept of deci-

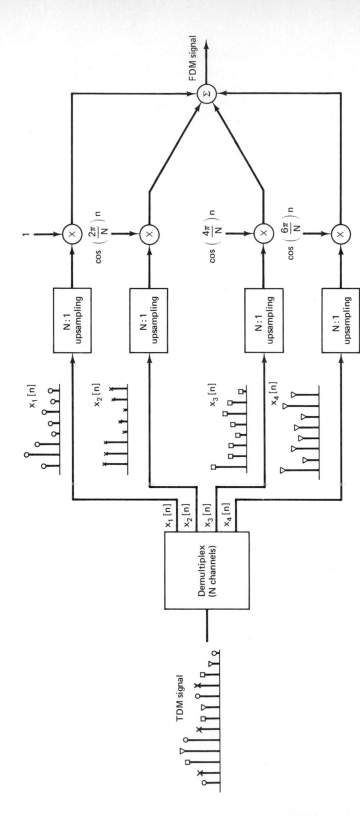

Figure 8.41 Block diagram for TDM to FDM transmultiplexing.

mation whereby the decimated sequence is obtained by extracting sequence values at periodic intervals. The difference between sampling and decimation lies in the fact that for the sampled sequence, values of zero lie in between the sample values, whereas in the decimated sequence these zero values are discarded, thereby compressing the sequence in time. The inverse of decimation is interpolation. The ideas of decimation and interpolation for sequences arise in a variety of important practical applications, including transmodulation systems.

PROBLEMS

8.1. In the system shown in Figure P8.1, two time functions $x_1(t)$ and $x_2(t)$ are multiplied and the product $w(t)$ is sampled by a periodic impulse train. $x_1(t)$ is bandlimited to ω_1 and $x_2(t)$ is bandlimited to ω_2, that is,

$$X_1(\omega) = 0, \qquad |\omega| > \omega_1$$
$$X_2(\omega) = 0, \qquad |\omega| > \omega_2$$

Determine the *maximum* sampling interval T such that $w(t)$ is recoverable from $w_p(t)$ through the use of an ideal lowpass filter.

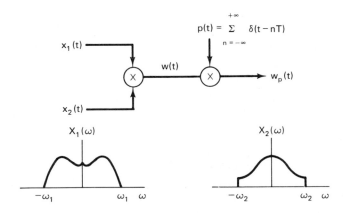

Figure P8.1

8.2. Shown in Figure P8.2 is a system in which the sampling signal is an impulse train with alternating sign. The Fourier transform of the input signal is as indicated in Figure P8.2.

(a) For $\Delta < \pi/2\omega_M$, sketch the Fourier transform of $x_p(t)$ and $y(t)$.

(b) For $\Delta < \pi/2\omega_M$, determine a system which will recover $x(t)$ from $x_p(t)$.

(c) For $\Delta < \pi/2\omega_M$, determine a system which will recover $x(t)$ from $y(t)$.

(d) What is the *maximum* value of Δ in relation to ω_M for which $x(t)$ can be recovered from either $x_p(t)$ or $y(t)$.

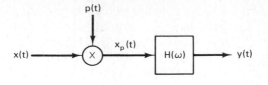

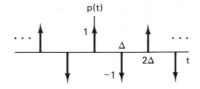

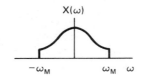

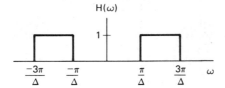

Figure P8.2

8.3. Shown in Figure P8.3 is a system in which the input signal is multiplied by a periodic square wave. The period of $s(t)$ is T. The input signal is bandlimited with $|X(\omega)| = 0$, $|\omega| \geq \omega_M$.

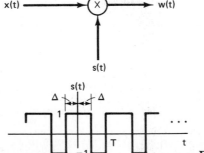

Figure P8.3

(a) For $\Delta = T/3$ determine in terms of ω_M the *maximum* value of T for which $x(t)$ can be recovered from $w(t)$. With this maximum value, determine a system to recover $x(t)$ from $w(t)$.

(b) For $\Delta = T/4$, determine in terms of ω_M the *maximum* value of T for which $x(t)$

can be recovered from $w(t)$. With this maximum value, determine a system to recover $x(t)$ from $w(t)$.

8.4. In Figure 8.5 and redrawn as Figure P8.4 is a sampler, followed by an ideal lowpass filter for reconstruction of $x(t)$ from its samples $x_p(t)$. From the sampling theorem, we know that if $\omega_s = 2\pi/T$ is greater than twice the highest frequency present in $x(t)$ and $\omega_c = \omega_s/2$, then the reconstructed signal $x_r(t)$ will exactly equal $x(t)$. If this condition on the bandwidth of $x(t)$ is violated, then $x_r(t)$ will *not* equal $x(t)$. However, as we show in this problem, if $\omega_c = \omega_s/2$, then for any choice of T, $x_r(t)$ and $x(t)$ will always be equal at the sampling instants, that is,

$$x_r(kT) = x(kT), \qquad k = 0, \pm 1, \pm 2, \ldots$$

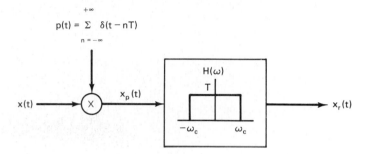

Figure P8.4

To obtain this result, consider eq. (8.10), which expresses $x_r(t)$ in terms of the samples of $x(t)$, specifically

[eq. (8.10)]
$$x_r(t) = \sum_{n=-\infty}^{+\infty} x(nT)T\frac{\omega_c}{\pi}\,\text{sinc}\left[\frac{\omega_c(t-nT)}{\pi}\right]$$

With $\omega_c = \omega_s/2$ this becomes

$$x_r(t) = \sum_{n=-\infty}^{+\infty} x(nT)\,\text{sinc}\left[\frac{(t-nT)}{T}\right] \tag{P8.4}$$

Using the properties of the sinc function [in particular, the values of α for which sinc $(\alpha) = 0$], show from eq. (P8.4) that without any restrictions on $x(t)$, $x_r(kT) = x(kT)$ for any integer value of k.

8.5. A signal limited in bandwidth to $|\omega| < W$ can be recovered from nonuniformly spaced samples as long as the average sample density is $2(W/2\pi)$ samples per second. This problem illustrates a particular example of nonuniform sampling. Assume that in Figure **P8.5-1**:

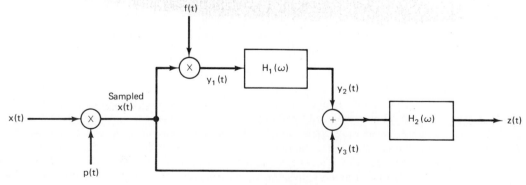

Figure P8.5-1

1. $x(t)$ is bandlimited; $X(\omega) = 0$, $|\omega| > W$.
2. $p(t)$ is a nonuniformly spaced periodic impulse train, as shown in Figure P8.5-2.

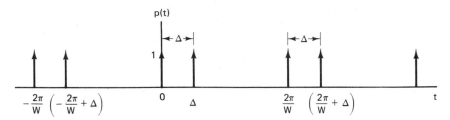

Figure P8.5-2

3. $f(t)$ is a periodic waveform with period $T = 2\pi/W$. Since $f(t)$ multiplies an impulse train, only its values, $f(0) = a$ and $f(\Delta) = b$, at $t = 0$ and $t = \Delta$ are significant.

4. $H_1(\omega)$ is a 90° phase shifter, that is,

$$H_1(\omega) = \begin{cases} j, & \omega > 0 \\ -j, & \omega < 0 \end{cases}$$

5. $H_2(\omega)$ is an ideal lowpass filter;

$$H_2(\omega) = \begin{cases} K, & 0 < \omega < W \\ K^*, & -W < \omega < 0 \\ 0, & |\omega| < W \end{cases}$$

where K is a (possibly complex) constant.

Find (and sketch where appropriate) the following:
(a) The Fourier transform, $P(\omega)$, of $p(t)$.
(b) The Fourier transform of the product, $p(t)f(t)$, in terms of the as-yet-unspecified parameters, a and b.
(c) An expression for the Fourier transform, $Y_1(\omega)$, of $y_1(t)$ valid in the interval $0 < \omega < W$.
(d) An expression for the Fourier transform, $Y_2(\omega)$, of $y_2(t)$ valid in the interval $0 < \omega < W$.
(e) An expression for the Fourier transform, $Y_3(\omega)$, of $y_3(t)$ valid in the interval $0 < \omega < W$.
(f) The values of the real parameters, a and b, and the complex gain K as functions of Δ such that $z(t) = x(t)$ for any bandlimited $x(t)$ and any Δ, $0 < \Delta < \pi/W$.

8.6. The sampling theorem as we have derived it states that a signal $x(t)$ must be sampled at a rate greater than its bandwidth (or equivalently a rate greater than twice its highest frequency). This implies that if $x(t)$ has a spectrum as indicated in Figure P8.6-1, then $x(t)$ must be sampled at a rate greater than $2\omega_2$. Since the signal has most of its energy concentrated in a narrow band, it would seem reasonable to expect that a sampling rate lower than twice the highest frequency could be used. A signal whose energy is concentrated in a frequency band is often referred to as a *bandpass signal*. There are a variety of techniques for sampling such signals, and these techniques are generally referred to as *bandpass sampling*.

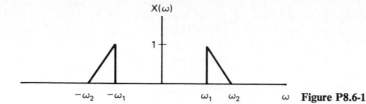

Figure P8.6-1

To examine the possibility of sampling a bandpass signal at a rate less than the total bandwidth, consider the system shown in Figure P8.6-2. Assuming that $\omega_1 > (\omega_2 - \omega_1)$, find the maximum value of T and the values of the constants A, ω_a, and ω_b such that $x_r(t) = x(t)$.

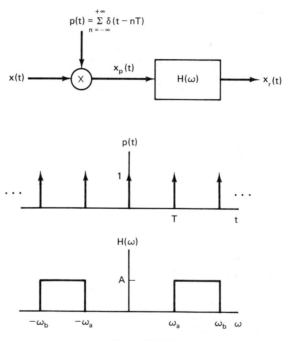

Figure P8.6-2

8.7. In Problem 8.6 we considered one procedure for bandpass sampling and reconstruction. Another procedure when $x(t)$ is real consists of using complex modulation followed by sampling. The sampling system is shown in Figure P8.7-1. With $x(t)$ real and with

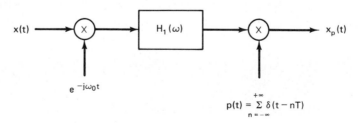

Figure P8.7-1

$X(\omega)$ nonzero only for $\omega_1 < |\omega| < \omega_2$, the modulating frequency ω_0 is chosen as $\omega_0 = \frac{1}{2}(\omega_1 + \omega_2)$ and the lowpass filter $H_1(\omega)$ has cutoff frequency $\frac{1}{2}(\omega_2 - \omega_1)$.

(a) For $X(\omega)$ as shown in Figure P8.7-2, sketch $X_p(\omega)$.

(b) Determine the maximum sampling period T such that $x(t)$ is recoverable from $x_p(t)$.

(c) Determine a system to recover $x(t)$ from $x_p(t)$.

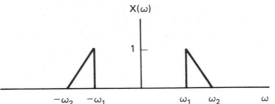

Figure P8.7-2

8.8. Consider a disc on which four cycles of a sinusoid are painted. The disc is rotated at approximately 15 revolutions/second, so that the sinusoid when viewed through a narrow slit has a frequency of 60 Hz.

The arrangement is indicated in Figure P8.8. Let $v(t)$ denote the position of the line seen through the slit. Then $v(t)$ is of the form

$$v(t) = A \cos(\omega_0 t + \phi), \qquad \omega_0 = 120\pi$$

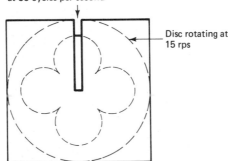

Position of line varies sinusoidally at 60 cycles per second

Disc rotating at 15 rps

Figure P8.8

For notational convenience, we will normalize $v(t)$ so that $A = 1$. At 60 Hz, the eye is not able to follow $v(t)$, and we will assume that this effect can be explained by modeling the eye as an ideal lowpass filter. The cutoff frequency of this filter will be taken to be 20 Hz.

Sampling of the sinusoid can be accomplished by illuminating the disc with a strobe light. Thus, the illumination $i(t)$ can be represented by an impulse train,

$$i(t) = \sum_{k=-\infty}^{+\infty} \delta(t - kT)$$

where $1/T$ is the strobe frequency in hertz. The resulting sampled signal is the product $r(t) = v(t) \cdot i(t)$. Let $R(\omega)$, $V(\omega)$, and $I(\omega)$ denote the Fourier transforms of $r(t)$, $v(t)$, and $i(t)$, respectively.

(a) Sketch $V(\omega)$, clearly indicating the effect of the parameters ϕ and ω_0.

(b) Sketch $I(\omega)$, indicating the effect of T.

(c) According to the sampling theorem, there is a maximum value for T in terms of ω_0 such that $v(t)$ can be recovered from $r(t)$ using a lowpass filter. Determine this value of T and the cutoff frequency of the lowpass filter. Sketch $R(\omega)$ when T is slightly less than that value.

If the sampling period T is made greater than the value determined in part (c), aliasing of the spectrum occurs. As a result of this aliasing, we perceive a lower-frequency sinusoid.

(d) Suppose that $2\pi/T = \omega_0 + 20\pi$. Sketch $R(\omega)$ for $|\omega| < 40\pi$. Denote by $v_a(t)$ the apparent position of the line as we perceive it. Assuming that the eye behaves as an ideal lowpass filter with 20-Hz cutoff and unity gain, express $v_a(t)$ in the form

$$v_a(t) = A_a \cos{(\omega_a t + \phi_a)}$$

where A_a is the apparent amplitude, ω_a the apparent frequency, and ϕ_a the apparent phase of $v_a(t)$.

(e) Repeat part (d) for $2\pi/T = \omega_0 - 20\pi$.

8.9. It is frequently necessary to display on an oscilloscope screen waveforms having very short time structures, for example, on the scale of thousandths of a nanosecond. Since the rise time of the fastest oscilloscope is longer than this, such displays cannot be achieved directly. If however, the waveform is periodic, the desired result can be obtained indirectly using an instrument called a sampling oscilloscope.

The idea, as shown in Figure P8.9-1, is to sample the fast waveform $x(t)$ once each period but at successively later points in successive periods. The increment Δ should be an appropriately chosen sampling interval in relation to the bandwidth of $x(t)$. If the resulting impulse train is then passed through an appropriate lowpass interpolating filter the output, $y(t)$, will be proportional to the original fast waveform slowed down or stretched out in time [i.e., $y(t)$ is proportional to $x(at)$, where $a < 1$].

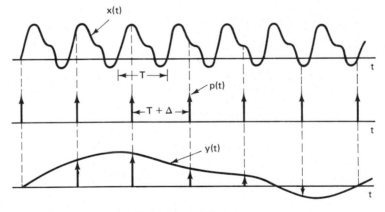

Figure P8.9-1

For $x(t) = A + B \cos{[(2\pi/T)t + \theta]}$ find a range of values of Δ so that $y(t)$ in Figure P8.9-2 is proportional to $x(at)$ with $a < 1$. Also, determine the value of a in terms of T and Δ.

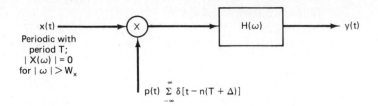

$$H(\omega) = \begin{cases} 1 & |\omega| < \dfrac{1}{2(T+\Delta)} \\ 0 & \text{elsewhere} \end{cases}$$

Figure P8.9-2

8.10. In Figure P8.10-1 is shown the overall system for filtering a continuous-time signal using a discrete-time filter. If $X_c(\omega)$ and $H(\Omega)$ are as shown in Figure P8.10-2 and with $1/T = 20$ kHz, sketch $X_p(\omega)$, $X(\Omega)$, $Y(\Omega)$, $Y_p(\omega)$, and $Y_c(\omega)$.

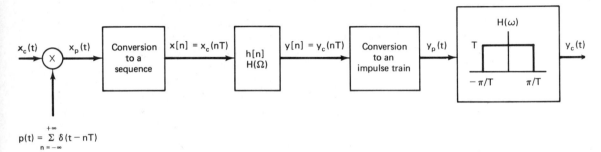

Figure P8.10-1

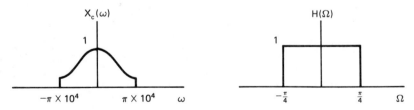

Figure P8.10-2

8.11. Figure P8.11-1 shows a system that converts a continuous-time signal to a discrete-time signal. The input $x(t)$ is periodic with a period of 0.1 second. The Fourier series coefficients of $x(t)$ are

$$a_k = (\tfrac{1}{2})^{|k|}, \qquad -\infty < k < +\infty$$

The lowpass filter $H(\omega)$ has the frequency response shown in Figure P8.11-2. The sampling period $T = 5 \times 10^{-3}$ second.
(a) Show that $x[n]$ is a periodic sequence and determine its period.
(b) Determine the Fourier series coefficients of $x[n]$.

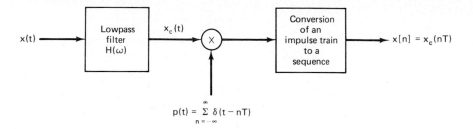

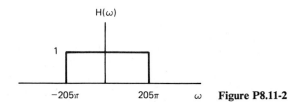

Figure P8.11-1

H(ω)

−205π 205π ω **Figure P8.11-2**

8.12. Consider a bandlimited signal $x_c(t)$ which is sampled at a rate greater than the Nyquist rate. The samples, spaced T seconds apart, are then converted to a sequence $x[n]$, as indicated in Figure P8.12.

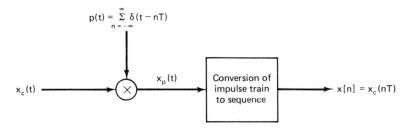

Figure P8.12

Determine the relation between the energy E_d of the sequence, the energy E_c of the original signal, and the sampling interval T. The energy of a sequence $x[n]$ is defined as

$$E_d = \sum_{n=-\infty}^{+\infty} |x[n]|^2$$

and the energy in a continuous-time function $x_c(t)$ is defined as

$$E_c = \int_{-\infty}^{+\infty} |x_c(t)|^2 \, dt$$

8.13. We wish to design a continuous-time sinusoidal signal generator that is capable of producing sinusoidal signals at any frequency satisfying

$$\omega_1 \leq \omega \leq \omega_2$$

where ω_1 and ω_2 are given positive numbers.

Our design is to take the following form. We have stored a discrete-time cosine wave of period N; that is, we have stored $x[0], \ldots, x[N-1]$, where

$$x[k] = \cos\left(\frac{2\pi k}{N}\right)$$

Every T seconds we output an impulse weighted by a value of $x[k]$, where we proceed through the values of $k = 0, 1, \ldots, N-1$ in a cyclic fashion. That is,

$$y(kT) = x(k \text{ modulo } N)$$

or equivalently

$$y(kT) = \cos\left(\frac{2\pi k}{N}\right)$$

and

$$y_p(t) = \sum_{k=-\infty}^{+\infty} \cos\left(\frac{2\pi k}{N}\right) \delta(t - kT)$$

(a) Show that by adjusting T, we can adjust the frequency of the cosine signal being sampled. Specifically, show that

$$y_p(t) = (\cos \omega_0 t) \sum_{k=-\infty}^{+\infty} \delta(t - kT)$$

where $\omega_0 = 2\pi/NT$. Determine a range of values for T so that $y(t)$ can represent samples of a cosine signal with a frequency that is variable over the full range

$$\omega_1 \leq \omega \leq \omega_2$$

(b) Sketch $Y_p(\omega)$.

The overall system for generating a continuous-time sinusoid is depicted in Figure P8.13-1. $H(\omega)$ is an ideal lowpass filter with unity gain in its passband

$$H(\omega) = \begin{cases} 1, & |\omega| < \omega_c \\ 0, & \text{otherwise} \end{cases}$$

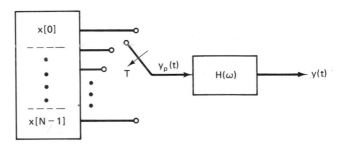

Figure P8.13-1

The parameter ω_c is to be determined so that $y(t)$ is a continuous-time cosine signal in the desired frequency band.

(c) Consider any value of T in the range determined in part (a). Determine the minimum value of N and some value for ω_c so that $y(t)$ is a cosine signal in the range $\omega_1 \leq \omega \leq \omega_2$.

(d) The amplitude of $y(t)$ will vary, depending upon the value of ω chosen between ω_1 and ω_2. Thus, we must design a system $G(\omega)$ that normalizes the signal, as shown in Figure P8.13-2. Find such a $G(\omega)$.

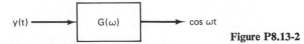

Figure P8.13-2

Sampling   Chap. 8

8.14. In Figure P8.14 is shown a system consisting of a continuous-time linear time-invariant system, followed by a sampler, conversion to a sequence, and a linear time-invariant discrete-time system. The continuous-time LTI system is causal and satisfies the LCCDE

$$\frac{dy_c(t)}{dt} + y_c(t) = x_c(t)$$

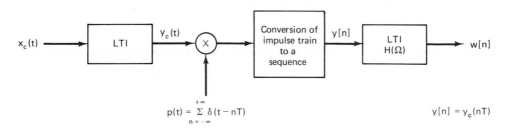

Figure P8.14

The input $x_c(t)$ is a unit impulse $\delta(t)$.

(a) Determine $y_c(t)$.

(b) Determine the frequency response $H(\Omega)$ and the impulse response $h[n]$ such that $w[n] = \delta[n]$.

8.15. Shown in Figure P8.15 is a system that processes continuous-time signals using a digital filter. The digital filter $h[n]$ is linear and causal with difference equation

$$y[n] = \tfrac{1}{2}y[n-1] + x[n]$$

For input signals which are bandlimited so that $X_c(\omega) = 0$ for $|\omega| > \pi/T$, the system of Figure P8.15 is equivalent to a continuous-time LTI system.

Determine the frequency response $H_c(\omega)$ of the equivalent overall system with input $x_c(t)$ and output $y_c(t)$.

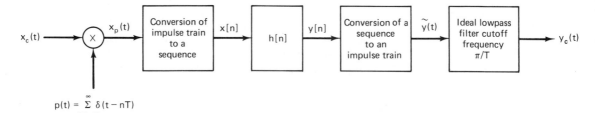

Figure P8.15

8.16. In Figure P8.16-1 is depicted a system for which the input and output are discrete-time signals. The discrete-time input $x[n]$ is converted to a continuous-time impulse train $x_p(t)$. The continuous-time signal $x_p(t)$ is then filtered by an LTI system to produce the output $y_c(t)$. $y_c(t)$ is then converted to the discrete-time signal $y[n]$. The LTI system with input $x_c(t)$ and output $y_c(t)$ is causal and is characterized by the linear constant-coefficient differential equation

$$\frac{d^2y_c(t)}{dt^2} + 4\frac{dy_c(t)}{dt} + 3y_c(t) = x_c(t)$$

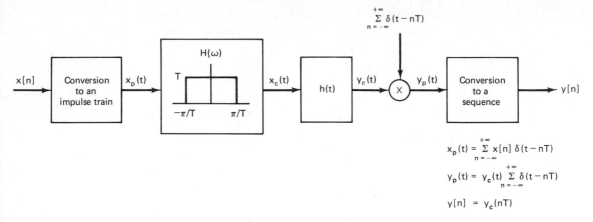

$$x_p(t) = \sum_{n=-\infty}^{+\infty} x[n]\, \delta(t-nT)$$

$$y_p(t) = y_c(t) \sum_{n=-\infty}^{+\infty} \delta(t-nT)$$

$$y[n] = y_c(nT)$$

Figure P8.16-1

The overall system is equivalent to a causal discrete-time LTI system, as indicated in Figure P8.16-2.

Determine the frequency response $H(\Omega)$ and the unit sample response, $h[n]$, of the equivalent LTI system.

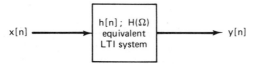

Figure P8.16-2

8.17. In the system shown in Figure P8.17, the input $x_c(t)$ is bandlimited with $X_c(\omega) = 0$, $|\omega| > 2\pi \times 10^4$. The digital filter $h[n]$ is described by the input–output relation

$$y[n] = T \sum_{k=-\infty}^{n} x[k] \tag{P8.17-1}$$

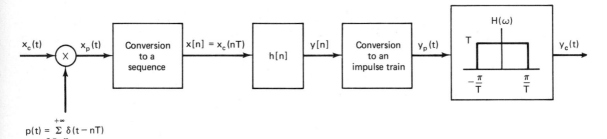

Figure P8.17

(a) What is the maximum value of T allowed if aliasing is to be avoided in the transformation from $x_c(t)$ to $x_p(t)$?

(b) With the discrete-time LTI system $h[n]$ specified through eq. (P8.17-1), determine its impulse response $h[n]$.

(c) Determine if there is any value of T for which

$$\lim_{n\to\infty} y[n] = \lim_{t\to\infty} \int_{-\infty}^{t} x_c(t)\, dt \tag{P8.17-2}$$

If so, determine the *maximum* value. If not, explain and specify how T would be chosen so that the equality in eq. (P8.17-2) is best approximated. (Think carefully about this part. It is easy to jump to the wrong conclusion.)

8.18. Consider a continuous-time signal

$$x_c(t) = s_c(t) + \alpha s_c(t - T_D)$$

Assume that the Fourier transform of $x_c(t)$ is bandlimited such that $X_c(\omega) = 0$ for $|\omega| > \pi/T$ and that $x_c(t)$ is sampled with a sampling period T to obtain the sequence

$$x[n] = x_c(nT)$$

We want to find the unit sample response, $h[n]$, of a discrete-time system such that

$$x[n] = \sum_{k=-\infty}^{\infty} s[k]h[n - k]$$

where $s[n] = s_c(nT)$.
(a) Find $h[n]$ when $T_D = T$.
(b) Find $h[n]$ when $T_D = T/2$.

8.19. In many practical situations a signal is recorded in the presence of an echo, which we would like to remove by appropriate processing. For example, in Figure P8.19-1,

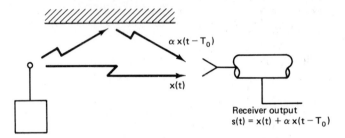

Figure P8.19-1

we illustrate a system in which a receiver simultaneously receives a signal $x(t)$ and an echo represented by an attenuated delayed replication of $x(t)$. Thus, the receiver output is $s(t) = x(t) + \alpha x(t - T_0)$, where $|\alpha| < 1$. The receiver output is to be processed to recover $x(t)$ by first converting to a sequence and using an appropriate digital filter $h[n]$ as indicated in Figure P8.19-2.

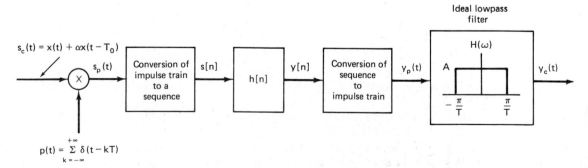

Figure P8.19-2

Assume that $x(t)$ is bandlimited [i.e., $X(\omega) = 0$ for $|\omega| > \omega_M$] and that $|\alpha| < 1$.

(a) If $T_0 < \pi/\omega_M$, and the sampling period is taken equal to T_0 (i.e., $T = T_0$), determine the difference equation for the digital filter $h[n]$ so that $y_c(t)$ is proportional to $x(t)$.

(b) With the assumptions of part (a), specify the gain A of the ideal lowpass filter so that $y_c(t) = x(t)$.

(c) Now suppose that $\pi/\omega_M < T_0 < 2\pi/\omega_M$. Determine a choice for the sampling period T, the lowpass filter gain A, and the frequency response for the digital filter $h[n]$ so that $y_c(t)$ is proportional to $x(t)$.

8.20. In discussing frequency-domain sampling in Section 8.5 we assumed that the interval over which $x(t)$ is nonzero is centered about $t = 0$. Then, with $x(t) = 0$ for $|t| > T_m$ we showed that $x(t)$ could be recovered from equally spaced samples of $X(\omega)$ provided that the sample spacing ω_0 satisfies the inequality in eq. (8.35), specifically

[eq. (8.35)]
$$\frac{2\pi}{\omega_0} > 2T_m$$

In this problem we generalize that result to apply to a time-limited signal that is zero outside *any* specified time interval of length $2T_m$.

Paralleling the discussion in Section 8.5, consider $x(t)$ as sketched in Figure P8.20, and let $X(\omega)$ denote its Fourier transform. Applying the system of Figure 8.30 to $X(\omega)$, we have in the frequency domain $\tilde{X}(\omega) = X(\omega)P(\omega)$, where $P(\omega) = \sum_{k=-\infty}^{+\infty} \delta(\omega - k\omega_0)$.

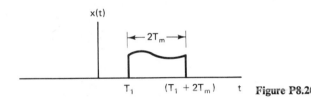

Figure P8.20

(a) For $2\pi/\omega_0 > 2T_m$, sketch $\tilde{x}(t)$, the inverse Fourier transform of $\tilde{X}(\omega)$.

(b) As shown in Section 8.5, for $2\pi/\omega_0 > 2T_m$, $x(t)$ can be recovered from $\tilde{x}(t)$ by windowing, that is,

$$x(t) = \tilde{x}(t)w(t)$$

Determine and sketch $w(t)$ for the more general case considered in this problem.

8.21. We have a continuous-time signal $x_c(t)$ with Fourier transform $X_c(\omega)$. We wish to evaluate samples of $X_c(\omega)$ by first sampling $x_c(t)$ to obtain a discrete-time sequence $x[n]$, where

$$x[n] = x_c(nT)$$

Samples of the discrete-time Fourier transform $X(\Omega)$ are then computed. If samples of $X(\Omega)$ are computed with a sample spacing in frequency of $\Delta\Omega$, this corresponds to computing samples of $X_c(\omega)$ with a sample spacing in frequency which we denote by $\Delta\omega$.

Assume that $x_c(t)$ is bandlimited to $10\,\text{kHz}$, that the sampling period $T = 50\,\mu\text{sec}$ and that $\Delta\Omega = \pi \times 10^{-3}$. Determine $\Delta\omega$.

8.22. In this problem we consider the discrete-time counterpart of the zero-order hold (ZOH) and first-order hold (FOH), which were discussed for continuous time in Sections 8.1.2 and 8.2.

We consider a sequence $x[n]$ to which discrete-time sampling as illustrated in Figure 8.32 has been applied. We assume that the conditions of the discrete-time sampling theorem are satisfied; that is, $\Omega_s > 2\Omega_M$ where Ω_s is the sampling frequency and $X(\Omega) = 0$, $\Omega_M < |\Omega| \le \pi$. The original sequence $x[n]$ is then exactly recoverable from $x_p[n]$ by ideal lowpass filtering, which, as discussed in Section 8.6, corresponds to bandlimited interpolation.

The ZOH represents an approximate interpolation whereby each sample value is repeated (or held) $N - 1$ successive times as illustrated in Figure P8.22-1 for the case of $N = 3$. The FOH represents a linear interpolation between samples as illustrated in Figure P8.22-1.

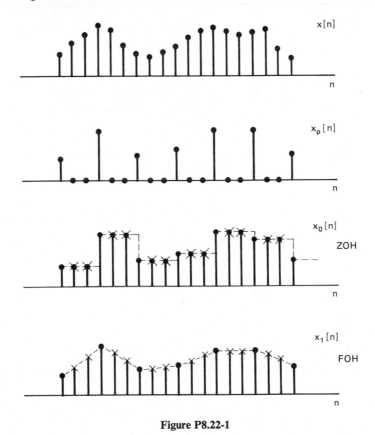

Figure P8.22-1

(a) The ZOH can be represented as an interpolation in the form of eq. (8.51) and the system in Figure P8.22-2. Determine and sketch $h_0[n]$ for the general case of a sampling period N.

(b) $x[n]$ can be exactly recovered from the ZOH sequence $x_0[n]$ using an appropriate LTI filter $H(\Omega)$ as indicated in Figure P8.22-3. Determine and sketch $H(\Omega)$.

(c) The FOH (linear interpolation) can be represented as an interpolation in the form of eq. (8.51) and equivalently the system in Figure P8.22-4. Determine and sketch $h_1[n]$ for the general case of a sampling period N.

(d) $x[n]$ can be exactly recovered from the FOH sequence $x_1[n]$ using an appropriate LTI filter with frequency response $H(\Omega)$. Determine and sketch $H(\Omega)$.

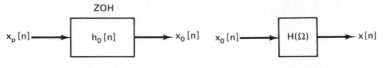

Figure P8.22-2   Figure P8.22-3

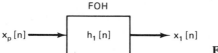

Figure P8.22-4

8.23. A signal $x[n]$ is sampled in discrete time as shown in Figure P8.23. $h_r[n]$ is an ideal low-pass filter with frequency response

$$H(\Omega) = \begin{cases} 1, & |\Omega| < \dfrac{2\pi}{N} \\ 0, & \dfrac{2\pi}{N} < |\Omega| < \pi \end{cases}$$

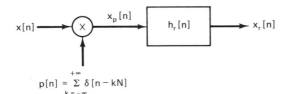

Figure P8.23

From eqs. (8.50) and (8.51), the filter output is expressible as

$$x_r[n] = \sum_{k=-\infty}^{+\infty} x[kN]h_r[n - kN] = \sum_{k=-\infty}^{+\infty} x[kN]\frac{N\Omega_c}{2\pi} \text{sinc}\left[\frac{\Omega_c}{\pi}(n - kN)\right]$$

where $\Omega_c = 2\pi/N$. Show that independent of whether the sequence $x[n]$ is sampled above or below the Nyquist rate, $x_r[mN] = x[mN]$, where m is any positive or negative integer.

8.24. Consider a discrete-time sequence $x[n]$ from which we form two new sequences, $x_p[n]$ and $x_d[n]$, where $x_p[n]$ corresponds to *sampling* $x[n]$ with a sampling period of 2 and $x_d[n]$ corresponds to *decimating* $x[n]$ by a factor of 2, so that

$$x_p[n] = \begin{cases} x[n], & n = 0, \pm 2, \pm 4, \ldots \\ 0, & n = \pm 1, \pm 3, \ldots \end{cases}$$

and

$$x_d[n] = x[2n]$$

(a) If $x[n]$ is as illustrated in Figure P8.24-1, sketch the sequences $x_p[n]$ and $x_d[n]$.
(b) If $X(\Omega)$ is as shown in Figure P8.24-2, sketch $X_p(\Omega)$ and $X_d(\Omega)$.

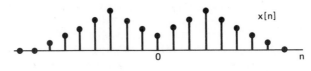

Figure P8.24-1

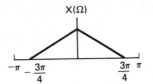

Figure P8.24-2

8.25. As discussed in Section 8.7 and illustrated in Figures 8.40 and P8.25-1, the procedure for interpolation or upsampling by an integer factor N can be thought of as the cascade of two operations. The first system, system A, corresponds to inserting $(N - 1)$ zero-sequence values between each sequence value of $x[n]$, so that

$$x_p[n] = \begin{cases} x_d\left[\dfrac{n}{N}\right], & n = 0, \pm N, \pm 2N, \ldots \\ 0, & \text{otherwise} \end{cases}$$

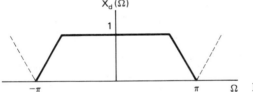

Figure P8.25-1

For exact bandlimited interpolation, $H(\Omega)$ is an ideal lowpass filter.

(a) Determine whether or not system A is linear.

(b) Determine whether or not system A is time-invariant.

(c) For $X_d(\Omega)$ as sketched in Figure P8.25-2 and with $N = 3$, sketch $X_p(\Omega)$.

(d) For $N = 3$, $X_d(\Omega)$ as in Figure P8.25-2, and $H(\Omega)$ appropriately chosen for exact bandlimited interpolation, sketch $X(\Omega)$.

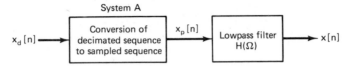

Figure P8.25-2

8.26. As shown in Figure 8.40 and discussed in Section 8.7, the procedure for interpolation or upsampling by an integer factor N can be thought of as a cascade of two operations. For exact bandlimited interpolation, the filter $H(\Omega)$ in Figure 8.40 is an ideal lowpass filter. In any specific application it would be necessary to implement an approximate lowpass filter. In this problem we explore some useful constraints that are often imposed on the design of these approximate lowpass filters.

(a) Let us first consider $H(\Omega)$ approximated by a zero phase FIR filter. The filter is to be designed with the constraint that the original sequence values $x_d[n]$ get reproduced *exactly*, that is,

$$x[n] = x_d\left[\frac{n}{L}\right], \qquad n = 0, \pm L, \pm 2L, \ldots \qquad (P8.26\text{-}1)$$

This guarantees that even though the interpolation between the original sequence values may not be exact, the original values are reproduced exactly in the interpolation. Determine the constraint on the impulse response $h[n]$ of the lowpass filter to guarantee that eq. (P8.26-1) will hold exactly for any sequence $x_d[n]$.

(b) Now suppose that the interpolation is to be carried out with a *linear-phase causal symmetric* FIR filter of length N, that is,

$$h[n] = 0, \qquad n < 0, n > N - 1 \tag{P8.26-2}$$

$$H(\Omega) = H_R(\Omega)e^{-j\alpha\Omega} \tag{P8.26-3}$$

where $H_R(\Omega)$ is real. The filter is to be designed with the constraint that the original sequence values $x_d[n]$ get reproduced *exactly* but with an integer delay α, where α is the negative of the slope of the phase of $H(\Omega)$, that is,

$$x[n] = x_d\left[\frac{n - \alpha}{L}\right], \qquad n - \alpha = 0, \pm L, \pm 2L, \ldots \tag{P8.26-4}$$

Determine whether this imposes any constraint on whether the filter length N is odd or even.

(c) Again, suppose that the interpolation is to be carried out with a linear-phase causal symmetric FIR filter, so that $H(\Omega)$ is of the form

$$H(\Omega) = H_R(\Omega)e^{-j\beta\Omega}$$

where $H_R(\Omega)$ is real. The filter is to be designed with the constraint that the original sequence values $x_d[n]$ get reproduced exactly, but with a delay M which is not necessarily equal to the negative of the slope of the phase, that is,

$$x[n] = x_d\left[\frac{n - M}{L}\right], \qquad n - M = 0, \pm L, \pm 2L, \ldots$$

Determine whether this imposes any constraint on whether the filter length N is odd or even.

In the preceding chapters we have seen that the tools of Fourier analysis are extremely useful in the study of many problems involving signals and LTI systems. This is due in large part to the fact that a broad class of signals can be represented by a linear combination of complex exponentials and that complex exponentials are eigenfunctions of linear time-invariant systems. The representation of continuous-time signals as linear combinations of complex exponentials of the form e^{st} with $s = j\omega$ formed the basis for the Fourier series and Fourier transform representation of signals and this, in combination with the eigenfunction property, led to the notions of frequency response, filtering, and so on. The eigenfunction property and many of its consequences still apply when s is not restricted to be pure imaginary, and in many cases there are important advantages to be gained in reformulating some of the ideas developed in Chapters 4 and 5 in the context of a more general set of complex exponentials. In this chapter we consider this generalization of the Fourier transform, utilizing a broader class of complex exponential signals. The resulting transform is referred to as the Laplace transform. In Chapter 10 a corresponding generalization is carried out for discrete-time signals.

9.1 THE LAPLACE TRANSFORM

For a linear time-invariant system, with impulse response $h(t)$, we saw in Chapter 4 that the response $y(t)$ of the system to a complex exponential input of the form e^{st} is given by

The Laplace Transform

$$y(t) = H(s)e^{st} \tag{9.1a}$$

where

$$H(s) = \int_{-\infty}^{+\infty} h(t)e^{-st}\, dt \tag{9.1b}$$

For s imaginary (i.e., $s = j\omega$), the integral in eq. (9.1b) corresponds to the Fourier transform of $h(t)$. For general values of the complex variable s, it is referred to as the *Laplace transform* of the impulse response $h(t)$. As with the Fourier transform, the Laplace transform plays an important role in representing not only the system impulse response for LTI systems, but also the input signals and output signals to such systems. The Laplace transform of a general signal $x(t)$ is defined as†

$$\boxed{X(s) \triangleq \int_{-\infty}^{+\infty} x(t)e^{-st}\, dt} \tag{9.2}$$

and we note in particular that it is a function of the independent variable s corresponding to the complex variable in the exponent of e^{-st}. The complex variable s is in general of the form $s = \sigma + j\omega$, with σ and ω the real and imaginary parts, respectively. For convenience we will sometimes denote the Laplace transform in operator form as $\mathcal{L}\{x(t)\}$ and denote the transform relationship between $x(t)$ and $X(s)$ as

$$x(t) \xleftrightarrow{\mathcal{L}} X(s) \tag{9.3}$$

When $s = j\omega$ eq. (9.2) becomes

$$X(j\omega) = \int_{-\infty}^{+\infty} x(t)e^{-j\omega t}\, dt \tag{9.4}$$

which corresponds to the *Fourier transform* of $x(t)$, that is,

$$X(s)|_{s=j\omega} = \mathcal{F}\{x(t)\} \tag{9.5}$$

The Laplace transform also bears a straightforward relationship to the Fourier transform when the complex variable s is not purely imaginary. To see this relationship, consider $X(s)$ as specified in eq. (9.2) with s expressed as $s = \sigma + j\omega$, so that

$$X(\sigma + j\omega) = \int_{-\infty}^{+\infty} x(t)e^{-(\sigma+j\omega)t}\, dt \tag{9.6}$$

or

$$X(\sigma + j\omega) = \int_{-\infty}^{+\infty} [x(t)e^{-\sigma t}]e^{-j\omega t}\, dt \tag{9.7}$$

We recognize the right-hand side of eq. (9.7) as the Fourier transform of $x(t)e^{-\sigma t}$; that is, the Laplace transform of $x(t)$ can be interpreted as the Fourier transform

†The transform defined by eq. (9.2) is often called the bilateral Laplace transform, to distinguish it from the unilateral Laplace transform, which we discuss in Section 9.8. As we are concerned primarily with the bilateral transform, we will omit the word "bilateral" except where it is needed to avoid ambiguity.

of $x(t)$ after mutiplication by a real exponential signal. The real exponential $e^{-\sigma t}$ may be decaying or growing in time, depending on whether σ is positive or negative.

To illustrate the Laplace transform and its relationship to the Fourier transform, let us consider the following example.

Example 9.1

Consider the signal $x(t) = e^{-at}u(t)$. From Example 4.7 the Fourier transform $X(j\omega)$ converges for $a > 0$ and is given by†

$$X(j\omega) = \int_0^\infty e^{-at}e^{-j\omega t}\, dt = \frac{1}{j\omega + a}, \qquad a > 0 \tag{9.8}$$

From eq. (9.2) the Laplace transform is

$$X(s) = \int_{-\infty}^{+\infty} e^{-at}e^{-st}u(t)\, dt$$

$$= \int_0^\infty e^{-(s+a)t}\, dt$$

or, with $s = \sigma + j\omega$,

$$X(\sigma + j\omega) = \int_0^\infty e^{-(\sigma+a)t}e^{-j\omega t}\, dt \tag{9.9}$$

By comparison with eq. (9.8) we recognize (9.9) as the Fourier transform of $e^{-(\sigma+a)t}u(t)$ and thus

$$X(\sigma + j\omega) = \frac{1}{(\sigma + a) + j\omega}, \qquad (\sigma + a) > 0$$

or equivalently,

$$X(s) = \frac{1}{s + a}, \qquad \mathcal{Re}\{s\} > -a \tag{9.10}$$

For $a = 0$, for example, $x(t)$ is the unit step with Laplace transform $X(s) = 1/s$ $\mathcal{Re}\{s\} > 0$.

We note, in particular, that just as the Fourier transform does not converge for all signals, the Laplace transform may converge for some values of $\mathcal{Re}\{s\}$ and not for others. In Example 9.1, the Laplace transform converges only for $\mathcal{Re}\{s\} > -a$. If a is positive, then $X(s)$ can be evaluated at $\sigma = 0$ to obtain

$$X(0 + j\omega) = \frac{1}{j\omega + a} \tag{9.11}$$

As indicated in eq. (9.5), for $\sigma = 0$ the Laplace transform is equal to the Fourier transform, as is evident for this example by comparing eqs. (9.8) and (9.11). If a is negative or zero, the Laplace transform still exists but the Fourier transform does not.

†In Chapter 4, in referring to the Fourier transform, we represented it notationally as $X(\omega)$. Because of the fact that the Fourier transform is equal to the Laplace transform for $s = j\omega$, so that for Laplace transform $X(s)$ the Fourier transform is $X(j\omega)$, it is somewhat more convenient at this point to denote the independent variable associated with the Fourier transform as $j\omega$ as an explicit reminder that it is equal to the Laplace transform for that specific set of values for s. Consequently, for the remainder of this book, we adopt this simple change of notation.

Example 9.2

For comparison with Example 9.1, let us consider as a second example the signal

$$x(t) = -e^{-at}u(-t)$$

Then

$$X(s) = -\int_{-\infty}^{+\infty} e^{-at}e^{-st}u(-t)\,dt$$

$$= -\int_{-\infty}^{0} e^{-(s+a)t}\,dt$$

or

$$X(s) = \frac{1}{s+a}$$

For this example, however, we require that for convergence $\mathcal{R}e\{s+a\} < 0$ or $\mathcal{R}e\{s\} < -a$, that is,

$$-e^{-at}u(-t) \quad \overset{\mathcal{L}}{\longleftrightarrow} \quad \frac{1}{s+a}, \qquad \mathcal{R}e\{s\} < -a \tag{9.12}$$

The algebraic expression for the Laplace transform is identical in both Example 9.1 and 9.2. However, the set of values for s for which the expression is valid is very different in the two examples. This serves to illustrate the fact that in specifying the Laplace transform of a signal, both the algebraic expression and the range of values of s for which this expression is valid are required. In general, the range of values of s for which the integral in eq. (9.2) converges is referred to as the *region of convergence* (which we abbreviate as ROC) of the Laplace transform. That is, the ROC consists of those values of $s = \sigma + j\omega$ for which the Fourier transform of $x(t)e^{-\sigma t}$ converges. We will have a great deal more to say about the ROC as we develop some insight into the properties of the Laplace transform.

A convenient way to display the ROC is shown in Figure 9.1. The variable s is a complex number and in Figure 9.1 we display the complex plane, generally referred to as the *s*-plane, associated with this complex variable. The coordinate axes are $\mathcal{R}e\{s\}$ along the horizontal axis and $\mathcal{I}m\{s\}$ along the vertical axis. The horizontal and vertical axes are sometimes referred to as the σ-axis and the $j\omega$-axis, respectively. The

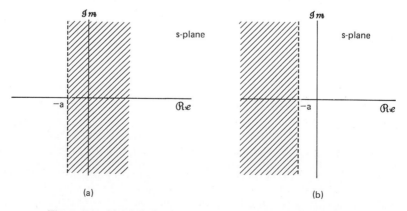

(a)   (b)

Figure 9.1 (a) ROC for Example 9.1; (b) ROC for Example 9.2.

shaded region in Figure 9.1(a) represents the set of points in the *s*-plane corresponding to the region of convergence for Example 9.1. The shaded region in Figure 9.1(b) indicates the region of convergence for Example 9.2.

Example 9.3

In this example we consider the signal which is the sum of two real exponentials, specifically

$$x(t) = e^{-t}u(t) + e^{-2t}u(t) \tag{9.13}$$

The algebraic expression for the Laplace transform is then

$$X(s) = \int_{-\infty}^{+\infty} [e^{-t}u(t) + e^{-2t}u(t)]e^{-st}\, dt$$

$$= \int_{-\infty}^{+\infty} e^{-t}e^{-st}u(t)\, dt + \int_{-\infty}^{+\infty} e^{-2t}e^{-st}u(t)\, dt$$

or

$$X(s) = \frac{1}{s+1} + \frac{1}{s+2} \tag{9.14}$$

To determine the ROC, we note that since $x(t)$ is a sum of two real exponentials and the Laplace transform operator is linear, as is evident in eq. (9.14), $X(s)$ is the sum of the Laplace transforms of each of the individual terms. The first term is the Laplace transform of $e^{-t}u(t)$ and the second term the Laplace transform of $e^{-2t}u(t)$. From Example 9.1 we know that

$$e^{-t}u(t) \quad \overset{\mathcal{L}}{\longleftrightarrow} \quad \frac{1}{s+1}, \qquad \mathcal{Re}\{s\} > -1$$

$$e^{-2t}u(t) \quad \overset{\mathcal{L}}{\longleftrightarrow} \quad \frac{1}{s+2}, \qquad \mathcal{Re}\{s\} > -2$$

Thus, the set of values of $\mathcal{Re}\{s\}$ for which the Laplace transform of both terms converge is $\mathcal{Re}\{s\} > -1$, and thus

$$e^{-t}u(t) + e^{-2t}u(t) \quad \overset{\mathcal{L}}{\longleftrightarrow} \quad \frac{1}{s+1} + \frac{1}{s+2}, \qquad \mathcal{Re}\{s\} > -1$$

or equivalently, combining the two terms on the right-hand side,

$$e^{-t}u(t) + e^{-2t}u(t) \quad \overset{\mathcal{L}}{\longleftrightarrow} \quad \frac{2s+3}{s^2+3s+2}, \qquad \mathcal{Re}\{s\} > -1 \tag{9.15}$$

In each of the three examples above, the Laplace transform is a ratio of polynomials in the complex variable *s*, that is, is of the form

$$X(s) = \frac{N(s)}{D(s)} \tag{9.16}$$

where $N(s)$ and $D(s)$ are the numerator polynomial and denominator polynomial, respectively. When $X(s)$ is of this form, it is referred to as *rational*. As suggested by Example 9.3, $X(s)$ will be rational whenever $x(t)$ is a linear combination of real or complex exponentials, and as we will see in Section 9.7.1, rational transforms also arise when we consider LTI systems specified in terms of linear constant-coefficient differential equations. Except for a scale factor, the numerator and denominator polynomials in a rational Laplace transform can be specified by their roots; thus, marking the location of the roots of $N(s)$ and $D(s)$ in the *s*-plane provides a convenient pictorial way of describing the Laplace transform. For example, we show in Figure 9.2 the s-plane representation of the Laplace transform of Example 9.3 with the location of each root of the denominator polynomial in eq. (9.15) indicated with an × and the location of the root of the numerator polynomial in eq. (9.15) indicated with an ○. The region of convergence for this example is also indicated as a shaded region. For rational Laplace transforms, the roots of the numerator polynomial are commonly referred to as the *zeros* of $X(s)$ since for those values of s, $X(s) = 0$. The roots of the denominator polynomial are referred to as the *poles* of $X(s)$, and for those values of s, $X(s)$ becomes unbounded. If the order of the denominator polynomial is greater than the order of the numerator polynomial, then $X(s)$ will become zero as s approaches infinity. Conversely, if the order of the numerator polynomial is greater than the order of the denominator polynomial, then $X(s)$ will become unbounded as s approaches infinity. This behavior can be interpreted as zeros or poles at infinity, an interpretation which will be useful in Chapter 11. However, it should be recognized that the poles and zeros of $X(s)$ in the finite s-plane are sufficient to completely characterize the algebraic expression $X(s)$ to within a scale factor since these represent all the roots of the polynomials $N(s)$ and $D(s)$ in eq. (9.16) but do not by themselves identify the ROC for the Laplace transform. The representation of $X(s)$ through its poles and zeros in the s-plane is referred to as the *pole–zero plot*. To within a scale factor, the pole–zero plot, together with the ROC, provides a complete specification of the Laplace transform.

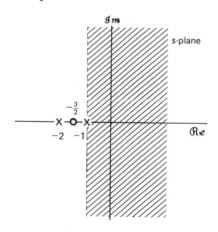

Figure 9.2 s-Plane representation of the Laplace transform for Example 9.3. × and ○ mark the locations of the roots of the denominator and the numerator, respectively. The shaded region indicates the ROC.

Example 9.4

Let

$$x(t) = \delta(t) - \tfrac{4}{3}e^{-t}u(t) + \tfrac{1}{3}e^{2t}u(t)$$

Then

$$X(s) = 1 - \frac{4}{3}\frac{1}{s+1} + \frac{1}{3}\frac{1}{s-2}, \qquad \mathcal{R}e\{s\} > 2 \tag{9.17a}$$

or

$$X(s) = \frac{(s-1)^2}{(s+1)(s-2)}, \qquad \mathcal{R}e\{s\} > 2 \tag{9.17b}$$

The pole–zero plot for this example is shown in Figure 9.3 together with the ROC.

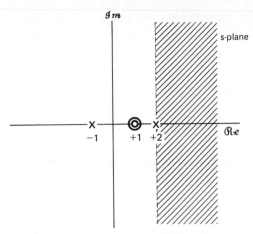

Figure 9.3 Pole–zero plot and ROC for Example 9.4.

Recall from eq. (9.5) that for $s = j\omega$ the Laplace transform corresponds to the Fourier transform. However, if the ROC of the Laplace transform does not include the $j\omega$-axis, [i.e., $\mathcal{R}e\{s\} = 0$], then the Fourier transform does not converge. As we see from Figure 9.3, this, in fact, is the case for Example 9.4. We observe also for this example that the two zeros in eq. (9.17) occur at the same value of s. In general, we will refer to the *order* of a pole or zero as the number of times it is repeated at a given location. Thus, in Example 9.4 there is a second-order zero at $s = 1$ and two first-order poles, one at $s = -1$, the other at $s = 2$. In this example the ROC lies to the right of the rightmost pole. In general, for rational Laplace transforms there is a close relationship between the locations of the poles and the possible ROCs that can be associated with a given pole–zero plot, and the specific constraints are closely associated with time-domain properties of $x(t)$. In the next section we explore some of these constraints and relationships.

9.2 THE REGION OF CONVERGENCE FOR LAPLACE TRANSFORMS

In the foregoing discussion, we have seen that a complete specification of the Laplace transform requires not only the algebraic expression for $X(s)$, but also the associated region of convergence, and as evidenced by Examples 9.1 and 9.2, two very different signals can have identical algebraic expressions for $X(s)$, so that their Laplace transforms are distinguishable *only* by the region of convergence. In this section we explore some specific constraints on the ROC for various classes of signals. As we will see, an understanding of these constraints often permits us to specify implicitly, or to reconstruct, the ROC from knowledge only of the algebraic expression for $X(s)$ and certain general characteristics of $x(t)$ in the time domain. In developing these properties we justify the statements intuitively rather than rigorously.

Property 1: The ROC of $X(s)$ consists of strips parallel to the $j\omega$-axis in the s-plane.

The validity of this property stems from the fact that the ROC of $X(s)$ consists of those values of $s = \sigma + j\omega$ for which the Fourier transform of $x(t)e^{-\sigma t}$ converges, and thus the ROC depends only on the real part of s.

> *Property 2:* For rational Laplace transforms, the ROC does not contain any poles.

This property is easily observed for all the examples studied thus far. Since $X(s)$ is infinite at a pole, the integral in eq. (9.2) clearly does not converge at a pole, and thus the ROC cannot contain those values of s.

> *Property 3:* If $x(t)$ is of finite duration and if there is at least one value of s for which the Laplace transform converges, then the ROC is the entire s-plane.

We verify Property 3 as follows. A finite-duration signal has the property that it is zero outside an interval of finite duration, as illustrated in Figure 9.4. Now, let us assume that $x(t)e^{-\sigma t}$ is absolutely integrable for some value of σ, say σ_0, so that

$$\int_{T_1}^{T_2} |x(t)| e^{-\sigma_0 t} \, dt < \infty \tag{9.18}$$

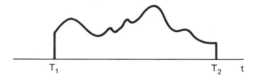

T_1 T_2 t   **Figure 9.4**   Finite-duration signal.

In this case, the line $\mathcal{R}e\{s\} = \sigma_0$ is in the ROC. For $\mathcal{R}e\{s\} = \sigma_1$ to also be in the ROC we require that

$$\int_{T_1}^{T_2} |x(t)| e^{-\sigma_1 t} \, dt = \int_{T_1}^{T_2} |x(t)| e^{-\sigma_0 t} e^{-(\sigma_1 - \sigma_0)t} \, dt < \infty$$

Let us assume that $\sigma_1 > \sigma_0$, so that $e^{-(\sigma_1 - \sigma_0)t}$ is a decaying exponential. Then, over the interval when $x(t)$ is nonzero, the maximum value of this exponential is $e^{-(\sigma_1 - \sigma_0)T_1}$, and thus we can write that

$$\int_{T_1}^{T_2} |x(t)| e^{-\sigma_1 t} \, dt < e^{-(\sigma_1 - \sigma_0)T_1} \int_{T_1}^{T_2} |x(t)| e^{-\sigma_0 t} \, dt \tag{9.19}$$

Since the right-hand side of eq. (9.19) is bounded, so is the left-hand side, and thus the s-plane for $\mathcal{R}e\{s\} > \sigma_0$ must also be in the ROC. By a similar argument, if $\sigma_1 < \sigma_0$, then

$$\int_{T_1}^{T_2} |x(t)| e^{-\sigma_1 t} \, dt < e^{-(\sigma_1 - \sigma_0)T_2} \int_{T_1}^{T_2} |x(t)| e^{-\sigma_0 t} \, dt \tag{9.20}$$

and again $x(t)e^{-\sigma_1 t}$ is absolutely integrable. Thus, the ROC includes the entire s-plane. The intuition behind this result is suggested in Figure 9.5. In Figure 9.5(a), we have shown $x(t)$ of Figure 9.4 multiplied by a decaying exponential, and in Figure 9.5(b) the same signal multiplied by a growing exponential. Since the interval is finite, the exponential weighting is never unbounded, and consequently it is reasonable that the integrability of $x(t)$ is not destroyed by this exponential weighting.

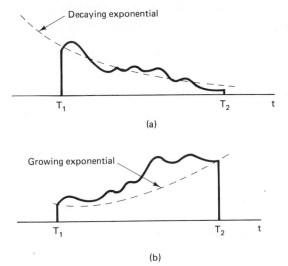

(a)

(b)

Figure 9.5 (a) Finite-duration signal of Figure 9.4 multiplied by a decaying exponential; (b) finite-duration signal of Figure 9.4 multiplied by a growing exponential.

It is important to recognize that to ensure that the exponential weighting is bounded over the interval in which $x(t)$ is nonzero, the discussion above relies heavily on the fact that $x(t)$ is finite length. In the next two properties we consider a modification of the result when $x(t)$ is of finite extent in only the positive-time or negative-time direction.

Property 4: If $x(t)$ is right-sided and if the line $\mathcal{R}e\{s\} = \sigma_0$ is in the ROC, then all values of s for which $\mathcal{R}e\{s\} > \sigma_0$ will also be in the ROC.

A right-sided signal is one for which $x(t) = 0$ prior to some finite time T_1, as illustrated in Figure 9.6. It is possible that for such a signal there is no value

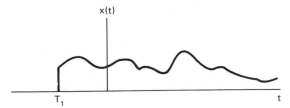

Figure 9.6 Right-sided signal.

of s for which the Laplace transform will converge. One example is the signal $x(t) = e^{t^2}u(t)$. However, suppose that the Laplace transform converges for some value of σ, which we denote by σ_0. Then

$$\int_{-\infty}^{+\infty} |x(t)| e^{-\sigma_0 t} \, dt < \infty$$

or equivalently, since $x(t)$ is right-sided,

$$\int_{T_1}^{+\infty} |x(t)| e^{-\sigma_0 t} \, dt < \infty \tag{9.21}$$

Then if $\sigma_1 > \sigma_0$, it must also be true that $x(t)e^{-\sigma_1 t}$ is absolutely integrable, since $e^{-\sigma_1 t}$ decays faster than $e^{-\sigma_0 t}$ as $t \to +\infty$, as illustrated in Figure 9.7. Formally,

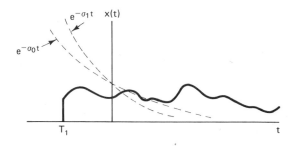

Figure 9.7 If $x(t)$ is right-sided and $x(t)e^{-\sigma_0 t}$ is absolutely integrable, then $x(t)e^{-\sigma_1 t}\sigma_1 > \sigma_0$ will be absolutely integrable also.

we can say that with $\sigma_1 > \sigma_0$,

$$\int_{T_1}^{\infty} |x(t)| e^{-\sigma_1 t} \, dt = \int_{T_1}^{\infty} |x(t)| e^{-\sigma_0 t} e^{-(\sigma_1 - \sigma_0)t} \, dt$$
$$\leq e^{-(\sigma_1 - \sigma_0)T_1} \int_{T_1}^{\infty} |x(t)| e^{-\sigma_0 t} \, dt \tag{9.22}$$

Since T_1 is finite, then from eq. (9.21) the right side of the inequality in eq. (9.22) is finite, and hence $x(t)e^{-\sigma_1 t}$ is absolutely integrable. Note that in the argument above we explicitly rely on the fact that $x(t)$ is right-sided so that although with $\sigma_1 > \sigma_0$, $e^{-\sigma_1 t}$ diverges faster than $e^{-\sigma_0 t}$ as $t \to -\infty$, $x(t)e^{-\sigma_1 t}$ cannot grow without bound in the negative-time direction since $x(t) = 0$ for $t < T_1$.

Property 5: If $x(t)$ is left-sided and if the line $\mathcal{R}e\{s\} = \sigma_0$ is in the ROC, then all values of s for which $\mathcal{R}e\{s\} < \sigma_0$ will also be in the ROC.

A left-sided signal is one for which $x(t) = 0$ after some finite time T, as illustrated in Figure 9.8, and the argument and intuition behind this property are exactly analogous to that for Property 4.

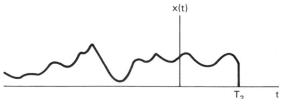

Figure 9.8 Left-sided signal.

Property 6: If $x(t)$ is two-sided and if the line $\mathcal{R}e\{s\} = \sigma_0$ is in the ROC, then the ROC will consist of a strip in the s-plane which includes the line $\mathcal{R}e\{s\} = \sigma_0$.

A two-sided signal is one that is of infinite extent for both $t > 0$ and $t < 0$, as illustrated in Figure 9.9(a). For such a signal, the ROC can be examined by choosing an arbitrary time T_0 and dividing $x(t)$ into the sum of a right-sided signal $x_R(t)$ and left-sided signal $x_L(t)$, as indicated in Figure 9.9(b) and (c). The Laplace transform

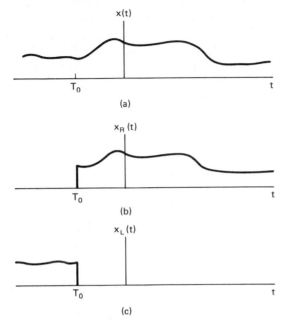

Figure 9.9 Two-sided signal divided into the sum of a right-sided and left-sided signal.

of $x(t)$ converges for values of s for which the transforms of *both* $x_R(t)$ and $x_L(t)$ converge. From Property 4, the ROC of $\mathcal{L}\{x_R(t)\}$ consists of a half-plane $\mathcal{R}e\{s\} > \sigma_R$ for some value σ_R, and from Property 5 the ROC of $\mathcal{L}\{x_L(t)\}$ consists of a half-plane $\mathcal{R}e\{s\} < \sigma_L$ for some value σ_L. The ROC of $\mathcal{L}\{x(t)\}$ is then the overlap of these two half-planes, as indicated in Figure 9.10. This assumes, of course, that $\sigma_R < \sigma_L$, so that there is some overlap. If this is not the case, then even if the Laplace transforms of $x_R(t)$ and $x_L(t)$ individually exist, the Laplace transform of $x(t)$ does not.

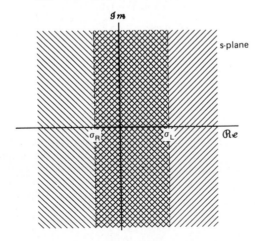

Figure 9.10 ROCs for $x_R(t)$ and for $x_L(t)$ assuming that they overlap. The overlap of the two ROCs is the ROC for $x(t) = x_R(t) + x_L(t)$.

Example 9.5

Let

$$x(t) = \begin{cases} e^{-at}, & 0 < t < T \\ 0, & \text{otherwise} \end{cases}$$

$$X(s) = \int_0^T e^{-at}e^{-st}\,dt = \frac{1}{s+a}[1 - e^{-(s+a)T}] \tag{9.23}$$

Since in this example, $x(t)$ is finite length, from Property 3, the ROC is the entire s-plane. In the form of eq. (9.23), $X(s)$ would appear to have a pole at $s = -a$, which from Property 2 would be inconsistent with an ROC that consists of the entire s-plane. In fact, in the algebraic expression in eq. (9.23), both numerator and denominator are zero at $s = -a$ and thus, to determine $X(s)$ at $s = -a$, we can use L'Hôpital's rule to obtain

$$\lim_{s \to -a} X(s) = \lim_{s \to -a}\left[\frac{\frac{d}{ds}(1 - e^{-(s+a)T})}{\frac{d}{ds}(s + a)}\right] = \lim_{s \to -a} Te^{-aT}e^{-sT}$$

so that

$$X(-a) = T \tag{9.24}$$

In particular, then, $X(s)$ in eq. (9.23) has no poles. There are an infinite number of zeros of the numerator corresponding to the values of s for which

$$1 - e^{-(s+a)T} = 0 \tag{9.25}$$

or equivalently,

$$e^{-(s+a)T} = 1 = e^{-j2\pi k}, \qquad k \text{ any integer} \tag{9.26}$$

Thus eq. (9.25) is satisfied whenever

$$(s + a)T = j2\pi k$$

or

$$s = -a + j\frac{2\pi k}{T}, \qquad k = 0, \pm1, \pm2, \dots \tag{9.27}$$

The resulting pole–zero plot for $X(s)$ is shown in Figure 9.11. Note that although

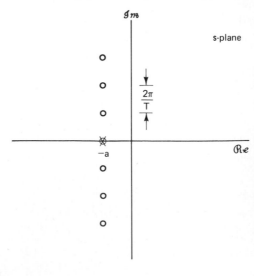

Figure 9.11 Pole–zero pattern for Example 9.5. As indicated, the pole, represented by the denominator term in eq. (9.23), is canceled by a zero of the numerator. Thus, the Laplace transform contains only zeros and the ROC is the entire s-plane.

eqs. (9.25) and (9.27) imply a zero at $s = -a$, this is, in effect, canceled by a pole represented by the denominator term in eq. (9.23), as indicated in Fig. 9.11, so that there is neither a pole nor a zero at $s = -a$. This, of course, is consistent with our conclusion in eq. (9.24).

Example 9.6

Let

$$x(t) = e^{-b|t|} \tag{9.28}$$

as illustrated in Figure 9.12 for both $b > 0$ and $b < 0$. Since this is a two-sided signal, let us divide it into the sum of a right-sided and left-sided signal, that is,

$$x(t) = e^{-bt}u(t) + e^{+bt}u(-t) \tag{9.29}$$

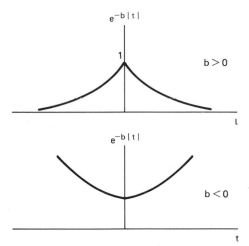

Figure 9.12 Signal $x(t) = e^{-b|t|}$ for both $b > 0$ and $b < 0$.

From Example 9.1,

$$e^{-bt}u(t) \quad \overset{\mathcal{L}}{\longleftrightarrow} \quad \frac{1}{s + b} \Re e\{s\} > -b \tag{9.30}$$

and from Example 9.2,

$$e^{+bt}u(-t) \quad \overset{\mathcal{L}}{\longleftrightarrow} \quad \frac{-1}{s - b} \Re e\{s\} < +b \tag{9.31}$$

Although the Laplace transforms of each of the individual terms in eq. (9.29) have a region of convergence, there is no *common* region of convergence if $b < 0$, and thus for those values of b, $x(t)$ has no Laplace transform. If $b > 0$, the Laplace transform of $x(t)$ is

$$e^{-b|t|} \quad \overset{\mathcal{L}}{\longleftrightarrow} \quad \frac{1}{s + b} - \frac{1}{s - b} = \frac{2b}{s^2 - b^2}, \qquad -b < \Re e\{s\} < +b \tag{9.32}$$

The corresponding pole–zero plot is shown in Figure 9.13 with the shading indicating the ROC.

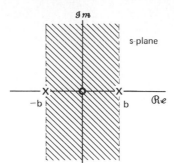

Figure 9.13 Pole–zero plot and ROC for Example 9.6.

Note that any signal either does not have a Laplace transform or it falls into one of the four categories covered by Properties 3 to 6. Thus, for any signal with a Laplace transform, the ROC *must* be either the entire *s*-plane (for finite-length signals), a left-half plane (for left-sided signals), a right-half plane (for right-sided signals), or a single strip (for two-sided signals). In all the examples that we have considered, the ROC has the additional property that in each direction (i.e., $\mathcal{R}e\{s\}$ increasing and $\mathcal{R}e\{s\}$ decreasing) it is bounded by poles or extends to infinity. In fact, this is *always* true for rational Laplace transforms. A formal argument is somewhat tedious, but its validity is essentially a consequence of the fact that a signal with a rational Laplace transform consists of linear combinations of exponentials and from Examples 9.1 and 9.2, the ROC for the transform of individual terms in this linear combination must have this property. As a consequence of this, together with Properties 1 and 4, the ROC for a right-sided signal is the region in the *s*-plane to the right of the rightmost pole. For a left-sided signal the ROC is the region in the *s*-plane to the left of the leftmost pole.

To illustrate how different ROCs can be associated with the same pole–zero pattern, let us consider the following example.

Example 9.7

Consider the algebraic expression

$$X(s) = \frac{1}{(s + 1)(s + 2)} \tag{9.33}$$

with the associated pole–zero pattern in Figure 9.14(a). As indicated in Figure 9.14(b)–(d), there are three possible ROCs that can be associated with this algebraic expression, corresponding to three distinct signals. The signal associated with the pole–zero pattern in Figure 9.14(b) is right-sided. Since the ROC includes the $j\omega$-axis, the Fourier transform of this signal converges. Figure 9.14(c) corresponds to a left-sided signal and 9.14(d) to a two-sided signal. Neither of these two signals have Fourier transforms since the ROC does not include the $j\omega$-axis.

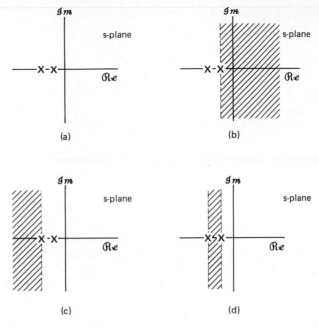

Figure 9.14   (a) Pole–zero pattern for Example 9.7; (b) ROC corresponding to a right-sided sequence; (c) ROC corresponding to a left-sided sequence; (d) ROC corresponding to a two-sided sequence.

9.3 THE INVERSE LAPLACE TRANSFORM

In Section 9.1 we discussed the interpretation of the Laplace transform of a time function as the Fourier transform of the function exponentially weighted; that is, with s expressed as $s = \sigma + j\omega$, where $\mathcal{R}e\{s\} = a$ is a line in the ROC, the Laplace transform $X(\sigma + j\omega)$ of a signal $x(t)$ is given by

$$X(\sigma + j\omega) = \mathcal{F}\{x(t)e^{-\sigma t}\} = \int_{-\infty}^{+\infty} x(t)e^{-\sigma t}e^{-j\omega t}\, dt \tag{9.34}$$

We can invert this relationship using the inverse Fourier transform as given in eq. (4.62), specifically

$$x(t)e^{-\sigma t} = \mathcal{F}^{-1}\{X(\sigma + j\omega)\} = \frac{1}{2\pi}\int_{-\infty}^{+\infty} X(\sigma + j\omega)e^{j\omega t}\, d\omega \tag{9.35}$$

or multiplying both sides by $e^{\sigma t}$, we obtain

$$x(t) = \frac{1}{2\pi}\int_{-\infty}^{+\infty} X(\sigma + j\omega)e^{(\sigma + j\omega)t}\, d\omega \tag{9.36}$$

Finally, if we change the variable of integration from ω to s, and use the fact that σ is constant so that $ds = j\,d\omega$, we obtain the basic inverse Laplace transform equation

$$x(t) = \frac{1}{2\pi j} \int_{\sigma - j\infty}^{\sigma + j\infty} X(s)e^{st}\, ds \tag{9.37}$$

This states that $x(t)$ can be represented as a weighted integral of complex exponentials. The contour of integration is a straight line in the complex plane, parallel to the $j\omega$-axis and determined by any value of σ so that $X(\sigma + j\omega)$ converges. The formal evaluation of this integral for a general $X(s)$ requires the use of contour integration in the complex plane, a topic that we will not consider here. However, for the class of rational transforms, the inverse Laplace transform can be determined without direct evaluation of eq. (9.37) by utilizing the partial fraction expansion in a manner similar to that used in Chapter 4 to determine the inverse Fourier transform. Basically, the procedure consists of expanding the rational algebraic expression into a linear combination of lower-order terms. For example, assuming no multiple-order poles and assuming that the order of the denominator polynomial is greater than the order of the numerator polynomial, $X(s)$ can be expanded in the form

$$X(s) = \sum_{i=1}^{m} \frac{A_i}{s + a_i} \tag{9.38}$$

From the ROC of $X(s)$, the ROC of each of the individual terms can be inferred. Since from Examples 9.1 and 9.2 we know the inverse Laplace transform of the individual first-order terms in eq. (9.38), the composite time function $x(t)$ representing the inverse transform of $X(s)$ is then obtained. The details of the procedure are best presented through a number of examples, which follow.

Example 9.8

Let

$$X(s) = \frac{1}{(s + 1)(s + 2)}, \qquad \mathcal{Re}\{s\} > -1 \tag{9.39}$$

To obtain the inverse Laplace transform, our first step is to perform a partial fraction expansion to obtain

$$X(s) = \frac{1}{(s + 1)(s + 2)} = \frac{A}{s + 1} + \frac{B}{s + 2} \tag{9.40}$$

As discussed in the Appendix, we can evaluate the coefficients A and B by multiplying both sides of (9.40) by $(s + 1)(s + 2)$ and then equating coefficients of equal powers of s on both sides. Alternatively, we can observe that

$$A = [(s + 1)X(s)]|_{s=-1} = 1 \tag{9.41}$$

$$B = [(s + 2)X(s)]|_{s=-2} = -1 \tag{9.42}$$

Thus, the partial fraction expansion for $X(s)$ is

$$X(s) = \frac{1}{s + 1} - \frac{1}{s + 2} \tag{9.43}$$

Next, we need to determine the ROC to associate with each of the individual first-order terms in eq. (9.43). This is done by reference to the properties of the ROC developed in Section 9.3. Since the ROC for $X(s)$ is $\mathcal{Re}\{s\} > -1$, the ROC for the individual terms in the partial fraction expansion (9.43) includes $\mathcal{R}e\{s\} > -1$. For each term, it can then be extended to the left and/or right to be bounded by a pole or infinity. This is illustrated in Figure 9.15. Figure 9.15(a) shows the pole–zero plot and ROC for $X(s)$ as specified in eq. (9.39). Figure 9.15(b) and (c) represent the individual terms in the partial fraction expansion in eq. (9.43). The ROC for the sum is indicated with diagonal shading. For the term represented by Figure 9.15(c), the common ROC can be extended to the left as shown.

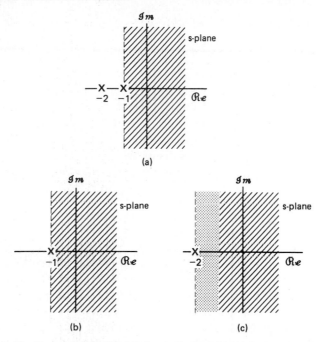

(a)

(b)   (c)

Figure 9.15 Construction of ROC for the individual terms in a partial fraction expansion: (a) pole–zero plot and ROC for $X(s)$; (b) pole at $s = -1$ and its ROC; (c) pole at $s = -2$ and its ROC.

Since both poles in eq. (9.39) are to the left of the ROC, both terms in eq. (9.43) correspond to right-sided signals. The inverse transform of the individual terms in eq. (9.43) is obtained by reference to Example 9.1, specifically,

$$e^{-t}u(t) \quad \overset{\mathcal{L}}{\longleftrightarrow} \quad \frac{1}{s+1}, \qquad \mathcal{R}e\{s\} > -1 \tag{9.44a}$$

$$e^{-2t}u(t) \quad \overset{\mathcal{L}}{\longleftrightarrow} \quad \frac{1}{s+2}, \qquad \mathcal{R}e\{s\} > -2 \tag{9.44b}$$

so that

$$[e^{-t} - e^{-2t}]u(t) \quad \overset{\mathcal{L}}{\longleftrightarrow} \quad \frac{1}{(s+1)(s+2)}, \qquad \mathcal{R}e\{s\} > -1 \tag{9.44c}$$

Continuing this example, let us now suppose that the algebraic expression for $X(s)$ is again that in eq. (9.39) but that the ROC is now $\mathcal{R}e\{s\} < -2$. The partial fraction expansion for $X(s)$ relates only to the algebraic expression and in particular eq. (9.43) is still valid. With this new ROC, however, both poles are to the *right* of the ROC, and thus both terms in (9.43) correspond to left-sided signals. From Example 9.2

we then have

$$x(t) = [-e^{-t} + e^{-2t}]u(-t) \quad \overset{\mathcal{L}}{\longleftrightarrow} \quad \frac{1}{(s+1)(s+2)}, \quad \mathcal{R}e\{s\} < -2 \quad (9.45)$$

Finally, suppose that the ROC of $X(s)$ is $-2 < \mathcal{R}e\{s\} < -1$. In this case, the pole at $s = -1$ is still to the right of the ROC (corresponding to a left-sided signal), while the pole at $s = -2$ is to the left (corresponding to a right-sided signal). Using Examples 9.1 and 9.2, we obtain

$$x(t) = -e^{-t}u(-t) - e^{-2t}u(t) \quad \overset{\mathcal{L}}{\longleftrightarrow} \quad \frac{1}{(s+1)(s+2)}, \quad -2 < \mathcal{R}e\{s\} < -1 \quad (9.46)$$

In Example 9.8, we used the partial fraction expansion technique to express the Laplace transform as a sum of terms whose inverse transform we could then recognize "by inspection." As discussed in the Appendix, when $X(s)$ has multiple-order poles, and/or the denominator is not of higher degree than the numerator, its partial fraction expansion will include other terms in addition to the first-order terms considered above. In Section 9.6, after discussing properties of the Laplace transform, we develop some other Laplace transform pairs which, in conjunction with the properties, are useful in extending the inverse transform method outlined in Example 9.8.

9.4 GEOMETRIC EVALUATION OF THE FOURIER TRANSFORM FROM THE POLE–ZERO PLOT

As we saw in Section 9.1, the Fourier transform of a signal is the Laplace transform evaluated on the $j\omega$-axis. In this section we discuss a procedure for geometrically evaluating the Fourier transform and more generally the Laplace transform at any set of values from the pole–zero pattern associated with the Laplace transform. To develop the procedure, let us first consider a Laplace transform with a single zero [i.e., $X(s) = (s - a)$], which we evaluate at a specific value of s, say $s = s_1$. The algebraic expression $(s_1 - a)$ is the sum of two complex numbers, s_1 and $-a$, each of which can be represented as a vector in the complex plane, as illustrated in Figure 9.16. The vector representing the complex number $(s_1 - a)$ is then the vector sum which we see in Figure 9.16 to be a vector from the zero at $s = a$ to the point s_1. The value of $X(s_1)$ then has a magnitude which is the length of this vector and an angle which is the angle of the vector relative to the real axis. If $X(s)$ has instead a single

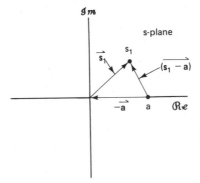

Figure 9.16 Complex plane representation of the vectors s_1, **a**, and $(s_1 - \mathbf{a})$ representing the complex numbers s_1, a, and $(s_1 - a)$ respectively.

The Laplace Transform   Chap. 9

pole at $s = a$ [i.e., $X(s) = 1/(s - a)$], then the denominator would be represented by the same vector as above and the value of $X(s_1)$ would have a magnitude that is the *reciprocal* of the length of the vector from the pole to $s = s_1$ and an angle that is the *negative* of the angle of the vector with the real axis. A more general rational Laplace transform consists of a product of pole and zero terms of the form discussed above; that is, it can be factored into the form

$$X(s) = M \frac{\prod_{i=1}^{R} (s - \beta_i)}{\prod_{j=1}^{P} (s - \alpha_j)} \tag{9.47}$$

To evaluate $X(s)$ at $s = s_1$, each term in the product is represented by a vector from the zero or pole to the point s_1. The magnitude of $X(s_1)$ is then the magnitude of the scale factor M, times the product of the lengths of the zero vectors (i.e., the vectors from the zeros to s_1) divided by the product of the lengths of the pole vectors (i.e., the vectors from the poles to s_1). The angle of the complex number $X(s_1)$ is the sum of the angles of the zero vectors minus the sum of the angles of the pole vectors. If the scale factor M in eq. (9.47) is negative, an additional angle of π would be included. Clearly, if $X(s)$ has a multiple pole and/or zero, corresponding to some of the α_j's being equal and/or some of the β_i's being equal, the lengths and angles of the vectors from these poles or zeros must be included a number of times equal to the order of the pole or zero.

Example 9.9

Consider $X(s)$ given by

$$X(s) = \frac{1}{s + \frac{1}{2}} \qquad \mathcal{Re}\{s\} > -\frac{1}{2} \tag{9.48}$$

The Fourier transform is $X(s)|_{s=j\omega}$. For this example, then, the Fourier transform is given by

$$X(j\omega) = \frac{1}{j\omega + \frac{1}{2}} \tag{9.49}$$

The pole–zero plot for $X(s)$ is shown in Figure 9.17. To determine the Fourier transform

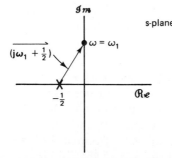

Figure 9.17 Pole–zero plot for Example 9.9. $X|(j\omega_1|$ is the reciprocal of the length of the vector shown and $\sphericalangle X(j\omega)$ is the negative of the angle of the vector.

graphically, we construct the pole vector as indicated. The magnitude of the Fourier transform at $\omega = \omega_1$ is the reciprocal of the length of the vector from the pole to the point $j\omega_1$ on the $j\omega$-axis. The phase of the Fourier transform is the negative of the angle of the vector. Geometrically, from Fig. 9.17 we can write that

$$|X(j\omega_1)|^2 = \frac{1}{\omega_1^2 + (\frac{1}{2})^2} \qquad (9.50a)$$

and

$$\sphericalangle X(j\omega_1) = -\tan^{-1} 2\omega_1 \qquad (9.50b)$$

Often, part of the value of the geometric determination of the Fourier transform lies in its usefulness in obtaining an approximate view of its overall characteristics. For example, in Fig. 9.17 it is readily evident that the length of the pole vector monotonically increases with increasing ω_1, and thus the magnitude of the Fourier transform will monotonically *decrease* with increasing ω_1. The ability to draw general conclusions about the behavior of the Fourier transform, particularly its magnitude, from the pole–zero plot is further illustrated by the consideration of general first- and second-order systems.

9.4.1 First-Order Systems

As a generalization of Example 9.9, let us consider the class of first-order systems that was discussed in some detail in Section 4.12. The impulse response for this class of systems is

$$h(t) = \frac{1}{\tau} e^{-t/\tau} u(t) \qquad (9.51)$$

and its Laplace transform is

$$H(s) = \frac{1}{s\tau + 1}, \qquad \mathcal{R}e\{s\} > -\frac{1}{\tau} \qquad (9.52)$$

The pole–zero plot is shown in Figure 9.18 and the Bode plot for the frequency

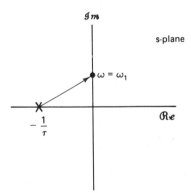

Figure 9.18 Pole–zero plot for the first-order system of eq. (9.51).

response is shown in Figure 9.19. From the behavior of the pole vector as ω_1 varies, it is clear that the magnitude of the frequency response monotonically decreases as ω_1 increases, which is, of course, consistent with the behavior in Figure 9.19. We also note that as the pole moves farther into the left-half plane, the effective cutoff frequency of the system increases. Also, from eq. (9.51) and from Figure 4.36, we see that this same movement of the pole to the left corresponds to a faster decay of the impulse response and correspondingly a faster rise time in the step response. This relationship between the real part of the pole locations and the speed of the system response holds more generally; that is, poles farther away from the $j\omega$-axis are associated with faster response terms in the impulse response.

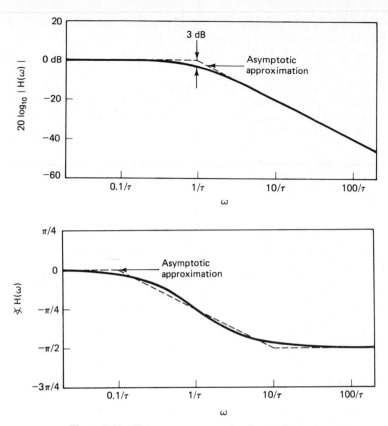

Figure 9.19 Frequency response for first-order system.

9.4.2 Second-Order Systems

Let us next consider the class of second-order systems, which was discussed in some detail in Section 4.12. The impulse response $h(t)$ and frequency response $H(j\omega)$ for the system are given in eqs. (4.175) and (4.171), respectively, which we repeat below for convenience:

[eq. (4.175)]
$$h(t) = M[e^{c_1 t} - e^{c_2 t}]u(t)$$

where

$$c_1 = -\zeta\omega_n + \omega_n\sqrt{\zeta^2 - 1}$$
$$c_2 = -\zeta\omega_n - \omega_n\sqrt{\zeta^2 - 1}$$
$$M = \frac{\omega_n}{2\sqrt{\zeta^2 - 1}}$$

[eq. (4.171)]
$$H(j\omega) = \frac{\omega_n^2}{(j\omega)^2 + 2\zeta\omega_n(j\omega) + \omega_n^2}$$

The Laplace transform of the impulse response is

$$H(s) = \frac{\omega_n^2}{s^2 + 2\zeta\omega_n s + \omega_n^2} = \frac{\omega_n^2}{(s - c_1)(s - c_2)} \tag{9.53}$$

For $0 < \zeta < 1$, c_1 and c_2 are complex, so that the pole–zero plot is that shown in Figure 9.20(a). Correspondingly, the impulse response and step response have oscillatory parts. We note that the two poles occur in complex conjugate locations. In fact, as considered in Problem 9.12, the complex poles (and zeros) for a real-valued signal always occur in complex conjugate pairs. For $\zeta > 1$, both poles lie on the real axis, as indicated in Figure 9.20(b). For this case and from our previous discussion we note that as ζ increases, one pole moves closer to the $j\omega$-axis, indicative of a term in the time domain with a slow speed of response and the other pole moves farther into the left-half plane, indicative of a term in the time domain with a more rapid speed of response. Thus, for large values of ζ it is the pole close to the $j\omega$-axis that dominates the time response for large time. Similarly, from a consideration of the pole vectors for $\zeta \gg 1$, as indicated in Figure 9.20(c), for low frequencies the length and angle of the pole vector for the pole close to the $j\omega$-axis is much more sensitive to changes in ω than those for the pole far from the $j\omega$-axis. Thus, we see that for low frequencies, the frequency-response characteristics are influenced principally by the pole close to the $j\omega$-axis.

Next, let us focus on the pole–zero plot in Figure 9.20(a), corresponding to $0 < \zeta < 1$. From the figure, particularly when ζ is small so that the poles are close to the $j\omega$-axis, as ω approaches $\omega_m\sqrt{1 - \zeta^2}$, the behavior of the frequency response is dominated by the pole vector in the second quadrant, and in particular, the length of that pole vector has a minimum at $\omega = \omega_n\sqrt{1 - \zeta^2}$. Thus qualitatively, we would expect the magnitude of the frequency response to exhibit a peak in the vicinity of that frequency. A careful sketch of the magnitude of the frequency response is shown in Figure 9.21, where the expected behavior in the vicinity of the poles is clearly evident. This, of course, is consistent with the Bode plot in Figure 4.40.

From Figure 9.20 we also can readily determine how the poles change as we vary ζ, keeping ω_n constant. Specifically, since $\cos\theta = \zeta$, the poles move along a semicircle with fixed radius ω_n. For $\zeta = 0$, the two poles are on the imaginary axis. Correspondingly, in the time domain, the impulse response is sinusoidal with no damping. As ζ increases from 0 to 1, the two poles remain complex and move into the left-half plane and the vectors from the origin to the poles maintain constant overall magnitude ω_n. As the real part of the poles becomes more negative, the associated time response will decay more quickly as $t \to \infty$.

9.4.3 All-Pass Systems

As a final illustration of the geometric evaluation of the frequency response let us consider a system for which the Laplace transform of the impulse response has the pole–zero plot shown in Figure 9.22(a). From this figure it is evident that for any value along the $j\omega$-axis, the pole and zero vectors have equal length and consequently the magnitude of the frequency response is constant, independent of frequency. Such a system is commonly referred to as an all-pass system, since it passes all frequencies with equal gain (or attenuation). The phase of the frequency response is $\theta_1 - \theta_2$ or, since $\theta_1 = \pi - \theta_2$,

$$\sphericalangle H(j\omega) = \pi - 2\theta_2(\omega) \tag{9.54}$$

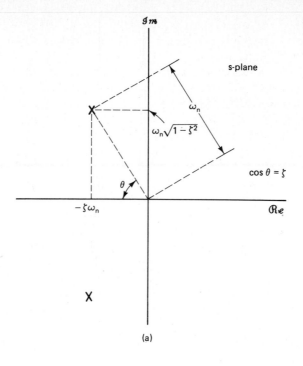

(a)

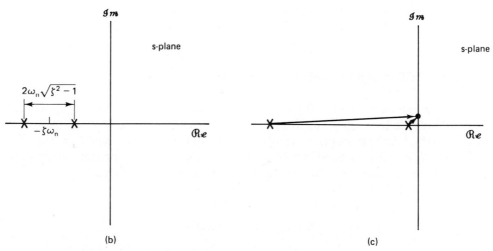

(b)   (c)

Figure 9.20 (a) Pole–zero plot for a second-order system with $0 < \zeta < 1$; (b) pole–zero plot for a second-order system with $\zeta > 1$; (c) pole vectors for $\zeta \gg 1$.

From Figure 9.22(a), $\theta_2 = \tan^{-1}(\omega/a)$ and thus

$$\measuredangle H(j\omega) = \pi - 2\tan^{-1}\left(\frac{\omega}{a}\right) \tag{9.55}$$

The magnitude and phase of $H(j\omega)$ are illustrated in Figure 9.22(b).

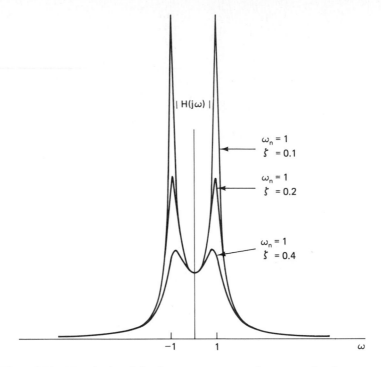

Figure 9.21 Magnitude of the frequency response for a second-order system with $0 < \zeta < 1$.

9.5 PROPERTIES OF THE LAPLACE TRANSFORM

In exploiting the Fourier transform, we relied heavily on the set of properties developed in Sections 4.6 to 4.8. In this section we consider the corresponding set of properties for the Laplace transform. The derivations for many of these results are analogous to those of the corresponding properties for the Fourier transform. Thus, we will not present the derivations in detail, some of which are left as exercises at the end of the chapter (see Problems 9.9 to 9.11).

9.5.1 Linearity of the Laplace Transform

If

$$x_1(t) \overset{\mathcal{L}}{\longleftrightarrow} X_1(s) \qquad \text{with a region of convergence that will be denoted as } R_1$$

and

$$x_2(t) \overset{\mathcal{L}}{\longleftrightarrow} X_2(s) \qquad \text{with a region of convergence that will be denoted as } R_2$$

then

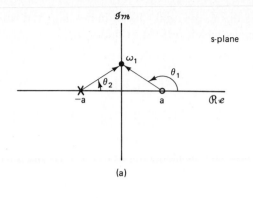

(a)

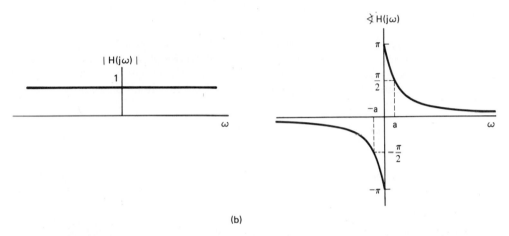

(b)

Figure 9.22 (a) Pole–zero plot for all-pass system; (b) magnitude and phase of all-pass frequency response.

$$ax_1(t) + bx_2(t) \overset{\mathcal{L}}{\longleftrightarrow} aX_1(s) + bX_2(s) \qquad \text{with ROC containing} \qquad R_1 \cap R_2 \tag{9.56}$$

As indicated, the region of convergence of $X(s)$ is at least the intersection of R_1 and R_2, which could be empty, in which case $X(s)$ has no region of convergence— i.e., $x(t)$ has no Laplace transform. For example, for $x(t)$ as in eq. (9.29) (Example 9.6), with $b > 0$ the ROC for $X(s)$ is the intersection of the ROC for the two terms in the sum. If $b < 0$, there are no common points in R_1 and R_2; that is, the intersection is empty and thus $x(t)$ has no Laplace transform. The ROC can also be larger than the intersection. As a simple example, for $x_1(t) = x_2(t)$ and $a = -b$ in eq. (9.56), $x(t)$ and thus $X(s) = 0$. The ROC of $X(s)$ is then the entire s-plane.

The ROC associated with a linear combination of terms can always be constructed by using the properties of the ROC developed in Section 9.2. Specifically, from the intersection of the ROCs for the individual terms (assuming that it is not empty), we can find a line or strip that is in the ROC of the linear combination.

We then extend this to the right ($\mathcal{R}e\{s\}$ increasing) and to the left ($\mathcal{R}e\{s\}$ decreasing) to the nearest pole (which may be at infinity).

Example 9.10

$$X_1(s) = \frac{1}{s+1}, \qquad \mathcal{R}e\{s\} > -1 \qquad (9.57\text{a})$$

$$X_2(s) = \frac{1}{(s+1)(s+2)}, \qquad \mathcal{R}e\{s\} > -1 \qquad (9.57\text{b})$$

$$x(t) = x_1(t) - x_2(t) \qquad (9.57\text{c})$$

The pole–zero plot, including the ROC for $X_1(s)$ and $X_2(s)$, is shown in Figure 9.23(a) and (b). From eq. (9.57c),

$$X(s) = \frac{1}{s+1} - \frac{1}{(s+1)(s+2)} = \frac{s+1}{(s+1)(s+2)} = \frac{1}{s+2} \qquad (9.58)$$

Thus, in the linear combination of $x_1(t)$ and $x_2(t)$, the pole at $s = -1$ is canceled by a zero at $s = -1$. The pole–zero plot for $X(s)$ is shown in Figure 9.23(c). The intersection of the ROCs for $X_1(s)$ and $X_2(s)$ is $\mathcal{R}e\{s\} > -1$. However, since the ROC is always bounded by a pole or infinity, for this example the ROC can be extended to the left to be bounded by the pole at $s = -2$, as a result of the pole–zero cancellation at $s = -1$.

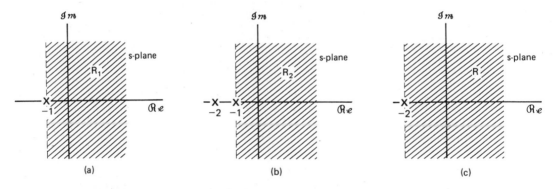

Figure 9.23 Pole–zero plots and ROCs for Example 9.10: (a) $X_1(s)$; (b) $X_2(s)$; (c) $X_1(s) - X_2(s)$. The ROC for $X_1(s) - X_2(s)$ includes the intersection of R_1 and R_2 which can then be extended to be bounded by the pole at $s = -2$.

9.5.2 Time Shifting

If

$$x(t) \quad \overset{\mathcal{L}}{\longleftrightarrow} \quad X(s) \qquad \text{with ROC} = R$$

then

$$\boxed{x(t - t_0) \quad \overset{\mathcal{L}}{\longleftrightarrow} \quad e^{-st_0}X(s) \qquad \text{with ROC} = R} \qquad (9.59)$$

9.5.3 Shifting in the s-Domain

If

$$x(t) \quad \overset{\mathcal{L}}{\longleftrightarrow} \quad \text{ROC} = R$$

then

$$e^{s_0 t}x(t) \quad \overset{\mathcal{L}}{\longleftrightarrow} \quad X(s - s_0) \qquad \text{with ROC } R_1 = R + \Re e\{s_0\} \qquad (9.60)$$

That is, the ROC associated with $X(s - s_0)$ is that of $X(s)$, shifted by $\Re e\{s_0\}$. Thus, for any value s that is in R, the value $s + \Re e\{s_0\}$ will be in R_1. This is illustrated in Figure 9.24.

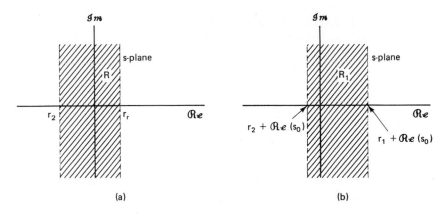

Figure 9.24 Effect on the ROC of shifting in the s-domain: (a) the ROC of $X(s)$; (b) the ROC of $X(s - s_0)$.

9.5.4 Time Scaling

If

$$x(t) \quad \overset{\mathcal{L}}{\longleftrightarrow} \quad X(s) \qquad \text{ROC} = R$$

then

$$x(at) \quad \overset{\mathcal{L}}{\longleftrightarrow} \quad \frac{1}{|a|}X\left(\frac{s}{a}\right) \qquad \text{with ROC } R_1 = \frac{R}{a} \qquad (9.61)$$

That is, for any value s in R, the value s/a will be in R_1, as illustrated in Figure 9.25.

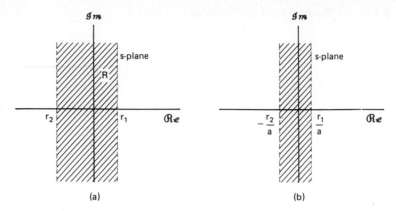

Figure 9.25 Effect on the ROC of time scaling: (a) ROC of $X(s)$; (b) ROC of $(1/|a|)X(s/a)$ for a positive.

9.5.5 Convolution Property

If

$$x_1(t) \overset{\mathcal{L}}{\longleftrightarrow} X_1(s), \qquad \text{ROC} = R_1$$

$$x_2(t) \overset{\mathcal{L}}{\longleftrightarrow} X_2(s), \qquad \text{ROC} = R_2$$

then

$$\boxed{x_1(t) * x_2(t) \overset{\mathcal{L}}{\longleftrightarrow} X_1(s)X_2(s) \qquad \text{with ROC containing } R_1 \cap R_2} \qquad (9.62)$$

Thus, in a manner similar to the linearity property 9.5.1, the ROC of $X_1(s)X_2(s)$ includes the intersection of the ROCs of $X_1(s)$ and $X_2(s)$ and may be larger if pole–zero cancellation occurs in the product. For example, if

$$X_1(s) = \frac{s+1}{s+2}, \qquad \mathcal{Re}\{s\} > -2$$

and

$$X_2(s) = \frac{s+2}{s+1}, \qquad \mathcal{Re}\{s\} > -1$$

then $X_1(s)X_2(s) = 1$ and its ROC is the entire s-plane.

As we saw in Chapter 4, the convolution property in the context of the Fourier transform played an important role in the analysis of linear time-invariant systems. In Section 9.7 we will exploit in some detail the convolution property for Laplace transforms for the analysis of LTI systems in general, and more specifically, for the class of systems represented by linear constant-coefficient differential equations.

9.5.6 Differentiation in the Time Domain

If

$$x(t) \quad \overset{\mathcal{L}}{\longleftrightarrow} \quad X(s), \qquad \text{ROC} = R$$

then

$$\boxed{\frac{dx(t)}{dt} \quad \overset{\mathcal{L}}{\longleftrightarrow} \quad sX(s) \qquad \text{with ROC containing } R} \tag{9.63}$$

This property follows by differentiating both sides of the inverse Laplace transform expression equation (9.38). Specifically, with

$$x(t) = \frac{1}{2\pi j} \int_{\sigma - j\infty}^{\sigma + j\infty} X(s)e^{st} \, ds$$

then

$$\frac{dx(t)}{dt} = \frac{1}{2\pi j} \int_{\sigma - j\infty}^{\sigma + j\infty} sX(s)e^{st} \, ds \tag{9.64}$$

Consequently, $dx(t)/dt$ is the inverse Laplace transform of $sX(s)$. The ROC of $sX(s)$ includes the ROC of $X(s)$ and may be larger if $X(s)$ has a first-order pole at $s = 0$ which is canceled by the multiplication by s. For example, if $x(t) = u(t)$, then $X(s) = 1/s$ with an ROC that is $\mathcal{R}e\{s\} > 0$. The derivative of $x(t)$ is an impulse with an associated Laplace transform which is unity and an ROC which is the entire s-plane.

9.5.7 Differentiation in the s-Domain

Applying differentiation to both sides of the Laplace transform equation (9.2), we have

$$X(s) = \int_{-\infty}^{+\infty} x(t)e^{-st} \, dt$$

$$\frac{dX(s)}{ds} = \int_{-\infty}^{+\infty} (-t)x(t)e^{-st} \, dt$$

Consequently,

$$\boxed{-tx(t) \quad \overset{\mathcal{L}}{\longleftrightarrow} \quad \frac{dX(s)}{ds}, \qquad \text{ROC} = R} \tag{9.65}$$

Let us consider an example of the use of this property.

Example 9.11

Consider obtaining the Laplace transform of

$$x(t) = te^{-at}u(t) \tag{9.66}$$

Since

$$e^{-at}u(t) \overset{\mathcal{L}}{\longleftrightarrow} \frac{1}{s+a}, \qquad \mathcal{Re}\{s\} > -a$$

it follows from eq. (9.65) that

$$te^{-at}u(t) \overset{\mathcal{L}}{\longleftrightarrow} -\frac{d}{ds}\left[\frac{1}{s+a}\right] = \frac{1}{(s+a)^2}, \qquad \mathcal{Re}\{s\} > -a \qquad (9.67)$$

In fact, by repeated application of eq. (9.65), it follows that

$$\frac{t^2}{2}e^{-at}u(t) \overset{\mathcal{L}}{\longleftrightarrow} \frac{1}{(s+a)^3}, \qquad \mathcal{Re}\{s\} > -a \qquad (9.68)$$

and, more generally,

$$\frac{t(n-1)}{(n-1)!}e^{-at}u(t) \overset{\mathcal{L}}{\longleftrightarrow} \frac{1}{(s+a)^n}, \qquad \mathcal{Re}\{s\} > -a \qquad (9.69)$$

This specific Laplace transform pair is particularly useful when applying the partial fraction expansion to the determination of the inverse Laplace transform of a rational function with multiple-order poles.

9.5.8 Integration in the Time Domain

If

$$x(t) \overset{\mathcal{L}}{\longleftrightarrow} X(s), \qquad \text{ROC} = R$$

then

$$\boxed{\int_{-\infty}^{t} x(\tau)\, d\tau \overset{\mathcal{L}}{\longleftrightarrow} \frac{1}{s}X(s), \qquad \text{ROC contains } R \cap \{\mathcal{Re}\{s\} > 0\}} \qquad (9.70)$$

This property follows by integrating both sides of the inverse Laplace transform expression and is the inverse of the differentiation property 9.5.6. It can also be interpreted through property 9.5.5. Specifically,

$$\int_{-\infty}^{t} x(\tau)\, d\tau = u(t) * x(t) \qquad (9.71)$$

From Example 9.1, with $a = 0$,

$$u(t) \overset{\mathcal{L}}{\longleftrightarrow} \frac{1}{s}, \qquad \mathcal{Re}\{s\} > 0 \qquad (9.72)$$

and thus, from property 9.5.5,

$$u(t) * x(t) \overset{\mathcal{L}}{\longleftrightarrow} \frac{1}{s}X(s) \qquad (9.73)$$

with an ROC that contains the intersection of the ROC of $X(s)$ and the ROC of the Laplace transform of $u(t)$, which corresponds to the ROC associated with Property 9.5.8.

9.5.9 The Initial and Final Value Theorems

Under the specific constraints that $x(t) = 0$, $t < 0$, and that $x(t)$ contains no impulses or higher-order singularities at the origin, one can directly calculate from the Laplace transform the initial value $x(0+)$, i.e., $x(t)$ as t approaches zero from positive values of t, and the final value, i.e., the limit as $t \to \infty$ of $x(t)$. Specifically, the *initial value theorem* states that

$$x(0+) = \lim_{s \to \infty} sX(s) \qquad (9.74)$$

while the *final value theorem* is

$$\lim_{t \to \infty} x(t) = \lim_{s \to 0} sX(s) \qquad (9.75)$$

The derivation of these results is considered in Problem 9.11.

9.5.10 Table of Properties

In Table 9.1 we summarize the properties developed in this section. In Section 9.7 many of these properties are used in applying the Laplace transform to the analysis and characterization of linear time-invariant systems.

TABLE 9.1 PROPERTIES OF THE LAPLACE TRANSFORM

Property	Signal	Transform	ROC
	$x(t)$	$X(s)$	R
	$x_1(t)$	$X_1(s)$	R_1
	$x_2(t)$	$X_2(s)$	R_2
9.5.1	$ax_1(t) + bx_2(t)$	$aX_1(s) + bX_2(s)$	At least $R_1 \cap R_2$
9.5.2	$x(t - t_0)$	$e^{-st_0}X(s)$	R
9.5.3	$e^{s_0 t}x(t)$	$X(s - s_0)$	Shifted version of R [i.e., s is in the ROC if $(s - s_0)$ is in R]
9.5.4	$x(at)$	$\dfrac{1}{\|a\|}X\left(\dfrac{s}{a}\right)$	"Scaled" ROC [i.e., s is in the ROC if (s/a) is in the ROC of $X(s)$]
9.5.5	$x_1(t) * x_2(t)$	$X_1(s)X_2(s)$	At least $R_1 \cap R_2$
9.5.6	$\dfrac{d}{dt}x(t)$	$sX(s)$	At least R
9.5.7	$-tx(t)$	$\dfrac{d}{ds}X(s)$	R
9.5.8	$\displaystyle\int_{-\infty}^{t} x(\tau)\, d\tau$	$\dfrac{1}{s}X(s)$	At least $R \cap \{\mathcal{R}e\{s\} > 0\}$

9.6 SOME LAPLACE TRANSFORM PAIRS

As we indicated in Section 9.3, the inverse Laplace transform can often be easily evaluated by decomposing $X(s)$ into a linear combination of simpler terms, the inverse transform of each of which can be recognized. Listed in Table 9.2 are a

TABLE 9.2 LAPLACE TRANSFORMS OF ELEMENTARY FUNCTIONS

Transform pair	Signal	Transform	ROC
1	$\delta(t)$	1	All s
2	$u(t)$	$\dfrac{1}{s}$	$\mathcal{R}e\{s\} > 0$
3	$-u(-t)$	$\dfrac{1}{s}$	$\mathcal{R}e\{s\} < 0$
4	$\dfrac{t^{n-1}}{(n-1)!}u(t)$	$\dfrac{1}{s^n}$	$\mathcal{R}e\{s\} > 0$
5	$-\dfrac{t^{n-1}}{(n-1)!}u(-t)$	$\dfrac{1}{s^n}$	$\mathcal{R}e\{s\} < 0$
6	$e^{-\alpha t}u(t)$	$\dfrac{1}{s+\alpha}$	$\mathcal{R}e\{s\} > \alpha$
7	$-e^{-\alpha t}u(-t)$	$\dfrac{1}{s+\alpha}$	$\mathcal{R}e\{s\} < -\alpha$
8	$\dfrac{t^{n-1}}{(n-1)!}e^{-\alpha t}u(t)$	$\dfrac{1}{(s+\alpha)^n}$	$\mathcal{R}e\{s\} > -\alpha$
9	$-\dfrac{t^{n-1}}{(n-1)!}e^{-\alpha t}u(-t)$	$\dfrac{1}{(s+\alpha)^n}$	$\mathcal{R}e\{s\} < -\alpha$
10	$\delta(t-T)$	e^{-sT}	All s
11	$[\cos \omega_0 t]u(t)$	$\dfrac{s}{s^2+\omega_0^2}$	$\mathcal{R}e\{s\} > 0$
12	$[\sin \omega_0 t]u(t)$	$\dfrac{\omega_0}{s^2+\omega_0^2}$	$\mathcal{R}e\{s\} > 0$
13	$[e^{-\alpha t}\cos \omega_0 t]u(t)$	$\dfrac{s+\alpha}{(s+\alpha)^2+\omega_0^2}$	$\mathcal{R}e\{s\} > -\alpha$
14	$[e^{-\alpha t}\sin \omega_0 t]u(t)$	$\dfrac{\omega_0}{(s+\alpha)^2+\omega_0^2}$	$\mathcal{R}e\{s\} > -\alpha$

number of useful Laplace transform pairs. Transform pair 1 follows easily from eq. (9.2). Transform pairs 2 and 6 follow directly from Example 9.1 with $a = 0$ and $a = \alpha$, respectively. Transform pair 4 was developed in Example 9.11 using the differentiation property. Transform pair 8 follows from 4 using Property 9.5.3. Transform pairs 3, 5, 7, and 9 are based on 2, 4, 6, and 8, respectively, together with Property 9.5.4. Similarly, transform pairs 10 through 14 can all be obtained from prior ones in the table using appropriate properties in Table 9.1 (see Problem 9.13).

9.7 ANALYSIS AND CHARACTERIZATION OF LTI SYSTEMS USING THE LAPLACE TRANSFORM

In previous sections we developed the Laplace transform and discussed some of its characteristics and properties. One of the important applications of the Laplace transform is in the analysis and characterization of linear time-invariant systems. Its role for this class of systems stems directly from the convolution property 9.5.5, from which it follows that the Laplace transforms of the input and output of an LTI system are related through multiplication by the Laplace transform of the system impulse response. Thus,

$$Y(s) = H(s)X(s) \qquad (9.76)$$

where $X(s)$, $Y(s)$, and $H(s)$ are the Laplace transforms of the system input, output, and impulse response, respectively. Equation (9.76) is the counterpart, in the context of Laplace transforms, of eq. (4.117) as developed for the Fourier transform. In fact, for $s = j\omega$, each of the Laplace transforms in eq. (9.76) reduces to the respective Fourier transforms and eq. (9.76) corresponds exactly to eq. (4.117).

For $s = j\omega$, $H(s)$ is the frequency response of the LTI system. In the broader context of the Laplace transform, $H(s)$ is commonly referred to as the *system function* or, alternatively, the *transfer function*. Many properties of LTI systems can be closely associated with the characteristics of the system function in the s-plane, and in particular with the pole locations and the region of convergence. For example, for a causal LTI system, the impulse response is zero for $t < 0$ and thus, in particular, is right-sided. Consequently, from the discussion in Section 9.2, the ROC associated with the system function for a *causal* system with a rational system function will be the entire region in the s-plane to the right of the rightmost pole (i.e., the pole with the most positive real part). Similarly, if the system is anticausal [i.e., $h(t) = 0$, $t > 0$], then the ROC for $H(s)$ will be the region in the s-plane to the left of the leftmost pole (i.e., the pole with the most negative real part). It should be stressed, however, that the reverse statements are not necessarily true. An ROC to the right of the rightmost pole does not guarantee that the system is causal, only that the impulse response is right-sided. Similarly, an ROC to the left of the leftmost pole only guarantees that the impulse response is left-sided, not that it is anticausal.

The ROC of $H(s)$ can also be related to the stability of the system. As we discussed in Chapter 4, the Fourier transform of the impulse response for a stable LTI system exists (converges). Thus, for a stable system, the ROC of $H(s)$ must include the $j\omega$-axis [i.e., $\mathcal{R}e\{s\} = 0$].

The relationship of the ROC to both causality and stability also leads to the conclusion that for an LTI system with a rational system function and which is *both* causal and stable, all the poles must lie in the left half of the s-plane (i.e., they must all have negative real parts). This is a straightforward consequence of the observation that due to causality the ROC is to the right of the rightmost pole and due to stability, the ROC must include the $j\omega$-axis. The relationship of the ROC to stability and causality is illustrated in the following examples.

Example 9.12

Consider a system with impulse response

$$h(t) = e^{-t}u(t) \qquad (9.77)$$

The system function is the Laplace transform of $h(t)$, which from Example 9.1 is

$$H(s) = \frac{1}{s + 1}, \qquad \mathcal{R}e\{s\} > -1 \qquad (9.78)$$

The ROC in eq. (9.78) is to the right of the rightmost pole and includes the $j\omega$-axis. Consequently, from our discussion above, the system is both stable and causal. Stability and causality of the system is also evident from the impulse response as given in eq. (9.77).

Example 9.13

Consider the system function

$$H(s) = \frac{e^s}{s+1}, \qquad \mathcal{R}e\{s\} > -1 \tag{9.79}$$

For this example, the ROC is again to the right of the rightmost pole. Therefore, the impulse response must be right-sided. To determine the impulse response, we first use the result of Example 9.1, in particular

$$e^{-t}u(t) \quad \overset{\mathcal{L}}{\longleftrightarrow} \quad \frac{1}{s+1}, \qquad \mathcal{R}e\{s\} > -1 \tag{9.80}$$

Next, from the time-shifting property 9.5.2, eq. (9.59), we see that the factor e^s in eq. (9.79) can be accounted for by a time shift in the time function in eq. (9.80). In particular, then

$$e^{-(t+1)}u(t+1) \quad \overset{\mathcal{L}}{\longleftrightarrow} \quad \frac{e^s}{s+1}, \qquad \mathcal{R}e\{s\} > -1 \tag{9.81}$$

so that the impulse response associated with the system is

$$h(t) = e^{-(t+1)}u(t+1) \tag{9.82}$$

which is zero for $t < -1$ but not for $t < 0$, and hence the system is not causal. This example serves as a reminder that in our previous discussion, we noted that causality implied that the ROC is to the right of the rightmost pole but that the reverse statement cannot in general be made.

Example 9.14

Let us consider the class of second-order systems that we have previously discussed in Sections 4.12 and 9.4.2. The impulse response and system function are, respectively,

$$h(t) = M[e^{c_1 t} - e^{c_2 t}]u(t) \tag{9.83}$$

and

$$H(s) = \frac{\omega_n^2}{s^2 + 2\zeta\omega_n + \omega_n^2} = \frac{\omega_n^2}{(s - c_1)(s - c_2)} \tag{9.84}$$

where

$$c_1 = -\zeta\omega_n + \omega_n\sqrt{\zeta^2 - 1} \tag{9.85a}$$

$$c_2 = -\zeta\omega_n - \omega_n\sqrt{\zeta^2 - 1} \tag{9.85b}$$

$$M = \frac{\omega_n}{2\sqrt{\zeta^2 - 1}} \tag{9.85c}$$

In Figure 9.20 we illustrated the pole locations for $\zeta > 0$. In Figure 9.26 we illustrate the pole locations for $\zeta < 0$. As is evident from Figure 9.26 and from eqs. (9.85), for $\zeta < 0$ both poles have positive real parts. Since in eq. (9.83) we have specified a causal impulse response, the ROC must lie to the right of the rightmost pole in Figure 9.26 and thus, in particular, cannot include the $j\omega$-axis. Consequently, for $\zeta < 0$ the causal second-order system cannot be stable. This is also evident in eq. (9.83), since with $\mathcal{R}e\{c_1\} > 0$ and $\mathcal{R}e\{c_2\} > 0$ each term will grow exponentially as t increases, and thus $h(t)$ cannot be absolutely integrable.

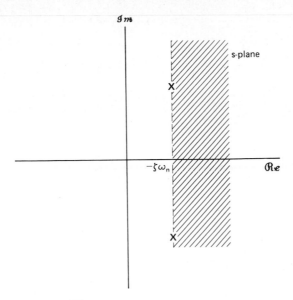

Figure 9.26 Pole locations and ROC for a causal second-order system with $\zeta < 0$.

Example 9.15

Let us consider an LTI system with system function

$$H(s) = \frac{s - 1}{(s + 1)(s - 2)} \tag{9.86}$$

Since the ROC has not been specified, we know from our discussion in Section 9.2 that there are several different ROCs and consequently several different system impulse responses that can be associated with the algebraic expression for $H(s)$ given in eq. (9.86). If, however, information about causality or stability is known, the appropriate ROC can be identified. For example, if the system is known to be *causal*, the ROC will be that indicated in Figure 9.27(a). If the system is known to be *stable*, it will be as indicated in Figure 9.27(b). For the ROC shown in Figure 9.27(c), the system is neither causal nor stable.

9.7.1 Systems Characterized by Linear Constant-Coefficient Differential Equations

In Section 4.11 we saw how the Fourier transform can be used to obtain the frequency response of a system characterized by a linear constant-coefficient differential equation without first solving for the impulse response or time-domain solution. In an exactly analogous manner, the properties of the Laplace transform can be exploited to obtain directly the system function for a system characterized by a linear constant-coefficient differential equation. To first illustrate the procedure, consider an LTI system for which the input $x(t)$ and output $y(t)$ satisfy the linear constant-coefficient differential equation

$$\frac{dy(t)}{dt} + 3y(t) = x(t) \tag{9.87}$$

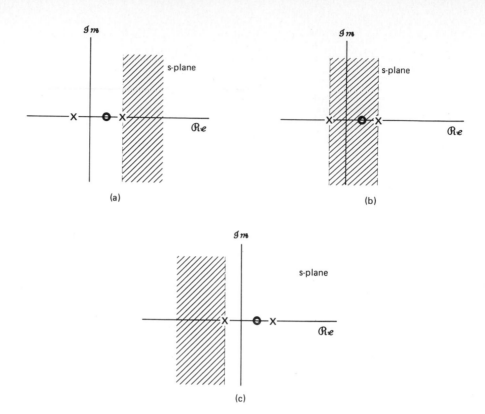

Figure 9.27 Possible ROCs for the system function of Example 9.15: (a) causal, unstable system; (b) noncausal, stable system; (c) noncausal, unstable system.

Applying the Laplace transform to both sides of eq. (9.87) and utilizing the derivative and linearity properties 9.5.6 and 9.5.1 [(eqs. (9.63) and (9.56)], we obtain the algebraic equation

$$sY(s) + 3Y(s) = X(s) \tag{9.88}$$

Since from eq. (9.62), the system function $H(s)$ is

$$H(s) = \frac{Y(s)}{X(s)}$$

we obtain for this example

$$H(s) = \frac{1}{s + 3} \tag{9.89}$$

This then provides the algebraic expression for the system function, but not the region of convergence. In fact, the differential equation by itself is not a complete specification of the LTI system, and there are in general different impulse responses all consistent with the differential equation, but depending for example on whether the equation is to be solved forward in time (causally) or backward in time (noncausally). If, in addition to the differential equation, we know that the system is causal, then the ROC can be inferred to be to the right of the rightmost pole, which for this example corresponds to $\mathcal{R}e\{s\} > -3$. If the system were known to be anticausal, then the ROC associated with $H(s)$ would be $\mathcal{R}e\{s\} < -3$. The corresponding impulse response in the causal case is

$$h(t) = e^{-3t}u(t) \tag{9.90}$$

whereas in the anticausal case it is

$$h(t) = -e^{-3t}u(-t) \tag{9.91}$$

It is easily verified by substituting into eq. (9.87) that both solutions are consistent with the differential equation.

The same procedure used to obtain $H(s)$ from the differential equation in this example can be applied more generally. Specifically, consider a general linear constant-coefficient differential equation of the form

$$\sum_{k=0}^{N} a_k \frac{d^k y(t)}{dt^k} = \sum_{k=0}^{M} b_k \frac{d^k x(t)}{dt^k} \tag{9.92}$$

Applying the Laplace transform to both sides and using the derivative property 9.5.6 [eq. (9.63)] and the linearity property 9.5.1 [eq. (9.56)] repeatedly, we obtain

$$\left\{ \sum_{k=0}^{N} a_k s^k \right\} Y(s) = \left\{ \sum_{k=0}^{M} b_k s^k \right\} X(s) \tag{9.93}$$

or

$$H(s) = \frac{\left\{ \sum_{k=0}^{M} b_k s^k \right\}}{\left\{ \sum_{k=0}^{N} a_k s^k \right\}} \tag{9.94}$$

Thus, the system function for a system specified by a differential equation is always rational, with zeros at the solutions of

$$\sum_{k=0}^{M} b_k s^k = 0 \tag{9.95}$$

and poles at the solutions of

$$\sum_{k=0}^{N} a_k s^k = 0 \tag{9.96}$$

Consistent with our previous discussion, eq. (9.94) does not include a specification of the region of convergence of $H(s)$ since the linear constant-coefficient differential equation by itself does not constrain the region of convergence. However, with additional information such as the stability or causality of the system, the region of convergence can be inferred.

9.7.2 System Function for Interconnections of LTI Systems

As we have seen, the use of the Laplace transform allows us to replace time-domain operations such as differentiation, convolution, time shifting, and so on, with algebraic operations. A further example of the usefulness of this is in analyzing and describing systems which are combined in series, parallel, and feedback interconnections. For example, for a parallel combination of two systems as shown in Figure 9.28(a), the

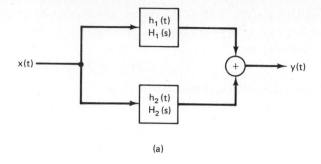

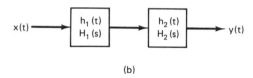

Figure 9.28 (a) Parallel connection of two LTI systems; (b) series combination of two LTI systems.

impulse response of the system is

$$h(t) = h_1(t) + h_2(t) \tag{9.97}$$

and from the linearity of the Laplace transform,

$$H(s) = H_1(s) + H_2(s) \tag{9.98}$$

Similarly, the impulse response of the system in Figure 9.28(b) is

$$h(t) = h_1(t) * h_2(t) \tag{9.99}$$

and the associated system function is

$$H(s) = H_1(s)H_2(s) \tag{9.100}$$

To illustrate the utility of the Laplace transform in representing combinations of linear systems through algebraic operations, consider the feedback interconnection of two systems as indicated in Figure 9.29. The design, applications, and analysis of such interconnections are treated in considerable detail in Chapter 11. For the moment let us simply focus on determining the overall system function from input $x(t)$ to

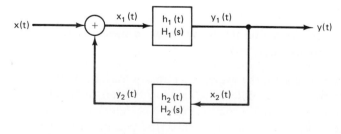

Figure 9.29 Feedback interconnection of two LTI systems.

output $y(t)$. Analysis of the system in the time domain is not particularly straightforward. However, let us consider the analysis in the Laplace transform domain. From Figure 9.29

$$Y_2(s) = H_2(s)Y(s) \tag{9.101}$$

$$Y(s) = H_1(s)[X(s) - Y_2(s)] \tag{9.102}$$

from which

$$Y(s) = H_1(s)X(s) - H_1(s)H_2(s)Y(s) \tag{9.103}$$

or

$$\frac{Y(s)}{X(s)} = H(s) = \frac{H_1(s)}{1 + H_1(s)H_2(s)} \tag{9.104}$$

Thus, through the use of the Laplace transform, the analysis of the system to obtain the overall system function reduces to a set of straightforward algebraic manipulations.

9.7.3 Butterworth Filters

As a further illustration of the usefulness of Laplace transforms in characterizing LTI systems, let us return to a discussion of the class of Butterworth filters, which was introduced in Section 6.5. From eq. (6.23), the magnitude squared of the frequency response of an Nth-order lowpass Butterworth filter is

$$|B(j\omega)|^2 = \frac{1}{1 + (j\omega/j\omega_c)^{2N}} \tag{9.105}$$

where N is the filter order. In rewriting eq. (6.23), we have modified it slightly by multiplying both ω and ω_c by j. From eq. (9.105) we would like to determine the system function $B(s)$ which gives rise to $|B(j\omega)|^2$. We first note that by definition,

$$|B(j\omega)|^2 = B(j\omega)B^*(j\omega) \tag{9.106}$$

and from Table 4.1, if we restrict the impulse response of the Butterworth filter to be real, then

$$B^*(j\omega) = B(-j\omega) \tag{9.107}$$

so that

$$B(j\omega)B(-j\omega) = \frac{1}{1 + (j\omega/j\omega_c)^{2N}} \tag{9.108}$$

Next, we note that $B(s)|_{s=j\omega} = B(j\omega)$ and consequently from eq. (9.108),

$$B(s)B(-s) = \frac{1}{1 + (s/j\omega_c)^{2N}} \tag{9.109}$$

The roots of the denominator polynomial corresponding to the combined poles of $B(s)B(-s)$ are at

$$s = (-1)^{1/2N}(j\omega_c) \tag{9.110}$$

Equation (9.110) is satisfied for any value $s = s_p$ for which

$$|s_p| = \omega_c \tag{9.111}$$

$$\angle s_p = \frac{\pi(2k + 1)}{2N} + \frac{\pi}{2}, \qquad k \text{ an integer} \tag{9.112}$$

that is,

$$s_p = \omega_c \exp \left\{ j \left[\frac{\pi(2k+1)}{2N} + \frac{\pi}{2} \right] \right\} \tag{9.113}$$

In Figure 9.30 we illustrate the positions of the poles of $B(s)B(-s)$ for $N - 1, 2,$ 3, and 6. In general, the following observations can be made about the poles of $B(s)B(-s)$:

1. There are $2N$ poles equally spaced in angle on a circle of radius ω_c in the s-plane.
2. A pole never lies on the $j\omega$-axis and occurs on the σ-axis for N odd but not for N even.
3. The angular spacing between the poles of $B(s)B(-s)$ is π/N radians.

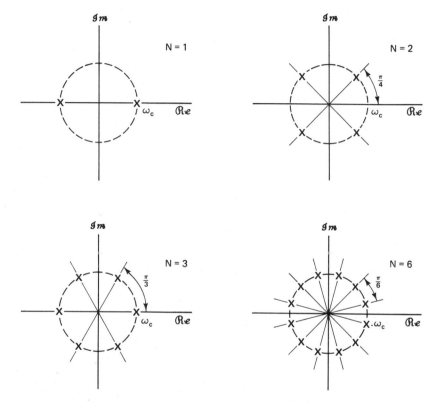

Figure 9.30 Position of the poles of $B(s)B(-s)$ for $N = 1, 2, 3$ and 6.

To determine the poles of $B(s)$ given the poles of $B(s)B(-s)$, we observe that the poles of $B(s)B(-s)$ occur in pairs, so that if there is a pole at $s = s_p$, then there is also a pole at $s = -s_p$. Consequently, to construct $B(s)$ we choose one pole from each pair. If we restrict the system to be stable and causal, then the poles that we associate with $B(s)$ are the poles along the circle in the left-half plane. The pole locations specify $B(s)$ only to within a scale factor. However, from eq. (9.109) we see that $B^2(s)|_{s=0} = 1$ or, equivalently, from eq. (9.105) the scale factor is chosen so that the magnitude squared of the frequency response has unity gain at $\omega = 0$.

The Laplace Transform Chap. 9

To illustrate the determination of $B(s)$, let us consider the cases $N = 1$, $N = 2$, and $N = 3$. In Figure 9.30 we had shown the poles of $B(s)B(-s)$ as obtained from eq. (9.113). In Figure 9.31 we show the poles associated with $B(s)$ for each of these three cases. The corresponding transfer functions are:

$N = 1$ $\qquad B(s) = \dfrac{\omega_c}{s + \omega_c}$ $\hfill$ (9.114)

$N = 2$ $\qquad B(s) = \dfrac{\omega_c^2}{(s + \omega_c e^{j(\pi/4)})(s + \omega_c e^{-j(\pi/4)})}$

$\qquad\qquad\qquad = \dfrac{\omega_c^2}{s^2 + \sqrt{2}\,\omega_c s + \omega_c^2}$ $\hfill$ (9.115)

$N = 3$ $\qquad B(s) = \dfrac{\omega_c^3}{(s + \omega_c)(s + \omega_c e^{j(\pi/3)})(s + \omega_c e^{-j(\pi/3)})}$

$\qquad\qquad\qquad = \dfrac{\omega_c^3}{(s + \omega_c)(s^2 + \omega_c + \omega_c^2)}$ $\hfill$ (9.116)

$\qquad\qquad\qquad = \dfrac{\omega_c^3}{s^3 + 2\omega_c s^2 + 2\omega_c^2 s + \omega_c^3}$

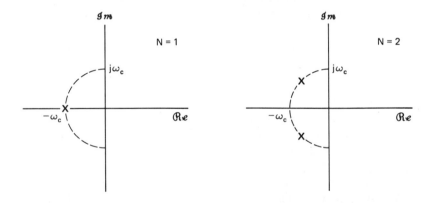

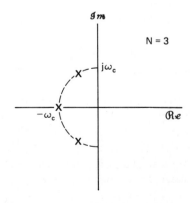

Figure 9.31 Position of the poles of $B(s)$ for $N = 1, 2, 3$.

Based on the discussion in Section 9.7.1, from $B(s)$ we can determine the associated linear constant-coefficient differential equation. Specifically, for the three cases considered above, the corresponding differential equations are:

$$N = 1: \qquad \frac{dy(t)}{dt} + \omega_c y(t) = \omega_c x(t) \tag{9.117}$$

$$N = 2: \qquad \frac{d^2 y(t)}{dt^2} + \sqrt{2}\,\omega_c \frac{dy(t)}{dt} + \omega_c^2 y(t) = \omega_c^2 x(t) \tag{9.118}$$

$$N = 3: \qquad \frac{d^3 y(t)}{dt^3} + 2\omega_c \frac{d^2 y(t)}{dt^2} + 2\omega_c^2 \frac{dy(t)}{dt} + \omega_c^3 = \omega_c^3 x(t) \tag{9.119}$$

9.8 THE UNILATERAL LAPLACE TRANSFORM

In the preceding sections of this chapter we have dealt with a form of the Laplace transform referred to as the bilateral Laplace transform. A somewhat different form of the Laplace transform, referred to as the *unilateral Laplace transform*, plays a particularly important role in analyzing causal systems specified by linear constant-coefficient differential equations with initial conditions (i.e., which are not initially at rest).

The unilateral Laplace transform $\mathfrak{X}(s)$ of a signal $x(t)$ is defined as

$$\mathfrak{X}(s) \triangleq \int_{0+}^{\infty} x(t)e^{-st}\, dt \tag{9.120}$$

From comparison of eqs. (9.120) and (9.2) we see that the difference in the definition of the unilateral and bilateral Laplace transforms lies in the lower limit on the integral. The bilateral transform depends on the entire signal from $t = -\infty$ to $t = +\infty$, whereas the unilateral transform depends only on the signal from $t = 0+$ to ∞. Consequently, two signals which differ for $t < 0$ but which are identical for $t > 0$ will have different bilateral Laplace transforms but identical unilateral transforms. We note also that since the unilateral transform does not include $t = 0$, it does not incorporate any impulses or higher-order singularity functions $u_n(t)$, $n > 0$. Basically, the unilateral transform should not be thought of as a new transform. It is the *bilateral transform* of a signal whose values for $t < 0+$ have been set to 0. Thus, using Property 4 in Section 9.2 for right-sided signals, we see that the ROC for eq. (9.120) is always a right-half plane. To illustrate the unilateral Laplace transform, let us consider two examples.

Example 9.16

Consider the signal

$$x(t) = \frac{t^{n-1}}{(n-1)!} e^{-at} u(t) \tag{9.121}$$

Since $x(t) = 0$, $t < 0$, and contains no singularities, the unilateral and bilateral transforms are identical. Thus, from Table 9.2,

$$\mathfrak{X}(s) = \frac{1}{(s+a)^n}, \qquad \mathcal{R}e\{s\} > -a \tag{9.122}$$

Example 9.17

Consider next

$$x(t) = e^{-a(t+1)}u(t + 1) \tag{9.123}$$

The *bilateral* transform $X(s)$ for this example can be obtained from Example 9.1 and the time-shifting property (Section 9.5.2). Specifically,

$$X(s) = \frac{e^s}{s + a}, \qquad \mathcal{R}e\{s\} > -a \tag{9.124}$$

By contrast, the unilateral transform is

$$\mathfrak{X}(s) = \int_{0+}^{\infty} e^{-a(t+1)}u(t + 1)e^{-st}\, dt$$

$$= \int_{0+}^{\infty} e^{-a}e^{-t(s+a)}\, dt \tag{9.125}$$

$$= e^{-a}\frac{1}{s + a}, \qquad \mathcal{R}e\{s\} > -a$$

Thus, for this example, the unilateral and bilateral Laplace transforms are distinctly different. In fact, we should recognize $\mathfrak{X}(s)$ as the bilateral transform not of $x(t)$ but of $x(t)u(t)$ consistent with our comment above that the unilateral transform *is* the bilateral transform of a signal whose values for $t < 0+$ have been set to zero.

Most of the properties of the unilateral transform are the same as for the bilateral transform. In fact, the initial and final value properties, eqs. (9.78) and (9.79), are more appropriately associated directly with the unilateral Laplace transform because they required for their validity that $x(t)$ be zero for $t < 0$ and contain no impulses or higher-order singularities. Under these conditions the bilateral Laplace transform is identical to the unilateral Laplace transform.

A particularly important difference between the properties of the unilateral and bilateral transforms is the differentiation property. Specifically, let $x(t)$ have unilateral Laplace transform $\mathfrak{X}(s)$. Then, integrating by parts, we find that the unilateral transform of $dx(t)/dt$ is given by

$$\int_{0}^{\infty} \frac{dx(t)}{dt} e^{-st}\, dt = x(t)e^{-st}\Big|_{0+}^{\infty} + s \int_{0+}^{\infty} x(t)e^{-st}\, dt \tag{9.126}$$

$$= s\mathfrak{X}(s) - x(0+)$$

Similarly, a second application of this would yield the unilateral Laplace transform of $d^2x(t)/dt^2$,

$$s^2\mathfrak{X}(s) - sx(0+) - x'(0+) \tag{9.127}$$

where $x'(0+)$ denotes the derivative of $x(t)$ evaluated at $t = 0+$. Clearly, we can continue the procedure to obtain the unilateral transform of higher derivatives.

A primary use of the unilateral Laplace transform is in obtaining the solution of linear constant-coefficient differential equations with nonzero initial conditions. For example, consider a causal system characterized by the equation

$$\frac{d^2y(t)}{dt} + 3\frac{dy(t)}{dt} + 2y(t) = x(t) \tag{9.128}$$

with the initial conditions

$$y(0+) = 3, \qquad \frac{dy(0+)}{dt} = -5 \tag{9.129}$$

Let $x(t) = 2u(t)$. Then, applying the unilateral transform to both sides of eq. (9.128), we obtain

$$s^2 \mathcal{Y}(s) - 3s + 5 + 3s\mathcal{Y}(s) - 9 + 2\mathcal{Y}(s) = \frac{2}{s} \tag{9.130}$$

or

$$\mathcal{Y}(s) = \frac{3s + 4}{(s + 1)(s + 2)} + \frac{2}{s(s + 1)(s + 2)} \tag{9.131}$$

where $\mathcal{Y}(s)$ is the unilateral Laplace transform of $y(t)$. To obtain $y(t)$, we can expand $\mathcal{Y}(s)$ in a partial fraction expansion to obtain

$$\mathcal{Y}(s) = \frac{1}{s} - \frac{1}{s + 1} + \frac{3}{s + 2} \tag{9.132}$$

Application of Example 9.16 to each term yields

$$y(t) = [1 - e^{-t} + 3e^{-2t}]u(t) \tag{9.133}$$

9.9 SUMMARY

In this chapter we have developed and studied the Laplace transform, which can be viewed as a generalization of the Fourier transform. It is particularly useful as an analytical tool in the analysis and study of LTI systems. Because of the properties of Laplace transforms, LTI systems, including those represented by linear constant-coefficient differential equations, can be characterized and analyzed in the transform domain by algebraic manipulations.

For signals and systems with rational Laplace transforms, the transform is often conveniently represented in the complex plane (s-plane) by marking the locations of the poles and zeros and indicating the region of convergence. From the pole–zero plot, the Fourier transform can be geometrically obtained. Causality, stability, and other characteristics are also easily identified from the pole locations and knowledge of the region of convergence.

This chapter was concerned principally with the bilateral Laplace transform. A somewhat different form of the Laplace transform, the unilateral transform, was also introduced. In effect, the unilateral transform can be interpreted as the bilateral transform of a signal whose values prior to $t = 0+$ have been set to zero. This form of the Laplace transform is particularly useful for obtaining the solution of linear constant-coefficient differential equations with nonzero initial conditions.

PROBLEMS

9.1. Determine the Laplace transform and the associated region of convergence and pole–zero plot for each of the following time functions.
  (a) $e^{-at}u(t)$, $\qquad a < 0$
  (b) $-e^{at}u(-t)$, $\qquad a > 0$

(c) $e^{at}u(t),$ $\qquad a > 0$

(d) $e^{-a|t|},$ $\qquad a > 0$

(e) $u(t)$

(f) $\delta(t - t_0)$

(g) $\sum\limits_{k=0}^{\infty} a^k \delta(t - kT)$ (put your answer in closed form before determining the pole–zero plot)

(h) $te^{-at}u(t),$ $\qquad a > 0$

(i) $\delta(at + b),$ $\qquad a, b$ real constants

(j) $\cos(\omega_0 t + \phi)u(t)$

9.2. For each of the following statements about $x(t)$ and for each of the four pole–zero plots in Figure P9.2, determine the corresponding constraint on the ROC.

1. The Fourier transform of $x(t)e^{-t}$ exists.
2. $x(t) = 0,$ $t > 10.$
3. $x(t) = 0,$ $t < 0.$

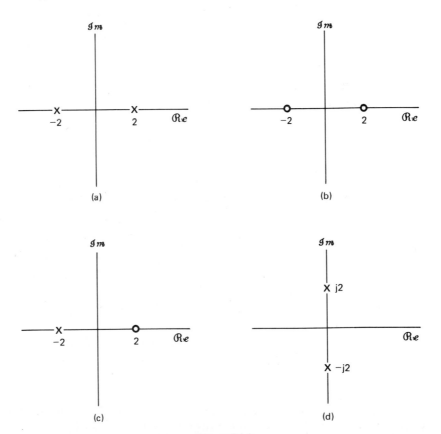

Figure P9.2

9.3. The Laplace transform is said to exist for a specific complex s if the magnitude of the transform is finite, that is, if $|X(s)| < \infty$.

Show that a *sufficient* condition for the existence of the transform $X(s)$ at

$s = s_0 = \sigma_0 + j\omega_0$ is given by

$$\int_{-\infty}^{+\infty} |x(t)|\, e^{-\sigma_0 t}\, dt < \infty$$

that is, that $x(t)$ exponentially weighted by $e^{-\sigma_0 t}$ is absolutely integrable. You will need to use the result that for a complex function $f(t)$,

$$\left| \int_a^b f(t)\, dt \right| \leq \int_a^b |f(t)|\, dt \qquad \text{(P9.3)}$$

Without rigorously proving eq. (P9.3), can you argue its plausibility?

9.4. Determine the time function $x(t)$ for each Laplace transform $X(s)$ and associated region of convergence listed below.

(a) $\dfrac{1}{s+1}$, $\qquad \mathcal{R}e\{s\} > -1$

(b) $\dfrac{1}{s+1}$, $\qquad \mathcal{R}e\{s\} < -1$

(c) $\dfrac{s}{s^2+4}$, $\qquad \mathcal{R}e\{s\} > 0$

(d) $\dfrac{s+1}{s^2+5s+6}$, $\qquad \mathcal{R}e\{s\} > -2$

(e) $\dfrac{s+1}{s^2+5s+6}$, $\qquad \mathcal{R}e\{s\} < -3$

(f) $\dfrac{s^2-s+1}{s^2(s-1)}$, $\qquad 0 < \mathcal{R}e\{s\} < 1$

(g) $\dfrac{s^2-s+1}{(s+1)^2}$, $\qquad -1 < \mathcal{R}e\{s\}$

(h) $\dfrac{s+1}{(s+1)^2+4}$, $\qquad \mathcal{R}e\{s\} > -1$ [*Hint:* Use result from part (c).]

9.5. The Laplace transform $X(s)$ of a signal $x(t)$ has four poles and an unknown number of zeros. $x(t)$ is known to have an impulse at $t = 0$. Determine what information, if any, this provides about the number of zeros and their locations.

9.6. Throughout this problem we will consider the region of convergence of the Laplace transforms to always include the $j\omega$-axis.

(a) Consider a signal $x(t)$ with Fourier transform $X(j\omega)$ and Laplace transform $X(s) = s + \frac{1}{2}$. Draw the pole–zero plot for $X(s)$. Also draw the vector whose length represents $|X(j\omega)|$ and whose angle with respect to the real axis represents $\sphericalangle X(j\omega)$ for a given ω.

(b) By examining the pole–zero plot and vector diagram in part (a), determine a different Laplace transform $X_1(s)$ corresponding to a time function $x_1(t)$ so that

$$|X_1(j\omega)| = |X(j\omega)|$$

but

$$x_1(t) \neq x(t)$$

Show the pole–zero plot and associated vectors that represent $X_1(j\omega)$.

(c) For your answer in part (b), determine, again by examining the related vector diagrams, the relationship between $\sphericalangle X(j\omega)$ and $\sphericalangle X_1(j\omega)$.

(d) Determine a Laplace transform $X_2(s)$ so that

$$\sphericalangle X_2(j\omega) = \sphericalangle X(j\omega)$$

but $x_2(t)$ is not proportional to $x(t)$. Show the pole–zero plot for $X_2(s)$ and the associated vectors that represent $X_2(j\omega)$.

(e) For your answer in part (d), determine the relationship between $|X_2(j\omega)|$ and $|X(j\omega)|$.

(f) Consider a signal $x(t)$ with Laplace transform $X(s)$ for which the pole–zero plot is as shown in Figure P9.6. Determine $X_1(s)$ such that $|X(j\omega)| = |X_1(j\omega)|$ and all poles and zeros of $X_1(s)$ are in the left-half of the s-plane [i.e., $\mathfrak{Re}\{s\} < 0$]. Also, determine $X_2(s)$ such that $\sphericalangle X(j\omega) = \sphericalangle X_2(j\omega)$ and all poles and zeros of $X_2(s)$ are in the left-half of the s-plane.

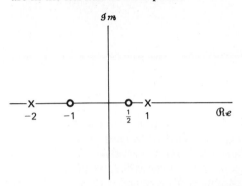

Figure P9.6

9.7. By considering the geometric determination of the Fourier transform, as developed in Section 9.4, sketch, for each of the pole–zero plots in Figure P9.7, the magnitude of the associated Fourier transform.

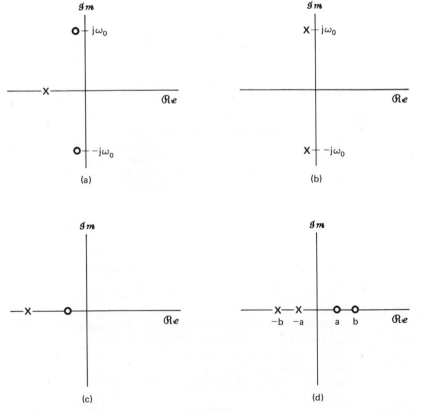

Figure P9.7

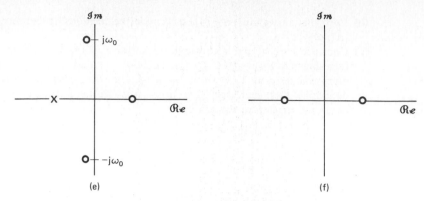

Figure P9.7 (cont.)

9.8. In long-distance telephone communication, an echo is sometimes encountered due to the transmitted signal being reflected at the receiver, sent back down the line, reflected again at the transmitter, and returned to the receiver. The impulse response for a system that models this effect is shown in Figure P9.8, where we have assumed that only one echo is received. The parameter T corresponds to the one-way travel time along the communication channel and the parameter α represents the attenuation in amplitude between transmitter and receiver.

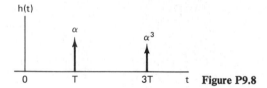

Figure P9.8

(a) Determine the system function $H(s)$ and associated region of convergence for the system.

(b) From your result in part (a), you should observe that $H(s)$ does not consist of a ratio of polynomials. Nevertheless, it is useful to represent it in terms of poles and zeros, where, as usual, the zeros are the values of s for which $H(s) = 0$ and the poles are the values of s for which $[1/H(s)] = 0$. For the system function determined in part (a), determine the zeros and demonstrate that there are no poles.

(c) From your result in part (b), sketch the pole–zero plot for $H(s)$.

(d) By considering the appropriate vectors in the s-plane, sketch the magnitude of the frequency response of the system.

9.9. As indicated in Section 9.5, many of the properties of the Laplace transform and their derivation are analogous to corresponding properties of the Fourier transform, as developed in Chapter 4. In this problem you are asked to outline the derivation for a number of the Laplace transform properties in Section 9.5.

By paralleling the derivation for the corresponding property in Chapter 4 for the Fourier transform, derive each of the following Laplace transform properties. Your derivation must include a consideration of the region of convergence.

(a) Time-shifting property (9.5.2)

(b) Shifting in the s-domain (9.5.3)

(c) Time scaling (9.5.4)

(d) Convolution property (9.5.5)

9.10. (a) If $x(t)$ is an even time function so that $x(t) = x(-t)$, show that this requires that $X(s) = X(-s)$.

(b) If $x(t)$ is an odd time function so that $x(t) = -x(-t)$, show that $X(s) = -X(-s)$.

(c) Determine which, if any, of the pole–zero plots in Figure P9.10 could correspond to an even time function. For those that could, indicate the required ROC.

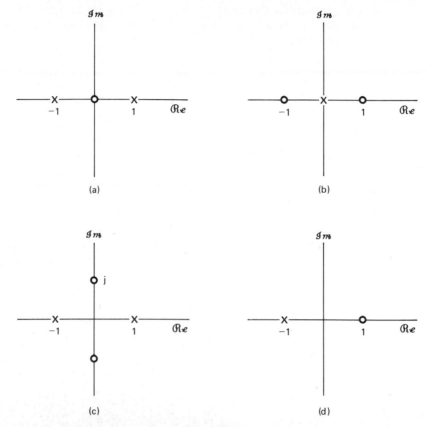

Figure P9.10

(d) Determine which, if any, of the pole–zero plots in Figure P9.10 could correspond to an odd time function. For those that could, indicate the required ROC.

9.11. As presented in Section 9.5.9, the initial value theorem states that for a signal $x(t)$ with Laplace transform $X(s)$ and for which $x(t) = 0$ for $t < 0$, the initial value of $x(t)$ [i.e., $x(0+)$] can be obtained from $X(s)$ through the relation

[eq. (9.74)]
$$x(0+) = \lim_{s \to \infty} s X(s)$$

First, we note that since $x(t) = 0$ for $t < 0$, $x(t) = x(t)u(t)$. Next, expanding $x(t)$ as a Taylor series at $t = 0+$,

$$x(t) = \left[x(0+) + x^{(1)}(0+)t + \ldots + x^{(n)}(0+)\frac{t^n}{n!} + \ldots \right] u(t) \qquad \text{(P9.11)}$$

where $x^{(n)}(0+)$ denotes the nth derivative of $x(t)$ evaluated at $t = 0+$.

(a) Determine the Laplace transform of an arbitrary term $x^{(n)}(0+)(t^n/n!)u(t)$ on the right-hand side of eq. (P9.11). (You may find it helpful to review Example 9.11.)

(b) From your result in part (a) and the expansion in eq. (P9.11), show that $X(s)$ can be expressed as

$$X(s) = \sum_{n=0}^{\infty} x^{(n)}(0+) \frac{1}{s^{n+1}}$$

(c) Demonstrate that eq. (9.74) follows from the result in part (b).

(d) By first determining $x(t)$, verify the initial value theorem for each of the following examples.

    1. $X(s) = \dfrac{1}{s + 2}$

    2. $X(s) = \dfrac{s + 1}{(s + 2)(s + 3)}$

(e) A more general form of the initial value theorem states that if $x^{(n)}(0+) = 0$ for $n < N$, then $x^{(N)}(0+) = \lim_{s \to \infty} s^{N+1} X(s)$. Demonstrate that this more general statement also follows from the result in part (b).

9.12. Consider a real-valued signal $x(t)$ with Laplace transform $X(s)$.

(a) By applying complex conjugation to both sides of eq. (9.37), show that $X(s) = X*(s*)$.

(b) From your result in (a), show that if $X(s)$ has a pole (zero) at $s = s_0$ it must also have a pole (zero) at $s = s_0^*$, i.e., for $x(t)$ real, the poles and zeros of $X(s)$ which are not on the real axis must occur in complex conjugate pairs.

9.13. In Section 9.6, Table 9.2, we listed a number of Laplace transform pairs and we indicated specifically how transform pairs 1 through 9 follow from Examples 9.1 and 9.11 together with various properties from Table 9.1.

By exploiting appropriate properties from Table 9.1, show how transform pairs 10 through 14 follow from transform pairs 1 through 9 in Table 9.2.

9.14. Consider an LTI system for which the system function $H(s)$ has the pole–zero pattern shown in Figure P9.14.

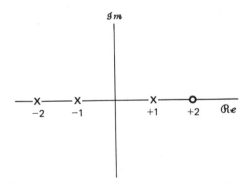

Figure P9.14

(a) Indicate all possible ROCs that can be associated with this pole–zero pattern.

(b) For each ROC identified in part (a), specify whether the associated system is stable and/or causal.

9.15. Consider an LTI system with input $x(t) = e^{-t}u(t)$ and impulse response $h(t) = e^{-2t}u(t)$.

(a) Determine the Laplace transform of $x(t)$ and $h(t)$.

(b) Using the convolution property, determine the Laplace transform $Y(s)$ of the output $y(t)$.

(c) From the Laplace transform of $y(t)$ as obtained in part (a), determine $y(t)$.

(d) Verify your result in part (b) by explicitly convolving $x(t)$ and $h(t)$.

9.16. A pressure gauge, which can be modeled as an LTI system, has a time response to a unit step input given by $(1 - e^{-t} - te^{-t})u(t)$. For a certain unknown input $x(t)$, the output is observed to be $(2 - 3e^{-t} + e^{-3t})u(t)$.

For this observed measurement, determine the true pressure input to the gauge as a function of time.

9.17. The inverse of an LTI system $H(s)$ is defined as a system which, when cascaded with $H(s)$, results in an overall transfer function of unity or equivalently an overall impulse response which is an impulse.

(a) If $H_1(s)$ denotes the transfer function of an inverse system for $H(s)$, determine the general algebraic relationship between $H(s)$ and $H_1(s)$.

(b) Shown in Figure P9.17 is the pole–zero plot for a stable, causal system $H(s)$. Determine the pole–zero plot for the associated inverse system.

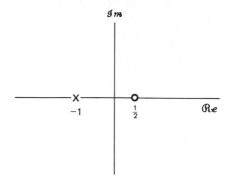

Figure P9.17

(c) Determine the impulse response $h_1(t)$ of the inverse system if it is assumed to be stable.

(d) By explicitly convolving $h(t)$ with $h_1(t)$, as determined in part (c), demonstrate that the impulse response of the two systems in cascade is an impulse.

9.18. A class of systems, referred to as minimum-delay or minimum-phase systems, are sometimes defined through the statement that they are causal and stable and that the inverse system is also causal and stable.

Develop an argument to demonstrate that based on the definition above, all poles *and* zeros of the transfer function of a minimum-delay system must be in the left half of the s-plane [i.e., $\Re e\{s\} < 0$].

9.19. Determine whether or not each of the following statements about LTI systems is true. If a statement is true, construct a convincing argument. If false, demonstrate a counter-example.

Statement 1 A stable continuous-time system must have all its poles in the left half of the s-plane [i.e., $\Re e\{s\} < 0$].

Statement 2 If the system function has more poles than zeros, and the system is causal, the step response will be continuous at $t = 0$.

Statement 3 If the system function has more poles than zeros, and the system is not restricted to be causal, the step response can be discontinuous at $t = 0$.

Statement 4 A stable, causal system must have all its poles and zeros in the left half of the s-plane.

9.20. The autocorrelation function $\phi_{xx}(\tau)$ of a signal $x(t)$ is defined as

$$\phi_{xx}(\tau) = \int_{-\infty}^{+\infty} x(t)x(t + \tau)\,dt$$

(a) Determine, in terms of $x(t)$, the impulse response $h(t)$ of an LTI system for which, when the input is $x(t)$, the output is $\phi_{xx}(t)$ (Figure P9.20-1).

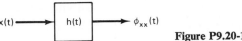

Figure P9.20-1

(b) From your answer in part (a), determine $\Phi_{xx}(s)$, the Laplace transform of $\phi_{xx}(\tau)$ in terms of $X(s)$. Also, express $\Phi_{xx}(j\omega)$, the Fourier transform of $\phi_{xx}(\tau)$, in terms of $X(j\omega)$.

(c) If $x(t)$ has the pole–zero pattern and ROC shown in Figure P9.20-2, sketch the pole–zero pattern and indicate the ROC for $\phi_{xx}(\tau)$.

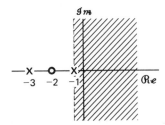

Figure P9.20-2

9.21. In a number of applications in signal design and analysis, the class of signals $\phi_n(t)$, $n = 0, 1, 2, \ldots$ are encountered, defined by the relations

$$\phi_n(t) = e^{-t/2}L_n(t)u(t) \qquad n = 0, 1, 2, \ldots \tag{P9.21a}$$

$$L_n(t) = \frac{e^t}{n!}\frac{d^n}{dt^n}(t^n e^{-t}) \tag{P9.21b}$$

(a) The functions $L_n(t)$ are, in fact, polynomials and are referred to as *Laguerre polynomials*. To verify that they have the form of polynomials, explicitly determine $L_0(t)$, $L_1(t)$, and $L_2(t)$.

(b) Using properties of the Laplace transform in Table 9.1 and Laplace transform pairs in Table 9.2, determine the Laplace transform $\Phi_n(s)$ of $\phi_n(t)$.

(c) The set of signals $\phi_n(t)$ can be generated by excting a network of the form in Figure P9.21 with an impulse. From your result in part (b), determine $H_1(s)$ and $H_2(s)$ so that the impulse responses along the cascade chain are the signals $\phi_n(t)$ as indicated.

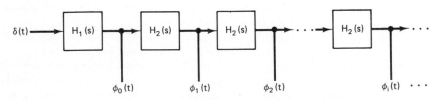

Figure P9.21

The Laplace Transform   Chap. 9

9.22. Consider an LTI system for which we are given the following information:

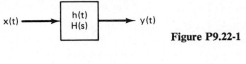

Figure P9.22-1

$$X(s) = \frac{s+2}{s-2}$$

$$x(t) = 0, \qquad t > 0$$

and

$$y(t) = -\tfrac{2}{3}e^{2t}u(-t) + \tfrac{1}{3}e^{-t}u(t)$$

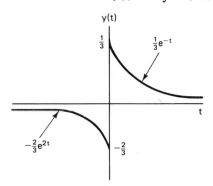

Figure P9.22-2

(a) Determine $H(s)$ and its region of convergence.

(b) Determine $h(t)$.

(c) Using the system function $H(s)$ found in part (a), determine the output $y(t)$ if the input $x(t)$ is

$$x(t) = e^{3t}, \qquad -\infty < t < +\infty$$

9.23. Consider a continuous-time LTI system for which the input $x(t)$ and output $y(t)$ are related by the differential equation

$$\frac{d^2 y(t)}{dt} - \frac{dy}{dt} - 2y(t) = x(t)$$

Let $X(s)$ and $Y(s)$ denote the Laplace transforms of $x(t)$ and $y(t)$, and let $H(s)$ denote the Laplace transform of $h(t)$, the system impulse response.

(a) Determine $H(s)$ as a ratio of polynomials in s. Sketch the pole–zero pattern of $H(s)$.

(b) Determine $h(t)$ for each of the following three cases:

  1. The system is stable.

  2. The system is causal.

  3. The system is *neither* stable nor causal.

9.24. A causal linear time-invariant system with impulse response $h(t)$ has the following properties:

  1. When the input to the system is $x(t) = e^{2t}$ for all t, the output is $y(t) = (\tfrac{1}{6})e^{2t}$ for all t.

2. The impulse response $h(t)$ satisfies the differential equation

$$\frac{dh(t)}{dt} + 2h(t) = (e^{-4t})u(t) + bu(t)$$

where b is an unknown constant.

Determine the system function $H(s)$ of the system consistent with the information above. There should be no unknown constants in your answer; that is, the constant b should *not* appear in your answer.

9.25. $H(s)$ represents the system function for a causal, stable system. The input to the system consists of the sum of three terms, one of which is an impulse $\delta(t)$ and another a complex exponential of the form $e^{s_0 t}$ where s_0 is a complex constant. The output is

$$y(t) = -6e^{-t}u(t) + \tfrac{4}{34}e^{4t}\cos 3t + \tfrac{18}{34}e^{4t}\sin 3t + \delta(t)$$

Determine $H(s)$ consistent with this information.

9.26. The signal

$$y(t) = u(t)e^{-2t}$$

is the output of a causal all-pass system for which the system function is

$$H(s) = \frac{s-1}{s+1}$$

(a) Find and sketch at least two possible inputs $x(t)$ that could produce this output.

(b) What is the input if it is known that

$$\int_{-\infty}^{+\infty} |x(t)|\, dt < \infty?$$

(c) What is the input $x(t)$ if it is known that a stable (but not necessarily causal) system exists which will have $x(t)$ as an output if $y(t)$ is the input? Find the impulse response $h(t)$ of this filter and show by direct convolution that it has the property claimed [i.e., that $y(t) * h(t) = x(t)$].

9.27. The system function, $H(s)$, of a causal LTI system, is given by

$$H(s) = \frac{s+1}{s^2 + 2s + 2}$$

Determine and sketch the response $y(t)$, when the input $x(t)$ is given by

$$x(t) = e^{-|t|}, \qquad -\infty < t < \infty$$

9.28. As we discussed in Chapter 6, in filter design it is often possible and convenient to transform a lowpass filter design to a highpass filter, and vice versa. With $H(s)$ denoting the transfer function of the original filter and $G(s)$ that of the transformed filter, one such commonly used transformation consists of replacing s by $1/s$, that is,

$$G(s) = H\!\left(\frac{1}{s}\right)$$

(a) For $H(s) = 1/(s + \tfrac{1}{2})$, sketch $|H(j\omega)|$ and $|G(j\omega)|$.

(b) Determine the linear constant-coefficient differential equation associated with $H(s)$ and with $G(s)$.

(c) Now consider a more general case in which $H(s)$ is the transfer function associated with a linear constant-coefficient differential equation in the general form

$$\sum_{k=0}^{N} a_k \frac{d^k y(t)}{dt^k} = \sum_{k=0}^{N} b_k \frac{d^k x(t)}{dt^k} \tag{P9.28}$$

Without any loss of generality, we have assumed that the number of derivatives N is the same on both sides of the equation, although in any particular case, some of the coefficients may then be zero. Determine $H(s)$ and $G(s)$.

(d) From your result in part (c), determine, in terms of the coefficients in eq. (P9.28), the linear constant-coefficient differential equation associated with $G(s)$.

9.29. Consider a general continuous-time system with input $x(t)$ and output $y(t)$. Listed below are a number of possible input–output properties.

1. $x(t - t_0) \longrightarrow y(t - t_0)$ for any t_0
2. $kx(t) \longrightarrow ky(t)$ for any k
3. $\dfrac{dx(t)}{dt} \longrightarrow \dfrac{dy(t)}{dt}$
4. $e^{st} \longrightarrow H(s)e^{st}$

For *each* of the four properties above, indicate whether it is true, in general, for:

(a) Every linear system   (b) Every time-invariant system
(c) Every LTI system   (d) Every system

9.30. $B(s)$ denotes the transfer function for an Nth-order Butterworth filter with cutoff frequency ω_c.
(a) Draw the pole–zero plot for $B(s)$ with $\omega_c = 2\pi \times 10^3$ and:

1. $N = 4$
2. $N = 5$

(b) Express $B(s)$ algebraically for the two cases in part (a).
(c) Determine the differential equation associated with $B(s)$ for $\omega_c = 2\pi \times 10^3$ and $N = 4$.

9.31. If $\mathfrak{X}(s)$ denotes the unilateral Laplace transform of $x(t)$, determine in terms of $\mathfrak{X}(s)$ the unilateral Laplace transform of:

(a) $x(t - 1)$ (b) $x(t + 1)$
(c) $\displaystyle\int_{-\infty}^{t} x(\tau)\, d\tau$ (d) $\dfrac{d^3 x(t)}{dt^3}$

9.32. (a) Determine the differential equation relating $v_i(t)$ and $v_0(t)$ for the *RLC* circuit of Figure P9.32.

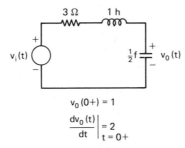

$$v_0(0+) = 1$$
$$\left.\frac{dv_0(t)}{dt}\right|_{t=0+} = 2$$

Figure P9.32

(b) Suppose that $v_i(t) = e^{-3t}u(t)$. Using the unilateral Laplace transform, determine $v_0(t)$ for $t > 0$.

In Chapter 9 we developed the Laplace transform as an extension of the continuous-time Fourier transform. This extension was motivated in part by the fact that it can be applied to a broader class of signals than the Fourier transform, since there are many signals for which the Fourier transform does not converge but the Laplace transform does. As with the Fourier transform, the Laplace transform arises naturally in considering the response of an LTI system to inputs that are complex exponentials. Furthermore, as we saw, the Laplace transform possesses many of the same properties as the Fourier transform and thus is a powerful tool in the analysis of continuous-time LTI systems.

In this chapter we consider the z-transform, which is the discrete-time counterpart of the Laplace transform, and which is the corresponding generalization of the discrete-time Fourier transform. As we will see, the motivations for and properties of the z-transform closely parallel those of the Laplace transform. Just as with the relationship between continuous- and discrete-time Fourier transforms, however, we will encounter some important distinctions between the z-transform and the Laplace transform that arise from the fundamental differences between continuous- and discrete-time signals and systems.

The z-Transform

10.1 THE z-TRANSFORM

As we saw in Section 5.1, for a discrete-time linear time-invariant system with impulse response $h[n]$, the response $y[n]$ of the system to a complex exponential input of the form z^n is

$$y[n] = H(z)z^n \qquad (10.1)$$

where

$$H(z) = \sum_{n=-\infty}^{+\infty} h[n]z^{-n} \qquad (10.2)$$

For $z = e^{j\Omega}$ with Ω real (i.e., with $|z| = 1$), the summation in eq. (10.2) corresponds to the discrete-time Fourier transform of $h[n]$. More generally, when $|z|$ is not restricted to unity, the summation in eq. (10.2) is referred to as the z-*transform* of $h[n]$. As with the Fourier transform for both continuous-time and discrete-time signals and the Laplace transform in the continuous-time case, the z-transform plays an important role as a transformation applied to sequences in general, whether or not they represent a system impulse response.

The z-transform of a sequence $x[n]$ is defined as

$$X(z) \triangleq \sum_{n=-\infty}^{+\infty} x[n]z^{-n} \qquad (10.3)$$

where z is a complex variable. The z-transform in this form is often referred to as the *bilateral* z-transform to distinguish it from the unilateral z-transform, which we discuss in Section 10.9. We will refer to $X(z)$ as defined in eq. (10.3) simply as the z-transform and use the term "bilateral" only where it is needed to avoid ambiguity. For convenience, the z-transform of $x[n]$ will sometimes be denoted as $\mathfrak{Z}\{x[n]\}$ and the relationship between $x[n]$ and its z-transform indicated as

$$x[n] \xleftrightarrow{\;\;\mathbf{Z}\;\;} X(z) \qquad (10.4)$$

In Chapter 9 we considered for continuous-time signals a number of important relationships between the Laplace transform and the Fourier transform. In a similar but not identical way, there are a number of important relationships between the z-transform and the Fourier transform. To explore these relationships, let us express the complex variable z in polar form as

$$z = re^{j\Omega} \qquad (10.5)$$

with r as the magnitude of z and Ω as the angle of z. In terms of r and Ω, eq. (10.3) becomes

$$X(re^{j\Omega}) = \sum_{n=-\infty}^{+\infty} x[n](re^{j\Omega})^{-n}$$

or, equivalently,

$$X(re^{j\Omega}) = \sum_{n=-\infty}^{+\infty} \{x[n]r^{-n}\}e^{-j\Omega n} \qquad (10.6)$$

From eq. (10.6) we see that $X(re^{j\Omega})$ is the Fourier transform of the sequence $x[n]$

multiplied by a real exponential r^{-n}, that is,

$$X(re^{j\Omega}) = \mathcal{F}\{x[n]r^{-n}\} \tag{10.7}$$

The exponential weighting r^{-n} may be decaying or growing with increasing n, depending on whether r is greater than or less than unity. We note in particular that for $r = 1$ or equivalently $|z| = 1$, the z-transform reduces to the Fourier transform, that is,

$$X(z)|_{z=e^{j\Omega}} = \mathcal{F}\{x[n]\} \tag{10.8}$$

The relationship between the z-transform and Fourier transform for discrete-time signals parallels closely the corresponding discussion in Section 9.1 for continuous-time signals, but with some important differences. In the continuous-time case, the Laplace transform reduces to the Fourier transform when the real part of the transform variable is zero. Interpreted in terms of the s-plane, this means that the Laplace transform reduces to the Fourier transform on the imaginary axis (i.e., for $s = j\omega$). By contrast, the z-transform reduces to the Fourier transform when the magnitude of the transform variable z is *unity* (i.e., for $z = e^{j\Omega}$). Thus, the z-transform reduces to the Fourier transform on the contour in the complex z-plane corresponding to a circle with a radius of unity as indicated in Figure 10.1. This circle in the z-plane

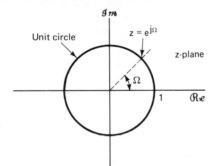

Figure 10.1 Complex z-plane. The z-transform reduces to the Fourier transform for values of z on the unit circle.

is referred to as the *unit circle*, and will play a role in the discussion of the z-transform similar to the role of the imaginary axis in the s-plane for the Laplace transform. Because of this relationship between the z-transform and the Fourier transform it is convenient at this point to make a simple change of notation in representing the discrete-time Fourier transform. Specifically, we will now denote the independent variable associated with the discrete-time Fourier transform as $e^{j\Omega}$ rather than simply Ω, to emphasize the fact that it is equal to the z-transform for $z = e^{j\Omega}$. With this change in notation, we can also express eq. (10.8) as

$$X(z)|_{z=e^{j\Omega}} = \mathcal{F}\{x[n]\} = X(e^{j\Omega}) \tag{10.9}$$

From eq. (10.7), we see that for convergence of the z-transform we require that the Fourier transform of $x[n]r^{-n}$ converge. For any specific sequence $x[n]$ we would expect this convergence for some values of r and not for others. In general there is associated with the z-transform of a sequence a range of values of z for which $X(z)$ converges. As with the Laplace transform, this range of values is referred to as the *region of convergence* (ROC). If the ROC includes the unit circle, then the Fourier transform also converges. To illustrate the z-transform and the associated region of convergence, let us consider several examples.

Example 10.1

Consider the signal $x[n] = a^n u[n]$. Then from eq. (10.3),

$$X(z) = \sum_{n=-\infty}^{+\infty} a^n u[n] z^{-n} = \sum_{n=0}^{\infty} (az^{-1})^n$$

For convergence of $X(z)$, we require that $\sum_{n=0}^{\infty} |az^{-1}|^n < \infty$. Thus, the region of convergence is the range of values of z for which $|az^{-1}| < 1$ or equivalently $|z| > |a|$. Then,

$$X(z) = \sum_{n=0}^{\infty} (az^{-1})^n = \frac{1}{1 - az^{-1}} = \frac{z}{z - a}, \qquad |z| > |a| \qquad (10.10)$$

Consequently the z-transform converges for any finite value of a. The Fourier transform of $x[n]$, on the other hand, only converges if $|a| < 1$. For $a = 1$, $x[n]$ is the unit step sequence with z-transform

$$X(z) = \frac{1}{1 - z^{-1}}, \qquad |z| > 1$$

We see that the z-transform in Example 10.1 is a rational function. Consequently, just as with rational Laplace transforms it can be characterized by its zeros (the roots of the numerator polynomial) and its poles (the roots of the denominator polynomial). For this example, there is one zero, at $z = 0$, and one pole, at $z = a$. The pole–zero plot and the region of convergence for Example 10.1 are shown in Figure 10.2. For $|a| > 1$, the ROC does not include the unit circle, consistent with the fact that for these values of a, the Fourier transform of $a^n u[n]$ does not converge.

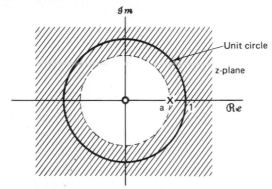

Figure 10.2 Pole–zero plot and region of convergence for Example 10.1.

Example 10.2

Now let $x[n] = -a^n u[-n - 1]$. Then

$$X(z) = -\sum_{n=-\infty}^{+\infty} a^n u[-n - 1] z^{-n} = -\sum_{n=-\infty}^{-1} a^n z^{-n} \qquad (10.11)$$

$$= -\sum_{n=1}^{\infty} a^{-n} z^n = 1 - \sum_{n=0}^{\infty} (a^{-1} z)^n$$

If $|a^{-1}z| < 1$ or equivalently $|z| < |a|$, the sum in eq. (10.11) converges and

$$X(z) = 1 - \frac{1}{1 - a^{-1}z} = \frac{1}{1 - az^{-1}} = \frac{z}{z - a} \qquad (10.12)$$

The pole–zero plot and region of convergence for this example are shown in Figure 10.3.

The z-Transform   Chap. 10

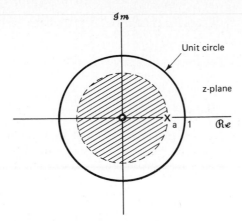

Figure 10.3   Pole–zero plot and region of convergence for Example 10.2.

Comparing eqs 10.10 and 10.12, and Figures 10.2 and 10.3, we see that the algebraic expression for $X(z)$ and the corresponding pole–zero plot are identical in Examples 10.1 and 10.2, and the z-transforms differ only in the region of convergence. Thus, as with the Laplace transform, specification of the z-transform requires both the algebraic expression and the region of convergence. Also, in both examples, the sequences were exponentials and the resulting z-transforms were rational. In fact, as further suggested by the next example, $X(z)$ will be rational whenever $x[n]$ is a linear combination of real or complex exponentials.

Example 10.3

Let us consider a signal that is the sum of two real exponentials:

$$x[n] = (\tfrac{1}{2})^n u[n] + (\tfrac{1}{3})^n u[n] \tag{10.13}$$

The z-transform is then

$$
\begin{aligned}
X(z) &= \sum_{n=-\infty}^{+\infty} \{(\tfrac{1}{2})^n u[n] + (\tfrac{1}{3})^n u[n]\}\, z^{-n} \\[2mm]
&= \sum_{n=-\infty}^{+\infty} (\tfrac{1}{2})^n u[n] z^{-n} + \sum_{n=-\infty}^{+\infty} (\tfrac{1}{3})^n u[n] z^{-n} \tag{10.14} \\[2mm]
&= \sum_{n=0}^{\infty} (\tfrac{1}{2} z^{-1})^n + \sum_{n=0}^{\infty} (\tfrac{1}{3} z^{-1})^n \\[2mm]
&= \frac{1}{1 - \tfrac{1}{2} z^{-1}} + \frac{1}{1 - \tfrac{1}{3} z^{-1}} = \frac{2 - (\tfrac{5}{6}) z^{-1}}{(1 - \tfrac{1}{2} z^{-1})(1 - \tfrac{1}{3} z^{-1})} \\[2mm]
&= \frac{z(2z - \tfrac{5}{6})}{(z - \tfrac{1}{2})(z - \tfrac{1}{3})} \tag{10.15}
\end{aligned}
$$

For convergence of $X(z)$, both sums in eq. (10.14) must converge, which requires that both $|\tfrac{1}{2} z^{-1}| < 1$ and $|\tfrac{1}{3} z^{-1}| < 1$ or equivalently $|z| > \tfrac{1}{2}$ and $|z| > \tfrac{1}{3}$. Thus, the region of convergence is $|z| > \tfrac{1}{2}$.

The z-transform for this example can also be obtained using the results of Example 10.1. Specifically, from the definition of the z-transform, eq. (10.3), we see that the z-transform is linear, that is, that if $x[n]$ is the sum of two terms, then $X(z)$ will be the sum of the z-transforms of the individual terms and will converge when both z-transforms converge. From Example 10.1 we see that

$$\left(\tfrac{1}{2}\right)^n u[n] \overset{\mathcal{Z}}{\longleftrightarrow} \frac{1}{1 - \tfrac{1}{2}z^{-1}}, \qquad |z| > \tfrac{1}{2} \tag{10.16}$$

$$\left(\tfrac{1}{3}\right)^n u[n] \overset{\mathcal{Z}}{\longleftrightarrow} \frac{1}{1 - \tfrac{1}{3}z^{-1}}, \qquad |z| > \tfrac{1}{3} \tag{10.17}$$

and consequently,

$$\left(\tfrac{1}{2}\right)^n u[n] + \left(\tfrac{1}{3}\right)^n u[n] \overset{\mathcal{Z}}{\longleftrightarrow} \frac{1}{1 - \tfrac{1}{2}z^{-1}} + \frac{1}{1 - \tfrac{1}{3}z^{-1}}, \qquad |z| > \tfrac{1}{2} \tag{10.18}$$

as we had determined above. The pole–zero plot and ROC for the z-transform of each of the individual terms and for the combined signal are shown in Figure 10.4.

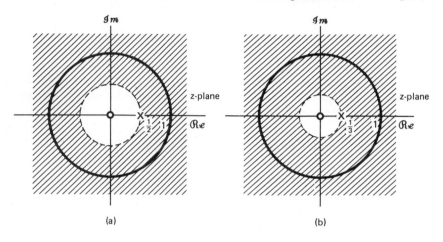

(a)   (b)

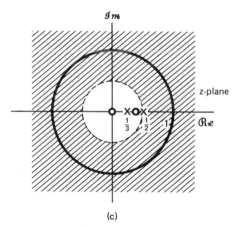

(c)

Figure 10.4 Pole–zero plot and region of convergence for the individual terms and the sum in Example 10.3: (a) $1/(1 - \tfrac{1}{2}z^{-1})$, $|z| > \tfrac{1}{2}$; (b) $1/(1 - \tfrac{1}{3}z^{-1})$, $|z| > \tfrac{1}{3}$; (c) $1/(1 - \tfrac{1}{2}z^{-1}) + 1/(1 - \tfrac{1}{3}z^{-1})$, $|z| > \tfrac{1}{2}$.

In each of the three examples above, we expressed the z-transform both as a ratio of polynomials in z and as a ratio of polynomials in z^{-1}. From the form of the definition of the z-transform as given in eq. (10.3), we see that for sequences that are zero for $n < 0$, $X(z)$ involves only negative powers of z. Thus for this class of signals it is particularly convenient for $X(z)$ to be expressed in terms of polynomials in z^{-1} rather than z, and when appropriate we will use that form in our discussion. When the z-transform is expressed in terms of factors of the form $(1 - az^{-1})$, it should be remembered that such a factor introduces both a pole and a zero, as evidenced in the algebraic expressions for the foregoing examples.

10.2 THE REGION OF CONVERGENCE FOR THE z-TRANSFORM

In Chapter 9 we saw that there were specific constraints on and properties of the region of convergence of the Laplace transform for different classes of signals and that understanding these led to further insights about the Laplace transform. In a similar manner, we explore a number of properties of the region of convergence for the z-transform. Each property discussed below and its justification closely parallels the corresponding property in Section 9.2.

> Property 1: The ROC of $X(z)$ consists of a ring in the z-plane centered about the origin.

This property is illustrated in Figure 10.5 and follows from the fact that the ROC consists of those values of $z = re^{j\Omega}$ for which $x[n]r^{-n}$ has a Fourier transform that converges. Thus, convergence is dependent only on $r = |z|$ and not on Ω. Consequently if a specific value of z is in the ROC, then all values of z on the same circle (i.e. with the same magnitude) will be in the ROC. This by itself guarantees that the ROC will consist of concentric rings. As we will see in Property 6, the ROC must in fact consist of only a single ring. In some cases, the inner boundary of the ROC may extend inward to the origin, thus reducing to a disk and/or the outer boundary may extend outward to infinity.

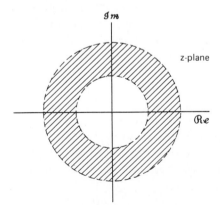

Figure 10.5 ROC as a ring in the z-plane. For specific cases the inner boundary can extend inward to the origin in which case the ROC becomes a disc. For specific cases the outer boundary can extend outward to infinity.

> Property 2: The ROC does not contain any poles.

As with the Laplace transform, this property is simply a consequence of the fact that at a pole, $X(z)$ is infinite and therefore by definition does not converge.

> **Property 3:** If $x[n]$ is of finite duration, then the ROC is the entire z-plane, except possibly $z = 0$ and/or $z = \infty$.

A finite-duration sequence has only a finite number of nonzero values, extending, say, from $n = N_1$ to $n = N_2$, where N_1 and N_2 are finite. Thus, the z-transform $X(z)$ is the sum of a finite number of terms, specifically

$$X(z) = \sum_{n=N_1}^{N_2} x[n]z^{-n} \tag{10.19}$$

For z not equal to zero or infinity, each term in the sum will be finite and consequently $X(z)$ will converge. If N_1 is negative and N_2 positive so that $x[n]$ has nonzero values both for $n < 0$ and $n > 0$, then the summation in eq. (10.19) includes terms with both positive powers of z and negative powers of z. As $|z| \rightarrow 0$, terms involving negative powers of z become unbounded, and as $|z| \rightarrow \infty$, those involving positive powers of z become unbounded. Consequently, for N_1 negative and N_2 positive, the ROC does not include $z = 0$ or $z = \infty$. If N_1 is zero or positive, there are only negative powers of z in eq. (10.19), and consequently the ROC includes $z = \infty$. If N_2 is zero or negative, there will be only positive powers of z in eq. (10.19) and consequently the ROC includes $z = 0$.

> **Property 4:** If $x[n]$ is a right-sided sequence and if the circle $|z| = r_0$ is in the ROC, then all finite values of z for which $|z| > r_0$ will also be in the ROC.

The justification of this property follows in a manner identical to that of Property 4 in Section 9.2. A right-sided sequence is zero prior to some value of n, say N_0. If the circle $|z| = r_0$ is in the ROC, then $x[n]r_0^{-n}$ is absolutely summable or equivalently the Fourier transform of $x[n]r_0^{-n}$ converges. Since $x[n]$ is right-sided, the term $x[n]$ multiplied by any real exponential sequence which, with increasing n, decays faster than r_0^{-n} will also be absolutely summable. Specifically, as illustrated in Figure 10.6, this more rapid exponential decay will further attenuate sequence values for positive values of n and cannot cause sequence values for negative values of n to become unbounded since $x[n]z^{-n} = 0$ for $n < N_0$. A more formal argument parallels very closely that used for the corresponding Property 4 for Laplace transforms in Section 9.2, and is left as an exercise (Problem 10.3).

For right-sided sequences in general, eq. (10.3) takes the form

$$X(z) = \sum_{n=N_1}^{\infty} x[n]z^{-n} \tag{10.20}$$

where N_1 is finite and may be positive or negative. If N_1 is negative, then the summation in eq. (10.20) includes terms with positive powers of z which become unbounded

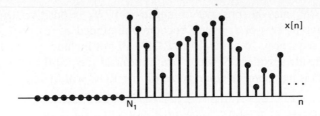

Figure 10.6 With $r_1 > r_0$, $x[n]r_1^{-n}$ decays faster with increasing n than $x[n]r_0^{-n}$. Since $x[n] = 0$, $n < N_1$, this implies that if $x[n]r_0^{-n}$ is absolutely summable, then $x[n]r_1^{-n}$ will be also.

as $|z| \rightarrow \infty$. Consequently, for right-sided sequences in general, the ROC will not include infinity. For the particular class of causal sequences,† however, N_1 will be nonnegative, and consequently the ROC will extend to infinity.

> **Property 5:** If $x[n]$ is a left-sided sequence and if the circle $|z| = r_0$ is in the ROC, then all values of z for which $0 < |z| < r_0$ will also be in the ROC.

Again, this property closely parallels the corresponding property for Laplace transforms and the argument and intuition are similar to those for Property 4. The formal justification is considered in Problem 10.3. In general, for left-sided sequences, from eq. (10.3) the summation for the z-transform will be of the form

$$X(z) = \sum_{n=-\infty}^{N_2} x[n]z^{-n} \qquad (10.21)$$

†A signal is often referred to as causal if it can correspond to the impulse response of a causal system, i.e., is zero for $t < 0$ (continuous time) or $n < 0$ (discrete time).

where N_2 may be positive or negative. If N_2 is positive, then eq. (10.21) includes negative powers of z, which become unbounded as $|z| \rightarrow 0$. Consequently, for left-sided sequences, in general the ROC will not include $z = 0$. For the particular class of left-sided sequences which are anticausal [i.e., $x[n] = 0$ $n \geq 0$ so that N_2 in eq. (10.21) is less than or equal to zero], the ROC will include $z = 0$.

Property 6: If $x[n]$ is two-sided and if the circle $|z| = r_0$ is in the ROC, then the ROC will consist of a ring in the z-plane which includes the circle $|z| = r_0$.

As with Property 6 in Section 9.2, the ROC for a two-sided signal can be examined by expressing $x[n]$ as the sum of a right-sided and left-sided signal. The ROC for the right-sided component is a region bounded on the inside by a circle and extending outward to (and possibly including) infinity. The ROC for the left-sided component is a region bounded on the outside by a circle and extending inward to, and possibly including, the origin. The ROC for the composite signal includes the intersection of these two. As illustrated in Figure 10.7, the overlap (assuming there is one) is a ring in the z-plane.

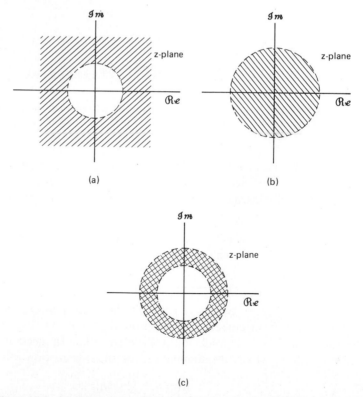

Figure 10.7 (a) ROC for a right-sided sequence; (b) ROC for a left-sided sequence; (c) intersection of the ROCs in (a) and (b) representing the ROC for a two-sided sequence that is the sum of a right-sided and left-sided sequence.

Let us illustrate the foregoing properties with several examples which closely parallel Examples 9.5 and 9.6.

Example 10.4

Consider the signal

$$x[n] = \begin{cases} a^n, & 0 \le n \le N - 1, \quad a > 0 \\ 0, & \text{otherwise} \end{cases}$$

Then

$$\begin{aligned} X(z) &= \sum_{n=0}^{N-1} a^n z^{-n} \\ &= \sum_{n=0}^{N-1} (az^{-1})^n \\ &= \frac{1 - (az^{-1})^N}{1 - az^{-1}} = \frac{1}{z^{N-1}} \frac{z^N - a^N}{z - a} \end{aligned} \tag{10.22}$$

Since $x[n]$ is finite length, from Property 3 the ROC includes the entire z-plane except possibly the origin and/or infinity. In fact, from our discussion of Property 3 we know that, since $x[n]$ is zero for $n < 0$, the ROC will extend to infinity. However, since $x[n]$ is nonzero from some positive values of n, the ROC will not include the origin. This is evident from eq. (10.22) where we see that there is a pole of order $N - 1$ at $z = 0$. The N roots of the numerator polynomial are at

$$z_k = ae^{j(2\pi k/N)}, \qquad k = 0, 1, \ldots, N - 1 \tag{10.23}$$

The root at $k = 0$ cancels the pole at $z = a$. Consequently there are no poles other than at the origin. The remaining zeros are at

$$z_k = ae^{j(2\pi k/N)}, \qquad k = 1, \ldots, N - 1 \tag{10.24}$$

The pole–zero pattern is shown in Figure 10.8.

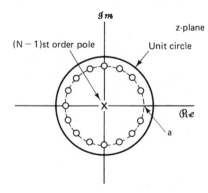

Figure 10.8 Pole–zero pattern for Example 10.4 with $N = 16$. The region of convergence for this example consists of all values of z except $z = 0$.

Example 10.5

Let

$$x[n] = b^{|n|}, \qquad b > 0 \tag{10.25}$$

This sequence is illustrated in Figure 10.9 both for $b < 1$ and $b > 1$. The z-transform for this two-sided sequence can be obtained by expressing it as the sum of a right-sided and left-sided sequence, specifically,

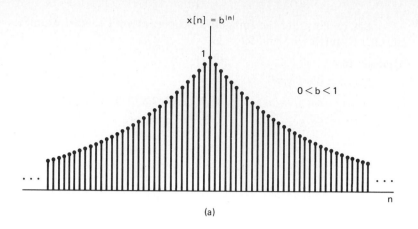

(a)

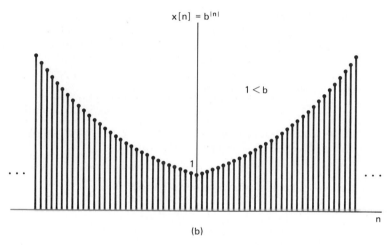

(b)

Figure 10.9 Sequence $x[n] = b^{|n|}$ for $|b| < 1$ and for $|b| > 1$: (a) $b = 0.95$; (b) $b = 1.05$.

$$x[n] = b^n u[n] + b^{-n} u[-n-1] \qquad (10.26)$$

From Example 10.1,

$$b^n u[n] \overset{\mathcal{Z}}{\longleftrightarrow} \frac{1}{1 - bz^{-1}}, \qquad |z| > b \qquad (10.27)$$

and, from Example 10.2,

$$b^{-n} u[-n-1] \overset{\mathcal{Z}}{\longleftrightarrow} \frac{-1}{1 - b^{-1}z^{-1}}, \qquad |z| < \frac{1}{b} \qquad (10.28)$$

In Figure 10.10(a)–(d) we show the pole–zero pattern and ROC for eqs. (10.27) and (10.28), both for $b > 1$ and $b < 1$. For $b > 1$ there is no common ROC and thus the sequence in eq. (10.25) will not have a z-transform, even though the right-sided and left-sided components do individually. For $b < 1$ the ROCs in eqs. (10.27) and (10.28) overlap, and thus the z-transform for the composite sequence is

$$X(z) = \frac{1}{1 - bz^{-1}} - \frac{1}{1 - b^{-1}z^{-1}}, \qquad b < |z| < \frac{1}{b} \qquad (10.29)$$

or, equivalently,

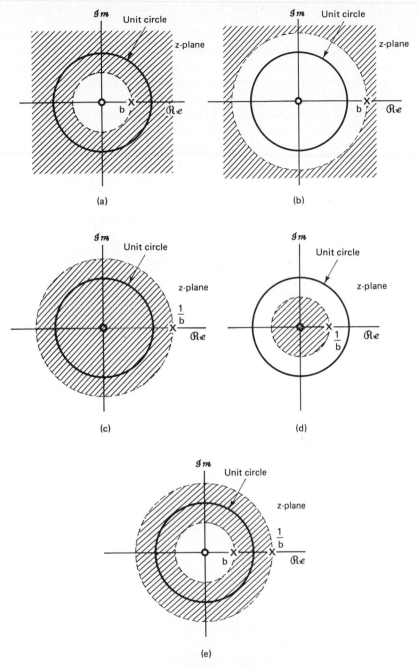

Figure 10.10 Pole–zero plots and ROCs for Example 10.5: (a) eq. (10.27) for $|b| < 1$; (b) eq. (10.27) for $|b| > 1$; (c) eq. (10.28) for $|b| < 1$; (d) eq. (10.28) for $|b| > 1$; (e) pole–zero plot and ROC for eq. (10.30) with $|b| < 1$. For $|b| > 1$, the z-transform does not converge for any value of z.

$$X(z) = \frac{b^2 - 1}{b} \frac{z}{(z - b)(z - b^{-1})}, \qquad b < |z| < \frac{1}{b} \tag{10.30}$$

The corresponding pole–zero pattern and ROC are shown in Figure 10.10(e).

In discussing the Laplace transform in Chapter 9 we remarked that for a rational Laplace transform, the ROC is always bounded by poles or infinity. We observe that in the foregoing examples a similar statement applies to the z-transform. In fact, as with the Laplace transform, it is always true that for any rational z-transform, the ROC will be bounded by poles or will extend to infinity. For a given pole–zero pattern, or equivalently a given rational algebraic expression $X(z)$, there are only a limited number of different ROCs that are consistent with the properties considered above. To illustrate how different ROCs can be associated with the same pole–zero pattern, we consider the following example, which closely parallels Example 9.7.

Example 10.6

Let us consider all of the possible ROCs that can be associated with the function

$$X(z) = \frac{1}{(1 - \frac{1}{3}z^{-1})(1 - 2z^{-1})} \tag{10.31}$$

The associated pole–zero pattern is shown in Figure 10.11(a). Based on our discussion

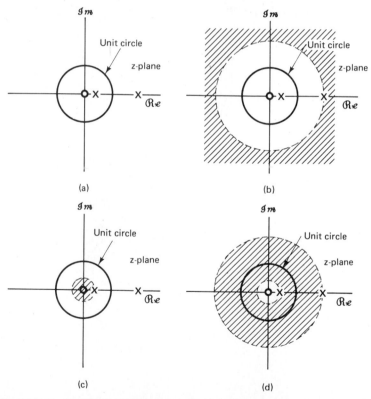

(a)

(b)

(c)

(d)

Figure 10.11 The three possible ROCs that can be associated with the z-transform expression in Example 10.6: (a) pole–zero pattern for $X(z)$; (b) pole–zero pattern and ROC if $x[n]$ is right-sided; (c) pole–zero pattern and ROC if $x[n]$ is left-sided; (d) pole–zero pattern and ROC if $x[n]$ is two-sided. In each case, the zero at the origin is a second-order zero.

in this section, there are three possible ROCs that can be associated with this algebraic z-transform expression. These are indicated in Figure 10.11(b)–(d). Each of the three corresponds to distinctly different sequences. Figure 10.11(b) is associated with a right-sided sequence, Figure 10.11(c) with a left-sided sequence, and Figure 10.11(d) with a two-sided sequence. Since Figure 10.11(d) is the only one for which the ROC includes the unit circle, the associated sequence is the only one of the three for which the Fourier transform converges.

10.3 THE INVERSE z-TRANSFORM

In this section we consider several procedures for obtaining a sequence when its z-transform is known. To begin, let us consider the formal relation expressing a sequence in terms of its z-transform. This expression can be obtained based on the interpretation, developed in Section 10.1, of the z-transform as the Fourier transform of an exponentially weighted sequence. Specifically, as expressed in eq. (10.8),

$$X(re^{j\Omega}) = \mathcal{F}\{x[n]r^{-n}\} \tag{10.32}$$

where $|z| = r$ is in the ROC. Applying the inverse Fourier transform to both sides of eq. (10.32),

$$x[n]r^{-n} = \mathcal{F}^{-1}\{X(re^{j\Omega})\}$$

or

$$x[n] = r^n \mathcal{F}^{-1}[X(re^{j\Omega})] \tag{10.33}$$

Using the inverse Fourier transform expression (5.43), we have

$$x[n] = r^n \frac{1}{2\pi} \int_{2\pi} X(re^{j\Omega})e^{j\Omega n} \, d\Omega$$

or, moving the exponential factor r^n inside the integral and combining it with the term $e^{j\Omega n}$, we have

$$x[n] = \frac{1}{2\pi} \int_{2\pi} X(re^{j\Omega})(re^{j\Omega})^n \, d\Omega \tag{10.34}$$

Let us now change the variable of integration from Ω to z. With $z = re^{j\Omega}$ and r fixed, then $dz = jre^{j\Omega} \, d\Omega = jz \, d\Omega$ or $d\Omega = (1/j)z^{-1} \, dz$. The integration in eq. (10.34) is over a 2π interval in Ω which, in terms of z, corresponds to one traversal around the circle $|z| = r$. Consequently, in terms of an integration in the z-plane, eq. (10.34) can be rewritten as

$$x[n] = \frac{1}{2\pi j} \oint X(z)z^{n-1} \, dz \tag{10.35}$$

where the symbol $\oint$ denotes a counterclockwise closed circular contour centered at the origin and with radius r. The value of r can be chosen as any value for which $X(z)$ converges. Equation (10.35) is the formal expression for the inverse z-transform and is the discrete-time counterpart of eq. (9.37) for the inverse Laplace transform. As in that case, formal evaluation of the inverse transform integral equation (10.35) requires the use of contour integration in the complex plane. There are, however, a number of alternative procedures for obtaining a sequence from its z-transform. As with Laplace transforms, one particularly useful procedure for rational z-transforms consists of expanding the algebraic expression into a partial fraction expansion and recognizing the sequence associated with the individual terms. Let us illustrate the procedure with a specific example.

Example 10.7

Consider the z-transform

$$X(z) = \frac{3 - \frac{5}{6}z^{-1}}{(1 - \frac{1}{4}z^{-1})(1 - \frac{1}{3}z^{-1})}, \qquad |z| > \frac{1}{3} \tag{10.36}$$

There are two poles, one at $z = \frac{1}{3}$ and one at $z = \frac{1}{4}$. Because the ROC lies outside the outermost pole, the inverse transform is a right-sided sequence. As described in the Appendix, $X(z)$ can be expanded in a partial fraction expansion. For this example, the partial fraction expansion, expressed in polynomials in z^{-1}, is

$$X(z) = \frac{1}{1 - \frac{1}{4}z^{-1}} + \frac{2}{1 - \frac{1}{3}z^{-1}} \tag{10.37}$$

Thus, $x[n]$ is the sum of two terms, one with z-transform $1/(1 - \frac{1}{4}z^{-1})$ and the other with z-transform $2/(1 - \frac{1}{3}z^{-1})$. In order to determine the inverse z-transform of each of these individual terms, we must specify the ROC associated with each. Since the ROC for $X(z)$ is outside the outermost pole, the ROC for the individual terms in eq. (10.37) must be outside the pole associated with each term. Thus,

$$x[n] = x_1[n] + x_2[n] \tag{10.38}$$

where

$$x_1[n] \xleftrightarrow{\;\mathcal{Z}\;} \frac{1}{1 - \frac{1}{4}z^{-1}}, \qquad |z| > \frac{1}{4} \tag{10.39a}$$

$$x_2[n] \xleftrightarrow{\;\mathcal{Z}\;} \frac{2}{1 - \frac{1}{3}z^{-1}}, \qquad |z| > \frac{1}{3} \tag{10.39b}$$

From Example 10.1 we can identify by inspection that

$$x_1[n] = (\tfrac{1}{4})^n u[n]$$

and

$$x_2[n] = 2(\tfrac{1}{3})^n u[n]$$

and thus

$$x[n] = (\tfrac{1}{4})^n u[n] + 2(\tfrac{1}{3})^n u[n] \tag{10.40}$$

Example 10.8

Let us consider the same algebraic expression for $X(z)$ as in Example 10.7 but now suppose that the ROC for $X(z)$ is $\frac{1}{4} < |z| < \frac{1}{3}$. Equation (10.37) is still a valid algebraic expansion for the algebraic expression for $X(z)$, but the ROC associated with the individual terms will now change. In particular, since the ROC for $X(z)$ is now outside the pole at $z = \frac{1}{4}$ and inside the pole at $z = \frac{1}{3}$, the z-transform for the individual components in eq. (10.38) becomes

$$x_1[n] \xleftrightarrow{\;\mathcal{Z}\;} \frac{1}{1 - \frac{1}{4}z^{-1}}, \qquad |z| > \frac{1}{4} \tag{10.41a}$$

$$x_2[n] \xleftrightarrow{\;\mathcal{Z}\;} \frac{2}{1 - \frac{1}{3}z^{-1}}, \qquad |z| < \frac{1}{3} \tag{10.41b}$$

Thus, $x_1[n]$ remains as before and from Example 10.2 we can identify $x_2[n]$ as

$$x_2[n] = -2(\tfrac{1}{3})^n u[-n-1] \tag{10.42}$$

so that

$$x[n] = (\tfrac{1}{4})^n u[n] - 2(\tfrac{1}{3})^n u[-n-1] \tag{10.43}$$

The foregoing examples illustrate the basic procedure in utilizing the partial fraction expansion to determine the inverse z-transform. As with the corresponding method for the Laplace transform, the procedure relies on expressing the z-transform as a linear combination of simpler terms. The inverse transform of each term is obtained by inspection. In general, the partial fraction expansion may include terms in addition to the first-order terms considered above. In Section 10.6 we develop a number of other z-transform pairs that can be used in conjunction with the z-transform properties to be developed in Section 10.5 to extend the inverse transform method outlined above.

Another very useful procedure for determining the inverse z-transform relies on a power-series expansion of $X(z)$. This procedure is motivated by the observation that the z-transform definition given in equation (10.3) can be interpreted as a power series involving both positive and negative powers of z. The coefficients in this power series are, in fact, the sequence values $x[n]$. To illustrate how a power-series expansion can be used to obtain the inverse z-transform, let us consider a simple example.

Example 10.9

Consider

$$X(z) = \frac{1}{1 - az^{-1}}, \qquad |z| > |a|$$

This expression can be expanded in a power series by long division. Specifically,

$$
\begin{array}{r}
1 + az^{-1} + a^2 z^{-2} + \cdots \\
1 - az^{-1} \enclose{longdiv}{1 \phantom{-az^{-1}}} \\
\underline{1 - az^{-1}} \\
az^{-1} \\
\underline{az^{-1} - a^2 z^{-2}} \\
a^2 z^{-2}
\end{array}
$$

or

$$\frac{1}{1 - az^{-1}} = 1 + az^{-1} + a^2 z^{-2} + \cdots \tag{10.44}$$

Comparing eq. (10.44) with the z-transform definition equation (10.3), we see by matching terms in powers of z that $x[n] = 0$, $n < 0$; $x[0] = 1$; $x[1] = a$; $x[2] = a^2$; and in general, $x[n] = a^n u[n]$, which is consistent with Example 10.1.

The power-series expansion of $1/(1 - az^{-1})$ in eq. (10.44) is correct for $|az^{-1}| < 1$ or, equivalently, $|z| > |a|$. If, instead, the ROC was specified as $|z| < |a|$ or, equivalently, $|az^{-1}| > 1$, then the power-series expansion for $1/(1 - az^{-1})$ is obtained by carrying out the long division as

$$
\begin{array}{r}
-a^{-1}z - a^{-2}z^2 - \cdots \\
-az^{-1} + 1 \enclose{longdiv}{1 \phantom{-a^{-1}z}} \\
\underline{1 - a^{-1}z} \\
a^{-1}z
\end{array}
$$

or

$$\frac{1}{1 - az^{-1}} = -a^{-1}z - a^{-2}z^2 - \cdots \qquad (10.45)$$

In this case, then, $x[n] = 0$, $n \geq 0$, and $x[-1] = -a^{-1}$, $x[-2] = -a^{-2} \ldots$; that is, $x[n] = -a^n u[-n - 1]$. This, of course, is consistent with Example 10.2.

The power-series expansion method for obtaining the inverse z-transform is particularly useful for nonrational z-transforms, which we illustrate with one additional example.

Example 10.10

Consider the z-transform

$$X(z) = \log (1 + az^{-1}), \qquad |z| > |a| \qquad (10.46)$$

With $|az^{-1}| < 1$, eq. (10.46) can be expanded in a power series using the Taylor's series expansion for $\log (1 + w)$ for $|w| < 1$ given by

$$\log (1 + w) = \sum_{n=1}^{\infty} \frac{(-1)^{n+1} w^n}{n}, \qquad |w| < 1 \qquad (10.47)$$

Applying this to $X(z)$ in eq. (10.46), we have

$$X(z) = \sum_{n=1}^{\infty} \frac{(-1)^{n+1} a^n z^{-n}}{n} \qquad (10.48)$$

from which we can identify $x[n]$ as

$$x[n] = \begin{cases} (-1)^{n+1} \dfrac{a^n}{n}, & n \geq 1 \\ 0, & n \leq 0 \end{cases} \qquad (10.49)$$

or, equivalently,

$$x[n] = \frac{-(-a)^n}{n} u[n - 1]$$

In Problem 10.10 we consider a related example with the region of convergence $|z| < |a|$.

10.4 GEOMETRIC EVALUATION OF THE FOURIER TRANSFORM FROM THE POLE–ZERO PLOT

In Section 10.1 we discussed the relationship between the z-transform and Fourier transform. We noted that the z-transform reduces to the Fourier transform for $|z| = 1$ (i.e., for the contour in the z-plane corresponding to the unit circle), provided that the ROC of the z-transform includes the unit circle so that the Fourier transform in fact converges. In a similar manner we saw in Chapter 9 that for continuous-time signals the Laplace transform reduces to the Fourier transform on the $j\omega$-axis in the s-plane. We also discussed in Section 9.4 the geometric evaluation of the Fourier transform from the pole–zero plot. In the discrete-time case the Fourier transform can again be evaluated geometrically by considering the pole and zero vectors in the z-

plane. However, since in this case the rational function is to be evaluated on the contour $|z| = 1$, we consider the vectors from the poles and zeros to the unit circle rather than to the imaginary axis. To illustrate the procedure, let us consider first- and second-order systems, as discussed in Section 5.12.

10.4.1 First-Order Systems

The impulse response of a first-order causal discrete-time system is of the general form

$$h[n] = a^n u[n] \tag{10.50}$$

and from Example 10.1 its z-transform is

$$H(z) = \frac{1}{1 - az^{-1}}, \qquad |z| > |a| \tag{10.51}$$

For $|a| < 1$, the ROC includes the unit circle and consequently its Fourier transform converges and is equal to $H(z)$ for $z = e^{j\Omega}$. Thus the frequency response for the first-order system is given by

$$H(e^{j\Omega}) = \frac{1}{1 - ae^{-j\Omega}} \tag{10.52}$$

The pole–zero plot, including the vectors from the pole and zero to the unit circle, is shown in Fig. 10.12. The magnitude of the frequency response at frequency

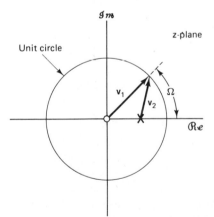

Figure 10.12 Pole and zero vectors for the geometric determination of the frequency response for a first-order system.

Ω is the ratio of the length of the vector $\mathbf{v}_1$ and the vector $\mathbf{v}_2$. The phase of the frequency response is the angle of $\mathbf{v}_1$ with respect to the real axis minus the angle of $\mathbf{v}_2$. First, we note that the vector $\mathbf{v}_1$ from the zero at the origin to the unit circle has a constant length of unity and thus has no effect on the magnitude of $H(e^{j\Omega})$. The phase contributed to $H(e^{j\Omega})$ by the zero is the angle of the zero vector with respect to the real axis, which we see is equal to Ω. For $0 < a < 1$, the pole vector has minimum length at $\Omega = 0$ and monotonically increases in length as Ω increases from zero to π.

Thus, the magnitude of the frequency response will be maximum at $\Omega = 0$ and will decrease monotonically as Ω increases from 0 to π. The angle of the pole vector begins at zero and increases monotonically but not linearly as Ω increases from zero to π. The resulting log magnitude and phase of $H(e^{j\Omega})$ were shown in Figure 5.31.

The magnitude of the parameter a in the discrete-time first-order system plays a role similar to that of the time constant τ for a continuous-time first-order system. As was discussed in Section 5.12 and illustrated in Figures 5.29 and 5.30, as $|a|$ decreases, the impulse response decays more sharply and the step response settles more quickly. With multiple poles, the speed of response associated with each pole is related to its distance from the origin, with those closest to the origin contributing the most rapidly decaying terms in the impulse response. This is further illustrated in the case of second-order systems, which we consider next.

10.4.2 Second-Order Systems

Next, let us consider the class of second-order systems as discussed in Section 5.12, with impulse response and frequency response given in eqs. (5.166) and (5.162):

[eq. (5.166)]
$$h[n] = r^n \frac{\sin (n + 1)\theta}{\sin \theta} u[n]$$

[eq. (5.162)]
$$H(e^{j\Omega}) = \frac{1}{1 - 2r \cos \theta e^{-j\Omega} + r^2 e^{-j2\Omega}}$$

where $0 < r < 1$ and $0 \leq \theta \leq \pi$. Since $H(e^{j\Omega}) = H(z)|_{z=e^{j\Omega}}$ we can infer from eq. (5.162) that the system function, corresponding to the z-transform of the system impulse response, is

$$H(z) = \frac{1}{1 - (2r \cos \theta)z^{-1} + r^2 z^{-2}} \tag{10.53}$$

The poles of $H(z)$ are located at

$$z_1 = re^{j\theta}, \qquad z_2 = re^{-j\theta} \tag{10.54}$$

and there is a double zero at $z = 0$. The pole–zero plot and the pole and zero vectors with $0 < \theta < \pi/2$ are illustrated in Figure 10.13(a). In Figure 10.13(b) we show the magnitude of the frequency response. We note in particular that, as we move along the unit circle, from $\Omega = 0$ toward $\Omega = \pi$, the length of the vector $\mathbf{v}_2$ first decreases and then increases, with a minimum length in the vicinity of the pole location. This is consistent with the fact that the magnitude of the frequency response peaks for Ω near θ when the length of the vector $\mathbf{v}_2$ is small. Based on the behavior of the pole vectors, it is also evident that as r increases toward unity, the minimum length of the pole vectors will decrease, causing the frequency response to peak more sharply with increasing r. From the form of the impulse response [eq. (5.166) and Figure 5.32] or the step response [eq. (5.171) and Figure 5.33], we see, as we did with the first-order system, that as the poles move farther from the unit circle, corresponding to r decreasing, the impulse response decays more rapidly and the step response settles more quickly.

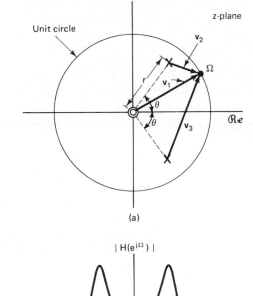

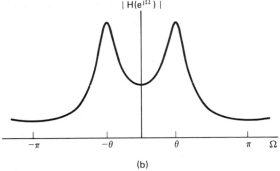

Figure 10.13 (a) Pole vectors $\mathbf{v}_1$ and $\mathbf{v}_2$ used in the geometric calculation of the frequency response for a second-order system; (b) magnitude of the frequency response corresponding to the reciprocal of the product of the lengths of the pole vectors.

10.5 PROPERTIES OF THE z-TRANSFORM

As with the other transforms we have developed, the z-transform possesses a number of properties that make it an extremely valuable tool in the study of discrete-time signals and systems. In this section we summarize many of these properties. The derivation of these results is analogous to the derivations of properties for the other transforms, and thus many of the derivations are left as exercises at the end of the chapter (see Problems 10.14–10.18).

10.5.1 Linearity

If

$$x_1[n] \overset{\mathbf{Z}}{\longleftrightarrow} X_1(z), \qquad \text{ROC} = R_1$$

and

$$x_2[n] \overset{\mathbf{Z}}{\longleftrightarrow} X_2(z), \qquad \text{ROC} = R_2$$

then

$$ax_1[n] + bx_2[n] \xleftrightarrow{\text{z}} aX_1(z) + bX_2(z), \qquad \text{ROC contains } R_1 \cap R_2 \qquad (10.55)$$

As indicated, the ROC of the linear combination is at least the intersection of R_1 and R_2. For sequences with rational z-transforms, if the poles of $aX_1(z) + bX_2(z)$ consist of all the poles of $X_1(z)$ and of $X_2(z)$ (i.e., there is no pole–zero cancellation), then the region of convergence will be exactly equal to the overlap of the individual regions of convergence. If the linear combination is such that some zeros are introduced which cancel poles, then the region of convergence may be larger. A simple example of this occurs when $x_1[n]$ and $x_2[n]$ are both of infinite duration but the linear combination is of finite duration. In this case the region of convergence of the linear combination is the entire z-plane with the possible exception of zero and/or infinity. For example, the sequences $a^n u[n]$ and $a^n u[n-1]$ both have a region of convergence defined by $|z| > |a|$, but the sequence corresponding to the difference $(a^n u[n] - a^n u[n-1]) = \delta[n]$ has a region of convergence which is the entire z-plane.

10.5.2 Time Shifting

If

$$x[n] \xleftrightarrow{\text{z}} X(z), \qquad \text{ROC} = R_x$$

then

$$x[n - n_0] \xleftrightarrow{\text{z}} z^{-n_0} X(z), \qquad \begin{array}{l} \text{ROC} = R_x \text{ except for the} \\ \text{possible addition or dele-} \\ \text{tion of the origin or infinity} \end{array} \qquad (10.56)$$

Because of the multiplication by z^{-n_0}, for $n_0 > 0$ poles will be introduced at $z = 0$ and will be deleted at infinity. Thus, whereas R_x may include the origin, the ROC of $x[n - n_0]$ may not. Similarly, if $n_0 < 0$, zeros will be introduced at $z = 0$ and poles at infinity. Consequently, although R_x may not include $z = 0$ because $X(z)$ contains poles at the origin, the ROC of $x[n - n_0]$ may include the origin. This is consistent with our discussion in Section 10.2 relating to the ROC for finite-duration sequences. In summary, then, the ROC of $x[n - n_0]$ is the same as the ROC of $x[n]$ except for the possible addition or deletion of the origin ($z = 0$) or infinity.

10.5.3 Frequency Shifting

If

$$x[n] \xleftrightarrow{\text{z}} X(z), \qquad \text{ROC} = R_x$$

then

$$e^{j\Omega_0 n} x[n] \xleftrightarrow{\text{z}} X(e^{-j\Omega_0} z), \qquad \text{ROC} = R_x \qquad (10.57a)$$

We recognize the left-hand side as corresponding to modulation by a complex exponential sequence. The right-hand side can be interpreted as a rotation in the z-plane; that is, all pole–zero locations rotate in the z-plane by an angle of Ω_0, as illustrated in Figure 10.14. This can be seen by noting that if $X(z)$ has a factor of the form

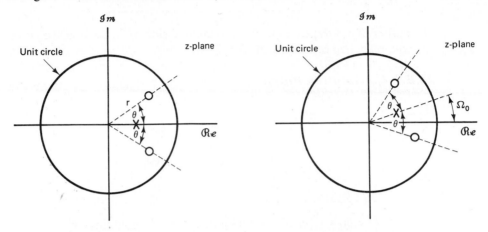

Figure 10.14 Effect on the pole–zero plot of time-domain modulation with a complex exponential sequence $e^{j\Omega_0 n}$.

$(1 - az^{-1})$, then $X(e^{-j\Omega_0}z)$ will have a factor $(1 - ae^{j\Omega_0}z^{-1})$, and thus a pole or zero at $z = a$ in $X(z)$ will become a pole or zero at $z = ae^{j\Omega_0}$ in $X(e^{-j\Omega_0}z)$. The behavior of the z-transform on the unit circle will then also shift by an angle of Ω_0. This is consistent with Property 5.5.4, where modulation with a complex exponential in the time domain was shown to correspond to a shift in frequency of the Fourier transform.

If $x[n]$ is real, then $e^{j\Omega_0 n}x[n]$ will not be real unless Ω_0 is an integer multiple of π. Correspondingly, if the poles and zeros of $X(z)$ are in complex conjugate pairs (for $x[n]$ real), they may no longer have this symmetry after a frequency shift. This is evident in Figure 10.14. More generally, the discrete-time counterpart to Property 9.5.3 is

$$z_0^n x[n] \stackrel{\mathcal{Z}}{\longleftrightarrow} X\left(\frac{z}{z_0}\right), \qquad \text{ROC} = z_0 R_x \tag{10.57b}$$

with $|z_0| = 1$ so that $z_0 = e^{j\Omega_0 n}$. This reduces to eq. (10.57a). With $z_0 = re^{j\Omega_0 n}$, the pole and zero locations are rotated in the z-plane by an angle of Ω_0 *and* scaled in position radially by a factor of r. The corresponding ROC is also scaled, so that if a point z_x is in the ROC for $x[n]$, then $z_0 z_x$ will be in the ROC for $z_0^n x[n]$.

10.5.4 Time Reversal

If

$$x[n] \stackrel{\mathcal{Z}}{\longleftrightarrow} X(z), \qquad \text{ROC} = R_x$$

then

$$x[-n] \xleftrightarrow{\ \mathbf{Z}\ } X\left(\frac{1}{z}\right), \qquad \text{ROC} = \frac{1}{R_x} \qquad\qquad (10.58)$$

The region of convergence for $x[-n]$ is an inversion of R_x; that is, if z_0 is in the ROC for $x[n]$, then $1/z_0$ is in the ROC for $x[-n]$.

10.5.5 Convolution Property

If

$$x_1[n] \xleftrightarrow{\ \mathbf{Z}\ } X_1(z), \qquad \text{ROC} = R_1$$

$$x_2[n] \xleftrightarrow{\ \mathbf{Z}\ } X_2(z), \qquad \text{ROC} = R_2$$

then

$$x_1[n] * x_2[n] \xleftrightarrow{\ \mathbf{Z}\ } X_1(z)X_2(z), \qquad \text{ROC contains } R_1 \cap R_2 \qquad (10.59)$$

Just as with the convolution property for the Laplace transform, the ROC of $X_1(z)X_2(z)$ includes the intersection of R_1 and R_2 and may be larger if pole–zero cancellation occurs in the product.

The convolution property can be derived in a variety of different ways. A formal derivation is developed in Problem 10.21. A development can also be carried out analogous to that used for the convolution property for the continuous-time Fourier transform in Section 4.7, which relied on the interpretation of the Fourier transform as the change in amplitude of a complex exponential through an LTI system. For the z-transform there is another often useful interpretation of the convolution property. Specifically, from the definition in equation (10.3) we recognize the z-transform as a polynomial in z^{-1} where the coefficient of z^{-n} is the sequence value $x[n]$. In essence, the convolution property equation (10.59) states that when two polynomials or power series $X_1(z)$ and $X_2(z)$ are multiplied, the coefficients in the polynomial representing the product are the convolution of the coefficients in the polynomials $X_1(z)$ and $X_2(z)$.

10.5.6 Differentiation in the z-Domain

If

$$x[n] \xleftrightarrow{\ \mathbf{Z}\ } X(z), \qquad \text{ROC} = R_x$$

then

$$nx[n] \xleftrightarrow{\ \mathbf{Z}\ } -z\frac{dX(z)}{dz}, \qquad \text{ROC} = R_x \qquad\qquad (10.60)$$

This property follows in a straightforward manner by differentiating both sides of the

z-transform expression given in eq. (10.3). As an example of the use of this property, let us apply it to determining the inverse z-transform considered in Example 10.10.

Example 10.11

If

$$X(z) = \log(1 + az^{-1}), \qquad |z| > |a| \tag{10.61}$$

then

$$nx[n] \overset{z}{\longleftrightarrow} -z\frac{dX(z)}{dz} = \frac{az^{-1}}{1 + az^{-1}}, \qquad |z| > |a| \tag{10.62}$$

By differentiating we have converted the z-transform to a rational expression. The inverse z-transform of the right-hand side of eq. (10.62) can be obtained by using Example 10.1 together with Property 10.5.2, eq. (10.56). Specifically, from Example 10.1 and the linearity property

$$a(-a)^n u[n] \overset{z}{\longleftrightarrow} \frac{a}{1 + az^{-1}}, \qquad |z| > |a| \tag{10.63}$$

and combining this with Property 10.5.2 yields

$$a(-a)^{n-1} u[n-1] \overset{z}{\longleftrightarrow} \frac{az^{-1}}{1 + az^{-1}}, \qquad |z| > |a|$$

Consequently,

$$x[n] = \frac{-(-a)^n}{n} u[n-1] \tag{10.64}$$

Example 10.12

As another example of the use of the differentiation property, consider determining the inverse z-transform for $X(z)$ given by

$$X(z) = \frac{az^{-1}}{(1 - az^{-1})^2}, \qquad |z| > |a| \tag{10.65}$$

From Example 10.1,

$$a^n u[n] \overset{z}{\longleftrightarrow} \frac{1}{1 - az^{-1}}, \qquad |z| > |a| \tag{10.66}$$

and hence

$$na^n u[n] \overset{z}{\longleftrightarrow} -z\frac{d}{dz}\left(\frac{1}{1 - az^{-1}}\right) = \frac{az^{-1}}{(1 - az^{-1})^2}, \qquad |z| > |a| \tag{10.67}$$

10.5.7 The Initial Value Theorem

If $x[n] = 0$, $n < 0$, then

$$x[0] = \lim_{z \to \infty} X(z) \tag{10.68}$$

This property follows by considering the limit of each term individually in the expression for the z-transform with $x[n]$ zero for $n < 0$. Specifically, with this constraint,

$$X(z) = \sum_{n=0}^{\infty} x[n]z^{-n}$$

As $z \to \infty$, $z^{-n} \to 0$ for $n > 0$, whereas for $n = 0$, $z^{-n} = 1$. Thus, eq. (10.68) follows.

As one consequence of the initial value theorem, we observe that for a causal sequence, if $x[0]$ is finite, then $\lim_{z \to \infty} X(z)$ is finite. Consequently, with $X(z)$

expressed as a ratio of polynomials in z, the order of the numerator polynomial cannot be greater than the order of the denominator polynomial; or, equivalently, the number of zeros of $X(z)$ cannot be greater than the number of poles.

10.5.8 Summary of Properties

In Table 10.1 we summarize these properties of the z-transform and list some additional properties.

TABLE 10.1 PROPERTIES OF THE z-TRANSFORM

Property	Sequence	Transform	ROC
	$x[n]$	$X(z)$	R_x
	$x_1[n]$	$X_1(z)$	R_1
	$x_2[n]$	$X_2(z)$	R_2
10.5.1	$ax_1[n] + bx_2[n]$	$aX_1(z) + bX_2(z)$	At least the intersection of R_1 and R_2
10.5.2	$x[n - n_0]$	$z^{-n_0} X(z)$	R_x except for the possible addition or deletion of the origin
10.5.3	$e^{j\Omega_0 n} x[n]$	$X(e^{-j\Omega_0}z)$	R_x
	$z_0^n x[n]$	$X\left(\dfrac{z}{z_0}\right)$	$z_0 R_x$
	$a^n x[n]$	$X(a^{-1}z)$	Scaled version of R_x (i.e., $\|a\| \cdot R_x =$ the set of points $\{\|a\| z\}$ for z in R_x)
10.5.4	$x[-n]$	$X(z^{-1})$	Inverted R_x (i.e., R_x^{-1} = the set of points z^{-1} where z is in R_x)
	$w[n] = \begin{cases} x[r], & n = rk \\ 0, & n \neq rk \text{ for some } r \end{cases}$	$X(z^k)$	$R_x^{1/k}$ (i.e., the set of points $z^{1/k}$ where z is in R_x)
10.5.5	$x_1[n] * x_2[n]$	$X_1(z)X_2(z)$	At least the intersection of R_1 and R_2
10.5.6	$nx[n]$	$-z\dfrac{dX(z)}{dz}$	R_x except for the possible addition or deletion of the origin
	$\displaystyle\sum_{k=-\infty}^{n} x[k]$	$\dfrac{1}{1 - z^{-1}} X(z)$	At least the intersection of R_x and $\|z\| > 1$

10.6 SOME COMMON z-TRANSFORM PAIRS

As with the inverse Laplace transform, the inverse z-transform can often be easily evaluated by expressing $X(z)$ as a linear combination of simpler terms, the inverse transforms of which are recognizable. In Table 10.2 we have listed a number of useful z-transform pairs. Each of these can be developed from previous examples in combination with the properties of the z-transform listed in Table 10.1. For example, transform pairs 2 and 5 follow directly from Example 10.1, and 7 follows from Example 10.12. These, together with Properties 10.5.4 and 10.5.2, then lead to transform pairs 3, 6, and 8. Transform pairs 9 and 10 can be developed using 2 together with Properties 10.5.1 and 10.5.3.

TABLE 10.2 SOME COMMON z-TRANSFORM PAIRS

Transform pair Signal	Transform	ROC
1. $\delta[n]$	1	All z
2. $u[n]$	$\dfrac{1}{1 - z^{-1}}$	$|z| > 1$
3. $u[-n - 1]$	$\dfrac{1}{1 - z^{-1}}$	$|z| < 1$
4. $\delta[n - m]$	z^{-m}	All z except 0 (if $m > 0$) or ∞ (if $m < 0$)
5. $\alpha^n u[n]$	$\dfrac{1}{1 - \alpha z^{-1}}$	$|z| > |\alpha|$
6. $-\alpha^n u[-n - 1]$	$\dfrac{1}{1 - \alpha z^{-1}}$	$|z| < |\alpha|$
7. $n\alpha^n u[n]$	$\dfrac{\alpha z^{-1}}{(1 - \alpha z^{-1})^2}$	$|z| > |\alpha|$
8. $-n\alpha^n u[-n - 1]$	$\dfrac{\alpha z^{-1}}{(1 - \alpha z^{-1})^2}$	$|z| < |\alpha|$
9. $[\cos \Omega_0 n]u[n]$	$\dfrac{1 - [\cos \Omega_0]z^{-1}}{1 - [2 \cos \Omega_0]z^{-1} + z^{-2}}$	$|z| > 1$
10. $[\sin \Omega_0 n]u[n]$	$\dfrac{[\sin \Omega_0]z^{-1}}{1 - [2 \cos \Omega_0]z^{-1} + z^{-2}}$	$|z| > 1$
11. $[r^n \cos \Omega_0 n]u[n]$	$\dfrac{1 - [r \cos \Omega_0]z^{-1}}{1 - [2 r \cos \Omega_0]z^{-1} + r^2 z^{-2}}$	$|z| > r$
12. $[r^n \sin \Omega_0 n]u[n]$	$\dfrac{[r \sin \Omega_0]z^{-1}}{1 - [2r \cos \Omega_0]z^{-1} + r^2 z^{-2}}$	$|z| > r$

10.7 ANALYSIS AND CHARACTERIZATION OF LTI SYSTEMS USING z-TRANSFORMS

The z-transform plays a particularly important role in the analysis and representation of discrete-time LTI systems. From the convolution property 10.5.5,

$$Y(z) = H(z)X(z) \tag{10.69}$$

where $X(z)$, $Y(z)$, and $H(z)$ are the z-transforms of the system input, output, and impulse response, respectively. $H(z)$ is referred to as the *system function* or *transfer function* of the system. For z evaluated on the unit circle (i.e., for $z = e^{j\Omega}$), $H(z)$

reduces to the frequency response of the system provided that the unit circle is in the ROC for $H(z)$.

Characteristics of the system such as stability and causality can be associated with constraints on the pole–zero pattern and the ROC of the system function. For example, if the system is causal, then from Properties 2 and 4 in Section 10.2 the ROC for $H(z)$ will be outside the outermost pole. If the system is stable, then the impulse response is absolutely summable, in which case the Fourier transform of $h[n]$ will converge and consequently the ROC of $H(z)$ must include the unit circle. For a system that is *both* stable and causal, the ROC must include the unit circle and be outside the outermost pole, from which it follows that for a causal, stable system, all the poles must be inside the unit circle. As an example, let us return to the class of second-order systems that we considered in Section 10.4.2.

Example 10.13

The system function for a second-order system with complex poles was given in eq. (10.53), specifically,

[eq. (10.53)] $$H(z) = \frac{1}{1 - (2r \cos \theta) z^{-1} + r^2 z^{-2}} \qquad (10.70)$$

with poles located at $z_1 = re^{j\theta}$ and $z_2 = re^{-j\theta}$. Assuming causality, the ROC is outside the outermost pole (i.e., $|z| > |r|$). The pole–zero plot and ROC are shown in Figure 10.15 for $r < 1$ and for $r > 1$. For $|r| < 1$, the poles are inside the unit circle, the ROC includes the unit circle, and therefore the system is stable. For $|r| > 1$, the poles are outside the unit circle, the ROC does not include the unit circle, and the system is unstable.

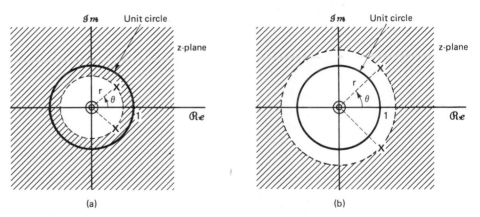

Figure 10.15 Pole–zero plot for second-order system with complex poles: (a) $r < 1$; (b) $r > 1$.

10.7.1 Systems Characterized by Linear Constant-Coefficient Difference Equations

For systems characterized by linear constant-coefficient difference equations the properties of the z-transform provide a particularly convenient procedure for obtaining the system function, frequency response, or time-domain response of the system. Let us first illustrate this with an example.

Example 10.14

Consider an LTI system for which the input $x[n]$ and output $y[n]$ satisfy the linear constant-coefficient difference equation

$$y[n] - \tfrac{1}{2}y[n-1] = x[n] + \tfrac{1}{3}x[n-1] \tag{10.71}$$

Applying the z-transform to both sides of eq. (10.71) and using the linearity property 10.5.1 and the time-shifting property 10.5.2, we obtain

$$Y(z) - \tfrac{1}{2}z^{-1}Y(z) = X(z) + \tfrac{1}{3}z^{-1}X(z)$$

or

$$Y(z) = X(z)\left[\frac{1 + \tfrac{1}{3}z^{-1}}{1 - \tfrac{1}{2}z^{-1}}\right] \tag{10.72}$$

From eq. (10.69), then,

$$H(z) = \frac{Y(z)}{X(z)} = \frac{1 + \tfrac{1}{3}z^{-1}}{1 - \tfrac{1}{2}z^{-1}} \tag{10.73}$$

This provides the algebraic expression for $H(z)$ but not the region of convergence. In fact, there are two distinct impulse responses that are consistent with the difference equation (10.71), one right-sided and the other left-sided. Correspondingly, there are two different choices for the ROC associated with the algebraic expression (10.73). One, $|z| > \tfrac{1}{2}$, is associated with the assumption that $h[n]$ is right-sided and the other, $|z| < \tfrac{1}{2}$, is associated with the assumption that $h[n]$ is left-sided.

For the more general case of an Nth-order difference equation, we proceed in a manner similar to Example 10.14, applying the z-transform to both sides of the equation and using the linearity and time-shifting properties. Thus, consider an LTI system for which the input and output satisfy a linear constant-coefficient difference equation of the form

$$\sum_{k=0}^{N} a_k y[n-k] = \sum_{k=0}^{M} b_k x[n-k] \tag{10.74}$$

Then

$$\sum_{k=0}^{N} a_k z^{-k} Y(z) = \sum_{k=0}^{M} b_k z^{-k} X(z)$$

or

$$Y(z) \sum_{k=0}^{N} a_k z^{-k} = X(z) \sum_{k=0}^{M} b_k z^{-k}$$

so that

$$H(z) = \frac{Y(z)}{X(z)} = \frac{\displaystyle\sum_{k=0}^{M} b_k z^{-k}}{\displaystyle\sum_{k=0}^{N} a_k z^{-k}} \tag{10.75}$$

We note in particular that the system function for a system satisfying a linear constant-coefficient difference equation is always rational. Consistent with our previous example and with the related discussion for the Laplace transform, the difference equation by itself does not provide information about the ROC to associate with the algebraic expression $H(z)$. An additional constraint such as causality or stability of the system, however, serves to specify the region of convergence. For example, if we know in addition that the system is causal, the ROC will be outside the outermost pole. If it is stable, the ROC must include the unit circle.

10.7.2 System Function for Interconnection of LTI Systems

Just as with the Laplace transform for continuous time, the z-transform for discrete time allows us to replace time-domain operations such as convolution and time shifting with algebraic operations. This was exploited in the previous subsection, where we were able to replace the difference-equation description of an LTI system with an algebraic description. The use of the z-transform to convert system descriptions to algebraic equations is also useful in analyzing interconnections of LTI systems, such as series, parallel, and feedback interconnections. For example, consider the feedback interconnection of two systems as shown in Figure 10.16. It is relatively difficult to

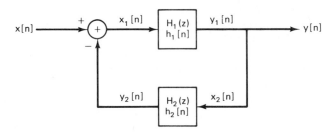

Figure 10.16   Feedback interconnection of two systems.

determine in the time domain the difference equation or impulse response for the overall system. However, with the systems and sequences expressed in terms of their z-transforms, the analysis involves only algebraic equations. The specific equations for the interconnection of Figure 10.16 exactly parallel eqs. (9.101)–(9.104), with the final result that the overall system function for the feedback system of Figure 10.16 is

$$\frac{Y(z)}{X(z)} = H(z) = \frac{H_1(z)}{1 + H_1(z)H_2(z)} \tag{10.76}$$

10.8 TRANSFORMATIONS BETWEEN CONTINUOUS-TIME AND DISCRETE-TIME SYSTEMS

In a variety of situations, it is of interest to transform a continuous-time system to a discrete-time system. For example, in Chapter 8 we considered the use of discrete-time systems for the processing of continuous-time signals. In such cases, the continuous-time signal is represented through periodic sampling by a discrete-time signal. After processing with an appropriate discrete-time system, the resulting sequence is converted back to a continuous-time signal. Thus, there is the issue of converting the desired continuous-time system to an appropriate discrete-time system. Another motivation for considering transformations from continuous-time to discrete-time systems is the fact that before the advent of high-density, high-speed digital technology essentially all filtering and signal processing systems were continuous-time systems. Consequently, there is a long history associated with the development of techniques for designing continuous-time systems to meet a given set of specifications in the time domain or frequency domain. Through the use of transformations such as those to be discussed in this section, system designs available for continuous-time systems can be transformed to designs for discrete-time systems with related specifications.

In this section we consider three transformations for mapping continuous-time systems with rational system functions to discrete-time systems with rational system functions. The first is a technique referred to as impulse invariance. In this technique a continuous-time system is transformed to a discrete-time system such that the impulse response of the discrete-time system corresponds to equally spaced samples of the impulse response of the continuous-time system. This is precisely the procedure that we considered in Chapter 8 when discussing discrete-time processing of continuous-time signals. In particular, since the discrete-time impulse response is obtained by sampling the continuous-time impulse response, it is a transformation whose utility is limited to continuous-time systems with a frequency response that is exactly or at least approximately bandlimited.

The second procedure we discuss corresponds to obtaining a difference equation for the discrete-time system by replacing derivatives in the differential equation representing the continuous-time system by backward differences. It is a procedure that is of somewhat limited utility, but its shortcomings help to illuminate the important considerations in transformations from continuous-time to discrete-time systems. The third technique we discuss is the use of the bilinear transformation in mapping from the Laplace transform to the z-transform. As with impulse invariance, we will see that this is an extremely useful transformation for certain classes of systems.

10.8.1 Impulse Invariance

If we are given a continuous-time impulse response $h_c(t)$, we can consider designing a discrete-time system with impulse response $h_d[n]$ consisting of equally spaced samples of $h_c(t)$ so that

$$h_d[n] = h_c(nT) \tag{10.77}$$

where T is a (positive) number to be chosen as part of the design procedure. The transformation for $h_c(t)$ to $h_d[n]$ can be viewed as shown in Figure 10.17. From the

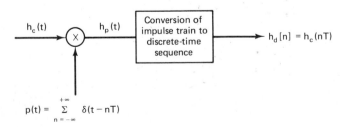

Figure 10.17 Transformation of a continuous-time impulse response to a discrete-time impulse response using impulse invariance.

discussion in Section 8.1.1 and specifically eq. (8.5), it follows that†

$$H_p(j\omega) = \frac{1}{T} \sum_{k=-\infty}^{+\infty} H_c\left(j\omega - j\frac{2\pi k}{T}\right) \tag{10.78}$$

Also, from eq. (8.20),

$$H_d(e^{j\Omega}) = H_p\left(j\frac{\Omega}{T}\right) = \frac{1}{T} \sum_{k=-\infty}^{+\infty} H_c\left(j\left(\frac{\Omega}{T} - \frac{2\pi k}{T}\right)\right) \tag{10.79}$$

Thus, for a discrete-time system obtained from a continuous-time system through impulse invariance, the discrete-time frequency response, $H_d(e^{j\Omega})$, is related to $H_c(j\omega)$ through a periodic replication of the continuous-time frequency response and linear scaling of the frequency axis. If $H_c(j\omega)$ is bandlimited and T is chosen so that aliasing is avoided, the discrete-time frequency response is then identical to that of the continuous-time frequency response except for a linear scaling in amplitude and frequency.

Let us explore further the properties of impulse invariance. From Figure 10.17, the sampled time signal $h_p(t)$ can be expressed as

$$\begin{aligned} h_p(t) &= h_c(t) \sum_{n=-\infty}^{+\infty} \delta(t - nT) \\ &= \sum_{n=-\infty}^{+\infty} h_c(nT)\delta(t - nT) \end{aligned} \tag{10.80}$$

Applying the Laplace transform to eq. (10.80),

$$\begin{aligned} H_p(s) &= \mathcal{L}\left\{ \sum_{n=-\infty}^{+\infty} h_c(nT)\,\delta(t - nT) \right\} \\ &= \sum_{n=-\infty}^{+\infty} h_c(nT)\mathcal{L}\{\delta(t - nT)\} \end{aligned}$$

or since

$$\mathcal{L}\{\delta(t - nT)\} = e^{-snT},$$

$$H_p(s) = \sum_{n=-\infty}^{+\infty} h_c(nT)e^{-snT} \tag{10.81}$$

On the other hand, $H_d(z)$, the z-transform of $h_d[n]$, is, by definition,

$$H_d(z) = \sum_{n=-\infty}^{+\infty} h_d[n]z^{-n}$$

or, using eq. (10.77),

$$H_d(z) = \sum_{n=-\infty}^{+\infty} h_c(nT)z^{-n} \tag{10.82}$$

Comparing eqs. (10.82) and (10.81), it follows that

$$H_d(z)|_{z=e^{sT}} = H_p(s) \tag{10.83a}$$

Furthermore, with $j\omega$ in eq. (10.78) replaced by the more general Laplace transform variable s, we have

$$H_p(s) = \frac{1}{T} \sum_{k=-\infty}^{+\infty} H_c\left(s - j\frac{2\pi k}{T}\right) \tag{10.83b}$$

or, equivalently,

$$H_d(z)|_{z=e^{sT}} = \frac{1}{T} \sum_{k=-\infty}^{+\infty} H_c\left(s - j\frac{2\pi k}{T}\right) \tag{10.83c}$$

---

†In keeping with our change of notation introduced at the beginning of Chapter 9, the Fourier transform is now expressed as a function of $j\omega$ rather than ω, as was the case in Chapter 8.

From eq. (10.83a) we note that impulse invariance corresponds to a transformation between $H_p(s)$ and $H_d(z)$ represented by the mapping $z = e^{sT}$ between the s-plane and z-plane. This mapping is illustrated in Figure 10.18. In particular, the shaded strip in Figure 10.18(a) maps into the entire z-plane. From eqs. (10.83), $H_d(z)$ can be viewed as being found by first superimposing or aliasing into the shaded strip the behavior of $H_c(s)$ in parallel strips of width $2\pi/T$ (eq. 10.83b) and then mapping the shaded strip to the entire z-plane according to the mapping $z = e^{sT}$ (eq. 10.83a). For this mapping, if $\mathcal{R}e\{s\} < 0$, then $|z| < 1$, so that points in the left half of the s-plane map to the interior of the unit circle. Likewise, points in the right half of the s-plane map to the exterior of the unit circle and points on the $j\omega$-axis map onto the unit circle.

When the Laplace transform of $h_c(t)$ is rational, the relationship between $H_d(z)$ and $H_c(s)$ can be expressed in a more convenient form than eq. (10.83). Let us consider the transfer function of a continuous-time system expressed in terms of a partial

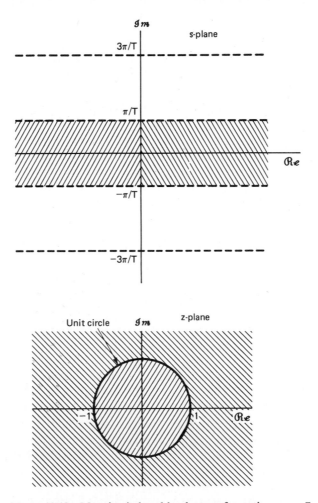

Figure 10.18 Mapping induced by the transformation $z = e^{sT}$.

fraction expansion. For convenience we will assume only first-order poles, although as illustrated in Problem 10.35, the argument is easily extended to multiple-order poles. With $H_c(s)$ expressed in the form

$$H_c(s) = \sum_{k=1}^{N} \frac{A_k}{s - s_k} \tag{10.84}$$

the corresponding impulse response is

$$h_c(t) = \sum_{k=1}^{N} A_k e^{s_k t} u(t) \tag{10.85}$$

and the unit sample response of the corresponding discrete-time system is

$$h_d[n] = h_c(nT) = \sum_{k=1}^{N} A_k e^{s_k nT} u[n]$$

$$= \sum_{k=1}^{N} A_k (e^{s_k T})^n u[n] \tag{10.86}$$

The transfer function of the discrete-time filter is then given by

$$H_d(z) = \sum_{k=1}^{N} \frac{A_k}{1 - e^{s_k T} z^{-1}} \tag{10.87}$$

Thus, $H_d(z)$ is rational, and in particular the residues A_k are preserved in the partial fraction expansion and a pole in $H_c(s)$ at $s = s_k$ is mapped to a pole in $H_d(z)$ at $z = e^{s_k T}$. One consequence of this mapping of the poles is that a stable continuous-time filter will always result in a stable discrete-time filter since, if $\mathcal{R}e\{s_k\} < 0$, then $|e^{s_k T}| < 1$. However, the zeros of $H_d(z)$ are a function of the poles and residues in eq. (10.87), and their locations do not in general correspond to a direct mapping of the zeros of $H_c(s)$ through the transformation $z = e^{sT}$.

In summary, the use of impulse invariance corresponds to converting the continuous-time impulse response to a discrete-time impulse response, through sampling. To avoid aliasing, the procedure is restricted to transforming bandlimited frequency responses. Except for aliasing, the discrete-time frequency response is a replication of the continuous-time frequency response linearly scaled in frequency. For rational transfer functions, the impulse invariance procedure can also be interpreted through eq. (10.87) as a relationship between the partial fraction expansion of $H_c(s)$ and $H_d(z)$, whereby the residues are retained and the poles of $H_c(s)$ are mapped to poles of $H_d(z)$ through the transformation $z = e^{sT}$.

10.8.2 Backward-Difference Approximation to Differential Equations

A second approach to transforming a continuous-time filter into a discrete-time one is to approximate derivatives in a differential equation representation of the continuous-time filter by finite differences. This is a common procedure in digital simulations of analog systems and can be motivated by the intuitive notion that the derivative of a continuous-time function can be approximated by the difference between consecutive samples of the signal to be differentiated. To illustrate the procedure, consider the first-order differential equation

$$\frac{dy(t)}{dt} + ay(t) = x(t) \tag{10.88}$$

The backward-difference method consists of replacing $x(t)$ by $x[n]$, $y(t)$ by $y[n]$, and the first derivative $dy(t)/dt$ by the first backward difference

$$\Delta^{(1)}y[n] = \frac{y[n] - y[n-1]}{T} \tag{10.89}$$

This yields the difference equation

$$\frac{y[n] - y[n-1]}{T} + ay[n] = x[n] \tag{10.90}$$

If T is sufficiently small, we would expect the solution $y[n]$ to yield a good approximation to the samples of $y(t)$.

To interpret the procedure in terms of a mapping of the continuous-time transfer function $H_c(s)$ to a discrete-time transfer function $H_d(z)$, we apply the Laplace transform to eq. (10.88) to obtain

$$sY(s) + aY(s) = X(s) \tag{10.91}$$

so that

$$H_c(s) = \frac{1}{s + a} \tag{10.92}$$

Alternatively, with the z-transform applied to eq. (10.90),

$$\left(\frac{1 - z^{-1}}{T}\right)Y(z) + aY(z) = X(z) \tag{10.93}$$

or

$$H_d(z) = \frac{1}{\left(\dfrac{1 - z^{-1}}{T}\right) + a} \tag{10.94}$$

Comparing $H_c(s)$ and $H_d(z)$, we see that

$$H_d(z) = H_c(s)\big|_{s=(1-z^{-1})/T} \tag{10.95}$$

Higher derivatives are approximated similarly. For example,

$$\frac{d^2 y(t)}{dt} = \frac{d}{dt}\left[\frac{dy(t)}{dt}\right] \tag{10.96}$$

and we approximate this second derivative as a backward difference of our approximation to the first derivative of $y(t)$. That is, we replace $d^2 y(t)/dt^2$ with

$$\Delta^{(2)}y[n] = \Delta^{(1)}(\Delta^{(1)}y[n])$$

$$= \frac{\dfrac{y[n] - y[n-1]}{T} - \dfrac{y[n-1] - y[n-2]}{T}}{T} \tag{10.97}$$

$$= \frac{y[n] - 2y[n-1] + y[n-2]}{T^2}$$

We see, then, that

$$\mathcal{L}\left\{\frac{d^2 y(t)}{dt^2}\right\} = s^2 Y(s) \tag{10.98}$$

$$\mathcal{Z}\{\Delta^{(2)}y[n]\} = \left(\frac{1 - z^{-1}}{T}\right)^2 Y(z) \tag{10.99}$$

Thus, the relationship (10.95), or equivalently the transformation

$$s = \frac{1 - z^{-1}}{T} \tag{10.100}$$

also holds for higher-order differential equations.

Let us now examine the transformation (10.100). We note first that if $H_c(s)$ is a rational function of s, then $H_d(z)$ will be a rational function of z. This is not surprising since by the very nature of the transformation, a linear constant-coefficient differential equation, which corresponds to a rational continuous-time system function, is transformed to a linear constant-coefficient difference equation, which we have shown will always correspond to a rational discrete-time system function. Next, let us invert the relationship in eq. (10.100) to express z in terms of s as

$$z = \frac{1}{1 - sT} \tag{10.101}$$

With $sT = \sigma + j\omega$,

$$|z| = \frac{1}{\sqrt{(1 - \sigma)^2 + \omega^2}} \tag{10.102}$$

and we see that if $\sigma < 0$, then $|z| < 1$, so that the left-half s-plane is mapped inside the unit circle. Thus, as with impulse invariance, poles in the left half of the s-plane will map to poles inside the unit circle so that a stable continuous-time system will always transform to a stable discrete-time filter.

The frequency response of the discrete-time system is examined by considering $H_d(z)$ on the unit circle, while for the continuous-time system it is examined by considering $H_c(s)$ on the $j\omega$-axis. Thus, to preserve the characteristics of the frequency response in mapping from $H_c(s)$ to $H_d(z)$, we would like the $j\omega$-axis in the s-plane to map to the unit circle in the z-plane, as was the case with impulse invariance.

For the transformation of eq. (10.101), let us consider how the $j\omega$-axis in the s-plane is mapped to the z-plane. For $s = j\omega$, from eq. (10.101)

$$z = \frac{1}{1 - j\omega T} = \frac{1}{2}\left[1 + \frac{1 + j\omega T}{1 - j\omega T}\right] \tag{10.103}$$

The complex number $1 + j\omega T$ can be expressed in polar form as

$$1 + j\omega T = [1 + (\omega T)^2]^{1/2} e^{j \tan^{-1}\omega T}$$

and similarly, the complex number $1 - j\omega T$ can be expressed in polar form as

$$1 - j\omega T = [1 + (\omega T)^2]^{1/2} e^{-j \tan^{-1}\omega T}$$

Thus, the ratio $(1 + j\omega T)/(1 - j\omega T)$ has unity magnitude and angle equal to $2 \tan^{-1} \omega T$, that is,

$$\frac{1 + j\omega T}{1 - j\omega T} = e^{j2 \tan^{-1}\omega T} \tag{10.104}$$

so that

$$z = \tfrac{1}{2}[1 + e^{j2 \tan^{-1}\omega T}] \tag{10.105}$$

From eq. (10.105), as ω varies from $\omega = -\infty$ to $\omega = +\infty$, the locus in the z-plane is a circle with center at $z = \frac{1}{2}$ and radius 1/2, as illustrated in Figure 10.19. It is only for very small ωT (i.e., only when the frequency, ω, of interest is much smaller

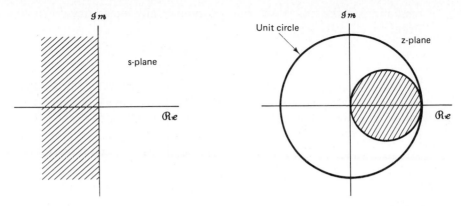

Figure 10.19 Transformation $z = 1/(1 - sT)$ associated with the substitution of backward differences for derivatives.

than the sampling rate $1/T$) that $|z|$ on this circle is near the unit circle. Thus, if $H_c(j\omega)$ is essentially bandlimited, we can choose T sufficiently small so that $H_d(e^{j\Omega})$ will approximate $H_c(j\omega)$ over the frequency band of interest. Said another way, if we wish to simulate a system specified by a differential equation by using the backward-difference approximation, the differential step size T must be chosen so that $1/T$ is much smaller than the bandwidth of the continuous-time system, corresponding to considerable oversampling of the signal.

In summary, although the backward-difference transformation avoids the aliasing problem encountered using impulse invariance, we have seen that it has problems of its own, specifically that the $j\omega$-axis does not even approximately map to the unit circle except in a very small region around $|z| = 1$.

In the preceding discussion, we developed in detail the transformation of a differential equation to a difference equation by replacing derivatives by backward differences. There are a variety of other ways in which derivatives can be approximated by differences, and each of these will correspond to different transformations between $H_c(s)$ and $H_d(z)$. One such alternative is the use of the forward difference rather than the backward difference, as explored in Problem 10.38.

10.8.3 The Bilinear Transformation

In this section we consider another mapping between the s-plane and the z-plane, called the *bilinear transformation* and specified by

$$s = \frac{2}{T} \frac{1 - z^{-1}}{1 + z^{-1}} \tag{10.106}$$

This transformation is invertible with the inverse mapping given by

$$z = \frac{1 + (T/2)s}{1 - (T/2)s} \tag{10.107}$$

In this case, then

$$H_d(z) = H_c(s)|_{s = (2/T)[(1-z^{-1})/(1+z^{-1})]} \tag{10.108}$$

As shown in Problem 10.40, this transformation also arises from a particular approximate method, the trapezoidal rule, for numerically integrating differential equations. From eq. (10.107), with $s = \sigma + j\omega$,

$$z = \frac{[1 + (T/2)(\sigma)] + j\omega T/2}{[1 - (T/2)(\sigma)] - j\omega T/2} \tag{10.109}$$

With $\sigma < 0$, $|z|$ is less than unity, so that the left half of the s-plane maps to the inside of the unit circle. Also, with $\sigma > 0$, $|z|$ is greater than unity, so that the right half of the s-plane maps to the outside of the unit circle. For $z = e^{j\Omega}$, from eq. (10.106),

$$s = \frac{2}{T} \frac{1 - e^{j\Omega}}{1 + e^{j\Omega}} = \frac{2}{T} \frac{j \sin (\Omega/2)}{\cos (\Omega/2)} = j\frac{2}{T} \tan\left(\frac{\Omega}{2}\right) \tag{10.110}$$

Thus, the unit circle maps onto the $j\omega$-axis with the relationship between the continuous-time and discrete-time frequency variables given by

$$\omega = \frac{2}{T} \tan \frac{\Omega}{2} \tag{10.111a}$$

$$\Omega = 2 \tan^{-1} \frac{\omega T}{2} \tag{10.111b}$$

The mapping characteristics of the bilinear transformation are shown in Figure 10.20. Since the entire $j\omega$-axis (ω from $-\infty$ to $+\infty$) maps into one circumference of the unit circle (Ω from $-\pi$ through 0 to $+\pi$) the use of the bilinear transformation also avoids aliasing.

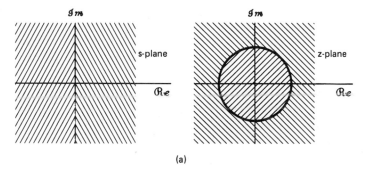

(a)

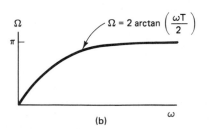

(b)

Figure 10.20 Mapping associated with the bilinear transformation: (a) the left half of the s-plane maps to the inside of the unit circle in the z-plane and the right half of the s-plane maps to the outside of the unit circle; (b) nonlinear mapping between the continuous-time frequency variable Ω and the discrete-time frequency variable ω.

With the use of the bilinear transformation, stable continuous-time systems will always map to stable discrete-time systems and the frequency response of the continuous-time system will be exactly replicated in the frequency response of the discrete-time system but with a nonlinear warping of the frequency axis as specified by eq. (10.111). For example, if the bilinear transformation were applied to a continuous-time equiripple filter with the frequency characteristic of Figure 6.33(c), the resulting discrete-time filter would also have an equiripple characteristic in both the passband and the stopband. This is a consequence of the fact that it is only the independent variable, i.e. the frequency axis, that is modified by the bilinear transformation. However, the frequency locations of the ripples and the passband and stopband edges are, of course, affected. If we are interested in using an analog design method and the bilinear transformation to design a digital filter with given specifications, we must "prewarp" these discrete-time specifications into specifications on $H_c(j\omega)$ so that the frequency warping caused by the bilinear transformation is taken into account. For example, as illustrated in Problem 10.41, in order to design a bandpass digital filter with given bandwidth and center frequency in the passband, we can design a continuous-time bandpass filter and use the bilinear transformation. However, because of the nonlinear frequency warping of eq. (10.111) the bandwidth needed in the continuous-time filter specification depends on both the bandwidth *and* the center frequency of the desired discrete-time system.

10.9 THE UNILATERAL z-TRANSFORM

The z-transform as it has been considered thus far in this chapter is in a form often referred to as the bilateral z-transform. As was the case with the Laplace transform, there is an alternative form, referred to as the unilateral z-transform. The unilateral z-transform is particularly useful in analyzing causal systems specified by linear constant-coefficient difference equations with initial conditions (i.e., which are not initially at rest). Our discussion of the unilateral z-transform closely parallels the discussion of the unilateral Laplace transform in Section 9.8.

The unilateral z-transform $\mathfrak{X}(z)$ of a sequence $x[n]$ is defined as

$$\mathfrak{X}(z) = \sum_{n=0}^{\infty} x[n]z^{-n} \tag{10.112}$$

and differs from the bilateral transform in that the summation is carried out only over nonnegative values of n whether or not $x[n]$ is zero for $n < 0$. Thus, in effect, the unilateral z-transform of $x[n]$ can be thought of as the bilateral transform of $x[n]u[n]$ (i.e., $x[n]$ multiplied by a unit step). In particular, then, for any sequence that is zero for $n < 0$, the unilateral and bilateral z-transforms will be identical. It should also be noted that since $x[n]u[n]$ is always a right-sided sequence, the region of convergence of $\mathfrak{X}(z)$ is always the exterior of a circle.

Example 10.15

Consider the signal

$$x[n] = a^n u[n] \tag{10.113}$$

Since $x[n] = 0$, $n < 0$, the unilateral and bilateral transforms are equal for this example and thus, in particular,

$$\mathcal{X}(z) = \frac{1}{1 - az^{-1}}, \qquad |z| > |a| \tag{10.114}$$

Example 10.16

Let

$$x[n] = a^{n+1}u[n + 1] \tag{10.115}$$

In this case the unilateral and bilateral transforms are *not* equal. In particular, the bilateral transform is obtained from Example 10.1 and the time-shifting property 10.5.2. Specifically,

$$X(z) = \frac{z}{1 - az^{-1}}, \qquad |z| > |a| \tag{10.116}$$

In contrast, the unilateral transform is

$$\mathcal{X}(z) = \sum_{n=0}^{\infty} x[n]z^{-n}$$

$$= \sum_{n=0}^{\infty} a^{n+1}z^{-n} \tag{10.117}$$

$$= \frac{a}{1 - az^{-1}}, \qquad |z| > |a|$$

As indicated previously, the importance of the unilateral z-transform lies in its application to analyzing systems characterized by linear constant-coefficient difference equations. In the case of the bilateral z-transform, we utilized the properties developed in Section 10.5. in applying the z-transform to the solution of difference equations. Most of the properties of the unilateral transform are the same as those for the bilateral transform, but there are several exceptions. The most important of these is the shifting property, which is of particular importance in solving difference equations. This property was developed in Section 10.5.2 for the bilateral transform. To develop the corresponding property for the unilateral transform, consider the signal

$$y[n] = x[n - 1] \tag{10.118}$$

Then,

$$\mathcal{Y}(z) = \sum_{n=0}^{\infty} x[n - 1]z^{-n}$$

$$= x[-1] + \sum_{n=1}^{\infty} x[n - 1]z^{-n}$$

$$= x[-1] + \sum_{n=0}^{\infty} x[n]z^{-(n+1)} \tag{10.119}$$

$$= x[-1] + z^{-1} \sum_{n=0}^{\infty} x[n]z^{-n}$$

so that

$$\mathcal{Y}(z) = x[-1] + z^{-1}\mathcal{X}(z)$$

Similarly, the signal

$$w[n] = x[n - 2] \tag{10.120}$$

has unilateral transform

$$\mathcal{W}(z) = x[-2] + x[-1]z^{-1} + z^{-2}\mathcal{X}(z) \tag{10.121}$$

and we can find corresponding expressions for the unilateral transform of $x[n-m]$. This property can be used in solving difference equations with initial conditions. To illustrate this, consider the following example.

Example 10.17

Consider the equation

$$y[n] + 3y[n-1] = x[n] \qquad (10.122)$$

with $x[n] = u[n]$ and the initial condition

$$y[-1] = 1 \qquad (10.123)$$

Applying the unilateral transform to both sides of (10.122), we obtain

$$\mathcal{Y}(z) + 3 + 3z^{-1}\mathcal{Y}(z) = \frac{1}{1 - z^{-1}} \qquad (10.124)$$

Solving for $\mathcal{Y}(z)$ we obtain

$$\mathcal{Y}(z) = -\frac{3}{1 + 3z^{-1}} + \frac{1}{(1 + 3z^{-1})(1 - z^{-1})} \qquad (10.125)$$

Performing a partial fraction expansion, we find that

$$\mathcal{Y}(z) = -\frac{9/4}{1 + 3z^{-1}} + \frac{1/4}{1 - z^{-1}} \qquad (10.126)$$

and application of Example 10.16 to each term yields

$$y[n] = [\tfrac{1}{4} - \tfrac{9}{4}(-3)^n]u[n] \qquad (10.127)$$

10.10 SUMMARY

In this chapter we have developed the z-transform for discrete-time signals. The discussion and development closely paralleled the corresponding treatment of the Laplace transform for continuous-time signals but with some important differences. For example, in the complex s-plane the Laplace transform reduces to the Fourier transform on the imaginary axis, whereas the z-transform reduces to the Fourier transform on the unit circle in the complex z-plane. For the Laplace transform the ROC consisted of a strip or half-plane (i.e., a strip extending to infinity in one direction), whereas for the z-transform the ROC is a ring, perhaps extending outward to infinity and/or inward to include the origin. As with the Laplace transform, time-domain characteristics such as the right-sided, left-sided, or two-sided nature of a sequence, and the causality or stability of an LTI system, can be associated with properties of the region of convergence. In particular for rational z-transforms, these time-domain characteristics can be associated with the pole locations in relation to the region of convergence.

In this chapter we also discussed a number of mappings from continuous-time to discrete-time systems. The use of such mappings is motivated by several considerations, including the fact that there is a rich history of design procedures for continuous-time systems, which, through the transformations discussed, can be exploited in designing discrete-time systems. An additional motivation lies in the fact that it is often useful to implement a continuous-time system by means of a discrete-time system. In such situations the desired continuous-time system must be transformed to an appropriate discrete-time system.

The chapter concluded with a brief discussion of the unilateral z-transform. This form of the z-transform is particularly useful for obtaining the response of systems characterized by linear constant-coefficient difference equations with non-zero initial conditions.

PROBLEMS

10.1. Determine the z-transform for each of the following sequences. Sketch the pole–zero plot and indicate the region of convergence. Indicate whether or not the Fourier transform of the sequence exists.

(a) $\delta[n]$ (b) $\delta[n-1]$

(c) $\delta[n+1]$ (d) $(\frac{1}{2})^n u[n]$

(e) $-(\frac{1}{2})^n u[-n-1]$ (f) $(\frac{1}{2})^n u[-n]$

(g) $\{(\frac{1}{2})^n + (\frac{1}{4})^n\} u[n]$ (h) $(\frac{1}{2})^{n-1} u[n-1]$

10.2. Determine the z-transform for the following sequences. Express all sums in closed form. Sketch the pole–zero plot and indicate the region of convergence. Indicate whether the Fourier transform of the sequence exists.

(a) $(\frac{1}{2})^n \{u[n] - u[n-10]\}$

(b) $(\frac{1}{2})^{|n|}$

(c) $7(\frac{1}{3})^n \cos\left[\dfrac{2\pi n}{6} + \dfrac{\pi}{4}\right] u[n]$

(d) $x[n] = \begin{cases} 0, & n < 0 \\ 1, & 0 \leq n \leq 9 \\ 0, & 9 < n \end{cases}$

10.3. In Section 10.2, Property 4, it was stated that if $x[n]$ is a right-sided sequence and if the circle $|z| = r_0$ is in the ROC, then all finite values of z for which $|z| < r_0$ will also be in the ROC. In the discussion an intuitive explanation was given. A more formal argument parallels closely that used for Property 4 in Section 9.2, relating to the Laplace transform. Specifically, consider a right-sided sequence $x[n]$ so that

$$x[n] = 0, \qquad n < N_1$$

and for which

$$\sum_{n=-\infty}^{+\infty} |x[n]| r_0^{-n} = \sum_{N_1}^{\infty} |x[n]| r_0^{-n} < \infty$$

Then if $r_0 \leq r_1$,

$$\sum_{N_1}^{\infty} |x[n]| r_1^{-n} \leq A \sum_{N_1}^{\infty} |x[n]| r_0^{-n} \tag{P10.3}$$

where A is a positive constant.

(a) Show that eq. (P10.3) is true and determine the constant A in terms of r_0, r_1, and N_1.

(b) From your result in part (a), show that Property 4, Section 10.2, follows.

(c) Develop an argument similar to that above to demonstrate the validity of Property 5, Section 10.2.

10.4. Using the method indicated, determine the sequence associated with each of the z-transforms below.

(a) Partial fractions:

$$X(z) = \frac{1 - 2z^{-1}}{1 - \frac{5}{2}z^{-1} + z^{-2}} \qquad \text{and } x[n] \text{ is absolutely summable}$$

(b) Long division:

$$X(z) = \frac{1 - \frac{1}{2}z^{-1}}{1 + \frac{1}{2}z^{-1}} \qquad \text{and } x[n] \text{ is right-sided}$$

(c) Partial fractions:

$$X(z) = \frac{3}{z - \frac{1}{4} - \frac{1}{8}z^{-1}} \qquad \text{and } x[n] \text{ is absolutely summable}$$

10.5. Listed here are several z-transforms. For each one, determine the inverse z-transform using both the method based on the partial fraction expansion and the Taylor's series method based on the use of long division.

$$X(z) = \frac{1}{1 + \frac{1}{2}z^{-1}}, \qquad |z| > \frac{1}{2}$$

$$X(z) = \frac{1}{1 + \frac{1}{2}z^{-1}}, \qquad |z| < \frac{1}{2}$$

$$X(z) = \frac{1 - \frac{1}{2}z^{-1}}{1 + \frac{3}{4}z^{-1} + \frac{1}{8}z^{-2}}, \qquad |z| > \frac{1}{2}$$

$$X(z) = \frac{1 - \frac{1}{2}z^{-1}}{1 - \frac{1}{4}z^{-2}}, \qquad |z| > \frac{1}{2}$$

$$X(z) = \frac{1 - az^{-1}}{z^{-1} - a}, \qquad |z| > \left| \frac{1}{a} \right|$$

10.6. Consider a right-sided sequence $x[n]$ with z-transform

$$X(z) = \frac{1}{(1 - \frac{1}{2}z^{-1})(1 - z^{-1})} \tag{P10.6}$$

(a) Carry out a partial fraction expansion of eq. (P10.6) expressed as a ratio of polynomials in z^{-1} and from this expansion determine $x[n]$.

(b) Rewrite eq. (P10.6) as a ratio of polynomials in z and carry out a partial fraction expansion of $X(z)$ expressed in terms of polynomials in z. From this expansion determine $x[n]$ and demonstrate that the sequence obtained is identical to that obtained in part (a).

10.7. Consider a left-sided sequence $x[n]$ with z-transform

$$X(z) = \frac{1}{(1 - \frac{1}{2}z^{-1})(1 - z^{-1})}$$

(a) Write $X(z)$ as a ratio of polynomials in z instead of z^{-1}.

(b) Using the partial fraction expansion, express $X(z)$ as a sum of terms, where each term represents a pole from your answer in part (a).

(c) Determine $x[n]$.

10.8. A right-sided sequence $x[n]$ has z-transform $X(z)$ given by

$$X(z) = \frac{3z^{-10} + z^{-7} - 5z^{-2} + 4z^{-1} + 1}{z^{-10} - 5z^{-7} + z^{-3}}$$

Determine $x[n]$ for $n < 0$.

10.9. (a) $x_1[n]$ is a left-sided sequence with z-transform $X_1(z) = e^z$. By expanding $X_1(z)$ in a Taylor's series about $z = 0$, determine $x_1[n]$.

(b) $x_3[n]$ is a right-sided sequence with z-transform $X_2(z) = e^{1/z}$. Determine $x_2[n]$.

10.10. By using the power-series expansion

$$\log (1 - w) = - \sum_{i=1}^{\infty} \frac{w^i}{i}, \qquad |w| < 1$$

determine the inverse of each of the following two z-transforms:
(a) $X(z) = \log (1 - 2z), \qquad |z| < \frac{1}{2}$
(b) $X(z) = \log (1 - \frac{1}{2}z^{-1}), \qquad |z| > \frac{1}{2}$

10.11. (a) Determine the z-transform of the sequence

$$x[n] = \delta[n] - 0.95\delta[n - 6]$$

(b) Sketch the pole–zero pattern for the sequence in part (a).
(c) By considering the behavior of the pole and zero vectors as the unit circle is traversed, develop an approximate sketch of the magnitude of the Fourier transform of $x[n]$.

10.12. By considering the geometric determination of the frequency response as discussed in Section 10.4, sketch, for each of the pole–zero plots in Figure P10.12, the magnitude of the associated Fourier transform.

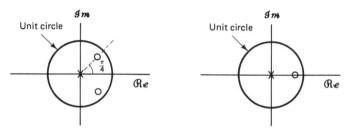

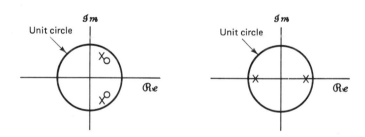

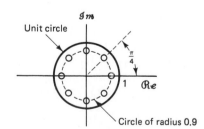

Figure P10.12

10.13. A discrete-time system with the pole–zero pattern shown in Figure P10.13-1 is referred to as a first-order all-pass system since the magnitude of the frequency response is constant independent of frequency.

(a) Demonstrate algebraically that $|H(e^{j\Omega})|$ is constant.

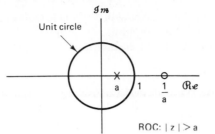

ROC: $|z| > a$ **Figure P10.13-1**

To demonstrate the same property geometrically, consider the vector diagram in Figure P10.13-2. We wish to show that the length of v_2 is proportional to the length of v_1 independent of Ω.

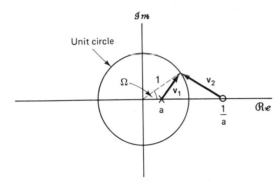

Figure P10.13-2

(b) Express the length of v_1 using the law of cosines and the fact that it is one leg of a triangle for which the other two legs are the unit vector and a vector of length a.

(c) In a manner similar to that in part (b), determine the length of v_2 and show that it is proportional in length to v_1 independent of Ω.

10.14. Consider a real-valued sequence $x[n]$ with rational z-transform $X(z)$.

(a) From the definition of the z-transform, show that

$$X(z) = X^*(z^*)$$

(b) From your result in part (a), show that if a pole (zero) of $X(z)$ occurs at $z = z_0$, then a pole (zero) must also occur at $z = z_0^*$.

(c) Verify the result in part (b) for each of the following sequences:

    1. $x[n] = (\frac{1}{2})^n u[n]$

    2. $x[n] = \delta[n] - \frac{1}{2}\delta[n-1] + \frac{1}{4}\delta[n-2]$

10.15. Consider a sequence $x_1[n]$ with z-transform $X_1(z)$ and a sequence $x_2[n]$ with z-transform $X_2(z)$, where $x_1[n]$ and $x_2[n]$ are related by

$$x_2[n] = x_1[-n]$$

Show that $X_2(z) = X_1(1/z)$ and, from this, show that, if $X_1(z)$ has a pole (or zero) at $z = z_0$, then $X_2(z)$ has a pole (or zero) at $z = 1/z_0$.

10.16. Consider an even sequence $x[n]$ (i.e., $x[n] = x[-n]$) with rational z-transform $X(z)$.

(a) From the definition of the z-transform show that

$$X(z) = X\left(\frac{1}{z}\right)$$

(b) From your result in part (a), show that if a pole (zero) of $X(z)$ occurs at $z = z_0$, then a pole (zero) must also occur at $z = 1/z_0$.

(c) Verify the result in part (b) for each of the following sequences:
   1. $\delta[n+1] + \delta[n-1]$
   2. $\delta[n+1] - \frac{5}{2}\delta[n] + \delta[n-1]$

(d) By combining your result in part (b) with the result of Problem 10.14(b), show that for a real, even sequence, if there is a pole (zero) of $H(z)$ at $z = \rho e^{j\theta}$, then there is also a pole (zero) of $H(z)$ at $z = (1/\rho)e^{j\theta}$ and at $z = (1/\rho)e^{-j\theta}$.

10.17. (a) Carry out the proof for each of the following properties in Table 10.1:
   1. Property 10.5.2
   2. Property 10.5.3
   3. Property 10.5.4

(b) With $X(z)$ denoting the z-transform of $x[n]$ and R_x its ROC, determine in terms of $X(z)$ and R_x the z-transform and associated ROC for each of the following sequences:
   1. $x^*[n]$
   2. $z_0^n x[n]$, where z_0 is a complex number

10.18. In Section 10.5.7 we stated and proved the initial value theorem for causal sequences.

(a) State and prove the corresponding theorem if $x[n]$ is anticausal (i.e., if $x[n] = 0$, $n > 0$).

(b) Show that if $x[n] = 0$, $n < 0$, then

$$x[1] = \lim_{z \to \infty} z(X(z) - x[0])$$

10.19. By first differentiating $X(z)$ and using the appropriate properties of the z-transform, determine the sequence for which the z-transform is each of the following:

(a) $X(z) = \log(1 - 2z)$, $\quad |z| < \frac{1}{2}$

(b) $X(z) = \log(1 - \frac{1}{2}z^{-1})$, $\quad |z| > \frac{1}{2}$

Compare your results for (a) and (b) with the results obtained in Problem 10.10, in which the power-series expansion was used.

10.20. Let $x[n]$ denote a causal sequence (i.e., $x[n] = 0$, $n < 0$) for which $x[0]$ is nonzero and finite.

(a) Using the initial value theorem, show that there are no poles or zeros of $X(z)$ at $z = \infty$.

(b) Show that as a consequence of your result in part (a), the number of poles of $X(z)$ in the finite z-plane equals the number of zeros in the finite z-plane. (The finite z-plane excludes $z = \infty$.)

10.21. In Section 10.5.5 we stated the convolution property for the z-transform. To prove this property, we begin with the convolution sum expressed as

$$x_3[n] = x_1[n] * x_2[n] = \sum_{k=-\infty}^{+\infty} x_1[k]x_2[n-k] \qquad \text{(P10.21)}$$

(a) By taking the z-transform of eq. (P10.21) using eq. (10.3), show that

$$X_3(z) = \sum_{k=-\infty}^{+\infty} x_1[k]\hat{X}_2(z)$$

where $\hat{X}_2(z) = \mathbb{Z}\{x_2[n-k]\}$.

(b) Using the result in part (a) and property 10.5.2, show that

$$X_3(z) = X_2(z) \sum_{k=-\infty}^{+\infty} x_1[k]z^{-k}$$

(c) From part (b), show that

$$X_3(z) = X_1(z)X_2(z)$$

as stated in eq. (10.59).

10.22. The autocorrelation sequence $\phi_{xx}[n]$ of a sequence $x[n]$ is defined as

$$\phi_{xx}[n] = \sum_{k=-\infty}^{\infty} x[k]x[n+k]$$

Determine the z-transform of $\phi_{xx}[n]$ in terms of the z-transform of $x[n]$.

10.23. Let $x[n]$ be a discrete-time signal with z-transform $X(z)$. For each of the following signals, determine the z-transform in terms of $X(z)$.

(a) $\Delta x[n]$, where Δ is the first backward difference operator defined by

$$\Delta x[n] = x[n] - x[n-1]$$

(b) $x_1[n] = \begin{cases} x\left[\dfrac{n}{2}\right], & n \text{ even} \\ 0, & n \text{ odd} \end{cases}$

(c) $x_1[n] = x[2n]$

10.24. **(a)** Determine the system function for the causal LTI system with difference equation

$$y[n] - \tfrac{1}{2}y[n-1] + \tfrac{1}{4}y[n-2] = x[n]$$

(b) Using z-transforms, determine $y[n]$ if

$$x[n] = (\tfrac{1}{2})^n u[n]$$

10.25. A sequence $x[n]$ is the output of an LTI system whose input is $s[n]$. This system is described by the difference equation

$$x[n] = s[n] - e^{-8\alpha}s[n-8]$$

where $0 < \alpha < 1$.

(a) Find the system function

$$H_1(z) = \frac{X(z)}{S(z)}$$

and plot its poles and zeros in the z-plane. Indicate the region of convergence.

(b) We wish to recover $s[n]$ from $x[n]$ with an LTI system. Find the system function

$$H_2(z) = \frac{Y(z)}{X(z)}$$

such that $y[n] = s[n]$. Find all possible regions of convergence for $H_2(z)$ and, for each, tell whether or not the system is causal and stable.

(c) Find all possible choices for the unit sample response $h_2[n]$ such that

$$y[n] = h_2[n] * x[n] = s[n]$$

10.26. A causal LTI system is described by the difference equation

$$y[n] = y[n-1] + y[n-2] + x[n-1]$$

(a) Find the system function $H(z) = Y(z)/X(z)$ for this system. Plot the poles and zeros of $H(z)$ and indicate the region of convergence.

(b) Find the unit sample response of this system.

(c) You should have found this to be an unstable system. Find a stable (noncausal) unit sample response that satisfies the difference equation.

10.27. Consider an LTI system with input $x[n]$ and output $y[n]$ for which
$$y[n-1] - \tfrac{5}{2}y[n] + y[n+1] = x[n]$$
The system may or may not be stable and/or causal.

By considering the pole–zero pattern associated with this difference equation, determine three possible choices for the unit sample response of the system. Show that each choice satisfies the difference equation.

10.28. Consider a linear discrete-time shift-invariant system with input $x[n]$ and output $y[n]$ for which
$$y[n-1] - \tfrac{10}{3}y[n] + y[n+1] = x[n]$$
The system is stable. Determine the unit sample response.

10.29. Given here are four z-transforms. Determine which ones could be the transfer function of a discrete-time linear system that is not necessarily stable but for which the unit sample response is zero for $n < 0$. Clearly state your reasons.

(a) $\dfrac{(1 - z^{-1})^2}{1 - \frac{1}{2}z^{-1}}$

(b) $\dfrac{(z - 1)^2}{z - \frac{1}{2}}$

(c) $\dfrac{(z - \frac{1}{4})^5}{(z - \frac{1}{2})^6}$

(d) $\dfrac{(z - \frac{1}{4})^6}{(z - \frac{1}{2})^5}$

10.30. Consider an LTI system with impulse response $h[n]$ and input $x[n]$ given by
$$h[n] = \begin{cases} a^n, & n \geq 0 \\ 0, & n < 0 \end{cases}$$
$$x[n] = \begin{cases} 1, & 0 \leq n \leq N - 1 \\ 0, & \text{otherwise} \end{cases}$$

(a) Determine the output $y[n]$ by explicitly evaluating the discrete convolution of $x[n]$ and $h[n]$.

(b) Determine the output $y[n]$ by computing the inverse z-transform of the product of the z-transforms of the input and the unit sample response.

10.31. In Problem 9.18 we considered minimum-phase continuous-time systems. Minimum-phase discrete-time systems are defined similarly. Specifically, a minimum-phase system is one that is causal and stable and for which the inverse system is also causal and stable.

Determine the necessary constraints on the location in the z-plane of the poles and zeros of the system function of a minimum-phase system.

10.32. Consider the digital filter structure shown in Figure P10.32.

(a) Find $H(z)$ for this causal filter. Plot the pole–zero pattern and indicate the region of convergence.

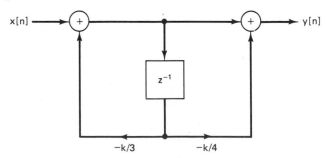

Figure P10.32

(b) For what values of k is the system stable?

(c) Determine $y[n]$ if $k = 1$ and $x[n] = (\frac{2}{3})^n$ for all n.

10.33. We would like to design a system for which, at each n, the output $y[n]$ is the average of the inputs at n, $n - 1, \ldots, n - M + 1$.

(a) Specify a difference equation relating $y[n]$ and $x[n]$ for this system.

(b) Determine $H(z)$ for this system.

(c) Draw a pole–zero plot for the case $M = 3$.

(d) For $M = 3$, sketch a realization of the system, using adders, coefficient multipliers, and unit delays. Use as few unit delay elements as possible.

(e) The system described in parts (a)–(d) often has excessive storage demands. As an alternative one often uses a recursive system of the form

$$y[n] = \alpha y[n - 1] + \beta x[n]$$

Find a relation between α and β so that this system has the same response to a *constant* input as the system described in parts (a)–(d).

(f) Suppose that the input $x[n]$ consists of two components, that is,

$$x[n] = c + w[n]$$

where c is a constant and $w[n]$ is contaminating noise at a frequency of approximately $\Omega = \pi$. We would like to design a system to estimate the value of c. To perform this task we must choose between the system described in parts (a)–(d) with $M = 3$ and a system of the form described in part (e) with $\alpha = 0.7$. Which system would you prefer to use? Justify your answer.

10.34. The following is known about a discrete-time LTI system with input $x[n]$ and output $y[n]$:

1. If $x[n] = (-2)^n$ for all n, then $y[n] = 0$ for all n.
2. If $x[n] = (\frac{1}{2})^n u[n]$ for all n, then $y[n]$ for all n is of the form

$$y[n] = \delta[n] + a(\tfrac{1}{4})^n u[n]$$

where a is a constant.

(a) Determine the value of the constant a.

(b) Determine the response $y[n]$ if the input $x[n]$ is

$$x[n] = 1 \qquad \text{for all } n$$

10.35. In discussing impulse invariance in Section 10.8.1 we considered $H_c(s)$ to be of the form of eq. (10.84) with only first-order poles. In this problem we consider how the presence of a second-order pole in eq. (10.84) would be reflected in eq. (10.87). Toward this end, consider $H_c(s)$ to be

$$H_c(s) = \frac{A}{(s - s_0)^2}$$

(a) By referring to Table 9.2, determine $h_c(t)$. (Assume causality.)

(b) Determine $h_d[n]$ defined as $h_d[n] = h_c(nT)$.

(c) By referring to Table 10.2, determine $H_d(z)$, the z-transform of $h_d[n]$.

(d) Determine the system function and pole–zero pattern for the discrete-time system obtained by applying impulse invariance to the continuous-time system

$$H_c(s) = \frac{1}{(s + 1)(s + 2)^2}$$

10.36. Let $h_c(t)$, $s_c(t)$, and $H_c(s)$ denote the impulse response, step response, and system function of a continuous-time linear time-invariant filter. Let $h[n]$, $s[n]$, and $H(z)$ denote the unit sample response, step response, and system function of a discrete-time linear shift-invariant digital filter.

(a) If $h[n] = h_c(nT)$, does $s[n] = \sum_{k=-\infty}^{n} h_c(kT)$?

(b) If $s[n] = s_c(nT)$, does $h[n] = h_c(nT)$?

10.37. Consider a continuous-time system with system function

$$H_c(s) = \frac{s+a}{(s+a)^2 + b^2}$$

and impulse response $h_c(t)$. Determine the system function $H_d(z)$ of the discrete-time system designed from this system on the basis of:

(a) Impulse invariance, that is,

$$h_d[n] = h_c[nT]$$

(b) Step invariance, that is,

$$s_d[n] = s_c[nT]$$

where

$$s_d[n] = \sum_{k=-\infty}^{n} h_d[k]$$

and

$$s_c(t) = \int_{-\infty}^{t} h_c(\tau)\, d\tau$$

10.38. Consider a continuous-time system with impulse response $h_c(t)$ and frequency response $H_c(j\omega)$ as depicted in Figure P10.38. Assume that this system may be represented by

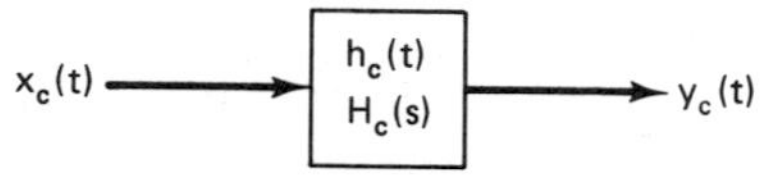

$$x_c(t) \longrightarrow \boxed{\begin{array}{c} h_c(t) \\ H_c(s) \end{array}} \longrightarrow y_c(t)$$

Figure P10.38

a linear constant-coefficient differential equation of the form

$$\sum_{k=0}^{N} a_k \frac{d^k y_c(t)}{dt^k} = \sum_{k=0}^{M} b_k \frac{d^k x_c(t)}{dt^k}$$

We wish to approximate this system through the use of a discrete-time system. The operation of differentiation will be approximated by forward differencing, that is, by making the approximation

$$\left. \frac{dx_a(t)}{dt} \right|_{t=nT} \simeq \frac{x_c(nT + T) - x_c(nT)}{T} = \frac{x[n+1] - x[n]}{T}$$

Thus, we define the first forward difference of $x[n]$ as

$$\nabla^{(1)}\{x[n]\} = \frac{x[n+1] - x[n]}{T}$$

and the kth forward difference of $x[n]$ as

$$\nabla^{(k)}\{x[n]\} = \nabla^{(k-1)}\{\nabla^{(1)}[x[n]]\}$$

where $\nabla^{(0)}\{x[n]\} = x[n]$. The difference equation for the discrete-time system will then be specified as

$$\sum_{k=0}^{N} a_k \nabla^{(k)}\{y[n]\} = \sum_{k=0}^{M} b_k \nabla^{(k)}\{x[n]\}$$

(a) If $H_c(s)$, the Laplace transform of $h_c(t)$, is given by

$$H_c(s) = \frac{s+2}{(s+1)(s+3)}$$

determine $H_d(z)$, the system function of the discrete-time system.

(b) In general, what is the relationship between $H_c(s)$ and $H_d(z)$?

(c) We know that sampling can be thought of as a mapping of the $j\omega$-axis in the s-plane to the unit circle in the z-plane. To what contour in the z-plane does the approximation by differences map the $j\omega$-axis of the s-plane?

(d) If the continuous-time filter $H_c(s)$ is stable, is the discrete-time filter $H_d(z)$ guaranteed to be stable? (Assume causality of both systems.)

10.39. Consider a continuous-time filter with input $x_c(t)$ and output $y_c(t)$ which is described by a linear constant-coefficient differential equation of the form

$$\sum_{k=0}^{N} a_k \frac{d^k y_c(t)}{dt^k} = \sum_{k=0}^{M} b_k \frac{d^k x_c(t)}{dt^k} \tag{P10.39}$$

The filter is to be mapped to a discrete-time filter with input $x[n]$ and output $y[n]$ by replacing derivatives with central differences. Specifically, let $\nabla^{(k)}\{x[n]\}$ denote the kth central difference of $x[n]$, defined as follows:

$$\nabla^{(0)}\{x[n]\} = x[n]$$

$$\nabla^{(1)}\{x[n]\} = \left[\frac{x[n+1] - x[n-1]}{2}\right]$$

$$\nabla^{(k)}\{x[n]\} = \nabla^{(1)}\{\nabla^{(k-1)}\{x[n]\}\}$$

The difference equation for the digital filter obtained from the differential equation (P10.39) is then

$$\sum_{k=0}^{N} a_k \nabla^{(k)}\{y[n]\} = \sum_{k=0}^{M} b_k \nabla^{(k)}\{x[n]\}$$

(a) If the transfer function of the continuous-time filter is $H_c(s)$ and if the transfer function of the corresponding discrete-time filter is $H_d(z)$, determine how $H_d(z)$ is related to $H_c(s)$.

(b) For the continuous-time frequency response $H_c(j\omega)$, as indicated in Figure P10.39,

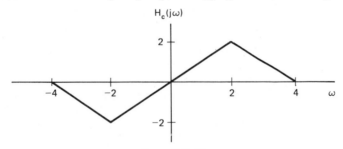

Figure P10.39

sketch the discrete-time frequency response $H_d(e^{j\Omega})$ that would result from the mapping determined in part (a).

(c) Assume that $H_c(s)$ corresponds to a causal stable filter. If the region of convergence of $H_d(z)$ is specified to include the unit circle, will $H_d(z)$ necessarily correspond to a *causal* filter?

10.40. As mentioned in Section 10.8.3, the bilinear transformation to map from the s-plane to the z-plane can be interpreted as arising from the use of the trapezoidal rule in numerically integrating differential equations.

(a) Let us consider a continuous-time system for which the differential equation is

$$\frac{dy(t)}{dt} = x(t) \tag{P10.40-1}$$

or equivalently,

$$y(t) = \int_{-\infty}^{t} x(\tau)\, d\tau \qquad\qquad \text{(P10.40-2)}$$

Determine the system function $H(s)$ for this continuous-time system.

In numerical analysis the procedure known as the trapezoidal rule for integration proceeds by approximating the continuous-time function as a set of contiguous trapezoids, as illustrated in Figure P10.40(a), and then adding their areas

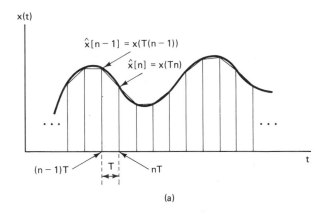

(a)

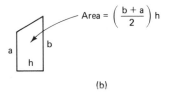

$$\text{Area} = \left(\frac{b+a}{2}\right) h$$

(b)

Figure P10.40

to compute the total integral. The area A of an individual trapezoid, with dimensions shown in Figure P10.40(b), is

$$A = \left(\frac{b+a}{2}\right) h$$

(b) What is the area A_n in the trapezoidal approximation, between $x((n-1)T)$ and $x(nT)$?

(c) From eq. (P10.40-2), $y(nT)$ denotes the area under $x(t)$ up to time $t = nT$. Let $\hat{y}[n]$ denote the approximation to $y(nT)$ obtained using the trapezoidal rule for integration, that is,

$$\hat{y}[n] = \sum_{k=-\infty}^{n} A_n$$

Show that

$$\hat{y}[n] = \hat{y}[n-1] + A_n$$

(d) With $\hat{x}[n]$ defined as $\hat{x}[n] = x(nT)$, show that the trapezoidal rule approximation to eq. (P10.40-2) becomes

$$\hat{y}[n] = \hat{y}[n-1] + \frac{T}{2}\{\hat{x}[n-1] + \hat{x}[n]\} \qquad\qquad \text{(P10.40-3)}$$

(e) Determine the system function corresponding to the difference equation in part (d). Demonstrate, in particular, that it is the same as would be obtained by applying the bilinear transformation to the continuous-time system function corresponding to eq. (P10.40-1).

10.41. The system function for a second-order continuous-time bandpass filter is of the form

$$H(s) = \frac{\omega_n^2}{s^2 + 2\zeta\omega_n s + \omega_n^2} \tag{P10.41}$$

The center frequency ω_c of such a filter is the frequency at which $|H(j\omega)|^2$ is maximum and the half-power frequencies are defined as the frequencies at which $|H(j\omega)|^2$ is one-half of its value at $\omega = \omega_c$. The bandwidth is defined as the width of frequency band between the half-power frequencies on either side of ω_c.

(a) For $0 < \zeta \ll 1$ and $\omega_n \gg 0$, draw the pole–zero plot of $H(s)$ and sketch $|H(j\omega)|$ by considering the vectors from the poles and zeros to the $j\omega$-axis. Show that ω_c is approximately ω_n and that the bandwidth is approximately $2\zeta\omega_n$.

(b) We would now like to design a second-order *discrete-time* bandpass filter by applying the bilinear transformation to the system of eq. (P10.41). The discrete-time bandpass filter is to have a center frequency of $\Omega_c = \pi/4$ and a bandwidth of 0.01π. Making reasonable approximations, determine the center frequency and bandwidth of the corresponding continuous-time filter.

(c) For the center frequency and bandwidth in part (a), and with $T = 2$ in eq. (10.106), determine ζ and ω_n in eq. (P10.41) so that when the bilinear transformation is applied to $H(s)$, the desired discrete-time filter will result.

10.42. A continuous-time highpass filter can be obtained from a continuous-time lowpass filter by replacing s by $1/s$ in the transfer function; that is, if $G_c(s)$ is the transfer function for a lowpass filter, then $H_c(s)$ is the transfer function for a highpass filter if

$$H_c(s) = G_c\left(\frac{1}{s}\right)$$

Assume that a discrete-time lowpass filter $G_d(z)$ and a discrete-time highpass filter $H_d(z)$ are obtained from $G_c(s)$ and $H_c(s)$, respectively, using the bilinear transformation. Show that $H_d(z)$ can be obtained from $G_d(z)$ by replacing z by some function of z, denoted by $m(z)$. Determine $m(z)$. This represents a discrete-time transformation to convert a lowpass filter to a highpass filter.

10.43. A discrete-time lowpass filter with frequency response $H(e^{j\Omega})$ is to be designed to meet the following specifications:

$$0.8 < |H(e^{j\Omega})| < 1.2 \quad \text{for } 0 \leq |\Omega| \leq 0.2\pi$$
$$|H(e^{j\Omega})| < 0.2 \quad \text{for } 0.8\pi \leq |\Omega| \leq \pi$$

The design procedure consists of applying the bilinear transformation to an appropriate continuous-time Butterworth lowpass filter.

(a) With $T = 2$ in eq. (10.106) and eq. (10.111), determine the necessary specifications on the continuous-time Butterworth filter which, when mapped through the bilinear transformation, will result in the desired discrete-time filter.

(b) Determine the lowest-order continuous-time Butterworth filter that meets the specifications in part (a).

(c) By applying the bilinear transformation to the filter in part (b), determine the transfer function of the discrete-time filter.

(d) Repeat parts (a)–(c) with $T = 1$.

(e) By first comparing the resulting discrete-time filter obtained in parts (c) and (d), discuss in general the effect of the parameter T in eqs. (10.106) and (10.111) on the discrete-time filter obtained by the procedure in parts (a)–(c).

10.44. A system frequently encountered in communications systems is the 90° phase splitter as depicted in its continuous-time form in Figure P10.44-1. The systems $H_1(j\omega)$ and

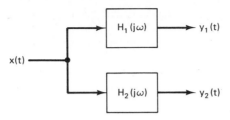

Figure P10.44-1

$H_2(j\omega)$ have the property that

$$H_1(j\omega) = e^{j\theta_1(\omega)}$$
$$H_2(j\omega) = e^{j\theta_2(\omega)}$$

where

$$\theta_1(\omega) - \theta_2(\omega) = \begin{cases} \dfrac{\pi}{2}, & \omega > 0 \\[2mm] -\dfrac{\pi}{2}, & \omega < 0 \end{cases}$$

A digital 90° phase splitter is depicted in Figure P10.44-2, where

$$G_1(e^{j\Omega}) = e^{j\phi_1(\Omega)}$$
$$G_2(e^{j\Omega}) = e^{j\phi_2(\Omega)}$$

where

$$\phi_1(\Omega) - \phi_2(\Omega) = \begin{cases} \dfrac{\pi}{2}, & 0 < \Omega < \pi \\[2mm] -\dfrac{\pi}{2}, & -\pi < \Omega < 0 \end{cases}$$

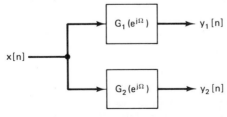

Figure P10.44-2

If $H_1(j\omega)$ and $H_2(j\omega)$ in the continuous-time 90° phase splitter are mapped to the digital filters $G_1(e^{j\Omega})$ and $G_2(e^{j\Omega})$, respectively, using the bilinear transformation, determine whether the result will be a digital 90° phase splitter.

10.45. Consider designing a digital filter with system function $H_d(z)$ from a continuous-time filter with rational system function $H_c(s)$ by the following transformation:

$$H_d(z) = H_c(s)\big|_{s=\beta\frac{1-z^{-\alpha}}{1+z^{-\alpha}}}$$

where α is a nonzero integer and β is real.

(a) If $\alpha > 0$, determine the range of values of β for which a stable, causal continuous-time filter with rational $H_c(s)$ will always lead to a stable, causal digital filter with rational $H_d(z)$.

(b) If $\alpha < 0$, determine the range of values of β for which a stable, causal continuous-time filter with rational $H_c(s)$ will always lead to a stable, causal digital filter with rational $H_d(z)$.

(c) For $\alpha = -1$, determine to what contour in the z-plane the $j\omega$-axis in the s-plane maps.

(d) Given $H_c(j\omega)$, as shown in Figure P10.45, sketch $H_d(e^{j\Omega})$ for $\beta = 1, \alpha = -1$.

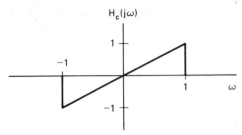

Figure P10.45

10.46. Determine the unilateral z-transform for each of the sequences in Problem 10.1.

10.47. If $\mathcal{X}(z)$ denotes the unilateral z-transform of $x[n]$, determine, in terms of $\mathcal{X}(z)$, the unilateral z-transform of:

(a) $x[n + 1]$ (b) $x[n - 3]$ (c) $\sum_{k=-\infty}^{n} x[k]$

10.48. For each of the following difference equations and associated input and initial conditions, determine the response $y[n]$ by using the unilateral z-transform.

(a) $y[n] + 3y[n - 1] = x[n]$
$$x[n] = (\tfrac{1}{2})^n u[n]$$
$$y[-1] = 1$$

(b) $y[n] - \tfrac{1}{2}y[n - 1] = x[n] - \tfrac{1}{2}x[n - 1]$
$$x[n] = u[n]$$
$$y[-1] = 0$$

(c) $y[n] - \tfrac{1}{2}y[n - 1] = x[n] - \tfrac{1}{2}x[n - 1]$
$$x[n] = u[n]$$
$$y[-1] = 1$$

It has long been recognized that in many situations there are particular advantages to be gained by using feedback, that is, by using the output of a system to control or modify the input. For example, it is common in electromechanical systems, such as a motor whose shaft position is to be maintained at a constant angle, to measure the error between the desired and true position and to use this error signal to turn the shaft in the appropriate direction. This is illustrated in Figure 11.1, where we have depicted the use of a dc motor for the accurate pointing of a telescope. In Figure 11.1(a) we have indicated pictorially what such a system would look like, where $v(t)$ is the input voltage to the motor and $\theta(t)$ is the angular position of the telescope platform. The block diagram for the motor-driven pointing system is shown in Figure 11.1(b). A feedback system for controlling the position of the telescope is illustrated in Figure 11.1(c), and a block diagram equivalent to this system is shown in Figure 11.1(d). The external or *reference* input to this feedback system is the desired shaft angle θ_D. A potentiometer is used to convert this desired angle into a voltage $K_1\theta_D$ proportional to θ_D. Similarly, a second potentiometer produces a voltage $K_1\theta(t)$ proportional to the actual platform angle. These two voltages are compared, producing an error voltage $K_1(\theta_D - \theta(t))$, which is amplified and then used to drive the electric motor.

Figure 11.1 suggests two different methods for pointing the telescope. One of these is the feedback system of Figure 11.1(c) and (d). Here the input that we must provide is the desired or reference angle θ_D. Alternatively, if the initial and desired angle and the detailed electrical and mechanical characteristics of the motor-shaft

Linear Feedback Systems

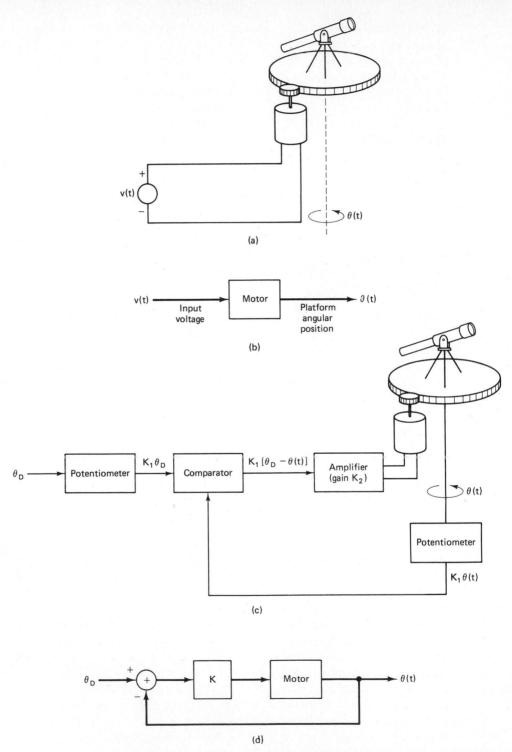

Figure 11.1 Use of feedback to control the angular position of a telescope: (a) dc motor-driven telescope platform; (b) block diagram of the system in (a); (c) feedback system for the pointing of the telescope; (d) block diagram of the system in (c) (here $K = K_1 K_2$).

assembly were known exactly, we could specify the precise time history of the input voltage $v(t)$ that would first accelerate and then decelerate the shaft, bringing the platform to a stop at the desired position without the use of feedback, as in Figure 11.1(a) and (b). A system operating as in Figure 11.1(a) and (b) is typically referred to as an *open-loop* system, in contrast to the *closed-loop* system of Figure 11.1(c) and (d). In a practical environment there are clear advantages to controlling the motor-shaft angle with the closed-loop system rather than with the open-loop system. For example, in the closed-loop system, when the shaft has been rotated to the correct position, any disturbance from this position will be sensed and the resulting error will be used to provide a correction. In the open-loop system there is no mechanism for providing a correction. As another advantage of the closed-loop system, consider the effect of errors in modeling the characteristics of the motor-shaft assembly. In the open-loop system a precise characterization of the system is required to design the correct input. In the closed-loop system the input is simply the desired shaft angle and does not require knowledge of the system. This insensitivity of the closed-loop system to disturbances and to imprecise knowledge of the system are two important advantages of feedback.

The control of an electric motor is just one of a great many examples in which feedback plays an important role. Similar uses of feedback can be found in a wide variety of applications, such as chemical process control, automotive fuel systems, household heating systems, and aerospace systems, to name just a few. In addition, feedback is also present in many biological processes and in the control of human motion. For example, when reaching for an object it is usual during the reaching process to monitor visually the distance between the hand and the object so that the velocity of the hand can be smoothly decreased as the distance (i.e., the error) between the hand and the object decreases. The effectiveness of using the system output (hand position) to control the input is clearly demonstrated by alternately reaching with and without the use of visual feedback.

In addition to its use in providing an error-correcting mechanism that can reduce sensitivity to disturbances and to errors in the modeling of the system that is to be controlled, another important characteristic of feedback is its potential for stabilizing a system that is inherently unstable. Consider the problem of trying to balance a broomstick in the palm of the hand [Figure 11.2(a)]. If the hand is held stationary, small disturbances (such as a slight breeze or inadvertent motion of the hand) will cause the broom to fall over. Of course, if one knows exactly what disturbances will occur and can control the motion of one's hand perfectly, it is possible to determine in advance how to move the hand to balance the broom. This is clearly unrealistic. However, by always moving the hand in the direction in which the broom is falling, it can be balanced. This of course *requires* feedback in order to sense the direction in which the broom is falling. A second example that is closely related to the balancing of a broom is the problem of controlling a so-called inverted pendulum, which is illustrated in Figure 11.2(b). As shown, an inverted pendulum consists of a thin rod with a heavier weight at the top. The bottom of the rod is mounted on a cart that can move in either direction along a track. Again, if the cart is kept stationary, the inverted pendulum will topple over. The problem of stabilizing the inverted pendulum is one of designing a feedback system that will move the cart to keep the pendu-

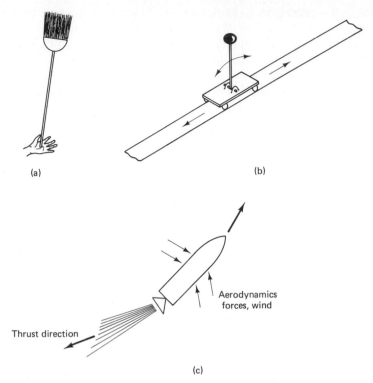

Figure 11.2 Three examples in which feedback is used to stabilize unstable systems: (a) the balancing of a broom; (b) an inverted pendulum; (c) control of the trajectory of a rocket.

lum vertical. This example is examined in Problem 11.6. In Figure 11.2(c) we have illustrated a third example, which again bears some similarity to the balancing of a broom. This is the problem of controlling the trajectory of a rocket. In this case, much as the movement of the hand is used to compensate for disturbances in the position of the broom, the direction of the thrust of the rocket is used to correct for changes in aerodynamic forces and wind disturbances that would otherwise cause the rocket to deviate from its course. Again feedback is important, because these forces and disturbances are never precisely known in advance.

The preceding examples provide some indication of why feedback may be useful. In the next two sections we introduce the basic block diagrams and equations for linear feedback systems and discuss in more detail a number of applications of feedback, both in continuous time and in discrete time. We also point out how feedback can have harmful as well as useful effects. These examples of the uses and effects of feedback will give us some insight into how changes in the parameters in a feedback system lead to changes in the behavior of the system. Understanding this relationship is essential in designing feedback systems that have certain desirable characteristics. With this as background, we will then develop in the remaining sections of the chapter several specific techniques that are of significant value in the analysis and design of continuous- and discrete-time feedback systems.

11.1 LINEAR FEEDBACK SYSTEMS

The general configuration of a continuous-time linear feedback system is shown in Figure 11.3(a) and that of a discrete-time linear feedback system in Figure 11.3(b). Because of the applications in which feedback is most typically applied, it is natural to restrict the systems in Figure 11.3(a) and (b) to be causal systems. This will be

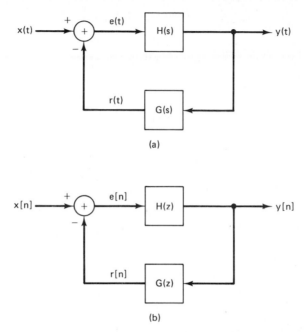

(a)

(b)

Figure 11.3 Basic feedback system configurations in (a) continuous time and (b) discrete time.

our assumption throughout the chapter. It should also be noted that the convention used in Figure 11.3(a) is that $r(t)$, the signal fed back, is subtracted from the input $x(t)$ to form $e(t)$. The identical convention is adopted in discrete time. Historically, this convention arose in tracking system applications, where $x(t)$ represented a desired command and $e(t)$ represented the error between the command $x(t)$ and the actual response $r(t)$. This was the case, for example, in our earlier discussion of the pointing of a telescope. In more general feedback systems, $e(t)$ and $e[n]$ may not correspond to or be directly interpretable as error signals.

The system function $H(s)$ in Figure 11.3(a) or $H(z)$ in Figure 11.3(b) is referred to as the *system function of the forward path* and $G(s)$ or $G(z)$ as the *system function of the feedback path*. The system function of the overall system of Figure 11.3(a) or (b) is referred to as the *closed-loop system function* and will be denoted by $Q(s)$ or $Q(z)$. In Sections 9.7.2 and 10.7.2 we derived expressions for the system functions of feedback interconnections of LTI systems. Applying these results to the feedback systems of Figure 11.3 we obtain

$$Q(s) = \frac{Y(s)}{X(s)} = \frac{H(s)}{1 + G(s)H(s)} \tag{11.1}$$

$$Q(z) = \frac{Y(z)}{X(z)} = \frac{H(z)}{1 + G(z)H(z)} \tag{11.2}$$

Equations (11.1) and (11.2) represent the fundamental equations for the study of linear feedback systems. In the following sections we use these equations as a basis for gaining insight into the properties of feedback systems and for developing several tools for their analysis.

11.2 SOME APPLICATIONS AND CONSEQUENCES OF FEEDBACK

In the introduction we provided a brief, intuitive look at some of the properties and uses of feedback systems. In this section we examine a number of the characteristics and applications of feedback in somewhat more quantitative terms using the basic feedback equations (11.1) and (11.2) as a starting point. Our purpose in this section is to provide an introduction to and an appreciation for the applications of feedback, rather than to develop any of these applications in detail. In the sections that follow we focus in more depth on several specific techniques for analyzing feedback systems that are useful in a wide range of problems, including many of the applications that we are about to describe.

11.2.1 Inverse System Design

In some applications one would like to synthesize the inverse of a given continuous-time system. Suppose that this system has system function $P(s)$, and consider the feedback system shown in Figure 11.4. Applying equation (11.1) with $H(s) = K$ and $G(s) = P(s)$ we find that the closed-loop system function is

$$Q(s) = \frac{K}{1 + KP(s)} \tag{11.3}$$

If the gain K is sufficiently large so that $KP(s) \gg 1$, then

$$Q(s) \simeq \frac{1}{P(s)} \tag{11.4}$$

so that the feedback system of Figure 11.4 approximates the inverse of the system with system function $P(s)$.

It is important to note that the result in eq. (11.4) requires that the gain K be sufficiently high, but is otherwise not dependent on the precise value of the gain.

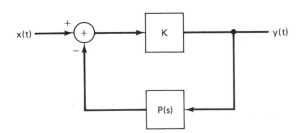

Figure 11.4 Form of a feedback system used in implementing the inverse of the system with system function $P(s)$.

A class of devices that provide this kind of gain are referred to as operational amplifiers and are widely used in feedback systems.† One common application of this general principle is in the implementation of integrators. A capacitor has the property that its current is proportional to the derivative of the voltage. By inserting a capacitor in the feedback path around an operational amplifier, this is inverted to provide integration. This specific application is explored in more detail in Problem 11.3.

Although our discussion is for the most part restricted to linear systems, it is worth pointing out that this same basic approach is commonly used in inverting a nonlinearity. For example, systems for which the output is the *logarithm* of the input are commonly implemented by utilizing the *exponential* current–voltage characteristics of a diode as feedback around an operational amplifier. This is explored in more detail in Problem 11.3.

11.2.2 Compensation for Nonideal Elements

Another common use of feedback is to correct for some of the nonideal properties of the open-loop system. For example, feedback is often used in the design of amplifiers to provide constant-gain amplification in a given frequency band. Specifically, consider an open-loop frequency response $H(j\omega)$ that provides amplification over the specified frequency band but is not constant over that range. If $G(s)$ in Figure 11.3(a) is chosen to be constant, $G(s) = K$, then the closed-loop frequency response $Q(j\omega)$ is

$$Q(j\omega) = \frac{H(j\omega)}{1 + KH(j\omega)} \tag{11.5}$$

If over the specified frequency range

$$|KH(j\omega)| \gg 1 \tag{11.6}$$

then

$$Q(j\omega) \simeq \frac{1}{K} \tag{11.7}$$

That is, the closed-loop frequency response is constant, as desired. This of course assumes that the system in the feedback path can be designed so that its frequency response $G(j\omega)$ has a constant gain K over the desired frequency band, which is precisely what we assumed we could not ensure for $H(j\omega)$. The difference between the requirements on $H(j\omega)$ and those on $G(j\omega)$, however, is that $H(j\omega)$ must provide amplification, whereas from eq. (11.7) we see that for the overall closed-loop system to provide a gain greater than unity, K must be less than 1. That is, $G(j\omega)$ must be an attenuator over the specified range of frequencies. In general, an attenuator with approximately flat frequency characteristics is considerably easier to realize than an amplifier with approximately flat frequency response (as an attenuator can be constructed from passive elements).

The use of feedback to flatten the frequency response incurs some cost, however.

---

†See J. K. Roberge, *Operational Amplifiers: Theory and Practice* (New York: John Wiley and Sons, Inc., 1975).

From eqs. (11.6) and (11.7) we see that

$$|H(j\omega)| \gg \frac{1}{K} \simeq Q(j\omega)$$

so that the closed-loop gain will be substantially less than the open-loop gain. The specific application of this result to extending the bandwidth of an amplifier is explored in Problem 11.1.

11.2.3 Stabilization of Unstable Systems

As we mentioned in the introduction, another extremely important application of feedback systems is in the stabilization of systems which without feedback are unstable. Examples of systems in which feedback stabilization is used include the control of the trajectory of a rocket, the regulation of nuclear reactions in a nuclear power plant, the stabilization of an aircraft, and the natural and regulatory control of animal populations.

To illustrate how feedback can be used to stabilize an unstable system, let us consider a simple first-order continuous-time system with

$$H(s) = \frac{b}{s - a} \tag{11.8}$$

With $a > 0$, the system is unstable. Choosing the system function $G(s)$ to be a constant gain K, the closed-loop system function $Q(s)$ in eq. (11.1) becomes

$$Q(s) = \frac{H(s)}{1 + KH(s)}$$

$$= \frac{b}{s - a + Kb} \tag{11.9}$$

The closed-loop system will be stable if the pole is moved into the left half of the s-plane. This will be the case if

$$K > \frac{a}{b} \tag{11.10}$$

Thus, we can stabilize this system with a constant gain in the feedback loop if that gain is chosen to satisfy eq. (11.10). This type of feedback system is referred to as a *proportional feedback system*, since the signal that is fed back is proportional to the output of the system.

As another example, consider the second-order system

$$H(s) = \frac{b}{s^2 + a} \tag{11.11}$$

If $a > 0$, the system is an oscillator [i.e., $H(s)$ has its poles on the $j\omega$-axis] and the impulse response of the system is sinusoidal. If $a < 0$, $H(s)$ has one pole in the right-half plane. Thus, in either case the system is unstable. In fact, as considered in Problem 11.6, the system function given in eq. (11.11) with $a < 0$ can be used to model the dynamics of the inverted pendulum which was described in the introduction.

Let us first consider the use of proportional feedback for this example; that is, we take

$$G(s) = K \tag{11.12}$$

In this case, substituting into eq. (11.1), we obtain

$$Q(s) = \frac{b}{s^2 + (a + Kb)} \tag{11.13}$$

In our discussion of second-order systems in Chapters 4 and 9, we considered a transfer function of the form

$$\frac{\omega_n^2}{s^2 + 2\zeta\omega_n s + \omega_n^2} \tag{11.14}$$

For such a system to be stable, ω_n must be real and positive (i.e., $\omega_n^2 > 0$) and ζ must be positive (corresponding to positive damping). Comparing eqs. (11.13) and (11.14), we see that with proportional feedback we can only influence the value of ω_n^2, and consequently we cannot stabilize the system because we cannot introduce any damping.

To suggest a type of feedback that can be used to stabilize this system, recall the mass–spring–dashpot mechanical system described in our examination of second-order systems in Section 4.12. For this system we saw that the damping in the system was the result of the inclusion of a dashpot, which provided a restoring force proportional to the *velocity* of the mass. This suggests, then, that we consider *proportional-plus-derivative* feedback, that is, $G(s)$ of the form

$$G(s) = K_1 + K_2 s \tag{11.15}$$

which yields

$$Q(s) = \frac{b}{s^2 + bK_2 s + (a + K_1 b)} \tag{11.16}$$

The closed-loop poles will be in the left-half plane and hence the closed-loop system will be stable as long as we choose K_1 and K_2 to guarantee that

$$bK_2 > 0, \qquad a + K_1 b > 0 \tag{11.17}$$

The preceding discussion illustrates how feedback can be used to stabilize continuous-time systems. The stabilization of unstable systems is an important application of feedback for discrete-time systems as well. A common example of discrete-time systems that are unstable in the absence of feedback are models of population growth. To illustrate how feedback can prevent the unimpeded growth of populations, let us consider a simple model for the evolution of the population of a single species of animal. Let $y[n]$ denote the number of animals in the nth generation, and assume that without the presence of any impeding influences, the birth rate is such that the population would double each generation. In this case the basic equation for the population dynamics of the species is

$$y[n] = 2y[n-1] + e[n] \tag{11.18}$$

where $e[n]$ represents any additions to or deletions from the population caused by external influences.

This population model is obviously unstable, with an impulse response that grows exponentially. However, in any ecological system there are a number of factors that will inhibit the growth of a population. For example, limits on the food supply for this species will manifest itself through a reduction in population growth when the number of animals becomes large. Similarly, if the species has natural enemies, it is often reasonable to assume that the population of the predators will grow when the population of the prey increases and consequently that the presence of natural enemies will retard population growth. In addition to natural influences such as these, there may also be effects introduced by man that are aimed at population control. For example, the regulation of the food supply or of the predator population will affect these natural influences. In addition, the stocking of lakes with fish or the importing of animals from other areas can be used to promote growth, and the control of hunting or fishing can also provide a regulating effect. As all of the regulating influences described in this paragraph depend on the size of the population (either naturally or by design), they represent feedback effects.

Based on the preceding discussion, we can separate $e[n]$ into two parts,

$$e[n] = x[n] - r[n] \tag{11.19}$$

where $r[n]$ represents the effect of the regulating influences described above and $x[n]$ incorporates any other external effects, such as the migration of animals, or the effect of singular events, such as a natural disaster or disease. Note that we have included a minus sign in eq. (11.19). This is consistent with our convention of using negative feedback, and here it also has the physical interpretation that, since the uninhibited growth of the population is unstable, the feedback term plays the role of a *retarding* influence. To see how the population can be controlled by the presence of this feedback term, suppose that the regulating influences account for the depletion of a fixed proportion β of the population in each generation. Since, according to our model, the surviving fraction of each generation will double in size, we find that

$$y[n] = 2(1 - \beta)y[n - 1] + x[n] \tag{11.20}$$

Comparing eq. (11.20) with eqs. (11.18) and (11.19), we see that this implies that

$$r[n] = 2\beta y[n - 1] \tag{11.21}$$

The factor of 2 here represents the fact that the depletion of the present population decreases the number of births in the next generation.

This example of the use of feedback is illustrated in Figure 11.5. Here the system function of the forward path is obtained from eq. (11.18) as

$$H(z) = \frac{1}{1 - 2z^{-1}} \tag{11.22}$$

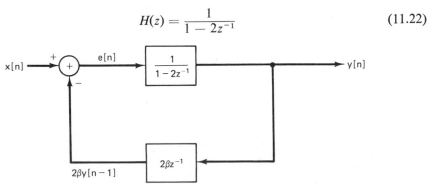

Figure 11.5 Block diagram of a simple feedback model of population dynamics.

while from eq. (11.21) the system function of the feedback path is

$$G(z) = 2\beta z^{-1} \qquad (11.23)$$

Consequently, the closed-loop-system function is

$$Q(z) = \frac{H(z)}{1 + G(z)H(z)} = \frac{1}{1 - 2(1 - \beta)z^{-1}} \qquad (11.24)$$

If $\beta < \frac{1}{2}$ the closed-loop system is still unstable, whereas it is stable if $\frac{1}{2} < \beta < 1$.

Clearly, this example of population growth and control is extremely simplified. For example, the feedback model of eq. (11.21) does not account for the fact that the part of $r[n]$ that is due to the presence of natural enemies depends upon the population of the predators, which in turn has its own growth dynamics. Such effects can be incorporated by making the feedback model more complex to reflect the presence of other dynamics in an ecological system, and the resulting models for the evolution of interacting species are extremely important in ecological studies. However, even without the incorporation of these effects, the simple model that we have described here does illustrate the basic ideas of how feedback can prevent the unlimited proliferation of a species or its extinction. In particular, we can see at an elementary level how human-induced factors can be used. For example, if a natural disaster or an increase in the population of natural enemies causes a drastic decrease in population of a species, a tightening of limits on hunting or fishing and accelerated efforts to increase the population can be used to decrease β in order to *destabilize* the system to allow for rapid growth until a normal-size population is again attained.

Note also that for this type of problem it is not usually the case that one wants strict stability. Specifically, if the regulating influences are such that $\beta = \frac{1}{2}$, and if all other external influences are zero (i.e., if $x[n] = 0$), then $y[n] = y[n-1]$. Therefore, as long as $x[n]$ is small and averages to zero over several generations, a value of $\beta = \frac{1}{2}$ will result in an essentially constant population. However, for this value of β the system is unstable, since in this case eq. (11.20) reduces to

$$y[n] = y[n-1] + x[n]$$

That is, the system is equivalent to an accumulator. Thus, if $x[n]$ is a unit step, the output grows without bound. Consequently, if a steady trend is expected in $x[n]$, caused, for example, by a migration of animals into a region, a value of $\beta > \frac{1}{2}$ would need to be used to stabilize the system and thus to keep the population within bounds and maintain an ecological balance.

In addition to problems such as the one just described, discrete-time feedback techniques are of great importance in a wide variety of applications involving continuous-time systems. Specifically, the extraordinary flexibility of digital systems has made the implementation of *sampled data feedback systems* an extremely attractive option. In such a system the output of a continuous-time system is sampled, some processing is done on the resulting sequence of samples, and a discrete sequence of feedback commands is generated. This sequence is then converted to a continuous-time signal which is fed back and subtracted from the external input to produce the actual input to the continuous-time system. Problem 11.14 investigates an example of the use of a sampled data feedback system to stabilize an unstable continuous-time system.

11.2.4 Tracking Systems

As mentioned in Section 11.1, one of the important applications of feedback is in the design of systems in which the objective is to have the output track or follow the input. There is a broad range of problems in which tracking is an important component. For example, the telescope pointing problem discussed in Section 11.0 is a tracking problem. Specifically, the feedback system of Figure 11.1(c) and (d) has as its input the desired pointing angle, and the purpose of the feedback loop is to provide a mechanism for driving the telescope to follow the commanded input. In airplane autopilots the input is the desired flight path of the vehicle, and the autopilot feedback system uses the aircraft control surfaces (rudder, ailerons, and elevator) in order to keep the aircraft on the prescribed course.

To illustrate some of the issues that arise in the design of tracking systems, consider the discrete-time feedback system depicted in Figure 11.6(a). The examination of discrete-time tracking systems of this form often arises in analyzing the characteristics of sampled data tracking systems for continuous-time applications (one example of such a system is a digital autopilot). In Figure 11.6(a) $H_p(z)$ denotes the system

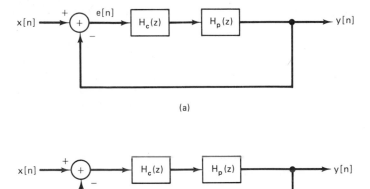

Figure 11.6 (a) Discrete-time tracking system; (b) tracking system of (a) with a disturbance $d[n]$ in the feedback path accounting for the presence of measurement errors.

function of the system whose output is to be controlled. This system is often referred to as the *plant*, terminology that can be traced to applications such as the control of power plants, heating systems, and chemical processing plants. The system function $H_c(z)$ represents a compensator which is the element to be designed. Here the input to the compensator is the tracking error, that is, the difference $e[n]$ between the input $x[n]$ and the output $y[n]$. The output of the compensator is the commanded input to the plant [for example, the actual voltage applied to the motor in the feedback system of Figure 11.1(c) and (d) or the actual physical input to the drive system of the rudder on an aircraft].

To simplify notation, let $H(z) = H_c(z)H_p(z)$. In this case the application of eq. (11.2) yields the relationship

$$Y(z) = \frac{H(z)}{1 + H(z)} X(z)$$

Also, since $Y(z) = H(z)E(z)$, we see that

$$E(z) = \frac{1}{1 + H(z)} X(z) \tag{11.25}$$

or, specializing to $z = e^{j\Omega}$,

$$E(e^{j\Omega}) = \frac{1}{1 + H(e^{j\Omega})} X(e^{j\Omega}) \tag{11.26}$$

Eq. (11.26) provides us with some insight into the design of tracking systems. Specifically, for good tracking performance we would like $e[n]$ or, equivalently, $E(e^{j\Omega})$ to be small. That is,

$$\frac{1}{1 + H(e^{j\Omega})} X(e^{j\Omega}) \simeq 0$$

Consequently, for that range of frequencies for which $X(e^{j\Omega})$ is nonzero we would like $|H(e^{j\Omega})|$ to be large. Thus we have one of the fundamental principles of feedback system design: Good tracking performance requires a large gain. This desire for a large gain, however, must typically be tempered for several reasons. One reason is that if the gain is *too* large, the closed-loop system may have undesirable characteristics (such as too little damping) or might in fact become unstable. This possibility is discussed in the next subsection and is also addressed by the methods developed in the remaining sections of this chapter.

In addition to the issue of stability, there are other reasons for wanting to limit the gain in a tracking system. For example, in implementing a tracking system we must measure the output $y[n]$ in order to compare it to the command input $x[n]$, and any measuring device used will have inaccuracies and error sources (such as thermal noise in the electronics of the measuring device). In Figure 11.6(b) we have included such error sources in the form of a disturbance input $d[n]$ in the feedback loop. Some simple system function algebra yields the following relationship between $Y(z)$ and the transforms $X(z)$ and $D(z)$ of $x[n]$ and $d[n]$:

$$Y(z) = \left[\frac{H(z)}{1 + H(z)} X(z) \right] - \left[\frac{H(z)}{1 + H(z)} D(z) \right] \tag{11.27}$$

From this expression we see that in order to minimize the influence of $d[n]$ on $y[n]$ we would like $H(z)$ to be small so that the second term on the right-hand side of eq. (11.27) is small.

From the preceding development we see that the goals of tracking and of minimizing the effect of measurement errors are conflicting, and one must take this into account in coming up with an acceptable system design. In general the design depends on more detailed information concerning the characteristics of the input $x[n]$ and the disturbance $d[n]$. For example, in many applications $x[n]$ has a significant amount of its energy concentrated at low frequencies, while measurement error sources such as thermal noise have a great deal of energy at high frequencies. Consequently one usually designs the compensator $H_c(z)$ so that $|H(e^{j\Omega})|$ is large at low frequencies and is small for Ω near $\pm\pi$.

There are a variety of other issues that one must consider in designing tracking systems, such as the presence of disturbances at other points in the feedback loop (for example, the effect of wind on aircraft motion must be taken into account in designing an autopilot). The methods of feedback system analysis introduced in this chapter provide the necessary tools for examining each of these issues. In Problem 11.9 we use some of these tools to investigate several other aspects of the problem of designing tracking systems.

11.2.5 Destabilization Caused by Feedback

As we mentioned in the introduction and again in the preceding subsection, as well as having many applications, feedback can also have undesirable effects and can in fact cause instability. For example, consider the telescope pointing system illustrated in Figure 11.1. From the discussion in the preceding section we know that it would be desirable to have a large amplifier gain in order to achieve good performance in tracking the desired pointing angle. On the other hand, as we increase the gain we are likely to obtain faster tracking response at the expense of a reduction in system damping, resulting in significant overshoot and ringing in response to changes in the desired angle. Furthermore, it is possible that instability would result if the gain is increased too much.

Another common example of the possible destabilizing effect of feedback is that of feedback in audio systems. Consider the situation depicted in Figure 11.7(a). Here a loudspeaker produces an audio signal that is an amplified version of the sounds picked up by a microphone. Note that in addition to other audio inputs, the sound coming from the speaker itself may be sensed by the microphone. How strong this particular signal is depends upon the distance between the speaker and the microphone. Specifically, because of the attenuating properties of air, the larger this distance is, the weaker the signal is that reaches the microphone. In addition, due to the finite propagation speed of sound waves, there is time delay between the signal produced by the speaker and that sensed by the microphone.

This audio feedback system is represented in block-diagram form in Figure 11.7(b). Here the constant K_2 in the feedback path represents the attenuation and T is the propagation delay. The constant K_1 is the amplifier gain. Also, in this example note that the output from the feedback path is *added* to the external input. This is an example of *positive feedback*. As discussed at the beginning of this section, the use of a negative sign in the definition of our basic feedback system of Figure 11.3 is purely convention, and positive and negative feedback systems can be analyzed using the same tools. For example, as illustrated in Figure 11.7(c), the feedback system of Figure 11.7(b) can be written as a negative feedback system by adding a minus sign to the feedback-path system function. From this figure and from eq. (11.1) we can determine the closed-loop system function as

$$Q(s) = \frac{K_1}{1 - K_1 K_2 e^{-sT}} \tag{11.28}$$

In Example 11.7 we will return to this example, and, using a technique that we will

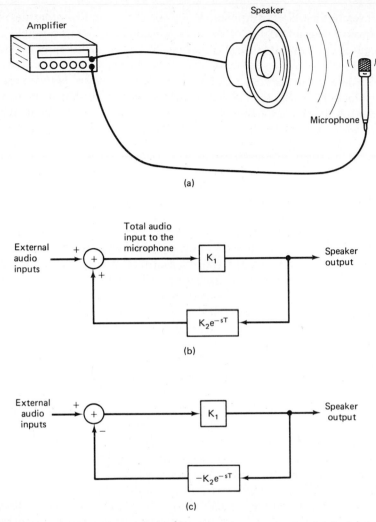

Figure 11.7 (a) Pictorial representation of the phenomenon of audio feedback; (b) block-diagram representation of (a); (c) block diagram (b) redrawn as a negative feedback system. *Note: e^{-sT} is the system function of a T-second time delay.*

develop in Section 11.4, we will show that the system of Figure 11.7 is unstable if

$$K_1 K_2 > 1 \qquad (11.29)$$

Since the attenuation due to the propagation of sound through the air decreases (i.e., K_2 *increases*) as the distance between the speaker and microphone decreases, we see that if the microphone is placed too close to the speaker so that eq. (11.29) is satisfied, the system will be unstable. The result of this instability is an excessive amplification and distortion of audio signals.

In this section we have described a number of the applications of feedback. These and others, such as the use of feedback in the implementation of recursive discrete-time filters (Problem 11.5), are considered in more detail in the problems. From our examination of the uses of feedback and the possible destabilizing effects that it can have, it is clear that some care must be taken in designing and analyzing feedback systems to ensure that the closed-loop system behaves in a desirable fashion. Specifically, in Sections 11.2.3 and 11.2.5, we have seen several examples of feedback systems in which the characteristics of the closed-loop system can be significantly altered by changing the values of one or two parameters in the feedback system. In the remaining sections of this chapter we develop several techniques for analyzing the effect of changes in such parameters on the closed-loop system.

11.3 ROOT-LOCUS ANALYSIS OF LINEAR FEEDBACK SYSTEMS

As we have seen in a number of the examples and applications that we have discussed, a useful and often encountered type of feedback system is that in which the system has an adjustable gain K associated with it. As this gain is varied, it is of interest to examine how the poles of the closed-loop system change, since the locations of these poles tell us a great deal about the behavior of the closed-loop system. For example, in stabilizing an unstable system, the adjustable gain is used to move the poles into the left-half plane for a continuous-time system or inside the unit circle for a discrete-time system. In addition, in Problem 11.1 we show that feedback can be used to broaden the bandwidth of a first-order system by moving the pole so as to decrease the time constant of the system. Furthermore, just as feedback can be used to relocate the poles to improve system performance, as we saw in Section 11.2.5, there is the potential danger that with an improper choice of feedback a stable system can be destabilized, which is generally undesirable.

In this section we discuss a particular method for examining the locus (i.e., the path) in the complex plane of the poles of the closed-loop system as an adjustable gain is varied. The procedure, referred to as the *root-locus method*, is a graphical technique for plotting the closed-loop poles of a rational system function $Q(s)$ or $Q(z)$ as a function of the value of the gain. The technique works in an identical manner for both continuous-time and discrete-time systems.

11.3.1 An Introductory Example

To illustrate the basic nature of the root-locus method for a feedback system, let us reexamine the discrete-time example considered in the preceding section, and specified by the system functions

[eq. (11.22)]
$$H(z) = \frac{1}{1 - 2z^{-1}} = \frac{z}{z - 2} \qquad (11.30)$$

[eq. (11.23)]
$$G(z) = 2\beta z^{-1} = \frac{2\beta}{z} \qquad (11.31)$$

where β now is viewed as an adjustable gain. Then, as we noted earlier, the closed-loop system function is

$$Q(z) = \frac{1}{1 - 2(1 - \beta)z^{-1}} = \frac{z}{z - 2(1 - \beta)} \tag{11.32}$$

In this example it is straightforward to identify the closed-loop pole as being located at $z = 2(1 - \beta)$. In Figure 11.8(a) we have plotted the locus of the pole for this system as β varies from 0 to $+\infty$. In part (b) of this figure we have plotted the locus as β varies from 0 to $-\infty$. In each plot we have indicated the point $z = 2$ which is the open-loop pole [i.e., it is the pole of $Q(z)$ for $\beta = 0$]. As β increases from 0 the pole moves to the left of the point $z = 2$ along the real axis, and we have indicated this by including an arrow on the thick line to show how the pole changes as β is increased. Similarly, for $\beta < 0$, the pole of $Q(z)$ moves to the *right* of $z = 2$, and the direction of the arrow in Figure 11.8(b) indicates how the pole changes as the *magnitude* of β increases. For $\frac{1}{2} < \beta < \frac{3}{2}$ the pole lies inside the unit circle and thus the system is stable.

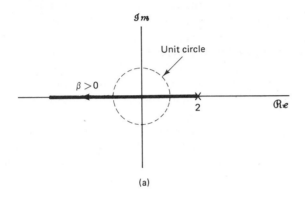

(a)

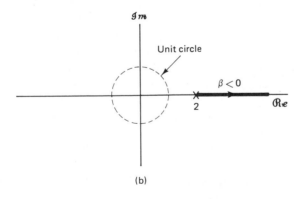

(b)

Figure 11.8 Root locus for the closed-loop system of eq. (11.32) as the value of β is varied: (a) $\beta > 0$; (b) $\beta < 0$. Note that we have marked the point $z = 2$ that corresponds to the pole location when $\beta = 0$.

As a second example, consider a continuous-time feedback system with

$$H(s) = \frac{s}{s - 2} \tag{11.33}$$

$$G(s) = \frac{2\beta}{s} \tag{11.34}$$

where β again represents the adjustable gain. Since $H(s)$ and $G(s)$ in this example are algebraically identical to the preceding example, with the exception that z is replaced by s, the same will be true for the closed-loop system function

$$Q(s) = \frac{s}{s - 2(1 - \beta)} \qquad (11.35)$$

and the locus of the pole as a function of β will be identical to the preceding example.

The relationship of these two examples stresses the fact that the locus of the poles is determined by the algebraic expressions for the system functions of the forward and feedback paths and is not inherently associated with whether the system is a continuous-time or discrete-time system. However, the interpretation of the result is intimately connected with its continuous-time or discrete-time context. In the discrete-time case, it is the location of the poles in relation to the unit circle that is important, whereas in the continuous-time case it is their location in relation to the imaginary axis. Thus, as we have seen for the discrete-time example in eq. (11.32), the system is stable for $\frac{1}{2} < \beta < \frac{3}{2}$, while the continuous-time system of eq. (11.35) is stable for $\beta > 1$.

11.3.2 Equation for the Closed-Loop Poles

In the simple example considered in the previous section the root locus was easy to plot, since we could first explicitly determine the closed-loop pole as a function of the gain parameter and then could plot the location of the pole as we changed the gain. For more complex systems, one cannot expect to find such simple closed-form expressions for the closed-loop poles. However, it is still possible to sketch accurately the locus of the poles as the value of the gain parameter is varied from $-\infty$ to $+\infty$, *without* actually solving for the location of the poles for any specific value of the gain. This technique for determining the root locus is extremely useful in gaining insight into the characteristics of a feedback system. Also, as we develop the method we will see that once we have determined the root locus, there is a relatively straightforward procedure for determining the value of the gain parameter that produces a closed-loop pole at any specified location along the root locus. We will phrase our discussion in terms of the Laplace transform variable s, with the understanding that it applies equally well to the discrete-time case.

Consider a modification of the basic feedback system of Figure 11.3(a), where either $G(s)$ or $H(s)$ is cascaded with an adjustable gain K. This is illustrated in Figure 11.9. In either of these cases the denominator of the closed-loop system function is $1 + KG(s)H(s)$.† Therefore, the equation for the poles of the closed-loop system are the solutions of the equation

$$1 + KG(s)H(s) = 0 \qquad (11.36)$$

Rewriting eq. (11.36), we obtain the basic equation determining the closed-loop poles as

---

†In the following discussion we assume for simplicity that there is no pole–zero cancellation in the product $G(s)H(s)$. The presence of such pole–zero cancellations does not cause any real difficulties, and the procedure that we will outline in this section is easily extended to this case (Problem 11.16). In fact, the simple example at the start of this section [eqs. (11.33) and (11.34)] *does* involve a pole–zero cancellation, at $s = 0$.

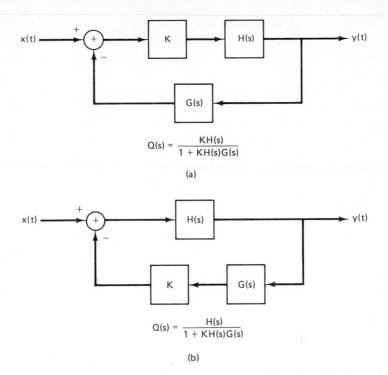

$$Q(s) = \frac{KH(s)}{1 + KH(s)G(s)}$$

(a)

$$Q(s) = \frac{H(s)}{1 + KH(s)G(s)}$$

(b)

Figure 11.9 Feedback systems containing an adjustable gain: (a) system in which the gain is located in the forward path; (b) system with the gain in the feedback path.

$$G(s)H(s) = -\frac{1}{K} \tag{11.37}$$

The technique for plotting the root locus is based on the properties of this equation and its solutions. In the remainder of this section we will discuss some of these properties and will indicate how they can be exploited in determining the root locus.

11.3.3 The End Points of the Root Locus: The Closed-Loop Poles for K = 0 and |K| = +∞

Perhaps the most immediate observation that one can make about the root locus is that obtained by examining eq. (11.37) for $K = 0$ and for $|K| = \infty$. In particular, note that for $K = 0$ the solution of eq. (11.37) must yield the poles of $G(s)H(s)$, since $1/K = \infty$. To illustrate this property, recall the example given by eqs. (11.33) and (11.34). If we let β play the role of K, we see that eq. (11.37) becomes

$$\frac{2}{s - 2} = -\frac{1}{\beta} \tag{11.38}$$

Therefore from the preceding observation, we conclude that for $\beta = 0$, the pole of the system will be located at the pole of $2/(s - 2)$ (i.e., at $s = 2$), which agrees with what we depicted in Figure 11.8.

Suppose now that $|K| = \infty$. In this case $1/K = 0$, so that the solutions of eq.

(11.37) must approach the *zeros* of $G(s)H(s)$. If the order of the numerator of $G(s)H(s)$ is smaller than the denominator, then some of these zeros, equal in number to the difference in order between the denominator and numerator, will be at infinity (see Chapter 9).

Referring again to the example given in eq. (11.38), since the order of the denominator of $2/(s-2)$ is 1, while the order of the numerator is zero, we conclude that in this example there is one zero at infinity and no zeros in the finite s-plane. Thus as $|\beta| \to \infty$, the closed-loop pole approaches infinity. Again this agrees with Figure 11.8, where we see that the magnitude of the pole increases without bound as $|\beta| \to \infty$ for either $\beta > 0$ or $\beta < 0$.

While the observations we have just made provide us with basic information as to the closed-loop pole locations for extreme values of K, it is the observation made in the following subsection that is the key to our being able to plot the root locus without actually solving for the closed-loop poles as explicit functions of the gain.

11.3.4 The Angle Criterion

Consider again eq. (11.37). Since the right-hand side of this equation is real, a point s_0 can be a closed-loop pole only if the left-hand side of eq. (11.37), $G(s_0)H(s_0)$, is also real. Writing

$$G(s_0)H(s_0) = |G(s_0)H(s_0)|e^{j\angle G(s_0)H(s_0)} \tag{11.39}$$

we see that for $G(s_0)H(s_0)$ to be real, it must be true that

$$e^{j\angle G(s_0)H(s_0)} = \pm 1$$

That is, for s_0 to be a closed-loop pole, we must have

$$\angle G(s_0)H(s_0) = \text{integer multiple of } \pi \tag{11.40}$$

Returning to the example of eq. (11.38), we see immediately that in order for $2/(s_0 - 2)$ to be real, it is necessary that s_0 be real. For more complex system functions, it is not *this* easy to determine the values of s_0 for which $G(s_0)H(s_0)$ is real. However, as we will see, the use of the angle criterion given by eq. (11.40), together with the geometric method described in Chapter 9 for evaluating $\angle G(s_0)H(s_0)$, greatly facilitates the determination of the root locus.

The angle criterion given by eq. (11.40) provides us with a direct method for determining if a point s_0 could possibly be a closed-loop pole for *some* value of the gain K. A further examination of eq. (11.37) provides us with a way in which to calculate the value of the gain corresponding to any point on the root locus. Specifically suppose that s_0 satisfies

$$\angle G(s_0)H(s_0) = \text{odd multiple of } \pi \tag{11.41}$$

Then $e^{j\angle G(s_0)H(s_0)} = -1$, and from eq. (11.39) we see that

$$G(s_0)H(s_0) = -|G(s_0)H(s_0)| \tag{11.42}$$

Substituting eq. (11.42) into eq. (11.37), we find that if

$$K = \frac{1}{|G(s_0)H(s_0)|} \qquad (11.43)$$

then s_0 is a solution of the equation and hence a closed-loop pole.

Similarly, if s_0 satisfies the condition

$$\sphericalangle G(s_0)H(s_0) = \text{even multiple of } \pi \qquad (11.44)$$

then

$$G(s_0)H(s_0) = |G(s_0)H(s_0)| \qquad (11.45)$$

Thus if

$$K = -\frac{1}{|G(s_0)H(s_0)|} \qquad (11.46)$$

then s_0 is a solution of eq. (11.37) and hence a closed-loop pole.

For the example given in eq. (11.38), if s_0 is on the real line and $s_0 < 2$, then

$$\sphericalangle \left(\frac{2}{s_0 - 2} \right) = -\pi$$

and from eq. (11.43) the value of β for which s_0 is the closed-loop pole is given by

$$\beta = \frac{1}{\left| \dfrac{2}{s_0 - 2} \right|} = \frac{2 - s_0}{2}$$

That is,

$$s_0 = 2(1 - \beta)$$

which agrees with eq. (11.35).

Summarizing the last two observations that we have made, we see that the *root locus* for the closed-loop system, that is, the set of points in the complex s-plane that are closed-loop poles for *some* value of K as K varies from $-\infty$ to $+\infty$, are precisely those points that satisfy the angle condition of eq. (11.40). Furthermore,

---

1. A point s_0 for which

$$\sphericalangle G(s_0)H(s_0) = \text{odd multiple of } \pi \qquad (11.47)$$

is on the root locus and is a closed-loop pole for some value of $K > 0$. The value of the gain that makes s_0 a closed-loop pole is given by eq. (11.43).

2. A point s_0 for which

$$\sphericalangle G(s_0)H(s_0) = \text{even multiple of } \pi \qquad (11.48)$$

is on the root locus and is a closed-loop pole for some value of $K < 0$. The value of the gain that makes s_0 a closed-loop pole is given by eq. (11.46).

Therefore, we have now reduced the problem of determining the root locus to that of searching for points that satisfy the angle requirements eqs. (11.47) and (11.48). These equations can be refined further into a set of properties that aid in sketching the root locus. Before discussing these, however, let us consider a simple example.

Example 11.1

Consider

$$H(s) = \frac{1}{s+1}, \qquad G(s) = \frac{1}{s+2} \qquad (11.49)$$

Recall that in Section 9.4 we discussed the geometric evaluation of Laplace transforms. Specifically, we saw that the angle of the rational Laplace transform

$$\frac{\prod\limits_{k=1}^{m} (s - \beta_k)}{\prod\limits_{k=1}^{n} (s - \alpha_k)}$$

evaluated at some point s_0 in the complex plane equals the sum of the angles of the vectors from each of the zeros to s_0 minus the sum of the angles from each of the poles to s_0. Applying this to the product of $G(s)H(s)$, where $G(s)$ and $H(s)$ are given in eq. (11.49), we can determine geometrically those points in the s-plane that satisfy the angle criteria, eqs. (11.47) and (11.48), and therefore can sketch the root locus.

In Figure 11.10 we have plotted the poles of $G(s)H(s)$ and have denoted by θ

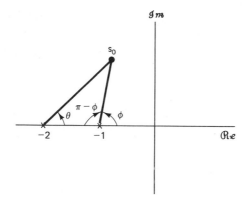

Figure 11.10 Geometric procedure for angle evaluation for Example 11.1.

and ϕ the angles from each of the poles to the point s_0. Let us first test the angle criterion for points s_0 on the real axis. Note first that the angle contribution from both poles is zero when s is on the real axis to the right of -1. Thus,

$$\sphericalangle G(s_0)H(s_0) = 0 = 0 \cdot \pi, \qquad s_0 \text{ real and greater than } -1$$

and by eq. (11.48) these points are on the root locus for $K < 0$. For points between the two poles, the pole at -1 contributes an angle of $-\pi$, and the pole at -2 contributes 0. Thus,

$$\sphericalangle G(s_0)H(s_0) = -\pi, \qquad s_0 \text{ real, } -2 < s_0 < -1$$

These points are on the locus for $K > 0$. Finally, each pole contributes an angle of $-\pi$ when s_0 is real and less than -2. Therefore, these points are on the locus for $K < 0$.

Let us now examine points in the upper half of the s-plane. Since we know that complex poles occur in conjugate pairs, we can immediately determine the poles in the lower half-plane after we have examined the upper half. Referring to Figure 11.10, the angle of $G(s_0)H(s_0)$ at the point s_0 is

$$\sphericalangle G(s_0)H(s_0) = -(\theta + \phi)$$

Also, it is clear that as s_0 ranges over the upper half-plane (but not the real axis), we have

$$0 < \theta < \pi, \quad 0 < \phi < \pi$$

Thus,

$$-2\pi < \sphericalangle G(s)H(s) < 0$$

Therefore, we see immediately that *no* point in the upper half-plane can be on the locus for $K < 0$ [since $\sphericalangle G(s)H(s)$ never equals an even multiple of π]. In addition, if s_0 is to be on the locus for $K > 0$, we must have

$$\sphericalangle G(s_0)H(s_0) = -(\theta + \phi) = -\pi$$

or

$$\theta = \pi - \phi$$

Examining the geometry of Figure 11.10, we see that this occurs only for those points located on the straight line that is parallel to the imaginary axis and that bisects the line joining the poles at -1 and -2. We have now examined the entire s-plane and have determined all those points on the root locus. In addition, we know that for $K = 0$, the closed-loop poles equal the poles of $G(s)H(s)$, and as $|K| \to \infty$, the closed-loop poles go to the zeros of $G(s)H(s)$, which in this case are both at infinity. Putting this together we can draw the entire root locus, depicted in Figure 11.11, where we have indicated the direction of increasing $|K|$, both for $K > 0$ and for $K < 0$.

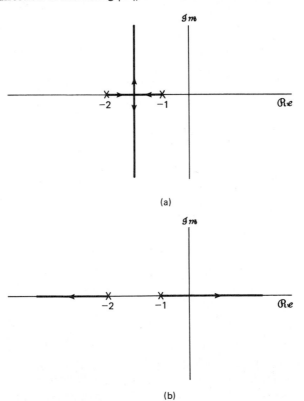

(a)

(b)

Figure 11.11 Root locus for Example 11.1: (a) $K > 0$; (b) $K < 0$. The poles of $G(s)H(s)$, which are located at $s = -1$ and $s = -2$, have been indicated.

Note from the figure that for $K > 0$ there are two branches of the root locus and that the same is true for $K < 0$. The reason for the existence of two branches is that for this example the closed-loop system is second-order and consequently has two poles for any specified value of K. Therefore, the root locus has two branches, *each* of which traces the location of *one* of the closed-loop poles as K is varied, and for any particular value of K there is one closed-loop pole on each branch. Again, if we wish to calculate the value of K for which a specific point s_0 on the locus is a closed-loop pole, we can use eqs. (11.43) and (11.46).

11.3.5 Properties of the Root Locus

The procedure outlined in the preceding section and example provides us, in principle, with a method for determining the root locus for any continuous- or discrete-time LTI feedback system. That is, we simply determine, graphically or otherwise, all those points that satisfy eq. (11.47) or (11.48). Fortunately, there are a number of other geometrical properties concerning root loci which make the sketching of a locus far less tedious. To begin our discussion of these properties, let us assume that we have placed $G(s)H(s)$ in the following standard form:

$$G(s)H(s) = \frac{s^m + b_{m-1}s^{m-1} + \ldots + b_0}{s^n + a_{n-1}s^{n-1} + \ldots + a_0} = \frac{\prod_{k=1}^{m}(s - \beta_k)}{\prod_{k=1}^{n}(s - \alpha_k)} \qquad (11.50)$$

where the β_k's denote the zeros and the α_k's denote the poles. In general, these may be complex. We are assuming that the leading coefficient in both the numerator and denominator in eq. (11.50) is $+1$. This can always be achieved by dividing the numerator and denominator by the denominator coefficient of s^n and absorbing the resulting numerator coefficient of s^m into the gain K. For example

$$K\frac{2s + 1}{3s^2 + 5s + 2} = K\frac{\frac{2}{3}s + \frac{1}{3}}{s^2 + \frac{5}{3}s + \frac{2}{3}} = (\tfrac{2}{3})K\frac{s + \frac{1}{2}}{s^2 + \frac{5}{3}s + \frac{2}{3}}$$

and the quantity

$$\tfrac{2}{3}K$$

is then regarded as the overall gain that is varied in determining the root locus.
To simplify the discussion somewhat, we also assume that

$$m \leq n \qquad (11.51)$$

Problem 11.17 considers the case $m > n$. The following are some properties that include earlier observations and that aid in our sketching of the root locus.

Property 1: For $K = 0$, the solutions of eq. (11.37) are the poles of $G(s)H(s)$. Since we are assuming n poles, the root locus has n branches, each one starting (for $K = 0$) at a pole of $G(s)H(s)$.

Property 1 is the general version of the property we noted in Example 11.1—that there is one branch of the root locus for each closed-loop pole. The next property is simply a restatement of one of our earlier observations.

Property 2: As $|K| \rightarrow \infty$, each branch of the root locus approaches a zero of $G(s)H(s)$. Since we are assuming that $m \leq n$, $(n - m)$ of these zeros are at infinity.

Property 3: Parts of the real s-axis that lie to the left of an *odd* number of real poles and zeros of $G(s)H(s)$ are on the root locus for $K > 0$. Parts of the real s-axis that lie to the left of an *even* number (possibly zero) of poles and zeros of $G(s)H(s)$ are on the root locus for $K < 0$.

We can see that Property 3 is true as follows. From our discussion in Example 11.1 and from Figure 11.12(a), we see that if a point on the real s-axis is to the right of a real pole or zero of $G(s)H(s)$, that pole or zero contributes zero to $\sphericalangle G(s_0)H(s_0)$. On the other hand, if s_0 is to the left of a zero, that zero contributes $+\pi$, whereas if it is to the left of a pole, we get a contribution of $-\pi$ (since we subtract the pole angles). Hence, if s_0 is to the left of an odd number of real poles and zeros, the total contribution of these poles and zeros is an odd multiple of π, whereas if it is to the left of an even number of real poles and zeros, the total contribution is an even mul-

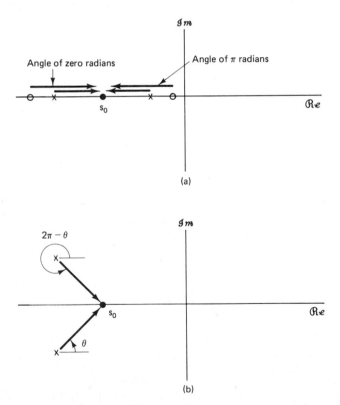

Figure 11.12 (a) Angle contribution from real poles and zeros to a point on the real axis; (b) total angle contribution from a complex-conjugate pole pair to a point on the real axis.

tiple of π. From eqs. (11.47) and (11.48), we will have the result stated in Property 3 if we can show that the total contribution from all poles and zeros with nonzero imaginary parts is an even multiple of π. The key here is that such poles and zeros occur in complex-conjugate pairs, and we can consider the contribution from each such pair, as illustrated in Figure 11.12(b). The symmetry in the picture clearly indicates that the sum of the angles from this pair to any point s_0 on the real axis is precisely 2π. Summing over all conjugate zero pairs and subtracting the sum over all conjugate pole pairs, we get the desired result. Thus, *any* segment of the real line between real poles or zeros is on the root locus either for $K > 0$ or for $K < 0$, depending on whether it lies to the left of an odd or an even number of poles and zeros of $G(s)H(s)$.

As one consequence of Properties 1 to 3, consider a segment of the real axis between two poles of $G(s)H(s)$ with no zeros between these poles. From Property 1, the root locus begins at the poles and from Property 3 the entire portion of the real axis between these two poles will lie on the root locus for a positive or negative range of values of K. Therefore, as $|K|$ increases from zero, the two branches of the root locus that begin at these poles move toward each other along the segment of the real axis between the poles. From Property 2, as $|K|$ increases toward infinity, each branch of the root locus must approach a zero. Since there are no zeros along that portion of the real axis, the only way that this can happen is if the branches break off into the complex plane for $|K|$ sufficiently large. This is illustrated in Figure 11.11, where the locus for $K > 0$ has a portion between two real poles. As K is increased, the root locus eventually leaves the real axis, forming two complex-conjugate branches. Summarizing this discussion, we have the following property of the root locus.

Property 4: Branches of the root locus between two real poles must break off into the complex plane for $|K|$ large enough.

Properties 1 to 4 serve to illustrate how characteristics of the root locus can be deduced form eqs. (11.37), (11.47), and (11.48). In many cases plotting the poles and zeros of $G(s)H(s)$ and then using these four properties suffices to provide a reasonably accurate sketch of the root locus (see Examples 11.2 and 11.3 to follow). In addition to these properties, however, there are numerous other characteristics of the root locus that allow one to obtain sketches of increasing accuracy. For example, from Property 2 we know that $(n - m)$ branches of the root locus approach infinity. In fact these branches approach infinity at specific angles that can be calculated, and therefore these branches are asymptotically parallel to lines at these angles. Moreover, it is possible to draw in the asymptotes and in particular to determine the point at which the asymptotes intersect. These two properties and several others are illustrated in Problems 11.21–25. A more detailed development of the root-locus method can be found in more advanced texts.†

In the remainder of this section we present two examples, one in continuous time and one in discrete time, that illustrate how the four properties that we have described

†See, for example, any of the texts on feedback listed in the Bibliography at the end of the book.

allow us to sketch the root locus and to deduce the stability characteristics of a feed-back system as the gain K is varied.

Example 11.2

Consider

$$G(s)H(s) = \frac{s-1}{(s+1)(s+2)}$$

In this example, from Properties 1 and 2, the root locus for both K positive and K negative starts at the points $s = -1$ and $s = -2$. One branch terminates at the zero at $s = 1$ and the other at infinity.

Let us first consider $K > 0$. The root locus in this case is illustrated in Figure 11.13(a). From Property 3 we can identify the regions of the real axis that are on the root locus, specifically $\Re e\{s\} < -2$ and $-1 < \Re e\{s\} < 1$. Therefore, one branch of the root locus for $K > 0$ originates at $s = -1$ and approaches $s = 1$ as $K \to +\infty$. The other branch begins at $s = -2$ and extends to the left toward $\Re e\{s\} = -\infty$ as $K \to +\infty$.

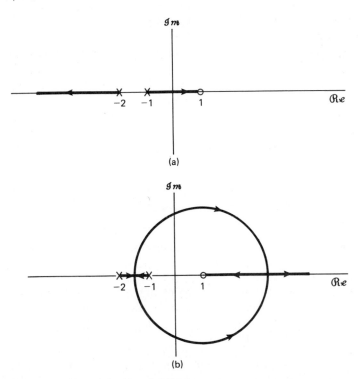

(a)

(b)

Figure 11.13 Root locus for Example 11.2: (a) $K > 0$; (b) $K < 0$. The poles of $G(s)H(s)$ at $s = -1$ and $s = -2$ and the zero of $G(s)H(s)$ at $s = 1$ are indicated in the figure.

Thus, we see that for $K > 0$, if K is sufficiently large, the system will become unstable, as one of the closed-loop poles moves into the right-half plane. The procedure that we have used for sketching the root locus does not, of course, indicate the value of K for which this instability develops. However, for this particular example, we see that the value of K for which the instability occurs corresponds to the root locus passing through $s = 0$. Consequently, from eq. (11.43) the corresponding value of K is

$$K = \frac{1}{|G(0)H(0)|} = 2$$

Thus the system is stable for $0 \leq K < 2$ but is unstable for $K \geq 2$.

For $K < 0$, the portions of the real axis lying on the root locus are $\mathcal{R}e\{s\} > 1$ and $-2 < \mathcal{R}e\{s\} < -1$. Thus, the root locus again starts at the points $s = -2$ and $s = -1$, moving into the region $-2 < \mathcal{R}e\{s\} < -1$. At some point it breaks off into the complex plane and follows a trajectory such that it returns to the real axis for $s > 1$. On returning to the real axis, one branch moves to the left toward the zero at $s = 1$ and the other to the right toward $s = \infty$, as indicated in Figure 11.13(b), where we have displayed an accurate plot of the root locus for $K < 0$.

Rules can be also developed to indicate the location at which the root locus leaves and enters the real axis. Even without that precise a description, however, we can sketch the general shape of the root locus in Figure 11.13(b) and can therefore deduce that for $K < 0$ the system also becomes unstable for $|K|$ sufficiently large.

Example 11.3

Consider the discrete-time feedback system illustrated in Figure 11.14. In this case

$$G(z)H(z) = \frac{z^{-1}}{(1 - \frac{1}{2}z^{-1})(1 - \frac{1}{4}z^{-1})} = \frac{z}{(z - \frac{1}{2})(z - \frac{1}{4})}$$

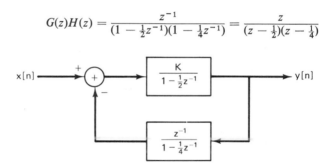

Figure 11.14 Discrete-time feedback system of Example 11.3.

As discussed at the beginning of this section, the techniques for sketching the root locus of a discrete-time feedback system are identical to the continuous-time case. Therefore, in a manner exactly analogous to that used in the preceding example, we can deduce the basic form of the root locus for this example, which is illustrated in Figure 11.15. In this case the portion of the real axis between the two poles of $G(z)H(z)$ (at $z = 1/4$ and $z = 1/2$) is on the root locus for $K > 0$, and as K increases, the locus breaks off into the complex plane and returns to the axis at some point in the left-half plane. From there one branch approaches the zero of $G(z)H(z)$ at $z = 0$ and the other approaches infinity as $K \longrightarrow \infty$. The form of the root locus for $K < 0$ consists of two branches on the real axis, one approaching 0 and the other infinity.

As we remarked earlier, while the form of the root locus does not depend on whether the system is a continuous- or discrete-time system, any conclusion regarding stability based on examining the locus certainly does. In particular, for this example we can conclude that for $|K|$ sufficiently large, the system is unstable, since one of the two poles has magnitude greater than 1.

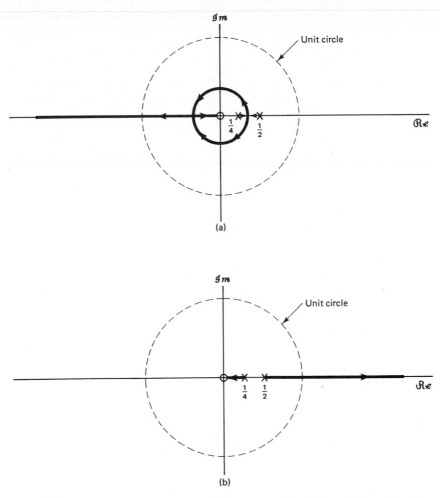

Figure 11.15 Root locus for Example 11.3: (a) $K > 0$; (b) $K < 0$. The poles of $G(z)H(z)$ at $z = 1/4$ and $z = 1/2$ and the zero of $G(z)H(z)$ at $z = 0$ are indicated in the figure.

11.4 THE NYQUIST STABILITY CRITERION

As developed in Section 11.3, the root-locus technique provides detailed information concerning the location of closed-loop poles as the system gain is varied. From such plots, one can determine the damping of the system and its stability characteristics as K is varied. Determination of the root locus requires the analytic description of the system functions of the forward and feedback paths and is applicable only when these transforms are rational. For example, it cannot be directly applied in situations in which our knowledge of these system functions is obtained purely from experimentation.

In this section we introduce another method for the determination of the stability of feedback systems as a function of an adjustable gain parameter. This technique, referred to as the Nyquist criterion, differs from the root-locus method in two basic ways. Unlike the root-locus method, the Nyquist criterion does *not* provide detailed information concerning the location of the closed-loop poles as a function of K but rather simply determines whether or not the system is stable for any specified value of K. On the other hand, the Nyquist criterion can be applied to nonrational system functions and in situations in which no analytic description of the forward and feedback path system functions is available.

Our objective in this section is to outline the basic ideas behind the Nyquist criterion for both continuous-time and discrete-time systems. As we will see, the discrete- and continuous-time Nyquist tests are both the result of the same fundamental concept, although, as with the root locus method, the actual criteria for stability differ because of the differences between continuous and discrete time. More detailed developments of the ideas behind the Nyquist criterion and its use in the design of feedback systems can be found in texts on the analysis and synthesis of feedback systems and automatic control systems.[†]

To introduce the method, let us recall that the poles of the closed-loop systems of Figure 11.9 and their discrete-time counterparts are the solutions of the equation

$$1 + KG(s)H(s) = 0 \qquad \text{(continuous time)} \tag{11.52}$$

and

$$1 + KG(z)H(z) = 0 \qquad \text{(discrete time)} \tag{11.53}$$

For discrete-time systems, we want to determine whether any of the solutions of eq. (11.53) lie outside the unit circle, and for continuous-time systems whether any of the solutions of eq. (11.52) lie in the right half of the s-plane. The Nyquist criterion determines this by examination of the values of $G(s)H(s)$ along the $j\omega$-axis and the values of $G(z)H(z)$ along the unit circle. The basis for these criteria is the encirclement property which we develop in the following subsection.

11.4.1 The Encirclement Property

Consider a general rational function $W(p)$, where p is a complex variable,[‡] and suppose that we plot $W(p)$ for values of p along a closed contour in the p-plane which we traverse in a clockwise direction. This is illustrated in Figure 11.16 for a function $W(p)$ that has two zeros and no poles. In Figure 11.16(a) we have shown a closed contour C in the p-plane, and in Figure 11.16(b) we have plotted the closed contour of the values of $W(p)$ as p varies around the contour C. In this example there is one zero of $W(p)$ inside the contour and one zero of $W(p)$ outside the contour. At any point p on the contour C, the angle of $W(p)$ is the sum of the angles of the two vectors $\mathbf{v}_1$ and $\mathbf{v}_2$ to the point p. As we traverse the contour once, the angle ϕ_1 of the vector from the zero *inside* the contour encounters a net change of -2π

[†]See the texts on feedback systems listed in the Bibliography at the end of the book.

[‡]Because we will use the property we are about to develop for both continuous-time and discrete-time feedback systems, we have chosen to phrase the general property in terms of a general complex variable p. In the next subsection we use this property to analyze continuous-time feedback systems where the complex variable is s. Following this, in Section 11.4.3 we use the encirclement property for discrete-time feedback systems in which context the complex variable is z.

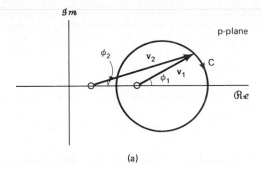

(a)

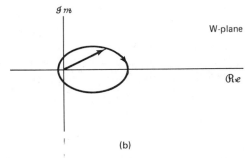

(b)

Figure 11.16 Basic encirclement property.

radians, whereas the angle ϕ_2 of the vector from the zero *outside* the contour encounters no net change. Thus, on the plot of $W(p)$, there is a net change in angle of -2π. Said another way, the plot of $W(p)$ in Figure 11.16(b) encircles the origin once in the clockwise direction. More generally, for an arbitrary rational $W(p)$, as we traverse a closed contour in the clockwise direction, any poles and zeros of $W(p)$ outside the contour will contribute no net change to the angle of $W(p)$, whereas each zero inside the contour will contribute a net change of -2π and each pole inside will contribute a net change of $+2\pi$. Since each net change of -2π in $W(p)$ corresponds to one clockwise encirclement of the origin in the $W(p)$ plot, we can state the following basic *encirclement property:*

> As a closed path C in the p-plane is traversed once in the clockwise direction, the plot of $W(p)$ along the contour encircles the origin in the clockwise direction a net number of times equal to the number of zeros minus the number of poles contained within the contour.

In applying this statement, a counterclockwise encirclement is interpreted as the negative of one clockwise encirclement. For example, if there is one pole and no zeros inside the contour there will be one counterclockwise or equivalently minus one clockwise encirclement.

Example 11.4

Consider the function

$$W(p) = \frac{p - 1}{(p + 1)(p^2 + p + 1)} \tag{11.54}$$

In Figure 11.17 we have depicted several closed contours in the complex p-plane and the corresponding plots of $W(p)$ along each of these contours. In Figure 11.17(a), the contour C_1 does not encircle any of the poles or zeros of $W(p)$, and consequently the plot of $W(p)$ has no net encirclements of zero. In Figure 11.17(b), only the pole at $p = -1$ is contained within the contour C_2, and the plot of $W(p)$ encircles the origin once in the counterclockwise direction. In Figure 11.17(c), C_3 encircles all three poles, and the plot of $W(p)$ encircles the origin three times in a counterclockwise direction. In Figure 11.17(d), C_4 encircles one pole and one zero, and therefore the plot of $W(p)$ has no net encirclements of the origin. Finally, in Figure 11.17(e), all of the poles and the one zero of $W(p)$ are contained within C_5, and thus the plot of $W(p)$ along this contour has two net counterclockwise encirclements of the origin.

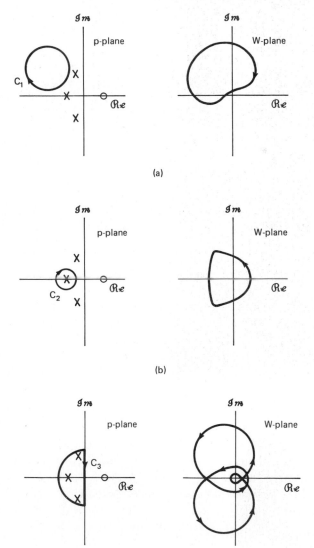

(a)

(b)

(c)

Figure 11.17 Basic encirclement property for Example 11.4.

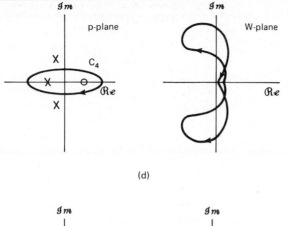

(d)

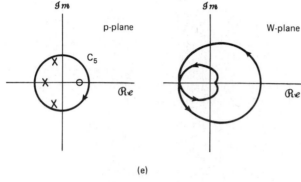

(e)

Figure 11.17 (cont.)

11.4.2 The Nyquist Criterion
for Continuous-Time LTI Feedback Systems

In this section we exploit the encirclement property in examining the stability of the continuous-time feedback system of Figure 11.9. Stability of this system requires that no zeros of $1 + KG(s)H(s)$ or equivalently of the function

$$R(s) = \frac{1}{K} + G(s)H(s) \tag{11.55}$$

lie in the right half of the s-plane. Thus, in applying the general result developed above, we can consider a contour as indicated in Figure 11.18. From the plot of $R(s)$ as s traverses the contour C, we can obtain a count of the number of zeros minus the number of poles of $R(s)$ contained within the contour by counting the number of clockwise encirclements of the origin. As M increases to infinity, this then corresponds to the number of zeros minus the number of poles of $R(s)$ in the right half of the s-plane.

Let us examine the evaluation of $R(s)$ along the contour in Figure 11.18 as M increases to infinity. Along the semicircular portion of the contour extending into the right-half plane, we must ensure that $R(s)$ remains bounded as M increases. Specifically, we will assume that $R(s)$ has at least as many poles as zeros. In this case

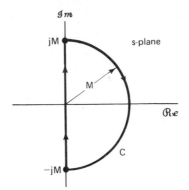

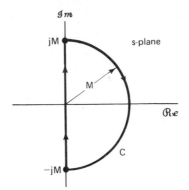

Figure 11.18 Closed contour containing a portion of the right-half plane; as $M \longrightarrow \infty$, the contour encloses the entire right-half plane.

$$R(s) = \frac{b_n s^n + b_{n-1} s^{n-1} + \ldots + b_0}{a_n s^n + a_{n-1} s^{n-1} + \ldots + a_0}$$

and

$$\lim_{|s| \to \infty} R(s) = \frac{b_n}{a_n} = \text{constant}$$

Therefore, as M increases to infinity, the value of $R(s)$ does not change as we traverse the semicircular part of the contour, and consequently the constant value along this part is equal to the value of $R(s)$ at the end points, [i.e., $R(j\omega)$ at $\omega = \pm\infty$].

Therefore, the plot of $R(s)$ along the contour of Figure 11.18 can be obtained by plotting $R(s)$ along the part of the contour that coincides with the imaginary axis, that is, the plot of $R(j\omega)$ as ω varies from $-\infty$ to $+\infty$. Since $R(j\omega)$ equals $1/K + G(j\omega)H(j\omega)$, $R(s)$ along the contour can be drawn from knowledge of $G(j\omega)$ and $H(j\omega)$. If both the forward and feedback path systems are stable, these are simply the frequency-response functions of these systems. However, the encirclement property for the general function $W(p)$ is simply a property of complex functions. It has nothing to do with whether this function arose as the Laplace or z-transform of any signal and consequently has nothing to do with regions of convergence. Thus, even if the forward and feedback path systems are unstable, if we examine the plot of the *function* $R(j\omega) = 1/K + G(j\omega)H(j\omega)$ for $-\infty < \omega < \infty$, we can use the encirclement property to count the number of zeros minus the number of poles of $R(s)$ that lie in the right-half plane.

Furthermore, from eq. (11.55) we see that the poles of $R(s)$ are simply the poles of $G(s)H(s)$, while the zeros of $R(s)$ are the closed-loop poles. In addition, since $G(j\omega)H(j\omega) = R(j\omega) - 1/K$, it follows that the plot of $G(j\omega)H(j\omega)$ encircles the point $-1/K$ *exactly* as many times as $R(j\omega)$ encircles the origin. The plot of $G(j\omega)H(j\omega)$ as ω varies from $-\infty$ to $+\infty$ is called the *Nyquist plot*. From the encirclement property we see that the net number of clockwise encirclements of the point $-1/K$ by the Nyquist plot equals the number of right-half-plane closed-loop poles minus the number of right-half-plane poles of $G(s)H(s)$. For the closed-loop system to be stable, we require no right-half-plane closed-loop poles. This yields the *Nyquist stability criterion:*

For the closed-loop system to be stable, the net number of clockwise encircle-ments of the point $-1/K$ by the Nyquist plot of $G(j\omega)H(j\omega)$ must equal *minus* the number of right-half-plane poles of $G(s)H(s)$. Equivalently, the net number of *counterclockwise* encirclements must *equal* the number of right-half-plane poles of $G(s)H(s)$.

For example, if the forward and feedback path systems are stable, then the Nyquist plot is simply the plot of the frequency response of the cascade of these two systems. In this case, since there are no poles of $G(s)H(s)$ in the right-half plane, the Nyquist criterion requires that for stability the net number of encirclements of the point $-1/K$ must equal zero.

Example 11.5

Let

$$G(s) = \frac{1}{s+1}, \qquad H(s) = \frac{1}{\frac{1}{2}s+1}$$

The Bode plot for $G(j\omega)H(j\omega)$ is shown in Figure 11.19. The Nyquist plot depicted in Figure 11.20 is constructed directly from these plots of the log-magnitude and phase of $G(j\omega)H(j\omega)$. That is, each point on the Nyquist plot has polar coordinates con-

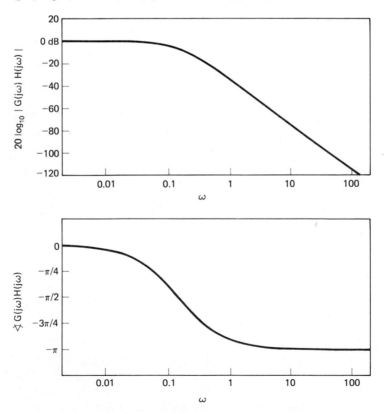

Figure 11.19 Bode plot for $G(j\omega)H(j\omega)$ in Example 11.5.

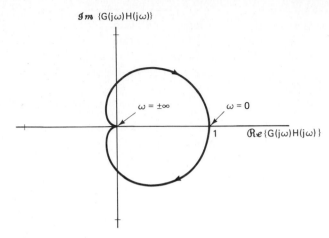

Figure 11.20 Nyquist plot of $G(j\omega)H(j\omega)$ for Example 11.5. The arrow on the curve indicates the direction of increasing ω.

sisting of the magnitude $|G(j\omega)H(j\omega)|$ and angle $\sphericalangle G(j\omega)H(j\omega)$ for some value of ω. The coordinates of $G(j\omega)H(j\omega)$ for $\omega < 0$ are obtained from the values for $\omega > 0$ through the use of the conjugate symmetry property of $G(j\omega)H(j\omega)$. This property manifests itself geometrically in a very simple way which facilitates the sketching of the Nyquist plot for any feedback system comprised of systems with real impulse responses. Specifically, since $|G(-j\omega)H(-j\omega)| = |G(j\omega)H(j\omega)|$ and $\sphericalangle G(-j\omega)H(-j\omega) = -\sphericalangle G(j\omega)H(j\omega)$, the Nyquist plot of $G(j\omega)H(j\omega)$ for $\omega \leq 0$ is a reflection about the real axis of the plot for $\omega \geq 0$. Note also that we have included an arrow on the Nyquist plot of Figure 11.20. This arrow indicates the direction of increasing ω. That is, it indicates the direction in which the Nyquist plot is traversed for the counting of encirclements in the application of the Nyquist criterion.

In this example there are no right-half-plane open-loop poles, and, consequently, the Nyquist criterion requires that, for stability, there be no net encirclements of the point $-1/K$. Thus, by inspection of Figure 11.20, the closed-loop system will be stable if the point $-1/K$ falls outside the Nyquist contour, that is, if

$$-\frac{1}{K} \leq 0 \qquad \text{or} \qquad -\frac{1}{K} > 1$$

which is equivalent to

$$K \geq 0 \qquad \text{or} \qquad 0 > K > -1$$

Combining these two conditions we obtain the result that the closed-loop system will be stable for any choice of K greater than -1.

Example 11.6

Consider now

$$G(s)H(s) = \frac{s+1}{(s-1)(\frac{1}{2}s+1)}$$

The Nyquist plot for this system is indicated in Figure 11.21. For this example, $G(s)H(s)$ has one right-half-plane pole. Thus, for stability we require one counterclockwise encirclement of the point $-1/K$, which in turn requires that the point $-1/K$ fall inside the contour. Thus, we will have stability if and only if $-1 < -1/K < 0$, that is, if $K > 1$.

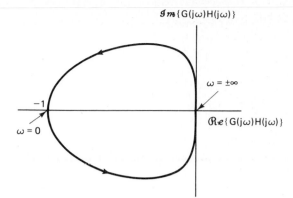

$\mathcal{I}m\{G(j\omega)H(j\omega)\}$

$\omega = \pm\infty$

-1

$\mathcal{R}e\{G(j\omega)H(j\omega)\}$

$\omega = 0$

Figure 11.21 Nyquist plot for Example 11.6. The arrow on the curve indicates the direction of increasing ω.

In the foregoing discussion we have presented a simplified version of the Nyquist criterion for determining stability. There are many refinements of the method. For example, as we have developed it, the Nyquist plot can be drawn without any difficulties for stable or unstable $G(s)H(s)$ as long as there are no poles of $G(s)H(s)$ exactly *on* the $j\omega$-axis. When such poles do occur, the value of $G(j\omega)H(j\omega)$ is infinite at these points. However, as considered in Problem 11.30, the Nyquist criterion can be modified to allow for poles of $G(s)H(s)$ on the $j\omega$-axis. In addition, as mentioned at the beginning of this section, the Nyquist criterion can also be extended to the case in which $G(s)$ and $H(s)$ are not rational. For example, if the forward and feedback path systems are both stable, it can be shown that the Nyquist criterion is the same when the system functions are nonrational as it is for the rational case. That is, the closed-loop system is stable if there are no net encirclements of the point $-1/K$. To illustrate the application of the Nyquist criterion for nonrational system functions, we present the following example.

Example 11.7

Consider the acoustic feedback example discussed in Section 11.2.5. Referring to Figure 11.7(b), let $K = K_1 K_2$ and

$$G(s)H(s) = -e^{-sT} = e^{-(sT + j\pi)}$$

where we have used the fact that $e^{-j\pi} = -1$. In this case

$$G(j\omega)H(j\omega) = e^{-j(\omega T + \pi)}$$

and as ω varies from $-\infty$ to $+\infty$, $G(j\omega)H(j\omega)$ traces out a circle of radius one in the clockwise direction, with one full revolution for every change of $2\pi/T$ in ω. This is illustrated in Figure 11.22. Since the forward and feedback path systems are stable [the cascade $G(s)H(s)$ is simply a time delay], the Nyquist stability criterion indicates that the closed-loop system will be stable if and only if $-1/K$ does not fall inside the unit circle. Equivalently, we require for stability that

$$|K| < 1$$

Since K_1 and K_2 represent an acoustic gain and attenuation, respectively, they are both positive, which yields the stability condition

$$K_1 K_2 < 1$$

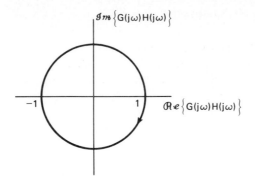

Figure 11.22 Nyquist plot for Example 11.7.

11.4.3 The Nyquist Criterion for Discrete-Time LTI Feedback Systems

As in the continuous-time case, the Nyquist stability criterion for discrete-time systems is based on the fact that the difference in the number of poles and zeros inside a contour, for a rational function, can be determined by examining a plot of the value of the function along the contour. The difference between the continuous- and discrete-time cases is the choice of the contour. For the discrete-time case, stability of the closed-loop feedback system requires that no zeros of

$$R(z) = \frac{1}{K} + G(z)H(z) \tag{11.56}$$

lie outside the unit circle.

Recall that the encirclement property relates to poles and zeros *inside* any specified contour. On the other hand, in examining the stability of a discrete-time system, we are concerned with the zeros of $R(z)$ *outside* the unit circle. Therefore, in order to make use of the encirclement property, we first make a simple modification. Specifically, let us consider the rational function

$$\hat{R}(z) = R\left(\frac{1}{z}\right)$$

obtained by replacing z by its reciprocal. As seen in Problem 10.15 if z_0 is a zero (pole) of $R(z)$, then $1/z_0$ is a zero (pole) of $\hat{R}(z)$. Since $1/|z_0|$ is less than 1 if $|z_0| > 1$, any zero or pole of $R(z)$ *outside* the unit circle corresponds to a zero or pole of $\hat{R}(z)$ *inside* the unit circle.

From the basic encirclement property we know that as z traverses the unit circle in a clockwise direction, the net number of clockwise encirclements of the origin by $\hat{R}(z)$ equals the difference between the number of its zeros and poles inside the unit circle. However, note from our discussion above that this equals the difference between the number of zeros and poles of $R(z)$ *outside* the unit circle. Furthermore, on the unit circle, $z = e^{j\Omega}$ and $1/z = e^{-j\Omega}$. Therefore,

$$\hat{R}(e^{j\Omega}) = R(e^{-j\Omega})$$

From this we see that evaluating $\hat{R}(z)$ as z traverses the unit circle in the clockwise direction is identical to evaluating $R(z)$ as z traverses the unit circle in the *counterclockwise* direction. In summary, as the unit circle is traversed once in the counterclockwise direction (e.g., as Ω increases from 0 to 2π), the plot of $R(e^{j\Omega})$ encircles the origin in a clockwise direction a net number of times equal to the number of zeros minus the number of poles of $R(z)$ outside the unit circle.

Much as in the continuous-time case, counting the encirclements of the origin by $R(e^{j\Omega})$ is equivalent to counting the number of encirclements of the point $-1/K$ by the plot of $G(e^{j\Omega})H(e^{j\Omega})$, again referred to as the Nyquist plot, which is plotted as Ω varies from 0 to 2π. Also, the poles of $R(z)$ are precisely the poles of $G(z)H(z)$ and the zeros of $R(z)$ are the closed-loop poles. Therefore, the encirclement property stated in the preceding paragraph implies that the net number of clockwise encirclements by the Nyquist plot of the point $-1/K$ equals the number of closed-loop poles outside the unit circle minus the number of poles of $G(z)H(z)$ outside the unit circle. In order that the closed-loop system be stable, we require no closed-loop poles outside the unit circle. This yields the discrete-time Nyquist stability criterion:

For the closed-loop system to be stable, the net number of clockwise encirclements of the point $-1/K$ by the Nyquist plot of $G(e^{j\Omega})H(e^{j\Omega})$ as Ω varies from 0 to 2π must equal *minus* the number of poles of $G(z)H(z)$ that lie outside the unit circle. Equivalently, the net number of *counterclockwise* encirclements must equal the number of poles of $G(z)H(z)$ outside the unit circle.

Example 11.8

Let

$$G(z)H(z) = \frac{z^{-2}}{1 + \frac{1}{2}z^{-1}} = \frac{1}{z(z + \frac{1}{2})}$$

The Nyquist plot for this example is shown in Figure 11.23. Since $G(z)H(z)$ has no poles outside the unit circle, for the stability of the closed-loop system there must be no encirclements of the $(-1/K)$ point. From the figure we see that this will be the case either if $(-1/K) < -1$ or if $(-1/K) > 2$. Thus, the system is stable for $-\frac{1}{2} < K < 1$.

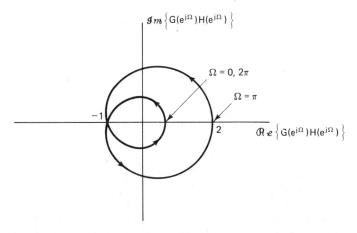

Figure 11.23 Nyquist plot for Example 11.8. The arrow on the curve indicates the direction in which the curve is traversed as Ω increases from 0 to 2π.

Just as in continuous time, if the forward and feedback path systems are stable, then the Nyquist plot can be obtained from the frequency responses $H(e^{j\Omega})$ and $G(e^{j\Omega})$ of these systems. If the forward and feedback path systems are unstable, then these frequency responses are not defined. Nevertheless, the *function* $G(z)H(z)$ can still be evaluated on this contour, and the Nyquist stability criterion can be applied.

As we have seen in this section, the Nyquist stability criterion provides a useful method for determining the range of values of the gain K for which a continuous- or discrete-time feedback system is stable (or unstable). This criterion and the root-locus method are extremely important tools in the design and implementation of feedback systems, and each has its own uses and limitations. For example, the Nyquist criterion can be applied to nonrational system functions, whereas the root-locus method cannot be. On the other hand, root-locus plots allow us to examine not only stability but also other characteristics of the closed-loop system response, such as damping, oscillation frequency, and so on, which are readily identifiable from the location of the poles of the closed-loop system. In the next section we introduce an additional tool for the analysis of feedback systems, which highlights another important characteristic of closed-loop system behavior.

11.5 GAIN AND PHASE MARGINS

In this section we introduce and examine the concept of the margin of stability in a feedback system. Specifically, it is often of interest not only to know *if* a feedback system is stable but also to determine how much the gain in the system can be perturbed and how much additional phase shift can be added to the system before it becomes unstable. Information such as this is of importance since in many applications the forward and feedback system functions are known only approximately or may change slightly during operation because of wear, the effect of high temperatures on components, or similar influences.

As an example, consider the telescope pointing system described in Section 11.0. This system consists of a motor, a potentiometer converting the shaft angle to a voltage, and an amplifier that is used to amplify the voltage representing the difference between the desired and the actual shaft angles. Assuming that we have obtained approximate descriptions of each of these components, we can set the amplifier gain so that the system will be stable *if these approximate descriptions are accurate*. However, the amplifier gain and the constant of proportionality that describes the angle-voltage characteristic of the potentiometer are never known exactly, and therefore the actual gain in the feedback system may differ from the nominal value assumed in designing the system. Furthermore, the damping characteristics of the motor cannot be determined with absolute precision, and thus the actual time constant of the motor response may differ from the approximate specification. For example, if the actual motor time constant is larger than the nominal value used in the design, the motor will respond more sluggishly than anticipated, thereby producing an effective time delay in the feedback system. As we have dis-

cussed in earlier chapters and as we will again in Example 11.11, time delays have
the effect of increasing the negative phase in the frequency response of a system, and
this phase shift can have a destabilizing influence. Because of the possible presence of
gain and phase errors such as those that we have just described, it is clearly desirable
to set the amplifier gain so that there is some margin for error, that is, so that the
actual system will remain stable even if it differs somewhat from the approximate
model used in the design process.

In this section we introduce one method for quantifying the margin of stability
in a feedback system. To do this, we consider a closed-loop system as depicted in
Figure 11.24, which has been designed to be stable assuming given, nominal values
for the forward and feedback path system functions. For our discussion here we let
$H(s)$ and $G(s)$ denote these nominal values. Also, since the basic concepts are identical
for both continuous- and discrete-time systems, we will again focus our development
on the continuous-time case, and at the end of the section we illustrate the application of these ideas to a discrete-time example.

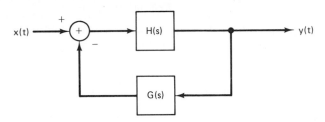

Figure 11.24 Typical feedback system designed to be stable assuming nominal
descriptions for $H(s)$ and $G(s)$.

To assess the margin of stability in our feedback system, suppose that the
actual system is as depicted in Figure 11.25, where we have allowed for the possibility
of a gain K and negative phase shift $-\phi$ in the feedback path. In our nominal system
K is unity and ϕ is zero, but in the actual system these quantities may have different
values. Therefore, it is of interest to know how much variation can be tolerated in
these quantities without losing closed-loop system stability. In particular, the *gain
margin* of the feedback system is defined as the minimum amount of additional gain
K, with $\phi = 0$, that is required so that the closed-loop system becomes unstable.
Similarly, the *phase margin* is the additional amount of phase shift, with $K = 1$,
that is required for the system to be unstable. By convention the phase margin is

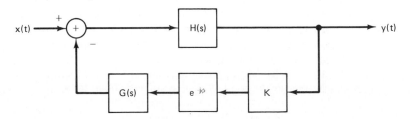

Figure 11.25 Feedback system containing possible gain and phase deviations
from the nominal description depicted in Figure 11.24.

expressed as a positive quantity. That is, it equals the magnitude of the additional negative phase shift for which the feedback system becomes unstable.

Since the closed-loop system of Figure 11.24 is stable, the system of Figure 11.25 can only become unstable if, as K and ϕ are varied, at least one pole of the closed-loop system crosses the $j\omega$-axis. If a pole of the closed-loop is on the $j\omega$-axis at, say, $\omega = \omega_0$, then at this frequency

$$1 + Ke^{-j\phi}G(j\omega_0)H(j\omega_0) = 0$$

or

$$Ke^{-j\phi}G(j\omega_0)H(j\omega_0) = -1 \tag{11.57}$$

Note that with $K = 1$ and $\phi = 0$, by our assumption of stability for the nominal feedback system of Figure 11.24, there is no value of ω_0 for which eq. (11.57) is satisfied. The gain margin of this system is the minimum value of $K > 1$ for which eq. (11.57) has a solution for *some* ω_0 with $\phi = 0$. That is, the gain margin is the smallest value of K for which the equation

$$KG(j\omega_0)H(j\omega_0) = -1 \tag{11.58}$$

has a solution. Similarly, the phase margin is the minimum value of ϕ for which eq. (11.57) has a solution for some value of ω_0 when $K = 1$. In other words, the phase margin is the smallest value of $\phi > 0$ for which the equation

$$e^{-j\phi}G(j\omega_0)H(j\omega_0) = -1 \tag{11.59}$$

has a solution.

To illustrate the calculation and graphical interpretation of gain and phase margins, we consider the following example.

Example 11.9

Let

$$G(s)H(s) = \frac{4(1 + \tfrac{1}{2}s)}{s(1 + 2s)(1 + 0.05s + (0.125s)^2)}$$

The Bode plot for this example is shown in Figure 11.26. Note that as discussed in Problem 4.47, the factor of $1/j\omega$ in $G(j\omega)H(j\omega)$ contributes $-90°$ ($-\pi/2$ radians) of phase shift and a 20-dB per decade decrease in $|G(j\omega)H(j\omega)|$. To determine the gain margin, we observe that with $\phi = 0$, the only frequency at which eq. (11.58) can be satisfied is that for which $\sphericalangle G(j\omega_0)H(j\omega_0) = -\pi$. At this frequency, the gain margin in decibels can be identified by inspection of Figure 11.26. Specifically, we first examine Figure 11.26(b) to determine the frequency ω_1 at which the angle curve crosses the $-\pi$ radians line. Locating the point at this same frequency in Figure 11.26(a) provides us with the value of $|G(j\omega_1)H(j\omega_1)|$. For eq. (11.58) to be satisfied for $\omega_0 = \omega_1$, K must equal $1/|G(j\omega_1)H(j\omega_1)|$. This value is the gain margin. As illustrated in Figure 11.26(a) the gain margin expressed in decibels can be identified as the amount the log-magnitude curve would have to be shifted up so that the curve intersects the 0-dB line at the frequency ω_1.

In a similar fashion we can determine the phase margin. Note first that the only frequency at which eq. (11.59) can be satisfied is that for which $20\log_{10}|G(j\omega_0)H(j\omega_0)| = 0$. To determine the phase margin we first find the frequency ω_2 in Figure 11.26(a) at which the log-magnitude curve crosses the 0-dB line. Locating the point at this same frequency in Figure 11.26(b) provides us with the value of $\sphericalangle G(j\omega_2)H(j\omega_2)$. For eq. (11.59) to be satisfied for $\omega_0 = \omega_2$ it must be true that the angle of the left-hand side

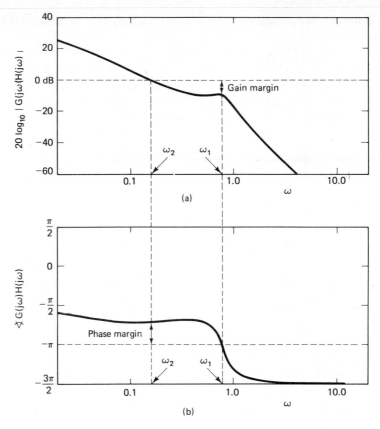

Figure 11.26 Use of Bode plots to calculate gain and phase margins for the system of Example 11.9.

of this equation is $-\pi$. The value of ϕ for which this is true is the phase margin. As illustrated in Figure 11.26(b), the phase margin can be identified as the amount the angle curve would have to be lowered so that the curve intersects the $-\pi$ line at the frequency ω_2.

In determining gain and phase margins, it is not always of interest to identify explicitly the *frequency* at which the poles will cross the $j\omega$-axis. Thus, it is more typical to identify gain and phase margins from a *log magnitude–phase diagram*. For example, the log magnitude–phase diagram for the example of Figure 11.26 is shown in Figure 11.27. In this figure we plot $20 \log_{10}|G(j\omega)H(j\omega)|$ versus $\angle G(j\omega)H(j\omega)$ as ω varies from 0 to $+\infty$. Therefore, because of the conjugate symmetry of $G(j\omega)H(j\omega)$, this plot contains the same information as the Nyquist plot, in which $\mathcal{R}e\{G(j\omega)H(j\omega)\}$ is plotted versus $\mathcal{I}m\{G(j\omega)H(j\omega)\}$ for $-\infty < \omega < \infty$. As we have indicated in the figure, the phase margin can be read off by locating the intersection of the log magnitude–phase plot with the 0-dB line. That is, the phase margin is the amount of additional negative phase shift required to shift the log magnitude–phase curve so that it intersects the 0-dB line with exactly 180° (π radians) of phase shift. Similarly, the gain margin is directly obtained from the intersection

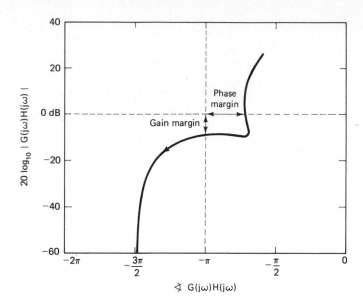

Figure 11.27 Log magnitude–phase plot for the system of Example 11.9.

of the log magnitude–phase curve with the $-\pi$ radians phase line, and this represents the amount of additional gain needed so that the curve crosses the $-\pi$ line with a magnitude of 0 dB.

The following examples provide several other elementary illustrations of log magnitude–phase diagrams.

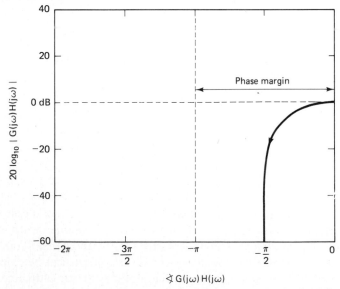

Figure 11.28 Log magnitude–phase plot for the first-order system of Example 11.10.

Example 11.10

Consider

$$G(s)H(s) = \frac{1}{\tau s + 1}, \qquad \tau > 0$$

In this case, we obtain the log magnitude–phase plot depicted in Figure 11.28. This has a phase margin of π and, since the curve does not intersect the $-\pi$ line, this system has infinite gain margin (i.e., we can increase the gain as much as we like and maintain stability). This is consistent with the conclusion that we can draw by examining the system illustrated in Figure 11.29(a). Specifically, in Figure 11.29(b) we have depicted the root locus for this system with $\phi = 0$ and $K > 0$. From this figure it is evident that the system is stable for any positive value of K. In addition, if $K = 1$ and $\phi = \pi$ so that $e^{j\phi} = -1$, the closed-loop system function for the system of Figure 11.29(a) is $1/\tau s$ so that the system is unstable.

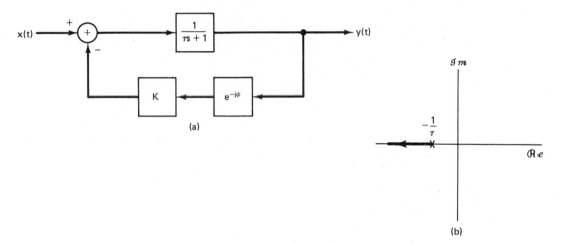

Figure 11.29 (a) First-order feedback system with possible gain and phase variations in the feedback path; (b) root locus for this system with $\phi = 0$, $K > 0$.

Example 11.11

Suppose we now consider the second-order system

$$H(s) = \frac{1}{s^2 + s + 1}, \qquad G(s) = 1 \tag{11.60}$$

The system $H(s)$ has an undamped natural frequency of 1 and a damping ratio of 0.5. The log magnitude–phase plot for this system is illustrated in Figure 11.30. Again we have infinite gain margin, but a phase margin of only $\pi/2$, since it can be shown by a straightforward calculation that $|H(j\omega)| = 1$ for $\omega = 1$, and at this frequency $\sphericalangle H(j\omega) = -\pi/2$.

With this example we can illustrate the type of problem that can be solved using the concepts of gain and phase margins. Specifically, suppose that the feedback system specified by eq. (11.60) cannot be realized. Rather, some unavoidable time delay is introduced in the feedback path. That is,

$$G(s) = e^{-s\tau}$$

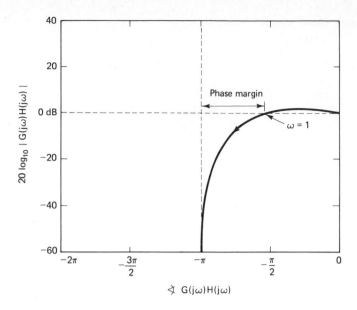

Figure 11.30 Log magnitude–phase plot for the second-order system of Example 11.11.

where τ is the time delay. What we would like to know is how small this time delay must be to ensure the stability of the closed-loop system.

The first point to note is that

$$|e^{-j\omega\tau}| = 1$$

so this time delay does not change the magnitude of $H(j\omega)G(j\omega)$. On the other hand,

$$\sphericalangle e^{-j\omega\tau} = -\omega\tau \text{ radians}$$

Thus, every point on the curve in Figure 11.30 is shifted to the *left*. The amount of the shift is proportional to the value of ω for each point on the log magnitude–phase curve.

From this discussion we see that instability will occur once the phase margin is reduced to zero, and this will occur when the phase shift introduced by the delay is equal to $-\pi/2$ at $\omega = 1$. That is, the critical value τ^* of the time delay satisfies

$$\sphericalangle e^{-j\tau} = -\tau = -\frac{\pi}{2}$$

or (assuming that the units of ω are radians/second)

$$\tau^* \simeq 1.57 \text{ seconds}$$

Thus, for any time delay $\tau < \tau^*$, the system remains stable.

Example 11.12

Consider again the acoustic feedback system discussed in Section 11.2.5 and in Example 11.7. Here we assume that the system of Figure 11.7 has been designed with $K_1K_2 < 1$, so that the closed-loop system is stable. In this case the log magnitude–phase plot for $G(s)H(s) = K_1K_2e^{-(sT+j\pi)}$ is illustrated in Figure 11.31. From this figure we see that the system has infinite phase margin and a gain margin in decibels of $-20\log_{10}(K_1K_2)$ (i.e., this is precisely the gain factor that, when multiplied by K_1K_2, equals 1).

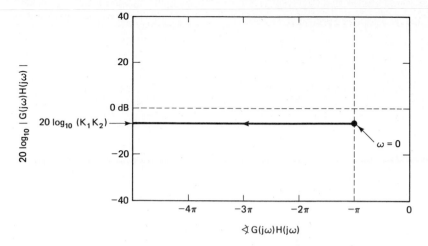

Figure 11.31 Log magnitude–phase plot for Example 11.12.

As indicated at the start of the section, the definitions of gain and phase margins are the same for discrete-time feedback systems. Specifically, if we have a stable discrete-time feedback system, the gain margin is the minimum amount of additional gain required in the feedback system so that the closed-loop system becomes unstable. Similarly, the phase margin is the minimum amount of additional negative phase shift required for the feedback system to be unstable. The following example illustrates the graphical calculation of phase and gain margins for a discrete-time feedback system. The procedure is essentially the same as for continuous-time systems.

Example 11.13

In this example we illustrate the concept of gain and phase margin for the discrete-time feedback system depicted in Figure 11.32. Here

$$G(z)H(z) = \frac{\dfrac{7\sqrt{2}}{4}z^{-1}}{1 - \dfrac{7\sqrt{2}}{8}z^{-1} + \dfrac{49}{64}z^{-2}}$$

and by direct calculation we can check that this system is stable for $K = 1$ and $\phi = 0$. In Figure 11.33 we have displayed the log magnitude–phase diagram for this system; that is, we have plotted $20\log_{10}|G(e^{j\Omega})H(e^{j\Omega})|$ versus $\angle G(e^{j\Omega})H(e^{j\Omega})$ as Ω varies from 0 to 2π. The system has a gain margin of 1.68 dB and a phase margin of 0.0685 radians (3.93°).

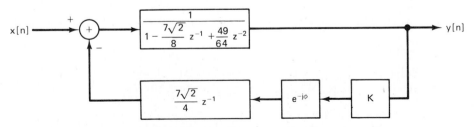

Figure 11.32 Discrete-time feedback system of Example 11.13.

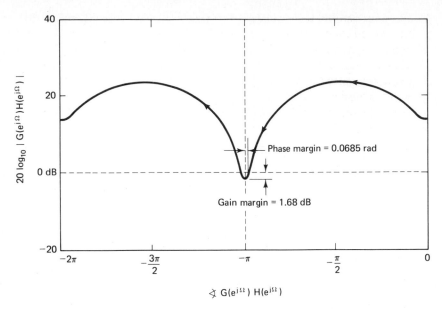

Figure 11.33 Log magnitude–phase diagram for the discrete-time feedback system of Example 11.13.

In concluding this section it should be stressed that the gain margin is the *minimum* value of gain that moves one or more of the closed-loop poles onto the $j\omega$-axis and consequently causes the system to become unstable. It is important to note, however, that this does *not* imply that the system is unstable for *all* values of gain above the value specified by the gain margin. For example, as illustrated in Problem 11.37, as K increases, the root locus may move from the left-half plane into the right-half plane and then cross back into the left-half plane. The gain margin provides us with the information about how much the gain can be increased until the poles *first* reach the $j\omega$-axis, but it tells us nothing about the possibility that the system may again be stable for even larger values of the gain. To obtain such information we must either refer to the root locus or use the Nyquist stability criterion (see Problem 11.37).†

11.6 SUMMARY

In this chapter we have examined a number of the applications and several techniques for the analysis of feedback systems. Specifically, we have seen how the use of Laplace and z-transforms allows us to analyze these systems algebraically and graphically. In Section 11.2 we indicated several of the applications of feedback, including the design of inverse systems, the stabilization of unstable systems, and the design of tracking systems. We also saw that feedback can destabilize as well as stabilize.

†For detailed discussions of this point and also of gain and phase margins and log magnitude–phase diagrams in general, see the texts on feedback listed in the Bibliography at the end of the book.

In Section 11.3 we described the root-locus method for plotting the poles of the closed-loop system as a function of a gain parameter. Here we found that the geometric evaluation of the phase of a rational Laplace or z-transform allowed us to gain a significant amount of insight into the properties of the root locus. These properties often allow us to obtain a reasonably accurate sketch of the root locus without performing complex calculations.

In contrast to the root-locus method, the Nyquist criterion of Section 11.4 is a technique for determining the stability of a feedback system, again as a function of a variable gain, *without* obtaining a detailed description of the location of the closed-loop poles. The Nyquist criterion is applicable to nonrational system functions and thus can be used when all that is available are experimentally determined frequency responses. The same is true of the gain and phase margins described in Section 11.5. These quantities provide a measure of the margin of stability in a feedback system and therefore are of importance to designers in allowing them to determine how robust a system is to discrepancies between estimates of the forward and feedback path system functions and their actual values.

PROBLEMS

11.1. In this problem we provide an illustration of how feedback can be used to increase the bandwidth of an amplifier. Consider an amplifier whose gain falls off at high frequencies. Specifically, suppose that the system function of this amplifier is

$$H(s) = \frac{Ga}{s + a}$$

(a) What is the dc gain of the amplifier (i.e., the magnitude of its frequency response at 0 frequency)?

(b) What is the system time constant?

(c) Suppose that we define bandwidth as the frequency at which the magnitude of the amplifier frequency response is $1/\sqrt{2}$ times its magnitude at dc. What is the bandwidth of this amplifier?

(d) Suppose that we place this amplifier in a feedback loop as depicted in Figure P11.1. What is the dc gain of the closed-loop system? What are the time constant and bandwidth of the closed-loop system?

(e) Find the value of K that leads to a closed-loop bandwidth that is exactly double

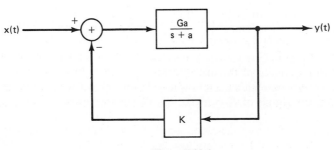

Figure P11.1

the bandwidth of the open-loop amplifier. What are the corresponding closed-loop system time constant and dc gain?

11.2. Consider the feedback system of Figure P11.2. Find the closed-loop poles and zeros of this system for the following values of K:

(i) $K = 0.1$
(ii) $K = 1$
(iii) $K = 10$
(iv) $K = 100$

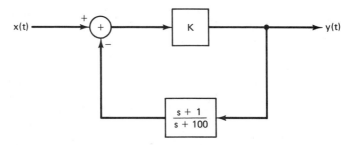

Figure P11.2

11.3. As mentioned in the text, an important class of devices used in the implementation of feedback systems are operational amplifiers. A model for such an amplifier is depicted in Figure P11.3-1. The amplifier's input is the difference between the two voltages $v_2(t)$ and $v_1(t)$, and the output voltage $v_o(t)$ is an amplified version of this input

$$v_o(t) = K[v_2(t) - v_1(t)] \qquad \text{(P11.3-1)}$$

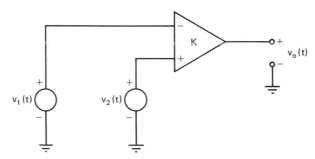

Figure P11.3-1

Consider the operational amplifier connection shown in Figure P11.3-2. In this figure $Z_1(s)$ and $Z_2(s)$ are impedances (that is, each is the system function of an LTI system whose input is the current flowing through the impedance element and whose output is the voltage across the element). Making the approximation that the input impedance of the operational amplifier is infinite and that its output impedance is zero, we obtain the following relationship between $V_1(s)$, $V_i(s)$, and $V_o(s)$, the Laplace transforms of $v_1(t)$, $v_i(t)$, and $v_o(t)$, respectively:

$$V_1(s) = \left[\frac{Z_2(s)}{Z_1(s) + Z_2(s)}\right] V_i(s) + \left[\frac{Z_1(s)}{Z_1(s) + Z_2(s)}\right] V_o(s) \qquad \text{(P11.3-2)}$$

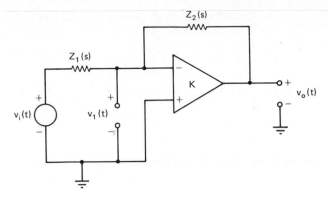

Figure P11.3-2

Also, from eq. (P11.3-1) and Figure P11.3-2, we see that

$$V_o(s) = -KV_1(s) \tag{P11.3-3}$$

(a) Show that the system function

$$H(s) = \frac{V_o(s)}{V_i(s)}$$

for the interconnection of Figure P11.3-2 is identical to the overall closed-loop system function for the system of Figure P11.3-3.

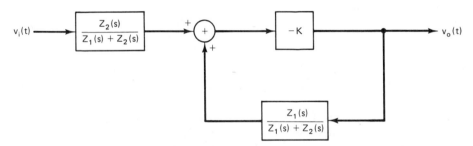

Figure P11.3-3

(b) Show that if $K \gg 1$, then

$$H(s) \simeq -\frac{Z_2(s)}{Z_1(s)}$$

(c) Suppose that $Z_1(s)$ and $Z_2(s)$ are both pure resistances, R_1 and R_2, respectively. A typical value for R_2/R_1 is in the range 1 to 10^3, while a typical value for K is 10^6. Calculate the actual system function for this value of K and for R_2/R_1 equal to 1 and to 10^3, and compare each resulting value to $-R_2/R_1$. This should give you some idea of how good the approximation of part (b) typically is.

(d) One of the important uses of feedback is in the reduction of system sensitivity to variations in parameters. This is particularly important for circuits involving operational amplifiers, which have high gains that may be known only approximately.

(i) Consider the circuit discussed in part (c), with $R_2/R_1 = 10^2$. What is the percentage change in the closed-loop gain of the system if K changes from 10^6 to 5×10^5?

(ii) How large must K be so that a 50% reduction in its value results in only a 1% reduction in the closed-loop gain? Again take $R_2/R_1 = 10^2$.

(e) Consider the circuit of Figure P11.3-4. In this case

$$Z_1(s) = R, \qquad Z_2(s) = \frac{C}{s}$$

Show that this system behaves essentially like an integrator. In what frequency range (expressed in terms of K, R, and C) does this approximation break down?

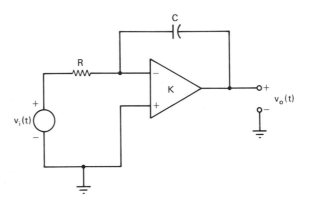

Figure P11.3-4

(f) Consider the circuit depicted in Figure P11.3-5, where the impedance $Z_1(s) = R$, while $Z_2(s)$ is replaced by a diode with an exponential current–voltage relationship. Assume that this relationship is of the form

$$i_d(t) = Me^{qv_d(t)/kT} \tag{P11.3-4}$$

where M is a constant depending upon the diode construction, q is the charge of an electron, k is Boltzmann's constant, and T is absolute temperature. Note that

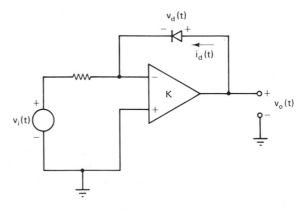

Figure P11.3-5

Linear Feedback Systems   Chap. 11

the idealized relationship of eq. (P11.3-4) assumes that there is no possibility of a negative diode current. Usually, there is some small maximum negative value of diode current, but we will neglect this possibility in our analysis.

(i) Again assuming that the input impedance of the operational amplifier is infinite and that its output impedance is zero, show that the following relations hold:

$$v_o(t) = v_d(t) + Ri_d(t) + v_i(t) \qquad \text{(P11.3-5)}$$

$$v_o(t) = -K[v_o(t) - v_d(t)] \qquad \text{(P11.3-6)}$$

(ii) Show that for K large, the relationship between $v_o(t)$ and $v_i(t)$ is essentially the same as in the feedback system of Figure P11.3-6, where the system in

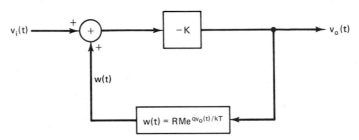

Figure P11.3-6

the feedback path is a nonlinear memoryless system with input $v_o(t)$ and output

$$w(t) = RMe^{qv_o(t)/kT}$$

(iii) Show that for K large

$$v_o(t) \simeq \frac{kT}{q} \ln\left(-\frac{v_i(t)}{RM}\right) \qquad \text{(P11.3-7)}$$

Note that eq. (P11.3-7) only makes sense for a negative $v_i(t)$, which is consistent with the requirement that the diode current cannot be negative. If a positive $v_i(t)$ is applied, the current $i_d(t)$ cannot balance the current through the resistor. Thus, a nonnegligible current is fed into the amplifier, causing it to saturate.

11.4. Consider the basic feedback systems of Figure 11.3. Determine the closed-loop system impulse response for each of the following specifications of the system functions in the forward and feedback paths.

(a) $H(s) = \dfrac{1}{(s + 1)(s + 3)}, \qquad G(s) = 1$

(b) $H(s) = \dfrac{1}{s + 3}, \qquad G(s) = \dfrac{1}{s + 1}$

(c) $H(s) = \tfrac{1}{2}, \qquad G(s) = e^{-s/3}$

(d) $H(z) = \dfrac{z^{-1}}{1 - \tfrac{1}{2}z^{-1}}, \qquad G(z) = \tfrac{2}{3} - \tfrac{1}{6}z^{-1}$

(e) $H(z) = \tfrac{2}{3} - \tfrac{1}{6}z^{-1}, \qquad G(z) = \dfrac{z^{-1}}{1 - \tfrac{1}{2}z^{-1}}$

11.5. (a) Consider the nonrecursive discrete-time LTI filter depicted in Figure P11.5-1. Through the use of feedback around this nonrecursive system, a recursive filter can be implemented. Specifically, consider the configuration shown in Figure

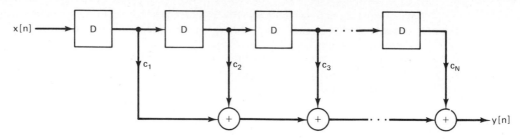

Figure P11.5-1

P11.5-2, where $H(z)$ is the system function of the nonrecursive LTI system of Figure P11.5-1. Determine the overall system function of this feedback system and also find the difference equation relating the input and output of the overall system.

(b) Now suppose that $H(z)$ in Figure P11.5-2 is the system function of a recursive LTI system. Specifically, suppose that

$$H(z) = \frac{\sum\limits_{i=1}^{N} c_i z^{-i}}{\sum\limits_{i=1}^{N} d_i z^{-i}}$$

Show how one can find values of the coefficients K, $c_1, \ldots, c_N$, and $d_0, \ldots, d_N$ so that the closed-loop system function is given by

$$Q(z) = \frac{\sum\limits_{i=0}^{N} b_i z^{-i}}{\sum\limits_{i=0}^{N} a_i z^{-i}}$$

where the a_i and b_i are specified coefficients.

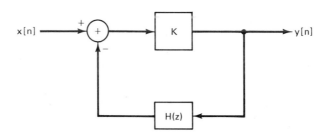

Figure P11.5-2

In this problem we have seen that the use of feedback provides us with alternative implementations of LTI systems specified by linear constant-coefficient difference equations. The implementation in part (a), consisting of feedback around a nonrecursive system, is particularly interesting, as some technologies are ideally suited to implementing tapped delay line structures (i.e., systems consisting of chains of delays with taps at each delay whose outputs are weighted and then summed).

11.6. Consider an inverted pendulum mounted on a movable cart as depicted in Figure P11.6. Here we have modeled the pendulum as consisting of a massless rod of length L, with a mass m attached at the end. The variable $\theta(t)$ denotes the pendulum's angular deflection from vertical, g is gravitational acceleration, $s(t)$ is the position of the cart

Linear Feedback Systems   Chap. 11

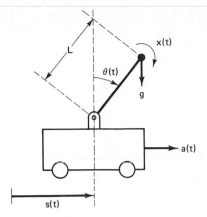

with respect to some reference point, $a(t)$ is the acceleration of the cart, and $x(t)$ represents the angular acceleration resulting from any disturbances, such as gusts of wind.

Our goal in this problem is to analyze the dynamics of the inverted pendulum and more specifically to investigate the problem of balancing the pendulum by judicious choice of the cart acceleration $a(t)$. The differential equation relating $\theta(t)$, $a(t)$, and $x(t)$ is

$$L \frac{d^2\theta(t)}{dt^2} = g \sin [\theta(t)] - a(t) \cos [\theta(t)] + Lx(t) \qquad \text{(P11.6-1)}$$

This relation merely equates the actual acceleration of the mass along a direction perpendicular to the rod to the applied accelerations [gravity, the disturbance acceleration due to $x(t)$, and the cart acceleration] along this direction.

Note that eq. (P11.6-1) is a nonlinear differential equation relating $\theta(t)$, $a(t)$, and $x(t)$. The detailed, exact analysis of the behavior of the pendulum therefore requires that we examine this nonlinear equation; however, we can obtain a great deal of insight into the dynamics of the inverted pendulum by performing a linearized analysis. Specifically, let us examine the dynamics of the pendulum when it is nearly vertical [i.e., when $\theta(t)$ is small]. In this case we can make the approximations

$$\sin [\theta(t)] \simeq \theta(t), \qquad \cos [\theta(t)] \simeq 1 \qquad \text{(P11.6-2)}$$

(a) Suppose that the cart is stationary [i.e., $a(t) = 0$] and consider the causal LTI system with input $x(t)$ and output $\theta(t)$, described by eq. (P11.6-1) together with the approximations given in eq. (P11.6-2). Find the system function for this system and show that it has a pole in the right-half plane, implying that the system is unstable.

What the result of part (a) indicates is that if the cart is stationary, any minor angular disturbance caused by $x(t)$ will lead to growing angular deviations from vertical. Clearly, at some point these deviations will become sufficiently large so that the approximations of eq. (P11.6-2) will no longer be valid. At this point the linearized analysis is no longer accurate, but the fact that it is accurate for small angular displacements allows us to conclude that the vertical equilibrium position is unstable, as small angular displacements will grow rather than diminish.

(b) We now wish to consider the problem of stabilizing the vertical position of the pendulum by moving the cart in an appropriate fashion. Suppose that we try proportional feedback,

$$a(t) = K\theta(t)$$

Assume that $\theta(t)$ is small so that the approximations in eq. (P11.6-2) are valid. Draw a block diagram of the linearized system with $\theta(t)$ as output, $x(t)$ as the external input, and $a(t)$ as the signal which is fed back. Show that the resulting closed-loop system is unstable. Find a value of K so that if $x(t) = \delta(t)$, the pendulum will sway back and forth in an undamped oscillatory fashion.

(c) Consider using proportional plus derivative (PD) feedback,

$$a(t) = K_1\theta(t) + K_2\frac{d\theta(t)}{dt}$$

Show that one can find values of K_1 and K_2 that do stabilize the pendulum. In fact, using the following values for g and L

$$g = 9.8 \text{ m/sec}^2$$
$$L = 0.5 \text{ m}$$

(P11.6-3)

choose values of K_1 and K_2 so that the damping ratio of the closed-loop system is 1 and the natural frequency is 3 rad/sec.

11.7. Consider the causal discrete-time system depicted in Figure P11.7.

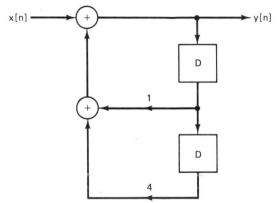

Figure P11.7

(a) Show that this is not a stable system.

(b) Suppose that we allow feedback with one unit delay. That is, suppose that

$$x[n] = x_e[n] - Ky[n-1]$$

where $x_e[n]$ is an externally applied signal now regarded as the input to the overall closed-loop system. Is it possible to stabilize the system with feedback of this form? If so, find the maximum range of values of K for which the system is stable.

(c) Suppose that instead of using feedback with one unit of delay as in part (b), we allow feedback with two units of delay so that

$$x[n] = x_e[n] - Ky[n-2]$$

Specify the full range of values of K (if any such values exist), for which this system is stable.

11.8. As we have seen, the system function algebra of Laplace and z transforms allows us to determine with relative ease the system functions of interconnections of LTI systems in terms of the system functions of the component systems. To illustrate this, consider the interconnected feedback systems shown in Figure P11.8. Determine the overall system function from input to output for each of these.

Linear Feedback Systems   Chap. 11

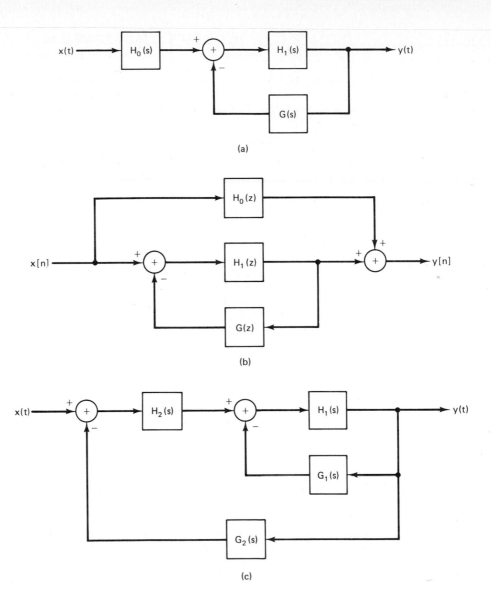

(a)

(b)

(c)

Figure P11.8

11.9. In this problem we consider several examples of the design of tracking systems. Consider the system depicted in Figure P11.9. Here $H_p(s)$ is the system whose output is to be controlled, and $H_c(s)$ is the compensator to be designed. Our objective in choosing $H_c(s)$ is that we would like the output $y(t)$ to follow the input $x(t)$. Specifically, in addition to stabilizing the system, we would also like to design the system so that the error $e(t)$ decays to zero for certain specified inputs.

(a) Suppose that

$$H_p(s) = \frac{\alpha}{s + \alpha}, \qquad \alpha \neq 0 \qquad \text{(P11.9-1)}$$

Show that if $H_c(s) = K$ (which is known as *proportional* or *P* control) we can

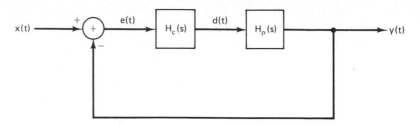

Figure P11.9

choose K to stabilize the system and so that $e(t) \to 0$ if $x(t) = \delta(t)$. Show that we *cannot* get $e(t) \to 0$ if $x(t) = u(t)$.

(b) Again let $H_p(s)$ be as in eq. (P11.9-1), and suppose that we use *proportional plus integral (PI)* control

$$H_c(s) = K_1 + \frac{K_2}{s}$$

Show that we can choose K_1 and K_2 to stabilize the system, and we can also get $e(t) \to 0$ if $x(t) = u(t)$. Thus, this system can track a step. In fact, this illustrates a basic and important fact in feedback system design: to track a step $[X(s) = 1/s]$, we need an integrator $(1/s)$ in the feedback system. An extension of this fact is considered in the next problem.

(c) Suppose that

$$H_p = \frac{1}{(s-1)^2}$$

Show that we *cannot* stabilize this system with a **PI** controller, but that we *can* stabilize it and have it track a step if we use *proportional-plus-integral-plus-differential (PID)* control.

$$H_c(s) = K_1 + \frac{K_2}{s} + K_3 s$$

11.10. In Problem 11.9 we discussed how the presence of an integrator in a feedback system can make it possible for the system to track a step input with zero error in steady state. In this problem we extend this idea. Specifically, consider the feedback system depicted in Figure P11.10, and assume that the overall closed-loop system is stable. Suppose also that $H(s)$ has the form

$$H(s) = \frac{K \displaystyle\prod_{k=1}^{m} (s - \beta_k)}{s^l \displaystyle\prod_{k=1}^{n-l} (s - \alpha_k)}$$

where the α_k and β_k are given, nonzero numbers and l is a positive integer. The feedback system of Figure P11.10 is often referred to as a *Type l* feedback system.

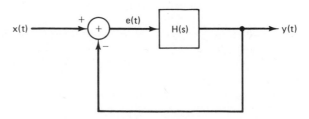

Figure P11.10

(a) Use the final value theorem (Section 9.5.9) to show that a Type 1 feedback system can track a step, that is, that

$$e(t) \longrightarrow 0 \qquad \text{if } x(t) = u(t)$$

(b) Similarly, show that a Type 1 system cannot track a ramp but rather that

$$e(t) \longrightarrow \text{a finite constant} \qquad \text{if } x(t) = u_{-2}(t)$$

(c) Show that for a Type 1 system unbounded errors result if

$$x(t) = u_{-k}(t)$$

with $k > 2$.

(d) More generally, show that for a Type l system
  (i) $e(t) \longrightarrow 0$ if $x(t) = u_{-k}(t)$ with $k \leq l$
  (ii) $e(t) \longrightarrow$ a finite constant if $x(t) = u_{-(l+1)}(t)$
  (iii) $e(t) \longrightarrow \infty$ if $x(t) = u_{-k}(t)$ with $k > l + 1$

11.11. (a) Consider the discrete-time feedback system of Figure P11.11. Suppose that

$$H(z) = \frac{1}{(z - 1)(z + \frac{1}{2})}$$

Show that this system can track a unit step in the sense that if $x[n] = u[n]$, then

$$\lim_{n \to \infty} e[n] = 0 \qquad\qquad \text{(P11.11-1)}$$

(b) More generally, consider the feedback system of Figure P11.11 and assume that the closed-loop system is stable. Suppose that $H(z)$ has a pole at $z = 1$. Show that the system can track a unit step. [*Hint:* Express the transform $E(z)$ of $e[n]$ in terms of $H(z)$ and the transform of $u[n]$; explain why all the poles of $E(z)$ are inside the unit circle.]

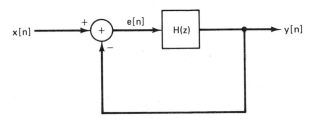

Figure P11.11

The results of parts (a) and (b) are discrete-time counterparts of the results for continuous-time systems discussed in Problems 11.9 and 11.10. In discrete time, we can also consider the design of systems that track specified inputs *perfectly* after a finite number of steps. Such systems are known as *deadbeat feedback systems*, several examples of which are illustrated in the remainder of this problem.

(c) Consider the discrete-time system of Figure P11.11 with

$$H(z) = \frac{z^{-1}}{1 - z^{-1}}$$

Show that the overall closed-loop system has the property that it tracks a step input exactly after one step. That is, if $x[n] = u[n]$, then $e[n] = 0$, $n \geq 1$.

(d) Show that the feedback system of Figure P11.11 with

$$H(z) = \frac{\frac{3}{4}z^{-1} + \frac{1}{4}z^{-2}}{(1 + \frac{1}{4}z^{-1})(1 - z^{-1})}$$

has the property that the output tracks a unit step perfectly after a finite number of steps. At what time step does the error $e[n]$ first settle to zero?

(e) More generally, consider the feedback system of Figure P11.11. Find $H(z)$ so that $y[n]$ perfectly tracks a unit step for $n \geq N$ and in fact so that

$$e[n] = \sum_{k=0}^{N-1} a_k \delta[n - k] \qquad \text{(P11.11-2)}$$

where the a_i are specified constants. [*Hint:* Use the relationship between $H(z)$ and $E(z)$ when the input is a unit step and $e[n]$ is given by eq. (P11.11-2).]

(f) Consider the system of Figure P11.11 with

$$H(z) = \frac{z^{-1} + z^{-2} - z^{-3}}{(1 + z^{-1})(1 - z^{-1})^2}$$

Show that this system tracks a ramp $x[n] = (n + 1)u[n]$ exactly after two time steps.

11.12. Sketch the root loci for $K > 0$ and $K < 0$ for each of the following.

(a) $G(s)H(s) = \dfrac{1}{s + 1}$

(b) $G(s)H(s) = \dfrac{1}{(s - 1)(s + 3)}$

(c) $G(z)H(z) = \dfrac{z - 1}{z^2 - \frac{1}{4}}$

(d) $G(z)H(z) = \dfrac{2}{z^2 - \frac{1}{4}}$

(e) $G(z)H(z) = \dfrac{z^{-1}(1 + z^{-1})}{1 - \frac{1}{4}z^{-2}}$

(f) $G(s)H(s) = \dfrac{1}{s^2 + s + 1}$

(g) $G(s)H(s) = \dfrac{s + 1}{s^2}$

(h) $G(z)H(z) = z^{-1} - z^{-2}$

(i) $G(z)H(z)$ is the system function of the causal LTI system described by the difference equation

$$y[n] - 2y[n - 1] = x[n - 1] - x[n - 2]$$

(j) $G(s)H(s) = \dfrac{(s + 1)^2}{s^3}$

(k) $G(s)H(s) = \dfrac{s^2 + 2s + 2}{s^2(s - 1)}$

(l) $G(s)H(s) = \dfrac{(s + 1)(s - 1)}{s(s^2 + 2s + 2)}$

(m) $G(s)H(s) = \dfrac{(1 - s)}{(s + 2)(s + 3)}$

11.13. Consider a feedback system with

$$G(s)H(s) = \frac{(s - a)(s - b)}{s(s + 3)(s + 6)}$$

Sketch the root locus for $K > 0$ and $K < 0$ for the following values of a and b.

(a) $a = 1, b = 2$　　　　(b) $a = -2, b = 2$　　　　(c) $a = -4, b = 2$
(d) $a = -7, b = 2$　　　　(e) $a = -1, b = -2$　　　　(f) $a = -4, b = -2$
(g) $a = -7, b = -2$　　　(h) $a = -5, b = -4$　　　　(i) $a = -7, b = -4$
(j) $a = -7, b = -8$

11.14. In this problem we investigate some of the properties of sampled data feedback systems and illustrate their use. Recall from Section 11.2.3 that in a sampled data feedback system the output of a continuous-time system is sampled. The resulting sequence of samples is processed by a discrete-time system the output of which is converted to a continuous-time signal which in turn is fed back and subtracted from the external input to produce the actual input to the continuous-time system.

Clearly, the constraint of causality for feedback systems imposes a restriction on the process of converting the discrete feedback signal to a continuous-time signal (e.g., ideal lowpass filtering or any noncausal approximation of it is not allowable). One of the most widely used conversion systems is the zero-order hold (introduced in Section 8.1.2). The structure of a sampled-data feedback system involving a zero-order hold is depicted in Figure P11.14(a). Here we have a continuous-time LTI system with system function $H(s)$, which is sampled to produce a discrete-time sequence

$$g[n] = y(nT) \qquad \text{(P11.14-1)}$$

The sequence $g[n]$ is then processed by a discrete-time LTI system with system function $G(z)$, and the resulting output is put through a zero-order hold to produce the continuous-time signal

$$z(t) = d[n] \qquad \text{for } nT \leq t < (n+1)T \qquad \text{(P11.14-2)}$$

This signal is subtracted from the external input $x(t)$ to produce $e(t)$.

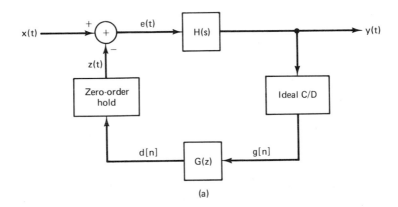

(a)

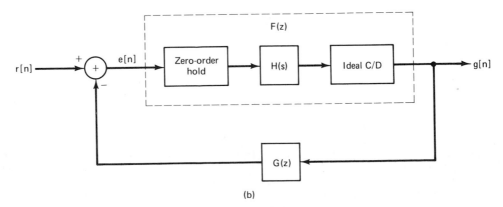

(b)

Figure P11.14

To simplify the subsequent analysis, we will assume that $x(t)$ is constant over intervals of length T. That is,

$$x(t) = r[n] \qquad \text{for } nT \leq t < (n+1)T \qquad \text{(P11.14-3)}$$

where $r[n]$ is a discrete-time sequence. This is an approximation that is usually valid in practice as the sampling rate is typically fast enough so that $x(t)$ does not change appreciably over time intervals of length T. Furthermore, in many applications the external input is itself actually generated by applying a zero-order hold operation to a discrete sequence. For example, in many systems (such as advanced aircraft) the external inputs represent human operator commands which are themselves first processed digitally and then converted back to continuous-time input signals. Because of the linearity of the zero-order hold, the feedback system of Figure P11.14(a) when $x(t)$ is given by eq. (P11.14-3) is equivalent to the system Figure P11.14(b).

(a) Consider the system within the dashed lines in Figure P11.14(b). This is a discrete-time system with input $e[n]$ and output $g[n]$. Show that this is an LTI system. As we have indicated in the figure, we will let $F(z)$ denote the system function of this system.

(b) Show that the discrete-time system with system function $F(z)$ is related to the continuous-time system with system function $H(s)$ by means of a *step-invariant* transformation. That is, if $s(t)$ is the step response of the continuous-time system and $q[n]$ is the step response of the discrete-time system, then

$$q[n] = s(nT) \qquad \text{for all } n$$

(c) Suppose that

$$H(s) = \frac{1}{s-1}, \qquad \mathcal{R}e\{s\} > 1$$

Show that

$$F(z) = \frac{(e^T - 1)z^{-1}}{1 - e^T z^{-1}}, \qquad |z| > e^T$$

(d) Suppose that $H(s)$ is as in part (c) and that $G(z) = K$. Find the range of values of K for which the closed-loop discrete-time system of Figure P11.14(b) is stable.

(e) Suppose that

$$G(z) = \frac{K}{1 + \frac{1}{2}z^{-1}}$$

Under what conditions on T can we find a value of K that stabilizes the overall system? Find a particular pair of values for K and T that yield a stable closed-loop system. (*Hint:* Examine the root locus and find the values of K for which poles enter or leave the unit circle.)

11.15. Consider the feedback system of Figure P11.15 with

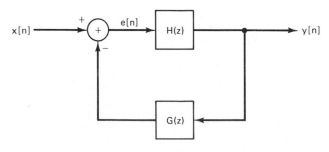

Figure P11.15

Linear Feedback Systems   Chap. 11

$$H(z) = \frac{1}{1 - \frac{1}{2}z^{-1}}, \qquad G(z) = K \qquad \text{(P11.15-1)}$$

(a) Plot the root locus for $K > 0$.

(b) Plot the root locus for $K < 0$. (*Note:* Be careful with this root locus. By applying the angle criterion on the real axis you will find that as K is decreased from zero, the closed-loop pole approaches $z = +\infty$ along the positive real axis and then returns along the negative real axis from $z = -\infty$. Check that this is in fact the case by explicitly solving for the closed-loop pole as a function of K. At what value of K is the pole at $|z| = \infty$?)

(c) Find the full range of values of K for which the closed-loop system is stable.

(d) The phenonenon observed in part (b) is a direct consequence of the fact that in this example the numerator and denominator of $G(z)H(z)$ have the same degree. When this occurs in a discrete-time feedback system it means that there is a delay-free loop. That is, the output at a given point in time is being fed back and in turn affects its own value at the same point in time. To see that this is the case in our example, write the difference equation relating $y[n]$ and $e[n]$. Then write $e[n]$ in terms of the input and output of the feedback system. Contrast this result with that for the feedback system with

$$H(z) = \frac{1}{1 - \frac{1}{2}z^{-1}}, \qquad G(z) = Kz^{-1} \qquad \text{(P11.15-2)}$$

The primary consequence of having delay-free loops is that such feedback systems cannot be implemented in the feedback form depicted. For example, for the system of eq. (P11.15-1) we cannot first calculate $e[n]$ and then $y[n]$, as $e[n]$ depends on $y[n]$. Note that we *can* perform this type of calculation for the system of eq. (P11.15-2) since $e[n]$ depends on $y[n-1]$.

(e) Show that the feedback system of eq. (P11.15-1) represents a causal system except for the value of K for which the closed-loop pole is at $|z| = \infty$.

11.16. (a) Consider the feedback system of Figure 11.9(b) with

$$H(s) = \frac{N_1(s)}{D_1(s)}, \qquad G(s) = \frac{N_2(s)}{D_2(s)} \qquad \text{(P11.16-1)}$$

Assume that there is no pole–zero cancellation in the product $G(s)H(s)$. Show that the zeros of the closed-loop system function consist of the zeros of $H(s)$ and the poles of $G(s)$.

(b) Use the result of part (a) together with the appropriate property of the root locus to confirm the fact that with $K = 0$ the closed-loop system zeros are the zeros of $H(s)$ and the closed-loop poles are the poles of $H(s)$.

(c) While it is usual for $H(s)$ and $G(s)$ in eq. (P11.16-1) to be in reduced form [i.e., the polynomials $N_1(s)$ and $D_1(s)$ have no common factors and the same is true of $N_2(s)$ and $D_2(s)$], it may happen that $N_1(s)$ and $D_2(s)$ have common factors and/or that $N_2(s)$ and $D_1(s)$ have common factors. To see what occurs when such common factors are present, let $p(s)$ denote the greatest common factor of $N_1(s)$ and $D_2(s)$, and let $q(s)$ denote the greatest common factor of $N_2(s)$ and $D_1(s)$. That is,

$$\frac{N_1(s)}{p(s)} \qquad \text{and} \qquad \frac{D_2(s)}{p(s)}$$

are both polynomials and have *no* common factors. Similarly,

$$\frac{N_2(s)}{q(s)} \qquad \text{and} \qquad \frac{D_1(s)}{q(s)}$$

are polynomials and have no common factors. Show that the closed-loop system function for the feedback system can be written as

$$Q(s) = \frac{p(s)}{q(s)}\left[\frac{\hat{H}(s)}{1 + K\hat{G}(s)\hat{H}(s)}\right] \qquad \text{(P11.16-2)}$$

where

$$\hat{H}(s) = \frac{N_1(s)/p(s)}{D_1(s)/q(s)}$$

and

$$\hat{G}(s) = \frac{N_2(s)/q(s)}{D_2(s)/p(s)}$$

Therefore, from eq. (P11.16-2) and part (a) we see that the zeros of $Q(s)$ are the zeros of $p(s)$, the zeros of $\hat{H}(s)$, and the poles of $\hat{G}(s)$, while the poles of $Q(s)$ are the zeros of $q(s)$ and the solutions of

$$1 + K\hat{G}(s)\hat{H}(s) = 0 \qquad \text{(P11.16-3)}$$

By construction there is no pole–zero cancellation in the product $\hat{G}(s)\hat{H}(s)$, and thus we can apply the root locus method described in Section 11.3 to sketch the location of the solutions of eq. (P11.16-3) as K is varied.

(d) Use the procedure outlined in part (c) to determine the closed-loop zeros, any closed-loop poles whose locations are independent of K, and the locus of the remaining closed-loop poles for $K > 0$, when

$$H(s) = \frac{s+1}{(s+4)(s+2)}, \qquad G(s) = \frac{s+2}{s+1}$$

(e) Repeat part (d) for

$$H(z) = \frac{1 + z^{-1}}{1 - \frac{1}{2}z^{-1}}, \qquad G(z) = \frac{z^{-1}}{1 + z^{-1}}$$

(f) Let

$$H(z) = \frac{z^2}{(z-2)(z+2)}, \qquad G(z) = \frac{1}{z^2}$$

(i) Sketch the root locus for $K > 0$ and for $K < 0$.
(ii) Find all values of K for which the overall system is stable.
(iii) Find the impulse response of the closed-loop system when $K = 4$.

11.17. Consider the feedback system of Figure 11.9(a) and suppose that

$$G(s)H(s) = \frac{\displaystyle\prod_{k=1}^{m} (s - \beta_k)}{\displaystyle\prod_{k=1}^{n} (s - \alpha_k)}$$

where $m > n$.† In this case $G(s)H(s)$ has $(m - n)$ poles at infinity (see Chapter 9), and we can adapt the root-locus rules given in the text simply by noting that (1) there are m branches of the root locus and (2) for $K = 0$ all branches of the root locus begin at poles of $G(s)H(s)$, $(m - n)$ of which are at infinity. Furthermore, as $|K| \rightarrow \infty$ these branches converge to the m zeros of $G(s)H(s)$, namely $\beta_1, \beta_2, \ldots, \beta_m$. Use

†Note that for a continuous-time system the condition $m > n$ implies that the system with system function $G(s)H(s)$ involves differentiation of the input [in fact, the inverse transform of $G(s)H(s)$ includes singularity functions up to order $m - n$]. In discrete time, if $G(z)H(z)$, written as a ratio of polynomials in z, has $m > n$, it is necessarily the system function of a noncausal system [in fact, the inverse transform of $G(z)H(z)$ has a nonzero value at time $n - m < 0$]. Thus, the case considered in this problem is actually only of interest for continuous-time systems.

these facts to assist you in sketching the root locus (for $K > 0$ and for $K < 0$) for each of the following.

(a) $G(s)H(s) = s - 1$ **(b)** $G(s)H(s) = (s + 1)(s + 2)$

(c) $G(s)H(s) = \dfrac{(s + 1)(s + 2)}{s - 1}$

11.18. Consider a feedback system with

$$H(s) = \frac{s + 2}{s^2 + 2s + 4}, \qquad G(s) = K$$

(a) Sketch the root locus for $K > 0$.

(b) Sketch the root locus for $K < 0$.

(c) Find the smallest positive value of K for which the closed-loop impulse response does not exhibit any oscillatory behavior.

11.19. Consider the feedback system of Figure 11.3, with

$$H(z) = \frac{Kz^{-1}}{1 - z^{-1}}$$

and

$$G(z) = 1 - az^{-1}$$

(a) Sketch the root locus of this system for $K > 0$ and for $K < 0$ when $a = \frac{1}{2}$.

(b) Repeat part (a) when $a = -\frac{1}{2}$.

(c) With $a = -\frac{1}{2}$, find a value of K for which the closed-loop impulse response is of the form

$$(A + Bn)\alpha^n$$

for some values of the constants A, B, and α, with $|\alpha| < 1$. (*Hint:* What must the denominator of the closed-loop system function look like in this case?)

11.20. Consider the discrete-time feedback system depicted in Figure P11.20. The system in the forward path is not very well damped and we would like to choose the feedback system function to improve overall damping. By using root-locus methods, show that this can be done with

$$G(z) = 1 - \tfrac{1}{2}z^{-1}$$

Specifically, sketch the root locus for $K > 0$ and specify a value of the gain K for which a significant improvement in damping is obtained.

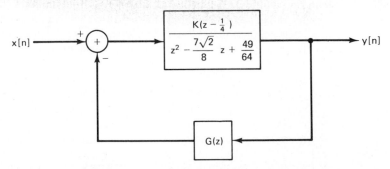

Figure P11.20

11.21. In Section 11.3 we derived a number of properties that can be of value in determining the root locus for a feedback system. In this problem we develop several additional properties. We derive these properties in terms of continuous-time systems, but as with all root-locus properties, they hold as well for discrete-time root loci. For our discussion of these properties we refer to the basic equation satisfied by the closed-loop poles:

$$G(s)H(s) = -\frac{1}{K} \qquad \text{(P11.21-1)}$$

where

$$G(s)H(s) = \frac{\prod_{k=1}^{m}(s - \beta_k)}{\prod_{k=1}^{n}(s - \alpha_k)} = \frac{\sum_{k=1}^{m} b_k s^k}{\sum_{k=1}^{n} a_k s^k} \qquad \text{(P11.21-2)}$$

Throughout this problem we assume that $m \leq n$.

(a) From Property 2 we know that $(n - m)$ branches of the root locus go to zeros of $G(s)H(s)$ located at infinity. In this first part we demonstrate that it is straightforward to determine the angles at which these branches of the root locus approach infinity. Specifically, consider searching the remote part of the s-plane [i.e., the region where $|s|$ is extremely large and far from any of the poles and zeros of $G(s)H(s)$]. This is illustrated in Figure P11.21. Use the geometry of this picture, together with the angle criterion for $K > 0$ and for $K < 0$, to deduce that:

- For $K > 0$, the $(n - m)$ branches of the root locus that approach infinity do so at the angles

$$\frac{(2k + 1)\pi}{n - m}, \qquad k = 0, 1, \ldots, (n - m - 1)$$

- For $K < 0$, the $(n - m)$ branches of the root locus that approach infinity do so at the angles

$$\frac{2k\pi}{n - m}, \qquad k = 0, 1, \ldots, (n - m - 1)$$

Thus, the branches of the root locus that approach infinity do so at specified angles that are arranged symmetrically. For example, for $n - m = 3$ and $K > 0$, we see that the asymptotic angles are $\pi/3$, π, and $5\pi/3$. The result of part (a) together with one additional fact allows us to draw in the asymptotes for the branches of the root locus that approach infinity. Specifically, all of the $(n - m)$ asymptotes intersect at a single point on the real axis. This is derived in the following part of this problem.

(b) (i) As a first step, consider a general polynomial equation

$$s_r + f_{r-1}s^{r-1} + \ldots + f_0 = (s - \xi_1)(s - \xi_2)\ldots(s - \xi_r) = 0$$

Show that

$$f_{r-1} = -\sum_{i=1}^{r} \xi_i$$

(ii) Perform long division on $1/G(s)H(s)$ to write

$$\frac{1}{G(s)H(s)} = s^{n-m} + \gamma_{n-m-1}s^{n-m-1} + \ldots \qquad \text{(P11.21-3)}$$

Show that

$$\gamma_{n-m-1} = a_{n-1} - b_{m-1} = \sum_{k=1}^{m} \beta_k - \sum_{k=1}^{n} \alpha_k$$

[see eq. (P11.21-2)].

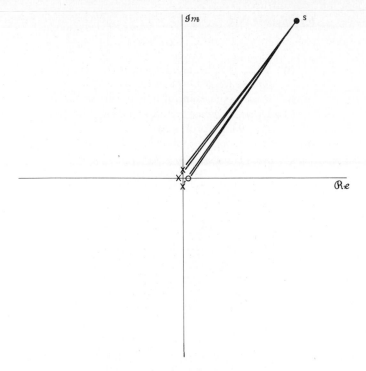

Figure P11.21

(iii) Argue that the solution of eq. (P11.21-1) for *large s* approximately solves the equation

$$s^{n-m} + \gamma_{n-m-1}s^{n-m-1} + \gamma_{n-m-2}s^{n-m-2} + \ldots + \gamma_0 + K = 0$$

(iv) Use the results of (i)–(iii) to deduce that the sum of the $(n - m)$ closed-loop poles that approach infinity is asymptotically equal to

$$b_{m-1} - a_{n-1}$$

Thus, the center of gravity of these $(n - m)$ poles is

$$\frac{b_{m-1} - a_{n-1}}{n - m}$$

which does not depend on K. Consequently, we have $(n - m)$ closed-loop poles that approach $|s| = \infty$ at evenly spaced angles and that have a center of gravity that is independent of K. From this we can deduce that:

The asymptotes of the $(n - m)$ branches of the root locus that approach infinity intersect at the point

$$\frac{b_{m-1} - a_{n-1}}{n - m} = \frac{\displaystyle\sum_{k=1}^{n} \alpha_k - \sum_{k=1}^{m} \beta_k}{n - m}$$

This point of asymptote intersection is the same for $K > 0$ and $K < 0$.

(c) Suppose that

$$G(s)H(s) = \frac{1}{(s+1)(s+3)(s+5)}$$

- (i) What are the asymptotic angles for the closed-loop poles that approach infinity for $K > 0$ and for $K < 0$?
- (ii) What is the point of intersection of the asymptotes?
- (iii) Draw in the asymptotes and use these to help you sketch the root locus for $K > 0$ and for $K < 0$.

(d) Repeat part (c) for each of the following.

- (i) $G(s)H(s) = \dfrac{s+1}{s(s+2)(s+4)}$

- (ii) $G(s)H(s) = \dfrac{1}{s^4}$

- (iii) $G(s)H(s) = \dfrac{1}{s(s+1)(s+5)(s+6)}$

- (iv) $G(s)H(s) = \dfrac{1}{(s+2)^2(s-1)^2}$

- (v) $G(s)H(s) = \dfrac{s+3}{(s+1)(s^2+2s+2)}$

- (vi) $G(s)H(s) = \dfrac{s+1}{(s+2)^2(s^2+2s+2)}$

- (vii) $G(s)H(s) = \dfrac{s+1}{(s+100)(s-1)(s-2)}$

(e) Use the result of part (a) to explain why the following statement is true: For any continuous-time feedback system, with $G(s)H(s)$ given by eq. (P11.21-2), if $(n-m) \geq 3$, we can make the closed-loop system unstable by choosing $|K|$ large enough.

(f) Repeat part (c) for the discrete-time feedback system specified by

$$G(z)H(z) = \frac{z^{-3}}{(1-z^{-1})(1+\frac{1}{2}z^{-1})}$$

(g) Explain why the following statement is true: For any discrete-time feedback system, with

$$G(z)H(z) = \frac{z^m + b_{m-1}z^{m-1} + \ldots + b_0}{z^n + a_{n-1}z^{n-1} + \ldots + a_0}$$

if $n > m$, we can make the closed-loop system unstable by choosing $|K|$ large enough.

11.22. (a) Consider a feedback system with

$$H(z) = \frac{z+1}{z^2+z+\frac{1}{4}}, \qquad G(z) = \frac{K}{z-1}$$

- (i) Write the closed loop system function explicitly as a ratio of two polynomials (the denominator polynomial will have coefficients that depend on K).
- (ii) Show that the sum of the closed-loop poles is independent of K.

(b) More generally, consider a feedback system with system function

$$G(z)H(z) = K\frac{z^m + b_{m-1}z^{m-1} + \ldots + b_0}{z^n + a_{n-1}z^{n-1} + \ldots + a_0}$$

Show that if $m \leq (n-2)$, the sum of the closed-loop poles is independent of K.

11.23. (a) Consider again the feedback system of Example 11.2:

$$G(s)H(s) = \frac{s-1}{(s+1)(s+2)}$$

The root locus for $K < 0$ is plotted in Figure 11.13b. For some value of K, the two closed-loop poles are on the $j\omega$-axis. Solve for this value of K and the corresponding locations of the closed-loop poles by examining the real and imaginary parts of the equation

$$G(j\omega)H(j\omega) = -\frac{1}{K}$$

which must be satisfied if the point $s = j\omega$ is on the root locus for any given values of K. Use this result plus the analysis in Example 11.2 to find the full range of values of K (positive and negative) for which the closed-loop system is stable.

(b) Note that this feedback system is unstable for $|K|$ sufficiently large. Explain why this is true in general for continuous-time feedback systems for which $G(s)H(s)$ has a zero in the right-half plane and for discrete-time feedback systems for which $G(z)H(z)$ has a zero outside the unit circle.

11.24. Consider again the discrete-time feedback system of Example 11.3:

$$G(z)H(z) = \frac{z}{(z - \frac{1}{2})(z - \frac{1}{4})}$$

The root loci for $K > 0$ and $K < 0$ are depicted in Figure 11.15.

(a) Consider the root locus for $K > 0$. In this case the system becomes unstable when one of the closed-loop poles is less than or equal to -1. Find the value of K for which $z = -1$ is a closed-loop pole.

(b) Consider the root locus for $K < 0$. In this case the system becomes unstable when one of the closed-loop poles is greater than or equal to 1. Find the value of K for which $z = 1$ is a closed-loop pole.

(c) What is the full range of values of K for which the closed-loop system is stable?

11.25. Consider a continuous-time feedback system with

$$G(s)H(s) = \frac{1}{s(s+1)(s+2)} \qquad \text{(P11.25-1)}$$

(a) Sketch the root locus for $K > 0$ and for $K < 0$. (*Hint:* The result of Problem 11.21 is useful here.)

(b) If you have sketched the locus correctly, you will see that for $K > 0$, two branches of the root locus cross the $j\omega$-axis, passing from the left-half plane into the right-half plane. Consequently, we can conclude that the closed-loop system is stable for $0 < K < K_0$, where K_0 is the value of the gain for which the two branches of the root locus intersect the $j\omega$-axis. Note that the sketch of the root locus does not by itself tell us what the value of K_0 is or the exact point on the $j\omega$-axis where the branches cross. As in Problem 11.23, determine K_0 by solving the pair of equations obtained as the real and imaginary parts of

$$G(j\omega)H(j\omega) = -\frac{1}{K_0} \qquad \text{(P11.25-2)}$$

Determine the corresponding two values of ω (which are the negatives of each other, since poles occur in complex-conjugate pairs).

From your root-locus sketches in part (a) you will note that there is a segment of the real axis between two poles that is on the root locus for $K > 0$, and a different segment is on the locus for $K < 0$. In both cases the root locus breaks off from the real axis at some point. In the next part of this problem we illustrate how one can calculate these breakaway points.

(c) Consider the equation defining the closed-loop poles

$$G(s)H(s) = -\frac{1}{K} \qquad \text{(P11.25-3)}$$

Using eq. (P11.25-1), show that an equivalent equation for the closed-loop poles is

$$p(s) = s^3 + 3s^2 + 2s = -K \qquad \text{(P11.25-4)}$$

Consider the segment of the real axis between 0 and -1. This segment is on the root locus for $K \geq 0$. For $K = 0$ two branches of the locus begin at 0 and -1 and approach each other as K is increased.

(i) Use the facts just stated, together with eq. (P11.25-4), to explain why the function $p(s)$ has the form shown in Fig. P11.25(a) for $-1 \leq s \leq 0$, and why the point s_+ where the minimum occurs is the breakaway point (i.e., it is the point where the two branches of the $K > 0$ root locus break from the segment of the real axis between -1 and 0).

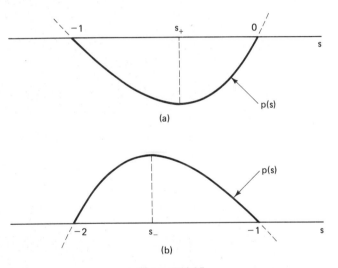

Figure P11.25

Similarly, consider the root locus for $K < 0$ and more specifically the segment of the real axis between -1 and -2, which is part of this locus. For $K = 0$ two branches of the root locus begin at -1 and -2 and as K is decreased, these poles approach each other.

(ii) In an analogous fashion to that used in (i), explain why the function $p(s)$ has the form shown in Figure P11.25(b) and why the point s_- where the maximum occurs is the breakaway point for $K < 0$.

Thus, the breakaway points correspond to the maxima and minima of $p(s)$ as s ranges over the negative real line.

(iii) The points at which $p(s)$ has a maximum or minimum are the solutions of the equation

Linear Feedback Systems   Chap. 11

$$\frac{dp(s)}{ds} = 0$$

Use this fact to find the breakaway points s_+ and s_-, and then use eq. (P11.25-4) to find the gains at which these points are closed-loop poles.

In addition to the method illustrated in part (c), there are other, partially analytical, partially graphical methods for determining breakaway points. It is also possible to use a procedure similar to the one illustrated in part (c) to determine "break-in" points, where two branches of the root locus merge onto the real axis. These methods plus the one just illustrated are described in advanced texts such as those listed in the Bibliography at the end of the book.

11.26. Consider a discrete-time feedback system with

$$G(z)H(z) = \frac{1}{z(z-1)}$$

(a) Sketch the root locus for $K > 0$ and for $K < 0$.

(b) If you have sketched the root locus correctly for $K > 0$, you will see that the two branches of the root locus cross and exit from the unit circle. Consequently, we can conclude that the closed-loop system is stable for $0 < K < K_0$, where K_0 is the value of the gain for which the two branches intersect the unit circle. At what points on the unit circle do the branches exit? What is the value of K_0?

11.27. Consider a feedback system, either in continuous time or discrete time, and suppose that the Nyquist plot for the system passes through the point $-1/K$. Is the feedback system stable or unstable for this value of the gain? Explain your answer.

11.28. Sketch the Nyquist plot for each of the following specifications of $G(s)H(s)$, and use the continuous-time Nyquist criterion to determine the range of values of K (if any such range exists) for which the closed-loop system is stable. [*Note:* In sketching the Nyquist plots you may find it useful to first sketch the corresponding Bode plots. It also is helpful to determine the values of ω for which $G(j\omega)H(j\omega)$ is real.]

(a) $G(s)H(s) = \dfrac{1}{s+1}$

(b) $G(s)H(s) = \dfrac{1}{s-1}$

(c) $G(s)H(s) = \dfrac{1}{(s+1)(s/10+1)}$

(d) $G(s)H(s) = \dfrac{1}{s^2-1}$

(e) $G(s)H(s) = \dfrac{1}{(s+1)^2}$

(f) $G(s)H(s) = \dfrac{1}{(s+1)^3}$

(g) $G(s)H(s) = \dfrac{1}{(s+1)^4}$

(h) $G(s)H(s) = \dfrac{1-s}{(s+1)^2}$

(i) $G(s)H(s) = \dfrac{s+1}{(s-1)^2}$

(j) $G(s)H(s) = \dfrac{s+1}{s^2-4}$

(k) $G(s)H(s) = \dfrac{1}{s^2+2s+2}$

(l) $G(s)H(s) = \dfrac{s+1}{s^2-2s+2}$

(m) $G(s)H(s) = \dfrac{s+1}{(s+100)(s-1)^2}$

(n) $G(s)H(s) = \dfrac{s^2}{(s+1)^3}$

11.29. Sketch the Nyquist plot for each of the following specifications of $G(z)H(z)$, and use the discrete-time Nyquist criterion to determine the range of values of K (if any such range exists) for which the closed-loop system is stable. [*Note:* In sketching the Nyquist plots you may find it useful to first sketch the magnitude and phase plots as a function of frequency or at least to calculate $|G(e^{j\Omega})H(e^{j\Omega})|$ and $\sphericalangle G(e^{j\Omega})H(e^{j\Omega})$ at several points. Also, it is helpful to determine the values of Ω for which $G(e^{j\Omega})H(e^{j\Omega})$ is real.]

(a) $G(z)H(z) = \dfrac{1}{z-\frac{1}{2}}$

(b) $G(z)H(z) = \dfrac{1}{z-2}$

(c) $G(z)H(z) = z^{-1}$

(d) $G(z)H(z) = z^{-2}$

(e) $G(z)H(z) = z^{-3}$

(f) $G(z)H(z) = \dfrac{1}{(z + \frac{1}{2})(z - \frac{3}{2})}$

(g) $G(z)H(z) = \dfrac{z - \sqrt{3}}{z(z + 1/\sqrt{3})}$

(h) $G(z)H(z) = \dfrac{1}{z^2 - z + \frac{1}{3}}$

(i) $G(z)H(z) = \dfrac{z - \frac{1}{2}}{z(z - 2)}$

(j) $G(z)H(z) = \dfrac{(z + 1)^2}{z^3}$

11.30. As mentioned in Section 11.4, the continuous-time Nyquist criterion can be extended to allow for poles of $G(s)H(s)$ on the $j\omega$-axis. In this problem we illustrate the general technique by means of several examples. Consider a continuous-time feedback system with

$$G(s)H(s) = \frac{1}{s(s + 1)} \qquad\qquad \text{(P11.30-1)}$$

When $G(s)H(s)$ has a pole at $s = 0$, we modify the contour of Figure 11.18 by avoiding the origin. To do this we indent the contour by adding a semicircle of infinitesimal radius ϵ into the right-half plane. This is illustrated in Figure P11.30-1. Thus, only a small part of the right-half plane is not enclosed by the modified contour, and this area goes to zero as we let $\epsilon \rightarrow 0$. Consequently, as $M \rightarrow \infty$, the contour will enclose the entire right-half plane. As in the text, $G(s)H(s)$ is constant (and in this case zero) along the circle of infinite radius. Thus, to plot $G(s)H(s)$ along the contour we need only plot it for the portion of the contour consisting of the $j\omega$-axis and the infinitesimal semicircle.

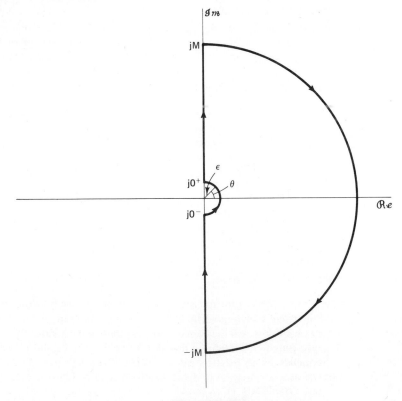

Figure P11.30-1

(a) Show that

$$\sphericalangle G(j0^+)H(j0^+) = -\frac{\pi}{2}$$

$$\sphericalangle G(j0^-)H(j0^-) = \frac{\pi}{2}$$

where $s = j0^-$ is the point where the infinitesimal semicircle meets the $j\omega$-axis just below the origin, and $s = j0^+$ is the corresponding point just above the origin.

(b) Use the result of part (a) together with eq. (P11.30-1) to verify that Figure P11.30-2 is an accurate sketch of $G(s)H(s)$ along the portions of the contour from $-j\infty$ to $j0^-$ and $j0^+$ to $j\infty$. In particular, check that $\sphericalangle G(j\omega)H(j\omega)$ and $|G(j\omega)H(j\omega)|$ behave in the manner depicted in the figure.

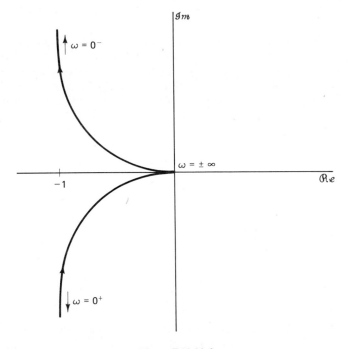

Figure P11.30-2

(c) All that remains to be done is to determine the plot of $G(s)H(s)$ along the small semicircle about $s = 0$. Note that as $\epsilon \to 0$ the magnitude of $G(s)H(s)$ along this contour goes to infinity. Show that as $\epsilon \to 0$, the contribution of the pole at $s = -1$ to $\sphericalangle G(s)H(s)$ along the semicircle is zero. Show then that as $\epsilon \to 0$,

$$\sphericalangle G(s)H(s) = -\theta$$

where θ is defined in Figure P11.30-1. Thus, since θ varies from $-\pi/2$ at $s = j0^-$ to $+\pi/2$ at $s = j0^-$ in a counterclockwise direction, $\sphericalangle G(s)H(s)$ must go from $+\pi/2$ at $s = j0^-$ to $-\pi/2$ at $s = j0^+$ in a clockwise direction. The result is the complete Nyquist plot depicted in Figure P11.30-3.

(d) Using the Nyquist plot of Figure P11.30-3, find the range of values of K for which the closed-loop feedback system is stable. (*Note:* The continuous-time Nyquist criterion as presented in the text states that for closed-loop system stability the

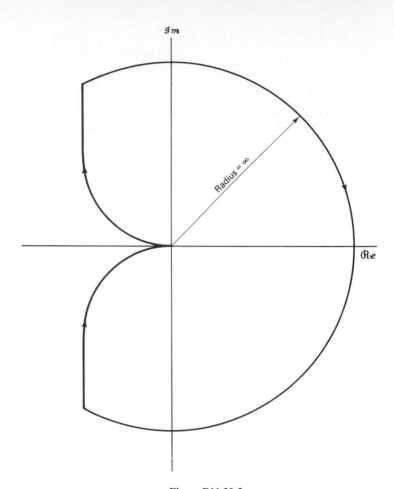

Figure P11.30-3

net number of clockwise encirclements of the point $-1/K$ must equal minus the number of right-half-plane poles of $G(s)H(s)$. In our present example note that the pole of $G(s)H(s)$ at $s = 0$ is *outside* the modified contour. Consequently, it is *not* included in counting the poles of $G(s)H(s)$ in the right-half plane [i.e., only poles of $G(s)H(s)$ *strictly inside* the right-half plane are counted in applying the Nyquist criterion]. Thus in this case, since $G(s)H(s)$ has no poles strictly inside the right-half plane, we must have *no* encirclements of the point $s = -1/K$ for closed-loop system stability.)

(e) Follow the steps outlined in parts (a)–(c) to sketch the Nyquist plots for each of the following.

(i) $G(s)H(s) = \dfrac{s/10 + 1}{s(s + 1)}$

(ii) $G(s)H(s) = \dfrac{1}{s(s + 1)^2}$

(iii) $G(s)H(s) = \dfrac{1}{s^2}$ [be careful in calculating $\sphericalangle G(s)H(s)$ along the infinitesimal semicircle]

(iv) $G(s)H(s) = \dfrac{s+1}{s(1-s)}$ [be careful in calculating $\sphericalangle G(j\omega)H(j\omega)$ as ω is varied; make sure to take the minus sign in the denominator into account]

(v) $G(s)H(s) = \dfrac{s+1}{s^2}$ [same remark as for (iii)]

In each case use the Nyquist criterion to determine the range of values of K (if there are any such values) for which the closed-loop system is stable. Also, use another method (root locus or direct calculation of the closed-loop poles as a function of K) to check your answer and thus to provide a partial check of the correctness of your Nyquist plot. [*Note:* In sketching the Nyquist plots you may find it useful to first sketch the Bode plots of $G(s)H(s)$. It may also be helpful to determine the values of ω for which $G(j\omega)H(j\omega)$ is real.]

(f) Repeat part (e) for:

(i) $G(s)H(s) = \dfrac{1}{s^2+1}$

(ii) $G(s)H(s) = \dfrac{s+1}{s^2+1}$

Note: In these cases there are *two* poles on the imaginary axis; you will need to modify the contour of Figure 11.18 to avoid each of these. Use infinitesimal semicircles as in Figure P11.30-1.

11.31. In this problem we illustrate the discrete-time counterpart of the technique described in Problem 11.30. Specifically, the discrete-time Nyquist criterion can be extended to allow for poles of $G(z)H(z)$ on the unit circle.

Consider a discrete-time feedback system with

$$G(z)H(z) = \frac{z^{-2}}{1-z^{-1}} = \frac{1}{z(z-1)} \qquad \text{(P11.31-1)}$$

In this case we modify the contour on which we evaluate $G(z)H(z)$ as illustrated in Figure P11.31-1.

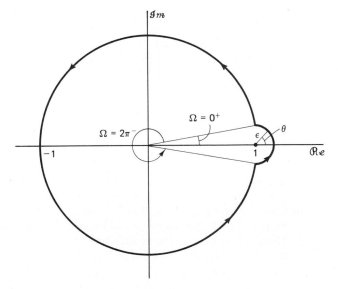

Figure P11.31-1

(a) Show that

$$\sphericalangle G(e^{j0^+})H(e^{j0^+}) = -\frac{\pi}{2}$$

$$\sphericalangle G(e^{j2\pi^-})H(e^{j2\pi^-}) = \frac{\pi}{2}$$

where $z = e^{j2\pi^-}$ is the point below the real axis where the small semicircle intersects the unit circle, and $z = e^{j0^+}$ is the corresponding point above the real axis.

(b) Use the result of part (a) together with eq. (P11.31-1) to verify that Figure P11.31-2 is an accurate sketch of $G(z)H(z)$ along the portion of the contour $z = e^{j\Omega}$ as Ω varies from 0^+ to $2\pi^-$ in a counterclockwise direction. In particular, verify that the angular variation of $G(e^{j\Omega})H(e^{j\Omega})$ is as indicated.

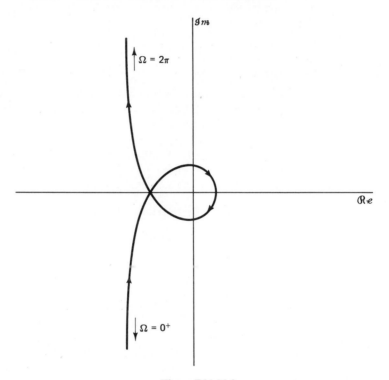

Figure P11.31-2

(c) Find the value of Ω for which $\sphericalangle G(e^{j\Omega})H(e^{j\Omega}) = -\pi$ and verify that

$$|G(e^{j\Omega})H(e^{j\Omega})| = 1$$

at this point. [*Hint:* Use the geometrical method for evaluating $\sphericalangle G(e^{j\Omega})H(e^{j\Omega})$ together with some elementary geometry to determine the value of Ω.]

(d) Consider next the plot of $G(z)H(z)$ along the small semicircle about $z = 1$. Note that as $\epsilon \to 0$, the magnitude of $G(z)H(z)$ along this contour goes to infinity. Show that as $\epsilon \to 0$, the contribution of the pole at $z = 0$ to $\sphericalangle G(z)H(z)$ along the semicircle is zero. Show then that as $\epsilon \to 0$,

$$\sphericalangle G(z)H(z) = -\theta$$

where θ is defined in Figure P11.31-1.

Thus, since θ varies from $-\pi/2$ to $+\pi/2$ in a counterclockwise direction, $\sphericalangle G(z)H(z)$ varies from $+\pi/2$ to $-\pi/2$ in a clockwise direction. The result is the complete Nyquist plot depicted in Figure P11.31-3.

(e) Using the Nyquist plot of Figure P11.31-3, find the range of values of K for which the closed-loop feedback system is stable. [*Note:* Since the pole of $G(z)H(z)$ at $z = 1$ is *inside* the modified contour, it is not included in counting the poles of $G(z)H(z)$ outside the unit circle. That is, only poles *strictly outside* the unit circle are counted in applying the Nyquist criterion. Thus in this case, since $G(z)H(z)$ has no poles strictly outside the unit circle, we must have no encirclements of the point $z = -1/K$ for closed-loop system stability.]

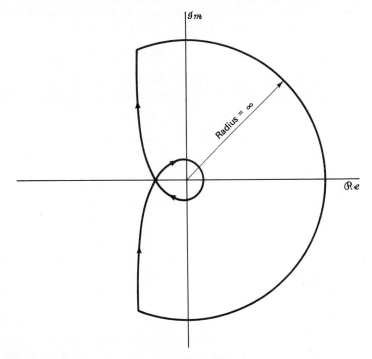

Figure P11.31-3

(f) Follow the steps outlined in parts (a), (b), and (d) to sketch the Nyquist plots for each of the following:

(i) $\dfrac{z + \frac{1}{2} + \sqrt{3}}{z - 1}$

(ii) $\dfrac{1}{(z - 1)(z + \frac{1}{2} + \sqrt{3})}$

(iii) $\dfrac{z + 1}{z(z - 1)}$

(iv) $\dfrac{z - 1/\sqrt{3}}{(z - 1)^2}$ [be careful in calculating $\sphericalangle G(z)H(z)$ along the infinitesimal semicircle]

For each of these use the Nyquist criterion to determine the range of values of K (if there are any such values) for which the closed-loop system is stable. Also use another method (root locus or direct calculations of the closed-loop poles as a function of K) to check your answers and thus to provide a partial check of the correctness of your Nyquist plots. [*Note:* In sketching the Nyquist plots you may find it useful to first sketch the magnitude and phase plots as a function of frequency or at least to calculate $|G(e^{j\Omega})H(e^{j\Omega})|$ and $\sphericalangle G(e^{j\Omega})H(e^{j\Omega})$ at several points. Also, it is helpful to determine the values of Ω for which $G(e^{j\Omega})H(e^{j\Omega})$ is real.]

(g) Repeat part (f) for

$$G(z)H(z) = \frac{1}{z^2 - 1}$$

In this case there are two poles on the unit circle and thus you must modify the contour around each of these by including an infinitesimal semicircle that extends outside the unit circle, thus placing the pole inside the contour.

11.32. Consider a system with system function

$$H(s) = \frac{1}{(s+1)(s-2)} \qquad (P11.32\text{-}1)$$

As this system is unstable, we would like to devise some method for its stabilization.

(a) Consider first a series compensation scheme as illustrated in Figure P11.32-1.

$x(t) \longrightarrow \boxed{C(s)} \longrightarrow \boxed{H(s)} \longrightarrow y(t)$

Figure P11.32-1

Show that the overall system of this figure is stable if the system function $C(s)$ is taken as

$$C(s) = \frac{s-2}{s+3}$$

In practice, this is *not* considered to be a particularly useful way to attempt to stabilize a system. Explain why this is so.

(b) Suppose that instead we use a feedback system as depicted in Figure P11.32-2. Is it possible to stabilize this system using a constant gain for the compensating element, that is,

$$C(s) = K$$

Justify your answer using Nyquist techniques.

(c) Show that the system of Figure P11.32-2 can be stabilized if $C(s)$ is a proportional plus derivative system:

$$C(s) = K(s+a)$$

Consider both the cases $0 < a < 1$ and $a > 1$.

(d) Suppose that

$$C(s) = K(s+2)$$

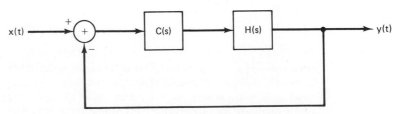

Figure P11.32-2

Choose the value of K so that the closed-loop system has a pair of complex poles with a damping ratio of $\zeta = \frac{1}{2}$. (*Hint*: In this case the denominator of the closed-loop system function must have the form

$$s^2 + \omega_n s + \omega_n^2$$

for some value of $\omega_n > 0$.)

(e) Pure derivative compensation is impossible to obtain and undersirable in practice, as the required amplification of arbitrarily high frequencies can neither be obtained nor is advisable, as all real systems are subject to some level of high-frequency disturbances. Thus, suppose that we consider a compensator of the form

$$C(s) = K\left(\frac{s + a}{s + b}\right) \qquad (a, b > 0) \tag{P11.32-2}$$

If $b < a$, this is called a *lag network*, as it can be checked that $\angle C(j\omega) < 0$ for all $\omega > 0$, so the phase of the output of this system lags the phase of the input. If $b > a$, $\angle C(j\omega) > 0$ for all $\omega > 0$, and this system is called a *lead network*.

(i) Show that it is possible to stabilize the system with the lead compensator

$$C(s) = K\frac{s + \frac{1}{2}}{s + 2} \tag{P11.32-3}$$

if K is chosen large enough.

(ii) Show that it is not possible to stabilize the feedback system of Figure P11.32-2 using the lag network

$$C(s) = K\frac{s + 3}{s + 2}$$

Hint: Use the results of Problem 11.21 in sketching the root locus. Then determine the points on the $j\omega$-axis which are on the root locus and the values of K for which each of these points is a closed-loop pole. Use this information to deduce that for no value of K are *all* of the closed-loop poles in the left-half plane.

11.33. One issue that must always be taken into account by the system designer is the possible effect of unmodeled aspects of the system one is attempting to stabilize or modify through feedback. In this problem we provide an illustration of why this is the case. Consider a continuous-time feedback system and suppose that

$$H(s) = \frac{1}{(s + 10)(s - 2)} \tag{P11.33-1}$$

and

$$G(s) = K \tag{P11.33-2}$$

(a) Use root locus techniques to show that the closed-loop system will be stable if K is chosen large enough.

(b) Suppose that the system we are trying to stabilize by feedback actually has a system function given by

$$H(s) = \frac{1}{(s + 10)(s - 2)(10^{-3}s + 1)} \tag{P11.33-3}$$

The added factor can be thought of as representing a first-order system in cascade with the system of eq. (P11.33-1). Note that the time constant of this system is extremely small and thus will appear to have a step response that is almost instantaneous. For this reason one often neglects such factors in order to obtain simpler and more tractable models that capture all of the important characteristics of the system. However, one must still keep these neglected dynamics in mind in obtaining a useful feedback design. To see why this is the case, show that if $G(s)$ is given by eq. (P11.33-2) and $H(s)$ is as in eq. (P11.33-3), then the closed-loop system will be unstable if K is chosen *too* large. (*Hint:* See Problem 11.21.)

(c) Use root-locus techniques to show that if

$$G(s) = K(s + 100)$$

then the feedback system will be stable for all values of K sufficiently large if $H(s)$ is given by eq. (P11.33-1) *or* eq. (P11.33-3).

11.34. Consider the continuous-time feedback system depicted in Figure P11.34(a).

(a) Use the straight-line approximations to Bode plots developed in Chapter 4 to obtain a sketch of the log magnitude–phase plot for this system. Determine estimates of the phase and gain margins from this plot.

(b) Suppose that there is an unknown delay within the feedback system so that the actual feedback system is as is shown in Figure P11.34(b). Approximately what is the largest delay τ that can be tolerated before the feedback system becomes unstable? Use your results from part (a) for this calculation.

(c) Calculate more precise values of the phase and gain margins and compare these to your results in part (a). This should give you some idea of the size of the errors that are incurred in using the approximate Bode plots.

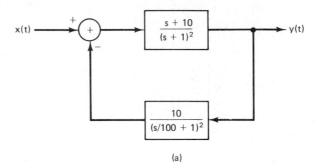

(a)

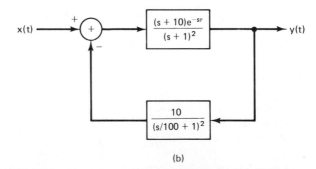

(b)

Figure P11.34.

11.35. Consider the basic continuous-time feedback system of Figure 11.3(a). Sketch the log magnitude–phase diagram and roughly determine the phase and gain margin for each of the following choices of $H(s)$ and $G(s)$. You may find it helpful to use the straight-line approximations to Bode plots developed in Chapter 4 to aid you in sketching the log magnitude–phase diagrams. Be careful, however, to take into account how the actual frequency response deviates from its approximation near break frequencies when there are underdamped second-order terms present (see Section 4.12).

(a) $H(s) = \dfrac{s + 1}{s^2 + s + 1}$, $\quad G(s) = 1$

(b) $H(s) = \dfrac{10s + 1}{s^2 + s + 1}$, $\quad G(s) = 1$

(c) $H(s) = \dfrac{s/10 + 1}{s^2 + s + 1}$, $\quad G(s) = 1$

(d) $H(s) = \dfrac{1}{(s + 1)^2(s + 10)}$, $\quad G(s) = 100$

(e) $H(s) = \dfrac{1}{(s + 1)^3}$, $\quad G(s) = \dfrac{1}{s + 1}$

(f) $H(s) = \dfrac{1 - s}{(s + 1)(s + 10)}$, $\quad G(s) = 1$

(g) $H(s) = \dfrac{1 - s/100}{(s + 1)^2}$, $\quad G(s) = \dfrac{10s + 1}{s/10 + 1}$

(h) $H(s) = \dfrac{1}{s(s + 1)}$, $\quad G(s) = \dfrac{1}{s + 1}$

[*Note:* Your sketch for part (h) should reflect the fact that for this feedback system $|G(j\omega)H(j\omega)| \rightarrow \infty$ as $\omega \rightarrow 0$; what is the phase of $G(j\omega)H(j\omega)$ for $\omega = 0^+$, i.e. for ω an infinitesimal amount larger than 0?]

11.36. Consider the basic discrete-time feedback system of Figure 11.3(b). Sketch the log magnitude–phase diagram and roughly determine the phase and gain margins for each of the following choices of $H(z)$ and $G(z)$. You may find it useful to determine those values of Ω for which either $|G(e^{j\Omega})H(e^{j\Omega})| = 1$ or $\measuredangle G(e^{j\Omega}) = -\pi$.

(a) $H(z) = z^{-1}$, $\quad G(z) = \dfrac{1}{2}$

(b) $H(z) = \dfrac{z^{-1}}{1 - \frac{1}{2}z^{-1}}$, $\quad G(z) = \dfrac{1}{2}$

(c) $H(z) = \dfrac{1}{(1 - \frac{1}{2}z^{-1})(1 + \frac{1}{2}z^{-1})}$, $\quad G(z) = z^{-2}$

(d) $H(z) = \dfrac{2}{z - 2}$, $\quad G(z) = 1$

(e) $H(z) = \dfrac{1}{z + \frac{1}{2}}$, $\quad G(z) = \dfrac{1}{z - \frac{3}{2}}$

(f) $H(z) = \dfrac{1}{z + \frac{1}{2}}$, $\quad G(z) = 1 - \dfrac{3}{2}z^{-1}$

(g) $H(z) = \dfrac{\frac{1}{2}}{z^2 - z + \frac{1}{3}}$, $\quad G(z) = 1$

(h) $H(z) = \dfrac{1}{z - 1}$, $\quad G(z) = \dfrac{1}{4}z^{-1}$

[*Note:* Your sketch for part (h) should reflect the fact that $G(z)H(z)$ has a pole at $z = 1$; what are the values of $\measuredangle G(e^{j\Omega})H(e^{j\Omega})$ for $e^{j\Omega}$ just on either side of the point $z = 1$?]

11.37. As mentioned at the end of Section 11.5, the concepts of phase and gain margin are *sufficient* conditions to ensure that a stable feedback system remains stable. Specifically, in the case of the gain margin, what we showed was that a stable feedback system will remain stable as the gain is increased up until we have reached a limit specified by the gain margin. This does *not* imply (a) that the feedback system cannot be made unstable by *decreasing* the gain or (b) that the system will be unstable for *all* values of gain larger than the gain margin limit. In this problem we illustrate these two points.

(a) Consider a continuous-time feedback system with

$$G(s)H(s) = \frac{1}{(s-1)(s+2)(s+3)}$$

Sketch the root locus for this system for $K > 0$. Use the properties of the root locus described in the text and in Problem 11.21 to help you draw the locus accurately. Once you do so you should see that for small values of the gain K the system is unstable, for larger values of K the system is stable, while for still larger values of K the system again becomes unstable. Find the range of values of K for which the system is stable. (*Hint:* Use the same method as is used in Example 11.2 and Problem 11.23 to determine the values of K at which branches of the root locus pass through the origin and cross the $j\omega$-axis.)

If we set our gain somewhere within the stable range that you have just found, we can increase the gain somewhat and maintain stability, but a large enough increase in gain causes the system to become unstable. This maximum amount of gain increase at which the closed-loop system just becomes unstable is the gain margin. Note also that if we *decrease* the gain too much, we can also cause instability.

(b) Consider the feedback system of part (a) with the gain K set at a value of 7. Show that the closed-loop system is stable. Sketch the log magnitude–phase plot for this system and show that there are two nonnegative values of ω for which $\angle G(j\omega)H(j\omega) = \pi$. Further show that for one of these $7|G(j\omega)H(j\omega)| < 1$ and for the other $7|G(j\omega)H(j\omega)| > 1$. The first of these provides us with the usual gain margin, that is, the factor $1/|7G(j\omega)H(j\omega)|$ by which we can increase the gain and just cause instability. The second of these provides us with the factor $1/|7G(j\omega)H(j\omega)|$, by which we can *decrease* the gain and just cause instability.

(c) Consider a feedback system with

$$G(s)H(s) = \frac{(s/100 + 1)^2}{(s+1)^3}$$

Sketch the root locus for $K > 0$. Show that two branches of the root locus begin in the left-half plane, and as K is increased move into the right-half plane and then back into the left-half plane. Do this by examining the equation

$$G(j\omega)H(j\omega) = -\frac{1}{K}$$

Specifically, by equating the real and imaginary parts of this equation, show that there are two values of $K \geq 0$ for which closed-loop poles lie on the $j\omega$-axis.

Thus, if we set the gain at a small enough value so that the system is stable, then we can increase the gain up until the point at which the two branches of the root locus intersect the $j\omega$-axis. For a range of values of gain beyond this point, the closed-loop system is unstable. However, if we *continue* to increase the gain, the system will again become stable for K large enough.

(d) Sketch the Nyquist plot for the system of part (c) and confirm the conclusions reached in part (c) by applying the Nyquist criterion (make sure to count the *net* number of encirclements of $-1/K$).

Systems such as that considered in parts (c) and (d) of this problem are often referred to as being *conditionally stable* systems, as the stability properties of such systems may change several times as the gain is varied.

APPENDIX
Partial Fraction Expansion

A.0 INTRODUCTION

The purpose of this appendix is to describe the technique of partial fraction expansion. This tool is of great value in the study of signals and systems, and in particular it is very useful in inverting Fourier, Laplace, or z-transforms and in analyzing LTI systems described by linear constant-coefficient differential or difference equations. The method of partial fraction expansion consists of taking a function that is the ratio of polynomials and expanding it as a linear combination of simpler terms of the same type. The determination of the coefficients in this linear combination is the basic problem to be solved in obtaining a partial fraction expansion. As we will see, this is a relatively straightforward problem in algebra which, with a bit of book-keeping, can be solved very efficiently.

To illustrate the basic idea behind and role of partial fraction expansion, consider the analysis performed in Section 4.12 for a second-order continuous-time LTI system specified by the differential equation

$$\frac{d^2y(t)}{dt^2} + 2\zeta\omega_n\frac{dy(t)}{dt} + \omega_n^2 y(t) = \omega_n^2 x(t) \tag{A.1}$$

Using the techniques developed in Chapter 4, we know that the frequency response of this system is

$$H(\omega) = \frac{\omega_n^2}{(j\omega)^2 + 2\zeta\omega_n(j\omega) + \omega_n^2} \tag{A.2}$$

or if we factor the denominator

$$H(\omega) = \frac{\omega_n^2}{(j\omega - c_1)(j\omega - c_2)} \tag{A.3}$$

where

$$c_1 = -\zeta\omega_n + \omega_n\sqrt{\zeta^2 - 1}$$
$$c_2 = -\zeta\omega_n - \omega_n\sqrt{\zeta^2 - 1}$$

(A.4)

Having $H(\omega)$, we are in a position to answer a variety of questions related to this system. For example, to determine the impulse response of the system, recall that for any number α with $\mathcal{R}e\{\alpha\} < 0$, the Fourier transform of

$$x_1(t) = e^{\alpha t}u(t)$$

(A.5a)

is

$$X_1(\omega) = \frac{1}{j\omega - \alpha}$$

(A.5b)

while if

$$x_2(t) = te^{\alpha t}u(t)$$

(A.6a)

then

$$X_2(\omega) = \frac{1}{(j\omega - \alpha)^2}$$

(A.6b)

Therefore, if we can expand $H(\omega)$ as a sum of terms of the form of eq. (A.5b) or (A.6b), we can determine the inverse transform of $H(\omega)$ by inspection. For example, in Section 4.12 we noted that when $c_1 \neq c_2$, $H(\omega)$ in eq. (A.3) could be rewritten in the form

$$H(\omega) = \left(\frac{\omega_n^2}{c_1 - c_2}\right)\frac{1}{j\omega - c_1} + \left(\frac{\omega_n^2}{c_2 - c_1}\right)\frac{1}{j\omega - c_2}$$

(A.7)

In this case the Fourier transform pair of eq. (A.5) allows us to write down immediately the inverse transform of $H(\omega)$ as

$$h(t) = \left[\frac{\omega_n^2}{c_1 - c_2}e^{c_1 t} + \frac{\omega_n^2}{c_2 - c_1}e^{c_2 t}\right]u(t)$$

While we have phrased the preceding discussion in terms of continuous-time Fourier transforms, similar concepts also arise in discrete-time Fourier analysis and in the use of Laplace and z-transforms. Specifically, in all of these cases we encounter the important class of *rational transforms*, that is, transforms that are ratios of polynomials in some variable. Also, in each of these contexts we find reasons for expanding such transforms as sums of simpler terms such as in eq. (A.7). In this section, in order to develop a general procedure for calculating such expansions, we consider rational functions of a general variable v. That is, we examine functions of the form

$$H(v) = \frac{\beta_m v^m + \beta_{m-1}v^{m-1} + \ldots + \beta_1 v + \beta_0}{\alpha_n v^n + \alpha_{n-1}v^{n-1} + \ldots + \alpha_1 v + \alpha_0}$$

(A.8)

For continuous-time Fourier analysis $(j\omega)$ plays the role of v, while for Laplace transforms that role is played by the complex variable s. In discrete-time Fourier analysis, v is usually taken to be $e^{-j\Omega}$, while for z-transforms, we can use either z^{-1} or z. After we have developed the basic technique of partial fraction expansion, we will illustrate its application to the analysis of both continuous-time and discrete-time LTI systems.

A.1 PARTIAL FRACTION EXPANSION
AND CONTINUOUS-TIME SIGNALS AND SYSTEMS

In our development it is convenient to consider rational functions in one of two standard forms. The second of these, which is often useful in the analysis of discrete-time signals and systems, will be discussed shortly. The first of the standard forms is given by

$$G(v) = \frac{b_{n-1}v^{n-1} + b_{n-2}v^{n-2} + \ldots + b_1 v + b_0}{v^n + a_{n-1}v^{n-1} + \ldots + a_1 v + a_0} \qquad (A.9)$$

In this form the coefficient of the highest-order term in the denominator is 1, and the order of the numerator is at least one less than the order of the denominator (the order of the numerator will be less than $n - 1$ if $b_{n-1} = 0$).

If we are given $H(v)$ in the form of eq. (A.8), we can obtain a rational function of the form of eq. (A.9) by performing two straightforward calculations. As a first step, we divide both the numerator and the denominator of $H(v)$ by α_n. This yields

$$H(v) = \frac{\gamma_m v^m + \gamma_{m-1}v^{m-1} + \ldots + \gamma_1 v + \gamma_0}{v^n + a_{n-1}v^{n-1} + \ldots + a_1 v + a_0} \qquad (A.10)$$

where

$$\gamma_m = \frac{\beta_m}{\alpha_n}, \quad \gamma_{m-1} = \frac{\beta_{m-1}}{\alpha_n}, \quad \ldots$$

$$a_{n-1} = \frac{\alpha_{n-1}}{\alpha_n}, \quad a_{n-2} = \frac{\alpha_{n-2}}{\alpha_n}, \quad \ldots$$

If $m < n$, $H(v)$ is called a *proper* rational function, and in this case $H(v)$ in eq. (A.10) is already of the form of eq. (A.9), with the identification that $b_0 = \gamma_0, b_1 = \gamma_1, \ldots,$ $b_m = \gamma_m$, with any remaining b's equal to zero. In most of the discussions in this book in which rational functions are considered, we are primarily concerned with proper rational functions. However, if $H(v)$ is not proper (i.e., if $m \geq n$), we can perform a preliminary calculation that allows us to write $H(v)$ as the sum of a polynomial in v and a proper rational function. That is,

$$H(v) = c_{m-n}v^{m-n} + c_{m-n-1}v^{m-n-1} + \ldots + c_1 v + c_0$$

$$+ \frac{b_{n-1}v^{n-1} + b_{n-2}v^{n-2} + \ldots + b_1 v + b_0}{v^n + a_{n-1}v^{n-1} + \ldots + a_1 v + a_0} \qquad (A.11)$$

The coefficients $c_0, c_1, \ldots, c_{m-n}$ and $b_0, b_1, \ldots, b_{n-1}$ can be obtained by equating eqs. (A.10) and (A.11) and then by multiplying through by the denominator. This yields

$$\gamma_m v^m + \ldots + \gamma_1 v + \gamma_0 = b_{n-1}v^{n-1} + \ldots + b_1 v + b_0$$

$$+ (c_{m-n}v^{m-n} + \ldots + c_0)(v^n + a_{n-1}v^{n-1} + \ldots + a_0)$$

$$(A.12)$$

By equating the coefficients of equal powers of v on both sides of eq. (A.12), we can determine the c's and b's in terms of the a's and γ's. For example, if $m = 2$ and $n = 1$ so that

$$H(v) = \frac{\gamma_2 v^2 + \gamma_1 v + \gamma_0}{v + a_1} = c_1 v + c_0 + \frac{b_0}{v + a_1}$$

then eq. (A.12) becomes

$$\gamma_2 v^2 + \gamma_1 v + \gamma_0 = b_0 + (c_1 v + c_0)(v + a_1)$$
$$= b_0 + c_1 v^2 + (c_0 + a_1 c_1)v + a_1 c_0$$

Equating the coefficients of equal powers of v, we obtain the equations

$$\gamma_2 = c_1$$
$$\gamma_1 = c_0 + a_1 c_1$$
$$\gamma_0 = b_0 + a_1 c_0$$

The first equation yields the value of c_1, which can then be used in the second to solve for c_0, which in turn can be used in the third to solve for b_0. The result is

$$c_1 = \gamma_2$$
$$c_0 = \gamma_1 - a_1 \gamma_2$$
$$b_0 = \gamma_0 - a_1(\gamma_1 - a_1 \gamma_2)$$

The general case of eq. (A.12) can be solved in an analogous fashion.

Our goal now is to focus on the proper rational function $G(v)$ in eq. (A.9) and to expand it into a sum of simpler proper rational functions. To see how this can be done, consider the case of $n = 3$, when eq. (A.9) reduces to

$$G(v) = \frac{b_2 v^2 + b_1 v + b_0}{v^3 + a_2 v^2 + a_1 v + a_0}$$

As a first step we factor the denominator of $G(v)$ so that we can write it in the form

$$G(v) = \frac{b_2 v^2 + b_1 v + b_0}{(v - p_1)(v - p_2)(v - p_3)} \tag{A.13}$$

Assuming for the moment that the roots p_1, p_2, and p_3 of the denominator are all distinct, we would like to expand $G(v)$ into a sum of the form

$$G(v) = \frac{A_1}{v - p_1} + \frac{A_2}{v - p_2} + \frac{A_3}{v - p_3} \tag{A.14}$$

The problem then is to determine the constants A_1, A_2, and A_3. One approach is to equate eqs. (A.13) and (A.14) and to multiply through by the denominator. In this case we obtain the equation

$$b_2 v^2 + b_1 v + b_0 = A_1(v - p_2)(v - p_3)$$
$$+ A_2(v - p_1)(v - p_3) \tag{A.15}$$
$$+ A_3(v - p_1)(v - p_2)$$

By expanding the right-hand side of eq. (A.15) and then equating coefficients of equal powers of v, we obtain a set of linear equations that can be solved for A_1, A_2, and A_3.

Although this approach always works, there is a much easier method. Consider eq. (A.14) and suppose that we would like to calculate A_1. Then, multiplying through by $(v - p_1)$, we obtain

$$(v - p_1)G(v) = A_1 + \frac{A_2(v - p_1)}{v - p_2} + \frac{A_3(v - p_1)}{v - p_3} \tag{A.16}$$

Since p_1, p_2, and p_3 are distinct, the last two terms on the right-hand side of eq.

(A.16) are zero for $v = \rho_1$. Therefore,

$$A_1 = [(v - \rho_1)G(v)]\big|_{v=\rho_1} \tag{A.17}$$

or, using eq. (A.13),

$$A_1 = \frac{b_2\rho_1^2 + b_1\rho_1 + b_0}{(\rho_1 - \rho_2)(\rho_1 - \rho_3)} \tag{A.18}$$

Similarly,

$$A_2 = [(v - \rho_2)G(v)]\big|_{v=\rho_2} = \frac{b_2\rho_2^2 + b_1\rho_2 + b_0}{(\rho_2 - \rho_1)(\rho_2 - \rho_3)} \tag{A.19}$$

$$A_3 = [(v - \rho_3)G(v)]\big|_{v=\rho_3} = \frac{b_2\rho_3^2 + b_1\rho_3 + b_0}{(\rho_3 - \rho_1)(\rho_3 - \rho_2)} \tag{A.20}$$

Suppose now that $\rho_1 = \rho_3 \neq \rho_2$, that is,

$$G(v) = \frac{b_2v^2 + b_1v + b_0}{(v - \rho_1)^2(v - \rho_2)} \tag{A.21}$$

In this case we look for an expansion of the form

$$G(v) = \frac{A_{11}}{v - \rho_1} + \frac{A_{12}}{(v - \rho_1)^2} + \frac{A_{21}}{v - \rho_2} \tag{A.22}$$

Here we need the $1/(v - \rho_1)^2$ term in order to obtain the correct denominator in eq. (A.22) when we collect terms over a least common denominator. We also need to include the $1/(v - \rho_1)$ term in general. To see why this is so, consider equating eqs. (A.21) and (A.22) and multiplying through by the denominator of eq. (A.21):

$$\begin{aligned}b_2v^2 + b_1v + b_0 &= A_{11}(v - \rho_1)(v - \rho_2) \\ &\quad + A_{12}(v - \rho_2) + A_{21}(v - \rho_1)^2\end{aligned} \tag{A.23}$$

Again if we equate coefficients of equal powers of v, we obtain three equations (for the coefficients of the v^0, v^1, and v^2 terms). If we omit the A_{11} term in eq. (A.22), we will then have three equations in two unknowns, which in general will not have a solution. By including this term, we can always find a solution. In this case also, however, there is a much simpler method. Consider eq. (A.22) and multiply through by $(v - \rho_1)^2$:

$$(v - \rho_1)^2G(v) = A_{11}(v - \rho_1) + A_{12} + \frac{A_{21}(v - \rho_1)^2}{v - \rho_2} \tag{A.24}$$

From the preceding example, we see immediately how to determine A_{12}:

$$A_{12} = [(v - \rho_1)^2G(v)]\big|_{v=\rho_1} = \frac{b_2\rho_1^2 + b_1\rho_1 + b_0}{\rho_1 - \rho_2} \tag{A.25}$$

As for A_{11}, suppose that we differentiate eq. (A.24) with respect to v:

$$\frac{d}{dv}[(v - \rho_1)^2G(v)] = A_{11} + A_{21}\left[\frac{2(v - \rho_1)}{v - \rho_2} - \frac{(v - \rho_1)^2}{(v - \rho_2)^2}\right] \tag{A.26}$$

It is then apparent that the final term in eq. (A.26) is zero for $v = \rho_1$, and therefore

$$\begin{aligned}A_{11} &= \left[\frac{d}{dv}(v - \rho_1)^2G(v)\right]\Big|_{v=\rho_1} \\ &= \frac{2b_2\rho_1 + b_1}{\rho_1 - \rho_2} - \frac{b_2\rho_1^2 + b_1\rho_1 + b_0}{(\rho_1 - \rho_2)^2}\end{aligned} \tag{A.27}$$

Finally, by multiplying eq. (A.22) by $(v - p_2)$, we find that

$$A_{21} = [(v - p_2)G(v)]|_{v=p_2} = \frac{b_2 p_2^2 + b_1 p_2 + b_0}{(p_2 - p_1)^2} \tag{A.28}$$

This example illustrates all of the basic ideas behind partial fraction expansion in the general case. Specifically, suppose that the denominator of $G(v)$ in eq. (A.9) has distinct roots $p_1, \ldots, p_r$ with *multiplicities* $\sigma_1, \ldots, \sigma_r$, that is,

$$G(v) = \frac{b_{n-1}v^{n-1} + \ldots + b_1 v + b_0}{(v - p_1)^{\sigma_1}(v - p_2)^{\sigma_2} \ldots (v - p_r)^{\sigma_r}} \tag{A.29}$$

In this case $G(v)$ has a partial fraction expansion of the form

$$\begin{aligned}
G(v) &= \frac{A_{11}}{v - p_1} + \frac{A_{12}}{(v - p_1)^2} + \ldots + \frac{A_{1\sigma_1}}{(v - p_1)^{\sigma_1}} \\
&+ \frac{A_{21}}{v - p_2} + \ldots + \frac{A_{2\sigma_2}}{(v - p_2)^{\sigma_2}} \\
&+ \ldots + \frac{A_{r1}}{v - p_r} + \ldots + \frac{A_{r\sigma_r}}{(v - p_r)^{\sigma_r}} \\
&= \sum_{i=1}^{r} \sum_{k=1}^{\sigma_i} \frac{A_{ik}}{(v - p_i)^k}
\end{aligned} \tag{A.30}$$

where the A_{ik} are computed from the equation†

$$\boxed{A_{ik} = \frac{1}{(\sigma_i - k)!} \left[\frac{d^{\sigma_i - k}}{dv^{\sigma_i - k}} [(v - p_i)^{\sigma_i} G(v)] \right]_{v=p_i}} \tag{A.31}$$

This result can be checked much as in the example: multiply both sides of eq. (A.30) by $(v - p_i)^{\sigma_i}$ and differentiate repeatedly until A_{ik} is no longer multiplied by a power of $(v - p_i)$. Then set $v = p_i$.

Example A.1

In Example 4.26 we examine an LTI system described by the differential equation

$$\frac{d^2 y(t)}{dt^2} + 4\frac{dy(t)}{dt} + 3y(t) = \frac{dx(t)}{dt} + 2x(t) \tag{A.32}$$

From Example 4.26 we have that the frequency response of this system is

$$H(\omega) = \frac{j\omega + 2}{(j\omega)^2 + 4j\omega + 3} \tag{A.33}$$

To determine the impulse response for this system, we expand $H(\omega)$ into a sum of simpler terms whose inverse transforms can be obtained by inspection. Making the substitution of v for $j\omega$, we obtain the following function:

$$G(v) = \frac{v + 2}{v^2 + 4v + 3} = \frac{v + 2}{(v + 1)(v + 3)} \tag{A.34}$$

The partial fraction expansion for $G(v)$ then is

$$G(v) = \frac{A_{11}}{v + 1} + \frac{A_{21}}{v + 3} \tag{A.35}$$

†Here we use the factorial notation $r!$ for the product $r(r - 1)(r - 2) \ldots 2 \cdot 1$. The quantity $0!$ is defined to be equal to 1.

where

$$A_{11} = [(v + 1)G(v)]\Big|_{v=-1} = \frac{-1 + 2}{-1 + 3} = \frac{1}{2} \tag{A.36}$$

$$A_{21} = [(v + 3)G(v)]\Big|_{v=-3} = \frac{-3 + 2}{-3 + 1} = \frac{1}{2} \tag{A.37}$$

Thus,

$$H(\omega) = \frac{\frac{1}{2}}{j\omega + 1} + \frac{\frac{1}{2}}{j\omega + 3} \tag{A.38}$$

and the impulse response of the system, obtained by inverting eq. (A.38), is

$$h(t) = \tfrac{1}{2}e^{-t}u(t) + \tfrac{1}{2}e^{-3t}u(t) \tag{A.39}$$

This example can also be analyzed using the techniques of Laplace transform analysis as developed in Chapter 9. The system function for this system is

$$H(s) = \frac{s + 2}{s^2 + 4s + 3} \tag{A.40}$$

and if we substitute v for s, we obtain the same $G(v)$ given in eq. (A.34). Thus, the partial fraction expansion proceeds exactly as in eqs. (A.35)–(A.37), with the result that

$$H(s) = \frac{\frac{1}{2}}{s + 1} + \frac{\frac{1}{2}}{s + 3} \tag{A.41}$$

Inverting this transform, we again obtain the impulse response as given in eq. (A.39).

Example A.2

We now illustrate the method of partial fraction expansion when there are repeated factors in the denominator. In Example 4.27 we consider the response of the system described by eq. (A.32) when the input is

$$x(t) = e^{-t}u(t) \tag{A.42}$$

From Example 4.27 we have that the Fourier transform of the output of the system is

$$Y(\omega) = \frac{j\omega + 2}{(j\omega + 1)^2(j\omega + 3)} \tag{A.43}$$

Substituting v for $j\omega$ we obtain the rational function

$$G(v) = \frac{v + 2}{(v + 1)^2(v + 3)} \tag{A.44}$$

The partial fraction expansion for this function is

$$G(v) = \frac{A_{11}}{v + 1} + \frac{A_{12}}{(v + 1)^2} + \frac{A_{21}}{v + 3} \tag{A.45}$$

where, from eq. (A.31),

$$A_{11} = \frac{1}{(2 - 1)!} \frac{d}{dv}[(v + 1)^2 G(v)]\Big|_{v=-1} = \tfrac{1}{4} \tag{A.46a}$$

$$A_{12} = \{(v + 1)^2\, G(v)\}|_{v=-1} = \tfrac{1}{2} \tag{A.46b}$$

$$A_{21} = \{(v + 3)G(v)\}|_{v=-3} = -\tfrac{1}{4} \tag{A.46c}$$

Therefore,

$$Y(\omega) = \frac{\frac{1}{4}}{j\omega + 1} + \frac{\frac{1}{2}}{(j\omega + 1)^2} - \frac{\frac{1}{4}}{j\omega + 3} \tag{A.47}$$

and taking inverse transforms,

$$y(t) = \{\tfrac{1}{4}e^{-t} + \tfrac{1}{2}te^{-t} - \tfrac{1}{4}e^{-3t}\}u(t) \tag{A.48}$$

Again this analysis could also have been performed using Laplace transforms, and the algebra would again be identical to that given in eqs. (A.44)–(A.46).

A.2 PARTIAL FRACTION EXPANSION AND DISCRETE-TIME SIGNALS AND SYSTEMS

As mentioned earlier, in performing partial fraction expansions for discrete-time Fourier transforms or for z-transforms, it is often more convenient to deal with a slightly different form for rational functions. Specifically, we will now assume that we have a rational function in the form

$$G(v) = \frac{d_{n-1}v^{n-1} + \ldots + d_1 v + d_0}{f_n v^n + \ldots + f_1 v + 1} \tag{A.49}$$

This form for $G(v)$ can be obtained from $G(v)$ in eq. (A.9) by dividing the numerator and denominator by a_0.

With $G(v)$ given as in eq. (A.49), the corresponding factorization of the denominator of $G(v)$ is of the form

$$G(v) = \frac{d_{n-1}v^{n-1} + \ldots + d_1 v + d_0}{(1 - \rho_1^{-1}v)^{\sigma_1}(1 - \rho_2^{-1}v)^{\sigma_2} \ldots (1 - \rho_r^{-1}v)^{\sigma_r}} \tag{A.50}$$

and the form of the partial fraction expansion that results is

$$G(v) = \sum_{i=1}^{r} \sum_{k=1}^{\sigma_i} \frac{B_{ik}}{(1 - \rho_i^{-1}v)^k} \tag{A.51}$$

The B_{ik} can be calculated in a manner that is similar to that used earlier:

$$\boxed{B_{ik} = \frac{1}{(\sigma_i - k)!}(-\rho_i)^{\sigma_i - k}\left[\frac{d^{\sigma_i - k}}{dv^{\sigma_i - k}}[(1 - \rho_i^{-1}v)^{\sigma_i}G(v)]\right]\bigg|_{v=\rho_i}} \tag{A.52}$$

As before, the validity of eq. (A.52) can be determined by multiplying both sides of eq. (A.51) by $(1 - \rho_i^{-1}v)^{\sigma_i}$, then by differentiating repeatedly with respect to v until B_{ik} is no longer multiplied by a power of $(1 - \rho_i^{-1}v)$, and finally by setting $v = \rho_i$.

Example A.3

Consider the causal LTI system characterized by the difference equation

$$y[n] - \tfrac{3}{4}y[n-1] + \tfrac{1}{8}y[n-2] = 2x[n] \tag{A.53}$$

This system is examined in Example 5.18. From this example we have that the frequency response of the LTI system described by eq. (A.53) is

$$H(\Omega) = \frac{2}{1 - \tfrac{3}{4}e^{-j\Omega} + \tfrac{1}{8}e^{-2j\Omega}} \tag{A.54}$$

For discrete-time transforms such as this, it is most convenient to substitute v for $e^{-j\Omega}$. Making this substitution, we obtain the rational function

$$G(v) = \frac{2}{1 - \tfrac{3}{4}v + \tfrac{1}{8}v^2} = \frac{2}{(1 - \tfrac{1}{2}v)(1 - \tfrac{1}{4}v)} \tag{A.55}$$

Using the partial fraction expansion specified by eqs. (A.50)–(A.52), we obtain

$$G(v) = \frac{B_{11}}{1 - \frac{1}{2}v} + \frac{B_{21}}{1 - \frac{1}{4}v} \qquad (A.56)$$

$$B_{11} = [(1 - \tfrac{1}{2}v)G(v)]|_{v=2} = \frac{2}{1 - \frac{1}{2}} = 4 \qquad (A.57)$$

$$B_{21} = [(1 - \tfrac{1}{4}v)G(v)]|_{v=4} = \frac{2}{1 - 2} = -2 \qquad (A.58)$$

Thus,

$$H(\Omega) = \frac{4}{1 - \frac{1}{2}e^{-j\Omega}} - \frac{2}{1 - \frac{1}{4}e^{-j\Omega}} \qquad (A.59)$$

and, taking the inverse transform of eq. (A.59), we obtain the unit impulse response

$$h[n] = 4(\tfrac{1}{2})^n u[n] - 2(\tfrac{1}{4})^n u[n] \qquad (A.60)$$

In Section 10.6 we develop the tools of z-transform analysis for the examination of discrete-time LTI systems specified by linear constant-coefficient difference equations. Applying those techniques to this example, we find that the system function can be determined by inspection from eq. (A.53):

$$H(z) = \frac{2}{1 - \frac{3}{4}z^{-1} + \frac{1}{8}z^{-2}} \qquad (A.61)$$

Then, substituting v for z^{-1}, we obtain $G(v)$ as in eq. (A.55). Thus, using the partial fraction expansion calculations in eqs. (A.56)–(A.58), we find that

$$H(z) = \frac{4}{1 - \frac{1}{2}z^{-1}} - \frac{2}{1 - \frac{1}{4}z^{-1}} \qquad (A.62)$$

which when inverted again yields the impulse response of eq. (A.60).

Example A.4

Suppose that the input to the system considered in Example A.3 is

$$x[n] = (\tfrac{1}{4})^n u[n]$$

Then from Example 5.19 we have that the Fourier transform of the output is

$$Y(\Omega) = \frac{2}{(1 - \frac{1}{2}e^{-j\Omega})(1 - \frac{1}{4}e^{-j\Omega})^2} \qquad (A.63)$$

Substituting v for $e^{-j\Omega}$, we obtain

$$G(v) = \frac{2}{(1 - \frac{1}{2}v)(1 - \frac{1}{4}v)^2} \qquad (A.64)$$

Thus, using eqs. (A.51) and (A.52), we obtain the partial fraction expansion

$$G(v) = \frac{B_{11}}{1 - \frac{1}{4}v} + \frac{B_{12}}{(1 - \frac{1}{4}v)^2} + \frac{B_{21}}{1 - \frac{1}{2}v} \qquad (A.65)$$

where

$$B_{11} = (-4)\left[\frac{d}{dv}(1 - \tfrac{1}{4}v)^2 G(v)\right]\Big|_{v=4} = -4 \qquad (A.66)$$

$$B_{12} = [(1 - \tfrac{1}{4}v)^2 G(v)]|_{v=4} = -2 \qquad (A.67)$$

$$B_{21} = [(1 - \tfrac{1}{2}v)G(v)]|_{v=2} = 8 \qquad (A.68)$$

Therefore,

$$Y(\Omega) = -\frac{4}{1 - \frac{1}{4}e^{-j\Omega}} - \frac{2}{(1 - \frac{1}{4}e^{-j\Omega})^2} + \frac{8}{1 - \frac{1}{2}e^{-j\Omega}} \qquad (A.69)$$

which can be inverted by inspection, using the Fourier transform pairs in Table 5.3:

$$y[n] = \{-4(\tfrac{1}{4})^n - 2(n+1)(\tfrac{1}{4})^n + 8(\tfrac{1}{2})^n\}u[n] \qquad \text{(A.70)}$$

Example A.5

Non-proper rational functions are often encountered in the analysis of discrete-time systems. To illustrate this and also to show how they can be analyzed using the techniques developed in this appendix, consider the causal LTI system characterized by the difference equation

$$y[n] + \tfrac{5}{6}y[n-1] + \tfrac{1}{6}y[n-2] = x[n] + 3x[n-1] + \tfrac{11}{6}x[n-2] + \tfrac{1}{3}x[n-3]$$

The frequency response of this system is given by

$$H(\Omega) = \frac{1 + 3e^{-j\Omega} + \tfrac{11}{6}e^{-j2\Omega} + \tfrac{1}{3}e^{-j3\Omega}}{1 + \tfrac{5}{6}e^{-j\Omega} + \tfrac{1}{6}e^{-j2\Omega}}. \qquad \text{(A.71)}$$

Substituting v for $e^{-j\Omega}$ we obtain

$$G(v) = \frac{1 + 3v + \tfrac{11}{6}v^2 + \tfrac{1}{3}v^3}{1 + \tfrac{5}{6}v + \tfrac{1}{6}v^2} \qquad \text{(A.72)}$$

This rational function can be written as the sum of a polynomial and a proper rational function:

$$G(v) = c_0 + c_1 v + \frac{b_1 v + b_0}{1 + \tfrac{5}{6}v + \tfrac{1}{6}v^2} \qquad \text{(A.73)}$$

Equating eqs. (A.72) and (A.73) and multiplying by $(1 + \tfrac{5}{6}v + \tfrac{1}{6}v^2)$ we obtain

$$1 + 3v + \tfrac{11}{6}v^2 + \tfrac{1}{3}v^3 =$$
$$(c_0 + b_0) + (\tfrac{5}{6}c_0 + c_1 + b_1)v + (\tfrac{1}{6}c_0 + \tfrac{5}{6}c_1)v^2 + \tfrac{1}{6}c_1 v^3 \qquad \text{(A.74)}$$

Equating coefficients we see that

$$\begin{aligned}
\tfrac{1}{6}c_1 &= \tfrac{1}{3} & &\Rightarrow c_1 = 2 \\
\tfrac{1}{6}c_0 + \tfrac{5}{6}c_1 &= \tfrac{11}{6} & &\Rightarrow c_0 = 1 \\
\tfrac{5}{6}c_0 + c_1 + b_1 &= 3 \Rightarrow b_1 = \tfrac{1}{6} \\
c_0 + b_0 &= 1 & &\Rightarrow b_0 = 0
\end{aligned} \qquad \text{(A.75)}$$

Thus

$$H(\Omega) = 1 + 2e^{-j\Omega} + \frac{\tfrac{1}{6}e^{-j\Omega}}{1 + \tfrac{5}{6}e^{-j\Omega} + \tfrac{1}{6}e^{-j2\Omega}} \qquad \text{(A.76)}$$

Also, we can use the method developed in the appendix to expand the proper rational function in eq. (A.73):

$$\frac{\tfrac{1}{6}v}{1 + \tfrac{5}{6}v + \tfrac{1}{6}v^2} = \frac{\tfrac{1}{6}v}{(1 + \tfrac{1}{3}v)(1 + \tfrac{1}{2}v)} = \frac{B_{11}}{(1 + \tfrac{1}{3}v)} + \frac{B_{21}}{(1 + \tfrac{1}{2}v)}$$

The coefficients B_{11} and B_{21} are given by

$$B_{11} = \left(\frac{\tfrac{1}{6}v}{1 + \tfrac{1}{2}v}\right)\Bigg|_{v=-3} = 1$$

$$B_{21} = \left(\frac{\tfrac{1}{6}v}{1 + \tfrac{1}{3}v}\right)\Bigg|_{v=-2} = -1$$

Therefore we find that

$$H(\Omega) = 1 + 2e^{-j\Omega} + \frac{1}{1 + \tfrac{1}{3}e^{-j\Omega}} - \frac{1}{1 + \tfrac{1}{2}e^{-j\Omega}} \qquad \text{(A.77)}$$

and by inspection we can determine the impulse response of this system:

$$h[n] = \delta[n] + 2\delta[n-1] + [(-\tfrac{1}{3})^n - (-\tfrac{1}{2})^n]u[n] \qquad \text{(A.78)}$$

Bibliography

B.0 INTRODUCTION

The purpose of this bibliography is to provide the reader with sources for additional and more advanced treatments of topics in signal and system analysis. This is by no means meant to be an exhaustive list but rather it is intended to indicate directions for further study and several references for each.

We have divided the bibliography into twelve different subject areas. Several of these deal with more thorough and specialized treatments of topics in signals and systems that are introduced in this text (filtering; sampling and digital signal processing; modulation and communications; and feedback and automatic control). Others deal specifically with the mathematical techniques used in signal and system analysis (background mathematics, including differential and difference equations and complex variables; the theory of Fourier series and of Fourier, Laplace, and z-transforms; and generalized functions). We have also included a list of other basic books on signals and systems. In addition, we have provided lists of references on three very important topics (state space methods, random signals, and nonlinear systems) which are natural subjects for more advanced study for those interested in expanding their knowledge of the methods of signals and systems. Finally, we have included a list of references that deal with specific applications of signal and system analysis. While this is just a very small sampling of the uses of these methods, it should provide the reader with a significant appreciation for the range of practical problems in which the ideas treated in this book find application.

B.1 BACKGROUND AND BASIC MATHEMATICS

BIRKHOFF, GARRETT, and ROTA, GIAN-CARLO. *Ordinary Differential Equations*. 3rd ed. New York: John Wiley, 1978.

CHURCHILL, RUEL V. *Complex Variables and Applications*. 3rd ed. New York: McGraw-Hill, 1974.

HILDEBRAND, FRANCIS B. *Advanced Calculus for Applications*. 2nd ed. Englewood Cliffs, N.J.: Prentice-Hall, 1976.

LEVY, H., and LESSMAN, F. *Finite Difference Equations*. New York: Macmillan, 1961.

THOMAS, GEORGE B., JR., and FINNEY, ROSS L. *Calculus and Analytic Geometry*. 5th ed. Reading, Mass.: Addison-Wesley, 1979.

B.2 FOURIER SERIES AND FOURIER, LAPLACE, AND z-TRANSFORMS

BRACEWELL, RONALD N. *The Fourier Transform and Its Applications*. 2nd ed. McGraw-Hill Electrical and Electronic Engineering Series. New York: McGraw-Hill, 1978.

DAVIS, HARRY F. *Fourier Series and Orthogonal Functions*. Boston: Allyn and Bacon, 1963.

DOETSCH, GUSTAV. *Introduction to the Theory and Application of the Laplace Transformation, with a Table of Laplace Transformations*. Translated by Walter Nader. Berlin and New York: Springer-Verlag, 1974.

DYM, H., and MCKEAN, H. P. *Fourier Series and Integrals*. New York: Academic Press, 1972.

FRANKS, L. E. *Signal Theory*. Prentice-Hall Electrical Engineering Series. Englewood Cliffs, N.J.: Prentice-Hall, 1969.

JURY, E. I. *Theory and Application of the z-Transform Method*. New York: John Wiley, 1964.

PAPOULIS, ATHANASIOS. *The Fourier Integral and Its Applications*. New York: McGraw-Hill, 1962.

RAINVILLE, EARL DAVID. *The Laplace Transform: An Introduction*. New York: Macmillan, 1963.

B.3 FILTERING

CHRISTIAN, ERICH, and EISENMANN, EGON. *Filter Design Tables and Graphs*. New York: John Wiley, 1966.

DANIELS, RICHARD WILLIAM. *Approximation Methods for Electronic Filter Design*. New York: McGraw-Hill, 1974.

HUELSMAN, LAWRENCE P., and ALLEN, P. E. *Introduction to the Theory and Design of Active Filters*. New York: McGraw-Hill, 1980.

HUMPHERYS, D. S. *The Analysis, Design, and Synthesis of Electrical Filters*. Englewood Cliffs, N.J.: Prentice-Hall, 1970.

JOHNSON, DAVID E. *Introduction to Filter Theory*. Englewood Cliffs, N.J.: Prentice-Hall, 1976.

JOHNSON, DAVID E.; JOHNSON, J. R.; and MOORE, H. P. *A Handbook of Active Filters*. Englewood Cliffs, N.J.: Prentice-Hall, 1980.

ZVEREV, ANATOL I. *Handbook of Filter Synthesis*. New York: John Wiley, 1967.

B.4 MODULATION AND COMMUNICATIONS

BLACK, HAROLD S. *Modulation Theory*. Bell Telephone Laboratory Series. New York: Van Nostrand, 1953.

CARLSON, A. BRUCE. *Communications Systems*. 2nd ed. New York: McGraw-Hill, 1975.

HAYKIN, SIMON S. *Communication Systems*. New York: John Wiley, 1978.

LINDSEY, WILLIAM C. *Synchronization Systems in Communication and Control*. Prentice-Hall Series in Information and System Sciences. Englewood Cliffs, N.J.: Prentice-Hall, 1972.

McMULLEN, C. W. *Communication Theory Principles*. New York: MacMillan, 1968.

SCHWARTZ, MISCHA. *Information Transmission, Modulation, and Noise: A Unified Approach to Communication Systems*. 3rd ed. McGraw-Hill Series in Electrical Engineering. New York: McGraw-Hill, 1980.

SIMPSON, RICHARD S., and HOUTS, RONALD C. *Fundamentals of Analog and Digital Communication Systems*. Boston: Allyn and Bacon, 1971.

STREMLER, FERREL G. *Introduction to Communication Systems*. Addison-Wesley Series in Electrical Engineering. Reading, Mass.: Addison-Wesley, 1977.

TAUB, HERBERT, and SCHILLING, DONALD L. *Principles of Communication Systems*. McGraw-Hill Series in Electrical Engineering. New York: McGraw-Hill, 1971.

THOMAS, JOHN BOWMAN. *An Introduction to Statistical Communication Theory*. New York: John Wiley, 1969.

WOZENCRAFT, JOHN M., and JACOBS, IRWIN MARK. *Principles of Communication Engineering*. New York: John Wiley, 1965.

B.5 DIGITAL SIGNAL PROCESSING

ANTONIOU, ANDREAS. *Digital Filters: Analysis and Design*. McGraw-Hill Series in Electrical Engineering. New York: McGraw-Hill, 1979.

GOLD, BERNARD, and RADER, CHARLES M. *Digital Processing of Signals*. Lincoln Laboratory Publications. New York: McGraw-Hill, 1969.

HAMMING, RICHARD WESLEY. *Digital Filters*. Prentice-Hall Signal Processing Series. Englewood Cliffs, N.J.: Prentice-Hall, 1977.

OPPENHEIM, ALAN VICTOR, and SCHAFER, RONALD W. *Digital Signal Processing*. Englewood Cliffs, N.J.: Prentice-Hall, 1975.

PELED, ABRAHAM, and LIU, B. *Digital Signal Processing: Theory, Design, and Implementation*. New York: John Wiley, 1976.

RABINER, LAWRENCE RICHARD, and GOLD, BERNARD. *Theory and Application of Digital Signal Processing*. Englewood Cliffs, N.J.: Prentice-Hall, 1975.

TRETTER, STEVEN A. *Introduction to Discrete-Time Signal Processing*. New York: John Wiley, 1976.

B.6 FEEDBACK AND AUTOMATIC CONTROL

ANDERSON, B. D. O., and MOORE, J. B. *Linear Optimal Control*. Englewood Cliffs, N.J.: Prentice-Hall, 1971.

BRYSON, ARTHUR EARL, and HO, YU-CHI. *Applied Optimal Control: Optimization, Estimation, and Control.* Waltham, Mass.: Blaisdell Publishing Company, 1969.

CLARK, R. N. *Introduction to Automatic Control Systems.* New York: John Wiley, 1962.

D'AZZO, JOHN J., and HOUPIS, CONSTANTINE H. *Linear Control System Analysis and Design: Conventional and Modern.* 2nd ed. New York: McGraw-Hill, 1981.

GUPTA, SOMESHWAR CHANDER, and HASDORFF, LAWRENCE. *Fundamentals of Automatic Control.* New York: John Wiley, 1970.

JURY, ELIAHU IBRAHIM. *Sampled-Data Control Systems.* New York: John Wiley, 1958.

KUO, BENJAMIN C. *Automatic Control Systems.* 2nd ed. Prentice-Hall Electrical Engineering Series. Englewood Cliffs, N.J.: Prentice-Hall, 1967.

MELSA, JAMES LOUIS, and SCHULTZ, DONALD G. *Linear Control Systems.* McGraw-Hill Series in Electronic Systems. New York: McGraw-Hill, 1969.

RAGAZZINI, JOHN RALPH, and FRANKLIN, GENE F. *Sampled-Data Control Systems.* McGraw-Hill Series in Control Systems Engineering. New York: McGraw-Hill, 1958.

ROBERGE, JAMES KERR. *Operational Amplifiers: Theory and Practice.* New York: John Wiley, 1975.

B.7 BASIC SIGNALS AND SYSTEMS

COOPER, GEORGE R., and McGILLEM, CLARE D. *Methods of Signal and System Analysis.* Holt, Rinehart and Winston Series in Electrical Engineering, Electronics, and Systems. New York: Holt, Rinehart and Winston, 1967.

CRUZ, JOSÉ BEJAR, and VAN VALKENBURG, M. E. *Signals in Linear Circuits.* Boston: Houghton Mifflin Company, 1974.

GABEL, ROBERT A., and ROBERTS, RICHARD A. *Signals and Linear Systems.* New York: John Wiley, 1973.

LATHI, BHAGWANDAS PANNALAL. *Signals, Systems and Communication.* New York: John Wiley, 1965.

LIU, CHUNG LAUNG, and LIU, JANE W. S. *Linear Systems Analysis.* New York: McGraw-Hill, 1975.

LYNN, PAUL A. *An Introduction to the Analysis and Processing of Signals.* New York: Mac-Millan, 1973.

McGILLEM, CLARE D., and COOPER, GEORGE R. *Continuous and Discrete Signal and System Analysis.* Holt, Rinehart and Winston Series in Electrical Engineering, Electronics, and Systems. New York: Holt, Rinehart and Winston, 1974.

PAPOULIS, ATHANASIOS. *Signal Analysis.* New York: McGraw-Hill, 1977.

PAPOULIS, ATHANASIOS. *Circuits and Systems.* Holt, Rinehart and Winston Series in Electrical Engineering, Electronics, and Systems. New York: Holt, Rinehart and Winston, 1980.

B.8 STATE SPACE MODELS AND METHODS

BROCKETT, ROGER W. *Finite Dimensional Linear Systems.* New York: John Wiley, 1970.

CHEN, CHI-TSONG. *Introduction to Linear System Theory.* Holt, Rinehart and Winston Series in Electrical Engineering, Electronics and Systems. New York: Holt, Rinehart and Winston, 1970.

DeRusso, Paul M.; Roy, Rob J.; and Close, Charles M. *State Variables for Engineers.* New York: John Wiley, 1965.

Gupta, S. C. *Transform and State Variable Methods in Linear Systems.* New York: John Wiley, 1966.

Kailath, Thomas. *Linear Systems.* Englewood Cliffs, N.J.: Prentice-Hall, 1980.

Zadeh, Lotfi Asker, and Desoer, Charles A. *Linear System Theory: The State Space Approach.* McGraw-Hill Series in System Science. New York: McGraw-Hill, 1963.

B.9 SYSTEMS AND RANDOM SIGNALS

Davenport, Wilbur B., Jr. *Probability and Random Processes: An Introduction for Applied Scientists and Engineers.* New York: McGraw-Hill, 1970.

Drake, Alvin W. *Fundamentals of Applied Probability Theory.* McGraw-Hill Series in Probability and Statistics. New York: McGraw-Hill, 1967.

Gelb, Arthur. *Applied Optimal Estimation.* Cambridge, Mass.: The M.I.T. Press, 1974.

Laning, J. Halcombe, Jr., and Battin, Richard H. *Random Processes in Automatic Control.* New York: McGraw-Hill, 1956.

Newton, George C.; Gould, Leonard A.; and Kaiser, James F. *Analytical Design of Linear Feedback Controls.* New York: John Wiley, 1957.

Papoulis, Athanasios. *Probability, Random Variables, and Stochastic Processes.* New York: McGraw-Hill, 1965.

Van Trees, Harry L. *Detection, Estimation, and Modulation Theory, Part 1. Detection, Estimation, and Linear Modulation Theory.* New York: John Wiley, 1968.

B.10 GENERALIZED FUNCTIONS

Arsac, J. *Fourier Transforms and the Theory of Distributions.* Translated by Allen Nussbaum and Gretchen C. Heim. Prentice-Hall Applied Mathematics Series. Englewood Cliffs, N.J.: Prentice-Hall, 1966.

Gel'fand, I. M., et al. *Generalized Functions.* 5 vols. Translated by E. Saletan, et al. New York: Academic Press, 1964–68.

Lighthill, M. J., Sir. *Introduction to Fourier Analysis and Generalized Functions.* Cambridge Monographs on Mechanics and Applied Mathematics. Cambridge, 1958.

Zemanian, A. H. *Distribution Theory and Transform Analysis: An Introduction to Generalized Functions, with Applications.* International Series in Pure and Applied Mathematics. New York: McGraw-Hill, 1965.

B.11 NONLINEAR SYSTEMS

Chua, Leon O. *Introduction to Nonlinear Network Theory.* McGraw-Hill Series in Electronic Systems. New York: McGraw-Hill, 1969.

Gelb, Arthur, and Vander Velde, Wallace E. *Multiple-input Describing Functions and Nonlinear System Design.* McGraw-Hill Electronic Sciences Series. New York: McGraw-Hill, 1968.

GRAHM, DUNSTAN, and MCRUER, DUANE. *Analysis of Nonlinear Control Systems.* New York: Dover, 1971.

HOLTZMAN, JACK M. *Nonlinear System Theory: A Functional Analysis Approach.* Prentice-Hall Electrical Engineering Series. Englewood Cliffs, N.J.: Prentice-Hall, 1970.

HSU, JAY C., and MEYER, ANDREW U. *Modern Control Principles and Applications.* New York: McGraw-Hill, 1968.

LEFSCHETZ, SOLOMON. *Stability of Nonlinear Control Systems.* Mathematics in Science and Engineering, no. 13. New York: Academic Press, 1965.

VIDYASAGER, M. *Nonlinear Systems Analysis.* Prentice-Hall Network Series. Englewood Cliffs, N.J.: Prentice-Hall, 1978.

B.12 APPLICATIONS

BATH, MARKUS. *Spectral Analysis in Geophysics.* Developments in Solid Earth Geophysics, vol. 7. New York: American Elsevier, 1974.

BOX, GEORGE E. P., and JENKINS, GWILYM M. *Time Series Analysis: Forecasting and Control.* Rev. ed. Holden-Day Series in Time Series Analysis. San Francisco: Holden-Day, 1976.

GONZALEZ, RAFAEL C., and WINTZ, PAUL. *Digital Image Processing.* Applied Mathematics and Computation, no. 13. Reading, Mass.: Addison-Wesley, 1977.

MUSLIM, J. H. *Biological Control Systems Analysis.* New York: McGraw-Hill, 1966.

OPPENHEIM, A. V., ed. *Applications of Digital Signal Processing.* Englewood Cliffs, N.J.: Prentice-Hall, 1978.

PAPOULIS, ATHANASIOS. *Systems and Transforms with Applications in Optics.* McGraw-Hill Series in Systems Science. New York: McGraw-Hill, 1968.

PRATT, WILLIAM. *Digital Image Processing.* New York: John Wiley, 1978.

RABINER, LAWRENCE R., and SCHAFER, RONALD W. *Digital Processing of Speech Signals.* Prentice-Hall Signal Processing Series. Englewood Cliffs, N.J.: Prentice-Hall, 1978.

ROBINSON, ENDERS A., and TREITEL, SVEN. *Geophysical Signal Analysis.* Prentice-Hall Signal Processing Series. Englewood Cliffs, N.J.: Prentice-Hall, 1980.

VAN TREES, HARRY L. *Detection, Estimation and Modulation Theory, Part 3. Radar-Sonar Signal Processing and Gaussian Signals in Noise.* New York: John Wiley, 1971.

WOODWARD, P. M. *Probability and Information Theory, with Applications to Radar.* 2nd ed. International Series of Monographs on Electronics and Instrumentation, vol. 3. Oxford, England: Pergamon Press, 1964.

Index

Type *l* feedback system, 742

Underdamped systems, 245, 357
Undersampling, 527
Unilateral Laplace transform, 614-16
Unilateral *z*-transform, 667-69
Unit circle, 631
      role in discrete-time stability, 656
Unit delay, 112
Unit doublet, 122, 156
Unit impulse:
    continuous-time, 22-25
    discrete-time, 26-27
    signal representation with:
        continuous-time, 72-75
        discrete-time, 70-72
Unit impulse response (*See* Impulse
    response)
Unit ramp, 123
Unit sample, 26-27 (*See also* Unit impulse)
Unit sample response, 78, 87
Unit step:
    continuous-time, 22, 123
    discrete-time, 26-27
    Fourier transform of (*See* Fourier
      transform)
Unit step response (*See* Step response)
Unstable system, 99-100
Upsampling, 550

Vibrating string example, 162-63
Visual system as lowpass filter, 523-25

Walsh functions, 145, 257
Wavelength, 165
Wideband frequency modulation, 483-85
Windowing, low-time, 542
Windowing in spectral analysis, 387
Window signal, 388

Zero-order hold, 519-23
    reconstruction filter, 519-21
    sampled data control systems
      use in, 746
    sampling with, 519-21

Zero-phase filter, 431
Zeros:
    Laplace transform, 578
    z-transform, 632
z-transform:
    definition, 830
    geometric evaluation of frequency
      responses, 646-49
    inverse, 643-46 (*See also* Inverse
      z-transform)
    for Nyquist plots, 722-24
    poles, 632
    pole-zero plot, 632
    properties of, 649-54
      convolution, 652
      differentiation, 652-53
      frequency shift, 650-51
      initial value theorem, 653-54
      linearity, 649-50
      time reversal, 651-52
      time shift, 650
    region of convergence (ROC):
      631-43 (*See also* Region of
      convergence)
      bounded by poles or infinity,
        642
      centered about origin, 635
      finite-duration sequence, 636
      left-sided sequence, 637
      right-sided sequence, 636
      two-sided sequence, 638
    relation to Fourier transform,
      646-49
    for root locus, 749-52, 700-701,
      712-13
    summary of properties, 654
    as system function, 655-58
      causal systems, 636
      difference equations, use with,
        656-58
      interconnection of systems,
        658
      stable systems, 656
    table of pairs, 655
    table of properties, 654
    zeros, 632